AF566920

OUTSKIRTS OF GALAXY CLUSTERS: INTENSE LIFE IN THE SUBURBS

IAU COLLOQUIUM No. 195

COVER ILLUSTRATION

The cover illustration is a drawing by Emmanuele Turatti. It depicts the Mole Antonelliana, the symbol of Torino. Designed by the architect Alessandro Antonelli and built between 1863 and 1889, it was originally conceived to be a synagogue. The Mole now hosts the Italian National Museum of Cinema. With its 167 m vertical size, it is one of the tallest stone buildings in Europe.

The shape of the Mole resembles the trumpet shape of the caustics observed in the redshift diagram of the infall regions of galaxy clusters. The Mole therefore became the natural logotype of this IAU Colloquium.

INTERNATIONAL ASTRONOMICAL UNION
UNION ASTRONOMIQUE INTERNATIONALE

OUTSKIRTS OF GALAXY CLUSTERS: INTENSE LIFE IN THE SUBURBS

PROCEEDINGS OF THE 195th COLLOQUIUM OF THE INTERNATIONAL ASTRONOMICAL UNION
TORINO, ITALIA
MARCH 12–16, 2004

Edited by

ANTONALDO DIAFERIO
Dipartimento di Fisica Generale "Amedeo Avogadro",
Università degli Studi di Torino,
Torino, Italia

CAMBRIDGE
UNIVERSITY PRESS

CAMBRIDGE UNIVERSITY PRESS
The Edinburgh Building, Cambridge CB2 2RU, UK
40 West 20th Street, New York, NY 10011 4211, USA
477 Williamstown Road, Port Melbourne, VIC 3207, Australia
Ruiz de Alarcón 13, 28014 Madrid, Spain
Dock House, The Waterfront, Cape Town 8001, South Africa

First published 2004

Printed in the United Kingdom at the University Press, Cambridge

Typeset in System LaTeX 2ε

A catalogue record for this book is available from the British Library

Library of Congress Cataloguing in Publication data

ISBN 0 521 84908 X hardback
ISSN 1743-9213

Table of Contents

Part 1. THE COSMIC WEB

Section A. The matter distribution

Section B. Theoretical tools

Section C. Clusters in action

Part 2. FORMATION AND EVOLUTION OF GALAXY CLUSTERS

Section A. Diffuse baryons

Section C. Theoreticians at work

Part 3. FORMATION AND TRANSFORMATION OF GALAXIES

Section A. Nurture and genetics of galaxies

Section B. Building galaxies

Preface

Clusters of galaxies lie at the nodes of a cosmic web. According to the gravitational instability hypothesis, which succeeds at reproducing the gross features of the large-scale structure, clusters form by accreting smaller clumps of dark matter along preferred directions and the cluster assembly is far from proceeding smoothly and isotropically.

Galaxies in clusters are substantially different from galaxies in the field. The strong correlation of optical and radio properties of galaxies with their environment indicates that the formation and evolution of galaxies is intimately linked to the formation of large-scale structure. Evidence of this link should be present in the regions intermediate between the central cores of clusters and the surrounding large-scale structure.

In the universe, clusters resemble frenetic cities on the Earth: galaxies move rapidly like people running in the streets of Milano (or New York City for the American reader), might experience violent interactions and might rapidly change their lifestyle. At a superficial look, metropolitan suburbs might appear more quite. However, much action does indeed take place there: people tend to know each other personally, and interactions occur in small groups. In the centres of town, there often happen events that provoke a lot of discussion but they might not be that relevant for the life of most human beings. Many people quietly grow in the countryside and in the suburbs, where they produce most of the goods consumed by the cities for their survival. Some people are eventually attracted by the city lights, they move there, change their appearence and never come back.

In October 2002, Margaret Geller and I, while enjoying the cozy hospitality of Massimo Ramella at the Osservatorio Astronomico di Trieste, discussed the current state of the scientific community understanding of galaxy and structure formation. We realized that the external regions of galaxy clusters, despite their relevance, were never the topic of a dedicated conference: such a meeting appeared timely because the traditional difficulties at collecting information in these regions had recently been overcome by extensive surveys of the external regions of nearby clusters and by the development of wide-field imaging and spectroscopic techniques that can be applied to clusters at moderate redshift.

Therefore it wasn't difficult to shape a draft of the program with some members of the community who promptly accepted to join the Scientific Organizing Committee (SOC): how are the properties of clusters and galaxies linked to the large-scale matter distribution? What is the role of the hot intracluster medium and the warm, diffuse distribution of baryons in the intergalactic medium? What specific questions will be addressed with the current and future multi-wavelength surveys?

The response of the community was enthusiastic and the meeting turned out to be more successful than we optimistically hoped. The scientific program was separated into three main topics: *The Cosmic Web*, *Formation and Evolution of Clusters*, and *Formation and Transformation of Galaxies*. The meeting took advantage of the spectacular observational data available, which came from X-ray (XMM, Chandra), radio (e.g. WSRT), submillimeter (SCUBA), infrared (ISO) and optical (HST, Keck, Subaru, VLT) telescopes. State-of-the-art numerical simulations and theoretical models showed the impressive progress we made at understanding cluster and galaxy formation.

The numerous and high-quality contributions at the meeting confirm its success: 26 invited oral presentations (indicated by an "(I)" in the table of contents: many thanks to the excellent work of the invited speakers), 42 contributed oral presentations and 55 poster presentations. I wish to warmly acknowledge all the SOC members for their relevant suggestions: their work shaped a conference that has now made apparent that the outskirts of clusters are the site of an intense life for both the galaxies living there and the astronomers working on understanding them.

Despite my optimism, I honestly expected a more modest participation. After all, in the international view of Italy, Torino does not belong to the top list of the Italian towns to visit, neither it is a beach or a ski resort. Indeed, the main reason for choosing Torino as the Colloquium location was to give the scientific community the opportunity of discovering a lively town rich of historic, artistic, and cultural treasures. A major satisfaction came from those participants who confessed to enjoy the unexpected beauties of Torino.

Making everything go smoothly in a conference, hosting almost twice as many participants as expected, of course requires the joint effort of a number of people. I would like to gratefully acknowledge the friendly support of Dr. Edoardo Trussoni at the Osservatorio Astronomico di Torino, Ing. Luigi Maria Quaglino at Alenia Spazio S.p.A., Dr. Stefano Morelli at the Città di Torino, and Dr. Aldo Timossi at the Regione Piemonte. I also wish to warmly thank the patience and good advice of Dr. Rosanna Cauda and Ms. Maria Salvatore at the Regione Piemonte, and the understanding of Ms. Roberta Ghiringhelli at Torino Incontra, who didn't loose her patience when I kept posing her unexpected problems that she had to (and did!) resolve on the shortest time scale. I would also like to gratefully acknowledge the logistic advice of Ms. Francesca Di Sero at Torino Convention Bureau at the early stages of the conference organization.

A particular thank to Giuseppe Murante for his patience in dealing with the videoprojectors that sometimes didn't work as expected; Silvia Ameglio, Sara Buttiglione and Laura Disabato for running restless around the conference hall with the microphones, without being frustrated by the fact that some participants appeared to be allergic to them (I mean the microphones). I also wish to thank Attilio Ferrari, Giuseppe Murante and especially Jörg Colberg for taking exquisite pictures at the conference. You can enjoy some of them in this book.

The meeting could become a reality only thanks to the experience and wisdom of Attilio Ferrari and the versatile skill and intrepid imagination of Claudio Casacci: I am extremely grateful to both of them. A special thank to our rare contralto Gabriella Ardizzoia, for her competent secretary work realized over a weekend when she was supposed to sing for the Holy Father. Next time, we will make the Pope change His schedule, so that she can both go to Roma and direct our meeting.

It would be silly not to expect unexpected problems. However, they do arrive unexpectedly and solving them requires lucidity and calmness. Luisa supported me during the worst difficulties with the most relevant suggestions, even when she was hundreds of kilometers away. I dedicate all my work and efforts poured into this conference to her.

Although all the contributions of the people mentioned above were no doubt crucial, the real artifices of the success of this conference were the 138 participants from 21 countries, with their enthusiasm and professionality. A major thank to all of them.

Astronomy is not an exclusive property of professional astronomers. I felt it as a duty to associate the Colloquium to a public lecture. Rashid Sunyaev kindly accepted to deliver a lecture that was inserted into the Giovedì Scienza program, a series of public lectures on scientific disciplines that has now been popular in Torino for 18 years, thanks also to the intense work of Dr. Pino Zappalà. That night it was a real satisfaction to see the theatre packed with more than 1500 people. In addition, interviews of Rashid Sunyaev and Alan Dressler by Dr. Silvia Rosa-Brusin were broadcasted on the national TV channels in April

2004, and an article of mine on the conference appeared on a major national newspaper in May 2004, following a kind request of Dr. Piero Bianucci. Many thanks to all of them for contributing once again to the popularity of Astronomy.

The very generous financial contributions of the public and private foundations and institutions listed on page xvii confirm that believing in the relevance of Astronomy for the advance of human knowledge is not a prerogative of astronomers alone. It is a real pleasure to acknowledge the support of the Compagnia di San Paolo, who contributed to the publication of this volume, and all the other sponsors that made possible the actual materialization of this conference out of our imagination.

For the next conference in Torino, I hope that we will be able to spot another topic as hot as the outskirts of galaxy clusters. It won't be a big deal if we miss the goal, however: the participants will have more time to enjoy Torino and will not pack the conference hall at every single session, as it happened this time.

Antonaldo Diaferio, chair SOC and LOC
Torino, June 16, 2004

THE ORGANIZING COMMITTEE

Scientific

A. Diaferio (chair, Italia)
G. Kauffmann (Deutschland)
D.G. Lambas (Argentina)
E. Salvador-Solé (España)
H.K.C. Yee (Canada)
M.J. Geller (USA)
T. Kodama (Japan)
A. Reisenegger (Chile)
R. van de Weygaert (Nederland)

Local

A. Diaferio (chair, Univ. di Torino)
A. Ferrari (Univ. di Torino)
G. Murante (INAF - Oss. Astron. di Torino)
C. Casacci (Alenia Spazio S.p.A. – CIFS)
S. Massaglia (Univ. di Torino)

Acknowledgements

The colloquium is sponsored and supported by the IAU Division VIII (Galaxies and the Universe); and by the IAU Commissions No. 28 (Galaxies) and No. 47 (Cosmology).

The colloquium was organized under the patronage of the
Consorzio Interuniversitario per la Fisica Spaziale (CIFS),
Alenia Spazio S.p.A.,
Osservatorio Astronomico di Torino,
Dip. di Fisica Generale "A. Avogadro", Università degli Studi di Torino,
Città di Torino,
Provincia di Torino
and
Regione Piemonte.

Funding by the
International Astronomical Union,
Alenia Spazio S.p.A.,
Camera di Commercio, Industria e Artigianato di Torino,
Compagnia di San Paolo,
Fondazione CRT,
Città di Torino
and
Regione Piemonte
is gratefully acknowledged.

CONFERENCE PHOTOGRAPH

(by G. Murante)

(by G. Murante)

Participants

Ummi **Abbas**, University of Pittsburgh, PA, USA ummi@phyast.pitt.edu
J. Alfonso L. **Aguerri**, Instituto de Astrofisica de Canarias, La Laguna, Tenerife, Spain jalfonso@ll.iac.es
Silvia **Ameglio**, Università di Torino, Italy ameglio@studenti.ph.unito.it
Stefano **Andreon**, INAF-Osservatorio Astronomico di Brera, Milano, Italy andreon@brera.mi.astro.it
Miguel Angel **Aragon Calvo**, Kapteyn, University of Groningen, The Netherlands miguel@astro.rug.nl
Pablo **Araya Melo**, Kapteyn Astronomical Institute, Groningen, The Netherlands pablo@astro.rug.nl
Fernando **Atrio-Barandela**, Universidad de Salamanca, Spain atrio@usal.es
Michael **Balogh**, Department of Physics, University of Durham, UK M.L.Balogh@Durham.ac.uk
Sandro **Bardelli**, INAF-Osservatorio Astronomico di Bologna, Italy bardelli@bo.astro.it
Andrew **Benson**, University of Oxford, Astrophysics, UK abenson@astro.ox.ac.uk
Andrea **Biviano**, INAF-Osservatorio Astronomico di Trieste, Italy biviano@ts.astro.it
Kris **Blindert**, University of Toronto, Canada blindert@astro.utoronto.ca
Hans **Böhringer**, Max-Planck-Institut für extraterrestrische Physik, Garching, Germany hxb@mpe.mpg.de
Massimiliano **Bonamente**, University of Alabama, Huntsville, USA bonamem@email.uah.edu
Michiel **Brentjens**, Kapteyn Institute, ASTRON, Groningen, The Netherlands m.a.brentjens@astro.rug.nl
Gianfranco **Brunetti**, Istituto di Radioastronomia, Bologna, Italy brunetti@ira.cnr.it
Sara **Buttiglione**, Università di Torino, Italy sarabutty@libero.it
Silvia **Caffi**, IASF, Sezione di Milano, Italy silvia@mi.iasf.cnr.it
Patricia Eiko **Campos**, IAG-USP, Sao Paulo, Brazil campos@astro.iag.usp.br
Conrado **Carretero**, Instituto de Astrofisica de Canarias, La Laguna, Tenerife, Spain cch@ll.iac.es
Claudio **Casacci**, Alenia Spazio S.p.A. – CIFS, Torino, Italy casacci.claudio@spacegate-altec.it
Rossella **Cassano**, Istituto di Radioastronomia, Bologna, Italy rcassano@ira.cnr.it
Laura **Ceccarelli**, Observatorio Astronomico de Cordoba, Cordoba, Argentina laura@oac.uncor.edu
Jörg **Colberg**, University of Pittsburgh, Pittsburgh, PA, USA astro@jmcolberg.com
Luca **Cortese**, Università degli Studi di Milano-Bicocca, Milano, Italy luca.cortese@mib.infn.it
Oliver **Czoske**, IAEF, Universität Bonn, Bonn, Germany oczoske@astro.uni-bonn.de
Håkon **Dahle**, Institute of Theoretical Astrophysics, Oslo, Norway hdahle@astro.uio.no
Sabrina **De Grandi**, INAF-Osservatorio Astronomico di Brera, Merate, Italy degrandi@mi.astro.it
Valerio **D'Elia**, INAF-Osservatorio Astronomico di Roma, Monte Porzio Catone, Italy delia@mporzio.astro.it
Ian **Dell'Antonio**, Brown University, Providence, RH, USA ian@het.brown.edu
Antonino **Del Popolo**, Bogazici University, Istanbul, Turkey adelpop@unibg.it
Gabriella **De Lucia**, Max-Planck-Institut für Astrophysik, Garching, Germany lucia@mpa-garching.mpg.de
Jelle **de Plaa**, SRON, Utrecht, The Netherlands j.de.plaa@sron.nl
Antonaldo **Diaferio**, Università di Torino, Italy diaferio@ph.unito.it
Jörg **Dietrich**, IAEF, Universität Bonn, Germany dietrich@astro.uni-bonn.de
Laura **Disabato**, Università di Torino, Italy aurape@hotmail.com
Alan **Dressler**, Carnegie Observatories, Pasadena, CA, USA dressler@ociw.edu
Pierre-Alain **Duc**, CEA-Saclay, Gif sur Yvette Cedex, France paduc@cea.fr
Erica **Ellingson**, University of Colorado, Boulder, CO, USA elling@casa.colorado.edu
Simon **Ellis**, Anglo-Australian Observatory, Epping, Australia sce@aaoepp.aao.gov.au
Stefano **Ettori**, European Southern Observatory, Garching bei München, Germany settori@eso.org
Siegfried **Falter**, Max-Planck-Institut für Astronomy, Heidelberg, Germany falter@mpia-hd.mpg.de
Attilio **Ferrari**, Università di Torino, Italy ferrari@ph.unito.it
Chiara **Ferrari**, Institut für Astrophysik, Innsbruck, Austria Chiara.Ferrari@uibk.ac.at
Alexis **Finoguenov**, Max-Planck-Institut für extraterrestrische Physik, Garching, Germany alexis@xray.mpe.mpg.de
Matthew **Fleenor**, University of North Carolina, Chapel Hill, NC, USA fleenor2@physics.unc.edu
Piotr **Flin**, Pedagogical University, Institute of Physics, Kielce, Poland sfflin@cyf-kr.edu.pl
Paolo **Franzetti**, INAF – IASF Milano, Italy paolo@mi.iasf.cnr.it
Anna **Gallazzi**, Max-Planck-Institut für Astrophysik, Garching, Germany gallazzi@mpa-garching.mpg.de
Simona **Giacintucci**, Dip. di Astronomia, Università di Bologna, Italy giacin@excalibur.bo.astro.it
David **Gilbank**, Dept. of Astronomy and Astrophysics, University of Toronto, Canada gilbank@astro.utoronto.ca
Stuart **Gill**, Swinburne University of Technology, Hawthorn, Australia sgill@astro.swin.edu.au
Guillermo **González-Casado**, Universidad Politécnica de Cataluña, Barcelona, Spain guillermo.gonzalez@upc.es
Antonio-Cesar **González-García**, Inst. de Astrofisica de Canarias, La Laguna, Tenerife, Spain cglez@ll.iac.es
Meghan **Gray**, University of Nottingham, United Kingdom Meghan.Gray@nottingham.ac.uk
Giorgio **Grosso**, Torino, Italy grosso.giorgio@iol.it
Roald **Guandalini**, Dipartimento di Fisica, Università di Perugia, Italy guandalini@fisica.unipg.it
Claire **Halliday**, Osservatorio Astronomico di Padova, Italy halliday@pd.astro.it
Takashi **Hamana**, IAP/NAOJ, Mitaka, Japan hamanatk@cc.nao.ac.jp
Craig **Harrison**, RSAA, Weston Creek, Australia harrison@mso.anu.edu.au
George **Hau**, European Southern Observatory, Garching bei München, Germany ghau@eso.org
Ilian **Iliev**, CITA, Toronto, Canada iliev@cita.utoronto.ca
Klaus **Jäger**, Universitäts-Sternwarte Göttingen, Germany jaeger@uni-sw.gwdg.de
Yipeng **Jing**, Shanghai Astronomical Observatory, Shanghai, China ypjing@shao.ac.cn
Melanie **Johnston-Hollitt**, Leiden Observatory, Leiden, The Netherlands johnston@strw.leidenuniv.nl
Henning **Jørgensen**, Astronomical Observatory, Copenhagen, Denmark henning@astro.ku.dk
Roman **Juszkiewicz**, Copernicus Astronomical Center, Warsaw, Poland roman@camk.edu.pl
Jelle **Kaastra**, SRON, Utrecht, The Netherlands J.Kaastra@sron.nk
Birgit **Kelm**, Dip. di Astronomia, Università di Bologna, Italy birgit@orion.bo.astro.it
Per **Kjærgaard Rasmussen**, Copenhagen University Observatory, Denmark per@astro.ku.dk
Tadayuki **Kodama**, National Astronomical Observatory of Japan, Mitaka, Japan kodama@th.nao.ac.jp
Yutaka **Komiyama**, Subaru Telescope, Hilo, HI, USA komiyama@subaru.naoj.org
Jessica **Krick**, University of Michigan, Ann Arbor, MI, USA jkrick@umich.edu
Diego **Lambas**, IATE Group, Observatorio Astronomico de Cordoba, Argentina dgl@oac.uncor.edu
Andrea **Lapi**, Dip. Fisica, Università di "Tor Vergata", Roma, Italy lapi@roma2.infn.it
Samuel **LaRoque**, University of Chicago, Chicago, IL, USA laroque@bigmuddy.uchicago.edu
I-hui **Li**, University of Toronto, Canada tornado@astro.utoronto.ca
Gary **Mamon**, IAP, Paris, France gam@iap.fr
Silvia **Marinoni**, INAF – Osservatorio Astronomico di Bologna, Italy nursothca@yahoo.it
Chiara **Marmo**, Dipartimento di Astronomia, Università di Padova, Italy marmo@pd.astro.it
Chiara **Mastropietro**, University of Zürich, Switzerland chiara@physik.unizh.ch
Claudia **Mendes de Oliveira**, Universidade de Sao Paulo, Brazil oliveira@astro.iag.usp.br
Silvano **Molendi**, IASF Sezione di Milano, Italy silvano@mi.iasf.cnr.it
Ben **Moore**, Institute for Theoretical Physics, Zürich, Switzerland moore@physik.unizh.ch
Sean **Moran**, Caltech, Pasadena, CA, USA smm@astro.caltech.edu
Volker **Müller**, Astrophysikalisches Institut Potsdam, Germany vmueller@aip.de
Giuseppe **Murante**, INAF – Osservatorio Astronomico Torino, Italy murante@to.astro.it
Daisuke **Nagai**, University of Chicago, IL, USA daisuke@oddjob.uchicago.edu
Masahiro **Nagashima**, University of Durham, United Kingdom masahiro.nagashima@durham.ac.uk
Fumiaki **Nakata**, University of Durham, United Kingdom fumiaki.nakata@durham.ac.uk
Fabrizio **Nicastro**, Harvard-Smithsonian Center for Astrophysics, Cambridge, MA, USA fnicastro@cfa.harvard.edu
Takashi **Okamoto**, University of Durham, United Kingdom Takashi.Okamoto@durham.ac.uk

Giorgio **Palumbo**, Dipartimento di Astronomia, Università di Bologna, Italy ggcpalumbo@bo.astro.it
Elena **Panko**, Nikolaev State University, Astronomical Observatory, Nikolaev, Ukraine tajgeta@sp.mk.ua
Ricardo **Pérez Martínez**, XMM-Newton SOC, ISO DC, ESA, Madrid, Spain rperez@iso.vilspa.esa.es
Rocco **Piffaretti**, Institute for Theoretical Physics, University of Zürich, Switzerland piff@physik.unizh.ch
Maximiliano **Pivato**, IATE, Astronomical Observatory of Cordoba, Argentina maxip@mail.oac.uncor.edu
Manolis **Plionis**, National Obs. of Athens and INAOE-Mx, Athens, Greece mplionis@astro.noa.gr
Bianca Maria **Poggianti**, Osservatorio Astronomico di Padova, Italy poggianti@pd.astro.it
Dominique **Proust**, Observatoire de Meudon, France dominique.proust@obspm.fr
Vicent **Quilis**, Dep. d'Astronomia i Astrofisica, Universitat de Valencia, Burjassot, Spain vicent.quilis@uv.es
Elena **Rasia**, Dipartimento di Astronomia, Università di Padova, Italy rasia@pd.astro.it
Paola **Re Fiorentin**, Università di Torino, Italy paola.re.fiorentin@to.infn.it
Kenneth **Rines**, Yale University, New Haven, CT, USA krines@astro.yale.edu
Mariachiara **Rossetti**, IASF Milano, Italy rossetti@mi.iasf.cnr.it
Rachida **Sadat**, Observatoire de Midi-Pyrenees (OMP), Toulouse, France rsadat@ast.obs-mip.fr
Rubén **Sánchez Janssen**, Instituto de Astrofísica de Canarias, La Laguna, Spain ruben@ll.iac.es
Peter **Schneider**, IAEF, Universität Bonn, Germany peter@astro.uni-bonn.de
Sergei **Shandarin**, University of Kansas, Lawrence, KS, USA sergei@ku.edu
Ravi **Sheth**, University of Pittsburgh, PA, USA rks12@pitt.edu
Debora **Sijacki**, Dipartimento di Astronomia, Università di Padova, Italy sijacki@pd.astro.it
Gert **Sikkema**, Kapteyn Institute, Groningen, The Netherlands sikkema@astro.rug.nl
Patrick **Simon**, IAEF University of Bonn, Germany psimon@astro.uni-bonn.de
José M. **Solanes**, Universitat de Barcelona, Spain jm.solanes@ub.edu
Volker **Springel**, Max-Planck-Institut für Astrophysik, Garching, Germany volker@mpa-garching.mpg.de
John **Stocke**, CASA, University of Colorado, Boulder, CO, USA stocke@casa.colorado.edu
Rashid **Sunyaev**, Max-Planck-Institut für Astrophysik, Garching, Germany sunyaev@mpa-garching.mpg.de
Masayuki **Tanaka**, University of Tokyo, Japan tanaka@astron.s.u-tokyo.ac.jp
Giuseppe **Tormen**, Dipartimento di Astronomia, Università di Padova, Italy tormen@pd.astro.it
Luca **Tornatore**, Dipartimento di Astronomia, Università di Trieste, Italy tornatore@ts.astro.it
Kim-Vy **Tran**, Institute for Astronomy, Zürich, Switzerland vy@phys.ethz.ch
Tommaso **Treu**, University of California, Los Angeles, CA, USA ttreu@astro.ucla.edu
Brent **Tully**, Institute for Astronomy, Honolulu, HI, USA tully@ifa.hawaii.edu
Edwin **Valentijn**, Kapteyn Institute, Groningen, The Netherlands valentyn@astro.rug.nl
Rien **van de Weygaert**, Kapteyn Astronomical Inst., Groningen, The Netherlands weygaert@astro.rug.nl
Eelco **van Kampen**, Inst. for Astrophysics, University of Innsbruck, Austria Eelco.v.Kampen@uibk.ac.at
Alexandre **Vazdekis**, Instituto de Astrofisica de Canarias, La Laguna, Spain vazdekis@ll.iac.es
Marc **Verheijen**, Astrophysikalisches Institut Potsdam, Germany verheyen@aip.de
Michael **Vogeley**, Drexel University, Department of Physics, Philadelphia, PA, USA vogeley@drexel.edu
Mark **Voit**, Michigan State University, East Lansing, MI, USA voit@pa.msu.edu
Q. Daniel **Wang**, University of Massachusetts, Amherst, MA, USA wqd@astro.umass.edu
Tracy **Webb**, Leiden Observatory, Leiden, The Netherlands webb@strw.leidenuniv.nl
Peter **Williams**, Shanghai Astronomical Observatory, China williams@shao.ac.cn
Toru **Yamada**, National Astronomical Observatory in Japan, Mitaka, Japan yamada@optik.mtk.nao.ac.jp
Howard **Yee**, University of Toronto, Canada hyee@astro.utoronto.ca
Gustavo **Yepes**, Universidad Autonoma de Madrid, Spain gustavo.yepes@uam.es
Stefano **Zibetti**, Max-Planck-Institut für Astrophysik, Garching, Germany zibetti@mpa-garching.mpg.de
Elena **Zucca**, INAF – Osservatorio Astronomico di Bologna, Italy zucca@bo.astro.it

Part 1

THE COSMIC WEB

Section A
The matter distribution

Outskirts of Galaxy Clusters: Intense Life in the Suburbs
Proceedings IAU Colloquium No. 195, 2004
A. Diaferio, ed.

DOI: 10.1017/S1743921304000043

Mapping the cosmic web with the Sloan Digital Sky Survey

Michael S. Vogeley, Fiona Hoyle, Randall R. Rojas, and David M. Goldberg

Department of Physics, Drexel University, Philadelphia, PA 19104, USA email: vogeley@drexel.edu

Abstract. Wide-angle, moderately deep redshift surveys such as that conducted as part of the Sloan Digital Sky Survey (SDSS) allow study of the relationship between the structural elements of the large-scale distribution of galaxies – including groups, cluster, superclusters, and voids – and the dependence of galaxy formation and evolution on these enviroments. We present a progress report on mapping efforts with the SDSS and discuss recently constructed catalogs of clusters, voids, and void galaxies, and evidence for a $420h^{-1}$Mpc supercluster or "Great Wall." Analysis of multi-band photometry and moderate-resolution spectroscopy from the SDSS reveals environmental dependence of the star formation history of galaxies that extends over more than a factor of 100 in density, from clusters all the way to the deep interiors of voids. On average, galaxies in the rarified environments of voids exhibit bluer colors, higher specific star formation rates, lower dust content, and more disk-like morphology than objects in denser regions. This trend persists in comparisons of samples in low vs. high-density regions with similar luminosity and morphology, thus this dependence is not simply an extension of the morphology-density relation. Large-scale modulation of the halo mass function and the temperature of the intergalactic medium might explain this dependence of galaxy evolution on the large-scale environment.

1. Introduction: cities, suburbs, and countryside of the universe

The varied geography and demography of human civilization provides a colorful analogy for the dependence of galaxy formation and evolution on the large-scale environment of the universe. The dense population of cities promotes a frenetic lifestyle and facilitates many fleeting interactions. Only the very richest get richer (cD's?), while many, drawn to the bright lights, are quickly stripped of all that they arrived with. At the other extreme lie the rural stretches, where interactions are few but which tend to be tightly-bound when they occur. Without the competition of others, a few can live surprisingly well off the steady, albeit sometimes meager, offerings of the land. This meeting has been called to study the suburbs of the universe, including the outskirts of clusters. "Intense life in the suburbs" results from living where interactions are frequent, but lasting enough to form families (groups?) and the city's pull is always felt. This environment is of particular interest because this is the lowest density in which halos with multiple galaxies are common.

Galaxy clusters, groups, and filaments surround the voids in a cosmic web that large galaxy redshift surveys reveal to be a ubiquitous pattern in the universe. Below we report on progress in mapping the nearby universe using the SDSS (§2) and describe results of objective identification of individual structures in the survey. We then discuss analyses of photometric and spectroscopic data for galaxy samples that are now large enough to statistically reveal dependence of galaxy evolution on environment from clusters all the way to voids (§3). We conclude by reviewing physical mechanisms that may explain this dependence (§4).

2. Large-scale structures in the SDSS

To study structure formation in this wide range of environments requires large statisically-controlled samples of galaxies over a wide area of sky. The SDSS was designed to accomplish this goal. It employs a special-purpose 2.5m telescope with a 3-degree field of view, a mosaic CCD camera, and dual fiber-fed spectrographs, to obtain five-band digital photometry and moderate-resolution spectroscopy over essentially the full range of optical wavelengths. The completed survey will cover approximately 10^4 square degrees. York et al. (2000) provides an overview of the SDSS and Stoughton et al. (2002) describes the early data release (EDR) and details about the photometric and spectroscopic measurements made from the data. Technical articles providing details of the SDSS include descriptions of the photometric camera (Gunn 1998), photometric analysis (Lupton et al. 2004), the photometric system (Fukugita et al. 1996; Smith et al. 2002), the photometric monitor (Hogg et al. 2001), astrometric calibration (Pier et al. 2003), and spectroscopic tiling (Blanton et al. 2003). A thorough analysis of possible systematic uncertainties in the galaxy samples is described in Scranton et al. (2002).

2.1. *SDSS galaxy redshift surveys*

The SDSS is now the largest galaxy redshift survey to date. When completed, it will include 10^6 galaxies over 10^4 deg^2 of sky. The second major public release of SDSS data (DR2: Abazajian et al. 2004; `www.sdss.org/dr2` and `skyserver.sdss.org`), in March 2004, includes 3324 deg^2 of imaging (88 million unique objects) and spectroscopy over 2627 deg^2 (including 260,490 galaxies).

The main spectroscopic galaxy sample of the SDSS (Strauss et al. 2002) includes objects selected to have Petrosian magnitudes of $r < 17.77$ after correction for Galactic extinction. The median depth of this sample is approximately $z = 0.1$. Analyses of these data include measurement of the power spectrum, correlation function, luminosity function, topology, galaxy properties, etc. (among 389 SDSS papers so far from collaboration participants; see `www.sdss.org/publications/`).

To obtain a deeper tracer of structure in the universe, a spectroscopic target sample of luminous red galaxies (LRG's), is selected by color and has $r < 19.5$ (Eisenstein et al. 2001). These criteria yield a nearly volume-limited sample out to $z = 0.45$. The LRG selection criteria ensure that this sample includes all brightest cluster galaxies, as well as slightly less luminous objects in clusters and groups. The LRG's are an efficient tracer of dense regions in the universe and have selection criteria that are quite simple compared to, e.g., cluster or group-finding algorithms. The ease of measuring redshifts to such galaxies (they have strong absorption lines) makes them an excellent choice for deep mapping of the universe.

2.2. *An even Greater Wall*

A novel approach to plotting the distribution of objects over large ranges of cosmic time and distance (Gott et al. 2003) clearly shows the largest structure seen to date, a "Sloan Great Wall" that extends over roughly $420h^{-1}$Mpc. This supercluster-scale structure may be compared with the Great Wall found in the Center for Astrophysics Redshift Survey (Geller & Huchra 1989), which has a linear scale of roughly $240h^{-1}$Mpc. The size of the CfA Great Wall was limited by the extent of that survey. In contrast, the SDSS is large enough that a more extensive structure could have been found.

The mapping method employed by Gott et al. is conformal, i.e., shapes of structures are preserved locally, thus we can compare great walls and other structures seen at different distance. The x-axis is right ascension in degrees, while the vertical axis plots $y = \ln r$ where r is the comoving coordinate distance from Earth. In such a plot, the entire universe, from the surface of the Earth to the epoch of the big bang, may be shown in one plot.

2.3. *Cluster catalogs*

Several algorithms have been applied to SDSS photometric and spectroscopic data to identify clusters of galaxies (see, e.g., Nichol 2003 and references therein). Cluster catalogs have been constructed from the SDSS imaging data alone using color criteria (maxBCG: Annis et al.), adaptive matched filtering (AMF; Kim et al. 2002), and a combination of these two methods (Bahcall et al. 2003). Perhaps the most useful for the purpose of placing clusters in the context of the cosmic web is the C4 method which identifies clusters within the volume sampled by the main galaxy spectroscopic sample. The C4 algorithm looks for density enhancements in the 7-dimensional space of four colors (generated by comparison of the five SDSS bands), right ascension, declination, and redshift. 800 clusters have been identified in the DR2 sample using this method (Miller et al. 2004).

3. Galaxy properties and environment

The environmental dependence of galaxy properties has typically been described in terms of the well-known morphology-density relation (Dressler 1980; Postman & Geller 1984). Study of the morphology-density relation using SDSS data (Goto et al. 2003) shows that this relation extends into the field population. Recent analyses of large spectroscopic samples of galaxies from the 2dFGRS and SDSS reveal variation of galaxy properties, in particular dependence of star formation rates, on density and/or clustercentric distance well beyond the virial radii of clusters. Analyses of galaxies in voids shows that this trend continues down to a small fraction of the mean density. Evidence suggests that the star formation rate depends not only on the local density (on scales of 1 Mpc or so) but also on the larger-scale environment. Here we highlight some recent results (also see Tanaka et al. 2004 in this proceedings) and describe some possible mechanisms for this large-scale variation.

3.1. *Star formation rates as a function of clustercentric distance*

Lewis et al. (2002) estimate star formation rates for galaxies in the 2dFGRS that lie near clusters. They find that the mean star formation rates of galaxies, as well as the fraction of galaxies with high star formation rates, increases with clustercentric distance, with a sharp rise out to $\sim 2r/R_{virial}$. Analysis of SDSS galaxies near clusters and groups found in SDSS (using the C4 method described above) clearly shows a rise of the fraction of high star-forming galaxies out to several virial radii. Gomez et al. (2003) find similar results from analysis of the SDSS EDR sample and show that a "break" appears in the density-SFR relation around a projected local density of $\sim 1h_{75}^{-1}$Mpc.

3.2. *Star formation rate as a function of density*

Balogh et al. (2004) examine the variation of Hα emission equivalent width as a function of density for galaxies in volume-limited samples of both 2dFGRS and SDSS ($M_r < -20.6$ for $h = 0.7$). They estimate galaxy density using both a projected density to the fifth-nearest neighbor and a Gaussian kernel smoothing and find that the fraction of galaxies with $W(H\alpha) > 4$ Åincreases with decreasing local density down to the lowest density that they can reliably examine. Further, when they examine the fraction of high $W(H\alpha)$ galaxies as a function of density estimated on both 1.1 and 5.5 Mpc scales, they find that, for fixed small-scale density, high star formation fractions are larger for galaxies that lie in large-scale underdensities. This result appears to disagree with Kauffmann et al. (2004), who find no discernable trend of star formation rate with large-scale environment. This discrepancy may result from the different range of densities probed (the lowest density bin probed by Kauffmann et al. includes over 30% of the galaxies). In the next section we discuss what happens at truly low densities.

3.3. *Void galaxy properties*

To examine whether the trends identified near clusters continue all the way into voids, we compare properties of void galaxies with those of galaxies in denser regions (Rojas 2004; Rojas et al. 2003, 2004; Hoyle et al. 2003; Hao et al. 2004). What happens to star formation rates and related galaxy properties at a small fraction of the mean density?

To identify void galaxies, we measure the distance from each galaxy in the flux-limited sample to the third nearest neighbor, and require $d_3 > 7h^{-1}$Mpc for void galaxies in a volume-limited sample ($M_r < -19.87 - 5\log h$). This yields 1010 void galaxies (6-8% of the sample after removing objects near the survey boundary) with $\delta\rho/\rho < -0.6$ on a $7h^{-1}$Mpc scale. Comparison of this criterion with results of the voidfinder algorithm (Hoyle & Vogeley 2002) show that it reliably identifies galaxies that lie in large voids. At small redshift, we use the nearby UZC and SSRS2 catalogs to map local voids, and identify an additional 194 fainter void galaxies. This is the first sample of true void galaxies that is large enough to allow division into statistically-significant sub-samples.

Statistical analyses of the distribution functions of color, surface brightness profile, and emission line equivalent widths, show that void galaxies are, on average, bluer, more disklike, and have higher specific (per unit stellar mass) star formation rates than objects in denser regions ("wall" galaxies hereafter). K-S tests typically yield a probability $P < 10^{-4}$ of the void and wall galaxies being drawn from the same parent population. These same trends obtain when we compare void galaxies with the lowest-density 20% of wall galaxies, thus these differences are apparent not only between "cluster" and "void" samples but also between true "field" and "void" populations.

Several analyses of SDSS photometry have shown that galaxies have a nearly bimodal distribution of properties, either red and bulge-dominated, or blue and disk-like (Blanton et al. 2003; Baldry et al. 2004). To test the possibility that void galaxies differ from wall galaxies simply because of a demographic shift in the relative fractions of objects in these two populations (fewer elliptical in low-density regions), we compare void and wall galaxies in sub-samples with similar morphology and luminosity. We find that void galaxies persist in having bluer average colors. Thus, the morphology-density relation does not explain this density-dependence.

In summary, we find that at fixed luminosity and surface brightness profile, galaxies in voids have average colors that are bluer and higher average specific star formation rates. These trends are generally consistent with predictions of semi-analytic modelling of galaxy formation (Benson et al. 2003).

4. Discussion: possible causes of large-scale environmental dependence

What could cause these large-scale variations? Is galaxy formation a purely local process? Or does the large-scale environment contribute to the fate of a galaxy? Environment can affect galaxy formation and evolution through a number of factors that vary with density:

(*a*) The dark halo mass function (shifts to lower masses at lower density);

(*b*) The formation epoch of halo assembly;

(*c*) The rate of galaxy-galaxy interactions/mergers (including tidal effects and induced turbulence);

(*d*) Temperature of the IGM.

The question of whether environment determines galaxy properties is mostly physical, but partly semantic. The first point makes this clear: large-scale environment affects the properties of the objects found at different large-scale density through the mass function. Likewise for the second point: for objects of fixed total mass today but in different large-scale environments, the merging history of halos is likely to be different. On the other hand, if we compare galaxies of fixed total mass in regions of fixed small-scale density, but in different large-scale environments, their properties are likely to be similar (as found by Kauffmann et al. 2004), but perhaps not identical, as shown by the evidence cited above.

4.1. *The mass function and environment*

Following the Press-Schechter formalism, it is straightforward to predict how the mass function of galaxies varies with large-scale environment. In voids, the number of small halos decreases in proportion to the lower overall matter density, while the exponential cutoff at the high mass end shifts to lower mass. This variation is consistent with the interior of voids acting like very low density universes (Goldberg & Vogeley 2004). In Goldberg et al. (2004), we study this dependence, compute the mass function for voids with varying δ_v, estimate the mass function of SDSS galaxies, and estimate δ_v.

Several paradoxical results relevant to the question of environmental dependence on large-scales arise from this study. The growth factor $D(z|\delta)$ in voids is larger at high redshift than in the background universe, suggesting that void galaxy halos should be relatively old. However, detailed analysis shows that, because the mass function of void galaxies is quite steep at the relevant mass scale, a small amount of growth at the present epoch yields a larger fractional growth rate at the high mass end than in denser regions. In other words, the few high mass galaxies in voids may have a high fraction of recently accretion. This may explain why massive void galaxies have relatively large star formation rates, while faint void galaxies exhibit smaller differences from higher-density objects.

4.2. *Temperature, environment, and star formation*

A consequence of the temperature dependence of the Jeans mass and cooling rate of baryons is that we expect star formation to be more efficient where the IGM temperature is lower. In trying to explain the scale-dependence of bias in N-body/hydrodynamical simulations, Blanton et al. (1999) examine how bias depends on gas temperature. They find that, at fixed dark matter overdensity, the overdensity of baryonic matter that can form stars is larger in region of cooler gas temperature.

Because the gas temperature reflects fluctuations in density on scales much larger than the galaxy mass scale, star formation may be modulated by the large-scale density field. When the gas is virialized, its temperature reflects the local gravitational potential. Poisson's equation implies that $\tilde{\delta}_T(k) \propto k^{-2}\tilde{\delta}(k)$ (Blanton et al. 1999). In other words, the extra two powers of k cause fluctuations in temperature to depend on much larger wavelength fluctuations in density than the density field itself. Thus, the large-scale environment strongly influences the IGM temperature, through which it may strongly affect the efficiency of star formation.

Acknowledgements

M.S.V. acknowledges support from NSF grant AST-0071201.

Funding for the creation and distribution of the SDSS Archive has been provided by the Alfred P. Sloan Foundation, the Participating Institutions, the National Aeronautics and Space Administration, the National Science Foundation, the U.S. Department of Energy, the Japanese Monbukagakusho, and the Max Planck Society. The SDSS Web site is http://www.sdss.org/.

The SDSS is managed by the Astrophysical Research Consortium (ARC) for the Participating Institutions. The Participating Institutions are The University of Chicago, Fermilab, the Institute for Advanced Study, the Japan Participation Group, The Johns Hopkins University, Los Alamos National Laboratory, the Max-Planck-Institute for Astronomy (MPIA), the Max-Planck-Institute for Astrophysics (MPA), New Mexico State University, University of Pittsburgh, Princeton University, the United States Naval Observatory, and the University of Washington.

References

Abazajian, K. et al. 2004, AJ, submitted
Annis, J. et al. http://home.fnal.gov/~annis
Bahcall, N.A. et al. 2003, ApJS, 148, 243
Baldry et al. 2004, ApJ, 600, 681
Balogh, M., et al. 2004, MNRAS, 348, 1355
Benson, A. J., Hoyle, F., Torres, F & Vogeley, M. S., 2003, MNRAS, 340, 160
Blanton, M., Cen, R., Ostriker, J.P., & Strauss 1999, ApJ, 522, 590
Blanton et al. 2003, ApJ, 594, 186
Blanton, M. R., Lin, H., Lupton, R. H., Maley, F. M., Young, N., Zehavi, I., & Loveday, J. 2003, AJ, 125, 2276
Dressler, A. 1980, ApJ, 236, 351
Eisenstein, D., et al. 2001, AJ, 122, 2267
Fukugita, M., Ichikawa, T., Gunn, J. E., Doi, M., Shimasaku, K., & Schneider, D. P. 1996, AJ, 111, 1748
Geller, M. J., & Huchra, 1989, Science, 246, 897
Goldberg, D. M., & Vogeley, M. S. 2004, ApJ, 605, 1
Goldberg, D. M., et al. 2004, in preparation
Gomez, P., et al. 2003, ApJ, 584, 210
Gott, J. R., et al. 2003, astro-ph/0310571
Goto, T., et al. 2003, MNRAS, 346, 601
Gunn, J. E., et al. 1998, ApJ, 116, 3040
Hao, L., Strauss, M. A., Rojas, R. R., & Vogeley, M. S. 2004, in preparation
Hogg, D. W., Finkbeiner, D. P., Schlegel, D. J., & Gunn, J. E. 2001, AJ, 122, 2129
Hoyle, F. & Vogeley, M. S., 2002, ApJ, 566, 641
Hoyle, F., Rojas, R., Vogeley, M.S., & Brinkmann, J. 2003, ApJ, submitted, astro-ph/0309728
Kauffmann, G., et al. 2004, MNRAS, submitted, astro-ph/0402030
Kim, R. S. J. et al. 2002, AJ, 123, 20
Lewis, I. et al. 2002, MNRAS, 673, 683
Lupton, R. H., et al. 2004, in preparation
Miller, C., et al. 2004, in preparation
Nichol, R.C. 2003, in Cluster of Galaxies, Carnegie Obs. Astro. Series 3, eds. J. Mulchaey, A. Dressler, & A. Oemler (Cambridge: Cambridge U. Press)
Pier, J. R., Munn, J. A., Hindsley, R. B., Hennessy, G. S., Kent, S. M., Lupton, R. H., & Ivezić, Ž. 2003, AJ, 125, 1559
Postman, M., & Geller, M. J. 1984, ApJ, 281, 95
Rojas, R., Vogeley, M. S., Hoyle, F. & Brinkmann, J., 2003, ApJ submitted, astro-ph/0307274

Rojas, R. 2004, Ph.D. thesis, Drexel University
Rojas, R., Vogeley, M. S., & Hoyle, F. 2004, in preparation
Scranton, R. et al. 2002, ApJS, 579, 48
Smith, J. A., et al. 2002, AJ, 123, 2121
Stoughton, C. et al. 2002, AJ, 123, 485
Strauss, M. A., et al. 2002, AJ, 124, 1810
Tanaka, M., Goto, T., Okamura, S., Shimasaku, K., & Brinkman, J. 2004, these proceedings
York, D. G. et al. 2000, AJ, 120, 1579

Outskirts of Galaxy Clusters: Intense Life in the Suburbs
Proceedings IAU Colloquium No. 195, 2004
A. Diaferio, ed.

DOI: 10.1017/S1743921304000055

X-ray galaxy clusters in the large-scale structure

H. Böhringer, P. Schuecker, N. Nowak, P. Popesso and M. Huber

Max-Planck-Institut für extraterrestrische Physik, D-85748 Garching, Germany

Abstract. In this contribution we investigate the connection between clusters of galaxies and their large-scale environment, with an emphasis on clusters which are well characterized by their X-ray emission. We show that this connection is so tight that clusters can be used as perfect tracers of the large-scale matter distribution and thus for cosmological tests. The correlation of the X-ray traced cluster mass and the optical luminosity of the galaxy content of clusters shows that the dark matter and galaxy distribution are tightly connected, but we also observe a scatter which is so far not well understood. We further explore the correlation of the galaxy population mix with the geometry of the large-scale structure features. On larger scales we also find correlations of the properties of galaxy clusters with the density of their large-scale structure environment.

1. Introduction

Galaxy clusters as the largest well defined cosmic building blocks have dynamical time scales of Gigayears not much smaller than the age of the Universe as a whole. Therefore many of their properties are still reminiscent of the density fluctuations which were set as initial conditions in the early Universe, triggering the formation of the large-scale structure (LSS). Galaxy clusters thus form an integral part of the LSS. As shown in several talks at this conference their internal properties are connected to the surrounding LSS and the cluster outskirts are controlled by the interplay between the clusters and their cosmic environment.

In this contribution we illustrate some of these connections. We will investigate how well galaxy clusters trace the underlying large-scale matter distribution in the Universe. In the following section we provide an account of our understanding how well the main mass component, the dark matter, follows the same distribution as the optically visible component, the galaxies. We find a good correlation of the light and mass distribution but also some scatter that is most probably larger than the uncertainties in the measurement.

The next degree of detail in the galaxy distribution then concerns the mixture of galaxy types. There is a well known morphology-density relation, with an increasing fraction of early type galaxies in the denser environments – a topic that was touched by several talks at this meeting. From a slightly different point of view we search for a correlation of the mixture of galaxy types with the geometry of the structure in which the galaxies are embedded. Is there a significant difference of galaxy types if one is looking for example into a filament, a sheet of galaxies, or into the field?

In the final section we make a similar study one step up in the structural hierarchy of the Universe: we investigate if there is a significant correlation of the average properties of galaxy clusters with the density of the LSS environment. In the contribution to this meeting by Manolis Plionis it was for example shown that there is a clear correlation of the position angles of cluster ellipticities and the orientation of LSS features. In our

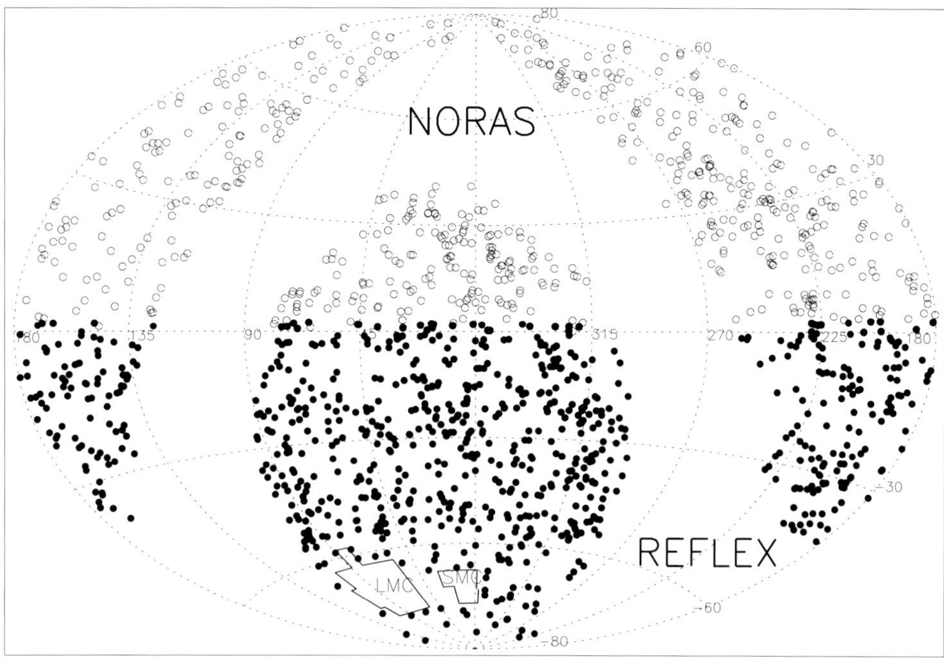

Figure 1. Sky distribution of the X-ray detected clusters in the ROSAT All-Sky Survey for a flux limit of 1.8×10^{-12} erg s^{-1} cm^{-2} in the South and 2×10^{-12} erg s^{-1} cm^{-2} in the North as far as they have been identified up to date.

study we find a correlation of the large-scale matter density with the frequency of cluster substructure and the X-ray luminosity function of galaxy clusters.

2. Galaxy clusters as tracers of the large-scale matter distribution

Structure formation theory predicts galaxy clusters to be ideal tracers of the large-scale matter distribution (e.g. Moscardini et al. 2000). As an additional advantage the clusters follow the amplitude variations in an amplified way (an effect called biasing) which facilitates the assessment of the LSS statistics in the presence of the unavoidable Poissonian noise. To make use of this very important property of the cluster population one needs to select clusters in a highly complete way above a given cluster mass threshold, which is an observational challenge. So far X-ray detections of clusters have been shown to be the best and most effective way for this selection (e.g. Henry et al. 1992). This is possibly due to the fact that the cluster mass and the X-ray luminosity are well correlated, however with enough scatter in the relation that it necessarily has to be taken into account (Reiprich & Böhringer 2002). The so far best basis for an X-ray cluster survey covering a large volume is the ROSAT X-ray All-Sky Survey (Trümper 1993). Within long term survey projects for the northern (NORAS Survey) and southern sky (REFLEX Survey) we have identified and obtained redshifts for more than 1400 X-ray galaxy clusters above a flux limit of 1.8×10^{-12} erg s^{-1} cm^{-2} in the South and 2×10^{-12} erg s^{-1} cm^{-2} in the North. Both samples are close to completion at these limits, but some further identification work is still necessary. The core parts of the REFLEX and NORAS survey

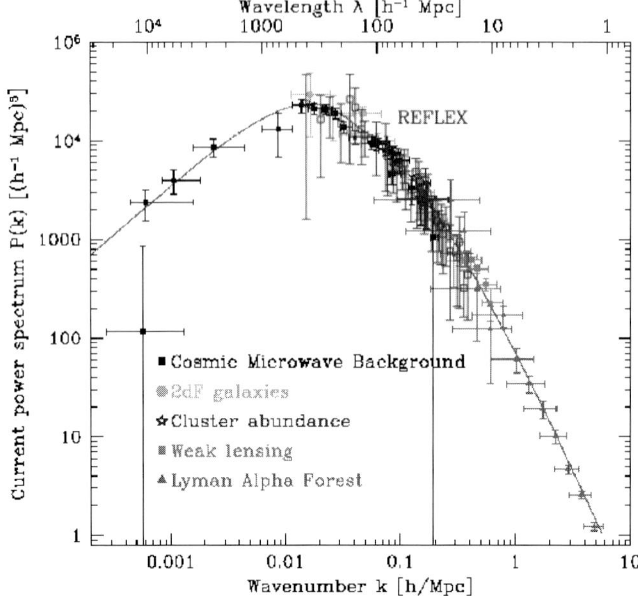

Figure 2. Power spectra for comoving coordinates of the mass density fluctuations as traced by various visible objects. Onto this plot, originally prepared by Max Tegmark, the REFLEX results from Schuecker et al. 2001a have been added.

are published (Böhringer et al. 2000, 2001, 2004). The sky distribution of the clusters from the combined catalogue at the stage of April 2004 is shown in Fig. 1.

This cluster sample, notably the core of the REFLEX Survey (REFLEX 1) has been used extensively for the assessment of the large-scale structure through the X-ray luminosity function (Böhringer et al. 2002), the two-point correlation function (Collins et al. 2001), and the power spectrum as well as Kurhunen-Loeve decomposition (Schuecker et al. 2001a, 2002a, 2002b, 2003). The results show that the X-ray luminosity function of clusters (a substitute of the mass function) and the spatial distribution of clusters (as described statistically by the density fluctuation power spectrum) is very closely bracketing the predicted statistical parameters describing the LSS in the concordance cosmology model (Turner 2001). The best constraint is obtained for the parameters $\Omega_m = 0.27 - 0.43$ and $\sigma_8 = 0.55 - 0.83$ (the amplitude parameter of the large-scale mass density fluctuations measured on a scale of $8h^{-1}$Mpc - Schuecker et al. 2002b).

Fig. 2 shows how the bias corrected cluster density fluctuation power spectrum fits to the other measures of LSS at different scales. There is a very good overall agreement in spite of the fact that the different measurements are assessing the LSS at different epochs and the proper cosmological model has to be applied to convert the different measurements into the correct comoving reference frame. Therefore, from these results we conclude that cluster trace the large scale structure in a perfect way as predicted by purely gravitationally driven CDM structure formation models.

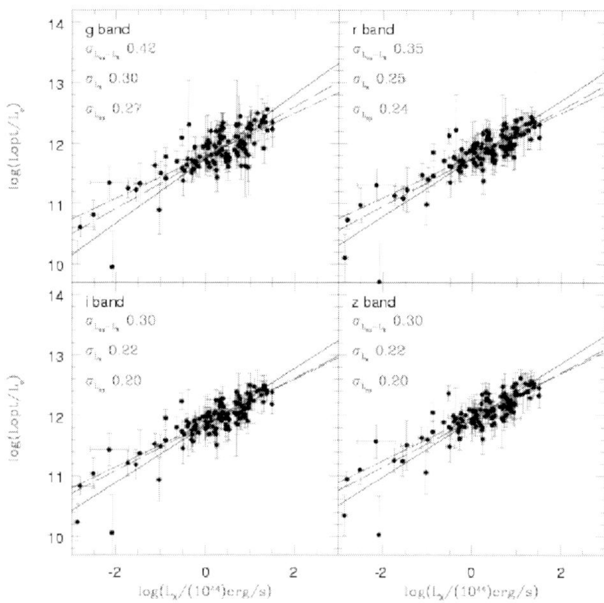

Figure 3. Correlation of the cluster X-ray luminosity with the optical luminosity measured in a $0.7h_{70}^{-1}$ Mpc aperture for the sample of ROSAT-SDSS clusters. The four panels show the results for four of the SDSS colour bands.

3. On the correlation of the galaxy and dark matter distribution

Large-scale galaxy redshift surveys, as the 2dF survey and the Sloan Digital Sky Survey (SDSS), have shown that the light is distributed in a very similar way as the dark matter and that the so-called biasing parameter for the galaxy distribution is close to one (e.g. Peacock et al. 2001). Is this also true for the galaxy and optical light content of galaxy clusters and their surroundings? This question was also raised in other talks at this conference, e.g. in the contribution by K. Rines. The size, precision, and homogeneity of the SDSS provides a new uniquely comprehensive approach to this quest. Based on the galaxy clusters detected in the NORAS and REFLEX surveys overlapping with the SDSS sky region we have performed a systematic comparison of the X-ray and optical properties of galaxy clusters added by a smaller sample of groups that have also been studied in X-rays with ASCA. A first important question to be answered by this study is: in which way can the measurement of the cluster light and the X-ray luminosity determination be combined such that the correlation of the two cluster properties shows the least scatter? That is, how can we best determine the properties in the optical to provide the highest accuracy in predicting the X-ray luminosity? The simplest and most straightforward way to determine an optical light measure is the integration of the background subtracted light of the cluster in a fixed metric aperture. A survey of different aperture parameters and different optical colors shows that the least scatter in the optical/X-ray correlation is produced by an aperture size of $0.7h_{70}^{-1}$ Mpc and the z-band (the reddest band) of the SDSS. The scatter in the i-band is similarly low (Popesso et al. 2004). Fig. 3 shows the correlation for the best correlating aperture for the 4 colors of the SDSS. The scatter in the relation in both directions is about 60% for both z-band and the i-band. We estimate that not much more than half of this scatter is due to uncertainties in the X-ray and optical luminosity measurements. Thus a significant fraction of this scatter is intrinsic.

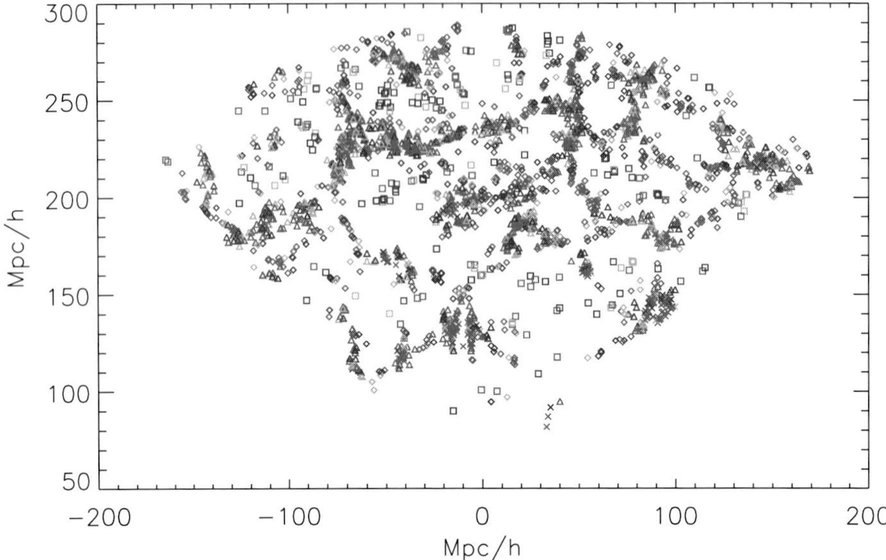

Figure 4. Distribution of galaxies with measured redshifts in a slice through the SDSS survey. The galaxies are sorted into scaling indices bins marked by symbols (crosses: $\alpha \leqslant 1.2$, triangles: $\alpha = 1.2 - 1.7$, diamonds: $\alpha = 1.7 - 2.4$, and squares: $\alpha \geqslant 2.4$) and the colors reflect the actual galaxy colors (blue: $g - r \leqslant 0.65$, green: $g - r = 0.65 - 0.8$, red: $g - r \geqslant 0.8$) from Huber et al. (2004).

First tests of a similar correlation based on primarily optically detected clusters give the surprising result that the scatter in the X-ray/optical correlation is much larger and in particular there are a lot of cases of optically luminous objects with low X-ray luminosity which fall outside the scattering range observed in the study of primarily X-ray selected clusters. This study is in progress and will hopefully shed more light on the not well understood difference between optically selected and X-ray selected cluster samples.

4. The galaxy population mix and large-scale geometry

While on the one hand we observe a tight correlation of the light and dark matter density, there is a surprisingly striking change in the galaxy population if we go from the low density field into the dense cluster environment (e.g. Dressler et al. 1997). Aspects of this morphology-density relation was one of the widely discussed topics at the meeting. Looking at the very pronounced bubbly and filamentary structure of our Universe one is led also to wonder if there is also a correlation of galaxy morphology and the type of structure the galaxies are embedded. Does it matter if the galaxies under investigation populate a cluster, a sheet, or a filament?

To determine the geometry of the local LSS we use scaling indices. These measures are equivalent to the determination of the local two-point correlation properties of a selected galaxy with the surrounding galaxies within a sphere with given radius. In addition a special radial weighting function may be included in the correlation analysis (Raeth et al. 2002). The scaling index $\alpha = 3 - \gamma$ (where γ is the power law index of the two-point correlation function) provides an indication of the dimension of the structure surrounding the selected galaxy, for example 1 for a filament, 2 for a sheet and less than 1 for a cluster (Huber et al. 2004). Fig. 4 illustrates the assignment of scaling indices to galaxies within the LSS. At present this method is tested on a slice through the SDSS survey -

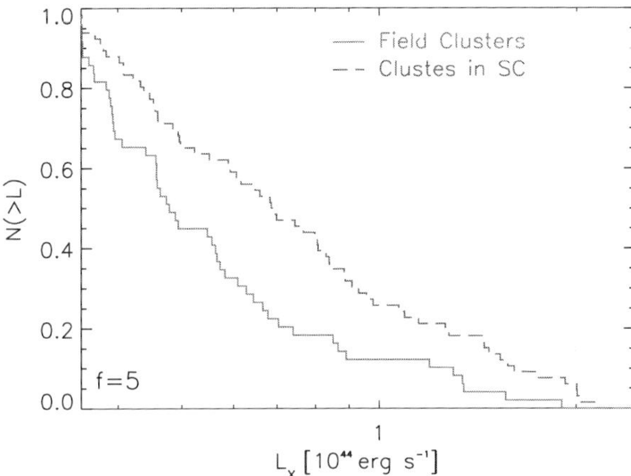

Figure 5. Cumulative X-ray luminosity distribution of galaxy clusters from the extended REFLEX sample assigned either to superclusters or to the field (from Nowak et al. 2004).

a pseudo-two-dimensional data set. There is a clear correlation between the geometry of the LSS features and the mean galaxy color, but most of this just reflects the well-known morphology density relation. However, we also have a first indication of a geometry - galaxy color correlation if the analysis is carried out for regions of similar density. A clearly significant result is expected if we apply this method to the bulk of the SDSS data in the near future.

5. The dependence of the cluster properties on the large-scale structure environment

While, as described above, it is well known that the galaxy population changes with the large-scale structure matter density, we know less if this is also true for the population of galaxy clusters. In the first study (Nowak et al. 2004) a supercluster catalogue for the 842 galaxy clusters of the extended REFLEX survey was constructed by means of the friends-of-friends algorithm. The most convincing supercluster structures were found for a linking length corresponding to a galaxy cluster overdensity in the range of 5 – 10. For this assignment of the galaxy clusters to belong either to the field or a supercluster, the luminosity distribution is investigated as shown in Fig. 5. There is a clear trend that clusters in superclusters are more luminous on average than clusters in the field. Detailed studies show that this result is significant and not an artefact of selection effects (Nowak 2004, Nowak et al. 2004). We attribute this luminosity excess either to a luminosity boost in merging clusters or to a more evolved luminosity function in the high density environment of superclusters.

In the second study the frequency of substructure was determined for a sample of clusters from the REFLEX and northern BCS sample (Ebeling et al. 1998). Three measures of substructure were used, the β-test (sensitive to deviations from mirror symmetry), the Lee statistic (most sensitive to bimodal structure), and the Fourier elongation analysis (sensitive to elongations and substructure - Schuecker et al. 2001b). As shown in Fig. 6 at least two of the three substructure measures show a clear signature that substructured clusters are more frequently observed in denser regions. That is, merging of clusters occurs with a higher frequency in denser environments.

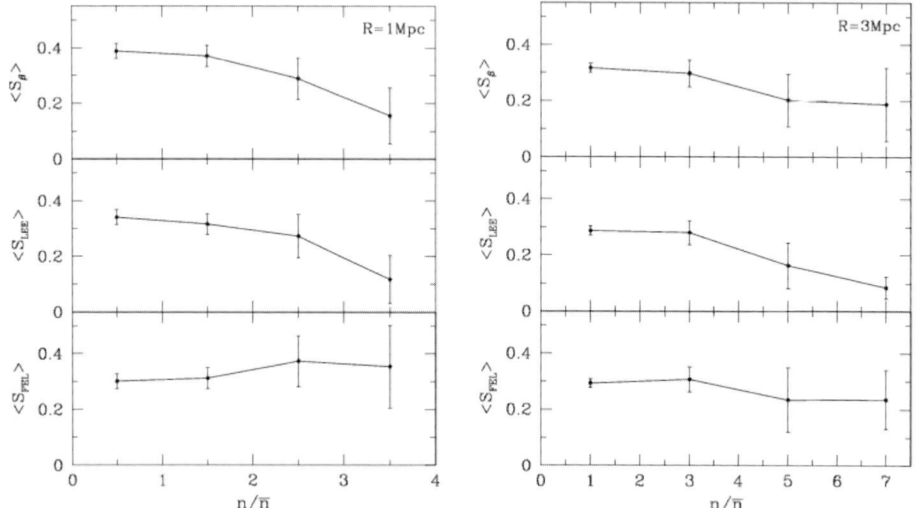

Figure 6. Average significance of clusters showing substructure for clusters from the REFLEX and BCS surveys binned according to the cluster density of the environment. Three measure of substructure are used, the β-test, Lee statistic, and the Fourier elongation test, respectively (Schuecker et al. 2001b).

Acknowledgements

We thank the ROSAT team and the REFLEX Survey for the collaboration which made several of these studies possible.

References

Böhringer, H., Voges, W., Huchra, et al. 2000, ApJS, 129, 435
Böhringer, H., Schuecker, P., Guzzo, L., et al., 2001, A&A, 369, 826
Böhringer, H., Collins, C.A., Guzzo, L., et al. 2002, ApJ, 566, 93
Böhringer, H., Schuecker, P., Guzzo, L., et al., 2004, A&A in press
Collins, C.A., Guzzo, L., Böhringer, et al., 2000, MNRAS, 319, 939
Dressler, A., Oemler, A., Couch, W.J., et al., 1997, ApJ, 490, 577
Ebeling, H., Edge, A.C., Böhringer, H., et al., 1998, MNRAS, 301, 881
Henry, J.P., Gioia, I.M., Maccacaro, T., 1992, ApJ, 386, 408
Huber, M., Schuecker, P., Böhringer, H., 2004, A&A (to be submitted)
Moscardini, L., Matarrese, S., Lucchin, F., Rosati, P., 2000, MNRAS, 316, 283
Nowak, N., 2003, Diplom Thesis, Ludwig-Maximilians-Universität München
Nowak, N., Schuecker, P., Böhringer, H., 2004, A&A (to be submitted)
Peacock, J.A., Cole, S., Norberg, P., et al., 2001, Nature, 410, 169
Popesso, P., Böhringer, H., Brinkmann, J., Voges, W., York, D.G., 2004, A&A in press – astro-ph/0403354
Raeth, C., Bunk, W., Huber, M.B., et al., 2002, MNRAS, 337, 413
Reiprich, T.H., Böhringer, H., 2002, ApJ, 567, 716
Schuecker, P., Böhringer, H., Guzzo, L., 2001a, A&A, 368, 86
Schuecker, P., Böhringer, H., Reiprich, T.H., Feretti, L., 2001b, A&A, 378, 408
Schuecker, P., Guzzo, L., Collins, C.A., Böhringer, H., 2002a, MNRAS, 335, 807
Schuecker, P., Böhringer, H., Collins, C.A., Guzzo, L., 2002b, A&A, 398, 867
Schuecker, P., Caldwell, R.R., Böhringer, H., et al., 2003, A&A, 402, 53
Trümper, J., 1993, Science, 260, 1769
Turner, M., 2001, astro-ph/0108103

Outskirts of Galaxy Clusters: Intense Life in the Suburbs
Proceedings IAU Colloquium No. 195, 2004
A. Diaferio, ed.

DOI: 10.1017/S1743921304000067

The large-scale environment of groups and clusters of galaxies

Manolis Plionis[1,2]

[1] Institute of Astronomy & Astrophysics, National Observatory of Athens, P.Penteli, 152 36 Athens, Greece

[2] Instituto Nacional de Astrofísica Óptica y Electrónica, AP 51 y 216, 72000, Puebla, Pue, México

Abstract. It appears that the dynamical status of clusters and groups of galaxies is related to the large-scale structure of the Universe. A few interesting trends have been established:
(1) *The Cluster Substructure - Alignment Connection* by which clusters show a strong correlation between their tendency to be aligned with their neighbors and their dynamical state (as indicated by the existence of significant substructres).
(2) *The Cluster Dynamics -Cluster Clustering Connection* by which dynamically young clusters are more clustered than the overall cluster population.
(3) *The Cluster- Supercluster Alignment Connection* by which clusters of galaxies show a statistical significant tendency to be aligned with the projected major axis orientation of their parent supercluster.
(4) *The Galaxy Alignment - Cluster Dynamics Connection* by which red-sequence cluster bright galaxies show a significant trend to be aligned with their parent cluster major axis, especially in dynamically young clusters.
(5) *The Group Richness - Shape Connection* by which groups of galaxies are flatter the poorer they are.

These are strong indications that clusters develop in a hierarchical fashion by anisotropic merging of smaller units along the large-scale filamentary structures within which they are embedded.

---

1. Introduction

In the framework of the hierarchical model for the formation of cosmic structures, galaxy clusters are supposed to form by accretion of smaller units (galaxies, groups, etc.). After the epoch of mass aggregation (which depend on the cosmological model), violent relaxation processes will tend to alter the cluster phase-space configuration producing 'regular', quasi-spherical, clusters having a smooth density profile.

Over the last decade, due to the increased spatial resolution in X-ray imaging (ROSAT, XMM-*Newton*, CHANDRA) and to the availability of wide-field cameras, many of the previously thought "regular" clusters have shown to be clumpy to some level, a fact that could have important consequences for structure formation theories, since the present fraction of dynamically young clusters, as well as the rate of cluster evolution (as measured for example by their luminosity and temperature functions and their morphology), are cosmology dependent (e.g., Richstone, Loeb & Turner 1992, Evrard et al. 1993, Lacey & Cole 1996). Detailed optical and X-ray studies have yielded evidence for significant amounts of subclustering within local and distant clusters (e.g. Buote & Tsai 1996; Plionis & Basilakos 2002; Jeltema et al. 2004) indicating that possibly a large fraction is still forming at the present time. Hints do exist for a very recent (within the last Gyr) dynamical evolution of the cluster population (Melott et al. 2001; Plionis 2002, see however

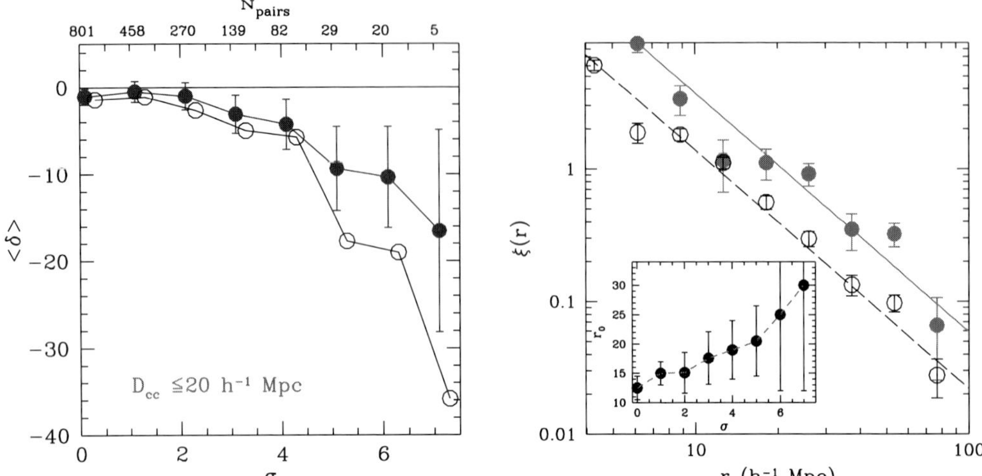

Figure 1. LEFT PANEL: Alignment signal as a function of substructure significance level. Filled circles correspond to nearest-neighbors and open circles to all neighbors within 20 h^{-1} Mpc. RIGHT PANEL: The APM cluster correlation function for all the APM clusters (open symbols) and for a sample with significant substructure (filled symbols). The lines represent the best $(r/r_\circ)^{-1.8}$ fit with $r_\circ \simeq 12$ and $\simeq 20\ h^{-1}$ Mpc respectively. In the insert we show the correlation length as a function of substructure significance level.

Floor et al. 2003). Note however that the existence of substructure does not necessarily mean that the corresponding cluster is dynamically young due to the ambiguity of cluster post-merging relaxation times (see discussion in Plionis 2001).

Furthermore, the dynamical evolution of member galaxies and of the ICM gas is also an open issue. Star formation seems to be active in clusters showing substantial substructure and velocity gradients, as expected if a recent merger has taken place. For example, the fraction of blue galaxies is strongly correlated with cluster ellipticity (Wang & Ulmer 1997), while ellipticity is strongly correlated with the dynamical state of the cluster (e.g. Kolokotronis et al. 2001). It appears that the violent merging events trigger star-formation, possibly through a multitude of different mechanisms; for example, the excess number of galaxy-galaxy interactions, the rapid variation of the cluster gravitational field (Bekki 1999), etc. An interesting observable that may be related to the dynamics of clusters is their tendency to be aligned with their nearest neighbor as well as with other clusters that reside in the same supercluster (e.g. Bingelli 1982; Plionis 1994). Analytical (Bond 1986) and numerical work (e.g. West et al. 1991, van Haarlem & van de Weygaert 1993, Splinter et al. 1999, Onuora & Thomas 2000; Faltenbacher et al. 2002; Knebe et al. 2004) have shown that such alignments occur in many hierarchical clustering models and are probably the result of an interesting property of Gaussian random fields which is the "cross-talk" between density fluctuations on different scales (e.g. West 1994).

2. Environmental trends

Our aim is to investigate whether there is any relation between the cluster and group large-scale environment and their internal dynamical state. To this end we have analysed APM and Abell clusters (in total more than 1200 clusters) as well as the UZC-SSRS2 (Ramella et al. 2002) and 2DFGRS (Eke et al. 2004) samples of groups of galaxies.

In order to realize our study we need to define, in an objective manner, the dynamical state of the clusters. Evrard et al. (1993) and Mohr et al. (1995) have suggested as

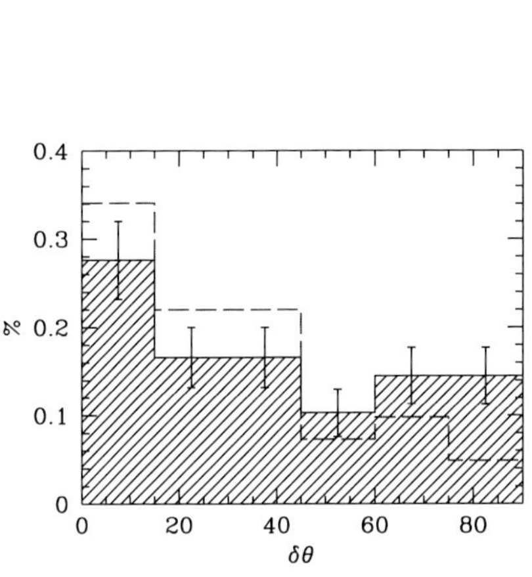

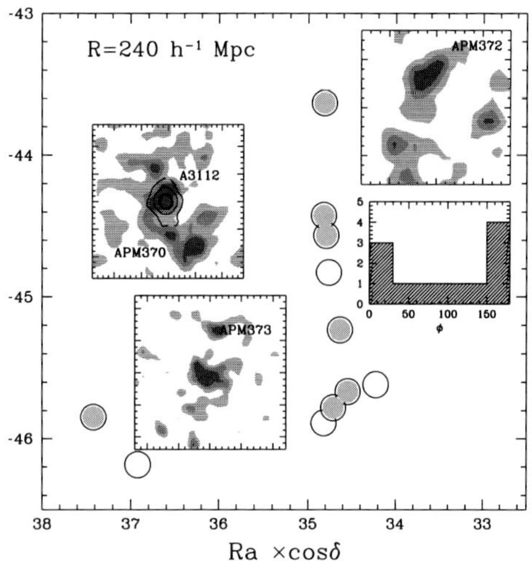

Figure 2. LEFT PANEL: Frequency distribution of the misalignment angle between cluster members and their parent supercluster orientations. The dashed line corresponds to superclusters defined with linking radius of 30 h^{-1} Mpc while the hatched distribution to that with linking radius of 20 h^{-1} Mpc. RIGHT PANEL: An example of a filamentary APM supercluster containing A3112, A3104, A3111 as well as 9 poorer APM clusters, identified using a linking radius of 12 h^{-1} Mpc. Filled dots represent clusters with significant substructure. In the inserts we show the smooth density distribution for three of the clusters (for the case of A3112 we also show its ROSAT-PSPC X-ray contours) as well as the frequency distribution of cluster position angles. The preferred cluster major axis alignment with the projected orientation of the supercluster is evident.

an indicator of cluster substructure the *centroid-shift* (*sc*) which is defined as the shift of the cluster center-of-mass position as a function of density threshold above which it is estimated. The significance, σ, of such centroid variations to the presence of background contamination, random density fluctuations and small galaxy number effects can be estimated using Monte Carlo cluster simulations in which, by construction, there is no substructure (see Plionis 2001). Kolokotronis et al. (2001) calibrated various substructure measures using APM galaxy data and ROSAT pointed observations for 22 Abell clusters and found that in most cases using either X–ray or optical data one can identify substructure unambiguously. Their conclusion was that a large and significant *centroid-shift* is a clear indication of cluster substructure.

Furthermore, in order to investigate the alignment between cluster, group or galaxy orientations, we use the relative position angle between pairs, $\phi_{i,j}$. A significant deviation from the isotropic expectation can be quantified by the measure: $\delta = \sum \phi_{i,j}/N - 45$ (Struble & Peebles 1995), which for an isotropic distribution, has a mean value $\langle\delta\rangle \simeq 0$, and a standard deviation $\sigma = 90/\sqrt{12N}$. A significantly negative (positive) value of δ would indicate alignment (misalignment).

Below we present the main results of a variety of studies that indicate a strong correlation between the dynamical state of galaxy structures (groups and clusters) and their large-scale environment.

2.1. *Cluster substructure - Alignment connection*

Plionis & Basilakos (2002), using a sample of $\sim$ 900 APM clusters, correlated the alignment signal between cluster neighbors, $\langle\delta\rangle$, with the existence of cluster substructure, as measured by the deviation from the expected density fluctuations in units of σ (e.g.

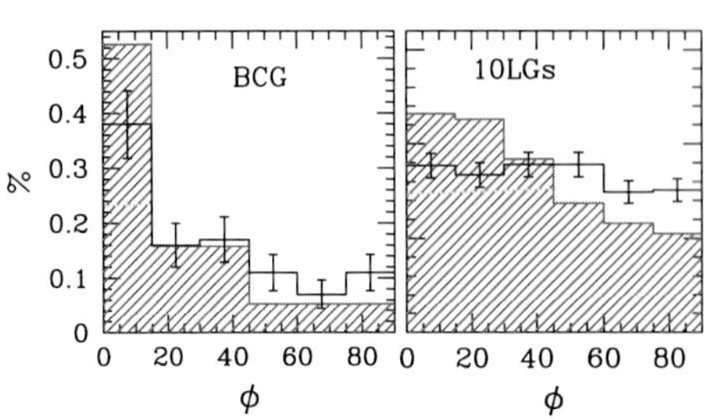

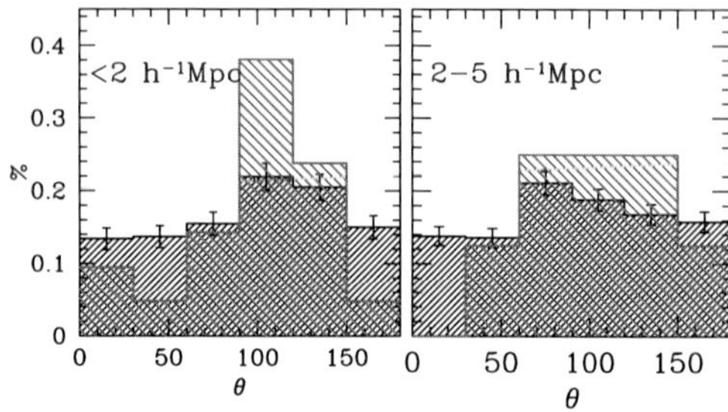

Figure 3. LEFT PANEL: The distribution of misalignment angles ϕ between the galaxy and cluster orientations within superclusters for the $v_\sigma < 900$ km/sec clusters (black lines) and for those with $v_\sigma \geqslant 900$ km/sec (red hatched histogram). RIGHT PANEL: Histogram of galaxy (black lines) and group (red lines) position angle distribution in Abell 521 for two indicated distance shells.

Kolokotronis et al. 2001; Plionis 2001). In the left panel of Fig. 1 we present $\langle\delta\rangle$ between cluster nearest-neighbors (filled dots) and between all pairs (open dots) with pair separations $< 20\ h^{-1}$ Mpc. Evidently, there is a strong correlation between the strength of the alignment signal and the substructure significance level. This result confirms statistically the analysis of West et al. (1995) based on EINSTEIN X-ray data of ~ 90 clusters.

2.2. *Cluster substructure - Clustering connection*

Where do clusters with significant substructure reside? The spatial 2-point cluster correlation function for cluster samples with different substructure significance levels show a strong trend. In the right panel of Fig. 1 we plot the correlation length, $r_\circ$, as a function of the cluster substructure significance level, which is clearly an increasing function of σ. The conclusion of this analysis is that clusters showing evidence of dynamical activity reside in high-density environments. Our results are in agreement with a similar environmental dependence found in the REFLEX and BCS X-ray cluster sample (Schuecker et al., 2001) and for cooling flow clusters with high mass accretion rates (Loken, Melott & Miller 1999).

2.3. *Cluster-supercluster alignment connection*

A further interesting question regarding environmental effects on large scales is whether clusters are also aligned with the orientation of their parent supercluster. We define APM superclusters by using a friends-of-friends algorithm and estimate for each supercluster the misalignment angle, $\delta\theta$, between its projected orientation and the mean position angle of its member clusters. In Fig. 2 we present the frequency distribution of $\delta\theta$ for two different supercluster catalogues (based on linking radii of 20 and 30 h^{-1} Mpc, respectively). The significant excess of small $\delta\theta$'s is evidence that indeed clusters do show preferential alignments with the orientation of their parent superclusters. As an example, in the right panel of Fig. 2 we present the projected distribution of clusters in a dense and filamentary APM supercluster. Embedded in the plot are the smoothed galaxy density distributions of a few member clusters. It is evident that the cluster position angle distribution reflects the projected orientation of the supercluster.

2.4. *The galaxy alignment - Cluster dynamics connection*

Analysing the DPOSS images of 303 Abell clusters (with $z \leqslant 0.12$), Plionis et al. (2003) found significant alignments between the projected major axis of the 10 brightest galaxies (excluding the BCG) and their parent cluster position angle, mostly in high-density

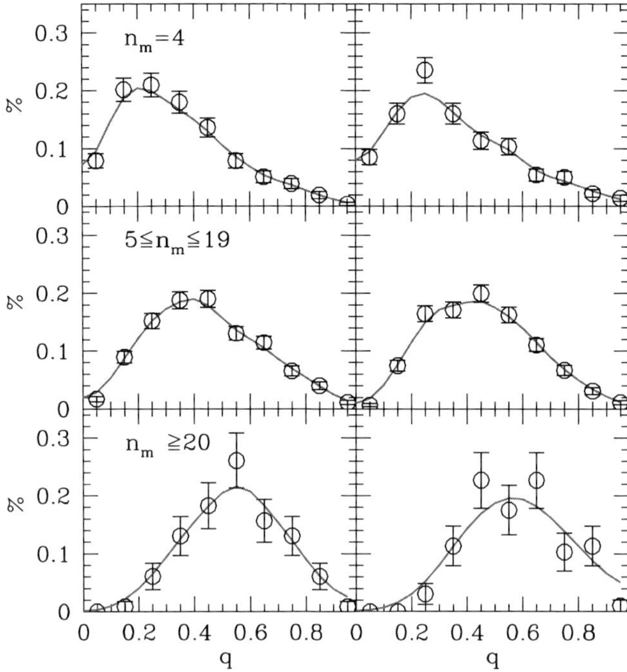

Figure 4. The apparent 2DFGRS group axial ratio distributions for different group membership, n_m, and for the northern (NGP) and southern (SGP) subsamples respectively The solid line is the smooth fit from the nonparametric kernel estimator.

environments. This alignment signal is stronger for higher velocity dispersion clusters, which in our case (based on a variety of criteria) appear also to be dynamically young merging clusters (see left panel of Fig. 3). This points strongly in the direction of galaxy alignments being correlated with their parent cluster dynamical youth and local density (see also West & Blakeslee 2000; Knebe et al. 2004). We have also studied in detail, using very deep WFI images, the highly unrelaxed Abell cluster (A521 at $z = 0.25$) and find that such alignments (see right panel of Fig. 3) are present also in fainter red-sequence galaxies as well as in the orientation of groups of galaxies in the extended environment around the cluster (see also Kitzbichler & Saurer 2003).

2.5. *Group shape-richness connection*

Plionis, Basilakos & Tovmassian (2004) and Plionis & Basilakos (2004) analysed two large samples of groups (the UZC-SSRS2 and the 2DFGRS-2PIGG samples) within a roughly volume-limited region ($cz \leqslant 5500$ and $cz \leqslant 30000$ km/sec for the two samples respectively). In Fig. 4 we present the projected, on the plane of the sky, axial ratio distribution for three group richness classes, while in Fig. 5 we show the corresponding inverted axial ratio distribution assuming that groups are intrinsically either prolate or oblate spheroids.

There is an obvious group richness-flatness relation, seen in both projected and intrinsic axial ratio distributions, with group flatness decreasing with richness. We also find a weak (Spearman correlation coefficient $\mathcal{R} = -0.13$) but significant correlation ($\mathcal{P} < 10^{-6}$), between the velocity dispersion of groups and their flattening, with flatter systems having lower velocity dispersions. Such a correlation as well as the increase of the group sphericity

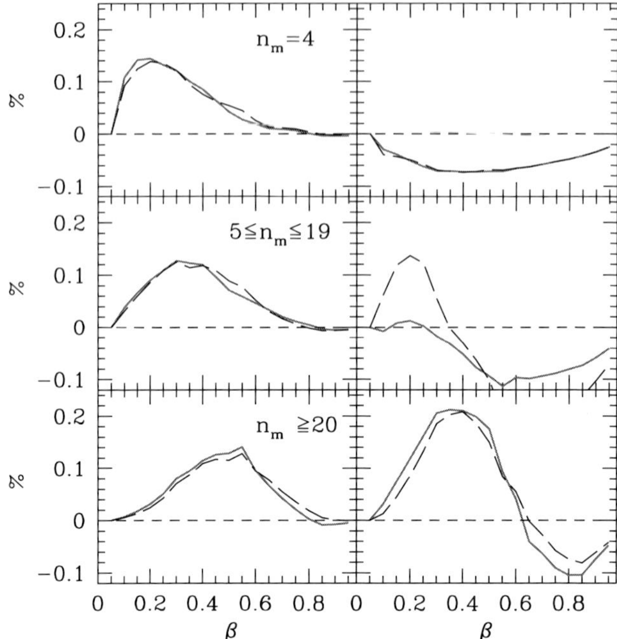

Figure 5. Right panel: The intrinsic distribution of group axial ratios for the NGP (continuous line) and SGP (dashed line) subsamples assuming that they are either prolate (left panel) or oblate (right panel) spheroids. The unphysical negative values of the latter model indicates its failure to represent the true three-dimensional shape of groups.

with richness, which extends also to clusters, could be explained as an indication of a higher degree of virialization, which is expected to be more rapid in more massive systems. Furthermore, we find that groups of galaxies are prolate-like systems which is in agreement with the shape of clusters and superclusters of galaxies (e.g. Cooray 2000; Basilakos, Plionis & Maddox 2000 and references therein). This is the expected shape for cosmic structures that form in quasi one-dimensional large-scale structures (filaments). Other interesting environmental effects that have been found is the decrease of the group velocity dispersion as their distance from nearby clusters increases (e.g. Ragone et al 2004).

3. Conclusions

We have presented evidence, based on the Apm and Abell cluster samples as well as the Uzc-Sssrs2 and 2dfgrs group samples, that there is a strong link between the dynamical state of clusters/groups of galaxies and their large-scale environment. Dynamically young clusters are significantly more aligned with their nearest neighbors and they are also much more spatially clustered. Clusters belonging to dense superclusters are preferentially aligned with their parent supercluster projected orientation; bright galaxies, in dynamically young clusters, are also orientated along their parent cluster major axis. These coherent orientation effects, from the scale of galaxies to that of superclusters, support the hierarchical formation scenario in which clusters form by anisotropic merging along the large-scale filamentary structures within which they are embedded.

Acknowledgements

This work has been partly funded within the framework of the program 'Promotion of Excellence in Technological Development and Research', project *'X-ray Astrophysics with ESA's mission XMM'*, as well as by the Mexican Government grant No CONACyT-2002-C01-39679.

References

Bingelli, B., 1982, A&A, **250**, 432
Bekki, K. 1982, ApJ, **510**, L15
Bond, J.R. 1986 *Galaxy Distances and Deviations from the Hubble Flow, eds. Madore, B.F., Tully, R.B., (Dordrecht: Reidel)*, 255
Buote, D. & Tsai, J. 1996, ApJ, **458**, 27
Cooray, R.A., 2000, MNRAS, **313**, 783
Eke, V.K., et al., 2004, MNRAS, **348**, 866
Evrard A.E., Mohr J.J., Fabricant D.G., Geller M.J.,1993, ApJ, 419, L9
Faltenbacher, A., Gottlöber, S., Kerscher, M., Müller, V., 2002, A&A, **395**, 1
Floor, S.N., Melott, A.L., Miller, C.J., Bryan, G.L., 2003, ApJ, **591**, 741
Jeltema, T.E., Canizares, C.R., Bautz, M.W., Buote, D.A., 2004, in *"Multiwavelength Cosmology"* proceedings, Ed. M.Plionis, Kluwer Academic Publishers
Knebe, A., Gill, S.P.D., Gibson, B.K., Lewis, G.F., Ibata, R.A., Dopita, M.M., 2004, ApJ, *in press*, astro-ph/0311202
Kitzbichler, M.G. & Saurer, W., 2003, ApJ, **590**, L9
Kolokotronis, V., Basilakos, S., Plionis, M., Georgantopoulos, I., 2001, MNRAS, **320**, 49
Lacey, C., Cole, S., 1996, MNRAS, **262**, 627
Loken, C., Melott, A.L., Miller, C.J., 1999, ApJ, **520**, L5
Mohr, J.J., Evrard, A.E., Fabricant, D.G., Geller, M.J., 1995, ApJ, **447**, 8
Onuora, L.I., Thomas, P.A., 2000, MNRAS, **319**, 614
Plionis M., 1994, ApJS, **95**, 401
Plionis M., 2001, in *Clusters and the High-Redshift Universe observed in X-rays* XXI$^{\rm th}$ Moriond Astrophysics Meeting, eds. Neumann et al., *in press*
Plionis M. & Basilakos, S., 2002, MNRAS, 329, L47
Plionis, M., Benoist, C., Maurogordato, S., Ferrari, C. & Basilakos, S., 2003, ApJ, **594**, 144
Plionis, M., Basilakos, S. & Tovamssian, H., 2004, MNRAS, *in press*
Plionis, M., Basilakos, S., 2004, *in preparation*
Richstone, D., Loeb, A., Turner, E.L., 1992, ApJ, **393**, 477
Ragone, CJ., Merchán, M., Muriel, H. & Zandivarez, A., 2004, astro-ph/0402155
Ramella, M., Geller, M. J., Pisani, A., & da Costa, L., 2002, AJ, **2976**, 123
Rizza E., Burns J. O., Ledlow M. J., Owen F. N., Voges, W., Blito M., 1998, MNRAS, **301**, 328
Sarazin, C.L., 2001, in *Merging Processes in clusters of Galaxies*, eds. Feretti, L., Gioia, M., Giovannini, G., (Dordrecht: Kluwer).
Schindler S., 1999, Vulcano Workshop 1999, *Multifrequency Behaviour of High Energy Cosmic Sources*, eds. Giovanelli F., Sabau-Graziati L.
Schuecker P., Böhringer H., Reiprich T. H., Feretti L., 2001, A&A, **378**, 408
Splinter, R.J., Melott, A.L., Linn, A.M., Buck, C., Tinker, J., 1997, ApJ, **479**, 632
Struble, M.F., Peebles, P.J.E., 1985, AJ, **90**, 582
van Haarlem, M., van de Weygaert, R., 1993, ApJ, **418**, 544
Wang, Q.D. & Ulmer, M.P., 1997, MNRAS, **292**, 920
West, M.J., Villumsen, J.V., Dekel, A., 1991, ApJ, **369**, 287
West, M.J., 1994, MNRAS, 268, 79
West, M.J., Jones C. & Forman W., 1995, ApJ, **451**, L5
West, M.J. & Blakeslee, J.P., 2000, ApJ, **543**, L27

Outskirts of Galaxy Clusters: Intense Life in the Suburbs
Proceedings IAU Colloquium No. 195, 2004
A. Diaferio, ed.

DOI: 10.1017/S1743921304000079

Superclusters and voids in the Sloan DSS

Volker Müller and Christian Maulbetsch

Astrophysikalisches Institut, Potsdam 14482, Germany email: vmueller@aip.de

Abstract. We analyze quasi-2-dimensional slices of the SDSS EDR. Gaussian smoothing with weighting by the inverse of the selection function provides 2D density fields across the full survey depth. Superclusters (SC) are characterized by a percolation algorithm in the large-scale smoothed field. Group candidates are identified with density maxima in the small-scale smoothed field. The group mass function depends on the SC environmental density. We derive the shape-dependent 3-point correlation function and the void size distribution. These are well reproduced by the galaxies identified in high-resolution ΛCDM simulations.

---

1. Cosmic density fields

Establishing large-scale cosmic density fields from the observed galaxy distribution was an early aim of analyzing galaxy redshift surveys (Saunders et al. 1991). Averaging over local random inhomogeneities, the density field provides genuine measures of the complex pattern of sheets and filaments in the universe comprising the 'cosmic web' (Bond, Kofman, & Pogosyan 1996). Our analysis is based on the ongoing SLOAN Redshift Survey intended to derive one million galaxy redshifts within one quarter of the sky. Here we analyze the first preliminary data release of 34 000 galaxies in two 2.5° thick stripes along the celestial equator in the Northern (top) and Southern (bottom) hemispheres. This part of the survey is 95% complete for red magnitudes $13.0 \leqslant r^* \leqslant 17.7$. We constrain the analysis to a depth of $z = 0.2$ that corresponds to a limiting distance of $570\ h^{-1}$Mpc $(h \approx 0.7)$.

The cosmic density field in Figure 1 was Gaussian smoothed with the galaxy selection function as inverse weight to get an approximately uniform coverage of the survey volume (Einasto et al. 2003). We identify 43 superclusters as the largest nonlinear structures in the survey. The most massive system in the North is the Supercluster 126 containing 7 Abell clusters and 6 X-ray clusters from the ROSAT bright survey (Schwope et al. 2000). Most superclusters form complex multi-branching filaments with one or two strong central concentrations. The richest system in the South is the Pegasus-Pisces supercluster 3 with 9 Abell clusters.

The high-resolution density field provides a detailed insight into the cosmic web. All supercluster concentrations are subdivided into interconnected filamentary branches with galaxies like pearls on a string. We identify about 5000 galaxy groups, for which we derive harmonic radii and the velocity dispersions to estimate the virial mass for gravitationally bound systems. We find that the upper limit of the group masses is about 5 times higher in the central supercluster regions. Furthermore, we find that the group masses scatter over a larger range in overdense regions due to a further evolution of the mass hierarchy. The environmental dependence of group and cluster properties was established in high-resolution numerical simulations of galaxy formation.

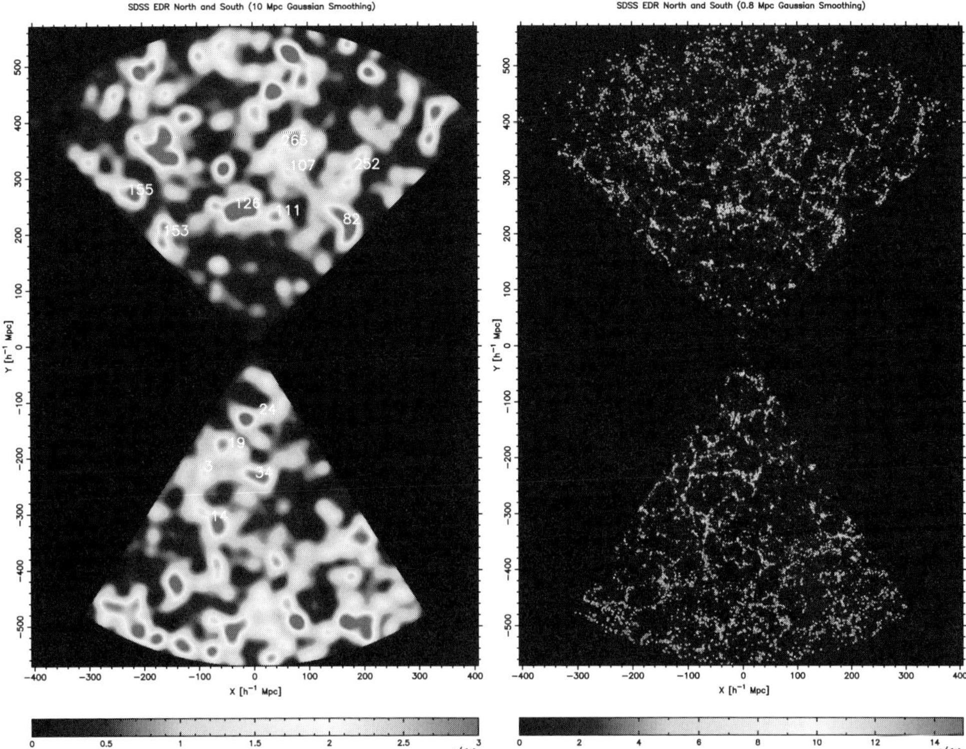

Figure 1. SDSS EDR density field with 10 h^{-1}Mpc (left panel) and 0.8 h^{-1}Mpc (right panel) smoothing. North is on top. Numbers correspond to superclusters from Einasto et al. (2001).

2. The 3-point correlation function and void size distribution

We quantify the filamentarity of the galaxy distribution in deriving the reduced 3-point correlation function Q shown in Figure 2, left panel. It is a reliable measure of hierarchical clustering, sensitive to the morphology of structures and accessible to gravitational perturbation theory. The four panels show the redshift space correlation function for galaxies sitting on the edges of triangles with sides s, us, and $(u+v)s$ ($u > 1$, $0 < v < 1$). The increase with v is characteristic for filaments. The solid lines show theoretical fits derived empirically for LCRS galaxies (Jing & Börner 1998) that are in agreement with our data (cf. also Kayo et al. 2004 for a recent analysis of the SDSS first data release). We compare our results with high-resolution DM-simulations (256^3 particles with spatial resolution of a few kpc) in a 60 h^{-1}Mpc box size for a concordance ΛCDM-model. There is a strong influence of redshift space corrections that remove almost all u-dependence of the reduced 3-point amplitude Q but well preserves the increase of Q with v (anisotropy of filamentarity of the clustering) predicted from gravitational perturbation theory. We find a weak but significant increase of this filamentarity effect for blue versus red galaxies. In the simulations we find a stronger anisotropy of Q for halos identified at high redshifts ($z = 3 \ldots 1$) than at the present epoch. Our results indicate that the amplitude of the 3-point function Q can be reproduced in high-resolution ΛCDM-simulations and is not overpredicted by a factor of two as claimed by Jing & Börner (1998).

We identify a sample of voids covering most of the survey area in seven volume-limited data sets from the SDSS-EDR with the 2D-void finder of Arbabi-Bidgoli & Müller (2002). The size distribution weighted with the void area is shown in Figure 2 as function of the

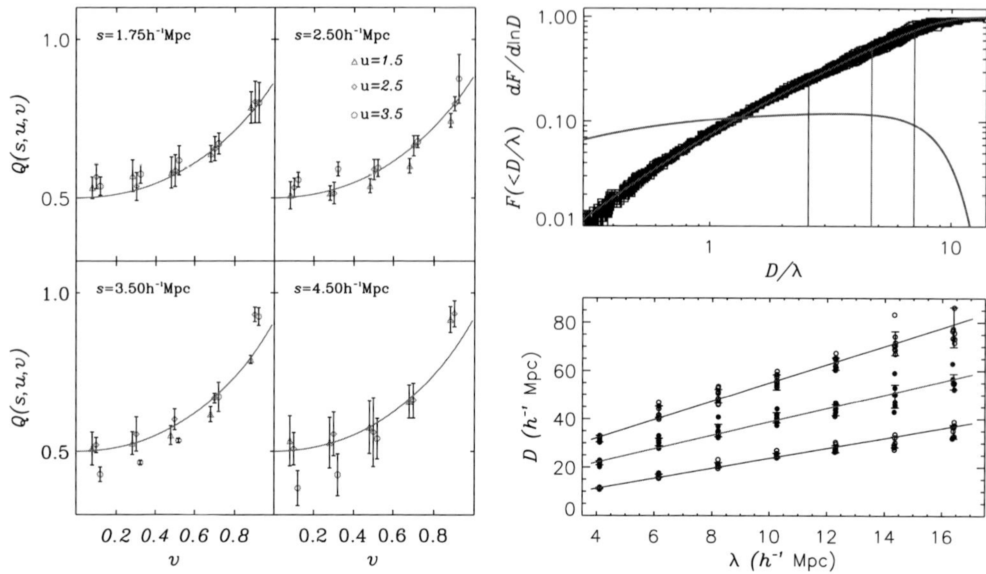

Figure 2. Left: The reduced redshift space correlation function in the strongly clustered regime. Right: Cumulative (histograms) and fitted differential (solid line) void size distribution. The lower panel shows the scaling of median and quartile void sizes for randomly diluted data.

ratio of the diameter D to the mean galaxy separation λ, $x = D/\lambda$, i.e. $dF/d\ln D \propto [x/(x+a)]^2 \exp[-(x/a)^4]$, where $a \approx 9$. The broad size distribution with a cut-off is in agreement with recent excursion set modeling by Sheth & van de Weygaert (2003), but it shows no abundance peak at that scale. The high abundance of small voids is typical for the galaxy void hierarchy. Voids among red galaxies (preferentially in clusters) are 15% larger than among blue galaxies. The lower panel of Figure 2 shows that the void sizes (median and quartiles) have a characteristic linear scaling with the mean galaxy separation λ if the galaxies are randomly diluted. This scaling law (cf. Müller et al. 2000) characterizes the void hierarchy in the universe and in simulations.

Acknowledgements

We want to thank our collaborators Maret and Jaan Einasto, Douglas Tucker, Sepehr Arbabi-Bidgoli, Enn Saar, and Pekka Heinämäki.

References

Arbabi-Bidgoli, S. & Müller, V. 2002 *MNRAS* **332**, 205.

Bond, J., Kofman, L. & Pogosyan D. 1996 *Nature* **380**, 603.

Einasto, M., Einasto, J., Tago, E., Müller & Andernach, H. 2001 *AJ* **122**, 2222.

Einasto, J., Hütsi, G., Einasto, M., Saar, E., Tucker, D., Müller, V., Heinämäki, P. & Allam, S. 2003 *A&A* **405**, 245.

Jing, Y. & Börner, G. 1998 *ApJ* **503**, 37.

Kayo, I., Suto, Y., Nichol, R. et al. 2004 *PASP* in press, asto-ph/0403638.

Müller, V., Arbabi-Bidgoli, S., Einasto, J. & Tucker, D. 2000 *MNRAS* **325**, 280.

Saunders, W., Frenk, C., Rowan-Robinson, M., Lawrence, A. & Efstathiou, G. 1991 *Nature* **349**, 32.

Schwope, A., Hasinger, G., Lehmann, I. et al. 2000 *Astron. N.* **321**, 1.

Sheth, R. & van de Weygaert, R. 2003 *MNRAS* submitted, astro-ph/0311260.

Outskirts of Galaxy Clusters: Intense Life in the Suburbs
Proceedings IAU Colloquium No. 195, 2004
A. Diaferio, ed.

DOI: 10.1017/S1743921304000080

Discovery of a large-scale clumpy structure of the Lynx supercluster at z∼1.27

Fumiaki Nakata[1,2,3], Tadayuki Kodama[2,3], Kazuhiro Shimasaku[3], Mamoru Doi[4], Hisanori Furusawa[5], Masaru Hamabe[6], Masahiko Kimura[7], Yutaka Komiyama[5], Satoshi Miyazaki[5], Sadanori Okamura[3], Masami Ouchi[3], Maki Sekiguchi[8], Masafumi Yagi[2] and Naoki Yasuda[8]

[1]Department of Physics, University of Durham, South Road, Durham DH1 3LE, UK
email: fumiaki.nakata@durham.ac.uk

[2]National Astronomical Observatory of Japan, Mitaka, Tokyo 181-8588, Japan

[3]Department of Astronomy, University of Tokyo, Bunkyo-ku, Tokyo 113-0033, Japan

[4]Institute of Astronomy, University of Tokyo, Mitaka, Tokyo 181-1500, Japan

[5]Subaru Telescope, National Astronomical Observatory of Japan, Hilo, HI 96720, USA

[6]Department of Mathematical and Physical Sciences, Faculty of Science, Japan Women's University, Bunkyo-ku, Tokyo 112-8681, Japan

[7]Department of Astronomy, Faculty of Science, Kyoto University, Sakyo-ku, Kyoto 606-8502, Japan

[8]Institute for Cosmic Ray Research, University of Tokyo, Kashiwa, Chiba 277-8582, Japan

Abstract. We report the discovery of a probable large-scale structure composed of many galaxy clumps around the known twin clusters at $z = 1.26$ and $z = 1.27$ in the Lynx region. Our analysis is based on deep, panoramic, and multi-colour imaging with the Suprime-Cam on the 8.2 m Subaru telescope. We apply a photometric redshift technique to extract plausible cluster members at $z \sim 1.27$ down to $\sim M^* + 2.5$. From the 2-D distribution of these photometrically selected galaxies, we newly identify seven candidates of galaxy groups or clusters where the surface density of red galaxies is significantly high ($>5\sigma$), in addition to the two known clusters, comprising the largest most distant supercluster ever identified.

---

1. Introduction

Superclusters are the largest systems of galaxies, composed of multiple clusters of galaxies, extending over 10-20Mpc (e.g., Postman, Geller, Huchra 1988; Small et al. 1998). According to the N-body simulations which successfully reproduce the observed filamentary structures at the local Universe (e.g., Peacock et al. 2001), clusters of galaxies at $z \sim 1$ are still in the process of formation (e.g., Moore et al. 1998). Therefore, clusters of galaxies at this cosmological distance, especially superclusters, are important sites where we can directly see the process of structure formation and evolution and mass assembly to cluster cores.

The advent of the Suprime-Cam (Miyazaki et al. 2002), a huge format optical camera with a 30′ field of view on the prime focus of the 8.2m Subaru telescope, has made it possible for us to view a >10Mpc region all at once at high redshifts. By utilizing the great light-collection power of the Subaru telescope, we now target the most distant supercluster ever firmly identified with spectroscopy: i.e., the Lynx supercluster region at $z \sim 1.27$ composed of two known clusters, RXJ 0848.9+4452 at $z = 1.26$ and RXJ 0848.6+4453 at $z = 1.27$ (Stanford et al. 1997; Rosati et al. 1999). Throughout this article, we use the

AB magnitude system. The adopted cosmological parameters are $\Omega_0 = 0.3$ and $\lambda_0 = 0.7$, which gives a physical scale of $11.7h_{50}^{-1}$ kpc arcsec^{-1} at the cluster redshift.

2. Observations, reduction and analysis

The Lynx field was observed in November 2000–March 2001 with the Suprime-Cam. Total exposure times of 96, 90, 60, and 54 mins were obtained in $VRi'z'$ respectively. The images are processed in a standard manner with IRAF and purpose-written software developed by us (Yagi et al. 2002). We register the V, R, and i' images with the z' image, and match their PSFs to a fixed value, $1''$ (FWHM). Taking an overlapped region for all the images in the four bands, the final area used in our analysis is restricted in a $26'.4 \times 24'.1$ region, which corresponds to $18.1 \times 17.5h_{50}^{-2}$ Mpc^2 at $z = 1.27$. We then constructed i'-band selected sample using SExtractor v.2.2.0 (Bertin, Arnouts 1996). After excluding 411 stellar-like objects, a total of 35143 objects, whose magnitudes are brighter than $i' = 26.15$ (5σ) in the ~ 590 arcmin^2 area, are contained in the final catalog.

Since spectroscopic measurements for the $\sim 35,000$ galaxies are not practical, we exploit photo-z technique as an observationally efficient method to largely subtract the foreground/background populations. We input the $VRi'z'$ magnitudes to the HYPERZ code (Bolzonella, Miralles & Pelló 2000) to get estimated redshifts for all the individual galaxies in our i'-band selected sample. By using the best estimate of photometric redshift of individual galaxies, we select the only galaxies within the range of $1 \leqslant z_{\rm phot} \leqslant 1.35$ to isolate the plausible members associated to the supercluster at $z \sim 1.27$. The number of these photometrically selected candidates for cluster members (hereafter photo-z selected candidates) is 2229.

3. Large-scale clumpy structure

Figure 1 shows the spatial distribution on the sky of the photo-z selected candidates at $z \sim 1.27$. The large and small dots indicate red and blue galaxies, redder and bluer respectively than a colour of the local Sab galaxies. The numbers of the red and blue galaxies are 575 and 1654, respectively. The contours indicate the local surface density of the 'red' photo-z selected candidates calculated from the 10 nearest neighbors, corresponding to 2, 3, 4 and 5σ. Here σ corresponds to the scatter (standard deviation) of the local surface density in the case of the random distribution of 575 galaxies over the 590 arcmin^2 field of view, i.e., 0.97 arcmin^{-2}. We do not include the blue galaxies here for the following reasons: (1) The accuracy of photo-z is relatively poor for the blue galaxies because of their intrinsically weak 4000Å feature. (2) The red galaxies trace clusters or groups of galaxies more neatly than the blue galaxies due to the morphology (colour)-density relation (Dressler 1980, Dressler et al. 1997).

In Figure 1, we mark by large circles the areas where the local surface densities of the red galaxies are higher than 5σ above the mean density. The cl1 and cl2 indicate the 2 known high redshift clusters at $z \sim 1.27$. We newly find 7 candidates of galaxy clusters or groups, where the local surface density is as large as those of the known clusters (gr1-gr7).

Figure 2 shows the $i' - z'$ versus z' diagram for each cluster/group candidate including the two known clusters. Galaxies within a circle of $1h_{50}^{-1}$ Mpc radius are plotted. The filled circles indicate the photo-z selected candidates and the small dots indicate the other galaxies. For comparison, we also show the galaxies randomly selected from the low density ($< 0\sigma$) regions in the bottom right panel of Figure 2. The dashed line in each panel shows a predicted colour-magnitude relation at $z = 1.27$ for passively evolving galaxies formed at $z_{\rm F} = 4.5$ (Kodama & Arimoto 1997). Not only in the two known

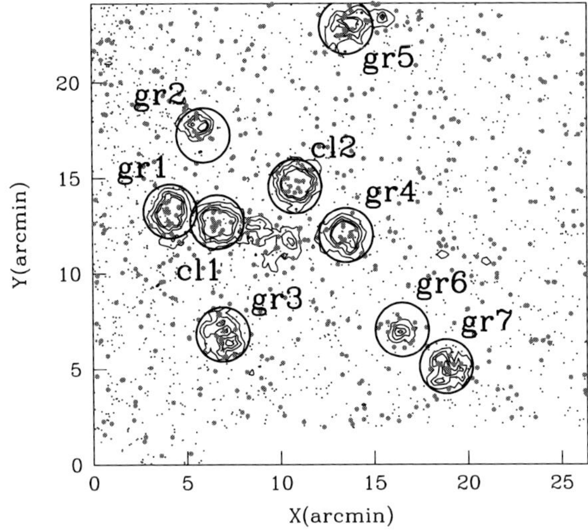

Figure 1. The 2-D distribution of photo-z selected candidates on the sky. North is up, and east is to the left. The large and small dots indicate red and blue galaxies respectively. The contour levels are 2, 3, 4 and 5σ. See the text for details. The radii of the large circles are $1h_{50}^{-1}$ Mpc at $z = 1.27$.

clusters (cl1 and cl2) but also in our new candidates (gr1-gr7), can we find many red galaxies consistent with passive evolution which comprise the red sequences in these diagrams.

We also show the radial profile of each cluster and cluster/group candidate in Figure 3. Only those galaxies brighter than $i' = 26$ are counted because of increasing incompleteness toward fainter magnitudes. Here we subtract the remaining field contamination using the Subaru Deep Field data (e.g., Kashikawa et al. 2003; Ouchi et al. 2003) from each bin after normalizing counts of field galaxies with referring areas. The averaged surface number densities of all galaxies and the red galaxies are 4.0 ± 3.6 and 0.20 ± 0.33 h_{50}^2Mpc^{-2}, respectively, to the same magnitude limit of $i' = 26$. We see a pronounced increase in galaxy density towards the center of most of our cluster/group candidates. The only exception is gr6 that shows no excess of galaxy density at the center. This may be due to the misidentification of the center of the group, or intrinsically non-axisymmetric structure. We combine the profiles of known clusters and those of our new candidates (groups) separately, and find that both of these combined profiles can be fitted very well with the King's law profile, $N(r) = N_0(1 + r^2/R_c^2)^{-1}$ (King 1966, 1972), with $R_c = 0.23 \pm 0.03$ h_{50}^{-1}Mpc (cluster) and $R_c = 0.32 \pm 0.04$ h_{50}^{-1}Mpc (group), which is consistent with estimations for local clusters (Bahcall 1975).

We estimate the richness of each cluster/group candidate, using the $N_{0.5}$ indicator introduced by Hill & Lilly (1991). This indicates the number of galaxies within a 0.5 Mpc radius from the cluster center and within the magnitude range between m_1 and $m_1 + 3$ measured in the R band at $z \sim 0.5$. Since the R band at $z \sim 0.5$ roughly corresponds to the z'-band at $z \sim 1.27$, we can make a direct comparison with their measurements. The cluster richness estimated based on $N_{0.5}$ is 1-2 for cl1 and cl2, while 0-1 for gr1-gr7 (See Table 4 of Hill & Lilly 1991). Therefore we find that the candidates we newly find are poorer systems than the two known clusters. This is also suggested by the deep (190 ks

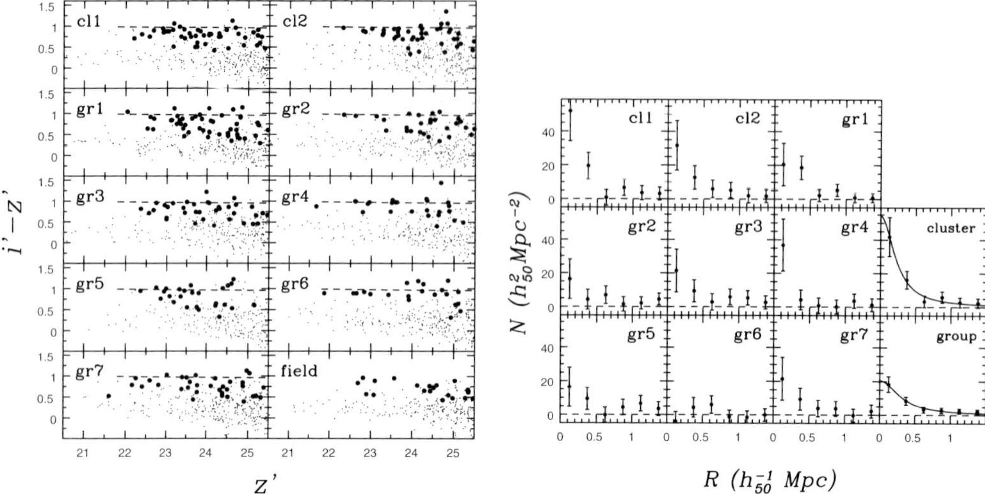

Figure 2 (left) : The $i' - z'$ versus z' diagrams of the galaxies in the known clusters (cl1 and cl2) and our new cluster/group candidates (gr1-gr7). See the text for details.
Figure 3 (right) : The surface number density of galaxies as a function of the distance from the cluster center. Cluster and group indicate the combined profiles of known clusters and those of our new candidates, respectively, and the best-fitted King's law profiles are also plotted by the solid lines.

exposure) X-ray data around the Lynx field taken by Chandra (Stanford et al. 2001; Stern et al. 2002). We compare the structures found in our optical data with those in the X-ray data. We cannot identify any counterparts (extended sources) in our cluster/group candidates except for the two known clusters. This may be because these candidates do not have dense or hot enough plasma to be detected in the Chandra data, whose limiting flux is $\approx 1.7 \times 10^{-16}$ ergs cm^{-2} s^{-1} in the 0.5-2 keV band. We argue therefore that our 'Suprime-Cam imaging + photo-z' method is a very powerful technique in finding clusters or groups in the distant Universe even beyond $z > 1$ out to $z \sim 1.3$. We note that this is practically the highest redshift structures of 'normal' galaxies that can be traced by optical images with our scheme (bracketing the 4000Å-break region in the rest frame).

4. Summary

We have presented deep panoramic imaging of the Lynx supercluster field at $z \sim 1.27$ taken with the Suprime-Cam on Subaru telescope. Our multicolor image covers a $26'.4 \times 24'.1$ ($18.1 \times 17.5 h_{50}^{-1}$ Mpc), allowing us for the first time to investigate the large-scale structures spreading around the known cluster regions at this high redshift. By applying photo-z technique, we have mapped out the spatial distributions of galaxies at $z \sim 1.27$ down to $\sim M^* + 2.5$.

With this technique, we have newly discovered seven cluster/group candidates. These candidates show red color-magnitude sequences and centrally concentrated profiles, similar to those of the two known clusters in this field, indicating that they are likely to be real clusters/groups of galaxies at $z \sim 1.27$, comprising a large-scale supercluster over $\sim$18 h_{50}^{-1}Mpc scale. The clumpiness of the structures suggests that its dynamical stage is young and they are in the process of assembly to a massive cluster.

References

Bahcall, N. A. 1975 ApJ, 198, 249
Bertin, E., Arnouts, S. 1996 A&AS, 177, 393
Bolzonella, M., Miralles, J.-M., Pelló, R. 2000 A&A, 363, 476
Dressler, A. 1980 ApJ, 236, 351
Dressler, A. et al. 1997 ApJ, 490, 577
Hill, G. J. Lilly, S. J. 1991 ApJ, 367, 1
Kashikawa, N. et al. 2003 AJ, 125, 53
King, I. R. 1966 AJ, 71, 1
King, I. R. 1972 ApJ, 174, L123
Kodama, T., Arimoto, N. 1997 A&A, 320, 41
Miyazaki, S. et al. 2002 PASJ, 54, 833
Moore, B., Governato, F., Quinn, T., Stadel, J., Lake, G. 1998 ApJ, 499, L5
Ouchi, M. et al. 2003 ApJ, 582, 60
Peacock, J. A. et al. 2001 Nature, 410, 169
Postman, M., Geller, M. J., Huchra, J. P. 1998 AJ, 95, 267
Rosati, P., Stanford, S. A., Eisenhardt, P. R., Elston, R., Spinrad, H., Stern, D., Dey, A. 1999 AJ, 118, 76
Small, T. A., Chung-Pei, M., Sargent, W. L. W., Hamilton, D. 1998 ApJ, 492, 45
Stanford, S. A., Elston, R., Eisenhardt, P. R., Spinrad, H., Stern, D., Dey A. 1997 AJ, 114, 2232
Stanford, S. A., Holden, B., Rosati, P., Tozzi, P., Borgani, S., Eisenhardt, P. R., Spinrad, H. 2001 ApJ, 552, 504
Stern, D. et al. 2002 AJ, 123, 2223
Yagi, M., Kashikawa, N., Sekiguchi, M., Doi, M., Yasuda, N., Shimasaku, K., Okamura, S. 2002 AJ, 123, 66

Outskirts of Galaxy Clusters: Intense Life in the Suburbs
Proceedings IAU Colloquium No. 195, 2004
A. Diaferio, ed.

DOI: 10.1017/S1743921304000092

Weak lensing evidence for a filament between A222/A223 †

Jörg P. Dietrich[1], Douglas Clowe[1], Peter Schneider[1], Jürgen Kerp[2] and Emilio Romano-Díaz[3]

[1]IAEF, University of Bonn, Auf dem Hügel 71, 53121 Bonn, Germany email: dietrich,clowe,peter@astro.uni-bonn.de

[2]Radioastronomisches Institut, University of Bonn, Auf dem Hügel 71, 53121 Bonn, Germany email: jkerp@astro.uni-bonn.de

[3]Kapteyn Institute, University of Groningen, P.O. Box 800, 9700, Av. Groningen, The Netherlands email: eromano@astro.rug.nl

Abstract. We present a weak lensing analysis and comparison to optical and X-ray maps of the close pair of massive clusters A222/223. Indications for a filamentary connection between the clusters are found and discussed.

---

1. Introduction

N-body simulations of cosmic structure formation predicts that matter in the universe should be concentrated along sheets and filaments and that clusters of galaxies form where these intersect (Kauffmann *et al.* 1999, Bond *et al.* 1996). This filamentary structure, often also dubbed "cosmic web", has been seen in galaxy redshift surveys (Vogeley *et al.* 1994) and X-rays (Zappacosta *et al.* 2002, Tittley & Henriksen 2001).

Because of the greatly varying mass–to–light (M/L) ratios between rich clusters and groups of galaxies (Tully & Shaya 1998) it is problematic to convert the measured galaxy densities to mass densities without further assumptions. Weak gravitational lensing, which is based on the measurement of shape and orientation parameters of faint background galaxies (FBG), is a model–independent method to determine the surface mass density of clusters and filaments. Due to the finite ellipticities of the unlensed FBG, every weak lensing mass reconstruction is unfortunately an inherently noisy process, and the expected surface mass density of a single filament is too low to be detected with current telescopes (Jain *et al.* 2000).

Cosmic web theory also predicts that the surface mass density of a filament increases towards clusters (Bond *et al.* 1996). Filaments connecting neighboring clusters should have surface mass densities high enough to be detectable with weak lensing (Pogosyan *et al.* 1998). Such filaments may have been detected in several recent weak lensing studies.

Kaiser *et al.* (1998) found a possible filament between two of the three clusters in the $z = 0.42$ super-cluster MS0302+17, but the detection remains somewhat uncertain because of a possible foreground structure overlapping the filament and possible edge effects due to the gap between two of the camera chips lying along the filament. Also, Gavazzi *et al.* (2004) could not confirm the detection of a filament in this system. Gray *et al.* (2002) claim to have found a filament extending between two of the three clusters of the Abell 901/902 super-cluster, but the significance of this detection is low and subject

† Based on observations made at ESO/La Silla under program Nos. 064.L-0248, 064.O-0248, 66.A-0165, 68.A-0269.

to possible edge effects, as again the filament is close to the intersection of four chips of the camera.

2. Data of the Abell 222/223 system

A222/223 are two Abell clusters at $z \approx 0.21$ separated by $\sim 14'$ on the sky, or $\sim 2600h_{70}^{-1}$ kpc, belonging to the Butcher *et al.* (1983) photometric sample. Both clusters are rich, having Abell richness class 3 (Abell 1958). While these are optically selected clusters, they have been observed by ROSAT (Wang & Ulmer 1997) and are confirmed to be massive clusters. Proust *et al.* (2000, P00) published a list of 53 spectra in the field of A222/223, 4 of them in region between the clusters (hereafter "intercluster region") and at the redshift of the clusters. Later, Dietrich *et al.* (2002, D02) reported on spectroscopy of 183 objects in the cluster field, 153 being members of the clusters or at the cluster redshift in the intercluster region. Taking the data of P00 and D02 together, 6 galaxies at the cluster redshift are known in the intercluster region, establishing this cluster system as a good candidate for a filamentary connection.

Imaging was performed with the Wide Field Imager (WFI) at the ESO/MPG 2.2 m telescope. In total, twenty 600 s exposures were obtained in R-band in October 2001 centered on A223, eleven 900 s R-band exposures were taken in December 1999 centered on A222. The images were taken with a dithering pattern filling the gaps between the chips in the co-added images of each field.

The R-band data used for the weak lensing analysis is supplemented with three 900 s exposures in the B and V-band centered on each cluster taken from November 1999 to December 2000. The final B- and V-band images have some remaining gaps and regions that are covered by only one exposure and – due to the dithering pattern – do not cover exactly the same region as the R-band images.

2.1. *Catalog creation*

Lensing catalogs were created from the R-band images using the KSB algorithm (Kaiser *et al.* 1995, corrections in Hoekstra *et al.* 1998). Stellar reflection rings and diffraction spikes were masked and the masked regions excluded from the catalogs. The catalogs from both pointings were merged and the ellipticities of objects contained in both catalogs were averaged.

From this merged catalog all objects with $R < 22$ mag, signal-to-noise $\nu < 7$, Gaussian radius $0.''4 < r_g < 1.''2$, and ellipticity $e > 0.8$ were deleted. All surviving objects fainter than $R > 23.5$ mag were kept as likely background galaxies, while objects brighter than this – for which colors were available – were selected according to the following criteria: Objects detected in B, V, and R matching colors of galaxies at $z < 0.5$, $-0.23 < (V - R) - 0.56 \times (B - V) < 0.67$, $0.5 < B - V < 1.6$, were deleted from the sample. Galaxies detected only in the V and R band were kept if $(V - R) > 1.0$. The final catalog contains 25583 objects, corresponding to a galaxy number density of 15.5 arcmin^{-2} if the masked regions are not taken into account for the computation of the total area.

3. Mass and light maps

Fig. 1 shows a reconstruction obtained from the catalog described in the previous section. The reconstruction was performed on a 214×200 points grid using the algorithm of Seitz & Schneider (2001) adapted to the field geometry. The smoothing scale of the shear data was set to $2'$. Galaxies in the catalog were weighted by the error estimate of their initial ellipticity measurement.

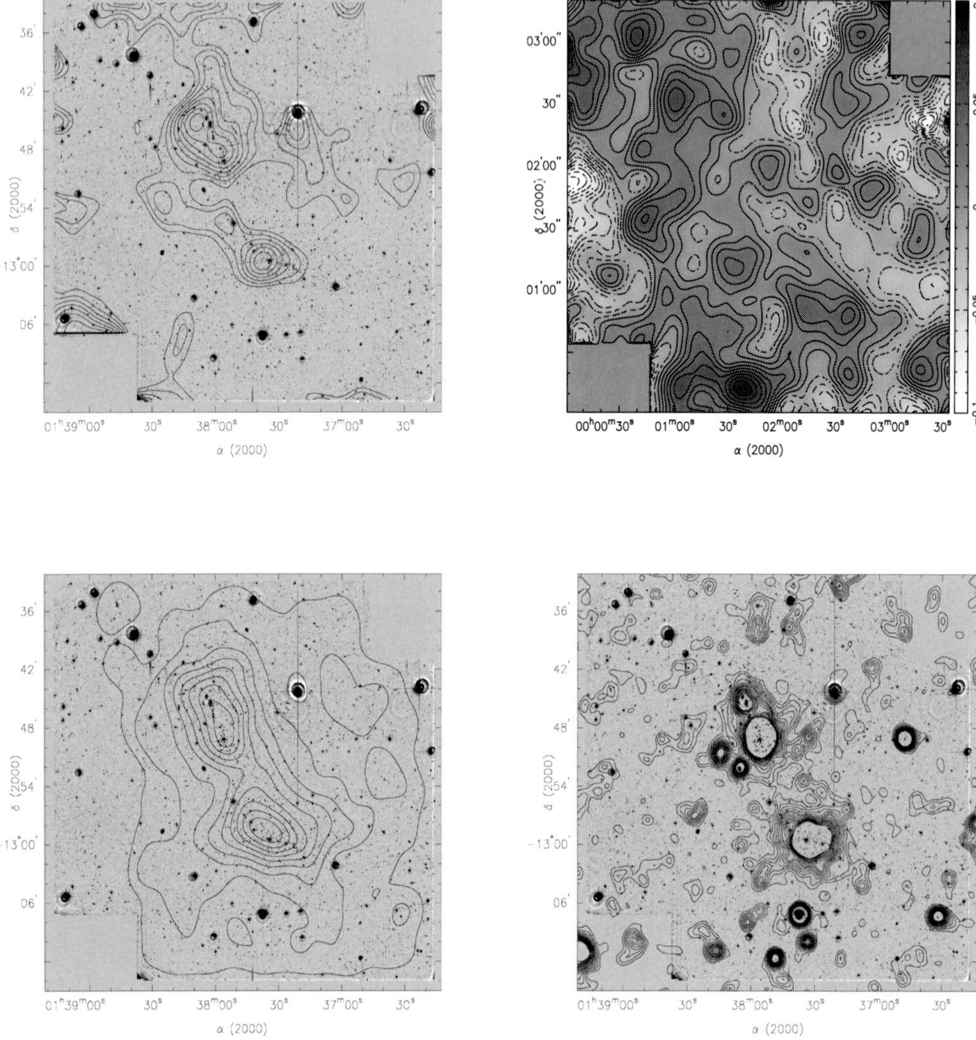

Figure 1. Various light and mass maps of the A222/223 system. A223 is the Northern cluster, A222 is South of it. *Top left*: Weak lensing reconstruction of A222/223. Each contour represents an increase in κ of 0.01 ($\sim 3.2 \times 10^{13}\ h_{70}\ M_{\odot}\ \mathrm{Mpc}^{-2}$, assuming $\overline{z}_{\mathrm{FBG}} = 1$) above the mean κ at the edge of the field. The peak $\sim 12'$ SE of A222 has an optical counter-part. *Top right*: Surface mass density map created from a catalog with randomly rotated galaxies used to estimate the noise level of the reconstruction on the right. *Bottom left*: Density distribution of galaxies matching the colors of the red cluster sequence ($0.78 < V - R < 0.98$) of the Abell clusters smoothed with a $2'$ Gaussian. The absence of galaxies in the Eastern part is caused by the very bad or missing V-band coverage. *Bottom right*: Contours in this plot are from ROSAT PSPC data in the energy range $0.5 - 2.4$ keV. The lowest contour is at the 6σ level. Higher contours rise in steps of 1σ. Note that the extended X-ray emission North of A223 is associated with the Northern extension of A223 in the surface mass density map.

Both clusters are well detected in the reconstruction. The two components of A223 are clearly visible. The strong mass peak West of A223 is most likely associated with the reflection ring around the bright $V = 7.98$ mag star at that position. Although the reflection ring and a large area around it were masked, diffuse stray light is visible extending beyond the masked region, well into A223, probably being the cause of the observed mass peak.

Also visible is a bridge in the surface mass density extending between A222 and A223. Although the signal of this possible filamentary connection between the clusters is very low, the feature is quite robust when the selection criteria of the catalog are varied and it never disappears.

The noise level in the reconstruction can be estimated by randomly rotating the galaxies while keeping their positions and ellipticity moduli fixed. Performing a reconstruction on this randomized catalog gives $\langle \kappa^2 \rangle \simeq 0.015$, suggesting that the intercluster connection is present at the $\sim 2\sigma$ level.

The galaxy density distribution and the X-ray contours in Fig. 1 both show a clear connection between the clusters. While the X-ray and galaxy density contours are aligned in the intercluster region, the surface mass density contours connect the clusters Eastwards of them.

4. Discussion

We presented weak lensing, optical galaxy density, and X-ray maps of the massive, close pair of galaxy clusters A222/223. All maps show a connection between the clusters in the intercluster region. The density distribution of color-selected objects, supplemented by the spectroscopic confirmation of galaxies at the cluster redshift in the intercluster region by P00 and D02, and the significant connection between both clusters in the $0.5-2.4$ keV band of ROSAT's PSPC establish a secure filamentary connection between the clusters. The weak lensing indications for a dark matter bridge are much less secure. The signal-to-noise of the structure extending between the clusters is low but the structure itself is very robust and never disappears when the selection criteria for the lensing catalog are varied.

We note that the luminosity density of color-selected galaxies in the intercluster region is only by a factor of ~ 2 lower than in the peaks of the galaxy clusters. If the early type galaxies trace the mass well and a constant M/L ratio is assumed, we would expect a clearly detectable lensing signal along the galaxy density contours. This is obviously not the case. A more detailled comparison of the observed and expected surface mass density, also addressing the apparent misalignment of mass and light in A223, will be done elsewhere (Dietrich *et al.* 2004, in preparation). We also note that a misalignment of mass and light is present in the filament candidate of Gray *et al.* (2002).

Acknowledgements

This work has been supported by the German Ministry for Science and Education (BMBF) through DESY under the project 05AE2PDA/8, and by the Deutsche Forschungsgemeinschaft under the project SCHN 342/3–1.

References

Abell, G. O. 1958, *ApJS* **3**, 211
Bond, H., Kofman, L., & Pogosyan, D. 1996, *Nature* **380**, 603
Butcher, H., Wells, D. C., & Oemler, A. 1983, *ApJS* **52**, 183
David, L. P., Forman, W., & Jones, C. 1999, *ApJ* **519**, 533
Dietrich, J. P., Clowe, D. I., & Soucail, G. 2002, *A&A* **394**, 395
Gavazzi, R., Mellier, Y., Fort, B., Cuillandre, J. & Dantel-Fort, M. 2004, *astro-ph/0401403*
Gray, M. E., Taylor, A. N., Meisenheimer, K., *et al.* 2002, *ApJ* **568**, 141

Hoekstra, H., Franx, M., Kuijken, K., & Squires, G. 1998, *ApJ* **504**, 636
Jain, B., Seljak, U., & White, S. 2000, *ApJ* **530**, 547
Kaiser, N., Squires, G., & Broadhurst, T. 1995, *ApJ* **449**, 460
Kaiser, N., Wilson, G., Luppino, G., *et al.* 1998, *astro-ph/99809268*
Kauffmann, G., Colberg, J. M., Diaferio, A., & White, S. D. M. 1999, *MNRAS* **303**, 188
Pogosyan, D., Bond, J. R., Kofman, L., & Wadsley, J. 1998, in Wide Field Surveys in Cosmology, 14th IAP meeting held May 26-30, 1998, Paris. Publisher: Editions Frontieres. ISBN: 2-8 6332-241-9, 61
Proust, D., Cuevas, H., Capelato, H. V., *et al.* 2000, *A&A* **355**, 443
Seitz, S. & Schneider, P. 2001, *A&A* **374**, 740
Tittley, E. R. & Henriksen, M. 2001, *ApJ* **563**, 673
Tully, R. B. & Shaya, E. J. 1998, *astro-ph/9810298*
Vogeley, M. S., Park, C., Geller, M. J., Huchra, J. P., & Gott, J. R. I. 1994, *ApJ* **420**, 525
Wang, Q. D. & Ulmer, M. P. 1997, *MNRAS* **292**, 920
Zappacosta, L., Mannucci, F., Maiolino, R., *et al.* 2002, *A&A* **394**, 7

Outskirts of Galaxy Clusters: Intense Life in the Suburbs
Proceedings IAU Colloquium No. 195, 2004
A. Diaferio, ed.

DOI: 10.1017/S1743921304000109

The faint X-ray source population near 3C 295

V. D'Elia[1], F. Fiore, M. Elvis, M. Cappi, S. Mathur, P. Mazzotta, E. Falco and F. Cocchia

[1]INAF - Osservatorio Astronomico di Roma - Via di Frascati, 33 I-00040 Monteporzio Catone (Rome), Italy email: delia@mporzio.astro.it

Abstract. We present a statistical analysis of the Chandra observation of the source field around the 3C 295 galaxy cluster ($z = 0.46$). Three different methods of analysis, namely a chip by chip LogN-LogS, a two-dimentional Kolmogorov-Smirnov (KS) test, and the angular correlation function (ACF) show a strong overdensity of sources in the North-East of the field, that may indicate a filament of the large scale structure of the universe towards 3C 295.

1. Introduction

N-body and hydrodynamical simulations show that high redshift clusters of galaxies lie at the nexus of several filaments of galaxies (see e.g. Peacock 1999). Such filaments map out the "cosmic web" of voids and filaments of the large scale structure of the universe. Thus, rich clusters represent good indicators of regions of sky where several filaments converge. The filaments themselves could be mapped out by Active Galactic Nuclei (AGNs), assuming that AGNs trace galaxies.

Cappi et al. (2001) studied the distribution of the X-ray sources around 3C 295 ($z = 0.46$). The cluster was observed with the ACIS-S CCD array for a short exposure time (18 ks). They found a high source surface density in the $0.5-2$ keV band which exceeds the ROSAT (Hasinger et al. 1998) and Chandra (Giacconi et al. 2002) LogN-LogS by a factor of 2, with a significance of 2σ.

In this work an analysis of a deeper (92 ks) Chandra observation of 3C 295 is performed, to check whether the overdensity of such a field in the $0.5-2$ keV band is real or not, to define the structure of the overdensity, and to extend the above considerations to the $2-10$ keV band.

2. Observation and data reduction

Chandra observed the $16' \times 16'$ field around the 3C 295 cluster with ACIS-I on May 18, 2001, for 92 ks. The data reduction was carried out using the Chandra Interactive Analysis of Observations software version 2.1.3. All the data analysis has been performed separately in the $0.5-2$ keV, in the $2-7$ keV and in the $0.5-7$ keV bands. An identical analysis was performed for the Chandra Deep Field South (CDFS) in the $0.5-2$ keV and $2-7$ keV bands for comparison, and to check our analysis methods. The source detection was carried out using the 'PWDetect' algorithm (Damiani et al. 1997). We identified 89 sources in the $0.5-2$ keV band, 71 sources in the $2-7$ keV band and 121 sources in the $0.5-7$ keV band (see fig. 1). The counts in the $0.5-2$ keV, $2-7$ keV and $0.5-7$ keV bands were converted in $0.5-2$ keV, $2-10$ keV and $0.5-10$ keV fluxes using conversion factors appropriate for a $\gamma = 1.8$ power law spectrum with a galactic absorption toward the 3C 295 field of $N_H = 1.33 \times 10^{20}$ cm^{-2}. Such values take into account the quantum efficiency degradation of the CCD.

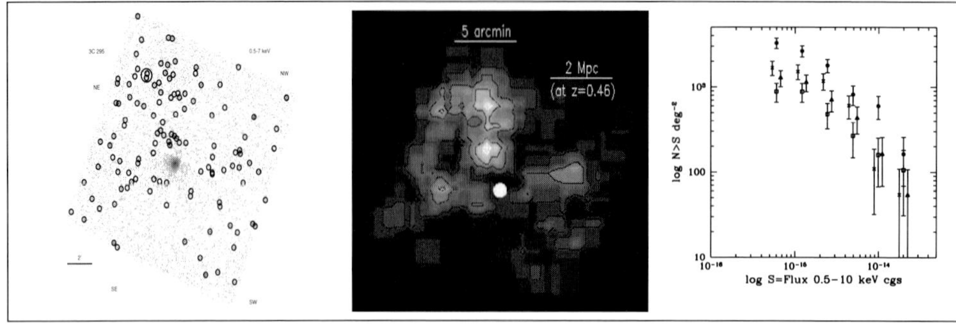

Figures 1, 2, 3 (left to right). Figure 1. - The *Chandra* 3C 295 field in the $0.5 - 7$ keV band. Circles represent the sources detected; the largest circle indicates an extended source, possibly a group of galaxies. The brightest source in the center of the field is the cluster of galaxies 3C 295. Figure 2. - The density map in the same band. The linear smoothing factor is 1.5 arcmin and the four contour levels indicate source densities of 1.3, 1.8, 2.2 and 2.7 sources per arcmin^2. Figure 3. - The mean 3C 295 LogN-LogS in the $0.5 - 10$ keV band, calculated for each ACIS-I chip separately. Filled circles represent counts for the NE chip, open triangles for the NW, open square for the SW and crosses for the SE.

3. Analysis

The following analysis has been perfomed. The sky coverage has been computed for 3C 295 and CDFS fields. The LogN-LogS in the soft and hard bands for the whole filelds have been produced. The LogN-LogS in the soft, hard and whole bands have been produced separately for each ACIS-I chip of the 3C 295 observation (see fig. 3 for the $0.5 - 10$ keV band). 3C 295 chip-to-chip and CDFS LogN-LogS have been fitted to evaluate slopes and normalizations (fig. 4 for the $0.5 - 10$ keV band). A two dimensional Kolmogorov-Smirnov test has been applied to check whether the sources were uniformly distributed or not. The angular correlation function (ACF) has been computed in order to estimate at which scales the sources in the two fields were clustered. The ACF has been evaluated also for a sample of 8 Chandra fields with an exposure time similar to that of 3C 295. The errors associated to the functions (figs. 5 and 6) have been calculated using both Poisson and bootstrap statistics (Barrow, Bhavsar & Sonoda 1984).

4. Main results

The following results have been achieved:

3C 295 and CDFS LogN-LogS are in very good agreement both in the soft and hard band, and with the CDFS LogN-LogS by Rosati et al. 2002. The 3C 295 LogN-LogS in the soft, hard and broad bands computed separately for each ACIS-I chip show an overdensity of sources in the North-East (NE) chip (fig. 3) which reflects the clustering of sources clearly visible in figs. 1 and 2. The discrepancy between the normalization of the LogN-LogS for the NE and SW chip is 3.2 σ, 3.3 σ and 4.0 σ in the soft, hard and broad band, respectively; fig. 4 shows the normalization vs. slope plot for the broad band.

The two-dimensional Kolmogorov-Smirnov test shows that there is a considerable probability that CDFS sources are uniformly distributed ($P \sim 15\%$ in the soft and hard bands), while the probability that the 3C 295 sources are uniformly distributed is only a few per cent, and drops below 1% if we consider the $0.5 - 7$ keV band.

The angular correlation function of the 3C 295 sources features a strong signal on scales of a few arcmins, and also on lower scales in the $0.5 - 7$ keV band (fig. 5). Moreover, the function is above the value found by Vikhlinin & Forman (1995) for a large sample of ROSAT sources. On the other hand, no signs of a similar behavior is featured by the CDFS and the sample of the 8 Chandra fields (fig. 6).

5. Discussion

In this work we studied in great quantitative details the excess of sources clearly visible in the upper left corner of the Chandra observation of the 3C 295 galaxy cluster field (see figs. 1 and 2). Since N-body and hydrodynamical simulations show that clusters of galaxies lie at the nexus of several filaments, this excess could represent a filament of the large scale structure of the universe.

Moreover, if the redshift of our sources were the same of 3C 295 ($z = 0.46$), we could compute the galaxy overdensity of the field assuming a spherical distribution. Under these assumption, we obtain a density contrast of ~ 70 which, despite large uncertainties, is intriguingly close to the expected galaxy overdensity of filaments $\sim 10 \div 10^2$, and much smaller than the overdensities of clusters of galaxies ($\sim 10^3 \div 10^4$).

Thus, in order to study this candidate filament, further Chandra observations and optical identifications are pursued to obtain the redshift of the sources, to map out the 3C 295 region, and delimiting the filament properties up to scales of 24 arcmin (i.e., ~ 6 Mpc) from the 3C 295 cluster.

More details on the present work can be found in D'Elia et al., 2004, A&A, submitted.

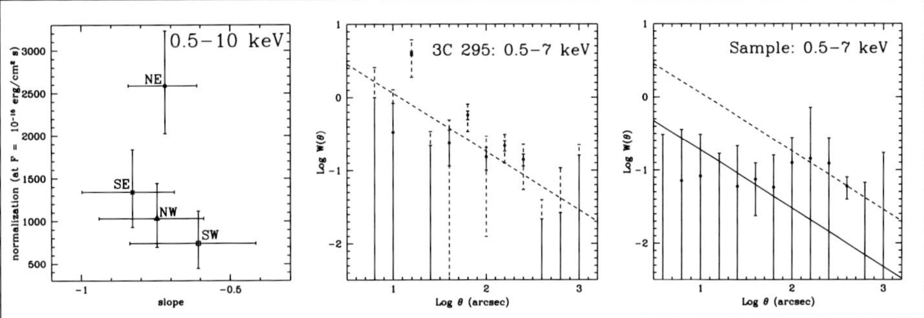

Figures 4, 5, 6 (left to right). Figure 4. - Results of the power law fits to the four LogN-LogS chips in the broad band; on the x axis is the slope of the power law, y axis the normalization. Symbols refer to chips as in fig. 3. Error bars are the 90% confidence limit. Figure 5. - The 3C 295 angular correlation function (ACF) in the $0.5 - 7$ keV band. The dashed line represents the best fit to the data. Solid error bars are Poisson; dashed error bars are bootstrap. Figure 6. - The ACF in the $0.5 - 7$ kev band. The solid line is the best fit to these data; the dashed line is the fit to the ACF of 3C 295 (fig. 5). Error bars are bootstrap.

References

J.D. Barrow, S.P. Bhavsar & D.H. Sonoda 1984, MNRAS, 210, 19

M. Cappi et al., 2001, ApJ, 548, 624

F. Damiani, A. Maggio, G. Micela & S. Sciortino, 1997, ApJ, 483, 350

G. Hasinger et al., 1998, A&A, 329, 482

J.A. Peacock, 1999, "Cosmological Physics" (Cambridge: Cambridge University Press)

P. Rosati et al., 2002, ApJ, 566, 667

A. Vikhlinin & W. Forman, 1995, ApJ, 455, L109

Alan Dressler shows to journalist Dr. Silvia Rosa-Brusin how much dark matter is there in the universe *(by J.M. Colberg)*.

Rachida Sadat and Antonaldo Diaferio at the welcome party *(by J.M. Colberg)*.

Section B

Theoretical tools

Outskirts of Galaxy Clusters: Intense Life in the Suburbs
Proceedings IAU Colloquium No. 195, 2004
A. Diaferio, ed.

DOI: 10.1017/S1743921304000110

Morphological statistics of the cosmic web

Sergei F. Shandarin

Department of Physics and Astronomy, University of Kansas, KS 66045, USA
email: sergei@ku.edu

Abstract. We report the *first* systematic study of the supercluster-void network in the ΛCDM concordance cosmology treating voids and superclusters on an equal footing. We study the dark matter density field in real space smoothed with the $L_s = 5\ h^{-1}$Mpc Gaussian window. Superclusters and voids are defined as individual members of over-dense and under-dense excursion sets respectively. We determine the morphological properties of the cosmic web at a large number of dark matter density levels by computing Minkowski functionals for every supercluster and void. At the adopted smoothing scale individual superclusters totally occupy no more than about 5% of the total volume and contain no more than 20% of mass if the largest supercluster is excluded. Likewise, individual voids totally occupy no more than 14% of volume and contain no more than 4% of mass if the largest void is excluded. The genus of individual superclusters can be ~ 5 while the genus of individual voids reaches ~ 55, implying significant amount of substructure in superclusters and especially in voids. Large voids are typically distinctly non-spherical.

1. Introduction

Redshift galaxy catalogues reveal a universe permeated by an interpenetrating network of superclusters and voids. It therefore becomes important to understand and quantify the geometrical and topological properties of large scale structure in an ΛCDM cosmology in a deep and integrated manner. The main aim of this talk is to study the supercluster-void network in ΛCDM cosmology with emphasis on the sizes, shapes and topologies of individual superclusters and voids. Concretely, we study the geometry and topology of isodensity surfaces $\delta(\mathbf{x}) \equiv \delta\rho(\mathbf{x})/\bar{\rho} = const.$ At a given threshold δ_{th} regions having higher then threshold density ($\delta > \delta_{th}$) will be called "superclusters", while regions with $\delta < \delta_{th}$ will be called "voids". We employ an elaborate surface modeling scheme, SURFGEN (short for 'surface generator'), that allows us to determine the geometry, morphology and topology of excursion sets defined on a density field in a very comprehensive manner (Sheth *et al.* 2003). Working with the density field also permits us to determine the morphological properties of the *full excursion set* describing the supercluster-void network. More detailed information is then gleaned at one particular threshold (usually associated with percolation) at which shapes and sizes of individual superclusters and voids yield rich information about properties of the cosmic web to which we belong.

We use dark matter distributions in a flat model with $\Omega_0 = 0.3$, $\Omega_\Lambda = 0.7$, $h = 0.7$ (ΛCDM). The initial spectrum was taken with the shape parameter $\Gamma = 0.21$. The amplitude ($\sigma_8 = 0.9$) of the power spectrum in the model is set so as to reproduce the observed abundance of rich galaxy clusters at the present epoch (see Jenkins *et al.* 1998 for details). The study of mock galaxy distributions was done by Sheth (2003).

SURFGEN operates on three-dimensional pixelized maps. Therefore we first generate the density field from the distribution of dark matter particles. This process was described in detail in Sheth *et al.* (2003); here we present a brief summary. The data consist of 256^3 particles in a box of size $239.5\ h^{-1}$Mpc. We fit a 128^3 grid to the box. We follow the smoothing technique used by Springel *et al.* (1998) which they adopted for their preliminary topological analysis of the Virgo simulations. We study the field smoothed with $L_s = 5\ h^{-1}$Mpc which is a fiducial smoothing scale in many studies of both density fields in N-body simulations and galaxy fields from redshift surveys; see for example, Grogin & Geller (2000). The Gaussian kernel for smoothing that we adopt here is $W(r) = \pi^{-3/2} L_s^{-3} exp(-r^2/L_s^2)$. Since the kernel is isotropic and uniform, it is likely to diminish the true extent of anisotropy in filaments, pancakes and voids. This effect could be minimized by considering adaptive kernels or smoothing techniques based on the wavelet transform. An even more ambitious approach is to reconstruct density fields using Delaunay tesselations using a technique reported by van de Weygaert (2002).

Four Minkowski functionals (MFs) are effective non-parametric descriptors of the morphological properties of surfaces in three dimensions Mecke *et al.* (1994). They are: *Volume* V enclosed by the surface, S; *Area* A of the surface; *Integrated mean curvature* C of the surface $C = \frac{1}{2} \oint_S \left(\frac{1}{R_1} + \frac{1}{R_2} \right) da$, where R_1 and R_2 are the principal radii of curvature at a given point on the surface; and finally, *the Euler characteristic* $\chi = \frac{1}{2\pi} \oint_S \left(\frac{1}{R_1 R_2} \right) da$. The genus is uniquely related to the Euler characteristic $G = 1 - \chi/2$ and thus carries exactly the same information. We measure the above parameters for every region in both over-dense and under-dense excursion sets at 99 density thresholds equispaced in the filling factor from $FF = 0.01$ to $FF = 0.99$ where FF is the fraction of the total volume occupied by the excursion set. As demonstrated in Sahni *et al.* (1998) and Sathyaprakash *et al.* (1998), particular ratios of the MFs called "Shapefinders" provide us with a set of non-parametric measures of sizes and shapes of objects. Therefore, in addition to determining MFs we shall also derive three quantities having the dimension of length that can be associated with three characteristic sizes: T (Thickness), B (Breadth) and L (Length) defined as follows: $T = 3V/A, \quad B = A/C, \quad L = C/4\pi$. The three Shapefinders describing an individual region bounded by one or several isolated surfaces of constant density have dimension of length and provide us with an estimate of the region extensions. The choice of the coefficients in the above equations results in a sphere having all three sizes equal to its radius $T = B = L = R$. It is worth noting that T, B and L are only the estimates of three basic sizes (semi-axes) of an object, which work quite well on objects such as a triaxial ellipsoid and a torus. An indicator of 'shape' is provided by a pair of dimensionless Shapefinder statistics $P = (B - T)/(B + T); \quad F = (L - B)/(L + B)$, where P and F are measures of Planarity and Filamentarity respectively ($P, F \leqslant 1$). A sphere has $P = F = 0$, an ideal filament has $P = 0, F = 1$, while $P = 1, F = 0$ for an ideal pancake.

2. Results

Understanding percolation is essential for understanding the morphology of the supercluster – void network. Percolation is important because the properties of superclusters and voids radically change at the percolation transitions (see Fig. 1). At relatively high thresholds $\delta_C > 1.8$, corresponding to small filling factors $FF_C < 0.07$, the largest supercluster has insignificant volume and mass compared to the total volume or mass contained in the over-dense excursion set, $\delta > \delta_{\rm TH}$. During the percolation transition at $FF_C \approx 0.07$, corresponding to $\delta_C \approx 1.8$, both volume and mass in the largest supercluster rapidly grow, overtaking the volume and mass in the entire excursion set, and completely dominating the entire sample from this point onwards. The largest void behaves in a qualitatively similar manner if plotted versus FF_V. At $FF_V < 0.22$, $\delta_V < -0.5$ its volume is small compared to the volume of the under-dense excursion set, $\delta < \delta_{\rm TH}$, but at the percolation transition $FF_V \approx 0.22$, $\delta_V \approx -0.5$, it takes over and from then on it remains

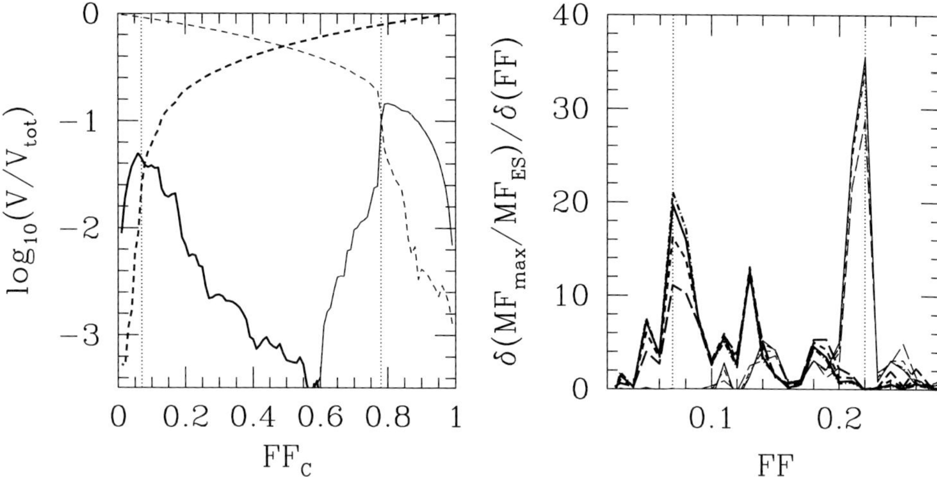

Figure 1. Left: the fractions of the total volume occupied by the largest supercluster (thick dashed line), all superclusters but the largest one (thick solid line), largest void (thin dashed line), and all voids but the largest one (thin solid line) are shown for the density field in the ΛCDM model smoothed with L_s=5 h^{-1}Mpc as a function of the filling factor, FF_C. Right: the estimates of the percolation thresholds. Thick lines show the rate of growth $\delta m^{(i)}/\delta(FF)$ for four estimators (Shandarin *et al.* 2003) for superclusters as a function of FF_C. All four curves consistently peak at $FF_C = 0.07$. Thin lines show similar quantities for voids with a distinct peak at $FF_C = 0.22$. Vertical dotted lines mark the percolation thresholds.

the dominant structure in the under-dense excursion set. Since $FF_C \equiv 1 - FF_V$ the void percolation transition takes place at $FF_C \approx 0.78$ as shown in Fig. 1 (left panel). Two obvious conclusions can be drawn from the above discussion. First, at percolation the object having the largest volume becomes very different from all the remaining objects, therefore it must be studied separately. Second, individual objects – both superclusters and voids – must be studied in the corresponding phase *before percolation occurs* in the corresponding phase. Both superclusters and voids reach their largest sizes, volumes and masses just before percolation sets in. The percolation thresholds can be accurately measured from the peaks in the rate of growth of the MFs of the largest supercluster or void as a function of FF (Fig. 1 right panel).

It is interesting to compare the percolation and genus statistics. Both were suggested as tests for the connectedness of the large-scale structure. First, Zel'dovich and Shandarin (Zel'dovich 1982, Shandarin 1983, Shandarin & Zel'dovich 1983) raised the question of topology of large-scale structure and suggested percolation statistic as a discriminator between models. The percolation test was first applied to a redshift catalog compiled by J. Huchra and Rood (Zel'dovich *et al.* 1982; Einasto *et al.* 1984) who found that the connectedness in this catalog was significantly stronger than for a Poisson distribution. In contrast, a non-dynamical model having approximately correct correlation functions up to the fourth order (Soneira & Peebles 1978) showed significantly weaker connectedness than in the observed catalog. Thus, percolation was able to detect connectedness in the galaxy distribution. It was also demonstrated that three lowest order correlation functions (two-, three- and four-point functions) are not sufficient to detect the connectedness in the galaxy distribution. Figure 2 illustrates the difference between the global genus (Fig. 2 left panel) and percolation statistic (Fig. 1 left panel and Fig. 2 right panel): at the percolation threshold many properties of the excursion set significantly change while the global genus curve remain smooth and featureless.

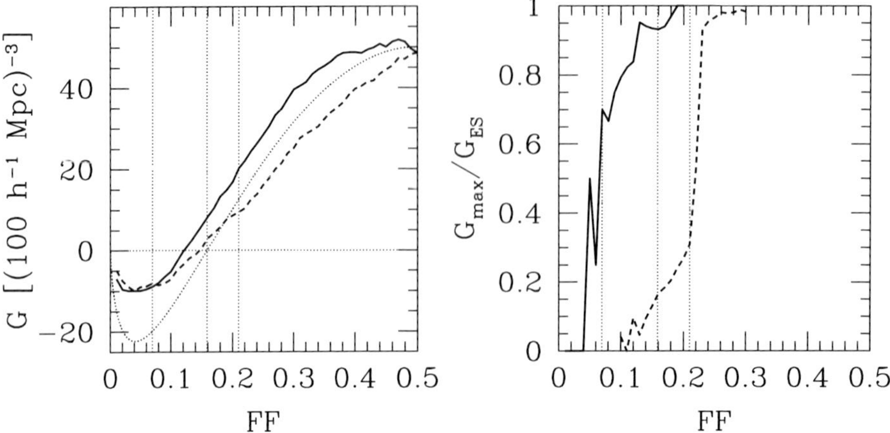

Figure 2. Left: the global genus is shown as a function of the filling factor for the density field smoothed with $L_s = 5\ h^{-1}$ Mpc. The half of the curve corresponding to high density thresholds is plotted as a function of FF_C (solid line) while the other half corresponding to low density thresholds is plotted as a function of FF_V (dashed line). For comparison, the dotted line shows the Gaussian genus curve having the same amplitude. Right: the percolation transitions in the same density field as indicated by the genus of the largest supercluster (solid line) and largest void (dashed line). The vertical dotted lines mark two percolation thresholds in the ΛCDM ($FF \approx 0.07$ and $FF \approx 0.22$) and Gaussian field ($FF \approx 0.16$) in both panels.

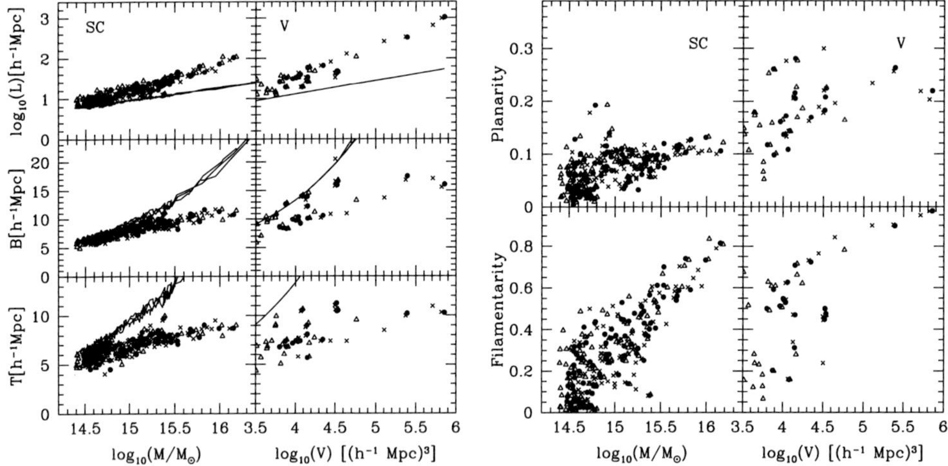

Figure 3. Left: length, breadth, and thickness versus mass for superclusters and versus volume for voids. Solid circles show the relation at percolation thresholds: $FF_C = 0.07$ for superclusters and $FF_V = 0.22$ for voids. Crosses show the parameters before percolation ($FF_C = 0.06$ for superclusters and $FF_V = 0.21$ for voids) and empty triangles after percolation ($FF_C = 0.08$ for superclusters and $FF_V = 0.23$ for voids). Solid lines show the radius of the sphere having the same volume as the supercluster or void. Right: planarity and filamentarity vs mass (for superclusters) and vs volume (for voids) at percolation. Notations are as in the left panel.

Three characteristic sizes and shapes of superclusters and voids can be estimated from the MFs. It is surprising that the thickness of superclusters depends on the threshold quite weakly; it is within $4\text{–}6h^{-1}$ Mpc interval for a range of thresholds between $0 < \delta < 6$. This may indicate that the actual thickness of superclusters is significantly smaller and the

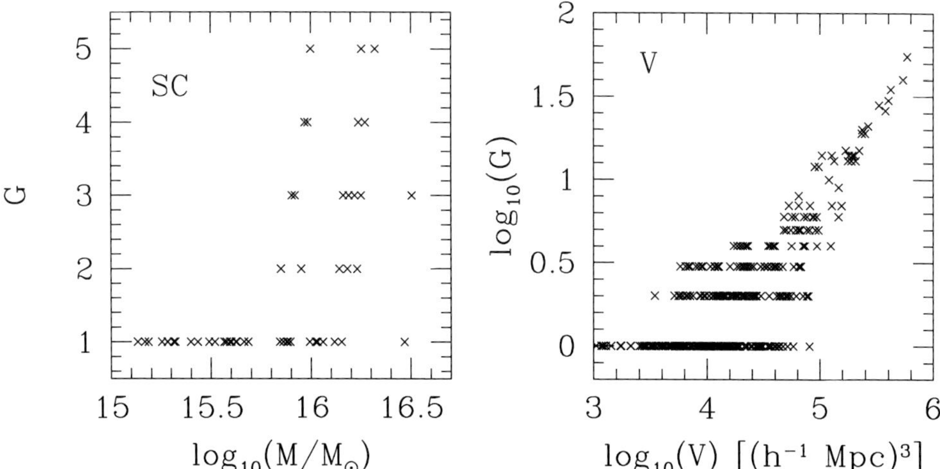

Figure 4. Left: genus-mass relation for superclusters. Right: genus-volume relation for voids. Every isolated supercluster and void having genus greater than zero at all density thresholds is shown.

measured values reflect the width of the smoothing window. The breadth of superclusters is not much greater than the thickness and it is likely that it is also affected by the width of the filtering window. Voids are a little fatter than superclusters and their thickness reaches about $9h^{-1}$ Mpc at the percolation threshold. Voids are also wider and longer than superclusters. Recalling that the size parameters are normalized to the radius of the sphere rather than to the diameter, we conclude that the longest 25% of superclusters are longer than about $50h^{-1}$ Mpc and 25% of voids are longer than about $60h^{-1}$ Mpc.

The correlation of the sizes (L, B, T) and shape parameters (P, F) of superclusters and voids with the masses (superclusters) and volumes (voids) are shown in Fig. 3. It was expected that all three sizes of the superclusters may correlate with their masses and the sizes of the voids correlate with their volumes. It is perhaps also not surprising that the filamentarity of superclusters increases with the mass. However, the strong correlation of the filamentarity (F) of voids with their volumes is quite unexpected and may require reconsidering some theoretical models of the void evolution. Figure 4 (left panel) demonstrates that the planarity of superclusters is quite small, indicating that pancakes are not typical structures in the ΛCDM cosmology. In addition, we have found a very strong indication of complex geometry and nontrivial topology in the largest superclusters (Fig. 4 left panel) and especially in voids (Fig. 4 right panel).

3. Summary

Individual superclusters totally occupy no more than about 5% of the total volume and comprise no more than 20% of mass if the largest (i.e. percolating) supercluster is excluded. The maximum of the total volume and mass comprised by all superclusters, except the largest one, is reached approximately at the percolation threshold: $\delta \approx 1.8$ corresponding to $FF_C \approx 0.07$. Individual voids totally occupy no more than 14% of volume and contain no more than 4% of mass if the largest void is excluded. The maximum of the total volume and mass comprised by all voids except the largest one is reached at about the void percolation threshold: $\delta \approx -0.5$ corresponding to $FF_V \approx 0.22$. Between these two percolation thresholds all superclusters and voids except the largest ones take up no more than about 10% of volume and mass. Both the largest supercluster and void span throughout the whole space and have a very large genus. Therefore neither of them has well defined sizes, volumes, masses or easily defined shapes.

The sizes of voids are significantly larger than those of superclusters even in the density field smoothed with $L_s = 5h^{-1}$ Mpc. The length of a quarter of the most massive superclusters exceeds $50h^{-1}$ Mpc. The most voluminous voids are even longer: 25% of them are longer than $60h^{-1}$ Mpc. The longest non-percolating supercluster is as long as $100h^{-1}$ Mpc and the longest non-percolating void is as long as $200h^{-1}$ Mpc. Both are comparable to the size of the box ($239.5h^{-1}$ Mpc) and therefore may be affected by the boundaries.

The genus value of individual superclusters can be ~ 5 while the genus of individual voids can reach ~ 55 (Fig. 4). This implies significant amount of substructure in superclusters and especially in voids. This is in a general agreement with other studies of voids (Kofman *et al.* 1992; Peebles 2001; Gottlöber *et al.* 2003).

One of our main results is that voids, as defined through the density field, can be distinctly non-spherical. Whether this result carries over to voids in galaxy surveys will depend upon the nature of the baryon-dark matter biasing and also on whether the density field is sampled in real or in redshift space.

The percolation thresholds as well as some other parameters depend on the smoothing scale and for smaller smoothing scales or adaptive filtering windows the supercluster percolation threshold must decrease ($FF_C^{perc.} < 0.07$) and the void percolation threshold increase ($FF_V^{perc.} > 0.22$).

Acknowledgements

I am thankful to Antonaldo Diaferio for a careful reading of the manuscript and useful critical comments.

References

Cole, S., Hatton, S., Weinberg, D.H., Frenk, C.S., 1998, MNRAS, 945
Einasto, J., Klypin, A.A., Saar, E., Shandarin, S.F., 1984, MNRAS, 206, 529
Gott, J.R., Melott, A.L. & Dickinson, M., 1986, ApJ, 306, 341
Gottlöber, S., Lokas, E.L., Klypin, A. Hoffman, Y., 2003, MNRAS, 344, 715
Jenkins, A.R. et al. (for the Virgo Constortium), 1998, ApJ, 499, 20
Kofman, L.A., Pogosyan, D., Shandarin, S.F. and Melott, A., 1992, ApJ, 393, 450.
Mecke, K.R., Buchert, T. & Wagner, H., 1994, A&A, 288, 697
Peebles, P.J.E., 2001, ApJ, 557,495
Sahni, V., Sathyaprakash, B.S. & Shandarin, S.F., 1998a, ApJL, 495, L5
Sathyaprakash, B.S., Sahni, V. & Shandarin, S.F., 1998b, ApJ, 508, 551
Sheth, J.V. 2003, astro-ph/0310755
Sheth, J.V., Sahni, V., Shandarin, S.F. & Sathyaprakash, B. 2003, MNRAS, 343, 22
Shandarin, S.F. 1983, Sov. Astron. Lett., 9, 104
Shandarin, S.F. & Zeldovich, Ya.B., 1983, Comments Astrophys. 10, 33
Shandarin, S.F. & Zeldovich, Ya.B., 1989, Rev. Mod. Phys.,61,185
Shandarin, S.F. Sheth, J., & Sahni, V. 2003, astro-ph/0312110
Soneira, R.M. & Peebles, P.J.E. 1978, Astron. J., 83, 845
Springel, V. et al. (for the Virgo Consortium), 1998, MNRAS, 298, 1169
van de Weygaert, R., 2002, astro-ph/0206427
Zel'dovich, Ya.B. 1982, Sov. Astron. Lett., 8, 102
Zeldovich, Ya.B., Einasto, J. and Shandarin, S.F., 1982, Nature, 300, 407

Outskirts of Galaxy Clusters: Intense Life in the Suburbs
Proceedings IAU Colloquium No. 195, 2004
A. Diaferio, ed.

DOI: 10.1017/S1743921304000122

Voids in a ΛCDM universe

Jörg M. Colberg

University of Pittsburgh, 3941 O'Hara Street, 100 Allen Hall, Pittsburgh PA 15260, USA
email: astro@jmcolberg.com

Abstract. We study the formation and evolution of voids in the dark matter distribution using various simulations of the popular Λ Cold Dark Matter Universe. We identify voids by requiring them to be spherical or elliptical regions of space with a mean overdensity of −0.8 or less. The distribution of void sizes in the different simulations shows good overlap. The size of a void is related to the depth of the smoothed density field at that position in the initial conditions. The rescaled mass profiles of voids in the different simulations agree remarkably well. We find a universal void mass profile of the form $\rho(<r)/\rho(r_{\rm eff}) \propto \exp[(r/r_{\rm eff})^{\alpha}]$ where $r_{\rm eff}$ is the effective void radius and $\alpha \sim 2$. The mass function of haloes in voids is steeper than that of haloes which populate denser regions.

---

1. Introduction

Galaxy redshift surveys show a complicated network of galaxies around large almost empty regions, so-called voids. One of the most famous voids, in the region of Boötes, has a diameter of $\sim 50h^{-1}$Mpc, and was found by Kirschner *et al.* in 1981.† Even though there have been some studies of properties of voids during the 1980s and 1990s (see e.g. Müller *et al.* 2000 and references therein), only recently have galaxy surveys become large enough to yield void samples which allow systematic studies (e.g. Rojas *et al.* 2003).

For similar reasons, voids in cosmological N–body simulations have also been somewhat less well-studied. Early simulations of Cold Dark Matter (CDM) universes showed that large empty regions were generic (Davis *et al.* 1985), and larger, more recent simulations (e.g., Jenkins *et al.* 1998) have provided a clearer picture of voids. Detailed studies of the properties of voids in the dark matter distribution are now becoming increasingly common (e.g. Gardner 2001, Schmidt *et al.* 2001, Gottlöber *et al.* 2003).

One of the problems with studies of voids is that there is little agreement on how to define a void in the galaxy distribution. Are voids regions which are completely devoid of galaxies? If no, how do void-galaxies differ from the galaxies which populate denser environments?

In models of galaxy formation within the context of hierarchical clustering, the galaxy distribution is determined by the underlying dark matter. Therefore, to understand void galaxies, it is important to define precisely what constitutes a void in the dark matter distribution. Dubinski *et al.* (1992) argued that the spherical evolution model (Gunn & Gott 1972) provides a useful guide. Here, we report on the results of a study which uses this definition of voids in the standard cosmology, ΛCDM.

† Throughout this work, we will express the Hubble constant in units of $H_0 = 100h$ km sec^{-1} Mpc^{-1}.

Simulation	Reference	n_p	l $[h^{-1}\text{Mpc}]$	m_p $[10^{10}h^{-1}M_\odot]$
GIF	Jenkins *et al.* 1998	256^3	141.3	1.4
VLS	Jenkins *et al.* 2001	512^3	479	6.9
GIF2	Gao *et al.* 2003	400^3	110	0.2
HV	Evrard *et al.* 2002	1000^3	3000	224.8

Table 1. Parameters of the simulations used in this work. All runs have $\Omega_m = 0.3$, $\Lambda = 0.7$, $n = 1$, $\sigma_8 = 0.9$, $h = 0.7$.

2. Finding voids in Cold Dark Matter universes

2.1. *The simulations*

For the study, we use a set of N–body simulations done by the Virgo Supercomputing Consortium‡. The simulations model regions of different sizes with different mass resolutions. Table 2.1 provides details about the simulations; in the following, we will refer to them as GIF, VLS, GIF2, and HV.

2.2. *The void finding algorithm*

A number of void-finding algorithms have been proposed (see e.g. Kauffmann & Fairall 1991, Aikio & Mähönen 1998, or Hoyle & Vogeley 2002). Most look for empty spherical or cubical regions, which are then usually merged together following some recipe. Our algorithm is a variant of the one advocated by Aikio & Mähönen (1998). It is based on the assumption that voids are primordial negative overdensity perturbations which grew gravitationally and have reached shell–crossing at present time. At shell-crossing, the comoving radius of a perturbation is 1.7 times larger than it was initially, so that the object has a density contrast of -0.8 (see Blumenthal *et al.* 1992, Dubinski *et al.* 1993). Strictly speaking, these numbers are correct for an Einstein de-Sitter cosmology. But the dependence on cosmology is weak, and we ignore it. Our algorithm looks for such regions in the simulations by finding spherical proto-voids first and then merging those so that final voids can be either spherical or have a more complex shape. However, we make sure that we never get configurations where two or more voids are joined through a thin tunnel. Thus, our void finder is analogous to the spherical–overdensity method for dark matter haloes (Lacey & Cole 1994). We compute the center of each void by taking the volume–weighted average of the centers of its constituent spheres. We compute an 'effective' radius by taking the radius of a sphere whose volume matches that of the void.

3. Voids in a ΛCDM universe

3.1. *Visual impression*

Figure 1 gives a visual impression of the voids in the VLS simulation volume at $z = 0$. The image was generated using a ray–tracing programme, rendering the proto–voids as spheres. From the image it is clear that the voids fill almost the entire volume.

3.2. *The void volume function*

On the left–hand side, Figure 2 shows the number density of voids larger than a given volume at $z = 0$ in the four simulations. Note that there is not a single very big void in the GIF simulations. The curves obtained from the different simulations show very good

‡ http://www.virgo.dur.ac.uk

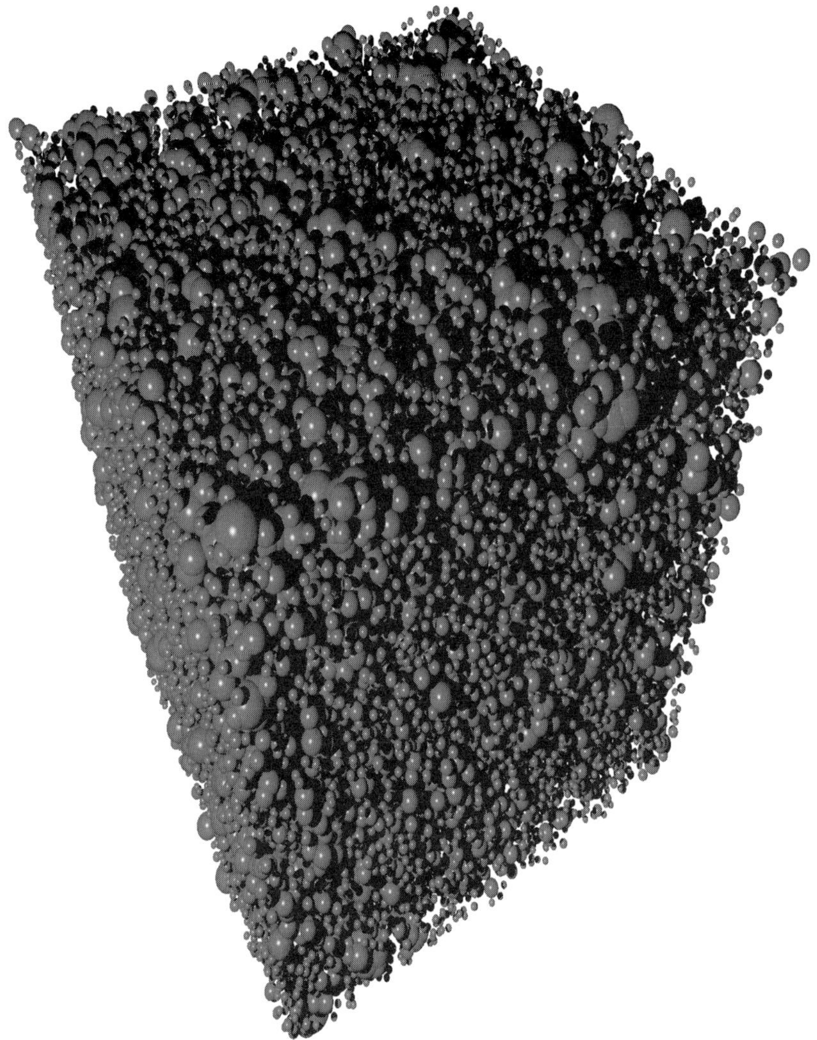

Figure 1. Voids in the VLS simulation at $z = 0$.

overlap. The steps visible at small V result from the discreteness of the grid. The largest void found in the HV simulation has an effective radius of $55\,\mathrm{Mpc}/h$.

On the right–hand side, Figure 2 shows how the cumulative volume function in the highest resolution simulation (GIF2) evolves. The evolution is smooth, and the volume functions at different redshifts cross each other.

3.3. *Voids in the initial density field*

Massive haloes in simulations are associated with higher peaks in the (smoothed) initial density field (Bardeen *et al.* 1986; Colberg *et al.* 2000). Voids are expected to form from initially underdense regions analogously to how clusters or haloes form from initially overdense regions. So one might wonder if a similar correlation exists between voids and minima in the initial density field. We used the GIF simulations to study this correlation.

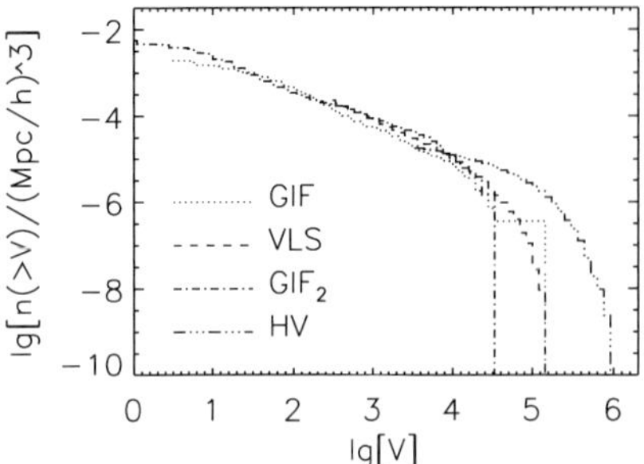

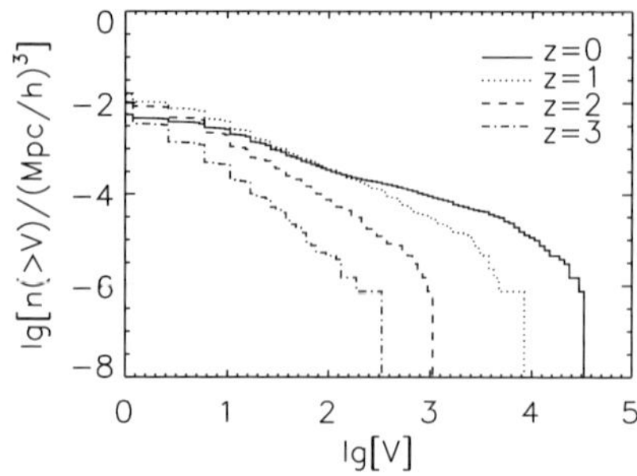

Figure 2. Left: Cumulative volume functions of voids at $z = 0$ in the GIF2 (dot–dashed), GIF (dotted), VLS (dashed), and HV simulations (dot-dot-dot-dashed). Right: Evolution of the the cumulative void volume fraction in the GIF2 simulation at $z = 0$ (solid), $z = 1$ (dotted), $z = 2$ (dashed), and $z = 3$ (dot–dashed).

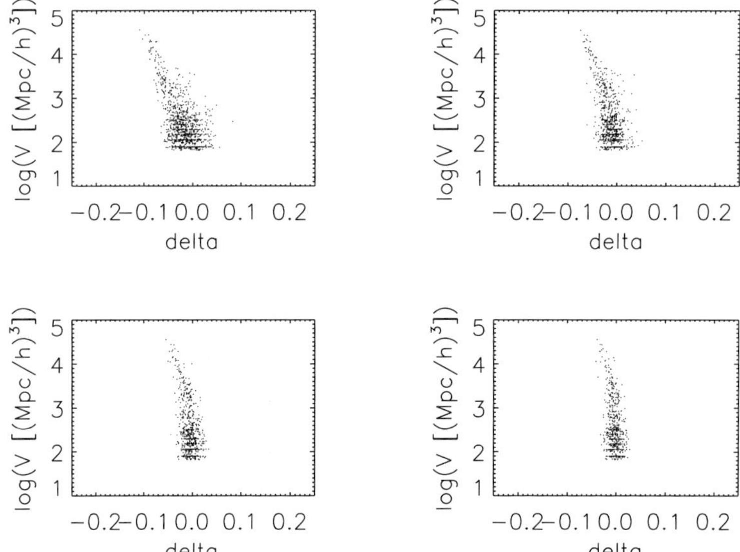

Figure 3. Voids in the smoothed GIF initial density field. There is a correlation between the void volume at $z = 0$, and the underdensity δ of the associated minimum in the initial field. Different panels show results for different smoothing scales: $5.0h^{-1}$Mpc (upper left), $7.5h^{-1}$Mpc (upper right), $10.0h^{-1}$Mpc (bottom left), and $12.5h^{-1}$Mpc (bottom right).

Since the voids in our sample enclose a large range of masses, we smoothed the initial ($z = 49$) density field using a set of four Top Hat filters: 5.0, 7.5, 10.0, and $12.5h^{-1}$Mpc. We identified the minima in each smoothed field and, where possible, associated a void with it. Where there are two or more minima for one void we pick the deepest one. Not all voids had associated troughs, but this is not very surprising, since Colberg *et al.* did not find a peak for each cluster either.

Figure 3 shows the results for the four smoothing scales. All four panels show the void volumes as a function of the overdensities of the associated troughs. Most of the voids can be associated with troughs in the initial density field. Larger voids form from somewhat deeper troughs than smaller voids, and there is a correlation between the size of a void

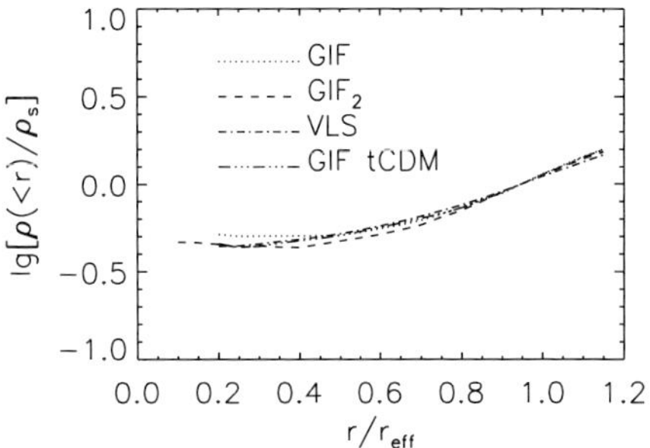

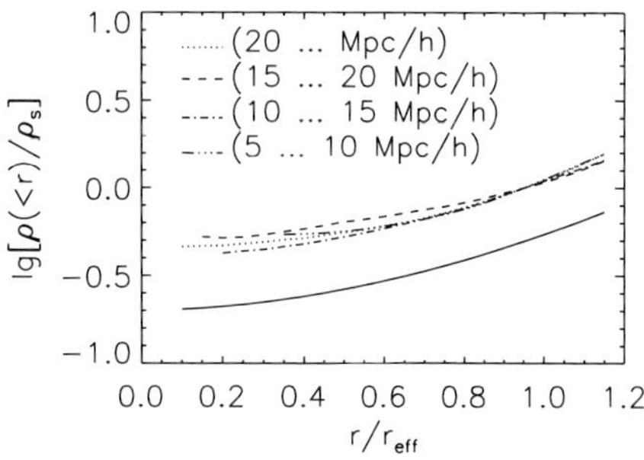

Figure 4. Left: Enclosed density in $z = 0$ voids as a function of radius for the GIF2 (dashed line), GIF (dotted line), and VLS (dot–dashed line) simulations (averaged over voids with $r_{eff} > 5\,\mathrm{Mpc}/h^{-1}$). In addition, the results from the $\Omega = 1$ τCDM GIF simulation are shown (three dots–dashed line). Curves are truncated at small radii because of the numerical resolution limits. Right: Enclosed density profiles in the VLS simulation, using four samples with void radii between 5 and $10h^{-1}$Mpc (dashed-three-dots), 10 and $15h^{-1}$Mpc (dot dashed), 15 and $20h^{-1}$Mpc (dashed), and voids with radii larger than $20h^{-1}$Mpc (dotted), averaged over all voids in each sample. Solid line shows equation (3.1), but shifted downwards by a factor of two.

and the depth of its corresponding trough, although this correlation is fairly weak for small voids.

3.4. *Void density profiles*

Navarro *et al.* (1996) have argued that CDM haloes have a universal density profile. We argue that the same holds true for voids.

Using our samples from the GIF, GIF2, and VLS simulations, we have computed the mass profiles of voids. On the left–hand side of Figure 4 we show the averaged enclosed density in $z = 0$ voids as a function of radius. For each void, we re–scaled the length scales (densities) by dividing by the effective radius (enclosed density at that radius). For almost the entire range, the average density profiles of voids in the three simulation sets agree very well. We also computed density profiles for the $\Omega = 1$ τCDM GIF simulation. These agree with the profiles of the ΛCDM simulations. This finding indicates that the form of void density profiles is indeed universal.

The cumulative profile shown in Figure 4 is quite well described by

$$\rho(< r)/\rho(r_{\rm eff}) = \frac{\exp[(r/r_{\rm eff})^{1.85}]}{2.5} \,. \tag{3.1}$$

3.5. *The void halo mass function*

We study the mass function by using the GIF2 simulation which has the highest mass resolution. We identify haloes using a friends–of–friends (fof) group finder with a linking length of 0.2 times the mean interparticle separation. At $z = 0$, we find void haloes by picking those haloes whose centres-of-mass lie within a void. We then mark those particles that are in a void at $z = 0$ and run the fof group finder on them at earlier redshifts.

Figure 5 compares the mass function of all haloes with that of haloes whose particles lie in a void at $z = 0$. The plot indicates that haloes which end up in a void at $z = 0$ – probably located at the very edges of a void – *at any fixed mass* undergo slightly more evolution than haloes with the same mass elsewhere. It also agrees with the more detailed investigation into this topic done by Gottlöber *et al.* (2003) who described the difference between the void and non–void haloes in much detail.

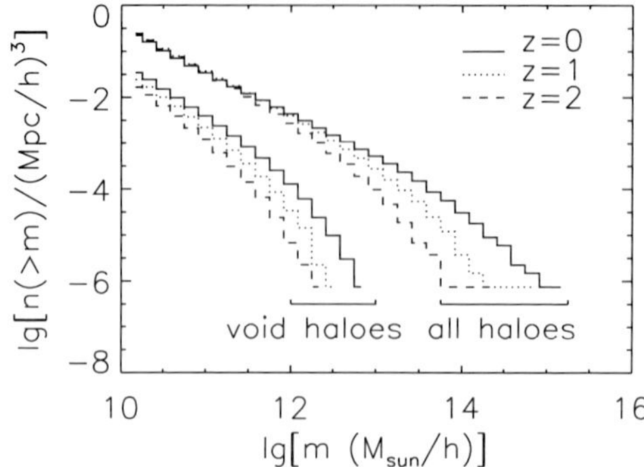

Figure 5. Mass function of haloes in the GIF2 simulation. Different curves show the mass function of all haloes, whatever their surrounding environment, and the mass function of those haloes whose particles lie in a void at $z = 0$.

4. Summary

We discussed the properties of voids in a set of large high-resolution N-body simulations of the ΛCDM cosmology, defining voids as spherical or elliptical regions of space with a mean overdensity of -0.8. The void volume functions of the different simulations agree well. There are more smaller voids at earlier times than at later times. Claims that CDM cosmologies do not form large enough voids can be put to rest given the sizes of the largest voids found in the Hubble Volume simulation. In addition, as our voids are defined through their mean overdensity we also show that CDM voids do not contain too much matter.

Voids correspond to troughs in the initial density field. There is a weak correlation of void size with initial density, larger voids form from somewhat deeper troughs than small voids.

When appropriately rescaled, voids appear to have a universal density profile. The void density profiles rise steeply at the edges of voids. Voids are thus very well defined in terms of their densities.

The mass function of haloes in voids is different from that in regions of average density, it is steeper in voids.

Acknowledgments

I thank my collaborators Ravi Sheth, Antonaldo Diaferio, Liang Gao, Naoki Yoshida, and Adrian Jenkins for allowing me to present the work. The simulations were carried out as part of the Virgo Consortium programme, at the Rechenzentrum of the Max–Planck-Gesellschaft in Garching, Germany and at the Edinburgh Parallel Computing Center.

I also want to thank the organizer of the conference, Antonaldo Diaferio, for enabling such a productive and stimulating workshop.

References

Aikio J. & Mähönen P., 1998 *ApJ* **497**, 534ff.

Blumenthal G. R., Da Costa L, N., Goldwirth D. S., Lecar M., Piran T., 1992 *ApJ* **388**, 234ff.

Colberg J. M., White S. D. M. MacFarland T. J., Jenkins A., Pearce F. R., Frenk C. S., Thomas P. A., Couchman H. M. P. (The Virgo Consortium), 2000 *MNRAS* **313**

Davis M., Efstathiou G., Frenk C.S., White S.D.M., 1985 *ApJ* **292**, 371ff.

Dubinski J., Da Costa L, N., Goldwirth D. S., Lecar M., Piran T., 1993 *ApJ* **410**, 458ff.

Evrard A. E., MacFarland T. J., Couchman H. M. P., Colberg J. M., Yoshida N., White S. D. M., Jenkins A., Frenk C. S., Pearce F. R., Peacock J. A., Thomas P. A., 2002 *ApJ* **573**, 7ff.

Gao L., White S. D. M. and The Virgo Consortium, 2004 *in preparation*

Gardner J.P, 2001 *ApJ* **557**, 616ff.

Gottlöber S., Lokas E., Klypin A., Hoffman Y. *astro-ph/0305393*

Hoyle F., Vogeley M. S., 2002 *ApJ* **566**, 641ff.

Jenkins A., Frenk C. S., Pearce F. R., Thomas P. A., Colberg J. M., White S. D. M., Couchman H. M. P., Peacock J. A., Efstathiou G., Nelson, A. H. (The Virgo Consortium), 1998 *ApJ* **499**, 20ff.

Jenkins A., Frenk C. S., White S. D. M., Colberg J. M., Cole S., Evrard A. E., Couchman H. M. P., Yoshida N., 2001 *MNRAS* **321**, 372ff.

Kauffmann G., Fairall A. P., 1991 *MNRAS* **248**, 313ff.

Kauffmann G., Colberg J. M., Diaferio A., White S. D. M., 1999 *MNRAS* **303**, 188ff.

Lacey C., Cole, S., 1994 *MNRAS* **271**, 676ff.

Müller V., Arbabi–Bidgoli S., Einasto J., Tucker D., 2000 *MNRAS* **318**, 280ff.

Navarro J. F., Frenk C. S., White S. D. M., 1996 *ApJ* **462**, 563ff.

Peebles P. J. E., 2001 *ApJ* **557**, 495ff.

Rojas R.R., Vogeley M.S., Hoyle F., Brinkmann J. *astro-ph/0307274*

Schmidt J.D., Ryden B.S., Melott A.L., 2001 *ApJ* **546**, 609ff.

Outskirts of Galaxy Clusters: Intense Life in the Suburbs
Proceedings IAU Colloquium No. 195, 2004
A. Diaferio, ed.

DOI: 10.1017/S1743921304000134

A hierarchy of voids

Rien van de Weygaert[1], Ravi Sheth[2,3] and Erwin Platen[1]

[1] Kapteyn Instituut, Rijksuniversiteit Groningen, the Netherlands

[2] Dept. of Physics and Astronomy, University of Pittsburgh, USA

[3] NASA/Fermilab Astrophysics Group, Batavia, IL 60510-0500, USA

Abstract. We present a model for the distribution of void sizes and its evolution within the context of hierarchical scenarios of gravitational structure formation. For a proper description of the hierarchical buildup of the system of voids in the matter distribution, not only the *void-in-void* problem should be taken into account, but also that of the *void-in-cloud* issue. Within the context of the excursion set formulation of an evolving void hierarchy is one involving a *two-barrier* excursion problem, unlike the *one-barrier* problem for the dark halo evolution. This leads to voids having a peaked size distribution at any cosmic epoch, centered on a characteristic void size that evolves self-similarly in time, in distinct contrast to the distribution of virialized halo masses in not having a small-scale cut-off.

---

1. Introduction: excursions

Hierarchical scenarios of structure formation have been very succesfull in understanding the formation histories of gravitationally bound virialized haloes. Particularly compelling has been the formulation of a formalism in which the collapse and virialization of overdense dark matter halos within the context of hierarchical clustering can be treated on a fully analytical basis. This approach was originally proposed by Press & Schechter (1974), which found a particularly useful and versatile formulation and modification in the *excursion set formalism* (Bond et al. 1991).

It is based on the assumption that for a structure to reach a particular nonlinear evolutionary stage, such as complete gravitational collapse, the sole condition is that its *linearly extrapolated primordial density* should attain a certain value. The canonic example is that of a spherical tophat overdensity collapsing once it reaches the collapse barrier $\delta_c \approx 1.69$. The successive contributions to the local density by perturbations on a (mass) resolution scale S_m may be represented in terms of a density perturbation random walk, the cumulative of all density fluctuations at a resolution scale smaller than S_m. By identifying the largest scale at which the density passes through the barrier δ_c it is possible (1) to infer at any cosmic epoch the mass spectrum of collapsed halos and (2) to reconstruct the merging history of each halo (see Fig. 3, lefthand). In this study we demonstrate that also the formation and evolution of foamlike patterns as a result of the gravitational growth of primordial density perturbations is liable to a succesfull description by the excursion set analysis. A slight extension and elaboration on the original formulation enables us to frame an analytical theory explaining how the characteristic observed weblike Megaparsec scale galaxy distribution, and the equivalent frothy spatial matter distribution seen to form in computer simulations of cosmic structure formation, are natural products of a hierarchical process of gravitational clustering. This is accomplished by resorting to a complementary view of clustering evolution in which we focus on the evolution of the *voids* in the Megaparsec galaxy and matter distribution, spatially *the* dominant component (see e.g. Müller & Maulbatsch, these proceedings).

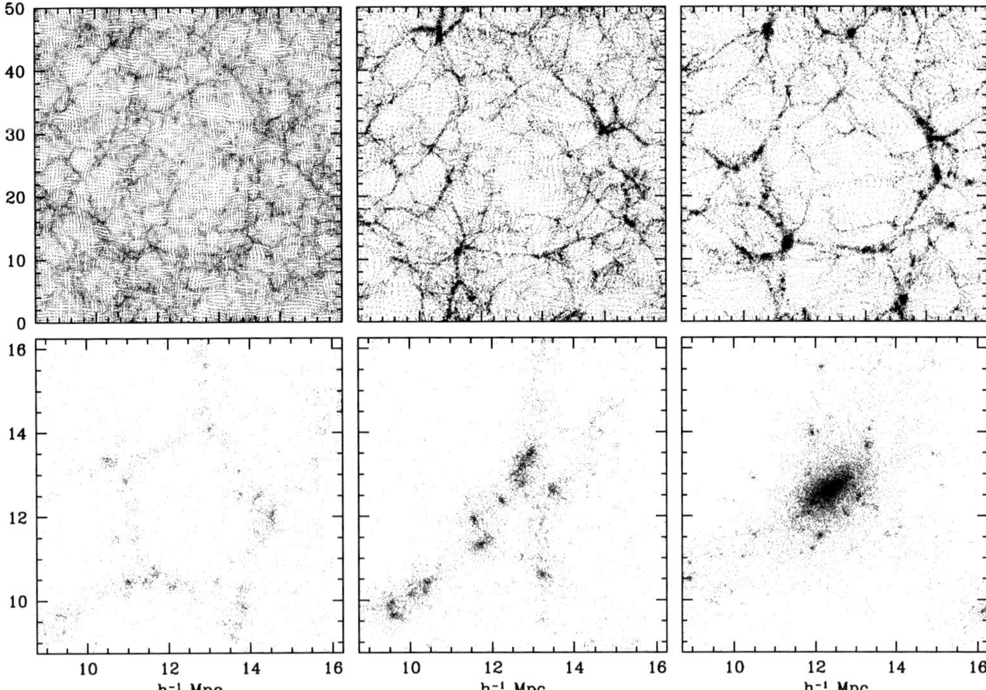

Figure 1. Illustration of the two essential "void hierarchy modes": (top) the *void-in-void* process, with a void growing through the merging of two or more subvoids; (bottom) the *void-in-cloud* process: a void demolished through the gravitational collapse of embedding region.

2. Void evolution: the two processes

Primordial underdensities are the progenitors of voids. Because underdensities are regions of suppressed gravitational attraction, representing an effective repulsive gravity, matter flows out of their interior and moves outward to the edges of these expanding *voids*. Voids *expand*, become increasingly *empty* and develop an increasingly *spherical* shape (Icke 1984). Matter from the void's interior piles up near the edge: usually a ridge forms around the void's rim and at a characteristic moment the void's interior shells take over the outer ones. At this *shell-crossing* epoch the void reaches maturity and becomes a nonlinear object expanding self-similarly, the implication being that the majority of observed voids is at or near this stage (Blumenthal et al. 1992). As voids develop from underdensities in the primordial cosmic density field, the interaction with internal substructure and external surrounding structures translates into a continuing process of hierarchical void evolution (Dubinski et al. 1993, van de Weygaert & van Kampen 1993, Colberg et al. 2004).

The evolution of voids resembles that of dark halos in that large voids form from mergers of smaller voids that have matured at an earlier cosmic time (Fig. 1, top row). However, in contrast to dark halos, the fate of voids is ruled by two processes. Crucial is the realization that the evolving void hierarchy does not only involve the *void-in-void* process but also an additional aspect, the *void-in-cloud* process. Small voids may not only merge into larger encompassing underdensities, they may also disappear through collapse when embedded within a larger scale overdensity (Fig. 1, bottom row). In terms of the *excursion set approach*, it means that the *one-barrier* problem for the halo population has to be extended to a more complex *two-barrier problem*. Voids not only should ascertain themselves of having decreased their density below the *void barrier* δ_v of the *shell-crossing*

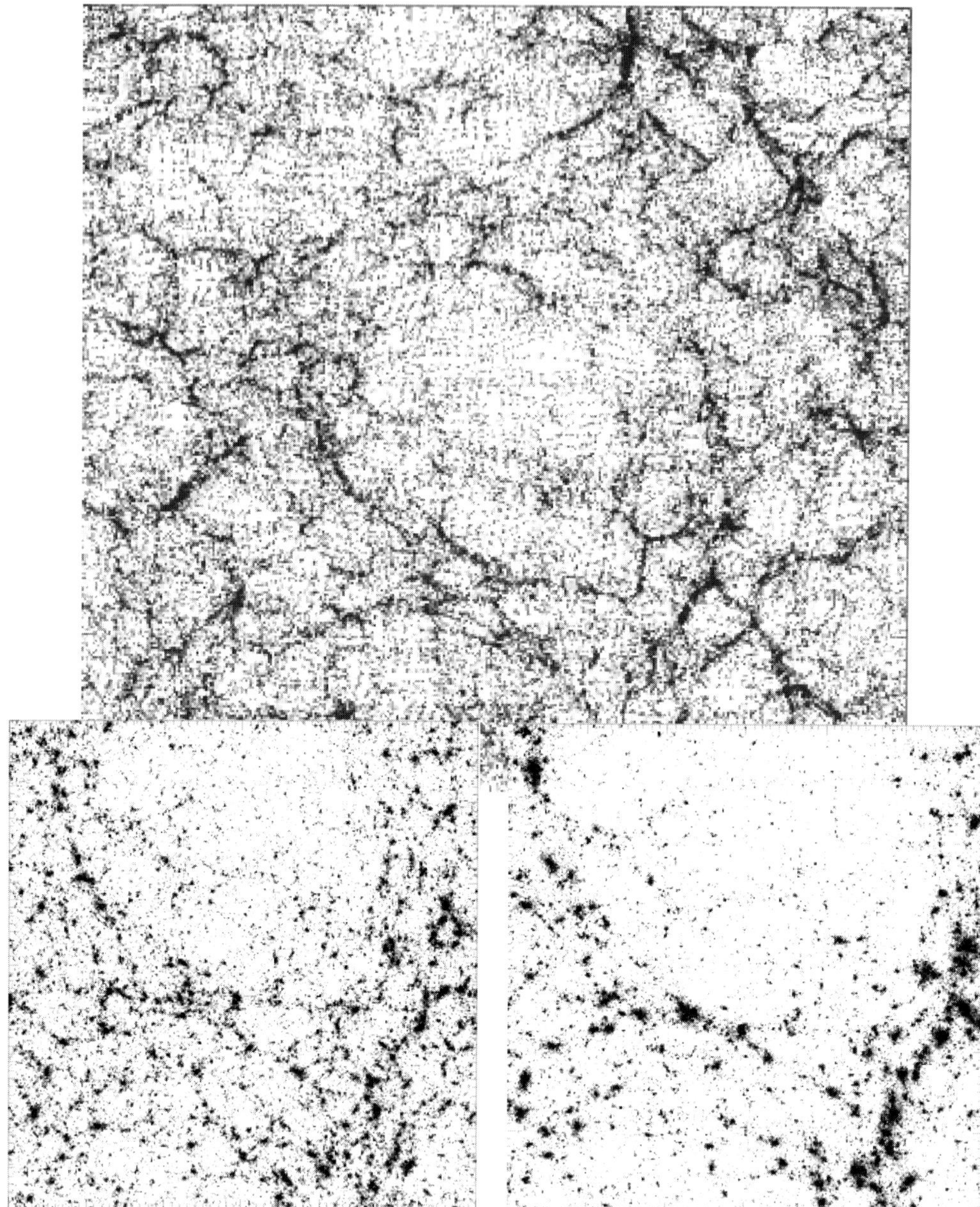

Figure 2. Identification of void collapse sites: near the boundary of an expanding voids small voids get squashed and sheared. Zoom-in on central lower region of large void (top image) at 2 different timesteps, showing the void compression process.

transition. For their survival and sheer existence it is crucial that they take into account whether they are not situated within a collapsing overdensity on a larger scale which crossed the collapse barrier δ_c. They should follow a random walk path like type "3" in Fig. 3, rather than the *void-in-cloud* path "4"). The repercussions of this are far-reaching and it leads to a major modification of the void properties and distribution.

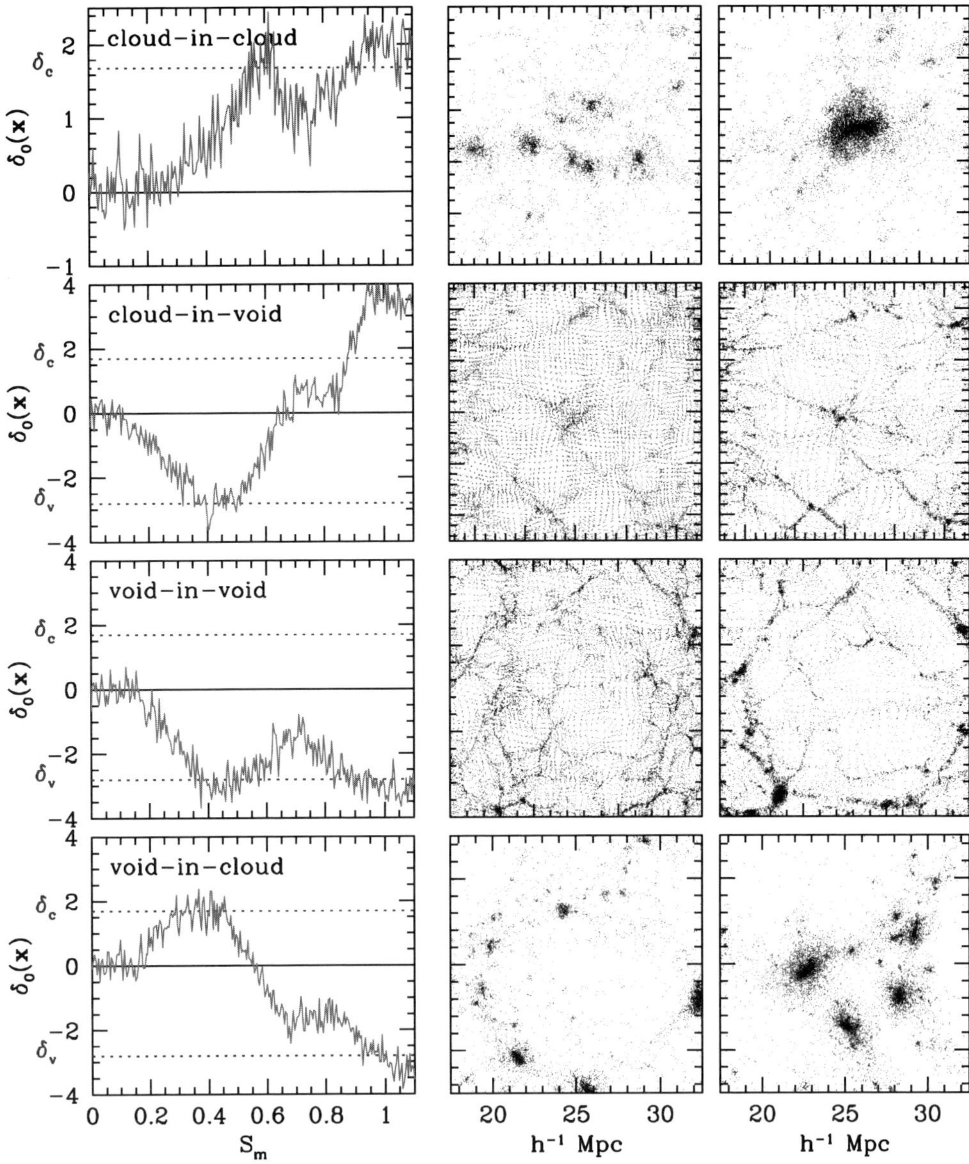

Figure 3. Four mode (extended) excursion set formalism. Each row illustrates one of the four basic modes of hierarchical clustering: the *cloud-in-cloud* process, *cloud-in-void* process, *void-in-void* process and *void-in-cloud* process (from top to bottom). Each mode is illustrated using three frames. Leftmost panels show 'random walks': the local density perturbation $\delta_0(\mathbf{x})$ as a function of (mass) resolution scale S_m at an early time in an N-body simulation of cosmic structure formation. In each graph, the dashed horizontal lines indicate the *collapse barrier* δ_c and the shell-crossing *void barrier* δ_v. The two frames on the right show how the associated particle distribution evolves. Whereas halos within voids may be observable (second row depicts a halo within a larger void), voids within collapsed halos are not (last row depicts a small void which will be squeezed to small size as the surrounding halo collapses). It is this fact which makes the calculation of void sizes qualitatively different from that usually used to estimate the mass function of collapsed halos.

3. Void distribution: universal, peaked and self-similar

Analytically, the resulting expression follows by evaluating the fraction of walks which first cross δ_v at S, and which do not cross δ_c until after they have crossed δ_v (see Fig. 3, cf. row 3 vs. row 4). This distribution $\mathcal{F}(S, \delta_v, \delta_c)$ is given by (Sheth & van de Weygaert 2003),

$$S\,\mathcal{F}(S, \delta_v, \delta_c) = \sum_{j=1}^{\infty} \frac{j^2\pi^2 D^2}{\delta_v^2/S}\,\frac{\sin(j\pi D)}{j\pi} \times \exp\left(-\frac{j^2\pi^2 D^2}{2\,\delta_v^2/S}\right), \tag{3.1}$$

in which the relative impact of void and halo evolution on the hierarchically evolving population of voids is parameterized through $D \equiv |\delta_v|/(\delta_c - \delta_v)$. A smaller value of D corresponds to a diminished importance of the *void-in-cloud* process. The above expression may be approximated by the more accessible expression, for values of $\delta_c/|\delta_v| \geqslant 1/4$, using the "self-similar" parameter $\nu \equiv \delta_v^2/S$

$$\nu f(\nu) \approx \sqrt{\frac{\nu}{2\pi}}\,\exp\left(-\frac{\nu}{2}\right)\,\exp\left(-\frac{|\delta_v|}{\delta_c}\,\frac{D^2}{4\nu} - 2\frac{D^4}{\nu^2}\right). \tag{3.2}$$

The resulting distributions, for various values of D, are shown in Figure 4. The void size distribution is clearly *peaked about a characteristic size*: the *void-in-cloud* mechanism is responsible for the demise of a sizeable population of small voids. The halo mass distribution diverges towards small scale masses, so that in terms of numbers the halo population is dominated by small mass objects. The void population, on the other hand, is "void" of small voids and has a sharp small-scale cut-off.

Four additional major observations readily follow from this analysis: (•) At any one cosmic epoch we may identify a *characteristic void size* which also explains why in the present-day foamlike spatial galaxy distribution voids of $\sim 20 - 30h^{-1}$Mpc are the predominant feature; (•) The *void distribution evolves self-similarly* and the *characteristic void size increases with time*: the larger voids present at late times formed from mergers of smaller voids which constituted the dominating features at earlier epochs (Fig. 4, top panels); (•) Volume integration shows that at any given time *the population of voids approximately fill space*, apparently squeezing the migrating high-density matter in between them; (•) As the size of the major share of voids will be in the order of that of the characteristic void size this observation implies that the *cosmic matter distribution* resembles a *foamlike packing of spherical voids of approximately similar size and excess expansion rate*.

To develop this generic picture into an encompassing view of the formation and evolution of the foamlike galaxy distribution various additional issues need to be addressed. One issue concerns the translation from matter into galaxy distribution, and thus the question of galaxy formation within voidlike environments. In this respect it should be noted that the suggested scale of the peaked matter distribution still falls short of that of the observed galaxy distribution (see e.g. Vogeley 2004). Another issue is the identification and location of the collapsing void population. Recent results (Fig. 2) have indicated that most of these events take place near and on the boundary of the large expanding voids: most small voids appear to get squashed and sheared (see van de Weygaert & Babul 1996) there where the matter currents out of voids collide with the surrounding overdensities.

In all, from this study an enticing image of cosmic structure evolution emanates. A continuously evolving hierarchy of voids produces a dynamical foamlike pattern whose characteristic dimension grows continuously along with the evolution of cosmic structure,

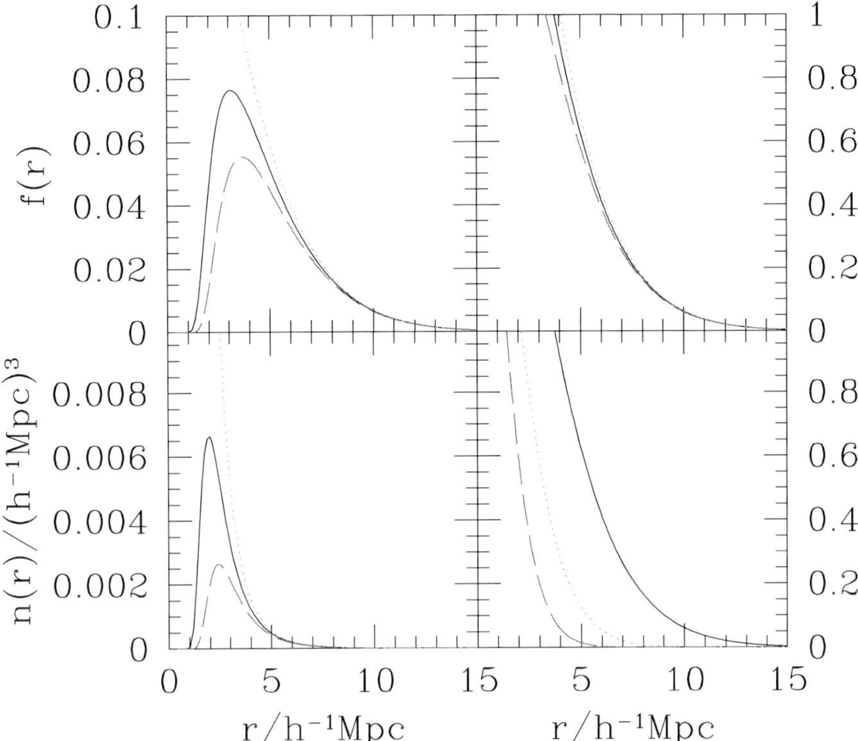

Figure 4. Distribution of predicted void radii predicted by, in an Einstein de-Sitter model with $P(k) \propto k^{-1.5}$, normalized to $\sigma_8 = 0.9$ at $z = 0$. Top left panel shows the mass fraction in voids of radius r. Bottom left panel shows the number density of voids of radius r. Note that the void-size distribution is well peaked about a characteristic size provided one accounts for the void-in-cloud process. Top right panel shows the cumulative distribution of the void volume fraction. Dashed and solid curves in the top panels and bottom left panel show the two natural choices for the importance of the void-in-cloud process discussed in the text: $\delta_c = 1.06$ and 1.686, with $\delta_v = -2.81$. Dotted curve shows the result of ignoring the *void-in-cloud* process entirely. Clearly, the number of small voids decreases as the ratio of $\delta_c/|\delta_v|$ decreases. Bottom right panel shows the evolution of the cumulative void volume fraction distribution. The three curves in this panel are for $\delta_c = 1.686(1+z)$, where $z = 0$ (solid), 0.5 (dotted) and 1 (dashed).

a Universe which at any one cosmic epoch is filled with bubbles whose size corresponds to the scale just reaching *maturity*.

References

Appel L., Jones B. J. T. 1990 *MNRAS* **245**, 522
Blumenthal G.R., et al. 1992 *ApJ* **388**, 234
Bond J. R., Cole S., Efstathiou G., Kaiser N. 1991 *ApJ* **379**, 440
Colberg J., et al. 2004, these proceedings
Dubinski J., et al. 1993 *ApJ* **410**, 458
Icke V. 1984 *MNRAS* **206**, 1
Müller V., Maulbetsch C. 2004, these proceedings
Press W., Schechter P. 1974, *ApJ* **187**, 425
Sheth R., van de Weygaert R. 2003, *MNRAS*, in the press
van de Weygaert R., Babul A. 1996, *unpublished*
van de Weygaert R., van Kampen E. 1993, *MNRAS* **263**, 481
Vogeley M. 2004, these proceedings

Outskirts of Galaxy Clusters: Intense Life in the Suburbs
Proceedings IAU Colloquium No. 195, 2004
A. Diaferio, ed.

DOI: 10.1017/S1743921304000146

Measuring the Mach number of the universe via the Sunyaev–Zel'dovich effect

F. Atrio-Barandela[1], A. Kashlinsky[2] and J.P. Mücket[3]

[1]Física Teórica. Facultad de Ciencias. 37008 Salamanca, Spain email: atrio@usal.es
[2]SSAI, Code 685, Goddard Space Flight Center, Greenbelt, MD 20771
[3]Astrophysikalisches Institut Potsdam, D-14482, Potsdam

Abstract. The cross-correlating cosmic microwave background (CMB) fluctuations caused by the Sunyaev-Zel'dovich effect from observed clusters of galaxies with their redshifts can be used to measure the mean squared cluster peculiar velocity with an error $\sigma_{C_S^2} \simeq (300\text{km/s})^2$. This can be done around $z > 0.3$ with clusters of flux above 200 mJy which will be detected by PLANCK, coupled with high resolution microwave images to eliminate the cosmological part of the CMB fluctuations. The latter can be achieved with observations by the planned ALMA array or the NSF South Pole telescope and other surveys. By measuring the rms peculiar velocity of clusters and their bulk flow in, e.g., 4 spheres of $\sim 100h^{-1}$Mpc at $z = 0.3$, we could have a direct measurement of the matter density $0.21 < \Omega_m < 0.47$ at 95% confidence level.

1. Introduction

The Mach number $\mathcal{M}$ characterizes the coldness of the velocity field. It was defined as $\mathcal{M}(r) = V_{\text{bulk}}(r)/C_S$ by Ostriker & Suto (1990) who introduce it as a test of cosmological models. The bulk flow component is the mean peculiar velocity over a region of characteristic scale r and C_S is the rms peculiar velocity in the reference frame of the flow. The cosmic Mach number effectively measures the *slope* of the power spectrum up to the scales corresponding to the size of the region and can be used to constrain cosmological models for structure formation. Hot gas in moving clusters produces redshift independent cosmic microwave background (CMB) anisotropies (Sunyaev & Zel'dovich, 1972, hereafter SZ). The thermal component (TSZ) is usually larger than the kinetic term (KSZ), but has a distinct spectral signature and can be removed using multifrequency observations, allowing the determination of the cluster peculiar velocities (Holzapfel et al. 1997; Mauskopf et al. 2000; LaRoque et al. 2002).

At present, the error bars on peculiar velocities of individual clusters, measured from KSZ (σ_{v_P}) are much larger than the estimated velocities v_P, preventing reliable measurement of the Mach number using CMB data alone. Measurements of the CMB temperature of clusters and *of their redshift* can be combined to determine $\mathcal{M}$ in some specific configurations. The bulk flow velocity can be evaluated directly from the KSZ signal by adding the velocity of all clusters in the sample (Benson et al. 2003) with uncertainty:

$$\sigma_{V_{\text{bulk}}} \simeq \frac{\sigma_{v_P}}{\sqrt{N_{cl}}} \sim 100\left(\frac{\sigma_{v_P}}{1000\ \text{km/s}}\right)\left(\frac{N_{cl}}{100}\right)^{-1/2} \text{km/s}\,. \tag{1.1}$$

The rms peculiar velocity along the line of sight of clusters located on a thin shell in redshift space can be reliably measured by cross-correlating the temperature field at cluster locations with their measured redshifts. The cross-correlation gives (Atrio-Barandela,

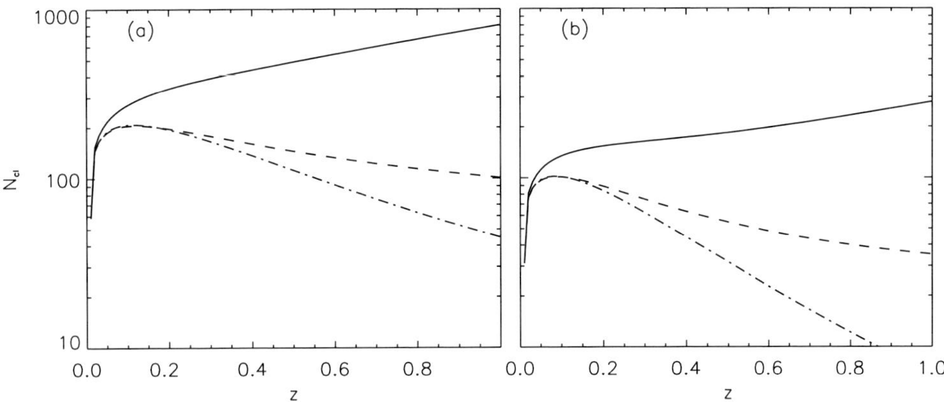

Figure 1. (a) Number of Clusters for three different XLF evolutionary histories: no evolution (solid line), dashed to $(A, B) = (-3, 0)$ and dot-dashed to $(-1, -2)$. The detection limit is 200 mJy. (b) Same as before, but the threshold limit is 400 mJy.

Kashlinsky & Mücket, 2004):

$$\frac{\langle c\delta T \cdot c\delta z \rangle}{T_o \langle \tau \rangle} = C_S^2 \pm \bar{\sigma}_{v_P} \frac{c\Delta z}{\sqrt{12 N_{cl}}} \pm \langle v_p \rangle \frac{(c\Delta z + 2C_S)}{\sqrt{12}}. \qquad (1.2)$$

The error bar of C_S depends linearly on Δz. To reduce it we compute C_S in shells with a small spread in redshift. For this geometry, the dominant contribution is

$$\sigma_{C_S^2} = (300\ \mathrm{km/s})^2 \Big(\frac{\bar{\sigma}_{v_P}}{1000\ \mathrm{km/s}} \Big) \Big(\frac{\Delta z}{0.01} \Big) \Big(\frac{100}{N_{cl}} \Big)^{1/2}. \qquad (1.3)$$

This represents an improvement of one order of magnitude on the current error bar of C_S.

2. Observational prospects

Observations of clusters to measure their SZ amplitude are already (almost) routinely carried out by several telescopes (e.g. Carlstrom et al. 2002) and exponential progress is expected with the construction of the ALMA array and the NSF funded South Pole microwave telescope. For the purpose of this project, high resolution observations of large samples of clusters with well measured redshifts are required. Blind surveys searching for clusters in large areas of the sky, like the cluster catalog expected to be obtained from the full sky CMB observations of the upcoming PLANCK ESA satellite mission will be most relevant. Most clusters detected by PLANCK will be unresolved. Those needed for computing the Mach number will have to be re-observed at high resolution with ALMA, where its high frequency coverage and its small projected noise of $\sim 2\mu$K rms after one hour of integration time for a beam of $1'$, could be very useful to subtract foreground and set the CMB contribution to its minimum.

To compute the number of clusters that will be detected by PLANCK, we modeled the X-ray Luminosity Function (XLF) as an evolving Schechter function: $\phi(L, z) = \phi_o (1 + z)^A L^{-\alpha} \exp(L/L_*)$, with $L_* = L_{*,0}(1 + z)^B$, where A, B are two evolutionary parameters, and L_*, α are the local XLF values. We took $L_* = 9.3 \times 10^{44} h^{-2}$ ergs s^{-1} and $\phi_o = 2.44 \times 10^{-7} h^3$Mpc^{-3} (Ebeling et al. 1997). In Fig. 1, solid lines correspond to no redshift evolution $(A, B) = (0, 0)$, dashed to $(-3, 0)$ and dot-dashed to $(-1, -2)$, obtained

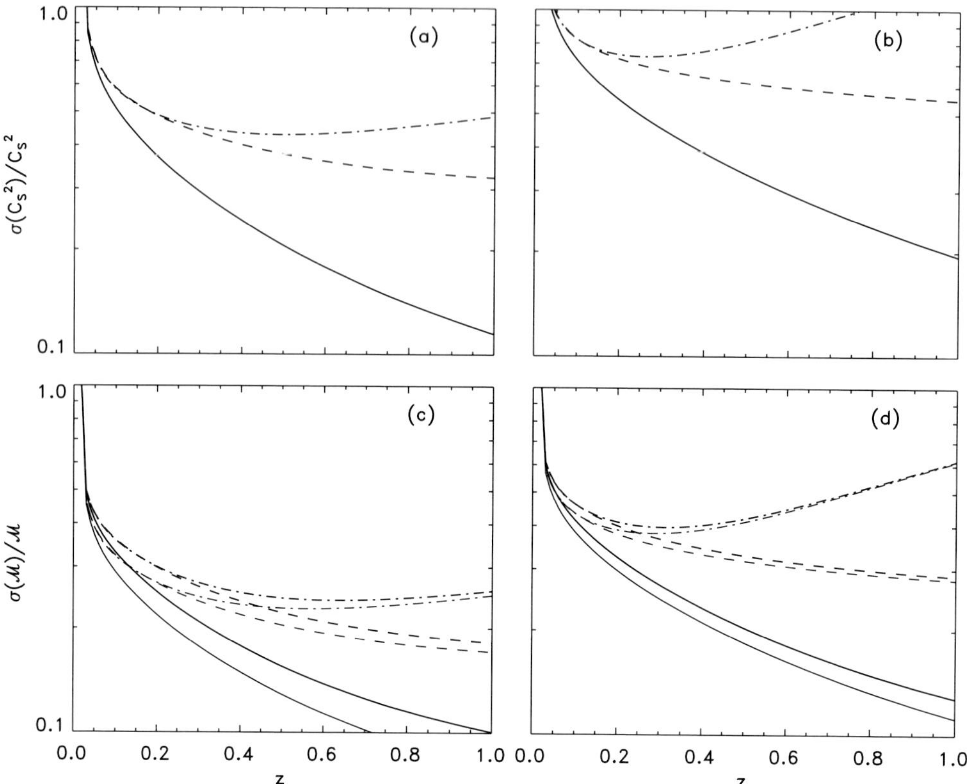

Figure 2. (a) Relative error of the sound speed for a flux detection limit of 200 mJy and (b) 400 mJy. (c) Relative error on the Mach number, for the same evolution histories as before. In every pair, the upper line corresponds to bulk flows on a sphere of $100h^{-1}$Mpc, and the lower line to a sphere of $150h^{-1}$Mpc. The flux limit is 200 mJy and (d) is 400 mJy.

from the analysis of the ROSAT Deep Cluster Survey and the Extended Medium Sensitivity Survey (EMSS), respectively. In (a) we show the number of clusters that produce a change in flux of 200 mJy relative to the mean flux of the CMB , and in (b) those producing a change of 400 mJy. For the 200 mJy limit (Fig. 1a) the RDCS evolution parameters predict about 14000 clusters out to $z = 1$, and for the 400 mJy, about 6000. Those numbers were 12000 and 4000 for the EMSS evolution parameters,

In Fig. 2(a,b) we show the relative errors of C_S^2 vs z. Lines correspond to the same evolution models as before. Spheres of radius $\simeq 100h^{-1}$Mpc will contain very few clusters in blind all-sky surveys such as PLANCK cluster catalog, and another type of survey is required. Romer et al. (2001) argue that an XMM serendipitous cluster survey of about $\simeq$800 deg^2 will detect more than 8000 clusters ranging from poor to very rich systems. In Fig. 2c and 2d we plot the error on the Mach number assuming that the sound speed was computed in shells of PLANCK detected clusters and the bulk velocity is computed using a top hat window on spheres of 100 (thick lines) and 150 h^{-1}Mpc (thin lines) radii.

Fig. 2c corresponds to PLANCK detecting all clusters with flux above 200 mJy and (d) to 400 mJy. We also considered $\bar{\sigma}_{v_P}$=1000 km s^{-1}, although now the bulk velocity and sound speed are computed over different samples.

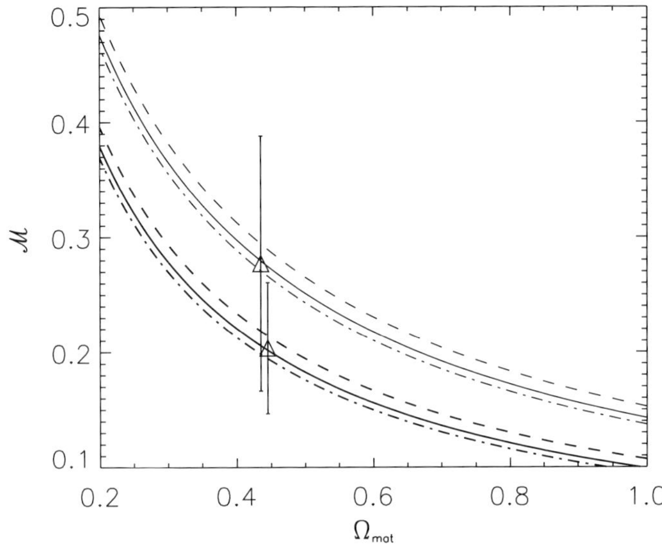

Figure 3. Mach number estimated by measuring bulk flows on spheres of 100 (upper set) or 150 h^{-1}Mpc (lower set). Dashed, solid and dot-dashed lines correspond to a spectral index $n = 0.95, 1.0$ and 1.03, in agreement with WMAP results. The triangles show the prediction for the WMAP concordance model. The 1σ error was computed assuming that there were 4 independent measurements of the bulk flow at that redshift. They are slightly shifted for a better display.

The Mach number can be used to estimate the mean matter density. For this purpose, let us assume that at redshift $z \simeq 0.3$ we have identified about $\sim$ 100 PLANCK clusters with fluxes above 200 mJy in a shell of Δz=0.01. The sample is used to compute C_S at that z. We also assume that the clusters identified by XMM in a sphere of radius $100h^{-1}$Mpc centered at z have data of similar quality. The Mach number would be the ratio of the bulk velocity in the sphere to the sound speed in the shell.

Fig. 3 plots $\mathcal{M}$ vs matter density for flat models with cosmological constant, $\Omega_\Lambda + \Omega_m$=1, and for three different values for the primordial spectral index $n = 0.95$, 1, 1.03 shown as dashed, solid and dot-dashed lines (all other cosmological parameters were chosen from the WMAP measurements, Bennett et al. 2003, Spergel et al. 2003). The matter power spectrum was normalized to σ_8=0.8. The upper set of lines corresponds to bulk flows on spheres of $100h^{-1}$Mpc and the lower set to 150 h^{-1}Mpc radius. For Ω_m=0.3 the triangle shows the expected Mach number and its 1-σ error bars. We assumed that the cluster XLF evolves with RDCS parameters. If the bulk flow is measured for 4 independent spheres of $100h^{-1}$Mpc radius, we can estimate $0.25(0.21) \leqslant \Omega_m \leqslant 0.37(0.47)$ at the 68 (95)% confidence level. Our method is insensitive to a running spectral index, quintessence or tilt: variations on these cosmological parameters lead to differences in the Mach number much smaller than $\sigma_{\mathcal{M}}$. In this respect, our measurement of the matter density would be robust.

3. Conclusions

We presented a new method to measure the cosmic sound speed of clusters of galaxies. Assuming that all clusters with SZ flux larger than 200 mJy will be identified by the

PLANCK mission, and that the XRLF evolution in the Rosat Deep Cluster Survey is representative of the overall cluster evolution, the relative error would be $\sigma_{C_S^2}/C_S^2 \sim 0.3$. If the XMM Serendipitous Cluster Survey detects all clusters above $kT_X = 4$ keV, by combining measurements of bulk flows in spheres of different sizes with the cosmic sound speed measured in shells at the same redshift, we can estimate the Cosmic Mach number with relative accuracy $\sigma_{\mathcal{M}}/\mathcal{M} \geqslant 0.2$. This accuracy can be achieved with high resolution microwave images in order to reduce the intrinsic CMB anisotropy with respect to the kinematic SZ signal, and is within the reach of the currently planned South Pole and ALMA telescopes. The Cosmic Mach number we can determine by our method is directly related to the matter density, and it can determine $0.21 \leqslant \Omega_m \leqslant 0.47$ at the 95% confidence level. This result, being independent of other mass density measurements, would provide an important self consistency check.

References

Atrio-Barandela, F. et al. 2004, ApJ, 601, L111
Bennett, C.L. et.al., 2003, ApJS, 148, 1
Benson, B.A. et al. 2003, astro-ph/0303510
Carlstrom, J. E., Holder, G.P. & Reese, E.D. 2002, ARA&A, 40, 643
Ebeling, H. et al. 1997, ApJ, 479, L101
Holzapfel, W.L. et al. 1997, ApJ, 481, 35
LaRoque, S.L. et al. 2002, astro-ph/0204134
Mauskopf, P.D. et al. 2000, ApJ, 538, 505
Ostriker, J.P. & Suto, Y. 1990, ApJ, 348, 378
Romer, A.K. et al. 2001, ApJ, 547, 594
Spergel, D.N. et.al., 2003, ApJS, 148, 175
Sunyaev, R.A. & Zel'dovich, Ya. B. 1972, Comments Astrophys. Space Phys., 4, 173

Section C

Clusters in action

Outskirts of Galaxy Clusters: Intense Life in the Suburbs
Proceedings IAU Colloquium No. 195, 2004
A. Diaferio, ed.

DOI: 10.1017/S1743921304000158

Interacting clusters and their environment

S. Bardelli

INAF-Osservatorio Astronomico di Bologna, via Ranzani 1, 40127 Bologna, Italy
e-mail: sandro.bardelli@bo.astro.it

Abstract. Central regions of superclusters are the ideal places where to study cluster merging phenomena: in fact the accretion activity is enhanced, as predicted by the cosmological simulations. In this paper I review the case-study of the Shapley Concentration, aimed to understand the effect of major mergings on the intracluster medium and the galaxy population of the involved clusters.

1. Introduction

Cluster mergings are known to be among the most energetic phenomena in the Universe, but until now studies at all wavelengths have not been extensively carried on: therefore it is still unclear how the collision energy is dissipated and which is the effect of merging on the emission properties of the galaxies and on the physics of the intracluster medium.

In cosmological N-body simulations the cluster accretion happens along specific directions defined by the density caustics and richer clusters form preferentially where the environment density is higher. Superclusters can be considered as the observational counterparts of these caustics and it is expected that at their centers the cluster accretion is still strongly active. Therefore, superclusters are the ideal places where to study merging phenomena, because the cross-section for cluster collisions is enhanced.

The best place for these studies is the central region of the Shapley Concentration supercluster, a huge concentration of clusters at $z \sim 0.05$ (Raychaudhury 1989, Plionis & Valdarnini 1991, Raychaudhury et al. 1991, Zucca et al. 1993). This region is anomalously rich in clusters, considering that it has 25 members, while the Great Attractor (with a similar mass overdensity) contains only 6 clusters (see Table 3 of Zucca et al. 1993): for some reasons, in this region the cluster formation efficiency has been enhanced also with respect to similar regions. Moreover, as noted by Raychaudhury et al. (1991), the fraction of clusters with substructures in this supercluster is higher than elsewhere, meaning that the process of cluster formation is still strongly active, suggesting that this region could be considered a "nursery" of rich clusters.

2. Large-scale structure

Various redshift surveys have been conducted in order to study the large scale distribution of galaxies in the Shapley Concentration (Quintana et al. 2000, Bardelli et al. 2000, Drinkwater et al. 2004) and its relation with the distribution of clusters. Up to now, there are known a few thousands of galaxy redshifts in the supercluster, allowing not only to determine its geometry but also its overdensity and mass (see Figure 1). The distribution of inter-cluster galaxies is well described by a plane tilted with respect to the line of sight: the distribution of galaxies around this plane is a Gaussian with a dispersion of 3.8 h^{-1} Mpc (Bardelli et al. 2000). The huge overdensity in number of

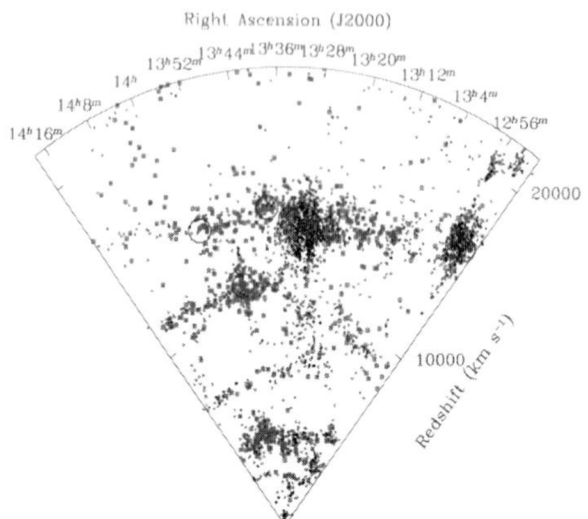

Figure 1. Central region of the Shapley Concentration supercluster (figure from Drinkwater et al. 2004; reproduced by permission of CSIRO Publishing, Melbourne, Australia; copyright Astronomical Society of Australia). Note the two structures (cluster complexes) at $\alpha = 13^h30^m$ and 12^h55^m, connected by a bridge of galaxies resembling the Great Wall. Circles indicate the position of clusters.

galaxies (~ 11 on a scale of ~ 10 h^{-1} Mpc) found for this supercluster is consistent only with ΛCDM or open CDM cosmological scenarios (Bardelli et al. 2000; for a more recent determination on larger scales see Drinkwater et al. 2004). The determination of the mass of the supercluster is more difficult because, as indicated by the overdensity, the region is far from virialization. Various methods have been applied (Ettori et al. 1997, Bardelli et al. 2000, Reisenegger et al. 2000), leading to an estimate of $\sim 10^{16}$ $M_\odot$.

All these properties are extreme also for ΛCDM models and for this reason it is quite difficult to find a supercluster like the Shapley Concentration in numerical simulations.

As can be seen in Figure 2, two main groups of clusters ("cluster complexes") dominate the central region of this supercluster. The A3558 complex (A3558 is the richest cluster in the region) is a structure elongated for ~ 7 h^{-1} Mpc in the East-West direction, comprising also A3562, A3556, SC1329-313 and SC1327-312. The A3528 complex extends for ~ 7 h^{-1} Mpc along the North-South direction and is formed by two pairs (actually A3528 is double) of interacting clusters, including also A3530 and A3532.

Dynamical studies of Bardelli et al. (2000) and Reisenegger et al. (2000) concluded that these structures (with mass of a few 10^{15} $M_\odot$ each) are in the collapse phase, while the entire central supercluster region already reached its turn-around radius and has started to collapse. In fact, the complexes represent a major merger at an advanced stage (the A3558 complex) and at an early stage (the A3528 complex). These two strucures are connected by a "bridge" of galaxies, resembling the Great Wall: within this wall the overdensity in number of galaxies is ~ 4, consistent with the overall overdensity of 3.3 obtained by Drinkwater et al. (2004) after having eliminated the cluster regions.

Formally, there is also another complex, dominated by A3571 (the cluster visible on the South-East in the lower panel of Figure 2) which is connected with A3572 and A3575 (two poor concentrations) in the optical image. However, in the X-ray this cluster appears well relaxed and this could indicate that the merging is "old", in the sense that the gas had

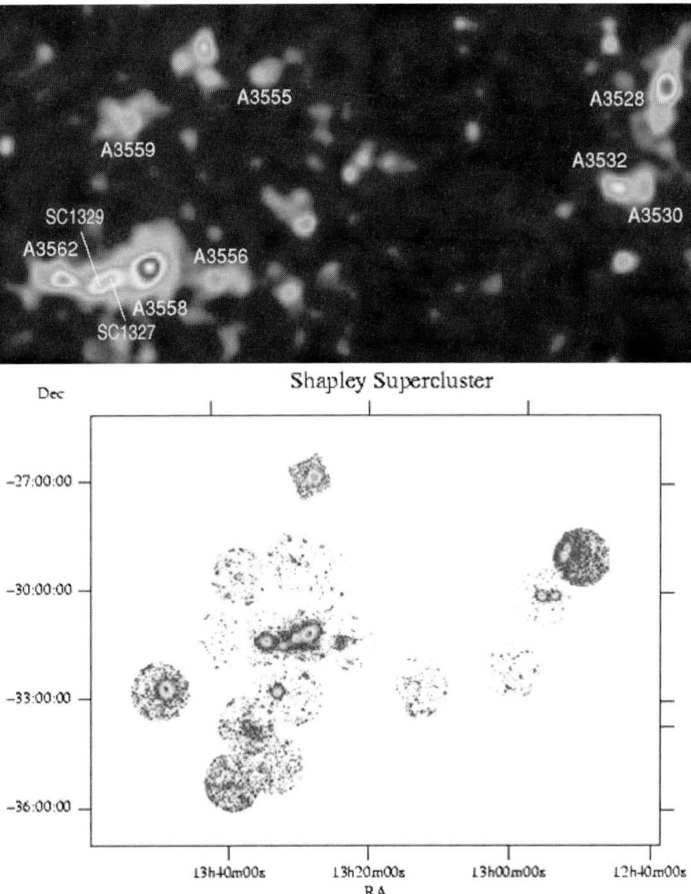

Figure 2. *Upper panel:* distribution of optical galaxies to b_J=19.5 in the central region of the Shapley Concentration. *Lower panel:* ROSAT-PSPC pointed observations mosaic of the same area. Note the presence of the two cluster complexes both in the optical and the X-ray bands.

time to reach dynamical equilibrium, while galaxies are still at the end of the relaxation process.

3. The A3558 complex

Clusters belonging to this structure are embedded in a continuous envelope of both hot gas (Kull & Böhringer 1999) and galaxies (Bardelli et al. 1998a) on a scale of ~ 7 h^{-1} Mpc , i.e. surrounding the entire structure (see Figure 3). The hot gas has not had origin from the cosmological filament (or "wall") seen in the redshift survey, but probably it is intracluster gas expelled from the clusters by the merging (see the spatial analysis of A3562 in Ettori et al. 2000). Also the galaxy envelope has had the same origin, being formed by the less bounded cluster objects, shared by the whole structure after the merging.

ROSAT, ASCA and Beppo-SAX (Bardelli et al. 2002, Hanami et al. 1999, Akimoto et al. 2004, Ettori et al. 2000) X-ray studies of this region did not detect shocks, although the gas distribution shows clear signs of disturbance. Only between A3562 and

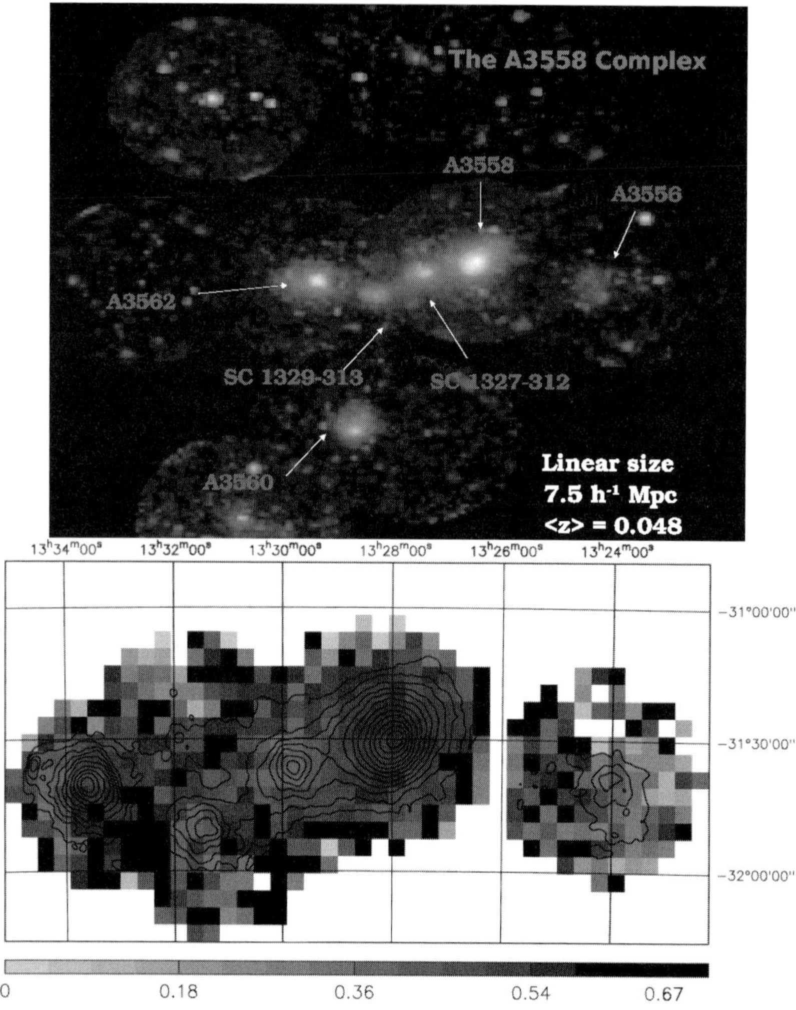

Figure 3. *Upper panel:* the A3558 cluster complex from the ROSAT-PSPC mosaic of Ettori et al. (1997). *Lower panel:* ASCA hardness ratio map of the A3558 complex from Akimoto et al. (2003). Note the "hot spot" between A3562 and SC1329-313 (reproduced with the permission of the American Astronomical Society).

SC1329-313, in the Eastern part of the structure, a hotter region is detected (see Figure 3). Furthermore, SC1329-313 has a gas distribution particularly disturbed with a comet-like shape (Bardelli et al. 2002): an ASCA analysis of its X-ray spectrum led Hanami et al. (1999) to claim the existence of significant turbulent motions or of a multiphase gas. This means that the merging is still at work and the clusters are far from equilibrium. As can be seen from Figure 4, most of the gas distribution features in the A3558 complex could be related with galaxy distribution substructures.

This spectacular major merging represents a unique opportunity to study the effect of merging at radio wavelengths (see also the contributions of Zucca and Giacintucci, this conference). In particular, a peculiar radio feature has been detected: it is formed by a radio halo and by a diffuse radiosource (see Figure 5); the only other known case is in the Coma cluster. The radio halo, detected at the center of A3562, is a head-tail radiogalaxy: we verified (Venturi et al. 2003) that this radiogalaxy furnished the electrons which, after

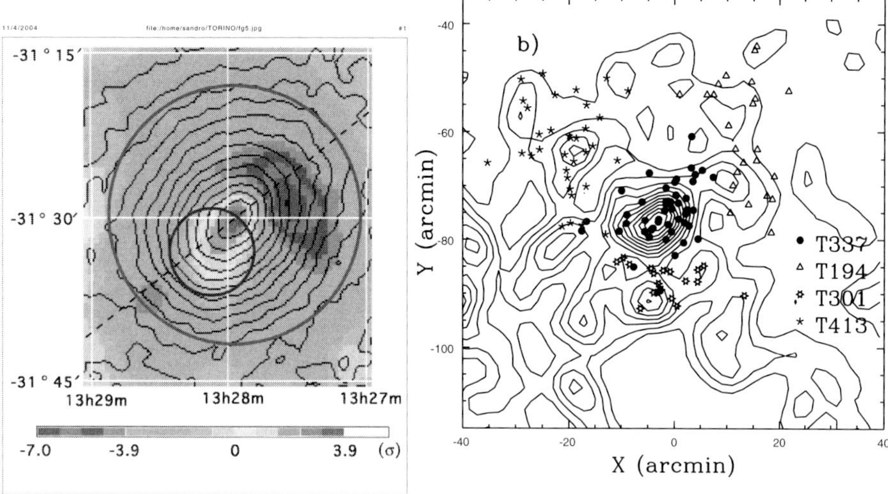

Figure 4. *Left panel:* ASCA surface brightness distribution of the cluster A3558, from Akimoto et al. (2003) (reproduced with the permission of the American Astronomical Society). Note the clumpy distribution. *Right panel:* optical isodensity contours; different symbols correspond to galaxies belonging to different substructures found in the α, δ and velocity space (see Bardelli et al. 1998b).

the reacceleration by a merging 0.4 Gyrs ago, are responsible for the halo emission. Also the radio spectrum of the halo is consistent with the last electron acceleration happened 0.4 Gyrs ago. Moreover, we found that there is a significant lack of radiosources in this structure (Venturi et al. 2000): this signal is coming mainly from the cluster A3558 and could indicate that merging could switch-off, at least for a period, the radiosource activity. Moreover, a relic radiosource has been found in the Westernmost part of the A3558 complex: a geometrical and dynamical reconstruction of this part of the structure lead to speculate that this relic had origin on the shock front (up to now undetected in the X-ray), caused by a small group in-falling onto the A3556 cluster (Venturi et al. 1998). A3556 itself presents peculiar characteristics, because of its very low X-ray surface brightness with respect to the optical richness: moreover, its optical luminosity function presents an unusual shape, with a pronouced excess of bright galaxies (Bardelli et al. 1998a).

4. The A3528 complex

A3528, the dominant cluster of the complex, is a double cluster formed by two twins subclumps separated by 0.9 h^{-1} Mpc and the other two clusters of the complex (A3530 and A3532) are a close pair, separated by ~ 1 h^{-1} Mpc .

Gastaldello et al. (2003) studied A3528 with XMM-Newton observations, obtaining surface brightness, temperature and abundance maps. Although a bridge of hot gas connecting the two clumps has been found, no shock is detected (see Figure 6): this fact is unexpected, given the estimated masses of the clumps ($\sim 8 \times 10^{13}$ $M_{\odot}$ each) and their relative distance. The most reasonable explanation is that the merging was not head-on but off-axis. After having subtracted a β model from the surface brightness of the two subclumps, we found emission excesses which can be used to determine the infalling direction (see right panel of Figure 6). The conclusion is that this system is in an off-axis post-merging phase, with the closest core encounter happened $\sim 1-2$ Gyrs ago.

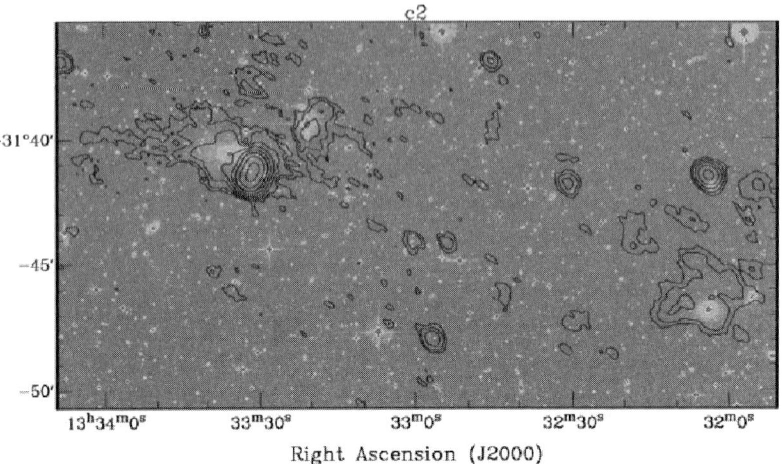

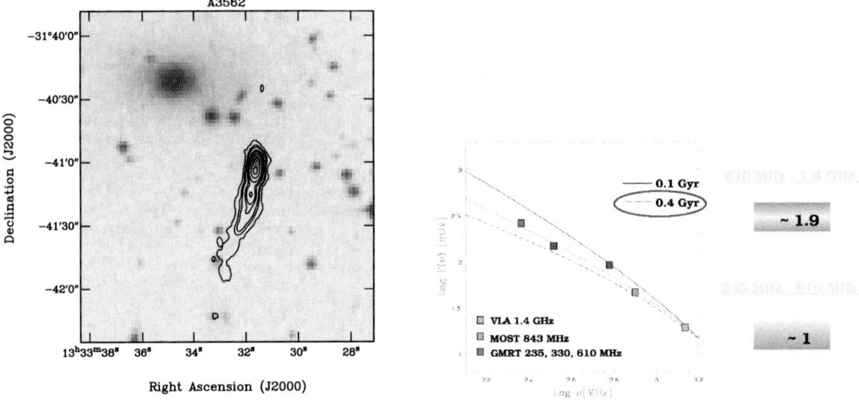

Figure 5. *Upper panel:* VLA 20cm isophotes overplotted to the Digital Sky Survey, in the A3562 region. On the left the system radio halo plus head-tail radiogalaxy (not resolved) is visible. Note the peculiar radiosource on the right. *Lower left panel:* the head-tail radiogalaxy seen at higher resolution. This object furnished the electrons to the halo. *Lower right panel:* radio spectrum of the halo. Solid lines are models with different reacceleration times.

The interesting point is that the optical blue luminosities of the two subclumps, which are twins for what regards the X-ray properties, differ by an order of magnitude. This could indicate that one of the two clumps suffered the galaxy "pealing" process more than the other, probably induced by a larger path through the large scale environment. XMM-Newton data on the couple A3530/A3532 are presently in the reduction phase.

Our general conclusion is that, although the two single pairs of clusters (the two clumps of A3528 and A3530/A3532) are mergings at an advanced state, the A3528 complex as a whole is at an earlier moment of collapse with respect to the A3558 complex, and the masses involved here are probably lower.

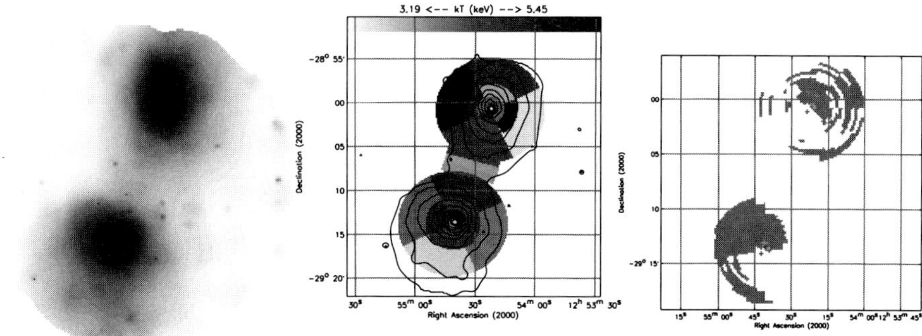

Figure 6. XMM-Newton observations of A3528. *Left panel:* X-ray surface brightness distribution; note the hot bridge connecting the two subclumps. *Middle panel:* temperature map. *Right panel:* surface brightness residuals, obtained after having subtracted a smoothed distribution.

5. Summary

Central regions of superclusters give the possibility to find cluster mergings at different phases and strengths and to study the consequences of cluster collisions on the intracluster medium and the galaxy population. The case study of the Shapley Concentration which I presented here shows that the multiwavelength approach is the best way to analyse merging clusters.

References

Akimoto, F., Kondou, K., Furuzawa, A., Tawara, Y., Yamashita, K. 2003 *ApJ* **596**, 170

Bardelli, S., Zucca, E., Zamorani, G., Vettolani, G., Scaramella, R. 1998a *MNRAS* **296**, 599

Bardelli, S., Pisani, A., Ramella, M., Zucca, E., Zamorani, G. 1998b *MNRAS* **300**, 589

Bardelli, S., Zucca, E., Zamorani, G., Moscardini, L., Scaramella, R. 2000 *MNRAS* **312**, 540

Bardelli, S., DeGrandi, S., Ettori, S., Molendi, S., Zucca, E., Colafrancesco, S. 2002 *A&A* **382**, 17

Drinkwater, M.J., Parker, Q.A., Proust, D., Slezak, E., Quintana, H. 2004 *PASA* **21**, 89

Ettori, S., Fabian, A.C., White, D.A. 1997 *MNRAS* **289** 787

Ettori, S., Bardelli, S., De Grandi, S., Molendi, S., Zamorani, G., Zucca, E. 2000 *MNRAS* **318**, 239

Gastaldello, F., Ettori, S., Molendi, S., Bardelli, S., Venturi, T., Zucca, E. 2003 *A&A* **411**, 21

Hanami, H., Tsuru, T., Shimasaku, K., Yamauchi, S., Ikebe, Y., Koyama, K. 1999 *ApJ* **521**, 90

Plionis, M. & Valdarnini, R. 1991 *MNRAS* **249**, 46

Quintana, H., Carrasco, E.R., Reisenegger, A. 2000 *AJ* **120**, 511

Raychaudhury, S., Fabian, A.C., Edge, A.C., Jones, C., Forman, W. 1991 *MNRAS* **248**, 101

Raychaudhury, S. 1989 *Nature* **342**, 251

Reisenegger, A., Quintana, H., Carrasco, E.R., Maze, J. 2000 *AJ* **120**, 523

Kull, A. & Böhringer, H. 1999 *A&A* **341**, 23

Venturi, T., Bardelli, S., Morganti, R., Hunstead, R.W. 1998 *MNRAS* **298**, 1113

Venturi, T., Bardelli, S., Morganti, R., Hunstead, R.W. 2000 *MNRAS* **314**, 594

Venturi, T., Bardelli, S., Dallacasa, D., Brunetti, G., Giacintucci, S., Hunstead, R.W., Morganti, R. 2003 *A&A* **402**, 913

Zucca, E., Zamorani, G., Scaramella, R., Vettolani, G. 1993 *ApJ* **407**, 470

Outskirts of Galaxy Clusters: Intense Life in the Suburbs
Proceedings IAU Colloquium No. 195, 2004
A. Diaferio, ed.

DOI: 10.1017/S174392130400016X

Galaxy cluster formation from the large-scale structure: a case study of the Abell 2125 complex at z=0.247

Q. Daniel Wang[1], Frazer Owen[2], Michael Ledlow[3] and William Keel[4]

[1]Department of Astronomy, University of Massachusetts, Amherst, MA, USA

[2]National Radio Astronomy Observatory, Scorro, NM, USA

[3]Gemini Observatory, La Serena, Chile

[4]Department of Astronomy, University of Alabama, Tuscaloosa, AL, USA

Abstract. The structure of the universe is believed to have formed by clustering hierarchically from small to large scales. Much of this evolution occurs very slowly but at a few special times more rapid, violent activity may occur as major subunits collide at high velocities. The Abell 2125 complex ($z = 0.247$) appears to be undergoing such an event as shown by modeling of the optical velocity field and by the detection with the VLA of an unusually large number of associated radio active galaxies. We present an 80 ksec *Chandra* imaging of Abell 2125, together with extensive complementary multi-wavelength data. We show direct evidence for galaxy transformation and destruction during the cluster formation. The *Chandra* data unambiguously separate the X-ray contributions from discrete sources and large-scale diffuse gas in the Abell 2125 complex, which consists of various merging clusters/groups of galaxies and low-surface brightness emission. This enables us to study processes affecting galaxy evolution during this special time from scales of Mpc down to a few kpc. The overall level of activity plus the special time for the cluster-cluster merger suggests that an important phase of galaxy evolution can take place during such events.

1. Introduction

We have identified a large-scale hierarchical complex (Fig. 1 left panel) that is well-suited for investigating the structure formation process and the galaxy/environment interactions. Revealed in a survey of 10 Butcher & Oemler clusters observed with the *ROSAT* PSPC, this complex contains various X-ray-emitting features, which are associated with concentrations of optical and radio galaxies (Fig. 1; Wang, Connolly, & Brunner 1997; Wang, Owen, & Ledlow 2004). The overall optical galaxy concentration of the region was classified previously as a cluster Abell 2125 (richness 4). In addition to its large blue galaxy fraction ($\sim 20\%$), Abell 2125 also contains an unusually high number of radio sources, a factor of a few higher than those in more-or-less relaxed clusters at about the same redshifts. The *ROSAT* image and follow-up optical observations have further shown that the complex contains three well-defined X-ray bright clusters (Wang et al. 1997). In addition, substantial amounts of unresolved low-surface brightness X-ray emission (LSBXE) are also present. The overall size of the complex is at least ~ 5 Mpc (assuming the now standard ΛCDM cosmology). The complex thus represents an X-ray-bright hierarchical filamentary superstructure, as predicted by numerical simulations of the structure formation (e.g., Cen & Ostriker 1996).

We have obtained an 80 ksec *Chandra* observation that covers part of the Abell 2125 complex (Fig. 1, right panel) to characterize its detailed X-ray properties and to study the interplay between galaxies with their environments. The high spatial resolution of

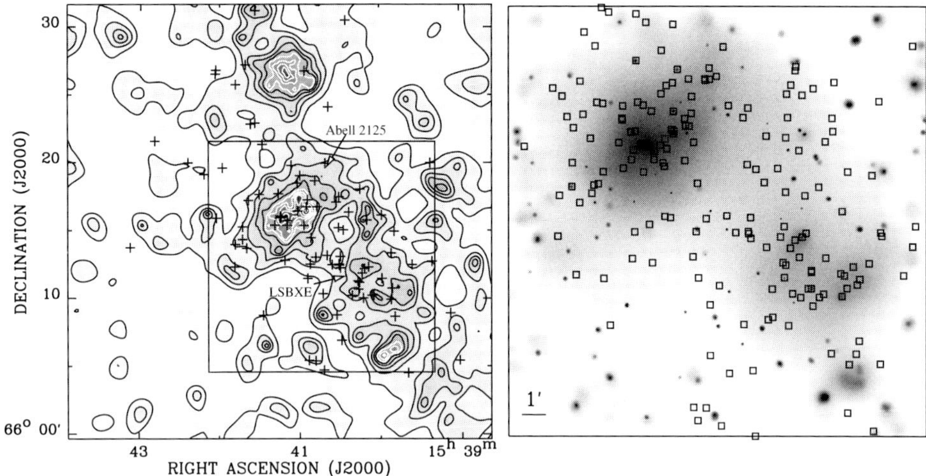

Figure 1. *ROSAT* PSPC X-ray image of the Abell 2125 complex in the 0.5-2 keV band (left panel; Wang et al. 1997). Point-like X-ray sources detected in the image have been excised. Spectroscopically confirmed, radio detected members of the Abell 2125 complex (Owen et al. 2004a) are marked by *pluses*. The box outlines the field covered by our *Chandra* ACIS-I observation shown in the right panel, where the intensity image is in the 0.5-2 keV band. *Squares* mark the positions of complex member galaxies.

Chandra enables us to detect point-like sources down to a limit of $\sim 1\times10^{-15}$ erg cm^{-2} s^{-1} in the 0.5-8 keV band and to examine diffuse X-ray structures down to a scale of ~ 3.8 kpc ($\sim 1''$). Here we summarize results from the analysis of the data and from the comparison with extensive multi-wavelength observations of the complex.

2. New results

We detect a total of 99 discrete sources in the field. Ten of these sources are identified with optical members of the complex. A statistical analysis shows that few (if any) of the remaining unidentified sources are likely to be members of the complex. The diffuse emission is substantially softer than typical discrete sources. There are substructures in the diffuse X-ray emission on all scales. The large-scale diffuse X-ray emission is correlated well with identified complex member galaxies (e.g., Fig. 1 right panel). Two relatively prominent concentrations are sampled by the ACIS-I observation: the Abell 2125 cluster and the southwest LSBXE patch. The redshifts obtained from our X-ray spectra of these features are consistent with the optical value ($z = 0.247$) of the complex.

2.1. *Nature of the LSBXE*

The LSBXE is distinctly different from the central cluster. Even the relatively X-ray-bright southwest patch does not seem to be centrally peaked, as would be expected from a more-or-less virialized intracluster medium (ICM). The kinematics (Miller et al. 2004) indicates that the galaxies in this patch are probably loosely bound, although their exact line-of-sight distribution is not clear. Fig. 2 shows a sub-field of the LSBXE. There are three bright ellipticals in the field; but most of the other galaxies appear to be spirals. X-ray emission from such galaxies are typically dominated by X-ray binaries and/or AGNs and is therefore expected to have a relatively hard spectrum. But the diffuse

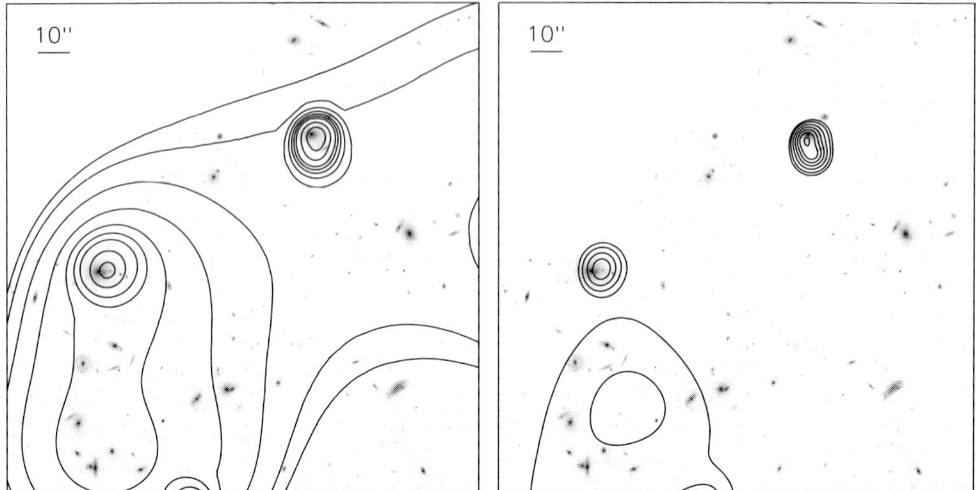

Figure 2. *HST* WFPC-2 V-band images of a field in the LSBXE and overlaid X-ray contours in the the 0.5-2 keV band (left panel) and the 2-8 keV band (right panel).

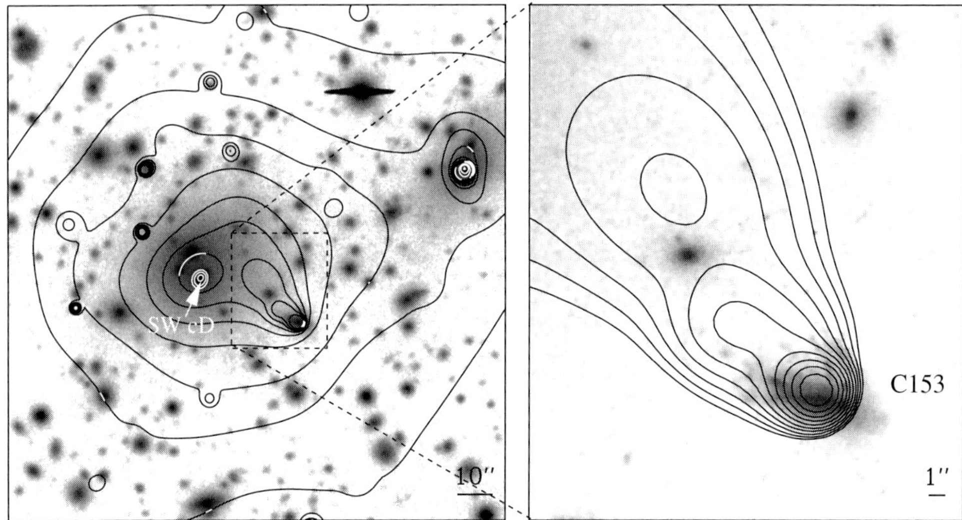

Figure 3. ACIS-I intensity images of the central cluster in the Kitt Peak 4-m mosaic V-band (red) as well as in the 0.5 - 2 keV (green) and 2 - 8 keV (blue) bands (left panel); (right panel) a close-up of C153 in an *HST* WFPC-2 V-band (red) and its diffuse X-ray trail.

X-ray enhancement associated with the galaxy concentration is soft, indicating an origin in diffuse hot gas around individual galaxies, in groups of galaxies, and/or in the intergroup IGM.

The heavy element abundance in the LSBXE ($< 10\%$ solar) appears to be substantially lower than that in the ICM of the central cluster ($\sim 1/4$ solar), although the abundance may be underestimated in the LSBXE if it contains multiple temperature components. If the low metallicity obtained from the single temperature plasma fit is real, there may be two possible explanations: 1) Metals (mainly irons) are largely locked in dust grains within or around galaxy groups; 2) The metals are still bound in individual galaxies. Dust grains may survive, and may even not mix well with, the X-ray-emitting gas in a quiescent, low density environment. But the mixing should likely occur during or after the merger of a galaxy or group with a cluster. The dust grains would then be destroyed rapidly by sputtering in the ICM of high density and temperature. This may explain the metal abundance difference in the X-ray-emitting gas between the LSBXE and the central cluster.

2.2. *Dynamical state of the central cluster*

The central cluster (Abell 2125; Fig. 1) represents the strongest enhancement of the diffuse X-ray emission in the complex. The relative values of our measured temperature (3 keV) and luminosity of the ICM ($\sim 7.9 \times 10^{43}$ ergs s^{-1} in the 0.5-2 keV band) are consistent with those for typical clusters. But, the presence of strong substructures in both the X-ray morphology and galaxy kinematics indicates that the cluster is undergoing a merger.

The cluster also represents the densest concentration of galaxies in the Abell 2125 complex. Although most of the associated galaxies are quiet in radio, the brightest radio galaxies in the field are found in the core of the cluster. While the galaxy C153 is a disturbed disk-like galaxy, the other three are all cD-like ellipticals (Wang et al. 2004; Owen et al. 2004b). Interestingly, the two galaxies (C153 and SW cD; Fig. 3) associated with enhanced X-ray emission are both moving at high velocities ($\sim 1.5 - 1.8 \times 10^3$ km s^{-1}) relative to the mean of the cluster.

The exact relationship between the LSBXE and the central cluster is not clear. The similar velocity centroids of the associated galaxy concentrations may indicate that they are about to merge along an axis that is perpendicular to the line of sight, although there is little direct evidence from both X-ray and galaxy kinematics.

2.3. *Galaxy-environment interaction*

Most of the 10 X-ray-bright member galaxies are not resolved by *Chandra* and are probably dominated by AGN activities. Two of the X-ray sources are apparently resolved and are associated with the disturbed disk-like galaxy C153 and the giant elliptical galaxy northwest of the central Abell 2125 cluster (Fig. 3 left panel). As our source detection algorithms are optimized to detect point-like sources, extended sources such as the one associated with the SW cD (Fig. 3 left panel) are not detected. These extended soft X-ray enhancements probably represent hot gas associated with individual massive galaxies or groups of galaxies, which may have entered the cluster for the first time. Outside the central cluster, the ambient density and relative velocity are typically low and the ram-pressure stripping is probably not important, at least for massive galaxies. The intergalactic gas around galaxies may even cool fast enough to replenish the gas consumed for star formation (Bekki et al. 2002). As they are plugging into a cluster, the surrounding gas may then be compressed by the high ram-pressure of the ICM, resulting in enhanced soft X-ray emission. The eventual stripping of the gas and dust may be important in both enriching the ICM and transforming the galaxies.

The most conspicuous soft X-ray feature in the core of the central cluster is a "trail" apparently attached to the radio galaxy C153 (Fig. 3 right panel). A similar, though shorter, trail in the same direction is also seen in [OII] line emission (Owen et al. 2004b). C153 probably represents an extreme case of the ram-pressure stripping. The observed steep X-ray spectral characteristics, as evidenced by the complete absence of the trail in the 2-8 keV band, suggests a thermal gas temperature of $\sim$ 0.7 keV, although the counting statistics of the data is not sufficient to give a tight constraint. We estimate the total luminosity of the trail as $\sim 5 \times 10^{41}$ ergs s^{-1} in the 0.5-2 keV band.

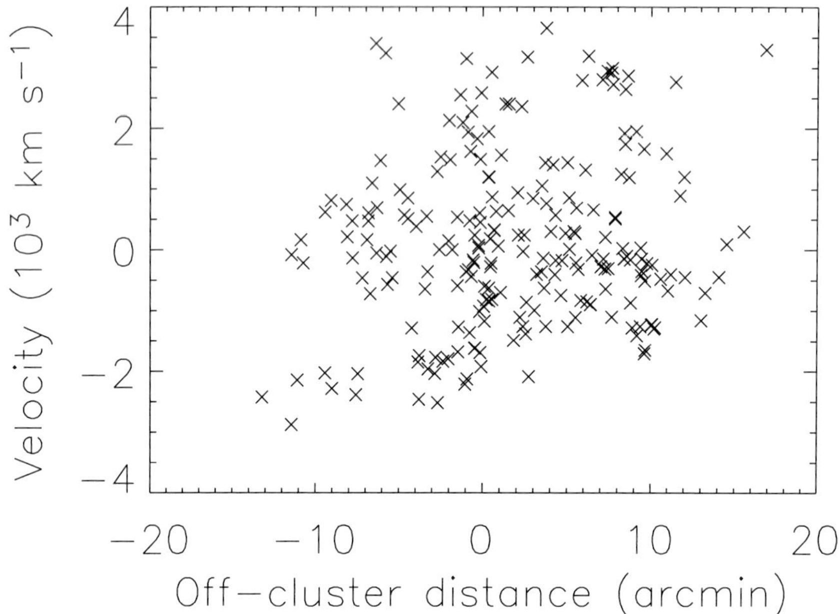

Figure 4. Radial velocities of galaxies versus the distance from the core cluster along the major axis ($\sim 40°$ east from the north) of the Abell 2125 complex.

2.4. *Abell 2125 as a large-scale hierarchical complex*

This distinct complex of galaxies and hot gas shows several remarkable characteristics:

• an unusually large galaxy velocity dispersion of $\sim 1.1 \times 10^3$ km s^{-1}, with multiple velocity components over a scale of ~ 5 Mpc (e.g., Fig. 4);

• an exceptionally large fraction of radio galaxies, which are located primarily outside the central Abell 2125 cluster (Owen et al. 2004a);

• the presence of multiple X-ray-emitting clusters and LSBXE features, each with substantially lower luminosity and temperature than expected from the overall galaxy richness and large velocity dispersion of the complex (Wang et al. 2004; Wang et al. 1997).

Comparisons with various numerical simulations indicate that the complex represents a projection of multiple components, which might be in a process of merging with each other (Miller et al. 2004; Owen et al. 2004a). Projection effects, together with enhanced activities during this process, may explain the observed characteristics.

References

Bekki, K., Couch, W. J., & Shioya, Y. 2002, ApJ, 577, 651
Cen, R., & Ostriker, J. P. 1999, ApJ, 514, 1
Miller et al. 2004, ApJ, submitted
Owen, F. N., et al. 2004a, in preparation
Owen, F. N., et al. 2004b, in preparation
Wang, Q. D., Connolly, A., & Brunner, R. 1997, ApJL, 487, 13
Wang, Q. D., et al. 2004, ApJ, submitted

Outskirts of Galaxy Clusters: Intense Life in the Suburbs
Proceedings IAU Colloquium No. 195, 2004
A. Diaferio, ed.

DOI: 10.1017/S1743921304000171

Radio activity in major mergings

E. Zucca[1], T. Venturi[2], S. Giacintucci[1,3], S. Bardelli[1] and D. Dallacasa[3]

[1]INAF-Osservatorio Astronomico di Bologna, via Ranzani 1, 40127 Bologna, Italy e-mail: elena.zucca@bo.astro.it

[2] INAF-Istituto di Radioastronomia, via Gobetti 101, 40129 Bologna, Italy

[3] Dipartimento di Astronomia, Università di Bologna, via Ranzani 1, 40127 Bologna, Italy

Abstract. Cluster mergings are expected to have an influence on the radio emission of the galaxy population. We present the results of a deep radio survey in the A3558 complex in the central region of the Shapley Concentration, in order to further explore our hypothesis of a dependence of the radio luminosity function on the age of the merging.

1. Introduction

During the huge energy release of a cluster merging, different mechanisms are at work on galaxies. On one side, the rapid changes in the tidal gravitational field can drive gas from outer to inner parts of galaxies, triggering star bursts and feeding the central AGN (Bekki 1999), while on the other side tidal stripping and ram pressure tend to devoid galaxies of their hydrogen.

Owen et al. (1999) found that A2125, a merging cluster, presents a factor 4 more radio galaxies than a relaxed one (A2645) at the same distance. In order to better study this phenomenon, we performed an extensive radio survey at 20cm with the ATCA telescope on the cluster complexes (Bardelli et al. 1998a, Bardelli et al. 1998b, Bardelli et al. 2000, Bardelli et al. 2002) located in the Shapley Concentration (Zucca et al. 1993).

The only way to estimate the radio activity of the clusters is to compute the bivariate radio-optical luminosity function (Figure 1), i.e. the number of radio galaxies in the considered structure (as confirmed by their redshift) over the total number of optical galaxies. This takes into account the different richness of the clusters. We found that the A3558 complex, which is a merging in an advanced phase, presents a significant lack of radiosources (Venturi et al. 2000, Venturi et al. 2001, Venturi et al. 2002). On the contrary the A3528 complex, which is at an early phase of interaction, presents a number of radiosources comparable to that of normal clusters. Also the A3571 complex, a merging at the end of the process, does not have pecularities in the number of radio objects.

This result is only apparently in contraddiction with Owen et al. (1999), because one has to take into account the spiral richness of interacting clusters and/or the merging phase. Our proposed scenario is the following: at very early stages of merging, the outer parts of clusters (which are spiral rich) begin to interact, triggering star formation activity and therefore enhancing the number of radiosources. After this episode, the radiosource excess disappears and the clusters return quiescent (right panel of Figure 1), until the ellipticals (with their AGNs) in the inner part of the clusters are reached by the merging effect, leading to a lowering of the AGN emission: as a result, there is a deficit of bright radiosources (left panel of Figure 1). Finally, the virialization re-establishes the starting situation.

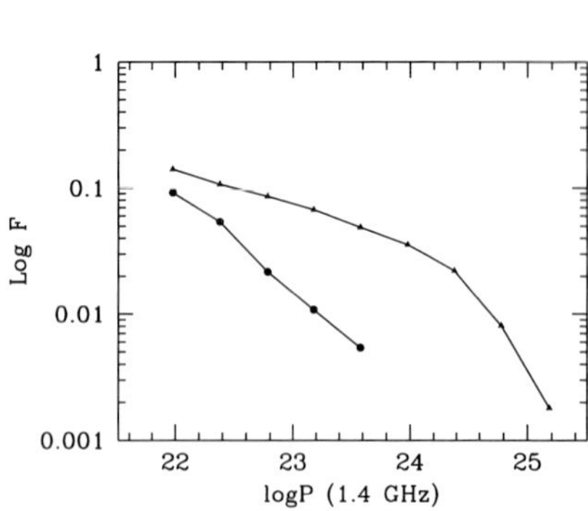

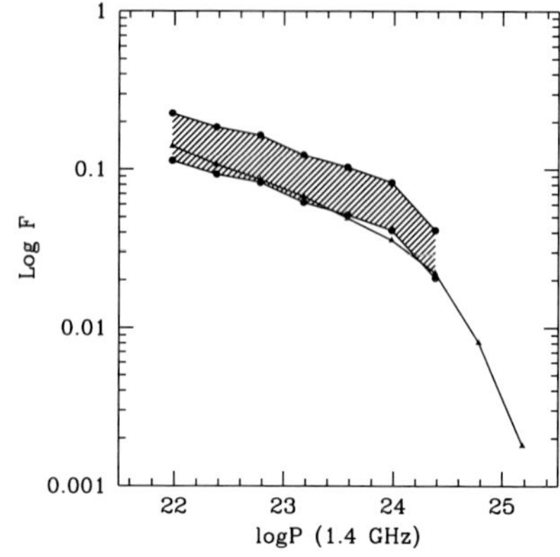

Figure 1. Bivariate radio-optical luminosity functions of the A3558 complex (left) and of the A3528 complex (right). Triangles represent the reference luminosity function of normal clusters from Ledlow & Owen (1996). The first complex is an advanced merging, while the second is a merging at early phases. For the A3528 complex, the radio luminosity function lies in the shadowed area.

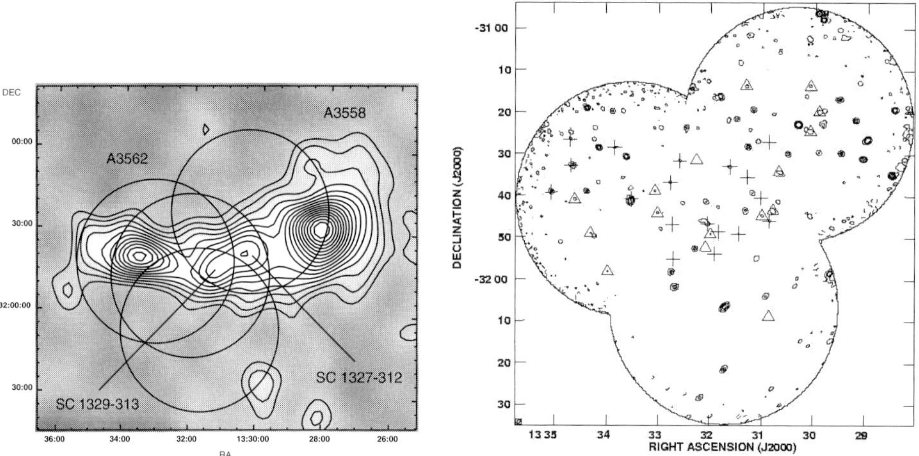

Figure 2. *Left panel:* circles represent the field of view of our VLA 20cm pointings, superimposed to the isodensities of the optical distribution of galaxies in the A3558 complex. *Right panel:* location of optical counterparts overplotted on the radio contours. Triangles are late type galaxies, crosses are early type galaxies.

2. The deep VLA survey

In order to better study the lack of radiosources in the A3558 complex, we performed a dedicated, deep ($\sim$ 0.25 mJy beam^{-1}) survey in the region between the two clusters A3558 and A3562 (see Figure 2, Giacintucci et al. 2004). This region is promising because it is highly substructured and presents features indicating an ongoing merging (see the contribution of Bardelli, this conference). Moreover, we discovered an interesting radio halo in the A3562 cluster (see the contribution of Giacintucci, this conference). The groups of this region tend to have bluer galaxies, indicating a possible ongoing starburst, and their hot gas distributions are highly disturbed (Bardelli et al. 2002).

Given the high statistics, we computed the radio luminosity function for both early and late type galaxies in this region, finding consistency with Ledlow & Owen (1996) for elliptical and S0 galaxies. Given the lack of bright radiosources that we found analysing the whole A3558 complex (see previous section), this result suggests that the deficit is entirely due to the A3558 cluster. Moreover, a population of faint ($\log P_{1.4\mathrm{GHz}}[\mathrm{W\ Hz}^{-1}] < 22$) radio galaxies has been found: half of these objects are blue, suggesting that starburst is the driving radio emission mechanism. Finally, we found 14 spiral galaxies, whose ratio between radio and optical emission is similar to that of galaxies located in rich and dynamically evolved clusters.

The next steps of this analysis will be the following:
1) extending the study to other systems in other superclusters: in particular we are analysing the cluster A2061 in the Corona Borealis supercluster;
2) extending the survey to other wavelengths: in particular we obtained 610 MHz observations of the A3558 complex with the GMRT telescope;
3) studying the HI emission of spiral galaxies in order to see whether the ongoing merging in the A3558 complex was able to strip gas from galaxies in between the dominant clusters.

References

Bardelli, S., Zucca, E., Zamorani, G., Vettolani, G., Scaramella, R. 1998a *MNRAS* **296**, 599
Bardelli, S., Pisani, A., Ramella, M., Zucca, E., Zamorani, G. 1998b *MNRAS* **300**, 589
Bardelli, S., Zucca, E., Zamorani, G., Moscardini, L., Scaramella, R. 2000 *MNRAS* **312**, 540
Bardelli, S., De Grandi, S., Ettori, S., Molendi, S., Zucca, E., Colafrancesco, S. 2002 *A&A* **382**, 17
Bekki, K. 1999 *ApJ* **510**, L15
Giacintucci, S., Venturi, T., Bardelli, S., Dallacasa, D., Zucca, E. 2004 *A&A* **419**, 71
Ledlow, M.J. & Owen, F.N. 1996 *ApJ* **112**, 9
Owen, F.N., Ledlow, M.J., Keel, W.C., Morrison, G.E. 1999 *AJ* **118**, 633
Venturi, T., Bardelli, S., Morganti, R., Hunstead, R.W. 2000 *MNRAS* **314**, 594
Venturi, T., Bardelli, S., Zambelli, G., Morganti, R., Hunstead, R.W. 2001 *MNRAS* **324**, 1131
Venturi, T., Bardelli, S., Zagaria, M., Prandoni, I., Morganti, R. 2002 *A&A* **385**, 39
Zucca, E., Zamorani, G., Scaramella, R., Vettolani, G. 1993 *ApJ* **407**, 470

Outskirts of Galaxy Clusters: Intense Life in the Suburbs
Proceedings IAU Colloquium No. 195, 2004
A. Diaferio, ed.

DOI: 10.1017/S1743921304000183

Radio halos in merging clusters of galaxies

S. Giacintucci[1,2], T. Venturi[3], S. Bardelli[2], G. Brunetti[3], D. Dallacasa[1,3], P. Rao[4] and E. Zucca[2]

[1]Dipartimento di Astronomia, Univ. Bologna, via Ranzani 1, I-40127 Bologna, Italy
[2] Osservatorio astronomico, via Ranzani 1, I-40127 Bologna, Italy
[3] Istituto di Radioastronomia - CNR, via Gobetti 101, I-40129 Bologna, Italy
[4] National Centre for Radio Astrophysics, Pune University Campus, India

Abstract. We present the preliminary results of 235 MHz, 327 MHz and 610 MHz observations of the galaxy cluster A3562 in the core of the Shapley Concentration (SC). The purpose of these observations, carried out with the Giant Metrewave Radio Telescope (GMRT, Pune, India) was to study the radio halo located at the centre of A3562 and determine the shape of its radio spectrum at low frequencies, in order to understand the origin of this source. In the framework of the re–acceleration model, the preliminary analysis of the halo spectrum suggests that we are observing a young source (few 10^8 yrs) at the beginning of the re–acceleration phase.

1. Introduction

A number of X–ray luminous galaxy clusters show large–scale synchrotron radio emission associated to the intracluster medium. These diffuse sources are know as radio halos, when they are located at the centre of the hosting cluster and show low or negligible polarization, and radio relics, when found in the cluster outskirts and highly polarized.

Both halos and relics have low surface brightness, large linear size (from ∼0.5 Mpc to more than 1 Mpc) and steep integrated spectra, i.e. $\alpha > 1$ ($S \propto \nu^{-\alpha}$).

Radio halos and relics represent the most striking evidence for μG magnetic fields on cluster scale and relativistic electrons diffused within the whole cluster volume (for a recent review see Giovannini & Feretti 2002).

The existence of this class of radio sources is believed to be connected to cluster mergers, since thus far they have only been found in clusters with significant signs of a current or recent merging event. In particular the leading hypothesis for the origin of the observed radio emission from these objects is a re–acceleration process, probably via turbolence powered by cluster mergers, of a population of relatively low energy ($\gamma \sim 10^3$) electrons, initially injected in the intracluster medium (two-phase model, Brunetti et al. 2001).

2. The radio halo in A3562

The cluster A3562 is the easternmost component of the A3558 complex, a chain of three clusters and two SC groups which form a single physically connected structure, whose complex dynamical stage of merging is supported by a large amount of observational data in all bands (see for instance Ettori et al. 1997, Ettori et al. 2000).

The existence of a radio halo at the A3562 centre was confirmed by deep radio observations at 1.4 GHz carried out with VLA (Venturi et al. 2003).

In the framework of the re–acceleration model, the halo in A3562 seems to play a very special role, since its largest linear size (LLS ∼ 600 kpc) and 1.4 GHz radio power

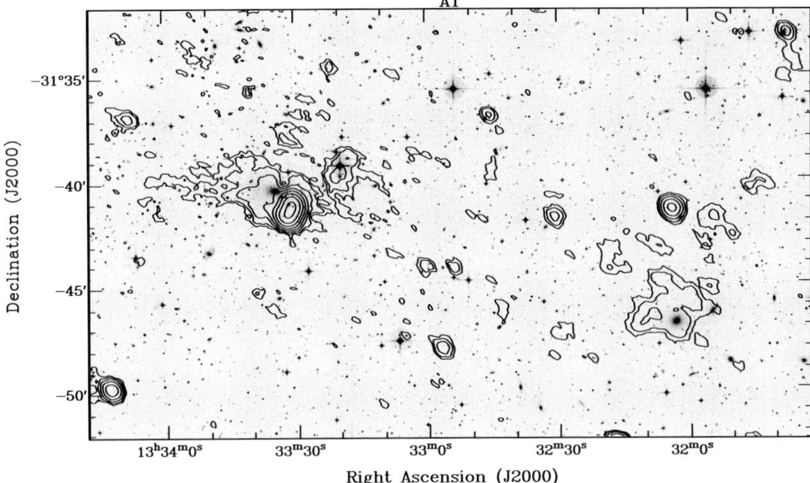

Figure 1. 327 MHz GMRT image of the radio halo in A3562 and the south–western extended emission in the direction of the SC 1329–313 group, overlaid on the optical DSS–2 image. The restoring beam is 41.98″ × 35.13″. The rms in the image is ∼ 0.25 mJy/beam and the lowest contour is 0.8 mJy/beam.

($P_{1.4GHz} = 2.14 \times 10^{23}$ W Hz^{-1}) are among the lowest values found for this class of radio sources.

The halo nicely fits the correlations existing between the halo radio power and the cluster X–ray luminosity and temperature (Bacchi et al. 2003), extending them to lower values for all quantities involved.

According to the two-phase model, the halo in A3562 could be interpreted either as the result of a low efficiency re-acceleration process or as a young source at the beginning of the re-acceleration phase. It has been established that the head–tail radio galaxy J1333–3141, located within the halo emission, has deposited in the intracluster medium a number of electrons high enough to feed the halo if it has been active for a considerable fraction of its crossing time ($t_{cross} \sim 6 \times 10^8$ yrs). This result seems to suggest that the halo at the centre of A3562 is young as compared to the other radio halos known to date (Venturi et al. 2003). However only low frequencies observations allow to discriminate between the two possibilities.

3. Low frequency observations of the halo in A3562

With the purpose of studing the morphology of the halo in A3562 and determine the shape of its synchrotron spectrum at low frequencies, we carried out observations at 235 MHz, 327 MHz and 610 MHz with the Giant Metrewave Radio Telescope (GMRT, Pune, India). Thanks to the u–v coverage of this instrument, a range of resolutions at all frequencies (from few arcsec to tens of arcsec) is allowed. This ensures an accurate determination of the total radio flux density of the halo, after a careful subtraction of the individual point sources embedded in the halo emission.

In Figure 1 we report the image of the radio halo in A3562 at 327 MHz overlaid on the optical DSS–2 image, and in Figure 2 we present the 610 MHz image of the halo overlaid on the X–ray XMM image.

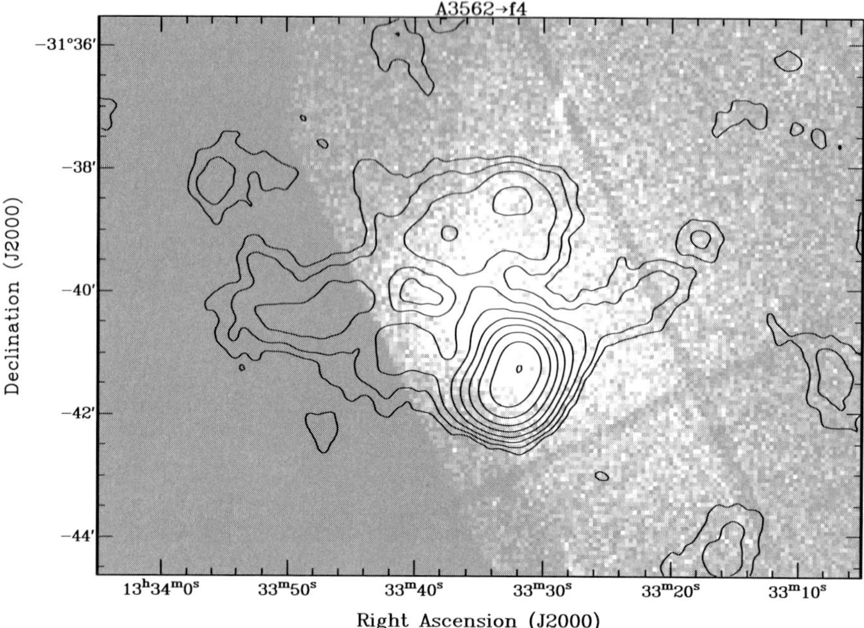

Figure 2. 610 MHz GMRT image of the radio halo in A3562 overlaid on the X–ray XMM image. The restoring beam is $41.98'' \times 35.13''$. The rms in the image is ~ 0.15 mJy/beam and the lowest contour is 0.45 mJy/beam.

An extended radio emission, located south–west of the halo, surrounds the radio source J1332–3146 (Figure 1), identified with the brightest galaxy in the SC 1329–313 group (Giacintucci et al. 2003). The nuclear component, detected at 1.4 GHz, disappears at low frequency. This seems to rule out any connection between the low brightness extended emission and an active nucleus, suggesting that its origin is most likely due to cluster merger re–acceleration.

The preliminary results of our low frequency data suggest that the spectrum of the radio halo is steep up to $\sim$ 610 MHz, with $\alpha_{610MHz}^{1.4GHz} \sim 1.9$, and it flattens beyond 610 MHz, with $\alpha_{235MHz}^{610MHz} \sim 1$. This seems to be in good agreement with a very recent re–acceleration phase, supporting the idea that the radio halo in A3562 is young and that its existence is strongly connected to the ongoing major merger in the A3558 complex (Venturi et al. in preparation).

References

Bacchi, M. Feretti, L. Giovannini, G. Govoni, F. 2003 *A&A* in press (astro-ph/0301296).

Brunetti, G. Setti, G. Feretti, L. Giovannini, G. 2001 *MNRAS* **320**, 365.

Ettori, S. Fabian, A.C. White, D.A. 1997 *MNRAS* **289**, 787.

Ettori, S. Bardelli, S. De Grandi, S. Molendi, S., Zamorani, G. Zucca, E. 2000 *MNRAS* **318**, 239.

Giacintucci, S. Venturi, T., Bardelli, S. Dallacasa, D. Zucca, E. 2003 *A&A* in press (astro-ph/0311251).

Giovannini, G. & Feretti, L. 2002 *Astroph. Sp. Sci. Lib.* **272**, 197.

Venturi, T. Bardelli, S. Dallacasa, D. Brunetti, G. Giacintucci, S. Hunstead, R.W., Morganti, R. 2003 *A&A* **402**, 913.

Outskirts of Galaxy Clusters: Intense Life in the Suburbs
Proceedings IAU Colloquium No. 195, 2004
A. Diaferio, ed.

DOI: 10.1017/S1743921304000195

Large-scale velocity flow in the Horologium-Reticulum supercluster

Matthew C. Fleenor[1], James A. Rose[1], Wayne A. Christiansen[1], Richard W. Hunstead[2], Melanie Johnston-Hollitt[3] and Michael J. Drinkwater[4]

[1] Department of Physics & Astronomy, University of North Carolina, Chapel Hill, North Carolina 27599-3255 USA email: fleenor2@physics.unc.edu

[2] School of Physics, University of Sydney, NSW 2006, Australia

[3] Leiden Observatory, NL-2300 RA Leiden, The Netherlands

[4] Department of Physics, University of Queensland, QLD 4072, Australia

Abstract. We present results from 547 optical galaxy redshifts obtained in the region of Horologium-Reticulum (HR) using the 6dF (six-degree field) multi-fiber spectrograph on the United Kingdom Schmidt Telescope. Concentrating upon the inter-cluster regions of the HR Supercluster (HRS), we derive three primary results from our 6dF observations. First, the HRS covers a redshift range from at least 16000 to 22500 km s^{-1}. Second, these HRS galaxies exhibit an overall gradient with increasing redshift along a spatial axis of SE-NW with high statistical probability. Third, along the best-fit line from the spatial-redshift linear regression, HRS galaxies comprising the redshift trend are distinctly split into a high- and low-redshift component. Finally, comparisons of the spatial-redshift distribution are made between the HRS and the Shapley supercluster.

---

1. The Horologium-Reticulum Supercluster (HRS)

Superclusters of galaxies provide a unique environment for investigating the growth of and inter-relationship between structures on all scales. These massive associations present the opportunity to assess whether structure is indeed assembling in a hierarchical fashion. Covering an area of more than 12° x 12° on the sky at approximately $\alpha = 03^h 19^m$ and $\delta = -50°02'$, the HRS is the second largest mass concentration within 300 Mpc, only exceeded in mass by the Shapley Supercluster (SSC).

To obtain a comprehensive picture of the variety of structures present in the HRS complex, we have utilized a combination of the AAT/2dF for investigations of Abell clusters A3125/A3128 and the UKST/6dF for observations of the inter-cluster regions. To probe the finer scale structures within the A3125/A3128 intra-cluster medium, we have also examined radio and X-ray imaging (Rose et al. 2002). Currently, our collaboration is involved in the analysis of extended radio emission in A3125 as it relates to the supercluster environment (Johnston-Hollitt et al. 2004). Here, we summarize several key results concerning the distribution of megaparsec-scale structure extracted from our 6dF inter-cluster observations.

1.1. *Kinematic extent of the HRS*

The redshift histograms in Figure 1 (left) show the connection between the 6dF galaxies and clusters of known redshift within the observing region. The majority of 6dF galaxies (53%) occupy the redshift range between 16000 and 22500 km s^{-1}. The histogram of cluster mean redshifts further supports the above kinematic boundaries for the HRS as most of the clusters fit between the above redshift range (Fig. 1c, left).

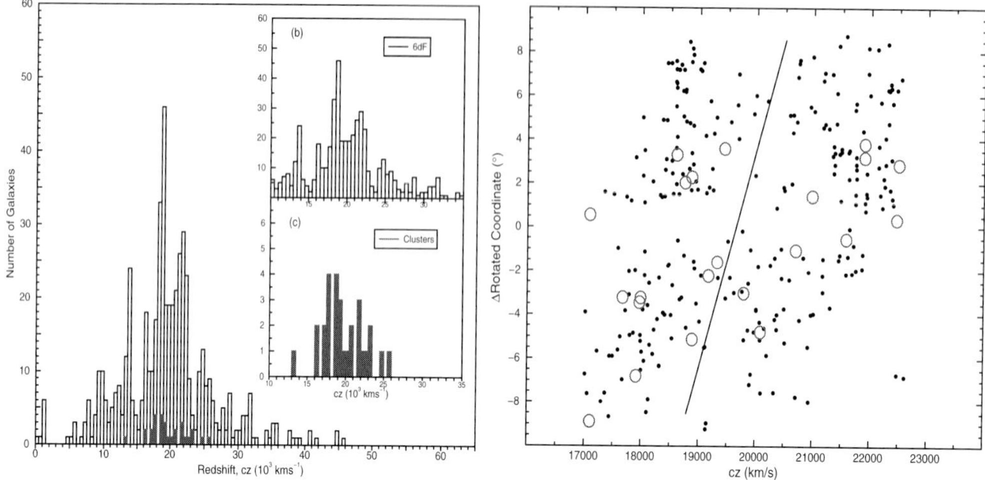

Figure 1. left: Histograms of the 6dF inter-cluster galaxies (solid lines) and of the clusters with known redshifts (filled histogram) are overlaid. In panels (b) and (c) we show an expanded view of the redshift histograms for the inter-cluster galaxies and the clusters respectively. The redshift bin size is 500 $\mathrm{km\,s^{-1}}$. **right:** Spatial-redshift plot for 6dF galaxies between 17000 and 22500 $\mathrm{km\,s^{-1}}$ (center located at $\alpha = 3^h 16^m$, $\delta = -52°$, PA $= -70°$). Open circles represent Abell clusters at an equivalent PA. Solid line is the best-fit line from the linear regression analysis.

1.2. *Large-scale redshift trend*

Qualitative examination of spatial plots of 6dF galaxies suggests an overall trend between spatial position and redshift, that is, there appears to be a trend of systematically increasing redshift along a principal axis extending from the SE to the NW end of the HRS. To quantify the notion of a systematic redshift trend for 6dF galaxies between 16000 and 22500 $\mathrm{km\,s^{-1}}$, we have conducted a linear correlation analysis between projected galaxy position and redshift. At consecutive 10° offsets in position angle (PA), we have projected the galaxy positions onto the principal axis through that PA. We find a significant correlation (with the null hypothesis of no correlation rejected at a high level) for a PA at $-20°$ (as measured east from north). However, when the spatially-compact galaxies between 16000 and 17000 $\mathrm{km\,s^{-1}}$ are excluded from the trend, the probability of correlation increases by three orders of magnitude ($> 99.9\%$). Furthermore, the clusters and galaxies from 17000 to 22500 $\mathrm{km\,s^{-1}}$ share the same best-fit line along the PA of $-70°$ after the low-redshift galaxies are excluded. The results of this analysis are summarized in Figure 1 (right) where the projected spatial coordinate is plotted against the redshift. In short, there is a large-scale kinematic trend in the HRS of increasing redshift for galaxies and clusters between 17000 and 22500 $\mathrm{km\,s^{-1}}$ along a SE-NW axis.

1.3. *Bi-modality of the HRS*

After fitting and removing the position-redshift tilt described above, a bi-modal redshift distribution of nearly equal populations emerges for the inter-cluster galaxies (Fig. 2, left). Both redshift components follow the overall trend and are separated by $\sim$2500 $\mathrm{km\,s^{-1}}$. The spatial-redshift trend of the inter-cluster galaxies (Fig. 1, right) displays a similar arrangement to that found in N-body simulations, whereby tidal encounters between equal-mass clusters create spatial-kinematic tilts only after the two clusters have passed through each other. Lastly, we note that the distribution of galaxy cluster mean

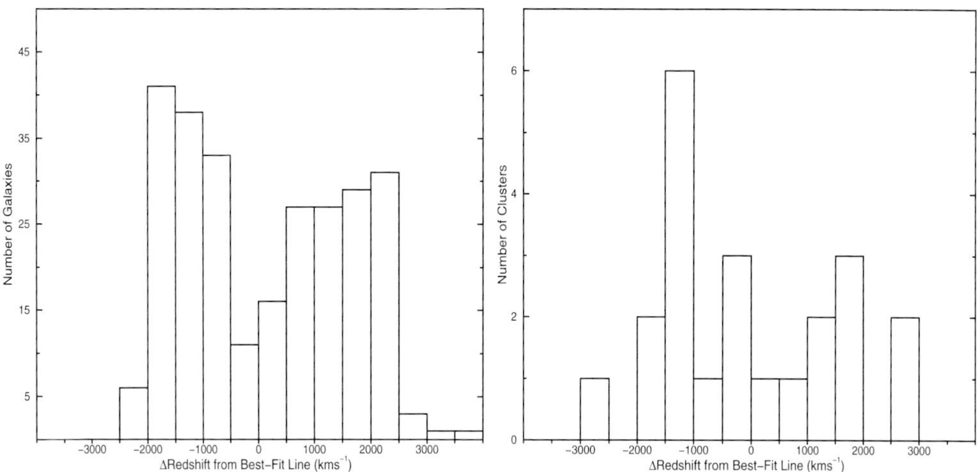

Figure 2. Redshift histograms for galaxies and clusters along the line of best-fit (PA$= -70°$, Fig. 1, right). **left:** 6dF galaxies. **right:** Clusters that are given as open circles in Figure 1, right.

redshifts in the HRS does not appear to reflect the bi-modality of the inter-cluster galaxies (Fig. 2, right).

2. Comparisons with the Shapley concentration

Since the SSC is both extremely well-studied and the most comparable supercluster in the local universe, we use it as a benchmark for assessing the properties of the HRS. The HRS redshift range from 16000 to 22500 km s^{-1} is comparable to the "Central Region" of the SSC at ~5500 km s^{-1} (Quintana et al. 2000). Bi-modality in redshift is also a feature emerging from inter-cluster studies in the SSC (c.f. Fig. 6 of Drinkwater et al. 1999). However, the redshift peaks in the HRS are nearly equal in amplitude (Fig. 2, left), while the higher redshift peak in the SSC has roughly three times the amplitude of the lower redshift component. Although Bardelli et al. (2000) have fit a plane in (α,δ,cz) space to their inter-cluster observations in the SSC, this result assumes the redshift is merely a distance and is not readily comparable to the spatial-redshift trend detected in the HRS. Comparisons like these mentioned serve to extricate the individual merging histories of specific superclusters from the dark-matter signatures present in all large-scale structures.

Acknowledgements

M.C.F. thanks the UNC Graduate School and the AAS for their generosity in providing travel funds. A portion of this work is supported by NSF grant, AST9900720.

References

Bardelli, S., Zucca, E., Zamorani, G., Moscardini, L., Scaramella, R. 2000, MNRAS, 312, 3

Drinkwater, M. J., Proust, D., Parker, Q. A., Quintana, H., Slezak, E. 1999, PASA, 16, 2

Johnston-Hollitt, M., Fleenor, M. C., Hunstead, R. W., Rose, J. A., Christiansen, W. A. 2004, Proc. IAU Coll. No. 195, A. Diaferio, ed.

Quintana, H., Carrasco, E. R., Reisenegger, A. 2000, AJ, 120, 511

Rose, J. A., Gaba, A. E., Christiansen, W. A., Davis, D. S., Caldwell, N., Hunstead, R. W., Johnston-Hollitt, M. 2002, AJ, 123, 3

Outskirts of Galaxy Clusters: Intense Life in the Suburbs
Proceedings IAU Colloquium No. 195, 2004
A. Diaferio, ed.

DOI: 10.1017/S1743921304000201

Multiple merging in the Abell cluster 1367

L. Cortese[1], G. Gavazzi[1] and A. Boselli[2]

[1]Università degli Studi di Milano-Bicocca, P.zza della Scienza 3, 20126 Milano, Italy email: Luca.Cortese@mib.infn.it; Giuseppe.Gavazzi@mib.infn.it

[2]Laboratoire d'Astrophysique de Marseille, BP8, Traverse du Siphon, F-13376 Marseille, France email: alessandro.boselli@oamp.fr

Abstract. We present a dynamical analysis of the central $\sim$ 1.3 square degrees of the galaxy cluster Abell 1367, based on 272 redshift (of which 118 are new measurements). From the analysis of the 146 confirmed cluster members Abell 1367 appears as a young cluster currently forming at the intersection of two filaments.

---

1. The cluster sample

The line of sight (LOS) velocity distribution of the 146 cluster members (4000 km s^{-1} < V < 10000 km s^{-1}) results significantly non gaussian with a mean location C_{BI} = 6484 ± 81 km s^{-1} and a scale $S_{BI} = 891 \pm 58$ km s^{-1} (Fig. 1, left). The structure of the cluster appears elongated from the North-West to the South-East with two main density peaks (Fig. 2). The North-West region has lower LOS and dispersion velocity than the South-East region and it is probably in the early phase of merging into the main cluster (Fig. 3). These results are in agreement with X-ray observations. The region between the two major galaxy density peaks coincides with a strong gap in the gas temperature observed for the first time by ASCA (Donnelly et al. 1998) and recently confirmed by *Chandra* (Sun & Murray 2002). This abrupt temperature change is strongly suggestive of a shock which has generated during a collision between two substructures, probably associated with the SE and the NW galaxy density peaks. We use the position of the gas temperature gap, shown by the straight line in Fig. 2 (right), to divide our sample into two regions and to study separately the dynamical properties of the two subclusters.

2. The North-West subcluster

The LOS velocity distribution of the North-West subcluster results consistent with a gaussian distribution, suggesting that this subcloud is a relaxed system (Fig. 1, center). However this subcluster presents some evidence that supports a multiple merging scenario. The Brightest Galaxy of this cloud CGCG97-095 (NGC3842), located 2 arcmin SE from the NW galaxy density peak, is a Narrow Angle Tail radio galaxy. The tail orientation suggests that this galaxy (and the associated substructure) is moving from north-west to south-east toward the main cluster core. Moreover the presence of a radio relic associated with two head-tail radio galaxies (CGCG97-073 and CGCG97-079) is consistent with a merging scenario in which the sub-group associated with these galaxies is falling into the NW cloud, compressing the plasma ejected from CGCG97-095 and originating the observed radio relic.

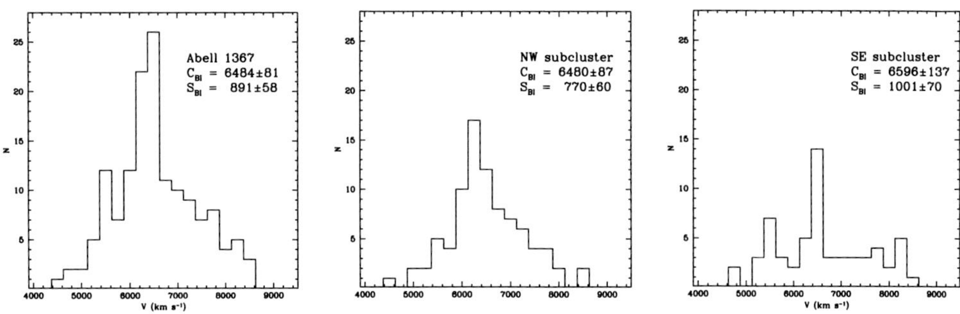

Figure 1. The LOS velocity distribution of the 146 cluster members (left), of galaxies belonging to the North-West subcluster (center) and of galaxies belonging to the South-East subcluster (right).

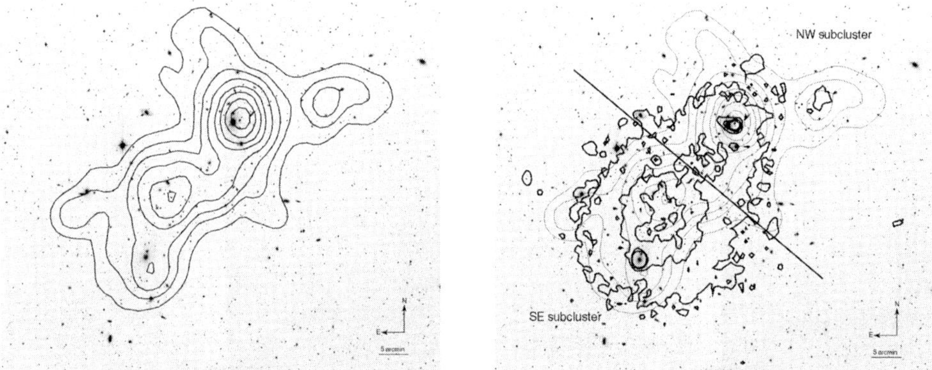

Figure 2. Left: Palomar DSS image of the central region (∼1.3 square degrees) of Abell 1367. In black are superimposed the iso-density contours (computed using the 10 nearest neighbors to each point) for the 146 confirmed cluster members. Right: The same as in the left figure. The ROSAT X-ray contours are superimposed in black. The straight line indicates the position of the abrupt gas temperature gap detected by ASCA (Donnelly et al. 1998) and used to divide our sample into two substructures.

3. The South-East subcluster

The LOS velocity distribution of the South-East subcluster results highly inconsistent with a gaussian distribution, supporting the idea that the cluster center is far from relaxation (Fig. 1, right). The velocity distribution presents three peaks at ~ 5500 km s^{-1}, ~ 6500 km s^{-1} and ~ 8200 km s^{-1} respectively, probably associated with three separate groups. The high-velocity group (V$\sim$ 8200 km s^{-1}) appears segregated in the northern part of the SE cloud and is associated with the infalling group of star-forming galaxies recently discovered by Sakai et al. (2002) and Gavazzi et al. (2003). Its spatial segregation and high star formation activity suggest that this group is a separate unit falling into the cluster, probably from the near side. It is remarkable that *Chandra* (Sun & Murray 2002) discovered a ridge-like structure around the cluster center, ∼6 arcmin south from the center of the high velocity group, probably associated with a compact subcluster penetrating the SE cluster core. The low-velocity group (V$\sim$ 5500 km s^{-1}) is partially segregated in the eastern part of the cloud, probably infalling from the eastern side into the cluster core. This scenario is supported by XMM observations (Forman et al. 2003), that detected cool gas streaming into the cluster core from the eastern side, probably

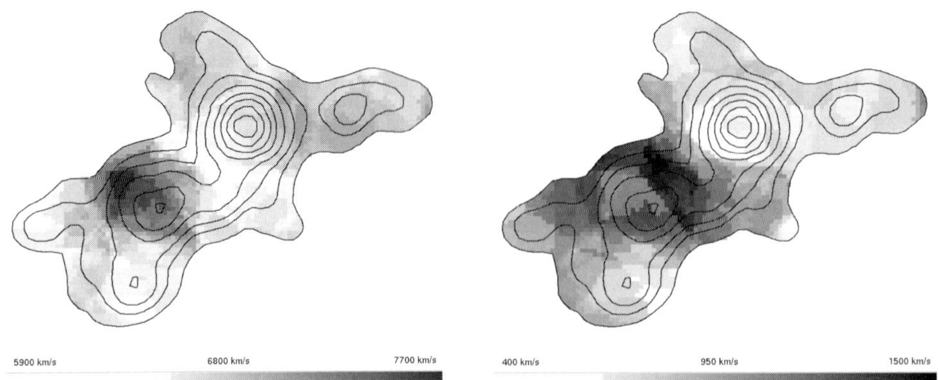

Figure 3. The LOS velocity field (left) and the dispersion velocity field (right) for the whole region studied in this work. The LOS velocity and the dispersion velocity values are calculated using the 10 nearest neighbors to each pixel of 36 arcsec size. The iso-density contours for the 146 confirmed cluster members are superimposed in black.

associated with this low velocity group of galaxies. Conversely the galaxies belonging to the main group (V$\sim$ 6500 km s^{-1}) are homogeneously distributed through the whole subcluster, representing its virialized galaxy population.

Finally the multiple merging scenario is also consistent with the particular position of Abell 1367. In fact this cluster lies at the intersection of two filaments, the first extending roughly 100 Mpc from A1367 toward Virgo (West & Blakeslee 2000) and the second extending between A1367 and Coma. Since the intersection of filaments are considered the natural place of cluster formation (Katz & White 1993), Abell 1367 might be at the first steps of its evolution into a rich relaxed cluster.

Acknowledgements

We thank Alessandro Donati for his help during the poster preparation.

References

Cortese L., Gavazzi G., Boselli A., Iglesias-Paramo J., & Carrasco L. 2004, submitted to A&A
Donnelly, R. H., Markevitch, M., Forman, W., et al. 1998 *ApJ* **500** 138
Forman, W., Churazov, E., David, L., et al. 2003 *astro-ph/0301476*
Gavazzi G., Cortese L., Boselli A., et al. 2003 *ApJ* **597** 210
Katz N. & White S. D. M. 1993 *ApJ* **412** 455
Sakay, S., Kennicutt, R. C., van der Hulst, J. M., & Moss, C. 2002 *ApJ* **578** 842
Sun, M., & Murray, S. S. 2002 *ApJ* **576** 708
West, M. J. & Blakeslee, J. P. 2000 *ApJ* **543** L27

Outskirts of Galaxy Clusters: Intense Life in the Suburbs
Proceedings IAU Colloquium No. 195, 2004
A. Diaferio, ed.

DOI: 10.1017/S1743921304000213

Wide-field spectroscopy of a galaxy cluster pair at z=0.4

David G. Gilbank[1], **F. J. Castander**[2], **M. L. Balogh**[3] and **R. G. Bower**[3]

[1]Department of Astronomy and Astrophysics, University of Toronto, 60 St George Street, Toronto, Ontario, Canada, M5S 3HE email: gilbank@astro.utoronto.ca

[2]Institut d'Estudis Espacials de Catalunya/CSIC Gran Capità 2-4, 08034, Barcelona, Spain.

[3]Institute for Computational Cosmology, University of Durham, South Road, Durham, DH1 3LE, UK

Abstract. We present preliminary results from a wide-field spectroscopic survey of two galaxy clusters at $z = 0.4$, separated by $<10\ h^{-1}$Mpc on the sky. Both clusters are similarly optically rich, have velocity dispersions ~ 700 km s^{-1}, but differ in X-ray luminosity by a factor of $\sim$20.

---

1. Introduction

While galaxy clusters are dominated by passively evolving galaxies, field galaxies at $z \sim 0.4$ generally have very high star formation rates. In the hierarchical merging model of galaxy formation (e.g. Cole et al. 2000 and references therein) clusters grow through the accretion of galaxies from surrounding filamentary structures, which implies that galaxy properties are altered as they drain into overdense regions. However, the contrast between the cluster and the field at large clustercentric radii, is extremely low and thus identifying cluster members in these outer regions is observationally expensive. To increase efficiency, we targeted the inter-cluster region between a pair of clusters at $z = 0.4$ separated by $\lesssim 10h^{-1}$ Mpc on the sky, identified via the selection of the colour-magnitude relation of early type galaxies (see method of Gladders & Yee 2000) from the X-ray Dark Cluster Survey (XDCS; Gilbank 2001, Gilbank et al. 2004). Furthermore, these systems have similar optical luminosities in early-type galaxies, $L_{\rm E}$, but very different (more than an order of magnitude) X-ray luminosities. Thus, this field provides the ideal sample for not only studying the evolution of galaxies in clusters out to large radii, but also the *differential* evolution of clusters with differing X-ray properties, observed at the same epoch. We present results from a wide-field spectroscopic survey tracing from one cluster core to the other across the lower density inter-cluster region, using LDSS2 on the 6.5-m Magellan telescope and FORS2 on the VLT.

2. Spectroscopic observations

The clusters were discovered in V- and I-band imaging taken with the 2.5-m INT. The field was re-imaged for spectroscopy in the R-band using FORS2 on the VLT, and galaxies brighter than $R < 20.5$ were selected for spectroscopy, adding fainter galaxies where space allowed. In total 626 galaxies were observed, resulting in 432 secure, 33 probable, 24 single-line and 36 lower confidence redshifts. 101 galaxies failed to yield a redshift. The mean redshifts of the two systems were measured to be $z_{\rm A} = 0.4127$ and $z_{\rm B} = 0.4203$, yielding a velocity difference of ≈ 1700 km s^{-1} in the mean rest frame, with velocity dispersions of $\sigma_{\rm A,rest} = 670 \pm 60$ km s^{-1} and $\sigma_{\rm B,rest} = 700 \pm 100$ km s^{-1} respectively. These implies virial radii of approximately $1.5h^{-1}$ Mpc (Girardi et al. 1998),

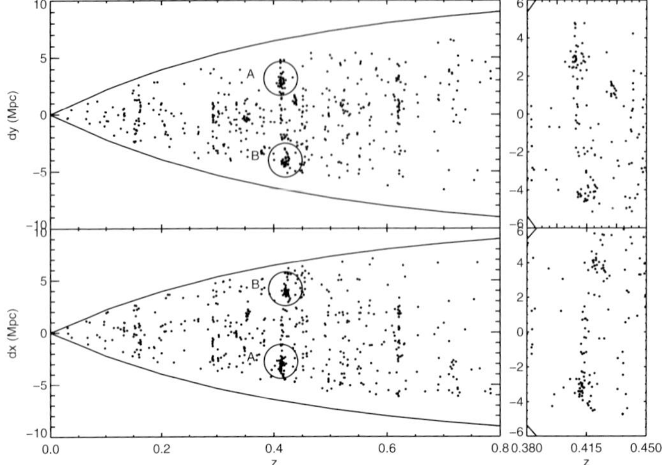

Figure 1. Declination (upper panels) and Right Ascension (lower panels) slices vs redshift for the field surveyed. The clusters A and B are circled and labelled in both wide field slices (left panels). A zoomed region around the cluster pair is shown to the right.

or ≈4.6 arcmins on the sky. The redshift distribution (shown as 'cone diagrams' in Fig. 1) shows evidence of filamentary structure connecting the clusters. We are unable to say if this connecting structure is due to a filament just linking the two clusters, or to an extended sheet of galaxies at the systems' redshift, because of our survey geometry: a line of masks from cluster core to cluster core.

In Fig. 2 we show the spectroscopically-selected colour-magnitude diagrams as a function of environment for each cluster in turn, and for the inter-cluster region. Both clusters show strong red-sequences of member galaxies (which is unsurprising, as this was the technique with which they were selected), and relatively low fractions of blue galaxies: 0.38 ± 0.11 and 0.25 ± 0.11 (although these fractions are possibly biased high, due to the greater ease of securing redshifts for blue, emission line galaxies over red, absorption line galaxies, at the faint end). The inter-cluster region also shows galaxies as red as those on the cluster red-sequences, but none as bright, and the fraction of blue galaxies is much higher (0.59 ± 0.13).

We attempt to characterise the environmental dependence of star-formation by examining the distribution of [OII] equivalent widths (EWs). Firstly we compare the distributions of each cluster with those of the X-ray luminous CNOC1 clusters (Balogh et al. 1999) and find that they are statistically indistinguishable. Secondly we compare the distribution in the inter-cluster region to that of a preliminary subsample of the CNOC2 field survey (R. Whitaker, priv. comm.). The distributions differ only at around the 1-σ level. Kodama et al. (2001) found that the transformation of galaxy properties occurs on a more local level, in groups in the cluster outskirts. This suggests that the lack of difference we see is due to the bulk of the galaxies in this large inter-cluster region being 'pristine' field galaxies, and that any signal of the truncation of star-formation is diluted over the large area we sample. Our next step is to examine the line-strengths of galaxies in more localised regions, in order to search for signatures of changes in star-formation properties.

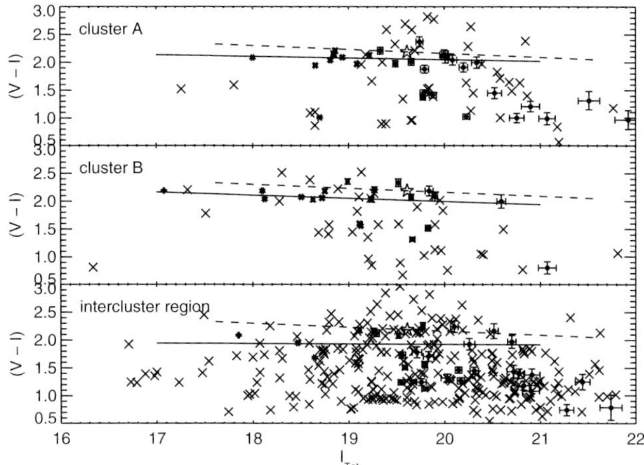

Figure 2. Colour-magnitude diagrams for the supercluster field, split by environment: top panel - cluster A; central panel - cluster B; lower panel - inter-cluster region. Filled points with error bars show spectroscopically confirmed members within the virial radius (∼4.6 arcmin) projected distance (or outside these radii for the inter-cluster area). Crosses are spectroscopically confirmed non-members. Overplotted are the biweight fit to the CMR (solid line) and the model CMR (broken line) for the clusters' redshift.

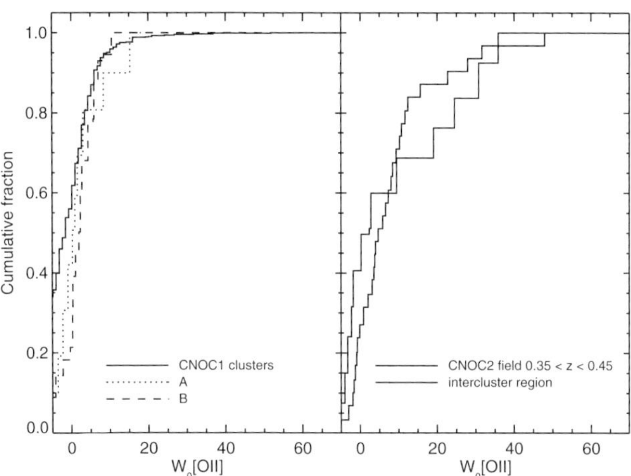

Figure 3. [OII] equivalent width distributions for our two $z = 0.4$ clusters and the CNOC1 X-ray luminous clusters (left panel) and for the inter-cluster region compared with a subsample of the CNOC2 field survey at comparable redshifts.

References

Balogh, M. L. and Morris, S. L. and Yee, H. K. C. and Carlberg, R. G. and Ellingson, E. 1999, ApJ, 527, 54

Cole, S., Lacey, C. G., Baugh, C. M., & Frenk, C. S. 2000, MNRAS, 319, 168

Gilbank, D. G. 2001, Ph.D. Thesis, University of Durham

Gilbank, D. G., Bower, R. G., Castander, F. J., & Ziegler, B. L. 2004, MNRAS, 348, 551

Girardi, M., Giuricin, G., Mardirossian, F., Mezzetti, M., & Boschin, W. 1998, ApJ, 505, 74

Gladders, M. D. & Yee, H. K. C. 2000, AJ, 120, 2148

Kodama, T. and Smail, I. and Nakata, F. and Okamura, S. and Bower, R. G., ApJL, 562, 9

Outskirts of Galaxy Clusters: Intense Life in the Suburbs
Proceedings IAU Colloquium No. 195, 2004
A. Diaferio, ed.

DOI: 10.1017/S1743921304000225

Structure and dynamics of the Shapley supercluster

Dominique Proust[1], Andreas Reisenegger[2], Hernán Quintana[2], Eric Slezak[3], Hernán Muriel[4], Laerte Sodré Jr.[5], Michael J. Drinkwater[6] and Quentin A. Parker[7]

[1] GEPI, Observatoire de Paris-Meudon F92195 Meudon CEDEX, France

[2] Departamento de Astronomìa y Astrofisica, Pontificia Universidad Católica de Chile, Casilla 306, Santiago 22, Chile

[3] Observatoire de Nice, 06304 Nice CEDEX4, France

[4] Grupo de Investigaciones en Astronomia Teórica y Experimental, Observatorio Astrónomico, Laprida 854, 5000 Córdoba, Argentina and CONICET, Buenos Aires, Argentina

[5] Instituto de Astronomia, Geofísica e Ciêncas Atmos., Universidade de São Paulo (IAG/USP), 01065-970 São Paulo/SP Brazil

[6] Department of Physics, University of Queensland, QLD 4072, Australia

[7] Department of Physics, Macquarie University, NSW 2109, Australia, and Anglo-Australian Observatory, PO Box 296, Epping NSW 1710, Australia

Abstract. We present results of our wide-field redshift survey of galaxies in a 285 square degree region of the Shapley Supercluster (SSC).

---

1. Introduction

The Shapley supercluster (SSC) is a remarkably rich concentration of galaxies centred around R.A.= 13^h25^m Dec. = $-30°$ which has been investigated by numerous authors since its discovery in 1930 (see Quintana *et al.* 1995). It consists of many clusters and groups of galaxies in the redshift range $0.04 < z < 0.055$, and lies in the general direction of the dipole anisotropy of the CMB. It is located $130\,h^{-1}$ Mpc beyond the Hydra-Centaurus supercluster. The SSC is one of the most massive concentrations of galaxies in the local universe, so it is also of particular interest to consider its effect on the dynamics of the Local Group. We present an analysis of the Shapley supercluster based on the most complete velocity catalogue obtained up to now. We discuss the completeness and overdensities of the velocity sample and the structure and dynamics of the supercluster using the spherical collapse model.

2. The velocity catalogue

We have made an investigation into the larger scale distributions of galaxies throughout the entire SSC region and close environs using data from wide-field multi-fibre spectrographs such as OPTOPUS and MEFOS at the 3.60m ESO telescope (Quintana *et al.* 1997), the fiber spectrograph at the 100" DuPont telescope at Las Campanas, Chile (Proust *et al.* 2004), the UKST/FLAIR-II system, Australia (Drinkwater *et al.* 1999, 2004), and more recently the UKST/6dF instrument (Jones *et al.* 2004). Combined with already published redshift sets from several surveys and papers (see Quintana *et al.* 1995, 2000, Bardelli *et al.* 2001, Kaldare *et al.* 2003) and data from the NED NASA/IPAC

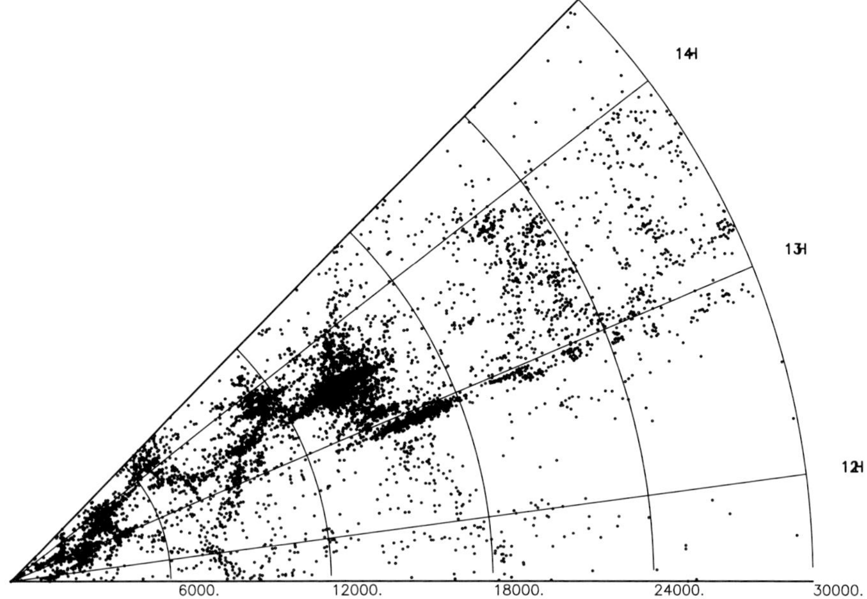

Figure 1. Wedge diagram in Right Ascension.

database, we obtain the most complete velocity catalogue for the SSC containing 10225 velocity measurements for 8341 galaxies (Figure 1). 4212 (50%) galaxies belong to the SSC with velocities in the range 9000–18000 km s^{-1} (a total depth of 90 h^{-1} Mpc).

3. Galaxy sample, completeness and overdensities

We compare our sample of observed galaxies to the total magnitude-limited galaxy distribution in the 225 square degrees survey region. We use the SuperCOSMOS sky survey to construct the parent galaxy catalogue for the region and analyse b_j magnitudes of all galaxies. Then, we determine the completeness of our survey (Drinkwater *et al.* 2004) as a function of 4 limiting magnitudes (Table 1) for both the full and the intercluster regions. It is highest for the brighter magnitude limits, peaking at 60 % for the whole sample at $b_j < 17.0$. In the velocity range ($9000 < cz < 18000\,\mathrm{km\,s^{-1}}$) of the SSC, we calculate galaxy overdensities of 7.0 ± 0.2 over the 225 square degree central region and 4.8 ± 0.2 in a 192 square degree region excluding rich clusters. Over the large region of our survey the inter-cluster galaxies make up 48 % (3705 objects) of all galaxies in the SSC region, and assuming a similar mass function they contribute a similar amount of mass to the cluster galaxies.

4. Velocity dispersions of ACO clusters in the Shapley supercluster

Velocity dispersions are an essential piece of information to study cluster dynamics, because they probe directly the cluster potential. To study the distribution of velocity dispersions of the Shapley clusters that are in the ACO catalog, we examined the velocity distribution of galaxies within 0.5 degree ($\sim$ Abell radius at $z = 0.05$ with $h = 0.7$) of each cluster center in the ACO catalog. The number of clusters with at least 6 galaxies in our velocity sample is 46. For a sample of 36 clusters in the best sampled region, the median velocity dispersion is 727 km s^{-1}. All clusters in the SSC with velocity dispersion larger than 1000 km s^{-1} present evidence of substructures.

Table 1. Completeness of the velocity catalogue for 4 b_j magnitude limits.

Field	b_j Mag lim.	Catalogue	Velocities	Differential
Full	17.0	2981	1796 (60%)	
Full	17.5	6331	2810 (44%)	[30%]
Full	18.0	13361	3931 (29%)	[16%]
Full	18.5	27177	4793 (18%)	[6%]
Inter-cluster	17.0	2404	1166 (49%)	
Inter-cluster	17.5	5105	1693 (33%)	[20%]
Inter-cluster	18.0	10862	2285 (21%)	[10%]
Inter-cluster	18.5	21917	2887 (13%)	[5%]

5. The spherical collapse model

We have applied the spherical collapse model (Regös & Geller 1989) to the SSC. The resulting mass enclosed by radius $r = 8h^{-1}$ Mpc lies between 2.0×10^{15} and $1.3 \times 10^{16} h^{-1} M_{\odot}$, corresponding to a density range $\rho/\rho_c \sim 3 - 20$ (Reisenegger *et al.* 2000). Diaferio's (1999) method gives similar results for the lower limit of the virialized mass in clusters. The mass required at the distance of the SSC to produce the observed motion of the Local Group with respect to the CMB is $M_{dipole} \approx (2 - 5) \times 10^{17} h^{-1} M_{\odot} \Omega^{0.4}$ (Reisenegger *et al.* 2000, Hoffman *et al.* 2001). The mass within $8h^{-1}$ Mpc can therefore produce at most $\simeq 5 - 10\%$ of the observed Local Group motion, which makes it unlikely that even the whole SSC would dominate its gravitational acceleration. On the other hand, consistent models of the density and velocity distribution on large scales (where density fluctuations are small) in the local Universe can now be built (e.g. Branchini *et al.* 1999). In these, the SSC figures prominently, although the Local Group motion originates from a combination of several "attractors".

Acknowledgements

This work was partially supported by the cooperative program ECOS/CONICYT C96U04 and the PICS-CNPq France-Brazil cooperation 910068/00-3.

References

Bardelli S., Zucca E., Baldi A.: 2001, *MNRAS* **320**, 387.
Branchini E. *et al.* : 1999, *MNRAS* **308**, 1.
Diaferio A: 1999, *MNRAS* **309**, 610.
Drinkwater M.J., Proust D., Parker Q.A., Quintana H., Slezak E.: 1999, *PASA* **16**, 113.
Drinkwater M.J., Parker Q.A., Proust D., Quintana H., Slezak E.: 2004, *PASA* **21**, 89.
Hoffman Y., Eldar A., Zaroubi S., Dekel A.: 2001, *astro-ph/0102190v1.*
Jones H., *et al.* (6dFGRS): 2004, *MNRAS* submitted.
Kaldare R., Colless M., *et al.* : 2003, *MNRAS* **339**, 652.
Proust D., Quintana H., Reisenegger A.: 2004, *A&A* in preparation.
Quintana H., Ramírez A., Melnick J., *et al.* : 1995, *AJ* **110**, 463.
Quintana H., Melnick J., Proust D., Infante L.: 1997, *A&AS* **125**, 247.
Quintana H., Carrasco E.R., Reisenegger, A: 2000, *AJ* **120**, 511.
Regös E., Geller M.J.: 1989, *AJ* **98**, 755.
Reisenegger A., Quintana H.., *et al.* : 2000, *AJ* **120**, 523.

Part 2

FORMATION AND EVOLUTION OF GALAXY CLUSTERS

Section A
Diffuse baryons

Outskirts of Galaxy Clusters: Intense Life in the Suburbs
Proceedings IAU Colloquium No. 195, 2004
A. Diaferio, ed.

DOI: 10.1017/S1743921304000237

The warm-hot intergalactic medium

Jelle S. Kaastra

Space Research Organization Netherlands, Sorbonnelaan 2, 3584 CA Utrecht, The Netherlands
email: J.Kaastra@sron.nl

Abstract. The warm-hot intergalactic medium contains a major part of all baryons in the universe. Despite this, it is very hard to observe in X-rays due to its relatively low temperature and the poor spectral resolution of most instruments. The high sensitivity and good spectral resolution of XMM-Newton allowed for the first time to show the presence of the warm-hot intergalactic medium in the outskirts of several nearby clusters of galaxies. In this contribution these results are presented. The physical state of the gas such as temperature, density, mass and chemical composition are discussed.

1. Introduction

Recent observational studies (e.g. using WMAP, Spergel et al. 2003) have brought us in the era of high accuracy cosmology. It has become now clear that matter constitutes only 27% of the mass density of the universe, and baryons even less: only 4%. At high redshift, all baryons are found predominantly in the Lyα forest (Rauch et al. 1998; Weinberg et al. 1997). However, at low redshifts, a census of the baryons (in stars of different types, atomic and molecular gas, dust, hot plasma in clusters and warm plasma in groups) finds only $\Omega = 0.0124$ (Fukugita et al. 1998), to be compared to the total baryon density $\Omega_b = 0.044$. Even accounting for the local Lyα forest (29% of all baryons, see the contribution by Stocke in these proceedings), these findings imply that currently we do not see a major fraction of all baryons.

Theoretical and numerical work (for example Cen & Ostriker 1999) has predicted that most of these "missing baryons" should be in a plasma phase with temperatures between 10^5 and 10^7 K, in filaments connecting the clusters and with on average higher density and temperature towards these clusters. They dubbed this component the Warm-Hot Intergalactic Medium (WHIM). Due to its low temperature and low density, it is difficult to observe in X-ray emission; observing it by means of absorption line studies requires high spectral resolution and sensitivity as well as bright background sources. In this paper I summarize the main results from observational studies of the WHIM using both emission and absorption techniques.

2. Emission (line) studies

2.1. *Cluster soft excess emission*

For a long time the hard X-ray spectra of clusters of galaxies have been modeled successfully with a thermal plasma emission model. The hot gas (10^7–10^8 K) is in hydrostatic equilibrium in the deep cluster potential well. It constitutes the major part of the baryonic mass contents of clusters (only the dark matter contribution is larger). With the advent of more sensitive low energy X-ray instrumentation (ROSAT PSPC, EUVE DS detector) it was found that some clusters contained soft excess emission above the predicted hard X-ray emission from the cluster (Virgo, Lieu et al. 1996a; Coma, Lieu et al.

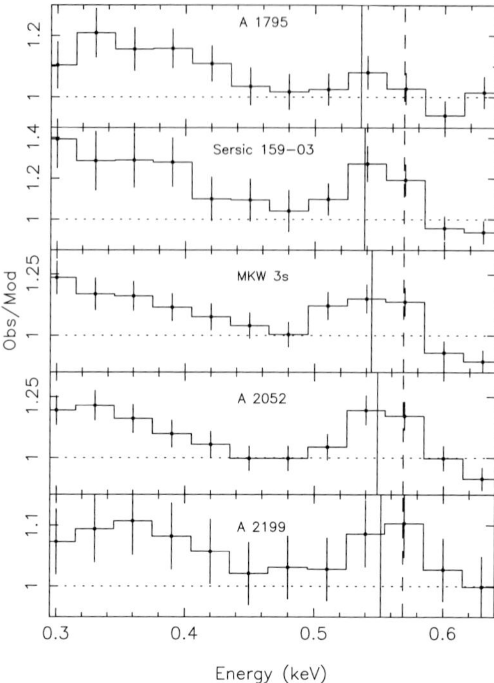

Figure 1. Fit residuals with respect to a two temperature model for the outer 4–12 arcmin part of six clusters. The position of the O VII triplet in the cluster restframe is indicated by a solid line and in our Galaxy's rest frame by a dashed line at 0.569 keV (21.80 Å). The instrumental resolution at 0.5 keV is ∼60 eV (FWHM).

1996b). Later a few other cases were found using either EUVE, ROSAT or BeppoSAX (see for instance Durret et al. 2002 for more references). Several different interpretations have been given, for example thermal emission from warm plasma (Lieu et al. 1996a, 1996b; Fabian 1997), or nonthermal emission, for instance inverse-Compton scattering between the cosmic microwave background (CMB) and intracluster relativistic electrons (Hwang 1997; Ensslin & Biermann 1998; Sarazin & Lieu 1998).

There has been a long debate in the literature and on conferences on this work, arguing about issues like proper background subtraction, modeling of the hot gas distribution, the physical implications etc. This is due to the fact that at the relevant low energies the above mentioned instruments have almost no spectral resolution, and calibration at these energies has always been a difficult task. The reality of the soft excess has been demonstrated recently, however, using the large sensitivity and moderate spectral resolution of XMM-Newton.

2.2. *XMM-Newton observations*

Using previous instrumentation, it was hard or impossible to demonstrate that the soft excess found in clusters has a thermal origin, due to the lack of spectral resolution. At the expected temperatures of the WHIM, the soft X-ray emission of a warm plasma is dominated by line radiation. A particular strong line blend is the O VII triplet near 22 Å (0.57 keV). This triplet is well isolated from other lines using CCD spectral resolution and is rather strong, since oxygen is the most abundant metal.

Table 1. XMM-Newton detection of the WHIM

Cluster	A 1795	Sérsic 159–03	MKW 3s	A 2052	A 2199
L_X [a]	4.7	0.78	0.83	0.71	1.5
Redshift	0.064	0.057	0.046	0.036	0.030
R (arcmin)	80	60	60	80	60
R (Mpc)	5.7	3.8	3.1	3.3	2.1
$n_{\rm H}(0)$ [b]	45	80	120	140	120
$N_{\rm H}(0)$ [c]	16	19	23	28	16
kT (keV) [d]	0.200±0.008	0.180±0.006	0.230±0.012	0.208±0.007	0.195±0.013
Abundance O	0.08±0.05	0.08±0.03	0.09±0.03	0.12±0.03	0.16±0.05
$N_{\rm O\,VII}(0)$ [e]	4	4	6	9	7
$I_{\rm triplet}$ [f]	1.3	2.8	3.4	7.7	7.1

[a] 2–10 keV luminosity of the cluster, in 10^{37} W (10^{44} erg s^{-1}), taken from David et al. (1993).
[b] Hydrogen density (assuming constant density sphere with radius R), in units of m^{-3} (10^{-6} cm^{-3}).
[c] Hydrogen column density through the core in units of 10^{24} m^{-2} (10^{20} cm^{-2}).
[d] Temperature of the WHIM component.
[e] O VII column density through the core in units of 10^{20} m^{-2} (10^{16} cm^{-2}).
[f] Total intensity of the O VII triplet (r+f+i line combined), in units of photons cm^{-2} s^{-1} sr^{-1}.

We have performed an extensive study of the presence of soft excess emission in a large sample of clusters using XMM-Newton data. Here we summarize the results of our work. A full account and details about the analysis are given in Kaastra et al. (2003a) for a sample of 14 clusters. This sample was extended to 20 clusters in Kaastra et al. (2003b), and the cluster A 2199 was added in Kaastra et al. (2004a). Our Table 1 is adapted from the last paper. In 7 out of 21 clusters there is evidence for a soft X-ray excess, in 5 of them there are signatures of a WHIM origin (the presence of O VII line emission plus the corresponding soft excess, Fig. 1), while in the remaining two (Coma, A 3112) the soft excess may have a non-thermal origin. The presence of soft excess in these last two clusters as well as in A 1795 was confirmed using a different analysis by Nevalainen et al. (2003).

The spatial extent of the soft excess emission is of the order of 1 degree in radius for these clusters (2–6 Mpc). The surface brightness drops only slowly toward the outer regions of the structure. In our study of Coma mentioned above we focussed on the central part. An extensive study of the outer parts was performed by Finoguenov et al. (2003). They found evidence for extended soft X-ray emission ($kT = 0.2$ keV), from plasma with 0.1 times solar metallicity, from a filamentary structure with a width of 3 Mpc and length 20 Mpc, seen in front of Coma (distance 90 Mpc). The presence of this filament is also evident from the redshift and spatial distribution of galaxies near Coma.

Stimulated by these findings, other studies using archival ROSAT PSPC data have been performed. Bonamente et al. (2002) did a statistical study of 38 clusters observed with ROSAT. Excess soft X-ray radiation, above the contribution from the hot intracluster medium, is evident in a large fraction of sources and is clearly detected with large statistical significance in the deepest observations. Zappacosta et al. (2004) studied archival PSPC data in the region around the Sculptor supercluster, and found a clear correlation between the 1/4 keV flux and the galaxy distribution, with indications for diffuse, filamentary structures sometimes connecting the known clusters of Sculptor.

3. Absorption line studies

3.1. *The Local Group ($z = 0$)*

X-ray absorption lines towards bright background sources have been reported now for several sources. A Chandra LETGS observation of PKS 2155-304 (Nicastro et al. 2002) showed unresolved O VII, O VIII and Ne IX lines (FWHM<800 km s^{-1}). Contrary, a FUSE observation showed O VI absorption consisting of a narrow (FWHM=100 km s^{-1}) and broader (FWHM=160 km s^{-1}) component. Nicastro et al. argue that the broad UV component as well as the X-ray component originate from low-density intergalactic plasma collapsing toward our Galaxy, consistent with the predictions of a warm-hot intergalactic medium from numerical simulations. The structure has a density of 6×10^6 cm^{-3}, overdensity δ of 60, temperature of 0.5 MK and ~3 Mpc size.

In several other Seyfert 1 galaxies and quasars, in particular the ones at higher redshift, it is now easy to see the $z = 0$ absorption component in long exposed grating spectra (3C 273, Fang et al. 2003; NGC 5548, Steenbrugge et al. 2003; NGC 4593, McKernan et al. 2003a). In other cases it is not clear if absorption is due to the WHIM or due to an ionized outflow of the background source, for example 3C 120 (McKernan et al. 2003b).

The number of lines of sight studied in X-rays up to now is very limited, due to the limited spectral resolution and sensitivity of the current X-ray spectrometers (but see Sect. 4). Using FUSE, Savage et al. (2003) studied 100 lines of sight in O VI (λ1032, λ1038 Å) absorption. Similar results using halo stars were obtained by Zsargó et al. (2003). However, with data from a single ion only it is hard to distinguish nearby absorption (Galactic halo) from distant absorption (Local Group). For example, Nicastro et al. (2003) argue, based upon the spatial distribution of high-velocity O VI absorbers, that this component originates from the Local Group and not the Galactic halo.

3.2. *Low redshift ($z > 0$) absorbers*

Absorption lines from $z > 0$ gas have now been found in FUSE spectra of a few bright quasars using O VI lines (H 1821+643 Tripp et al. 2000, Oegerle et al. 2000; PG 0353+415, Savage et al. 2002; PHL 1811, Jenkins et al. 2003). In the X-ray band, the results sofar have not been very convincing.

Fang et al. (2002) report the 4.5σ detection of O VIII Lyα absorption at $z = 0.055$ toward PKS 2155-304, utilizing a Chandra LETG/ACIS observation. Since there is also a small group of spiral galaxies at the same redshift, Fang et al. identified this with the first X-ray detection of O VIII from the WHIM.

However, these results have not been confirmed in both XMM-Newton RGS spectra with much longer exposure time and sensitivity (Paerels et al. 2003, Cagnoni et al. 2003), neither in a long observation using the LETG/HRC-S configuration.

Mathur et al. (2003) claim the detection of O VII and O VIII X-ray absorption lines at the redshifts of 6 intervening UV absorption systems (in particular at $z = 0.12$ and $z = 0.24$) toward the quasar H 1821+643 in a 500 ks Chandra observation. However, the significance of these detections is rather low and should best be regarded as upper limits until more sensitive data become available.

An exception to this lack of significance is the recent discovery of absorbers at $z = 0.011$ and $z = 0.027$ toward Mrk 421 (see Nicastro, these proceedings).

4. New X-ray instrumentation

The problem with detecting the WHIM in X-ray emission is its weakness in the filaments between the clusters. This is due to the fact that the X-ray emission scales with n^2

and the density is low ($< 10\ \mathrm{m}^{-3}$). Moreover, this emission is seen against the extragalactic background and a strong Galactic X-ray foreground. In particular the spectrum of the foreground is due to plasma at the same temperatures as the WHIM and therefore with similar spectra. As we showed in Sect. 2, with the present day CCD spectral resolution it is possible but non-trivial to disentangle foreground from (super)cluster emission near the brightest parts of the WHIM, where it connects to the clusters. Going beyond this requires high spectral resolution, high sensitivity and large field of view.

This last instrumental property (large field of view) is not present in the currently planned high resolution imaging X-ray missions (Astro-E2, Constellation-X, XEUS). These instruments will be able to investigate the properties of the WHIM both using absorption and emission studies, but only in "pencil-beam" regions of the sky. Good overviews of the capabilities of XEUS, the most powerfull of these instruments, are given by Paerels et al. (2003), Barcons (2003), and Kaastra & Paerels (2004b).

In order to map the WHIM, wide-field imaging is needed. Currently there is a Japanese proposal for such a mission, named DIOS (Yoshikawa et al. 2003; see also Suto et al. 2004). A similar proposal with smaller grasp (MBE, see Fang et al. 2004) was not awarded funding by NASA in 2003. Also our group is currently studying a concept for such a WHIM mission (Den Herder et al. 2004). More details will be given in that paper.

5. Conclusions

The presence of X-ray emission from the WHIM is now well established using XMM-Newton observations near bright clusters of galaxies. For the WHIM component in the filaments between the clusters, we currently rely upon a few UV and X-ray absorption measurements. In the coming years this situation is expected to improve, due to the launch of more sensitive X-ray observatories as well as dedicated WHIM missions.

References

Barcons, X. 2003 *MPE Report* **281**, 77

Bonamente, M., Lieu, R., Joy, M.K., & Nevalainen, J. 2002 *ApJ*, **576**, 688

Cagnoni, I., Nicastro, F., Maraschi, L., Treves, A., & Tavecchio, F. 2003 *New Astr. Rev.*, **47**, 561

Cen, R., & Ostriker, J.P. 1999 *ApJ*, **514**, 1

David, L.P., Slyz, A., Jones, C., Forman, W., Vrtilek, S.D., & Arnaud, K.A. 1993 *ApJ*, **412**, 479

Den Herder, J.W.A., Paerels, F.B.S., Kaastra, J.S., et al. 2004 *Proc. SPIE*, submitted

Durret, F., Slezak, E., Lieu, R., Dos Santos, S., & Bonamente, M. 2002 *A&A*, **390**, 397

Ensslin, T. A., & Biermann, P. L. 1998 *A&A*, **330**, 90

Fabian, A. C. 1997 *Science*, **275**, 48

Fang, T., Marshall, H.L., Lee, J.C., Davis, D.S., & Canizares, C.R. 2002 *ApJ*, **572**, L127

Fang, T., Sembach, K.R., & Canizares, C.R. 2003 *ApJ*, **586**, L49

Fang, T., Croft, R.A.C., Sanders, W.T., et al. 2004 *ApJ*, submitted (astro-ph/0311141)

Finoguenov, A., Briel, U.G. & Henry, J.P. 2003 *A&A*, **410**, 777

Fukugita, M., Hogan, C., & Peebles, P.J.E. 1998 *ApJ*, **503**, 518

Hwang, C.-Y. 1997 *Science*, **275**, 48

Jenkins, E.B., Bowen, D.V., Tripp, T.M., et al. 2003 *AJ*, **125**, 2824

Kaastra, J.S., Lieu, R., Tamura, T., Paerels, F.B.S., & den Herder, J.W.A., 2003a, *A&A*, **397**, 445

Kaastra, J.S., Lieu, R., Tamura, T., Paerels, F.B.S., & den Herder, J.W.A., 2003b, *Adv. Space Research*, in press.

Kaastra, J.S., Lieu, R., Tamura, T., Paerels, F.B.S., & den Herder, J.W.A. 2004a In *Soft X-ray emission from clusters of galaxies and related phenomena*, (ed. R. Lieu), Kluwer, in press (astro-ph/0305424)

Kaastra, J.S., & Paerels, F.B.S. 2004a In *Soft X-ray emission from clusters of galaxies and related phenomena*, (ed. R. Lieu), Kluwer, in press

Lieu, R., Mittaz, J.P.D., Bowyer, S. 1996a *ApJ*, **458**, L5

Lieu, R., Mittaz, J.P.D., Bowyer, S. 1996b *Science*, **274**, 1335

Mathur, S., Weinberg, D.H., & Chen, X. 2003 *ApJ*, **582**, 82

McKernan, B., Yaqoob, T., George, I.M., & Turner, T.J. 2003a *ApJ*, **593**, 142

McKernan, B., Yaqoob, T., Mushotzky, R., George, I.M., & Turner, T.J. 2003b *ApJ* **598**, L83

Nevalainen, J., Lieu, R., Bonamente, M., & Lumb, D. 2003 *ApJ*, **584**, 716

Nicastro, F., Zezas, A., Drake, J., et al. 2002 *ApJ*, **573**, 157

Nicastro, F., Zezas, A., Elvis, M., et al. 2003 *Nature*, **421**, 719

Oegerle, W.R., Tripp, T.M., Sembach, K.R., et al. 2000 *ApJ*, **538**, L23

Paerels, F.B.S., Rasmussen, A., Kahn, S., den Herder, J.W., de Vries, C.P. 2003 *MPE Report* **281**, 57

Rauch, M. 1998 *ARA&A*, **36**, 267

Sarazin, C. L., & Lieu, R. 1998 *ApJ*, **494**, L177

Savage, B.D., Sembach, K.R., Tripp, T.M., & Richter, P. 2002 *ApJ*, **564**, 631

Savage, B.D., Sembach, K.R., Wakker, B.P., et al. 2003 *ApJS*, **146**, 125

Spergel, D.N., Verde, L., Peiris, H.V., et al. 2003 *ApJS*, **148**, 175

Steenbrugge, K.C., Kaastra, J.S., de Vries, C.P., & Edelson, R. 2003 *A&A*, **402**, 477

Suto, Y., Yoshikawa, K., Yamasaki, N.Y., et al. 2004 In *proc. VI International Conference on Gravitation and Astrophysics of Asian-Pacfic Countries* (astro-ph/0402389)

Tripp, T.M., Savage, B.D., & Jenkins, E.B. 2000 *ApJ*, **534**, L1

Yoshikawa, K., Yamasaki, N.Y., Suto, Y., et al. 2003 *PASJ*,**55**, 879

Weinberg, D.H., Miralda-Escude, J., Hernquist, L., & Katz, N. 1997 *ApJ*, **490**, 564

Zappacosta, L., Maiolino, R., Mannucci, F., Gilli, R., & Schuecker, P. 2004 *MNRAS*, submitted (astro-ph/0402575)

Zsargó, J., Sembach, K.R., Howk, J.C., & Savage, B.D. 2003 *ApJ*, **586**, 1019

Outskirts of Galaxy Clusters: Intense Life in the Suburbs
Proceedings IAU Colloquium No. 195, 2004
A. Diaferio, ed.

DOI: 10.1017/S1743921304000249

First detection of WHIM filaments at cosmological distances

Fabrizio Nicastro

Harvard-Smithsonian Center for Astrophysics, 60 Garden Street, Cambridge, MA 02138, USA
email: fnicastro@cfa.harvard.edu

Abstract. In this contribution we present the detection of two WHIM filaments in the *Chandra*-LETG spectrum of Mkn 421. This spectrum has been obtained following 2 of our pre-approved Target of Opportunity requests to observe Blazars in outburst to efficiently 'X-ray' the IGM at high spectral resolution. These observations caught the source at the unprecedented levels of 60 and 40 mCrab in the soft X-ray (0.5-2 keV) band. We detect, for the first time, two WHIM filaments at redshifts of z=0.011 and z=0.027, respectively. Based on these two detections and on the upper limits on associated HI and OVI absorption as inferred from the HST and FUSE spectra of Mkn 421, we estimate a number of WHIM filaments per unit redshift, with He-like ion columns $N_X \gtrsim 8 \times 10^{14}$ cm^{-2} of $dN/dz = 67^{+88}_{-43}$, and a baryon mass density of $\Omega_b = 0.021^{+0.028}_{-0.014}$, virtually all of the 'missing baryons' at $z < 1$.

1. Introduction

Despite recent progress in cosmology, very little is still known about the location of baryons in the local Universe. The number of baryons detected in the "virialized Universe" at $z \lesssim 2$ (i.e. stars, neutral hydrogen associated with galaxies, and X-ray emitting gas in clusters) is more than a factor of 2 smaller than the concordance cosmology value of $\Omega_b = (4.4 \pm 0.4) h_{70}^{-2}$ % (Bennett *et al.* 2003), i.e. $\Omega_b < 2h_{70}^{-2}$ % (2σ; e.g. Fukugita, Hogan & Peebles 1998).

Hydrodynamical simulations for the formation of structures in the Universe, offer a possible solution to this puzzle. They predict that in the present epoch ($z \lesssim 1-2$), half of the normal baryonic matter in the Universe (the "missing baryons") is in a tenuous (overdensities $\delta \simeq 5-50$, relative to the mean baryon density in the Universe) warm-hot ($T = 10^{5.5} - 10^7$ K) phase, the so called 'Warm Hot Intergalactic Medium' (WHIM, e.g. Hellsten *et al.* 1998; Cen & Ostriker, 1999; Davé *et al.* 2001). This filamentary web of gas should imprint a 'forest' of metal absorption lines in the soft X-ray spectra of background sources. However, the vast majority of WHIM filaments is expected to be very thin, with metal ion columns typically much lower than $N_X \sim 10^{16}$ cm^{-2} (e.g. Fang, Bryan & Canizares, 2002).

A Local Group system of WHIM-like gas has been detected in X-ray species (Nicastro *et al.* 2002, Fang, Sembach & Canizares 2003, Rasmussen, Kahn & Paerels 2003, Cagnoni *et al.* 2004) and in high-velocity UV OVI (Nicastro *et al.* 2003). No secure detection of intervening WHIM absorption systems at cosmological distances has been claimed so far. We present the first high significance ($\gtrsim 5.8\sigma$ and $\gtrsim 11\sigma$) detection of two $z > 0$ WHIM filaments, in a high signal-to-noise *Chandra*-LETG spectrum of the blazar Mkn 421.

Throughout the paper we assume $H_0 = 70$ km s^{-1} Mpc^{-1}. Errors are quoted at 1σ for interesting parameters, unless otherwise stated.

2. Detecting the WHIM

With the exception of the very high-density tail of the intergalactic medium mass distribution (i.e. $\delta \gtrsim 1000$), in proximity of the most complex nodes of the WHIM web (i.e. outskirts of clusters of galaxies: Kaastra this conference), the vast majority of WHIM filaments should give rise to line emissions way below the sensitivity threshold of current X-ray instruments (e.g. Yoshikawa et al. 2003). Metal absorption lines from the WHIM, instead, can now in principle be detected in the *Chandra*-LETG or *Newton*-XMM RGS spectra of bright extragalactic sources, but only if the number of counts per resolution element ($\Delta\lambda = 50$ mÅ) in the continuum adjacent the line is larger than about 500. This number should be compared with the average 10-50 counts per resolution element in the archival *Chandra*-LETG spectra of the brightest Seyfert 1s (about 0.5-1 mCrab in the 0.5-2 keV band), collected, on average, with 100 ks exposures. So, either multi-megasecond integrations of the brightest nearby AGNs or transiently brighter sources are needed to detect the WHIM in absorption.

The public *Rossi* XTE All Sky Monitor (ASM) light curves show that several Blazars whose normal 2-10 keV flux states range between 0.1 and 5 mCrab have factor 10-50 outbursts, reaching 2-10 keV fluxes larger than 5-20 mCrab at least once per year. This flares usually last for a relatively short period of time (normally few days), but during these events these objects are, by far, the brightest extragalactic objects in the X-ray sky (with the exceptions of GRB afterglows that, however, typically are brighter than 1 mCrab only for ~ 1 day). We have pursued the less time-consuming of the two possible observational strategies: to detect the WHIM by waiting for one of these Blazars to flare, and then trigger a fast (i.e. less than 1 day from the trigger) Target of Opportunity Observation (TOO) with either the *Chandra* LETG or the *Newton*-XMM RGS.

We triggered our first two TOOs with the *Chandra*-ACIS-LETG and -HRC-LETG on 2002, October, 26-27, and 2003, July 1-2, on the blazar Mkn 421 ($z = 0.03$, De Vaucouleurs *et al.* 1991). These observations caught the source at 60 and 40 mCrab in the 0.5-2 keV band, and allowed us to collect a total of 6000 counts per resolution element at 21.6 Å. This is sufficient to detect He-like and H-like ion columns of $N_X \gtrsim 8 \times 10^{14}$ cm^{-2} at a significance level $\geqslant 3\sigma$.

3. WHIM absorption in the *Chandra*-LETG spectrum of Mkn 421

Figure 1 shows six portions of the LETG spectrum of Mkn 421. Three different ionized absorption systems are clearly identified in the spectrum: $z \sim 0$ (made up of both ISM, i.e. Savage *et al.* 2000, and Local-Group IGM lines, Nicastro *et al.* 2002, 2003), $z = 0.011$ (hereinafter WHIM System-I) and $z = 0.027$ (hereinafter WHIM System-II). The strongest line from WHIM System-I is the OVII$_{K\alpha}$ ($N_{OVII} = (1.0 \pm 0.3) \times 10^{15}$ cm^{-2}), while the two strongest lines of WHIM System-II are NVII$_{K\alpha}$ ($N_{NVII} = (1.5 \pm 0.4) \times 10^{15}$ cm^{-2}) and NVI$_{K\alpha}$ ($N_{NVI} = (0.7 \pm 0.2) \times 10^{15}$ cm^{-2}). We caution that, although clearly seen in the spectrum (Fig. 1), the NeIX$_{K\alpha}$ and OVII$_{K\beta}$ from System-II, as well as the OVIII$_{K\alpha}$ line from the System-I, lie very close to chip-gaps edges, which may induce an over-estimate of their EWs. We therefore consider the EWs of these lines only as upper-limits.

4. WHIM absorption in the UV spectra of Mkn 421

Mkn 421 has been observed with the HST-GHRS on 1995 February 1. An intervening HI Lyα system at $cz = (3046 \pm 2)$ km s^{-1} was discovered and reported by Shull *et al.*

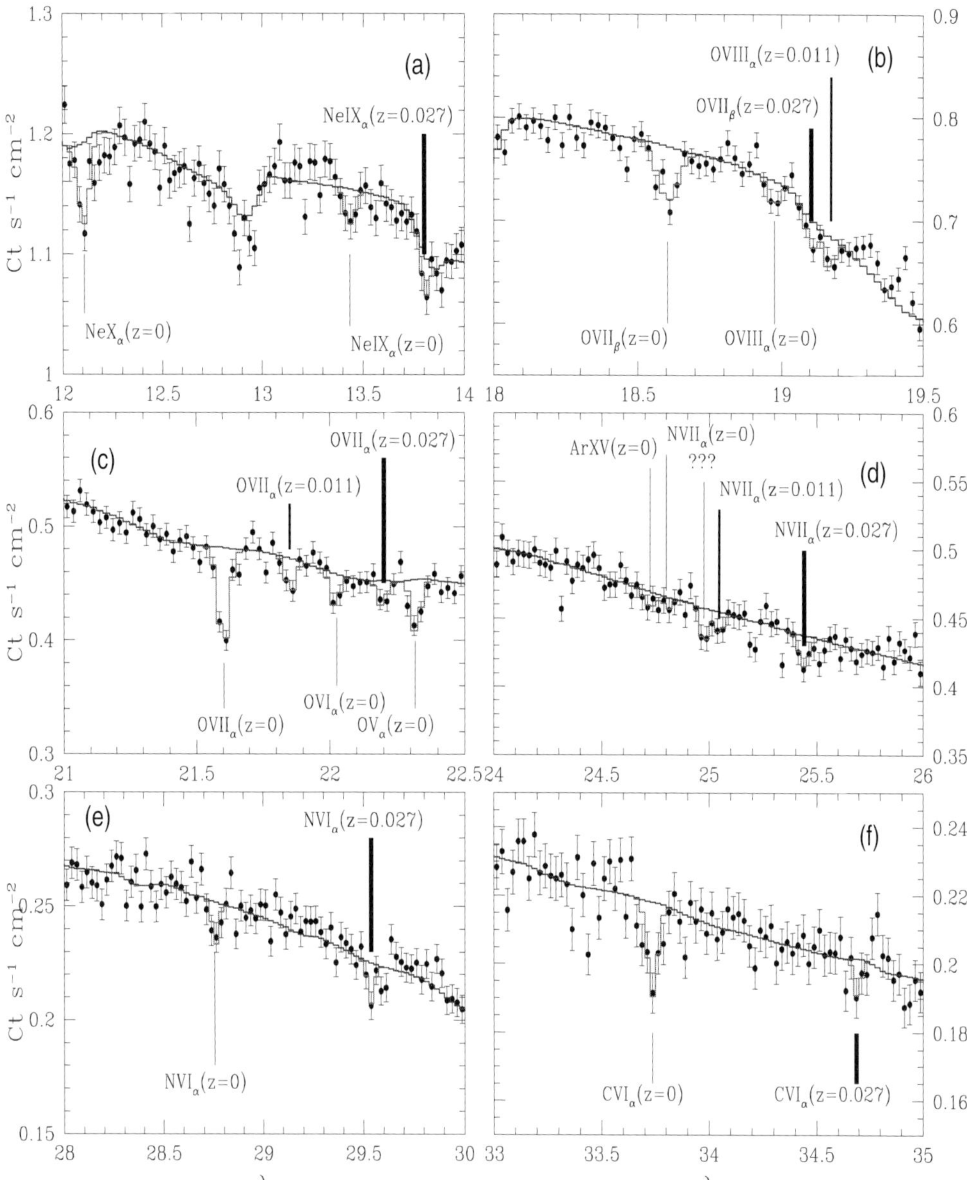

Figure 1. Six portions of the *Chandra*-LETG spectrum of Mkn 421 along with its best fitting continuum plus narrow absorption model (solid line: see details in Nicastro *et al.* 2004, ApJ, submitted), centered around the rest-wavelengths of the (a) NeXI-X$_{K\alpha}$, (b) OVIII$_{K\alpha}$, (c) OVII$_{K\alpha}$, (d) NVII$_{K\alpha}$, (e) NVI$_{K\alpha}$ and (f) CVI$_{K\alpha}$ transitions.

(1996). We retrieved this GHRS spectrum of Mkn 421 from the public HST archive, and re-analyzed the data. We fitted the 1220-1250 Å HST spectrum with a combination of broad continuum components and three narrow absorption Gaussians to model the intervening HI Lyα and two strong Galactic Si lines visible in the spectrum. The redshift of the HI Lyα line ($N_{HI} = (1.65 \pm 0.07) \times 10^{13}$ cm^{-2}) is consistent with that of our

System	T^{a}	N$_{H}$^{b}	D^{c}	[N/O]^{d}	[O/H]^{e}	n_b^{f}
I ($z = 0.011$)	(0.4-5.7)	(0.9-1.7)	(0.3-0.5)		$\geqslant -0.96$	$\gtrsim 1$
II ($z = 0.027$)	(0.7-1.7)	(1.5-3.0)	(0.5-1)	$\geqslant 0.2$	$\geqslant -0.3$	$>> 0.15$

Table 1. Physical parameters of the WHIM filaments. Units are as follows: ^{a}In 10^6 K. ^{b}In $10^{19}10^{-[O/H]}$ cm^{-2}. ^{c}In $10^{-[O/H]}n_b^{-1}$ Mpc. ^{d}[N/O] = $\mathrm{Log}\left(\frac{(N/O)}{(N/O)_\odot}\right)$. ^{e}[H/O] = $\mathrm{Log}\left(\frac{(O/H)}{0.1(O/H)_\odot}\right)$. ^{f}In 10^{-5} cm^{-3}.

X-ray WHIM System-I within the large systematic X-ray uncertainties ($\Delta z = 0.001$†). However, the average position of the X-ray system lies at ~ 5 Mpc from the accurate ($\Delta z = 7 \times 10^{-6}$) position of the HI system. Moreover the measured HI Lyα FWHM=0.28 Å gives an upper limit of $T < 2 \times 10^5$ K to the kinetic temperature of the HI, which is inconsistent with the temperature estimated for the X-ray WHIM System-I (see §5). If the systems are related, a multiphase WHIM is required. We also computed the 3σ HI column upper limit of a putative HI Lyα at the average redshift of the X-ray WHIM Systems. For System-I we found $N_{HI} < 3.9 \times 10^{12}$ cm^{-2}. Instead, for System-II we found a line at λ1248.80 Å at a significance of 4.7σ, which we identified with a HI Lyα at $z = 0.0273$ with column of $N_{HI} = (3.3 \pm 0.7) \times 10^{12}$ cm^{-2}. The FWHM of this line is $\leqslant$ 1.0 Å, implying $T \lesssim 2 \times 10^6$ K, consistent with the range of estimated HI kinetic temperatures for System-II (see §5).

Mkn 421 was also observed by FUSE on 2003, January 19-21, with a final total exposure time of 62 ks, following our TOO request based on a period of strong X-ray activity as recorded by the RXTE-ASM. We searched the FUSE spectrum of Mkn 421 for OVI$_{2s\to 2p}$ candidate intervening absorption line in the ~ 1032 to ~ 1062 Å wavelength range (i.e. from $z = 0$ up to to $z = 0.03$, the redshift of the source). With the exception of the strong Galactic Low-Velocity OVI (Savage et al. 2001) and the much weaker and probably extragalactic High Velocity OVI (Williams et al. 2004, in preparation), we did not detect any $z > 0$ OVI absorption down to a 3σ sensitivity limit of $N_{OVI} = 2 \times 10^{13}$ cm^{-2}.

5. Diagnostics analysis of WHIM System-I and -II

To infer the physical properties of the gas in these two WHIM filaments we first applied the equivalent width ratios technique to get direct information on the ionization balance in the gas and its metallicity, and then compared these quantities with predictions by gas ionization models to estimate the temperature and the equivalent H column density of the gas (see e.g. Nicastro et al. 2002, for details). Table 1 lists the physical parameters of the WHIM System-I and -II.

Limits on the baryon density n_b, in Table 1 are derived from the lower limits on [O/H] and from the fact that the total extent D of the filament along our line of sight has to be less than 5 Mpc for System-I (the distance from the average redshift of the X-ray system and that of the cooler HI system) and much lower than 13 Mpc for System-II (the distance of the filament from Mkn 421). We note that, although the ranges of uncertainty are large, physical and geometrical parameters for System-I and -II are all consistent with theoretical predictions for the WHIM. We also note that System-II ($z = 0.027$) tends to

† Errors associated with X-ray line wavelengths are largely dominated by systematics in the LETG wavelength calibrations. We estimated the amplitude of these systematics in our data looking for differences between expected and observed relative positions of two strong lines from the same ion, and then added a 20 mÅ systematic uncertainty to all our X-ray line wavelengths.

be somewhat bigger and more metal rich than System-I ($z = 0.011$), which may reflect a difference in the galaxy-environment in which the two systems lie.

6. dN/dz and Ω_b

Based on these two detections of WHIM filaments, we can now make a first, crude, estimate of the number of filaments per unit redshift with He-like ion columns larger than 8×10^{14} cm^{-2} (our 3σ detection threshold). We obtain $dN/dz(N_X > 8 \times 10^{14}) = 67^{+88}_{-47}$ (1σ, Gehrels 1986), in gratifying agreement with theoretical predictions of $dN/dz(N_X > 8 \times 10^{14}) = 65$.

Similarly, we can estimate the cosmological mass density of X-ray WHIM filaments. Following Savage et al. (2002) we measure $\Omega_b = (\frac{1}{\rho_c})(\frac{\mu m_p \sum_i N_H^i}{d_{Mkn421}}) = 0.021^{+0.028}_{-0.014}$, virtually all the 'missing' baryons at low redshift.

7. Conclusions

We have presented the first detection of two X-ray WHIM filaments at redshifts of $\langle z_1 \rangle = 0.011$ and $\langle z_2 \rangle = 0.027$, in the very high quality *Chandra*-LETG spectrum of the blazar Mkn 421. For System-I we estimate a temperature of $T = (0.4 - 5.7) \times 10^6$ K, an equivalent H column density of $N_H = (0.9 - 1.7) \times 10^{-[O/H]} \times 10^{19}$ cm^{-2}, a metallicity of [O/H]$\geqslant -0.96$, and a baryon density of $n_b \gtrsim 10^{-5}$ cm^{-3}. For System-II we obtain: $T = (0.7 - 1.7) \times 10^6$ K, $N_H = (1.5 - 3.0) \times 10^{-[O/H]} \times 10^{19}$ cm^{-2}, [O/H]$\geqslant -0.3$, and $n_b \gg 1.5 \times 10^{-6}$ cm^{-3}. The number of X-ray WHIM filaments per unit redshift and with He-like ion columns larger than 8×10^{14} cm^{-2}, inferred from these two detections is $dN/dz(N_X > 8 \times 10^{14}) = 67^{+88}_{-47}$, in gratifying agreement with theroretical predictions and so sufficient to account for all the 'missing' baryons in the local Universe. The cosmological mass density of X-ray WHIM filaments that we measure is $\Omega_b = 0.021^{+0.028}_{-0.014}$.

Acknowledgements

The author acknowledges the *Chandra* grant GO2-3122A.

References

Bennett, C.L., *et al.* 2003 *ApJS* **148**, 1-27.
Cagnoni, I. *et al.*, 2004 *ApJ* **603**, 449-455.
Cen, R. & Ostriker, J.P., 1999 *ApJ* **514**, 1-6.
Davé, R., *et al.*, 2001 *ApJ* **522**, 473-483.
De Vaucouleurs, G. *et al.*, 1991 *Springer-Verlag Berlin Heidelberg New York* **1-3, XII**, 1-2069.
Fang, T., Bryan, G.L. & Canizares, C.R., 2002 *ApJ* **564**, 604-623.
Fang, T., Sembach, K.R. & Canizares, C.R., 2003 *ApJ* **586**, L49-L52.
Fukugita, M., Hogan, C.J. & Peebles, P.J.E., 1998 *ApJ* **503**, 518-530.
Gehrels, N., 1986 *ApJ* **303**, 336-346.
Hellsten, U., Gnedin, N.Y. & Miralda-Escudé, J., 1998 *ApJ* **509**, 56-61.
Nicastro, F., *et al.*, 2003 *Nature* **421**, 719-721.
Nicastro, F., *et al.*, 2002 *ApJ* **573**, 157-167.
Rasmussen, *et al.* 2003 *ASSL Conference Proceedings* **281**, 109.
Savage, B.D., *et al.* 2002 *ApJ* **564**, 631-649.
Savage, B.D., *et al.* 2000 *ApJ* **538**, L27-L30.
Shull, J.M, Stocke, J.T. & Penton, S., 1996 *AJ* **111**, 72-77.
Yoshikawa, K., *et al.*, 2003 *PASJ* **55**, 879-890.

Outskirts of Galaxy Clusters: Intense Life in the Suburbs
Proceedings IAU Colloquium No. 195, 2004
A. Diaferio, ed.

DOI: 10.1017/S1743921304000250

Absorbers in the suburbs: HST views the local Lyα forest

John T. Stocke, J. M. Shull, S. V. Penton, B. A. Keeney and J. L. Rosenberg

CASA, Univ. of Colorado, Boulder CO 80309-0389, USA email: stocke@casa.colorado.edu

Abstract. We describe recent discoveries of low column density ($N_{HI} = 10^{12.5-16.0}$ cm^{-2}) H I Lyα absorbers made with the *Hubble Space Telescope* (HST) which have allowed us a first look at gas in local intergalactic space; i.e., between us and the "Great Wall". These absorbing clouds account for 29 ± 4% of all baryons at z=0 and are, in general, only loosely related to individual galaxies. Owing to the proximity of these absorbers to the Earth, the 187 absorbers in our combined GHRS + STIS sample provide our best view of the relationship between Lyα absorbers and galaxies, voids, and supercluster filaments. While only a few of the very strongest absorbers in our sample appear associated with individual galaxies, the majority (78%) are associated with large-scale filamentary structures of galaxies, while 22% are found in galaxy "voids". Amongst the highest column density absorbers (with $N_{HI} \approx 10^{16}$ cm^{-2}) in our sample, we find a low-ionization, metal-bearing absorber associated with a post-starburst dwarf galaxy $\sim$ 70h_{70}^{-1} kpc away. The detailed properties of the absorber and the galaxy provide strong evidence that these two are causally related. We argue that many/most weak metal absorption systems could be due to outflowing "superwinds" from dwarf galaxies.

1. The HST + STIS & GHRS Lyα absorber sample

Unlike virtually all other astronomical objects, Lyα absorbing "clouds" were first discovered at great distances ($z \geqslant$2) due to cosmological redshifts and the near-UV atmospheric cutoff. It has only been with the advent of the *Hubble Space Telescope* (HST) that nearby examples have been found and studied (Bahcall et al. 1991, 1993; Morris et al. 1991; and subsequent HST QSO Absorption Line Key Project papers by Jannuzi et al. 1998 and Weymann et al. 1998). While these absorbers are abundant enough to account for all baryons at $z \geqslant$2, their still substantial numbers at z=0 imply that ∼30% of all baryons remain in these clouds locally (Penton et al. 2000a,b; 2002, 2004; Papers I - IV hereafter, and Davé et al. 1998, 2000). Thus, any account of the present-day distribution of baryons must include an accurate census of these clouds and their masses, as inferred from their column densities, ionized fractions and physical extents.

Surprisingly and luckily for local IGM research, the modest HST aperture is competitive with the 10 m aperture of the *Keck* Telescope (+HIRES spectrograph) in detecting Lyα absorbers because much brighter targets can be observed. Figure 1 shows an HST+STIS (Space Telescope Imaging Spectrograph) far-UV spectrum of the bright BL Lac Object PKS 2005-489, which detects Lyα absorbers as low in column density, $N_{HI} \geqslant 10^{12.5}$ cm^{-2}, as the best *Keck* HIRES data (e.g., Hu et al. 1995), but these HST-discovered absorbers are within 20,000 km s^{-1} of the Earth. This allows us an unprecedented opportunity to search for faint galaxies that could be associated with these absorbers.

All results reported here are based on data toward 15 targets studied with the Goddard High Resolution Spectrograph (GHRS) (Papers I - III) and 16 targets observed using STIS (Paper IV). In the entire survey, we have detected 187 Lyα absorbers at $\geqslant 4\sigma$

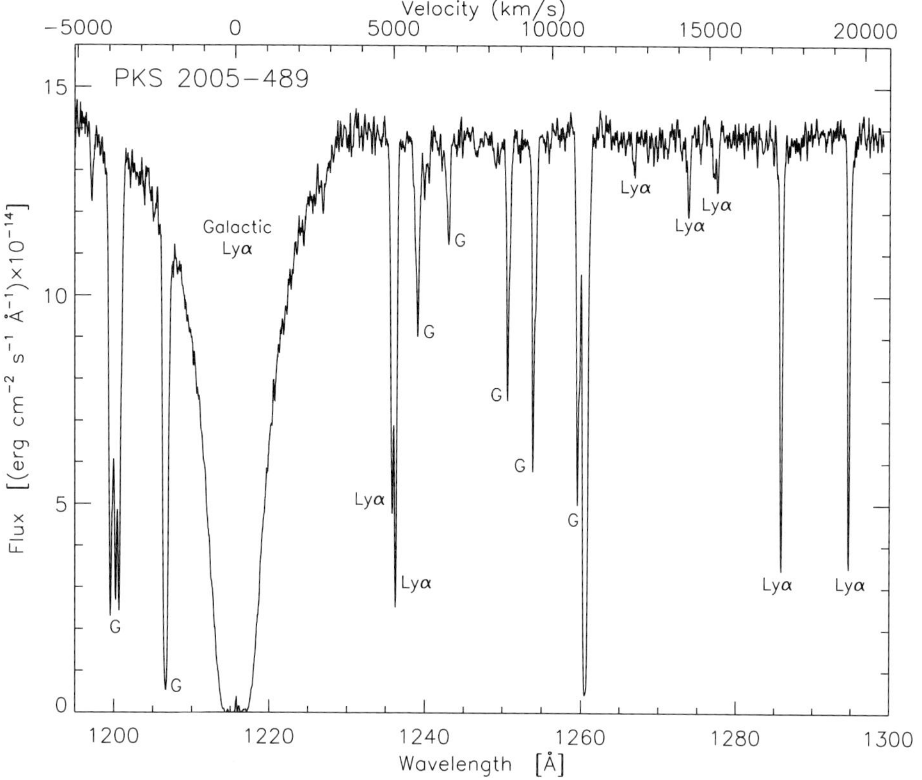

Figure 1. An HST/STIS medium resolution (20 km s^{-1}) spectrum (Paper IV) of the bright BL Lac Object PKS 2005-489 illustrates the best quality data obtained for this project. The deep, broad absorption at 1216 Å is the damped Lyα absorption due to the Milky Way. Other Galactic metal lines (S II, Si II, Mg II, N I, N V and Si III) are marked with a "G". The weakest Lyα absorbers have column densities, $N_{HI} = 10^{12.5}\,cm^{-2}$. The heliocentric velocity scale at top is for the Lyα absorbers only.

significance over a total unobscured path length $\Delta z = 1.16$. This yields $dN/dz \sim$300 per unit redshift at $N_{HI} \geqslant 10^{12.5}\,cm^{-2}$ or one "cloud" every $15\,h_{70}^{-1}$ Mpc. Using two different methods (Paper II and Schaye 2001) we estimate that the local, photoionized IGM contains 29±4% of all baryons at z=0 (see Paper IV for details). Since Lyα becomes extremely difficult to detect in our spectra at T$\geqslant$150,000°K, the baryon fraction above does not include much of the "warm-hot" medium predicted to account for another ∼10-30% of all baryons (Cen et al. 1994; Davé et al. 2000).

For a subset of 79 absorbers in our sample which lie in regions of the local universe surveyed for galaxies down to at least L* (based upon the CfA redshift survey catalog; Huchra et al. 1992, 2004 version), we find that 78% of all low column density absorbers lie in galaxy filaments and 22% lie in galaxy voids (nearest neighbor galaxy distances $> 3h_{70}^{-1}$ Mpc). The stronger absorbers ($N_{HI} \geqslant 10^{13.1}$ cm^{-2}) have much smaller nearest galaxy neighbor distances (median distance $\approx 400h_{70}^{-1}$kpc) than the weaker absorbers (median distance $\approx 1\ h_{70}^{-1}$Mpc). And, while the weaker absorbers cluster only very weakly with galaxies, they are not randomly distributed in space (7σ difference).

The two-point correlation function (TPCF; see Figure 2) for all "local" Lyα absorbers suggests a similar difference between the stronger and weaker absorbers. The full sample has a 7σ excess at Δv$\leqslant$ 250 km s^{-1}, but this excess is due entirely to the stronger half

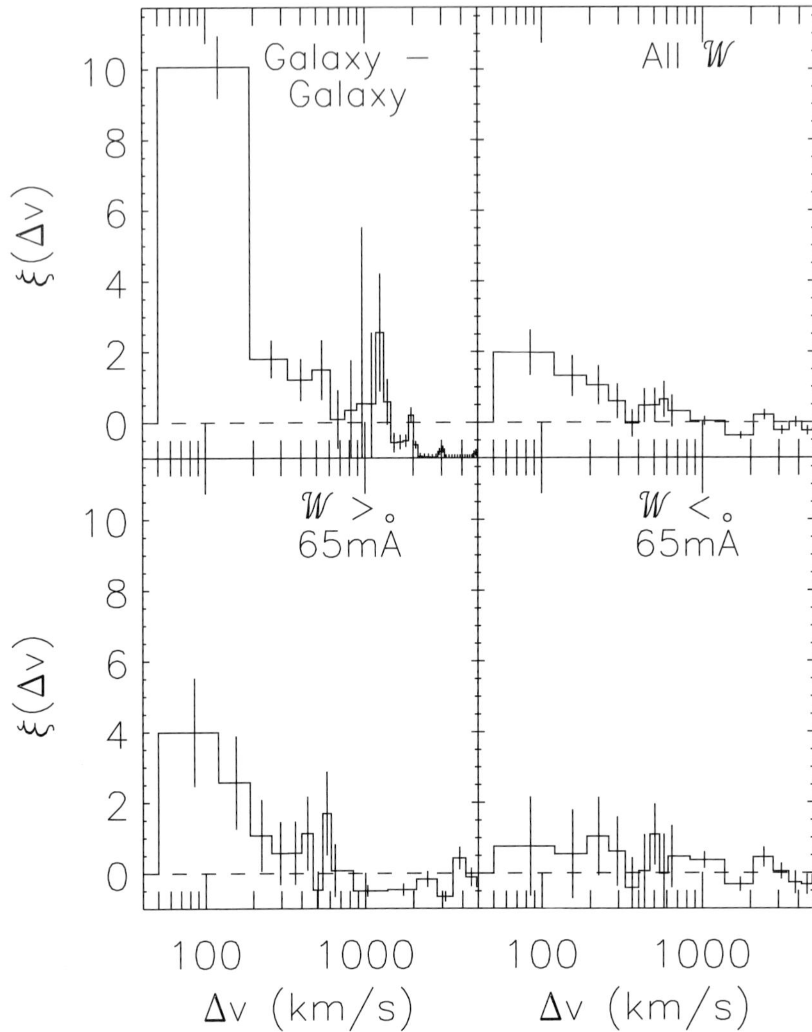

Figure 2. The two-point auto-correlation function amplitude (TPCF) for: galaxies in the CfA redshift survey region (upper left), our entire 187 Lyα sample (upper right), the stronger half ($N_{HI} \geqslant 10^{13.1}$ cm^{-2}) of our Lyα sample (lower left) and the weaker half sample (lower right). The galaxy-galaxy TPCF shows both an extremely large excess power at Δv$\leqslant$ 600 km s^{-1} due to galaxies within the same filament and a large deficit at larger Δv due to galaxy voids. Note that the stronger half-sample shows both the excess at small Δv and a *hint* of a deficit at larger Δv, while the weaker half sample shows no evidence for clustering on any scale.

of the sample (compare upper left and lower left panels). The weaker absorbers show no evidence for clustering on any velocity scale, indicating a relatively random distribution in space (Paper IV). Thus, it is generally the weaker Ly α absorbers which are in galaxy voids. These differences are suggestive of two populations of absorbers (Sargent, 1978), the stronger of which are somehow associated with galaxies (and may contain metals) while the weaker may be pristine gas.

2. Lyα absorbers in galaxy voids

When corrected for total path length surveyed, we find that 22% of all low column density Lyα absorbers are in galaxy voids, more than 3h^{-1} Mpc from the nearest galaxy. The number of Lyα absorbers we find in voids constitutes 4.5$\pm$1.5% of the total baryon density, a number which agrees quite well with the total mass density in voids found in ΛCDM simulations by Gottlöber et al. (2003). Additionally, we have conducted a deep optical search (McLin et al. 2002) and a sensitive H I 21cm search (Zamojski et al., in preparation) for galaxies and H I clouds near four very nearby void absorbers at $cz = 2000-3000$ km s^{-1}. No optical galaxies were found down to very faint limits (M$_B \leqslant$ -14 within 250h^{-1} kpc of the absorber) but one H I emitting cloud (but no stars) was found 0.4h^{-1} Mpc away from one absorber. The detection of this one $\sim$10^{8}M$_\odot$ object is in approximate agreement with the number of dark matter halos expected in voids (Gottlöber et al. 2003).

Finally, we have searched the strongest Lyα absorbers in voids for metals, and have not detected any Si III, C IV or O VI absorptions down to levels of $[Z] \leqslant -2.0$ solar. While these limits are already significantly below what would be expected from current-epoch "superwinds" from galaxies, they do not eliminate the possibility of "Population III level" metallicities suggested to be present throughout the high-z IGM (e.g., Cowie & Songaila 1998; but see Simcoe, Sargent & Rauch 2004).

3. Weak metal-line absorbers and starburst winds

Our low-column density Lyα sample provides a unique opportunity to study very nearby examples of Lyα forest absorbers. This is of particular importance for the weak metal-line systems because the origins of their metals is not well-understood. Two of the best examples of weak metal-line systems are found in the sightline pair 3C273/ RXJ 1230.8+0115, which is separated by 0.91° (350h$_{70}^{-1}$ kpc apart at cz=1600 km s^{-1}) on the sky. Both sightlines have Lyα absorbers at N$_{HI} \approx 10^{16}$ cm^{-2}) at similar redshift (cz=1586 and 1666 km s^{-1} respectively) and metallicity (6% and 10% solar; Tripp et al. 2002; Rosenberg et al. 2003). Recently, we (Stocke et al. 2004) have discovered a dwarf (M$_B = -13.9$), post starburst galaxy $70h_{70}^{-1}$ kpc away from the 3C 273 1586 km s^{-1} absorber mentioned above (see Figure 3). Not only do the absorber and galaxy redshifts match to within their combined errors but the absorber metallicity (6% solar) and the mean stellar metallicity of the galaxy ($\sim$10%) also approximately match. Further, the absorber has an overabundance of silicon to carbon indicative of recent supernova type II enrichment. The galaxy is a pure disk system whose optical spectrum has both strong Balmer and metal absorption lines, with no evidence for dust or gas (no emission lines and M$_{HI} \leqslant 5$ x 10^6 M$_\odot$; van Gorkom et al. 1993). From ratios of Lick absorption line indices, we estimate that the mean stellar age in this galaxy is 3.5$\pm$1.5 Gyrs.

Taken together, this information provides a consistent picture in which a massive ($\geqslant$ 0.3 M$_\odot$ yr^{-1}) starburst $\sim$ 3 Gyrs ago created enough supernovae to blow away the remaining gas from this galaxy into the IGM. Because the dwarf is quite low mass, this wind can easily escape from the galaxy and move $\sim$10^8 M$_\odot$ of gas to $\sim$100h_{70}^{-1} kpc at the required 20 - 30 km s^{-1} to create the metal-line absorber that we see between us and 3C 273 (Stocke et al. 2004).

But because this is the nearest weak metal-line system known, many other similar absorbers could be due to starburst winds from dwarf galaxies, which are too faint to find and observe (e.g., at $z \approx$0.1, a galaxy like this dwarf would have m$_B$ $\sim$24). And because dwarf galaxies are so numerous (1-3 Mpc^{-3}), if each dwarf galaxy had at least

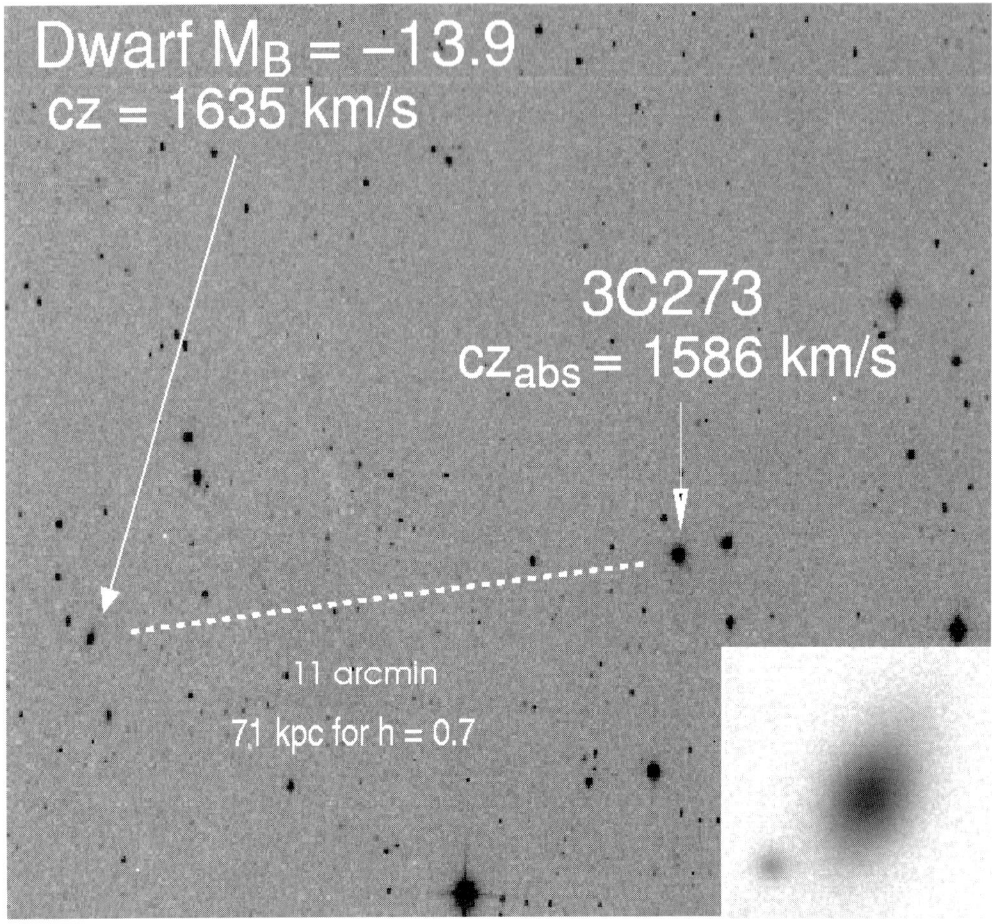

Figure 3. Digitized Sky Survey image of the region around 3C 273 (north up, east to left) showing the dwarf galaxy located $\sim$ 70 h_{70}^{-1} kpc away from the absorber on the sky and at the same recession velocity to within the errors. The R-band image of the dwarf galaxy taken with the ARC 3.5m telescope at Apache Point is shown as an insert at lower right. The surface brightness profile of this galaxy is well-fit by a pure exponential disk.

one massive starburst which ejected most or all of the gas from the galaxy to a distance of $\sim 100 h_{70}^{-1}$ kpc, this process would be sufficient to create several hundred weak metal-line absorbers per unit redshift. This number is comparable to the line density of all metal-line absorbers at z=2. While this dwarf galaxy would have been much more luminous ($M_B \approx -16.5$) when it was "starbursting", the present day absence of gas means that it will no longer form stars and will eventually fade to the approximate luminosity of a Local Group dwarf spheroidal. Thus, the dwarf in Figure 3 is the expected intermediate stage between the "faint blue galaxies" seen at intermediate redshift and the dwarf spheroids (Babul & Rees 1992). That such "Cheshire Cat" galaxies might be responsible for a large portion of the Lyα forest was predicted by Salpeter (1993) and Charlton (1995). Discovery of a similar dwarf near the RXJ 1230.8+0115 1666 km s^{-1} absorber would solidify this model. A galaxy survey of the RXJ 1230.8+0115 region is now in progress.

Acknowledgements

Financial support for "local" Lyα forest work at the University of Colorado comes from grants provided through HST GO program, #s 6593, 8182, 8125, 9221 & 9506, and NASA Theory Grant NAG5-7262. We thank Ray Weymann for the faint galaxy redshift survey work and his inspirational leadership in this field. We also thank Mark Giroux for photo- and collisional-ionization modeling and John Hibbard, Jacqueline van Gorkom and collaborators for the H I work at the VLA.

References

Babul, A. & Rees, M.J. 1992, MNRAS, 255, 346
Bahcall, J.N. et al. 1991,ApJ, 377, L5
Bahcall, J.N. et al. 1993, ApJS, 87, 1
Cen, R. et al. 1994, ApJ, 427, L9
Charlton, J.C. 1995, in QSO Absorption Lines, ed. G. Meylan (Berlin:Springer) , 405
Cowie, L.L. & Songaila, A. 1998, Nature, 394, 44
Davé, R., Hernquist, L., Katz, N. & Weinberg, D.H. 1999, ApJ, 511, 521
Davé, R., et al. 2001, ApJ, 552, 473
Gottlöber, S., Lokas, E.L. Klypin, A. & Hoffman, Y. 2003, MNRAS, 344, 715
Hu, E.M., Kim, T.-S., Cowie, L.L., Songaila, A. & Rauch, M. 1995 AJ, 110, 1526
Huchra, J. P., et al. 1992 Bull. C.D.S., 41, 31 (2004 electronic version)
Jannuzi, B.T. et al. 1998, ApJS, 118, 1
McLin, K.M., Stocke, J.T., Weymann, R.J. & Shull, J.M. 2001, ApJ, 574, L115
Morris, S.L., Weymann, R.J., Savage, B.D. & Gilliland, R.L. 1991, ApJ, 377, L21
Morris, S.L., et al. 1993, ApJ, 419, 524
Penton, S.V., Stocke, J.T. & Shull, J.M. 2000a, ApJS, 130, 121 (Paper I)
Penton, S.V., Shull, J.M. & Stocke, J.T. 2000b, ApJ, 544, 150 (Paper II)
Penton, S.V., Stocke, J.T. & Shull, J.M. 2002, ApJ, 565, 720 (Paper III)
Penton, S.V., Stocke, J.T. & Shull, J.M. 2004, ApJS, in press (astro-ph/0401036; Paper IV)
Rosenberg, J.L., Ganguly, R., Giroux, M.L. & Stocke, J.T. 2003, ApJ, 591, 6 77
Sargent, W.L.W. 1988, Proceedings of the QSO Absorption Line Meeting, ed. J.C . Blades, D. Turnshek & C.A. Norman (Cambridge U. Press), 1
Salpeter, E.E. 1993, AJ, 106, 1265
Schaye, J. 2001, ApJ, 559, 507
Simcoe, R.A., Sargent, W.L.W. & Rauch, M. 2004, ApJ, 606, 92
Stocke, J.T., Keeney, B.A., McLin, K.M., Rosenberg, J.L., Weymann, R.J. & Giroux, M.L. 2004, ApJ, in press (astro-ph/0403042)
Tripp, T. M., et al. 2002 ApJ, 575, 697
van Gorkom, J.H., Bahcall, J.N., Jannuzi, B.T. & Schneider, D.P., 1993, AJ, 106, 2213
Weymann, R.J. et al. 1998, ApJ, 506, 1

Outskirts of Galaxy Clusters: Intense Life in the Suburbs
Proceedings IAU Colloquium No. 195, 2004
A. Diaferio, ed.

DOI: 10.1017/S1743921304000262

Clusters outskirts at X-ray wavelengths: current status and future prospects

Silvano Molendi

Istituto di Astrofisica e Fisica Cosmica, Sezione di Milano, Via Bassini 15, I-20133, Milano, Italy email: silvano@mi.iasf.cnr.it

Abstract. A solid observational characterization of cluster regions around the virial radius would allow us to improve considerably our understanding of the physics of galaxy clusters as a whole: sadly current and planned experiments will not allow us to study these regions. Unbeknownst to most but not all, the development of an experiment sensitive to cluster outer regions could be achieved with currently available technology, with no need for breakthroughs. Amongst the major factors that will decide whether and when such an experiment will be flown are the awareness of the cluster community of the importance of such a mission and its determination at supporting it.

---

1. Why studying cluster outskirts?

In the hierarchical Universe clusters form at the intersection of cosmic filaments through accretion (see for example the picture on the the conference poster taken from Borgani et al. 2004). Shocks are expected to occur where the free-falling gas collides with the ICM (e.g. Tozzi et al. 2000). These shocks convert the bulk of the kinetic energy of the free-falling gas into thermal energy, a fraction of the kinetic energy may be retained by the shocked gas (e.g. Schuecker et al. 2004), playing a non trivial role in shaping the cluster. The boundary between cluster and free falling gas is likely not in a spherical form, as accretion occurs mainly from a few preferential directions. The thermodynamic status of the pre-shocked gas will likely determine the strength of the shock; if, for example, the gas has been pre-shocked in filaments, as envisaged in the pre-heating scenario, the shock will be weaker than if it has not been pre-shocked (e.g. Tozzi et al. 2000).

Relatively massive in-falling substructures, such as groups, are likely to retain at least part of their structure all the way down to the cluster central regions (e.g. A3667, Vikhlinin et al. 2001). Shocks generated by these structures will likely leave turbulence in their wakes, which will eventually be dissipated at smaller scales (e.g. Sunyaev these proceedings, Schuecker et al. 2004). Less massive structures such as galaxies will suffer severe modifications already in the cluster outskirts. The compression associated to the shock will amplify the strength of the magnetic fields which are frozen in the high conductivity plasma, and particles will likely be accelerated at the site of the shock.

Ram pressure stripping of the metal rich gas from the in-falling galaxies by the ICM (e.g. Treu, Moore these proceedings) will enrich the ICM in heavy elements. Direct accretion of pristine gas, never processed in galaxies, will have the opposite effect of diluting the metal content of the ICM.

These simple considerations illustrate how important the regions around the virial radius, the cluster outskirts, are. The rather obvious reason for the occurrence of the wide variety of phenomena sketched above is that these are the regions where virialization occurs. Viewing the issue from an observer's point of view we may say that as long as we do not have a solid observational characterization of cluster outer regions our understanding of clusters as a whole will necessarily be a limited one.

2. How much do available observations tell us?

Let us then see how much of this extremely important observational work we can carry out with currently available data. Let us consider the three principle X-ray observables: the surface brightness, the temperature and the metal abundance.

Measurements of the surface brightness extend out to about 0.6 r_{180} (Mohr et al. 1999, Ettori & Fabian 1999, Ettori et al. 2004). The detector that has been most successful in these measurements, at low redshifts, i.e. $z < 0.2$, is not part of the payload of the latest generation of X-ray satellites, indeed it is the ROSAT PSPC, which was launched 14 year ago.

ASCA (Markevitch et al. 1998), BeppoSAX (De Grandi & Molendi 2002) and XMM-Newton (Pratt & Arnaud 2002, Zhang et al. 2004) allow us to extend temperature measurements out to about 0.5-0.6 r_{180}. Is is rather interesting that XMM-Newton, with an effective area more than ten times larger than ASCA or BeppoSAX is not capable of extending temperature measurements beyond regions already explored by the latter satellites. It is also a matter of some concern that the first comparisons between the XMM-Newton and BeppoSAX temperature points to a non-negligible disagreement (e.g. De Grandi & Molendi these proceedings).

Finally, BeppoSAX, ASCA and XMM-Newton measurements of the metal abundance extend out to about 0.4 r_{180}, (e.g. De Grandi & Molendi 2001). This is indeed the most difficult observable to measure from low surface regions, at least with the current CCD resolution of a few percent.

Thus we are forced to conclude that current data do not go out to the virial radius, as a matter of fact they do not even come close to it. The answer to this section question is that *we know nothing from direct X-ray observations of the extremely important region surrounding the virial radius.*

3. How difficult would it be to design experiments capable of investigating these regions?

Having learned that current experiments are not sensitive to cluster outer regions we ask ourselves how difficult it would be to design experiments capable of investigating these regions. The answer to this question is that it would not be particularly difficult, no major technological breakthrough is required. Indeed the answer lies in keeping the background under control. Let us see how experiments could be optimized to do that.

3.1. *NXB < CXB*

The first thing to do is to reduce the non X-ray background (NXB) to values smaller than the Cosmic X-ray background (CXB) in the whole energy band covered by our experiment. Given the greater difficulty in focusing higher energy X-rays and the flatness of the NXB spectrum this typically means having NXB < CXB at the highest energies. It is worth remarking that none of the imaging experiments sensitive in the medium energy band (2-10 keV), flown so far, operate in the NXB < CXB regime. Neither the first generation experiments (BeppoSAX, ASCA), nor the second generation experiments (Chandra, XMM-Newton). As a matter of fact, the CXB/NXB ratio for the latter is worse than for the former and this is why XMM-Newton temperature profiles do not extend further out than ASCA or Beppo-SAX temperature profiles.

Achieving the NXB < CXB requirement would allow us to go about a factor 10 fainter in surface brightness. This is not really impossible: in the case of XMM-Newton for example, simply placing the satellite in an equatorial Low Earth Orbit would do more

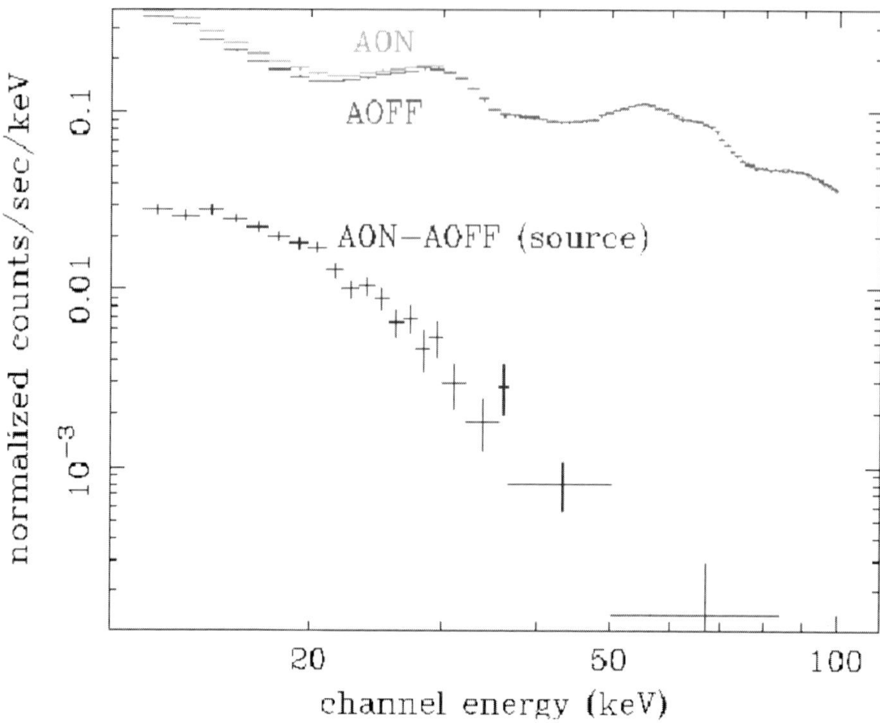

Figure 1. ON-spectrum, OFF-spectrum and source spectrum (difference between ON and OFF) for collimator row A during a PDS pointing.

than half of the job. The rest, a factor of about 2, could probably be achieved by working on the shielding of the detectors.

3.2. *Observing strategy*

The other thing to do is to improve the observing strategy. Typically temperature measurements can be made out to radii where the source counts, cts_{sou}, are equal to the background counts, cts_{bkg}. Measurements at larger radii cannot be made because the background is only poorly known. In other words we need to improve measurements in the $cts_{sou} < cts_{bkg}$. This is where differential techniques can be very effective. The most striking example comes from radio wavelengths, indeed WMAP detects CMB anisotropies of 1 part in 10^6! In the X-ray band the most successful experiment is the BeppoSAX PDS. This experiment makes use of 2 rows of rocking collimators: one is trained on the source, the other on the background; the collimators are switched every $\sim$ 100 s to allow the detectors, viewed by both collimators, to observe source and background fields.

In Fig. 1 we show the ON-spectrum, the OFF-spectrum and the source spectrum (difference between ON and OFF) for collimator row A during a PDS pointing. Extensive analysis (e.g. Rossetti 2003, Rossetti & Molendi 2004) have shown that the source spectrum measurement is reliable down to where source counts are about 1% of the background counts. The strategy adopted by the PDS could easily be implemented in future imaging missions, simply by substituting collimators with telescopes.

Combining the two strategies (i.e. reducing the NXB intensity below the CXB intensity and using differential measurements to keep the background under control) we could easily go down by a factor of 100 in surface brightness reaching the required sensitivity to map outer regions of galaxy clusters.

4. What about already planned missions?

What about already planned X-ray missions? Will they do better than current ones in investigating cluster outskirts? Probably not, missions such as ASTRO-E2, Con-X and XEUS are not optimized for the study of low surface brightness emission. The general approach is to achieve a high sensitivity through a large effective area, paying relatively little attention to background issues. This works well for point-sources where the background can be kept under control through the size of the PSF, but is of limited use for extended sources, where the instrumental background intensity per unit solid angle increases with the telescope focal length.

Of course we are not trying to imply that planned missions will be useless for galaxy clusters. They will certainly allow us to study important cluster properties; as an example consider the measurements of gas motions and turbulence that will be possible with ASTRO-E2 (e.g. Sunyaev these proceedings). However these studies will be limited to the brighter regions of galaxy clusters. In a sense our investigations will be restricted to the tip of the proverbial iceberg.

5. Summary

Observational characterization of outer regions is a fundamental step in understanding clusters. Unfortunately current and planned experiments will not allow us to study these regions. Fortunately no major technological breakthroughs are required to design an experiment sensitive to cluster outskirts. Whether and when such an experiment will be designed and flown will depend upon many factors. One of the most important will be the awareness of the cluster community of the need for such a mission and its determination at supporting it.

References

Borgani, S., Murante, G., Springel, et al. 2004, MNRAS, 348, 1078.

De Grandi S., & Molendi, S. *these proceedings*.

De Grandi, S. & Molendi S. 2002, ApJ, 567, 163.

De Grandi, S. & Molendi S. 2001, ApJ, 551, 153.

Ettori, S., Tozzi, P., Borgani, S., Rosati, P. 2004, A&A, 417, 13.

Ettori, S., Fabian, A. C. 1999, MNRAS, 305, 834.

Markevitch, M., Forman, W. R., Sarazin, C. L., & Vikhlinin, A. 1998, ApJ, 503, 77.

Mohr, J.J., Mathiesen, B., Evrard, A.E. 1999, ApJ, 517, 627.

Moore, B. *these proceedings*.

Pratt, G. W., Arnaud, M. 2002 A&A, 394, 375.

Rossetti, M., Molendi, S 2004, A&A, 414, L41.

Rossetti, M. *Laurea Thesis* 2003, Milano.

Schuecker, P., Finoguenov, A., Miniati, F., Boehringer, H., Briel, U.G. 2004, A&A sub. astro-ph/0404132.

Sunyaev, R. *these proceedings*.

Tozzi, P., Scharf, C. & Norman, C. 2000, ApJ, 542, 106.

Treu, T. *these proceedings*.

Vikhlinin, A., Markevitch, M., Forman, W., Jones, C 2001, ApJ, 555, L87.

Zhang, Y.-Y., Finoguenov, A., Boehringer, et al. 2004, A&A, 413, 49.

Outskirts of Galaxy Clusters: Intense Life in the Suburbs
Proceedings IAU Colloquium No. 195, 2004
A. Diaferio, ed.

DOI: 10.1017/S1743921304000274

XMM-Newton view on cluster outskirts

Alexis Finoguenov

Max-Planck-Institut für extraterrestrische Physik, Giessenbachstraße, 85748 Garching, Germany email: alexis@mpe.mpg.de

Abstract. We present an overview on the state of the gas in cluster outskirts. We show that the entropy of the gas in the outskirts often exhibit strong deviations from predictions based on the cluster scaling. We identify the cause of these deviations with incomplete shock propagation observed in clusters with ongoing major merging, survival of substructure and incomplete thermalization, and emission from the surrounding large-scale structure seen in projection.

1. Introduction

X-ray observations of outskirts of groups and clusters of galaxies are largely driven by an interest to the gravitational mass measurement out to the virial radius. Modelling the state of the gas in cluster outskirts implies a dominant role of gravitational heating, suggesting a simpler picture with respect to cluster centers, where entropy levels are much lower and are affected by various feedback effects, such as preheating and cooling.

Analysis of combined ROSAT and ASCA data on the entropy profiles for groups revealed a very high entropy level in the outskirts, up to 400 keV cm^2 (Finoguenov et al. 2002). After comparison with a larger sample of groups and clusters (Finoguenov et al. 2000, 2001), it was shown that there is a variation in the entropy ($kT_e/n_e^{2/3}$) between $\sim$ 100 keV cm^2 and $\sim$ 400 keV cm^2 within every system due to non-gravitational heating. The observed entropy levels are extremely high in the outer parts of groups and this suggests that there has been an energy release into a low-density environment (Finoguenov et al. 2002). In Finoguenov et al. (2003a) we introduce such a phenomenological model for the spatial distribution of the feeback into the GADGET code and were able to recover the observed entropy profiles from groups to clusters of galaxies. The entropy can be further increased by subsequent shock heating associated with cluster formation (Ponman, Sanderson, Finoguenov 2003; Voit et al. 2003).

Thus, these new observations revealed the crucial importance of cluster outskirts in understanding the state of the gas prior to its accretion to the cluster. In this contribution, we have picked out various examples on how the process of cluster formation is recorded in the entropy of the gas located in cluster outskirts.

2. RGH 80 / NRGs241

This group is selected from a sample of Ramella-Geller groups detected in X-rays, as cataloged in Mahdavi et al. (2000) and followed up with XMM-Newton (Finoguenov et al., Mahdavi et al., in preparation).

Fig. 1 shows a clear condensation in both the X-ray and the optical to the south. Fig. 1 also displays a velocity histogram of the group. It is double-peaked, and the high-redshift peak corresponds to the southern condensation. The entropy map shows that both the central clump and the southern clump are embedded in a low-entropy zone. The pressure map includes the southern clump, and we can show that it is more likely that the pressure enhancement corresponds to a separate dark matter blob, associated with the potential

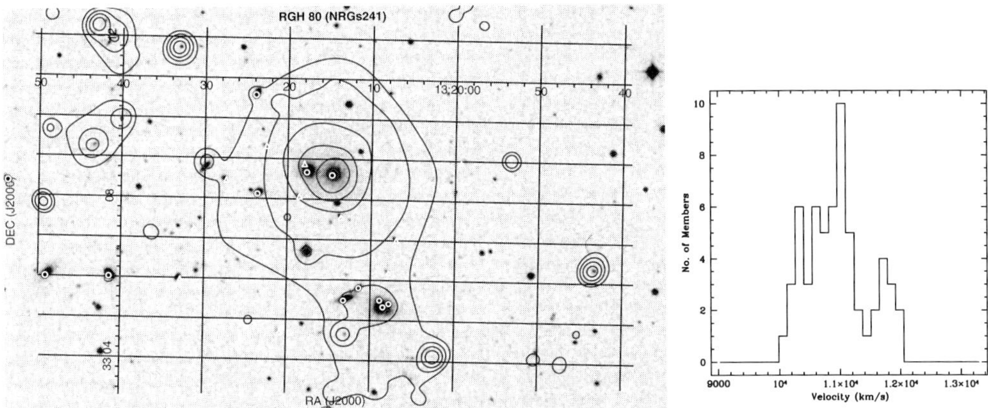

Figure 1. Left panel: optical (B-band) image of the RGH 80 with X-ray contours overlaid. Right panel: velocity histogram of the group.

of the infalling group, and less likely that it corresponds to a shock. The spectral fits shows that there are two components, which are consistent with being projected on top of each other. In the northern source, we observe a rise of the temperature from 0.9 keV to 1.2 keV, and then in the outer regions back down again to 0.7 keV. The metallicity is enhanced in the southern peak compared to the rest of the image — it is distinct from its surroundings. So, while the infalling group exhibits a low-entropy tail, indicative of a ram-pressure stripping, little disturbance is seen in the main group, indicating that the infalling group has a separation along the line-of-sight. The observed difference in the velocities between the groups is sufficient for the stripping to take place even in the outskirts of the group.

3. XMM cluster mosaics

XMM-Newton obtained a number of cluster mosaics as a part of the GTO program of the XMM telescope scientist. Analysis of the mosaics on A3667, A754, A3562, A3266, A2256 is completed (Briel et al. 2004; Henry et al. 2004; Finoguenov et al. 2004c; Jones et al., in preparation; Briel et al., in preparation). Advantages of mosaicing for cluster outskirts are many, in particular better control over the instrumental background that could be achieved; larger radii of the cluster that could be covered. Coma cluster takes a special place in cluster mosaics as the best-covered nearby cluster, yielding a rich collection of publications (Briel et al. 2001; Arnaud et al. 2001; Neumann et al. 2001; Neumann et al. 2003; Finoguenov et al. 2003b; Finoguenov et al. 2004a; Finoguenov & Miniati 2004; Schuecker et al. 2004).

For the purpose of this presentation we have selected three points clearly seen with these data: incomplete shock propagation, observed in A754; intra-supercluster medium observed in A3562 and projected emission from filaments, observed in the Coma cluster.

3.1. *A754*

Explanation of the observed properties of the A754 core has long been known to require a major merger (Roettiger et al. 1998). As a part of this process, one expects to see a forward shock propagating outward from the cluster center. Krivonos et al. (2001) has detected a sharp edge in the ROSAT data almost half a degree east from the A754 center. Our data provide a temperature measurement across the ROSAT edge, clearly revealing that the edge is a shock with a Mach number of 1.9 ± 0.4 and a temperature jump from

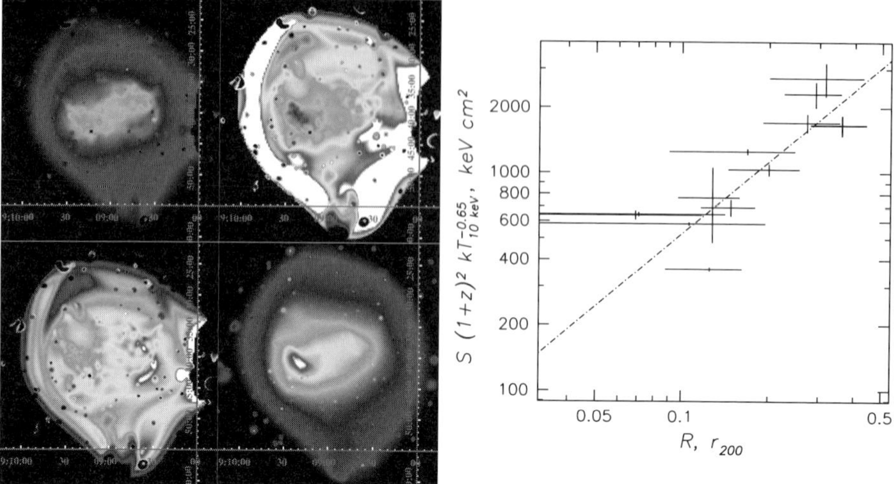

Figure 2. Left panel: pressure, entropy, temperature and surface brightness maps of A754 (from upper left to lower right). Right panel: scaled entropy profile of A754. The universal entropy scaling relation of Ponman et al. (2003) is shown as a dot-dashed line. Bold cross represents the pre-shock state of the gas in the outskirts of A754.

5.4 ± 0.6 keV to 9.6 ± 0.3 keV. As shown in Fig. 2, the material preceding the shock reveals a lower entropy compared to adjacent regions closer to the cluster center. Thus, the presence of low temperature, low entropy gas in the outskirts of clusters could simply reflect the state of the cluster before the cluster merger.

3.2. *A3562*

Using the mosaic of six XMM-Newton observations, we study the hydrodynamic state of A3562, a cluster in the center of the Shapley supercluster. The X-ray image reveals a sharp surface brightness gradient within the core of A3562, a 200 kpc ridge extending to the south-west. A nearby group, SC1329-313, imaged within this program also exhibits a tail aligned with the ridge. Study of the pressure and entropy identified the ridge with a 40% pressure enhancement. An associated Mach number of 1.15 relative to a polytropic sound speed for a 5 keV plasma requires the velocity of the group to be 1400 km/s, while the projected velocity difference between the cluster and the group amounts to 1200–1500 km/s. Apparent disruption of the group, if attributed to the ram pressure, also requires the velocity of the group in the cluster frame to amount to 1700 ± 150 km/s. The sharp surface brightness gradient at the center is identified with a contact discontinuity, which, together with dove tails in the the large-scale entropy distribution, reveals a sloshing of the BCG in response to the passage of the SC1329-313 group. By using the extent of the low-entropy tails in A3562 we estimate the amplitude of the sloshing motion to be $200h_{70}^{-1}$ kpc and the oscillation frequency 1 Gyr.

A filamentary X-ray structure on a several degree scale was reported by ROSAT (Kull & Böhringer 1999). In order to provide some insights into the origin of the large-scale emission, we have constructed two regions, which reach $25'$ distance from the A3562 center. Background subtraction becomes an issue for these regions and we have tried several background accumulations to test the robustness of our results. In addition to the level of instrumental background, as revealed by the data in the 5–12 keV band, we find a need to include an additional 0.2 keV component with solar element abundance typical to

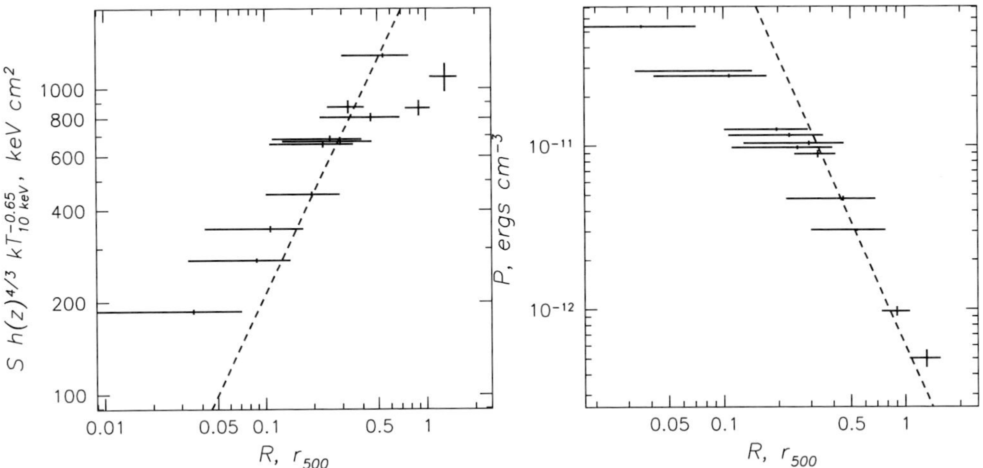

Figure 3. Left panel: entropy profile of the main zones in A3562. Universal entropy scaling relation of Ponman et al. (2003) is shown with a dashed line.Right panel: pressure profile of the main zones in A3562. A pressure slope of -2.5, typical of relaxed clusters, is shown with a dashed line. Normalization is done using the outmost point.

galactic halo emission. The main component, attributed to the Shapley supercluster, has a temperature of 2 keV, which is colder compared to the 3 keV temperature observed on the western part of A3562. The state of the gas in the outskirts of A3562, shown in Fig. 3 is characteristic of virial regions of clusters with already maintained pressure balance, but the lower entropy level indicates either a survival of low-entropy gas or an incomplete thermalization.

3.3. *Coma*

XMM-Newton observations of the outskirts of the Coma cluster of galaxies confirm the existence of warm X-ray gas claimed previously and provide a robust estimate of its temperature (~ 0.2 keV) and oxygen abundance (~ 0.1 solar). Associating this emission with a 20 Mpc infall region in front of Coma, seen in the skewness of its galaxy velocity distribution, yields an estimate of the density of the warm gas of $\sim 50\rho_{critical}$. Our measurements of gas mass associated with the warm emission strongly support its non-virialized nature, suggesting that we are observing the warm-hot intergalactic medium (WHIM). Our measurements provide a direct estimate of the O, Ne and Fe abundance of the WHIM. Differences with the reported Ne/O ratio for some OVI absorbers hint to different origins of the OVI absorbers and the Coma filament. Further details on this work could be found in Finoguenov et al. (2003).

Acknowledgements

The XMM-Newton project is supported by the Bundesministerium für Bildung und Forschung/Deutsches Zentrum für Luft- und Raumfahrt (BMBF/DLR), the Max-Planck-Gesellschaft (MPG) and the Heidenhain-Stiftung, and also by PPARC, CEA, CNES, and ASI. AF acknowledges support from BMBF/DLR under grant 50 OR 0207 and MPG. AF thanks his collaborators on various projects reported in this contribution, in particular Uli Briel, Pat Henry, Hans Böhringer, Andisheh Mahdavi, and Mark Henriksen. It is a pleasure to thank Antonaldo Diaferio for organizing such a productive meeting.

References

Arnaud, M., Aghanim, N., Gastaud, R., et al. 2001, A&A, 365, L67
Briel U.G., Henry J.P., Lumb D.H., et al. 2001, A&A, 365, L60
Briel, U.G., Finoguenov, A., Henry, J.P. 2003, A&A, submitted
Finoguenov, A., David, L. P., Ponman, T. J. 2000, ApJ, 544, 188
Finoguenov, A., Arnaud, M., David, L.P. 2001, ApJ, 555, 191
Finoguenov, A., Jones, C., Böhringer, H., Ponman, T.J. 2002, ApJ, 578, 74
Finoguenov, A., Borgani, S., Tornatore, L., Böhringer, H. 2003a, A&A, 398, L35
Finoguenov, A., Briel, U.G. and Henry, J.P. 2003b, A&A, 410, 777
Finoguenov, A., Briel, U.G., Henry, P.J., et al. 2004a, A&A, in press [arXiv:astro-ph/0403216]
Finoguenov, A., Miniati, F. 2004, A&A, 418, L21
Finoguenov, A., Henriksen, M.J., Briel, U.G., et al. 2004b, A&A, submitted
Henry, P.J., Finoguenov, A., Briel, U.G. 2004, ApJ, submitted
Krivonos, R.A., Vikhlinin, A.A., Markevitch, M.L., Pavlinsky, M.N. 2003, AstL, 29, 425
Kull, A., Böhringer, H. 1999, A&A, 341, 23
Mahdavi, A., et al. 2000, ApJ, 534, 114
Neumann, D.M., Arnaud, M., Gastaud, R., et al. 2001, A&A, 365, L74
Neumann, D.M., Lumb, D.H., Pratt, G.W., Briel, U.G. 2003, A&A, 400, 811
Ponman, T.J., Sanderson, A.J.R., Finoguenov, A. 2003, MNRAS, 343, 331
Roettiger, K., Stone, J.M., Mushotzky, R.F. 1998, ApJ, 493, 62
Schuecker, P., Finoguenov, A., Miniati, F., Böhringer, H., Briel, U.G. 2004, A&A submitted [arXiv:astro-ph/0404132]
Voit, G. M., Balogh, M.L., Bower, R.G., Lacey, C.G., Bryan, G.L. 2003, ApJ, 593, 272

Discussion

ETTORI: What are the implications of the entropy and pressure structure you reported to the mass measurements using the equation of hydrostatic equilibrium?

FINOGUENOV: The reported features could be segregated on those exhibiting pressure variations as due to the shock propagation and on those that exhibit only the entropy variation. In the former case, the formal mass measurements returns different values inside and outside the shock. E.g. in A754 the temperature in the outskirts experiences a transition from 5 keV to 10 keV. The smaller total mass that would be measured using the 5 keV temperature corresponds to a pre-merger state of the cluster, while the 10 keV temperature would correspond to a post-merger state of the cluster. An important note is that X-rays do not trace the total mass, but the virialized mass. As it becomes evident from other talks presented at this conference, the difference in the definition of a cluster is especially important in the comparison between clusters selected with weak lensing and X-rays. In the latter case, and in general, the use of the pressure to measure the total mass has certain advantages over using an X-ray image. However, an additional measure of the cluster virial temperature is required in the mass estimates and this task, as demonstrated in Silvano Molendi's contribution, is not trivial for current X-ray observatories.

WANG: What about the use of metallicity to study the substructure?

FINOGUENOV: Using metallicity as a tracer suffers from a poor statistics available at the moment and compared to the use of entropy achieves a much cruder spatial resolution. In most cases where it becomes possible, the entropy structure and metallicity structure are highly correlated. The level of chemical enrichment, on the other hand, sheds light on the origin of the low-entropy gas.

Outskirts of Galaxy Clusters: Intense Life in the Suburbs
Proceedings IAU Colloquium No. 195, 2004
A. Diaferio, ed.

DOI: 10.1017/S1743921304000286

Thermal structure of cooling clusters

Rocco Piffaretti[1,2], J. Kaastra[3], T. Tamura[4] and J. R. Peterson[5]

[1]Institute for Theoretical Physics, University of Zurich, Winterthurerstrasse 190, CH-8057 Zurich, Switzerland email: piff@physik.unizh.ch

[2]Laboratory for Astrophysics, Paul Scherrer Institute, CH-5232 Villigen PSI, Switzerland

[3]SRON, Sorbonnelaan 2, 3584 CA Utrecht, The Netherlands

[4]ISAS, 3-1-1 Yoshinodai, Sagamihara, Kanagawa 229-8510, Japan

[5]KIPAC, Stanford University, PO Box 90450, MS 29, Stanford, CA 94039, USA

Abstract. We present spatially resolved X-ray spectra taken with the EPIC cameras of XMM-Newton of a sample of 17 cooling clusters and three non-cooling clusters for comparison. The deprojected spectra are analyzed with a single-temperature model. All cooling clusters show a central decrement of the average temperature, most of them of a factor of ~2. Three clusters show a weak temperature decrement, while two others have a very strong temperature decrement. We investigate the role of heat conduction by electrons and find that the theoretically predicted conductivity rates are not high enough to balance radiation losses.

---

1. Introduction

The first high-resolution X-ray spectra of clusters of galaxies taken with the Reflection Grating Spectrometers (RGS) of XMM-Newton showed the presence of cooler gas in the cores of several clusters. However, the amount of cool gas at lower temperatures was much smaller than predicted by the isobaric cooling flow model (Sérsic 159−3, Kaastra et al. 2001; A1835, Peterson et al. 2001; A1795, Tamura et al. 2001a). The lack of relatively cool gas has been confirmed by the RGS spectra of other clusters, like A496, Tamura et al. (2001b) and Virgo, Sakelliou et al. (2001b) and also by Chandra observations. In order to be able to distinguish between the various theoretical models that have been proposed, it is important to measure the temperature structure for each radius. We report on a study of a large sample of clusters and use the spatially resolved XMM-Newton/EPIC spectra to investigate the temperature structure at each radius. We present here only results from single temperature fits and discuss heat conduction on cluster scales. More details are given in Kaastra et al. (2004).

2. Cluster sample and single temperature fits

The data analysis is described extensively by Kaastra et al. (2004). Here we summarize the results. Our sample of clusters consists of 17 cooling clusters (NGC 533, Virgo, A262, A1837, Sérsic 159−3, MKW 9, 2A 0335+096, MKW 3s, A2052, A4059, Hydra A, A496, A3112, A1795, A399, Perseus and A1835) and for comparison 3 non-cooling clusters (A3266, Coma and A754). Temperatures range between 1 and 15 keV, and most clusters have intermediate redshift (0.02–0.07). In our analysis, we assume spherical symmetry and we use deprojected spectra. Data obtained by the MOS and pn detectors were fitted simultaneously. The temperature profiles that we derive from our single temperature fits

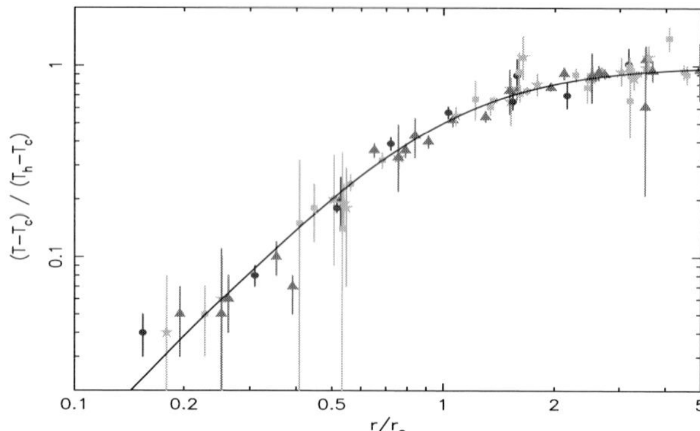

Figure 1. Scaled temperature profiles versus scaled radius. Triangles: clusters cooler than 3 keV; Squares: clusters with temperatures from 3 to 4 keV; Stars: clusters with temperatures from 4 to 6 keV; Circles: clusters with temperatures above 6 keV. The non-cooling clusters Coma, A754 and A3266 have been omitted from the plot. The solid line is the scaling curve Eq. (2.1).

for the cooling parts of the clusters are consistent with:

$$T(r) = T_c + (T_h - T_c)\frac{r^2}{r_c^2 + r^2}. \tag{2.1}$$

In this equation, r_c is a scale parameter for which we find best-fit values between 20 and 100 kpc. The parameter T_c corresponds to the temperature at the center of the cluster, while T_h is the asymptotic temperature at large radii. We find that the relative temperature decrement, given by the ratio $(T_h - T_c)/T_h$ has values between 0.15 and 0.70 for our cluster sample. We show the scaled temperature profiles in Fig. 1. There exist some scaling laws between the parameters of our fit to Eq. (2.1). For instance, there is a correlation between characteristic radius r_c and asymptotic temperature T_h of the form: $r_c \sim T_h^{1.84\pm0.14.}$ However, the virial radius scales proportionally to $T_h^{0.5}$. This implies that $r_c/r_{\rm vir}$ is not constant from cluster to cluster.

3. Heat conduction by electrons

Thermal conduction by electrons might play an important role in cooling clusters. We derive conduction coefficients assuming that heat conduction by electrons balances radiative losses, i.e.: $\int_{\rm V} n^2\Lambda(T)dV = \int_{\rm S} \kappa(\nabla T)dS$, where n, Λ, T and κ are the gas density, cooling function, temperature and thermal conductivity, respectively and V and S are the volume and surface area of the X-ray emitting region. Using data binned into shells, the conductivity coefficients at the boundary of each shell can be computed. The temperature gradient between shells is evaluated by fitting the temperature profile with the function (2.1). Our estimates of the conduction coefficients κ are shown in Fig. 2. The estimated conduction coefficients must be compared to theoretical calculations to see whether heat conduction from the outer regions can totally balance radiative losses. For a highly ionised plasma such as the ICM, the maximum rate is expected to be the Spitzer conductivity $\kappa_{\rm S}$. In the presence of a homogeneous magnetic field, the conductivity is the Spitzer rate only along the field, but severely decreased in the transverse direction. Thermal conduction in a tangled magnetic field has been studied by Chandran & Cowley (1998). They conclude that in cooling flows, thermal conductivity is below the Spitzer level by a factor of order 10^2 to 10^3. Conductivity is less severely decreased if the magnetic field behaves chaotically over a wide range of scales: Narayan & Medvedev (2001) estimate that in this case conductivity is only a factor ~5 below the Spitzer rate. Therefore,

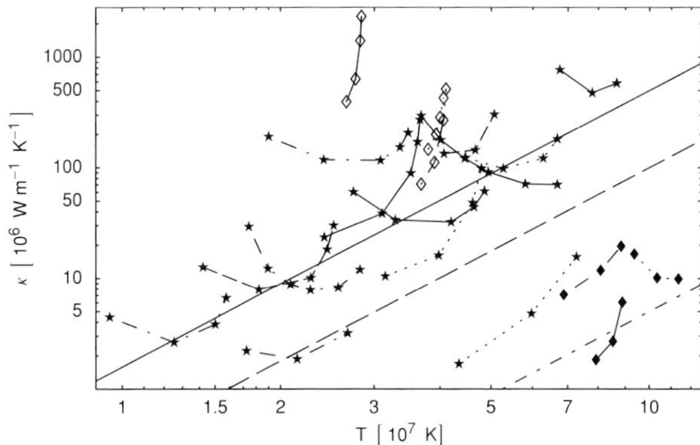

Figure 2. The conduction coefficients κ required for heat conduction to balance radiation losses as a function of temperature. The solid line is the Spitzer conductivity, the dashed lines are one fifth and one hundredth of the Spitzer conductivity, respectively. Filled diamonds: three clusters without cooling flow; Open diamonds: clusters with a shallow temperature profile (MKW 3s, Sérsic 159-3 and Hydra A); Filled stars: remaining clusters. Values for the same cluster are joined by a line and values are given only for bins with cooling times less than ~30 Gyr.

the Spitzer conductivity κ_s, $\kappa_s/5$ and $\kappa_s/100$ are shown in Fig. 2 for comparison. For the three non cooling clusters we find very low values for the conductivity coefficients: the absence of significant cooling allows even inefficient heat conduction to remove temperature inhomogeneities. The cooling clusters with shallow temperature profile show a very different trend and relatively high conductivity due to the small temperature gradients. Most important, we find conductivities larger than one fifth of the Spitzer rate in almost all the cooling clusters and therefore conclude that heat conduction alone is insufficient to balance radiative losses in cooling clusters.

Although we demonstrated that heat conduction on a global scale cannot maintain the large scale temperature gradients in cooling clusters, heat conduction may still play an important role on smaller spatial scales. In Kaastra et al. (2004) we considered a class of magnetic loop models as a possible mechanism in which heat conduction on smaller scales may occur.

Acknowledgements

This work is based on observations obtained with XMM-Newton, an ESA science mission with instruments and contributions directly funded by ESA Member States and the USA (NASA). The Space Research Organisation of the Netherlands (SRON) is supported financially by NWO, the Netherlands Organisation for Scientific Research.

References

Chandran, B.D.G., & Cowley 1998 *Phys. Rev. Lett.* **80**, 3077.
Kaastra, J.S. et al. 2001 *A & A* **365**, L99.
Kaastra, J.S. et al. 2004 *A & A* **413**, 415.
Narayan R. & Medvedev M. 2001 *ApJ* **562**, L129.
Peterson, J.R. et al. 2001 *A & A* **365**, L104.
Sakelliou, I. et al. 2002 *A & A* **391**, 903.
Tamura, T. et al. 2001a *A & A* **365**, L87.
Tamura, T. et al. 2001b *A & A* **379**, 107.
Voigt, L.M., & Fabian, A.C. 2004 *MNRAS* **347**, 1130.

Outskirts of Galaxy Clusters: Intense Life in the Suburbs
Proceedings IAU Colloquium No. 195, 2004
A. Diaferio, ed.

DOI: 10.1017/S1743921304000298

A note on temperature profiles of rich clusters of galaxies

Sabrina De Grandi[1] **and Silvano Molendi** [2]

[1] INAF OAB, via E. Bianchi 46, 23807 Merate (LC), Italy email: degrandi@mi.astro.it

[2] IASF, Sez. di Milano, Via Bassini 15, I-20133 Milano, Italy email: silvano@mi.iasf.cnr.it

Abstract. We derive here the mean temperature profile for a sample of hot, medium distant clusters recently observed with XMM-Newton, whose profiles are available from the literature, and compare it with the mean temperature profile found from BeppoSAX data. The XMM-Newton and BeppoSAX profiles are in good agreement between 0.05 and 0.25 r_{180}. From 0.25 to about 0.5 r_{180} both profiles decline; however the BeppoSAX profile does so much more rapidly than the XMM-Newton profile.

---

1. Introduction

Temperature profiles of galaxy clusters are of great importance for two main reasons: firstly, they allow us to measure the total mass of these systems through the hydrostatic equilibrium equation and secondly, they provide informations on the thermodynamic state of the Intra Cluster Medium (hereafter ICM).

Measurements of temperature profiles have been performed with the the first generation of X-ray satellites carrying telescopes operating in the medium energy band (2-10 keV), namely ASCA and BeppoSAX. A detailed description of results obtained with these experiments may be found in De Grandi & Molendi (2002) and references therein.

Over the last 3 years various authors have published XMM-Newton temperature profiles of individual clusters. Arnaud et al. (2003) with a sample of 7 objects comprising 5 clusters and 2 groups find that the temperature profiles are essentially isothermal within 0.5 r_{200}, and possibly declining at larger radii, where the statistics is rather limited. Recently, Zhang et al. (2003) have published temperature profiles for a sample of 9 clusters whose outer regions have both flat and strongly decreasing profiles. However they do not provide a mean temperature profiles of their sample. In this note we derive the mean temperature profile for a sample of hot, medium distant clusters whose profiles are available from the literature and compare it with our mean BeppoSAX profile.

2. XMM-Newton sample selection from the literature

We have selected from the literature all hot (i.e. kT> 3 keV) clusters in the redshift range between 0.1 and 0.3 with an available projected radial temperature profile. The resulting sample comprises a total of 15 clusters: 9 REFLEX clusters at redshift ~ 0.3 (Zhang et al. 2003), A1413 at $z = 0.143$ (Pratt & Arnaud 2002), A2163 at $z = 0.201$ (Pratt et al. 2001), A1835 at $z = 0.250$ (Majerovitz et al. 2002), PKS 0745$-$191 at $z = 0.1028$ (Chen et al. 2003), ZW 3146 at $z = 0.291$ and E1455+223 at $z = 0.258$ both taken from Mushotzky (2003). The adopted redshift range allows us to compare the temperature profiles measured from XMM-Newton data with those derived from BeppoSAX observations. The profiles published for the clusters in the XMM-Newton sample extend out to $\sim 9' - 10'$ from the core, corresponding to physical radii of $\sim 1.5 - 3$

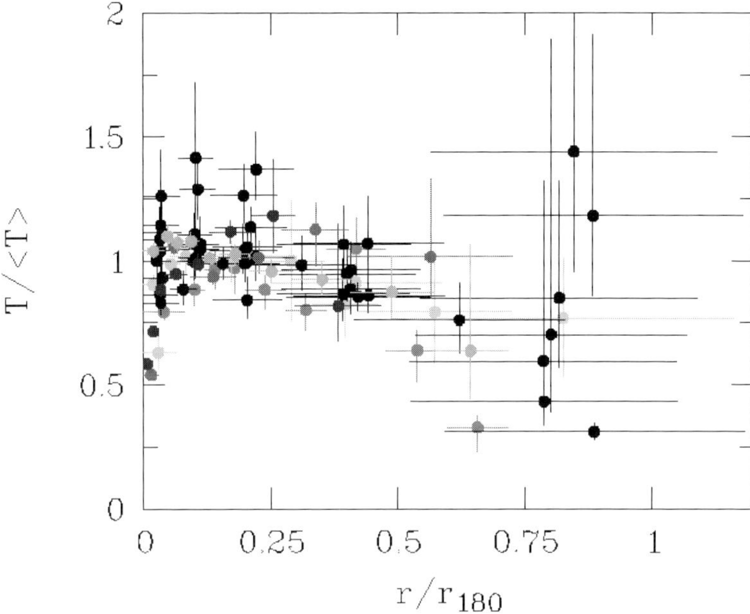

Figure 1. Projected temperature profile for 15 clusters observed with EPIC, radius in units of r180. Profiles are from the literature (references in text). Errors are at the 68% c.l.

Mpc (H_0 = 50 km s^{-1} Mpc^{-1}). The same physical radii are reached by BeppoSAX observations of clusters lying in the $\sim 0.05 - 0.08$ redshift range and detected out to $\sim 20'$ (see De Grandi & Molendi 2002). The average cluster temperature of the XMM-Newton sample is ~ 7 keV, very similar to that of the BeppoSAX sample (~ 6 keV). We have converted all temperature uncertainties at the 90% confidence level into errors at the 68% c.l. by dividing them by the scaling factor 1.65. For each cluster we have computed the virial radius r_{vir} ($= r_{180}$) from the relation derived by Evrard et al. (1996): $r_{vir} = 3.9\ (T/10\ \mathrm{keV})^{1/2}\ (1+z)^{-3/2}$ Mpc, using published cluster mean temperatures and redshifts.

Visual inspection of Fig.1, where we report the individual XMM-Newton temperature profiles, shows that the profiles are about isothermal from 0.10 to $0.3 - 0.5\ r_{180}$; beyond $0.3 - 0.5\ r_{180}$ there seems to be a decline. It is also clear that points at $r_{180} > 0.6 - 0.7$ are heavily fluctuating and tend to have larger errors.

We have modelled the XMM-Newton profiles in Fig. 1 with a power law and converted the slope into the polytrophic gas index accordingly to the calculation reported in the appendix of De Grandi & Molendi (2002). In the range $0.2 < r_{180} < 0.5$ we have obtained a polytrophic index $\gamma = 1.09 \pm 0.03$, which is close to the isothermal value 1. Whereas, for 0.5 $<\mathrm{r}_{180} < 0.75$ we have found $\gamma = 2.4 \pm 0.5$, which is formally above the adiabatic limit of 5/3, although consistent at the 68% confidence level with values below 5/3. The drop in the profiles at $\sim 0.55\ \mathrm{r}_{180}$ (clearly visible in Fig. 2) highlights a possible problem in the XMM-Newton profiles.

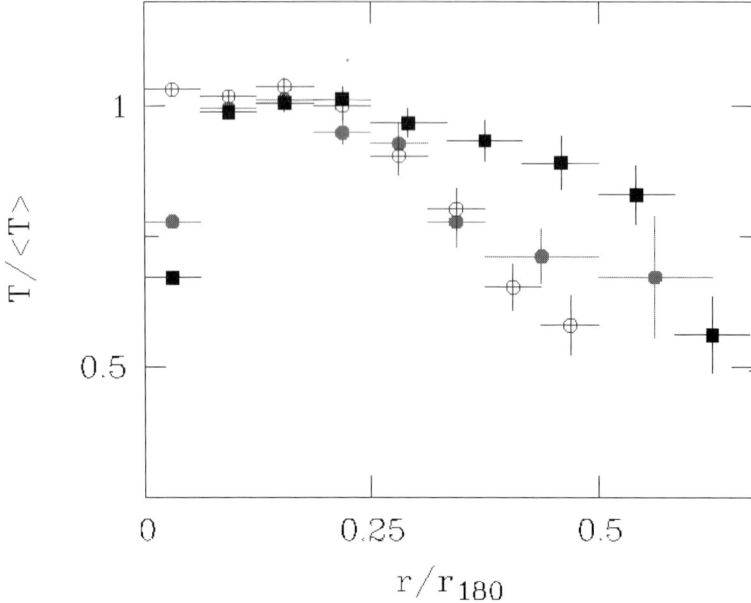

Figure 2. Mean temperature profile from BeppoSAX (De Grandi & Molendi 2002) (full and empty circles are respectively cool core and non cool core clusters) and from XMM-Newton (squares).

3. The mean profile

The binned error-weighted average temperature profile computed from the 15 clusters is shown in Fig. 2, where we also plot the BeppoSAX cool core and non cool core clusters profiles (see De Grandi & Molendi 2002 for details). We have added here to the BeppoSAX cool core clusters the Ophiuchus cluster.

The XMM-Newton and BeppoSAX profiles are in good agreement between 0.05 and 0.25 r_{180}. From 0.25 to about 0.5 r_{180} both profiles decline; however the BeppoSAX profile does so more rapidly than the XMM-Newton profile. The presence of a temperature jump, at the limit of convective stability, at the largest radii explored with XMM-Newton, hints to a problem possibly related to the low signal-to-noise ratio of the outermost cluster regions and to the difficulties in operating a correct background subtraction (see Molendi these proceedings for a discussion of this issue).

References

Arnaud, M., Pratt, G. W. & Pointecouteau, E. 2004, Mem. S.A.It. astro-ph/0312398.
Chen, Y., Ikebe, Y. & Boehringer, H. 2003, A&A, **407**, 41.
De Grandi, S. & Molendi, S. 2002, ApJ, **567**, 163.
Majerowicz, S., Neumann, D. M. & Reiprich, T. H. 2002, A&A, **394**, 77.
Molendi, S. 2004, these proceedings.
Mushotzky, R. F. 2003, astro-ph/0311105.
Pratt, G. W. & Arnaud, M. 2002, A&A, **394**, 375.
Pratt, G. W. et al. 2001, Proc. of XXIth Moriond Astroph. Meeting, p.38.
Zhang, Y.-Y., Finoguenov, A., Boehringer, H., et al. 2004, A&A, **413**, 49.

Outskirts of Galaxy Clusters: Intense Life in the Suburbs
Proceedings IAU Colloquium No. 195, 2004
A. Diaferio, ed.

DOI: 10.1017/S1743921304000304

Temperature and metallicity maps in cool core clusters of galaxies

Silvia Caffi, Fabio Gastaldello, Simona Ghizzardi and Silvano Molendi
IASF-CNR Sez. Milano

1. Observation and data analysis

We have analized XMM-Newton data of A2199, a nearby ($z = 0.030$), relaxed, rich cluster of galaxies, whose X-ray emission is strongly peaked on the central galaxy, NGC6166, suggesting the existence of a cluster cool core (Peres *et al.* 1998).
The background subtraction is performed using an observation characterized by features similar to those of the A2199 observation. We have selected, in the XMM-Newton archive, a field observation presenting the following properites:

- duration of about 90 ksec;
- performed with the same filter as the A2199 observation;
- characterized by a galactic HI column density and a soft X-ray background, the latter being calculated from the ROSAT All-Sky survey diffuse background maps, which differ from those of the A2199 observation ($N_H = 8.7 \times 10^{19}$ cm^{-2}, $R_{45} = 208.2 \times 10^{-6}$ cts/arcmin^2) by 3% and 30% respectively.

We have accumulated spectra from 9 concentric annuli, out to 12 arcmin, centered on the X-ray emission peak and have fit them with a single temperature model (mekal * wabs model in XSPEC) in the $1.5 - 10.0$ keV spectral band, well suited for temperature measurements in the range $kT \sim [3, 5]$ keV.

We have then deprojected the radial profiles using the technique described in Ettori *et al.* (2002).

2. Results

2.1. *Projected profiles*

Here we present the projected profiles of temperature and iron abundance (Fig. 1) plotted together with Chandra and BeppoSAX profiles derived from Johnstone *et al.* (2002) and De Grandi & Molendi (2001, 2002).

While the temperature profiles derived from XMM-Newton and Chandra data show a remarkable agreement, the comparison between BeppoSAX and our data points toward a systematic difference of about 10%, already noted in the Coma cluster: 0.8 keV on a temperature of 8.2 keV (De Grandi & Molendi, 2002).

There is a clear difference in the abundance profiles of XMM-Newton and Chandra. This difference could be due to the proximity of the $Fe - K_\alpha$ line at 6.7 keV, used to mesure the metallicity, to the spectral upper limit in Johnstone *et al.*'s analysis (7.0 keV).

2.2. *Deprojected profiles*

We have deprojected radial profiles using the method indicated by Ettori *et al.* (2002), obtaining three-dimensional profiles for temperature, abundance (see Fig. 2), electron density (n_e), emissivity (ϵ), cooling time (t_{cool}), pressure (see Fig. 3 left panel), entropy, gravitational mass, iron mass and iron mass excess (see Fig. 3 right panel).

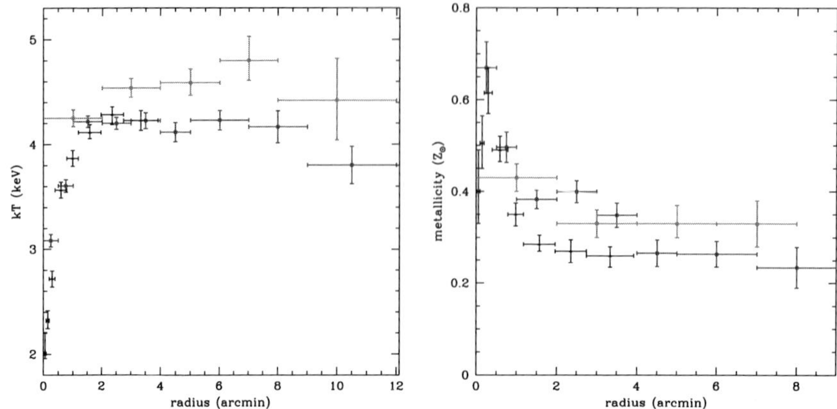

Figure 1. Temperature (left) and iron abundance (right) projected profiles: blue, red and black points are, respectively, for XMM-Newton, BeppoSAX and Chandra data.

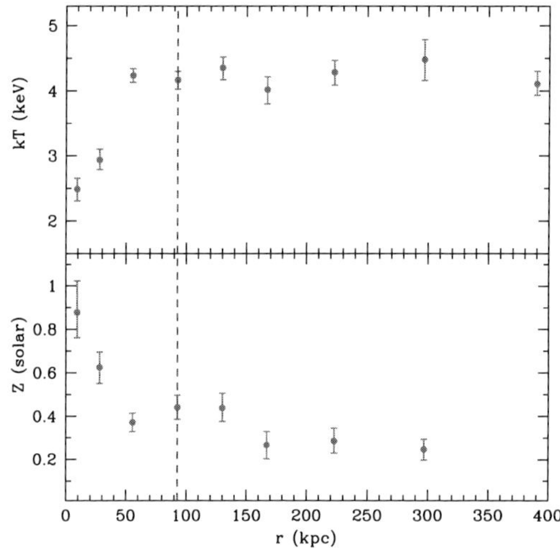

Figure 2. Temperature and iron abundance deprojected profiles.

The radius within which the temperature (abundance) profile decreases (increases) is about a factor of 2 smaller than the cooling radius, estimated from the t_{cool} profile. The value of r_{cool} is about 95 kpc, in good agreement with previous estimates by Perez *et al.* (1998) and Johnstone *et al.* (2002).

The total iron mass excess is 6.3×10^8 $M_\odot$, a factor of 2 smaller than previously estimated in De Grandi *et al.* (2003) with BeppoSAX data. The discrepancy could be due to the differences between the projected metallicity profiles (see Fig. 1).

References

Peres C. B. *et al.* 1998 MNRAS **298** 416
De Grandi S. & Molendi S. 2001 ApJ **551** 153
De Grandi S. & Molendi S. 2002 ApJ **567** 163
De Grandi S. *et al.* 2003 astro-ph/0310828, A&A, in press

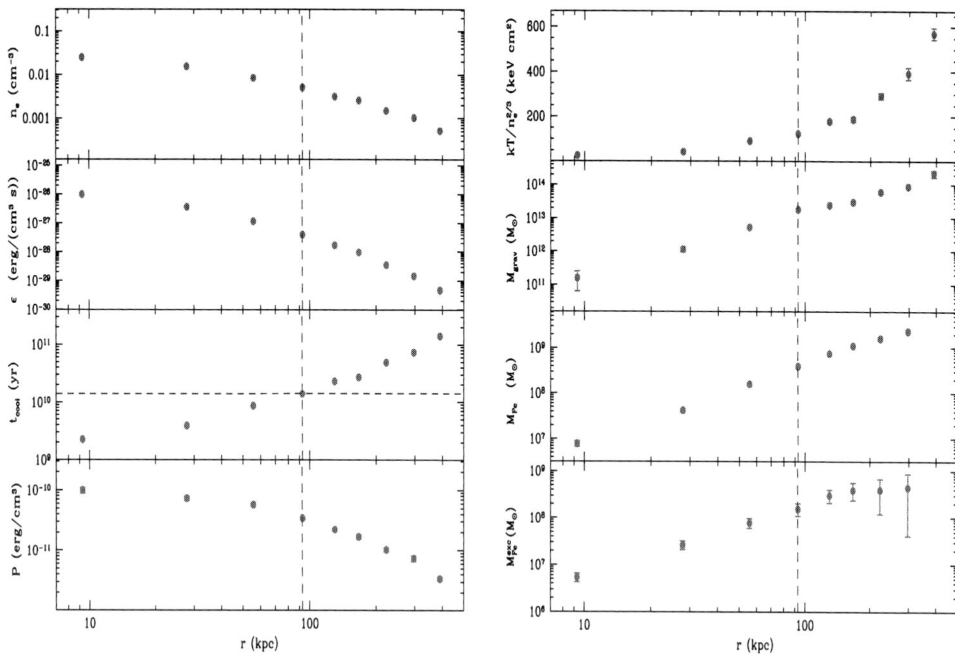

Figure 3. Deprojectd quantities of A2199: n_e, ϵ, t_{cool}, P in the left panel, entropy (defined as $kT/n_e^{2/3}$), M_{grav}, M_{Fe} and M_{Fe}^{exc} in the right panel.

Ettori S. *et al.* 2002 A&A **391** 841

Johnstone R. M. *et al.* 2002 MNRAS **336** 299

Outskirts of Galaxy Clusters: Intense Life in the Suburbs
Proceedings IAU Colloquium No. 195, 2004
A. Diaferio, ed.

DOI: 10.1017/S1743921304000316

Warm gas in the outskirts of galaxy clusters – The cluster soft excess phenomenon

Massimiliano Bonamente [1,2], Richard Lieu [1] and Jonathan P. D. Mittaz [1]

[1]Department of Physics, University of Alabama in Huntsville, Huntsville, AL
[2]National Space Science and Technology Center, NASA/MSFC, Huntsville, AL

Abstract. The cluster soft excess emission indicates the presence of large amounts of warm gas ($T \sim 10^6$ K) in the neighborhood of galaxy clusters. Among the clusters that display this phenomenon is the Coma cluster, the nearest rich galaxy cluster. The excess emission is more prominent at the cluster's outskirts than at its center. Detailed studies of its large-scale emission – up to ~ 2.6 Mpc from the cluster's center – reveal that these warm baryons are as massive as, or possibly more massive than, the well-known hot intra-cluster medium ($T \sim 10^8$ K). A possible interpretation of the excess emission from the Coma cluster is radiation from low-density filaments located in the neighborhood of the cluster. In this case, the filaments would extend for much larger distances, or feature higher density, than predicted by current cosmological simulations.

1. Introduction to the soft excess emission

Clusters of galaxies are strong emitters of X-rays, which originate from a hot and diffuse intra-cluster medium (ICM) at temperatures of a few 10^7 K. The soft X-ray band below ~ 1 keV often feature the 'soft excess' emission phenomenon, i.e., radiation in excess of that expected from the hot ICM. The excess emission could originate from Inverse-Compton scattering of cosmic microwave background (CMB) photons against a population of relativistic electrons in the intra-cluster medium (Hwang 1997, Sarazin and Lieu 1998, Ensslin and Biermann 1998, Lieu et al. 1999). Alternatively, warm gas at T$\sim 10^6$ K could be responsible for the soft emission (e.g., Lieu et al. 1996a,b, Nevalainen et al. 2003, Kaastra et al. 2003, Bonamente, Joy and Lieu 2003). Warm gas may reside inside the cluster, or in very diffuse filamentary structures outside the cluster, as seen in large scale hydrodynamic simulations (e.g., Cen and Ostriker 1999, Davé et al. 2001, Cen et al 2001). The warm gas scenario appears to be favored by the current X-ray data (e.g., Bonamente, Lieu and Mittaz 2001a, Buote 2001, Kaastra et al. 2003).

2. Spatial and spectral features of the soft excess emission

The soft excess emission is present in $\sim 50\%$ of the clusters. Two surveys by Bonamente et al. (2002) and Kaastra et al. (2003) reported, respectively, detection of soft excess emission in 18/38 clusters and in 7/21 clusters that were investigated. Also, the excess is normally detected when the observations have a high S/N, such as in the nearby Virgo (Bonamente, Lieu and Mittaz 2001a) and Coma clusters (Bonamente, Joy and Lieu 2003).

The relative importance of the soft excess component – with respect to the hot ICM component – normally increases with radial distance from the cluster center. This is the case of the Coma cluster (Fig. 1; Bonamente, Joy and Lieu 2003), of A1795 (Mittaz et al. 1998), A2199 (Lieu, Bonamente and Mittaz 1999), Virgo (Bonamente, Lieu and

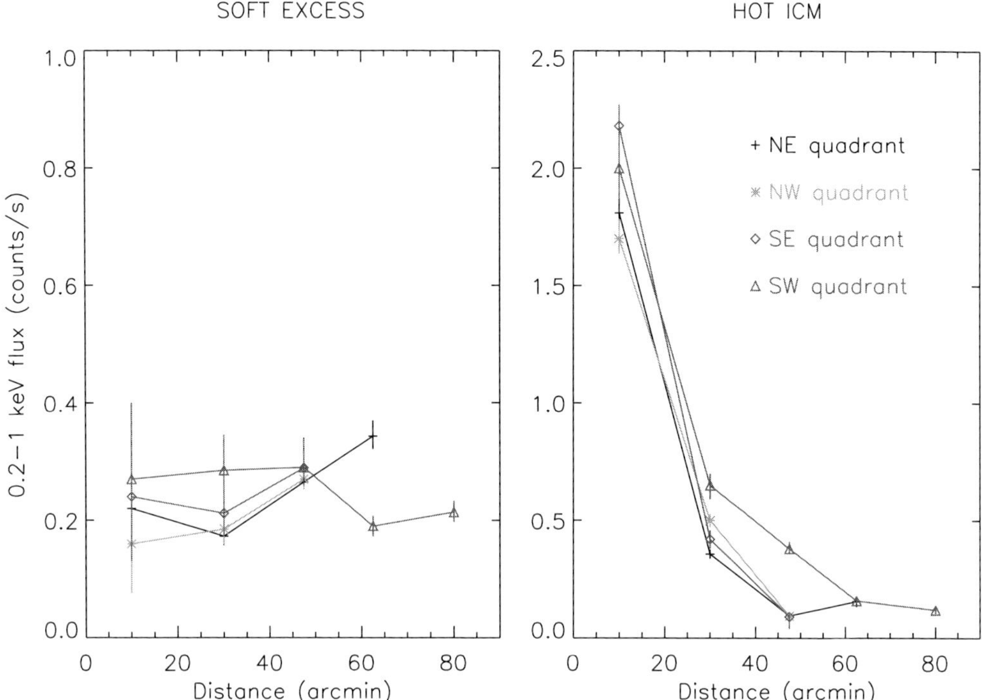

Figure 1. The radial distribution of the soft excess emission and of the hot ICM emission of Coma from ROSAT PSPC data. The data was analyzed in concentric annuli, further divided into quadrants.

Mittaz 2001a; Lieu et al. 1996a), MKW3s and A2052 (Kaastra et al. 2003). An important exception is AS1101 (Bonamente, Lieu and Mittaz 2001a), where the soft excess is sharply peaked at the cluster's center.

Thermal modelling of the soft excess component is normally favored by the goodness-of-fit analysis. Among the highest S/N data analyzed to date, the Coma cluster PSPC data favors the thermal interpretation for the majority of its regions (Fig. 2; Bonamente, Joy and Lieu 2003). The detection of OVII emission lines in the soft X-ray spectra of a few clusters, reported by Kaastra et al. (2003) from XMM data, would be the unmistakable signature of the thermal nature of the soft excess. A confirmation of those lines with higher S/N is however required before the detection be regarded as definitive.

The non-thermal interpretation remains a plausible explaination for the excess emission, and it cannot be formally rejected in its entirety. In order to explain the detected excess emission as non-thermal radiation, some clusters require that the cosmic rays be near or above pressure equipartition with the hot gas (Lieu et al. 1999; Bonamente, Lieu and Mittaz 2001a).

3. Thermal interpretation of the soft excess emission

If the excess emission is of thermal origin, it is possible to envisage two scenarios: (1) The 'warm' gas coexists with the 'hot' gas. In this case, it is possible to estimate the density of the warm gas from the emission integral $I = \int n^2 dV$; in Coma the warm gas

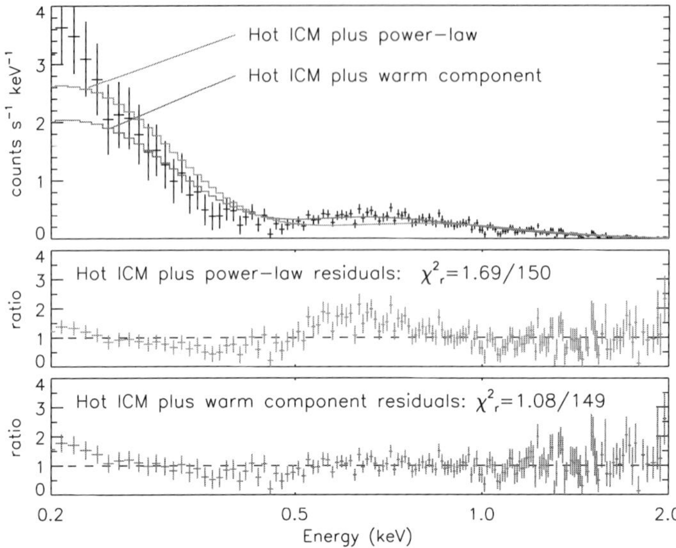

Figure 2. *ROSAT* PSPC spectrum of the 50-70' north-eastern quadrant of Coma. In green is the hot ICM plus power law model, in red the hot ICM model plus a low-energy thermal component.

would have a density of 9×10^{-4} cm^{-3} to $\sim 8\times10^{-5}$ cm^{-3} (Bonamente, Joy and Lieu 2003).

If the soft excess emission originates from a warm phase of the intra-cluster medium, the ratio of the emission integral ($I = \int n^2 dV$) of the hot ICM and the emission integral of the warm gas can be used to measure the relative mass of the two phases. The emission integral is readily measured by fitting the X-ray spectrum. For the PSPC data of the Coma cluster, we calculated M_{warm}/M_{hot}=0.75 within a radius of 2.6 Mpc (Bonamente, Joy and Lieu 2003).

(2) The 'warm' gas resides in low density filamentary structures outside the clusters, the warm-hot intergalactic medium (WHIM). This scenario follows several cosmological simulations (e.g., Cen and Ostriker 1999) which identify the majority of the low-redshift baryons in a tenuous network of WHIM filaments. Following this interpretation, the gas should feature overdensities of $\delta \sim 3-300$, corresponding to $\sim 10^{-4} - 10^{-6}$ cm^{-3} (for $\Omega_b \sim 0.05$), and a median overdensity of $\sim 10-30$ (Davé et al. 2001). We employ a simple model where constant-density filaments are directed towards the observer. In this case, the emission integral of the soft phase becomes $I = n^2 AL$, where A is the projected area of the filaments and L their length along the line of sight. In order to explain the detected I, filaments should extend for several Mpc in front of the cluster (see Table 1). Consider the case of filaments with density $n = 10^{-5}$ cm^{-3} ($\delta \sim 30$): the filaments should then extend for several hundred mega-parsecs! This figure is at odds with typical results from cosmological simulations, and with the fact that the Coma cluster is located at a distance of ~ 95 Mpc from the Galaxy. Filaments of higher density ($\sim 10^{-4}$ cm^{-3}) will require shorter L (Table 1); in this case the scenario becomes tenable (Bonamente, Joy and Lieu 2003), although still at odds with the results from cosmological simulations, which predict an average filament density considerably lower than 10^{-4} cm^{-3}. This result is consistent with a detailed analysis of the soft X-ray emission predicted by the WHIM filaments seen in some recent simulations (e.g., Cen et al. 2001). The analysis reveals that

Table 1. Lenght of filaments for the Coma PSPC soft excess emission

Region (arcmin)	I^*	L (Mpc) $[n = 10^{-5}\ \text{cm}^{-3}]$	L (Mpc) $[n = 10^{-4}\ \text{cm}^{-3}]$
0-20	0.057	1,800	18
20-40	0.039	350	3.5
40-55	0.066	610	6.1
55-70	0.019	250	2.5
70-90	0.007	220	2.2

(*) The emission integral I is in units of $10^{14}[4\pi(1+z)^2 D^2]$, as in the XSPEC optically-thin MEKAL code. D is distance to the Coma cluster in cm and z=0.023 the redshift. Description of the PSPC data used to obtain the emission integrals I can be found in Bonamente, Joy and Lieu (2003).

the WHIM filaments predict several times ($\geqslant 10$) lower soft X-ray fluxes than those of the typical soft excess emission (Mittaz et al. 2004) as, e.g., that of the Coma cluster. Similar results apply to the soft excess of the AS1101 cluster, where a strong centrally-peaked soft excess emission was detected (Bonamente, Lieu and Mittaz 2001a).

Following this scenario, and for a filament density of $n = 10^{-4}$ cm^{-3}, the PSPC data of Coma yields the conclusion that M_{fil}/M_{hot}=3 within a radius of 2.6 Mpc (Bonamente, Joy and Lieu 2003). The warm gas would therefore be more massive than the hot ICM if it is distributed in low-density filaments.

As the total mass of the Coma cluster within 14 Mpc is $1.6 \pm 0.4 \times 10^{15} M_\odot$ (Geller, Diaferio and Kurtz 1999) and the mass of the hot ICM is $\sim 4.3 \times 10^{14} M_\odot$ within 2.6 Mpc (Mohr, Mathiesen and Evrard 1999), the soft excess emission could account for a significant fraction of the low-redshift Ω_b. Kaastra et al. (2003) resported similar results from XMM observations of several clusters including AS1101, MKW3s and A2052.

4. Conclusions

Detection of soft X-ray excess emission from galaxy clusters is commonplace. The radiation is more prominent at the cluster outskirts, and it indicates that clusters may be a significant reservoir of 'warm' baryons. Precise mass estimates for the 'soft' component require knowledge of the exact location of the emitter. If the gas is located in filaments outside the clusters, the detection of soft excess in the Coma cluster indicates that the 'warm' gas is more massive than the hot ICM. Filaments of considerably higher density than predicted by the current cosmological simulations are needed to explain the detected excess emission from the Coma cluster.

References

Bonamente, M., Lieu, R., Joy, M.K. and Nevalainen, J.H. 2002, Astrophys. J., 576, 688
Bonamente, M., Lieu, R. and Mittaz, J.P.D. 2001a, Astrophys. J. Letters, 561, 63
Bonamente, M., Lieu, R. and Mittaz, J.P.D. 2001b, Astrophys. J., 547, 7
Buote, D.A., 2001, Astrophys. J., 548, 652
Cen, R. and Ostriker, J.P. 1999, Astrophys. J. Letters, 514, L1
Cen, R., Tripp, T.M., Ostriker, J.P. and Jenkins, E.B. 2001, Astrophys. J. Letters, 559,L5
Davé, R., Cen, R., Ostriker, J.P., Bryan, G.L., Hernquist, L., Katz, N., Weinberg, D.H.,
Ensslin, T. and Biermann, P. 1998, Astron. and Astrophys., 330, 90
Geller, M.J., Diaferio, A. and Kurtz, M.J. 1999, Astrophys. J., 517, L26
Hwang, C.-Y. 1997, Science, 278, 1917

Kaastra, J.S., Lieu, R., Tamura, T., Paerels, F.B.S. and den Herder, J.W. 2003, Astron. and Astrophys., 397, 445

Lieu, R., Ip, W.-I., Axford, W.I. and Bonamente, M. 1999b, Astrophys. J. Letters, 510,L25

Lieu, R., Bonamente, M. and Mittaz, J.P.D. 1999, Astrophys. J. Letters, 517, L91

Lieu, R., Mittaz, J.P.D., Bowyer, S., Lockman, F.J., Hwang, C. -Y., Schmitt, J.H.M.M. 1996a, Astrophys. J. Letters, 458, L5

Lieu, R., Mittaz, J.P.D., Bowyer, S., Breen, J.O., Lockman, F.J., Murphy, E.M. & Hwang, C. -Y. 1996b, Science, 274,1335

Mittaz, J.P.D., Lieu, R. and Lockman, F.J. 1998, Astrophys. J. Letters, 498, L17

Mittaz, J.P.D. et al. 2004, in preparation.

Mohr, J.J., Mathiesen, B and Evrard, A. E. 1999, Astrophys. J., 517, 627

Nevalainen, J.H., Lieu, R., Bonamente, M. and Oosterbroeck, T. 2003, Astrophys. J., in press

Sarazin, C.L. and Lieu, R. 1998, Astrophys. J. Letters, 494, L177

Outskirts of Galaxy Clusters: Intense Life in the Suburbs
Proceedings IAU Colloquium No. 195, 2004
A. Diaferio, ed.

DOI: 10.1017/S1743921304000328

Soft X-ray excess in the cluster of galaxies Sérsic 159-03

J. de Plaa[1,2] †, J. S. Kaastra[1], M. Méndez[1], T. Tamura[3], J. A. M. Bleeker[1], J. Peterson[4], F. B. S. Paerels[5], M. Bonamente[6] and R. Lieu[6]

[1]SRON, Utrecht, The Netherlands, email: j.de.plaa@sron.nl
[2]Astronomical Institute, Utrecht University, The Netherlands
[3]JAXA, Japan
[4]Stanford University, CA, USA
[5]Columbia University, NY, USA
[6]University of Alabama, AL, USA

Abstract. We present the results from a new 120 ks XMM-Newton observation of Sérsic 159-03. A previous XMM-Newton observation of this cluster shows the presence of a soft X-ray excess in the outer parts of the cluster, which is possibly connected to the interaction between the cluster and the gas from the surrounding filaments. We exploit the long exposure time to constrain the excess emission and discuss the relation to the warm-hot intergalactic medium. Furthermore, we show a high-resolution RGS spectrum of the core of the cluster and radial profiles from EPIC, which allow us to constrain the internal temperature structure and elemental abundances.

---

1. Introduction

The Sérsic 159-03 cluster of galaxies, also known as ACO S 1101, was discovered by Sérsic (1974). Since then it was studied in X-rays as part of several cluster samples: e.g. EXOSAT (Edge & Steward 1991) and ROSAT (Allen & Fabian 1980). Kaastra et al. (2001) reported results from an XMM-Newton observation with a useful exposure time of about 35 ks showing a radial temperature profile which peaks at a temperature of $kT \sim 2.7$ keV at a radius of about $2'$ from the core. The temperature drop in the core is relatively modest, while the temperature outside the $2'$ radius drops rapidly to values around 0.5 keV. Sérsic 159-03 shows a large soft X-ray excess in a re-analysis of this observation by Kaastra et al. (2003), as well as from archival ROSAT PSPC observations (Bonamente, Lieu & Mittaz 2001).

In this paper we show the results from a much longer 120 ks XMM-Newton observation of Sérsic 159-03. We exploit the very high statistics to obtain accurate temperature and abundance profiles with the EPIC instrument. We also make two-temperature model fits to account for the contribution of the Warm Hot Intergalactic Medium (WHIM). Together with high-resolution spectra from the RGS instrument we can constrain the temperature structure and abundances in the core.

2. Summary

We confirm the presence of a significant soft X-ray excess in the cluster of galaxies Sérsic 159-03. The excess can be reasonably fitted with a warm (0.2 keV) temperature

† Present address: SRON, Sorbonnelaan 2, 3584 CA Utrecht, The Netherlands.

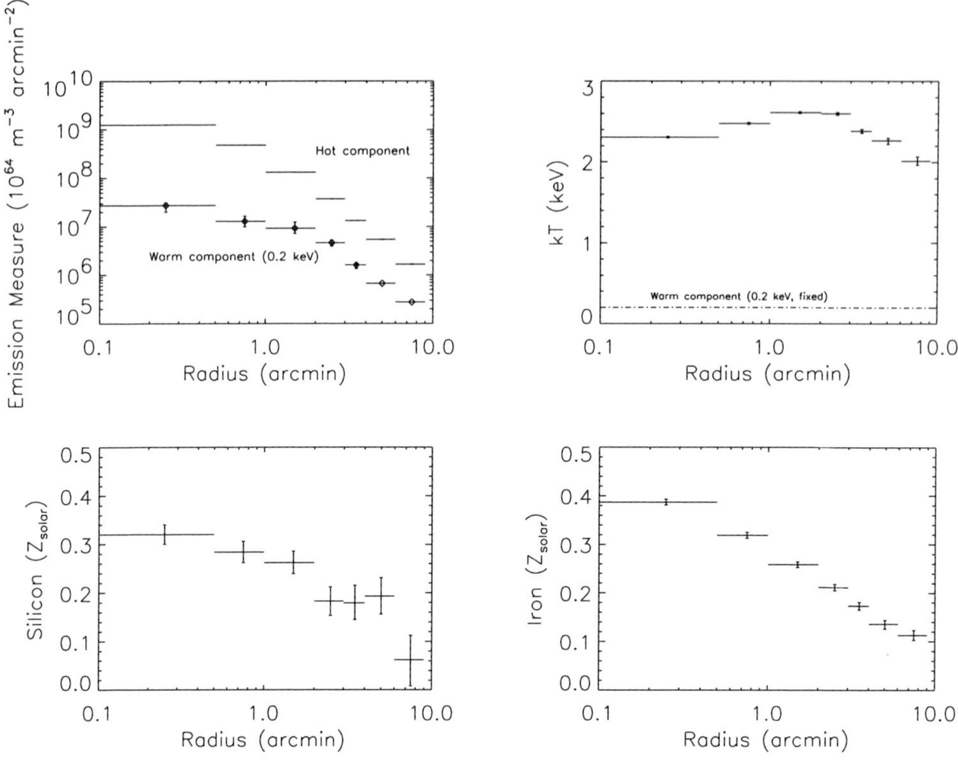

Figure 1. Radial profiles for a model with two-temperature components. *Top left:* Emission measure profiles for the hot and warm components. *Top right:* Temperature profile of the hot component. The warm component was fixed to 0.2 keV. *Lower left:* Silicon abundance profile. *Lower right:* Iron abundance profile. Both with respect to solar abundances.

Parameter	Value	Parameter	Value
$kT_{\rm hot}$ (keV)	2.89 ± 0.05	Mg	0.16 ± 0.05
O	0.216 ± 0.014	Fe	0.388 ± 0.010
Ne	0.43 ± 0.04		

Table 1. RGS fit results for a model with two temperature components.

component in combination with a hot component (see Fig. 1). In this figure we also see that the emission measure profile of the soft component shows a more gradual decrease toward the outer parts of the cluster than the emission measure of the hot component. In the outer parts the warm component grows to be relatively more important, while the hot component decreases in temperature. The iron and silicon abundances show an increase toward the core, consistent with the findings of Tamura et al. (2004) in a sample of clusters.

The high-resolution RGS spectrum (Fig. 2) shows line emission from Fe-L, Ne, Mg and O. In Table 1. the values for the hot temperature component are listed. Our RGS fits show that the ratio of the oxygen and iron abundance (O/Fe) in the core is 0.56 $\pm$ 0.04, which is consistent with other soft X-ray excess clusters measured by Tamura et al. (2004). This value is in between the theoretical values of < 0.05 and 1.5–4.0 for supernova types Ia and II, respectively.

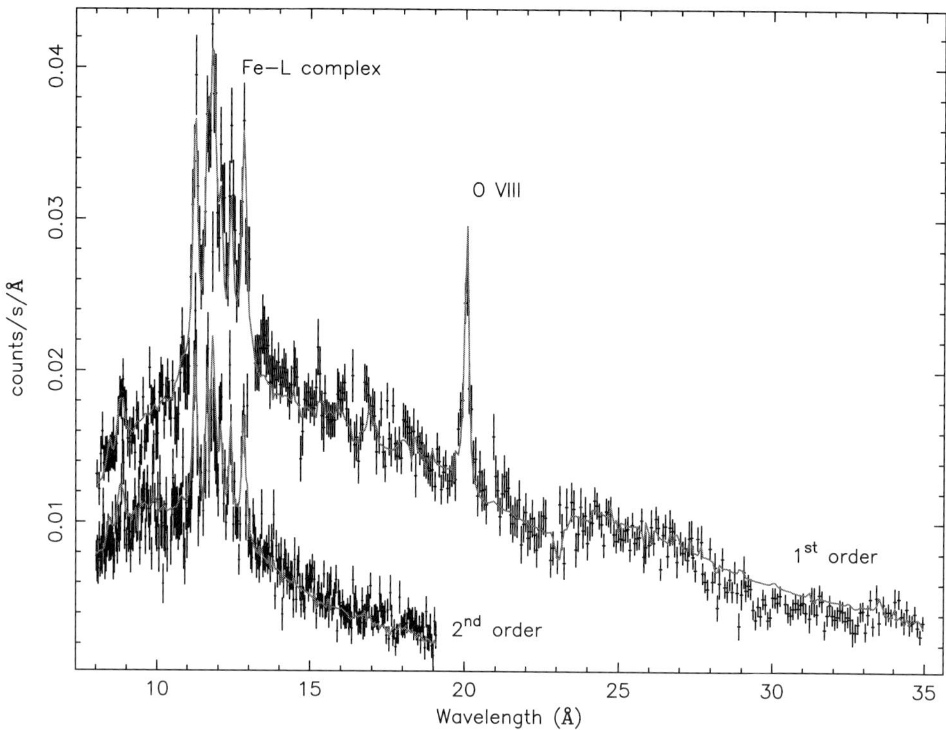

Figure 2. 1st and 2nd order RGS spectrum of Sérsic 159-03.

Acknowledgements

This work is based on observations obtained with XMM-Newton, an ESA science mission with instruments and contributions directly funded by ESA member states and the USA (NASA). The Space Research Organization of the Netherlands (SRON) is supported financially by NWO, the Netherlands Organization for Scientific Research.

References

Allen, S. W. & Fabian A. C. 1997 *MNRAS* **286**, 583.
Bonamente, M., Lieu, R. & Mittaz, J. P. D. 2001 *Ap J* **561**, L63.
Edge, A. C. & Steward, G. C. 1991 *MNRAS* **252**, 414.
Kaastra, J. S., Ferrigno, C., Tamura, T., et al. 2001 *A&A* **365**, L99.
Kaastra, J. S., Lieu, R., Tamura, T., et al. 2003 *A&A* **397**, 445.
Sérsic, J. L. 1974 *APSS* **28**, 365.
Tamura, T., Kaastra, J. S., den Herder, J. W. A., et al. 2004 *A&A*, in press.

Outskirts of Galaxy Clusters: Intense Life in the Suburbs
Proceedings IAU Colloquium No. 195, 2004
A. Diaferio, ed.

DOI: 10.1017/S174392130400033X

Non-thermal phenomena in galaxy clusters

Gianfranco Brunetti

Istituto di Radioastronomia INAF/CNR, via P.Gobetti, 101, I–40129 Bologna, Italy

Abstract. The discovery of diffuse synchrotron radio emission and, more recently, of the hard X-ray (HXR) tails have triggered a growing interest about non-thermal phenomena in galaxy clusters. After a brief review of the most important evidences for non-thermal emission, I will focus on the origin of the emitting particles and of the hadronic component. In particular I will describe the particle-injection and -acceleration mechanisms at work in the intra-cluster medium (ICM) and, at the same time, discuss the possibility to test current modellings of these phenomena with future radio, HXR, and gamma ray observatories.

1. Introduction

There is now firm evidence that the ICM is a mixture of hot gas, magnetic fields and relativistic particles. While the hot gas results in thermal bremsstrahlung X-ray emission, relativistic electrons and positrons generate non-thermal radio (synchrotron) and hard X-ray radiation (inverse Compton). In principle, the amount of the energy budget of the intracluster medium in the form of high energy hadrons can be large, due to the confinement of cosmic rays over cosmological time scales (Völk et al. 1996; Berezinsky, Blasi & Ptuskin 1997). The collisions between thermal and relativistic hadrons in the ICM generate π^o and secondary $e^{\pm}$. Both these species radiate a relevant fraction of their energy into the gamma band via π^o decay and via $e^{\pm}$-inverse Compton scatter of the photons of the cosmic microwave background. However, such gamma radiation that would allow us to constrain the energetics of relativistic hadrons in clusters has not been detected as yet (Reimer et al. 2003).

2. Observations

The most important evidence for relativistic electrons in clusters of galaxies comes from the diffuse synchrotron radio emission observed in a growing number of massive clusters (e.g., Feretti 2003). The diffuse emissions are referred to as *radio halos* (Fig. 1a) and/or radio *mini–halos* when they appear confined to the center of the cluster, while they are called *radio relics* when they are found in the cluster periphery. Giovannini, Tordi and Feretti (1999) found that ~5% of clusters in a complete X-ray flux limited sample have a diffuse radio source. The detection rate of diffuse radio sources shows a abrupt increase with the X-ray luminosity of the host clusters: about 30-35% of the galaxy clusters with X-ray luminosity larger than 10^{45} erg s^{-1} show diffuse radio emission (Feretti 2003). Interestingly, there is a correlation between the non–thermal diffuse radio emission and the presence of merger activity in the host clusters of galaxies (Buote 2001; Schuecker et al. 2001): this suggests a link between the process of formation of galaxy clusters and the origin of the non-thermal activity.

A second important evidence for relativistic electrons comes from the hard X-ray (HXR) excess emissions detected in a few galaxy clusters by the BeppoSAX and RXTE satellites (Fusco-Femiano et al. 2003a; Rephaeli et al. 2003). HXRs may be explained in

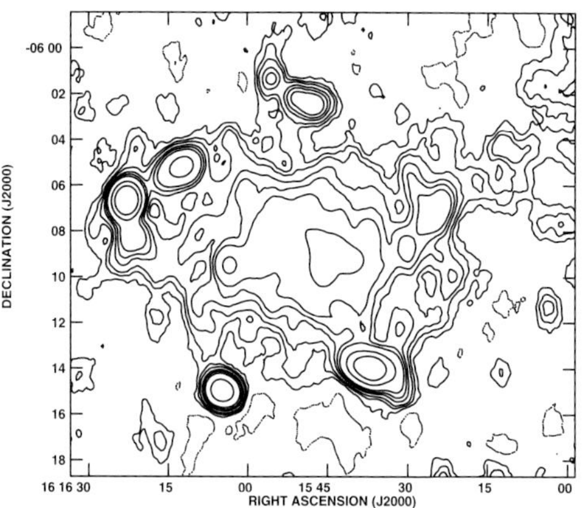

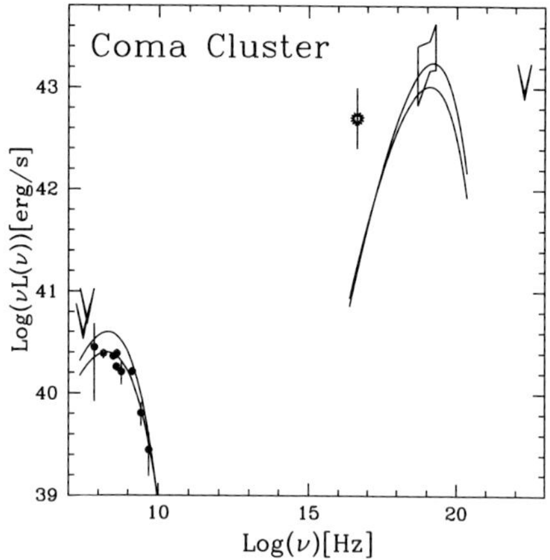

Figure 1. Left Panel: 327 MHz image of the giant radio halo in A 2163 (Feretti et al. 2004). Right Panel: Broad band non-thermal spectrum of the Coma radio halo: the fit to the radio and HXR data is provided by synchrotron and IC emission from a reaccelerated electron population.

terms of IC scattering of relativistic electrons off the photons of the cosmic microwave background (e.g., Fusco–Femiano et al. 2003a) and thus they represents a unique tool to unambiguously disentangle the energy of the relativistic electrons from that of the magnetic field in the ICM. Unfortunately the poor sensitivity of the present and past HXR-facilities does not allow to obtain an iron-clad detection of HXR excesses and thus future observatories (e.g. ASTRO-E2, NEXT) are necessary to definitely confirm these excesses (see Rossetti & Molendi 2004; Fusco-Femiano et al. 2004).

Additional evidence for non-thermal emission may also come from the extreme ultraviolet (EUV) excesses discovered in a number of galaxy clusters (e.g., Bowyer 2003). EUV excesses, and their origin, are however still matter of debate (see however Bonamente and Kaastra, these proceedings).

In Fig. 1b a compilation of the non-thermal spectrum of the Coma cluster is reported.

3. Modelling the origin of the emitting leptons

So far relativistic leptons are the unique non-thermal component detected in the ICM and thus our present understanding of the non-thermal activity is mostly based on the modelling of this component. The most spectacular example of non-thermal emission in galaxy clusters is given by the giant *radio halos* (Fig.1a). The difficulty in explaining these sources arises from the combination of their ~Mpc size, and the relatively short radiative lifetime of the radio emitting electrons (about 10^8yrs): the diffusion time necessary to these electrons to cover such distances is orders of magnitude larger than their radiative lifetime. Thus the emitting electrons cannot be injected by clusters' Galaxies and/or AGN and then simply diffuse into the ICM.

i) primary models: as proposed first by Jaffe (1977), a solution to this puzzle would be provided by continuous *in situ* reacceleration of the relativistic electrons on their way out; this possibility was then studied more quantitatively by Schlickeiser et al. (1987).

ii) secondary models: an alternative to the *in situ* reacceleration scenario was put forward by Dennison (1980), who suggested that relativistic electrons may be continuously injected in the ICM by inelastic proton-proton collisions through production and decay of charged pions.

3.1. *Broad band spectrum*

A first possibility to constrain the spectrum and origin of relativistic leptons is given by the study of the broad band spectrum of the non-thermal emission (e.g., Fig. 1b).

i) HXRs and γ-rays: If the HXR excess in the Coma cluster is of IC origin, then the number of relativistic electrons in the ICM should be large and this can be combined with the EGRET upper limit to reject the hypothesis of a secondary origin of the emitting particles. Indeed in this case a large number of relativistic hadrons is required and the gamma ray flux produced via π^o decay should overproduce the EGRET upper limit (e.g., Blasi & Colafrancesco 1999). Thus *secondary models* should admit a different origin for the HXRs. One possibility is given by supra-thermal bremsstrahlung emission (Blasi 2000; Dogiel 2000) which however would require a too large energy budget to maintain the HXR excess for more than 10^8 yr (Petrosian 2001).

ii) EUV and HXR excesses: A second possibility is given by the combination of the EUV and HXR excesses. If HXRs are due to IC scattering by the same electrons which emit the synchrotron radiation, then a low frequency (e.g., < 10 keV) flattening of the photon spectrum is required. The presence of this flattening indicates a corresponding flattening of the spectrum of the electrons at the energies responsible for IC emission below the HXR band; this flattening can result from Coulomb losses if the emitting electrons are in the cluster core, or it would be also naturally obtained if particles are accelerated by stochastic processes (e.g., via turbulent acceleration, Fig. 1b).

It is clear that the HXR fluxes are pivot points in the modelling of the non-thermal spectra of galaxy clusters and thus that the advent of the future observatories (ASTRO-E2, NEXT) will be crucial to confirm the above issues.

3.2. *Detailed properties of the radio emission*

Additional possibilities to constrain the origin of the emitting particles derive from the study of the detailed radio properties of a few well studied *radio halos*:

i) Broad synchrotron radio profiles: Giant *radio halos* are very extended sources with an extension up to 2-2.5 Mpc. The radial synchrotron profiles of these giant *radio halos* are found to be broader than that of the X-rays emitted by the hot gas (e.g., Govoni et al. 2001). This basically means that the synchrotron emissivity ($j \propto K_e B^{1+\alpha}$ for a spectrum of electrons $N(\gamma) = K_e \gamma^{-(1+2\alpha)}$) decreases with distance from the cluster center less rapidly than the bremsstrahlung emissivity $j \propto n_{th}^2$ (n_{th} is the density of the thermal plasma), and thus that the spatial distribution of the relativistic electrons is broader than that of the thermal particles. If radio electrons are of *secondary origin* then a very large number of relativistic hadrons is required in the clusters outskirts (since the production rate of secondary particles depends on n_{th}) and this causes a serious energetic problem at least if μG central fields and a relatively steep decrease of the field strength from the cluster center (as theoretically expected, e.g. Dolag et al. 2002) are assumed (Brunetti 2003). Obviously the energetic problem may be alleviated in some cases (e.g. Coma) by assuming stronger central fields and a slower radial decrease of the field strength (Pfrommer & Ensslin 2004).

ii) Observed synchrotron spectra: The integrated synchrotron spectrum of a few *radio halos* steepens at high frequencies (e.g., Giovannini et al. 1993; Fusco-Femiano et al. 2003b). Although, in some cases, the observed steepening may be mitigated by taking into account the SZ effect (Ensslin 2002), these observations point out to the presence of a high energy break or a cut-off in the spectrum of the emitting electrons. A second point is that the 0.3-1.4 GHz spectral index maps of an increasing number of *radio halos* indicate a progressive steepening of the radio spectrum from the center to the periphery of the clusters (Fig. 2a); the radial steepening in the Coma cluster, originally discovered with interferometric observations, is also required by single dish observations (Deiss et al. 1997). Radial spectral steepenings are theoretically explained in terms of the presence of a high energy break or cut-off in the electron spectrum combined with a radial decrease of the ICM-field strength (Brunetti et al. 2001; Kuo et al. 2003). Such high energy cut-offs should be at energy $\gamma \sim 10^3 - 10^5$ and are naturally expected if the electrons are re-accelerated by some kind of mechanism (e.g., Brunetti et al. 2001; Petrosian 2001).

iii) Synchrotron spectral and brightness variations: A more recent observational hint is that most *radio halos* show synchrotron brightness increments and complex spectral features in the spectral-index maps in coincidence with dynamically disturbed regions of the clusters and with temperature patches (Markevitch et al. 2002; Feretti et al. 2004). These observations suggest a close link between the dynamical status of the ICM and the spectrum of the particles (and/or the magnetic field) on 100-200 kpc scales.

Although future observations are obviously required to confirm all the above findings, it is clear that when the broad radial profiles of *radio halos* are *consistently* combined with the observed radial spectral steepenings, then *secondary models* have inescapable problems.

4. Cluster mergers and particle acceleration: a general view

In this Section we release the "historical" dichotomy between *primary* and *secondary* origin of the emitting particles and give a more general (but brief) theoretical overview of the mechanisms which are believed to drive non-thermal processes; a simple scheme is given in Fig.2b.

4.1. *Injection processes*

The first step in our understanding of non-thermal activity is given by the modelling of the injection and of the spatial diffusion of relativistic particles in the ICM (Fig.2b, bottom-left), and of the energy released by cluster mergers into shock waves and turbulence which will be responsible for particle acceleration (Fig.2b, top-center).

i) Injection of relativistic particles in the ICM: clusters contains galaxies and AGN which would inject relativistic leptons and hadrons which will remain confined in the cluster volume (Völk et al. 1996; Berezinsky et al. 1997). We don't know what is the fraction of energy channeled by these sources into relativistic leptons and hadrons; relativistic leptons, however, radiate their energy on a time-scale much shorter than a Hubble time, and thus it is very likely that the non-thermal energy stored in galaxy clusters is dominated by the hadronic component. Relativistic hadrons in the ICM continuously generate secondary leptons due to collisions with thermal protons and these particles contribute to the population of relativistic leptons in the ICM.

ii) Cluster mergers: injection of shocks and turbulence: accretion of matter at the virial radius is likely to form strong shocks, while cluster mergers should drive shock waves in the internal regions of the clusters (e.g., Miniati et al. 2000 and references therein). There is still some debate on the typical Mach number of the shocks developed during mergers. Some numerical simulations suggest that a relatively large fraction of these shocks have a high Mach number (Miniati et al. 2000, 2001). On the other hand semi–analytical calculations (Gabici & Blasi 2003; Berrington & Dermer 2003) find Mach numbers of order unity as also observed by *Chandra* (e.g., Markevitch et al. 2003). More recent numerical simulations (Ryu et al. 2003) seem to find weaker shocks than in Miniati et al.; however, the comparison with analytical calculations appears difficult because of a different classification of the shocks in the two approaches. Fluid turbulence is expected to be injected in galaxy clusters during cluster mergers. Numerical simulations find that an energy budget of 10-30% of the thermal energy of the ICM can be associated to

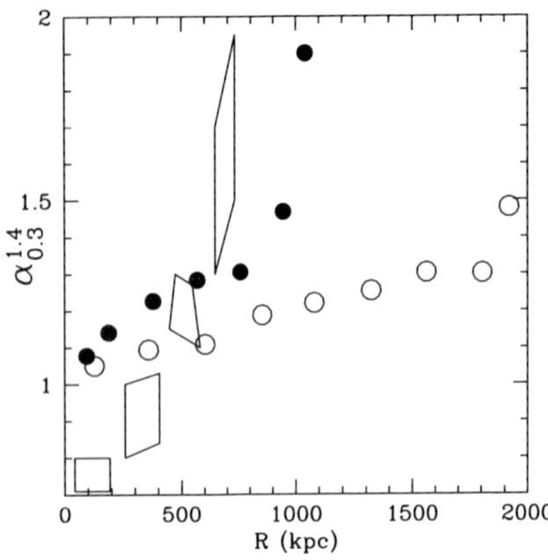

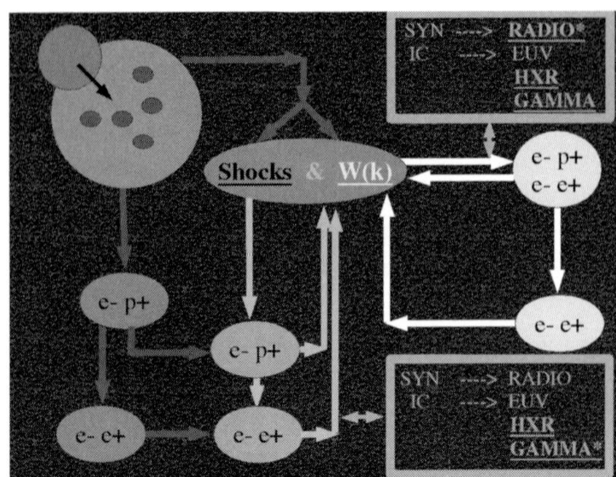

Figure 2. Left Panel: Spectral index (327 – 1400 MHz) vs. distance from the cluster center of Coma (boxes, Giovannini et al. 1993), A665 (filled circles) and A2163 (open circles, Feretti et al. 2004). Right Panel: Scheme of the processes in galaxy clusters (see text).

cluster turbulence (Sunyaev et al. 2003); this may be tested directly with future X-ray observatories (ASTRO-E2). In addition, first evidence for cluster turbulence has been found by the analysis of a very recent Newton-XMM observation of the Coma cluster (Schuecker et al. 2004).

4.2. *Particle acceleration*

The second point related to the generation of non-thermal phenomena is the interplay between particles and acceleration mechanisms. The idea is that this interplay is particularly efficient during cluster merger events. Two mechanism of interplay are reported in Fig.2b: shock-particle coupling (*shock loop*, Fig.2b bottom-center) and turbulence-particle coupling (*turbulence loop*, Fig.2b center-right).

i) Shock loop: Collisionless shocks are generally recognized as efficient particle accelerators through the so-called "diffusive shock acceleration" (DSA) process (Blandford & Eichler 1987). This mechanism has been invoked several times as acceleration process in clusters of galaxies (Takizawa & Naito 2000; Blasi 2001; Miniati et al. 2001; Fujita & Sarazin 2001). Although the scale and morphology of the synchrotron emission of *radio halos* are not naturally accounted for by shock-accelerated particles, evidence for shock acceleration in galaxy clusters may come from the so called *radio relics* which indeed can be produced through the combination of shock acceleration and adiabatic compression of *ghost radio plasma* by shock waves propagating into the ICM (e.g., Ensslin 2003). Shocks in the ICM may accelerate particles from the thermal pool and/or reaccelerate seed relativistic particles in the ICM with a power law in momentum with a slope $= 2(\mathcal{M}^2+1)/(\mathcal{M}^2-1)[+1]$ ($\mathcal{M}$ is the Mach number and "[+1]" should be added to calculate the spectrum of particles in the ICM under the assumption of a stationary continuous particle-injection by shock waves). If shocks related to major mergers are weak as suggested by semi-analytical calculations (Sect. 4.1), they accelerate very steep particle-spectra and thus they are not the dominant process for the injection of relativistic particles in the ICM. However, the maximum energy of particles accelerated at a shock wave is very high:

$$E^e_{max}(GeV) \sim 2.5 \times 10^4 B^{1/2}_{\mu G} v_8 \qquad \text{and} \qquad E^p_{max}(GeV) \sim 4.5 \times 10^8 B_{\mu G} v_8^2 \quad (4.1)$$

for leptons and hadrons, respectively (v_8 is the speed of the shock in units of 10^8km/s,

$B_{\mu G}$ is the field in units of μG and Bohm diffusion coefficient is assumed). The decay of π^o generated in hadronic collisions together with the IC emission from *secondary* leptons injected by the same collisions and from *primary* electrons accelerated at shock waves will produce gamma ray radiation above 100 MeV (e.g., Blasi 2001). The amount of radiation depends on the energy budget of relativistic hadrons and on their spectrum. Gamma rays emitted from the outskirts of galaxy clusters are expected to be dominated by IC emission from primary electrons accelerated at accretion shocks, these shocks are indeed expected to be very strong and thus the spectrum of the accelerated electrons should be flat ($\propto E^{-2}$). On the other hand, given the different results on the Mach number of merger shocks obtained by different approaches (Sect.4.1), there is no general agreement on the gamma ray flux and spectrum expected from the central regions of galaxy clusters where the contributions from the products of hadronic collisions should be important (e.g., Miniati 2002). Hopefully, future gamma ray observations will unambiguously clarify this issue and will allow us to constrain the energetics and spectrum of relativistic hadrons in galaxy clusters (Gabici & Blasi 2004).

ii) Turbulence loop: The origin of the giant *radio halos* and possibly of the HXR tails is most likely associated with the turbulence loop. Indeed, it has been shown that re–acceleration of a population of relic electrons by turbulence powered by major mergers is suitable to explain the very large scale of the observed radio emission and is also a promising possibility to account for the complex spectral behaviour observed in the diffuse radio sources and for the HXR tails (Brunetti et al. 2001a,b; Petrosian 2001; Ohno, Takizawa and Shibata 2002; Fujita, Takizawa and Sarazin 2003).

iii) Self-consistent modellings: Very recently, the problem of particle-Alfvén wave interactions has been investigated in the most general situation in which relativistic electrons, thermal protons and relativistic protons exist within the cluster volume (Brunetti et al. 2004). In this modelling the interaction of all these components with the waves, as well as the turbulent cascading and damping processes of Alfvén waves, have been treated in a fully time-dependent way in order to calculate the spectra of electrons, protons and waves at any fixed time. This work has provided a first investigation of the importance of the presence of the relativistic protons in the ICM for the electron acceleration. The most important result of this work is that Alfvénic electron acceleration can produce the observed phenomena provided that relativistic protons are not dynamically important in the ICM (less than few percents of the thermal energy). Of course additional MHD waves which do not interact with particles via resonant-acceleration (e.g., magnetosonic waves) may increase the efficiency of the electron acceleration without being very sensitive to the presence of relativistic hadrons. In addition, if reaccelerated, the spectrum of the hadronic component becomes harder and this may increase the efficiency in producing secondary particles in the ICM. These particles can be reaccelerated interacting with the MHD waves as in the case of the primary seed electrons, so that the leptonic acceleration can efficiently generate non-thermal phenomena also in the case in which the number of primary leptons in the ICM is negligible (Brunetti & Blasi, in preparation).

Acknowledgements

We acknowledge partial financial support from INAF (Istituto Nazionale di Astrofisica) through grant D4/03/IS. It is a pleasure to thank my collaborators and colleagues P. Blasi, R. Cassano, L. Feretti, S. Gabici and G. Setti for useful discussions.

References

Berezinsky V.S., Blasi P., Ptuskin V.S., 1997, ApJ 487, 529
Berrington R.C., Dermer C.D., 2003, ApJ 594, 709

Blandford R., Eichler D., 1987, PhR 154, 1B
Blasi P., 2000, ApJ 532, L9
Blasi P., 2001, APh 15, 223
Blasi P., Colafrancesco S., 1999, APh 12, 169
Bowyer S., 2003, in 'Matter and Energy in Clusters of Galaxies', ASP Conf. Series, 301, p.125,
Brunetti G., 2002, in 'Matter and Energy in Clusters of Galaxies', ASP Conf. Series, 301, p.349, eds. S. Bowyer and C.-Y. Hwang
Brunetti G., Setti G., Feretti L., Giovannini G., 2001a, MNRAS 320, 365
Brunetti G., Setti G., Feretti L., Giovannini G., 2001b, New Astronomy 6, 1
Brunetti G., Blasi P., Cassano R., Gabici S., 2004, MNRAS, in press; astro-ph/0312482
Deiss B.M., Reich W., Lesch H., Wielebinski R., 1997, A&A 321, 55
Dennison B., 1980, ApJ 239, L93
Dogiel V.A., 2000, A&A 357, 66
Dolag K., Bartelmann M., Lesch H., 2002, A&A 387, 383
Ensslin T.A., 2002, A&A 396, L17
Ensslin T.A., 2003, in 'Matter and Energy in Clusters of Galaxies', ASP Conf. Series, 301, p.159, eds. S. Bowyer and C.-Y. Hwang
Feretti L., 2003, in 'Matter and Energy in Clusters of Galaxies', ASP Conf. Series, 301, p.143, eds. S. Bowyer and C.-Y. Hwang
Feretti L., Orru' E., Brunetti G., et al., 2004, A&A in press; astro-ph/0404283
Fujita Y., Sarazin C.L., 2001, ApJ 563, 660
Fujita Y., Takizawa M., Sarazin C.L., 2003, ApJ 584, 190
Fusco-Femiano, R., Dal Fiume, D., Orlandini, et al., 2003a, in 'Matter and Energy in Clusters of Galaxies', ASP Conf. Series, 301, p.109, eds. S. Bowyer and C.-Y. Hwang
Fusco-Femiano R., Orlandini M., De Grandi S., et al., 2003b, A&A 398, 441
Fusco-Femiano R., Orlandini M., Brunetti G., et al., 2004, ApJ 602, L73
Gabici S., Blasi P., 2003, ApJ 583, 695
Gabici S., Blasi P., 2004, APh 20, 579
Giovannini G., Feretti L., Venturi T., Kim K.-T., Kronberg P.P., 1993, ApJ 406, 399
Giovannini G., Tordi M., Feretti L., 1999, NewA 4, 141
Govoni F., Ensslin T.A., Feretti L., Giovannini G., 2001, A&A 369, 441
Jaffe W.J., 1977, ApJ 212, 1
Kuo P-H., Hwang C-Y., Ip W-H., 2003, ApJ 594, 732
Markevitch M., Gonzalez A.H., David L., et al., 2002, ApJ 567, L27
Markevitch M., Vikhlinin A., Forman W.R., 2003, in 'Matter and Energy in Clusters of Galaxies', ASP Conf. Series, 301, p.37, eds. S. Bowyer and C.-Y. Hwang
Miniati F., Ryu D., Kang H., et al., 2000, ApJ 542, 608
Miniati F., Jones T.W., Kang H., Ryu D., 2001, ApJ 562, 233
Miniati F., 2002, Mem. S.A.It. 73, 23
Ohno H., Takizawa M., Shibata S., 2002, ApJ 577, 658
Petrosian V., 2001, ApJ 557, 560
Pfrommer C., Ensslin T.A., 2004, MNRAS in press; astro-ph/0404119
Reimer, O., Pohl, M., Sreekumar, P., Mattox, J.R., 2003, ApJ 588, 155
Rephaeli Y., Gruber D., 2003, ApJ 595, 137
Rossetti, M., Molendi, S., 2004, A&A 414, L41
Ryu D., Kang H., Hallman E., Jones T.W., 2003, ApJ 593, 599
Schlickeiser R., Sievers A., Thiemann H., 1987, A&A 182, 21
Schuecker P., Bhringer H., Reiprich T.H., Feretti L., 2001, A&A 378, 408
Schuecker P., Finoguenov A., Miniati F., Briel U.G., 2004, astro-ph/0404132
Sunyaev R.A., Norman M.L., Bryan G.L., 2003, AstL 29, 783
Takizawa M., Naito T., 2000, ApJ 535, 586
Völk H.J., Aharonian F.A., Breitschwerdt D., 1996, SSRv 75, 279

Outskirts of Galaxy Clusters: Intense Life in the Suburbs
Proceedings IAU Colloquium No. 195, 2004
A. Diaferio, ed.

DOI: 10.1017/S1743921304000341

Cluster mergers, radio halos and hard X-ray tails: a statistical magneto-turbulent model

Rossella Cassano[1,2] **and Gianfranco Brunetti**[2]

[1]Dip. di Astronomia, Università di Bologna
[2]Istituto di Radioastronomia del CNR di Bologna

Abstract. There is now firm evidence that the ICM consists of a mixture of hot plasma, magnetic fields and relativistic particles. The most important pieces of evidence for non-thermal phenomena in galaxy clusters come from the diffuse Mpc-scale synchrotron radio emission (radio halos), observed in a growing number of massive clusters (Feretti 2003), and from hard X-ray (HXR) excess emission (detected in a few cases), which can be explained in terms of IC scattering of relativistic electrons off the cosmic microwave background photons (Fusco-Femiano et al. 2003). There is now growing evidence that giant radio halos may be naturally accounted for by synchrotron emission from relativistic electrons reaccelerated by some kind of turbulence generated in the cluster volume during merger events (Brunetti 2003). With the aim of investigating the connection between thermal and non-thermal properties of the ICM, we have developed a statistical magneto-turbulent model which describes the evolution of the thermal and non-thermal emission from clusters. We calculate the energy and spectrum of the magne-!tosonic waves generated during cluster mergers, the acceleration and evolution of relativistic electrons and thus the resulting synchrotron and inverse Compton spectra. Here we give a brief description of the main results, while a more detailed discussion will be presented in a forthcoming paper (Cassano & Brunetti, in preparation). Einstein-De Sitter cosmology, $H_0 = 50$ km s^{-1} Mpc^{-1}, $q_0 = 0.5$, is assumed.

---

1. Introduction

Giovannini, Tordi and Feretti (1999) found that ∼5% of clusters from a complete X-ray flux limited sample have a radio halo source. The detection rate of radio halos shows an abrupt increase with increasing X-ray luminosity and mass of the host clusters: about 30-35% of the galaxy clusters with X-ray luminosity larger than 10^{45} erg s^{-1} show diffuse non-thermal radio emission (Feretti 2003). Recent papers (Enßlin & Röttgering 2003; Kuo, Hwang & Ip 2003) have investigated the statistics of the formation of radio halos from a more theoretical point of view. In these works, however, the expected statistics are not derived from formation models of radio halos, but simply from the observed luminosity-mass correlations and mass thresholds.

We model the formation of radio halos and HXR tails in a self–consistent approach which follows, at the same time, the evolution of the thermal properties of the ICM and the triggering and evolution of the non–thermal phenomena by assuming magneto-turbulent re–acceleration of relativistic particles. In particular, we follow the formation and evolution of clusters of galaxies, the generation of merger-driven turbulence and magnetosonic waves in the cluster volume, the acceleration and time–evolution of the relativistic particles, and of the related non–thermal emission.

2. The statistical magneto-turbulent model

The major steps of the model can be sketched as follows:

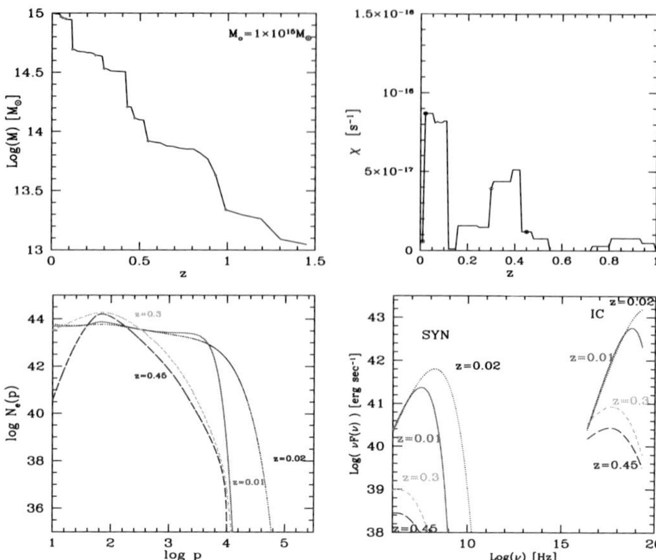

Figure 1. a) *Merger-tree* for a cluster with a present-day mass $M_o = 1 \times 10^{15} M_\odot$; b) evolution of the *electron acceleration coefficient* ($\chi_{acc}=\tau_{acc}^{-1}$, τ_{acc} is the acceleration time) ; c) *electron spectra* at four different z; d) correspondent *radio* (synchrotron) and *hard-X ray* (IC) emission.

i) Cluster formation: The evolution and formation of galaxy clusters is computed making use of a relatively simple semi–analytic procedure based on the hierarchical Press & Schechter (1974) theory of cluster formation. Given a present day mass and temperature of the parent clusters, the cosmological evolution (back in time) of the cluster properties (merger trees) are obtained making use of Montecarlo simulations. A suitable large number of trees allows us to describe the statistical cosmological evolution of galaxy clusters.

ii) Turbulence in galaxy clusters: The turbulence in the ICM is supposed to be injected during cluster mergers. The energetics of the turbulence is estimated from standard recipes based on the *Ram Pressure Stripping* theory and it is $E_{tur} \sim \rho v_i^2 \pi R_s^2 R_v$ where ρ is the density of the ICM, v_i is the velocity between the two colliding clusters, R_v is the virial radius of the most massive cluster and R_s is the stripping radius (Fujita, Takizawa & Sarazin 2003). We assume that a fraction η ($E_{MS} = \eta E_{tur}$) of the energy of the turbulence is associated to *fast magneto–acoustic waves* (MS waves). We use these waves since their damping rate and time evolution basically depend on the properties of the thermal plasma which are provided by our merger trees for each simulated cluster. The spectrum of the MS waves (following the injection of turbulence) is then calculated at each time step solving a turbulent-diffusion equation in the wave-number space (Eilek 1979) and assuming that the turbulence injected in the cluster volume for each merger event is injected, and thus dissipated, in a dynamical crossing time.

iii) Particle acceleration: We assume the presence of relativistic electrons in the ICM which are injected by AGNs, galactic winds, and/or merger shocks. Given the calculated spectrum of MS waves (*ii*) and the physical conditions in the ICM (*i*), we compute the time evolution of relativistic electrons at each time step by solving a Fokker-Planck equation including synchrotron, IC and Coulomb losses and the effect of the electron acceleration due to the coupling between MS waves and particles.

In Fig. 1 we report an example of the results in the case of a massive cluster. The calculations are obtained assuming that the energy injected in relativistic electrons is

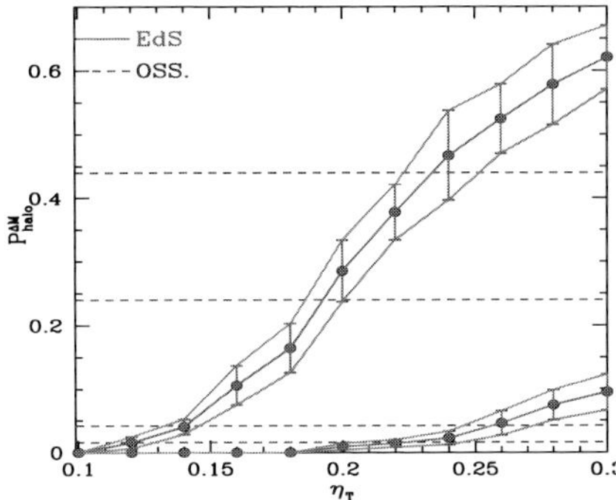

Figure 2. Formation rate of radio halos as a function of the parameter η_T in two different mass bins (solid lines with error bars): binA= $[1.8 - 3.6] \times 10^{15} M_{\odot}$ and binB= $[0.9 - 1.8] \times 10^{15} M_{\odot}$. The two bottom dashed lines mark the observed probabilities for radio halos in binB while the two top dashed lines mark the observed probabilities in binA.

of the order of $\sim$ 1% of the energy of the thermal plasma and a volume averaged $< B >\sim 0.5\mu$G. The first important result is that, given these reasonable assumptions, the radio (L_R) and hard-X ray (L_{HX}) luminosities are in agreement with the observed values (typically, $L_R \in [10^{40} - 10^{41}]$erg s^{-1} and $L_{HX} \in [10^{42} - 10^{44}]$erg s^{-1}).

3. Statistics: rate of radio halos with mass

Given a population of galaxy clusters with present day mass and temperature, combining *i)-iii)*, we follow the cosmological evolution of the non-thermal emission and the properties of the thermal ICM. We have calculated the formation rate of radio halos with cluster mass (at $z \leqslant 0.2$) in two bins consistent with those considerated in the observational studies: binA= $[1.8 - 3.6] \times 10^{15} M_{\odot}$ and binB= $[0.9 - 1.8] \times 10^{15} M_{\odot}$. In Fig. 2 we report the formation rate of radio halos in the two mass bins as a function of the parameter η ($\eta = E_{MS}/E_{tur}$, Sect. 2). The shaded regions mark the observed range of formation rate of radio halos (Giovannini et al. 1999) in binA (top shaded region) and binB (bottom shaded region). We find that our predictions match the observations in both the mass ranges for reasonable values of $\eta \in [0.16 - 0.22]$. In particular, in agreement with observations, we find that 20-30% of clusters in binA can form a radio halo and that only 2-3% of galaxy clusters in binB have a radio halo.

References

Brunetti, G. 2003, in *Matter and Energy in Clusters of Galaxies,* p.349, ASP, Conf. Ser. v.301, eds: S.Bowyer & C-Y.Hwang.

Eilek, A.J. 1979, *ApJ* **230**, 373.

Enßlin T.A., Röttgering H. 2002, *A&A* **396**, 83.

Feretti L. 2003, in *Matter and Energy in Clusters of Galaxis,* p.143, ASP Conf. Ser. v.301, eds: S.Bowyer & C-Y.Hwang.

Fujita Y., Takizawa M., Sarazin C.L. 2003, *ApJ* **584**, 190.

Fusco-Femiano R., Dal Fiume D., Orlandini M., et al. 2003, in *Matter and Energy in Clusters of Galaxies,* p.109, ASP, Conf. Ser. v.301, eds: S.Bowyer & C-Y.Hwang.

Giovannini G., Tordi M., Feretti L. 1999, *New Astronomy* **4**, 141.

Kuo P-H., Hwang C-Y., Ip W-H. 2004, *ApJ* **604**,108.

Press W.H., Schechter P. 1974, *ApJ* **187**, 425.

Outskirts of Galaxy Clusters: Intense Life in the Suburbs
Proceedings IAU Colloquium No. 195, 2004
A. Diaferio, ed.

DOI: 10.1017/S1743921304000353

The *XMM*- Ω project: gas mass fraction shape in high redshift clusters

R. Sadat[1] and the *XMM*- Ω collaboration

[1] Laboratoire d'astrophysique de l'observatoire Midi-Pyrénées, CNRS, UMR 5572, UPS, 14, Av. E. Belin, 31400 Toulouse, France

Abstract. We present the study of the gas mass fraction, $f_{\rm gas}$, properties of distant galaxy clusters observed within the *XMM*- Ω project, based on a sample of distant ($z \sim 0.5$) clusters of median luminosity. We first study the $f_{\rm gas}$ profiles traced up to the virial radius in both Einstein de Sitter (EdS) and concordance cosmologies. We find that the mean gas fraction profile of our high-z sample is similar to the local mean profile and in good agreement with numerical simulation predictions. This result confirms the self-similarity of the shape predicted by the simple scenario of structure formation up to high redshifts. We observe $f_{\rm gas}$ in the central part of distant clusters to lie below the local ones in the case of an EdS model, consistently with the observed evolution in the $L_{\rm X}$–T relation, which is indeed found to be weaker than expected in a self-similar model. We investigate quantitatively this departure from scaling laws by comparing $f_{\rm gas}$ in our distant *XMM* clusters with those derived from the local sample at fixed scaled radii. We show that the internal structure is very complex, i.e. the variation of the gas mass fraction with radius, with temperature and with redshift reveals differences that cannot be described in a simple scaling scheme. The observed variations in the central parts are a clear indication that, in these regions, the argument based on the non-evolving $f_{\rm gas}$ with redshift cannot be used to infer the cosmological parameters, given the present (lack of) understanding of the gas physics in clusters. At the virial radius, where this argument could be valid, we find that the ΛCDM model is excluded at more than 3-σ.

---

1. Introduction

Clusters of galaxies are unique cosmological probes whose statistical properties represent major sources of information for understanding the history of structure formation as well as for the determination of the cosmological parameters. In the simplest picture of purely gravitationally driven formation of virialized systems like galaxy clusters, it is expected that such objects exhibit self-similarity (Kaiser 1986). In this model physical properties of galaxy clusters obey scaling laws. The mass-temperature relation is consistent with the expected $M \propto T^{3/2}$ scaling while deviations from simple scaling predictions are observed in the $L_{\rm X}$–T, $M_{gas} - T$ relations for example. Moreover, self-similar models predict that the radial profile of any physical quantity should be similar independently of the cluster mass and redshift. Numerical simulations indicate that haloes of different masses follow a universal density profile, the so-called NFW profile (Navarro, Frenk & White 1996). The X-ray emissivity profiles in hot galaxy clusters are also found to be very similar at least in the clusters outer parts (Ponman et al. 1999). More recently, it has been shown that $f_{\rm gas}$ profiles are remarkably similar and seem to follow a universal shape (Roussel et al. 2000; hereafter RSB00). The self-similar model also makes predictions on the evolution of cluster properties. In particular the $L_{\rm X}$–T relation should scale as $(z+1)^{3/2}$ in an Einstein de Sitter (EdS) universe. Several studies have found evidence of a weak evolution in the $L_{\rm X}$–T relation (Sadat et al. 1998, Lumb et al. 2004, Vikhlinin et al. 2002). Using non-evolving physical properties of clusters as distance indicators can provide a new promising method for constraining cosmological parameters (Pen 1998, Sasaki 1996). The aim of the present study is to better understand the redshift evolution of the gas mass fraction scaling. We will discuss how the complexity of the internal structure can affect the application of the $f_{\rm gas}$ test and will present our result based on eight high redshift clusters observed with *XMM*.

2. The sample

We base our analysis on the *XMM* data obtained on a sample of eight distant SHARC clusters observed as part of the *XMM*- Ωproject (Bartlett et al. 2002). This sample represents a homogeneous sample of eight bona fide clusters with median luminosities, between 2 and 15 × 10^{44} erg/s (in the EdS cosmology with $h = 0.5$) with redshifts in a relatively restricted range, between 0.45 and 0.65. This sample is therefore expected to be fairly representative of the cluster population at high redshift. The detailed data reduction and analysis of this sample is presented in Lumb et al. (2004).

3. The shape of $f_{\rm gas}$ profiles

We compute the gas mass fraction $f_{\rm gas} = M_{gas}/M$ for each cluster of our high-z sample at different radii R_Δ (the radius enclosing a density contrast Δ which depends on cosmology). We assume a β-model for the X-ray gas distribution: the central electron density being computed from the Mekal constant. The total mass is computed from the M–T relation calibrated from numerical simulations (Bryan and Norman, 1998). Inside a given radius R_Δ, the binding mass is derived by assuming the so-called NFW profile. The temperature, β, the core radius, r_c and the Mekal constant are given in (Lumb et al. 2004). We estimate the apparent gas mass fraction within two different cosmological models (an $\Omega_M = 1$ and a flat $\Lambda = 0.7$ cosmology and $h = 0.5$). The $f_{\rm gas}$ radial distribution in the local sample is derived using the published gas mass measured up to the X-ray limiting radius $R_{\rm X\,lim}$ (RSB00); upward this radius we use the gas masses computed at two radii R_{1000} and R_{2000} as published by Vikhlinin et al 1999 (hereafter VFJ99). For this comparison we do not correct for the clumping (Mathiesen et al. 1999), as the radial variation of this quantity is unknown. Furthermore, if the scaling actually holds, the emission of both local and distant clusters should be biased by this effect by the same amount, and therefore the lack of correction does not affect our comparison. In figure 1 we compare the $f_{\rm gas}$ profile of the high-z clusters sample with nearby clusters in both cosmologies.

While a noticeable difference on the amplitude appears depending on the cosmological model, the shape of the observed profiles of distant clusters are globally in good agreement with what has been inferred for clusters at low redshift: the apparent $f_{\rm gas}$ profiles of our distant clusters increase from the center to the outer shells following the universal gas fraction shape. These measured $f_{\rm gas}$ profiles still exibit a dispersion which is consistent with what has been previously found (RSB00). In the Einstein-de Sitter model the most central values of $f_{\rm gas}$ appear to be smaller in distant clusters. Such deviation is consistent with the measured evolution of the $L_{\rm X}$–T relation, weaker than expected if the scaling was actually strictly holding. Conversely, in the low density flat model, $f_{\rm gas}$ values in the central parts of distant clusters seem to agree with the standard scaling, again in agreement with the evolution of the $L_{\rm X}$–T relation in this cosmology.

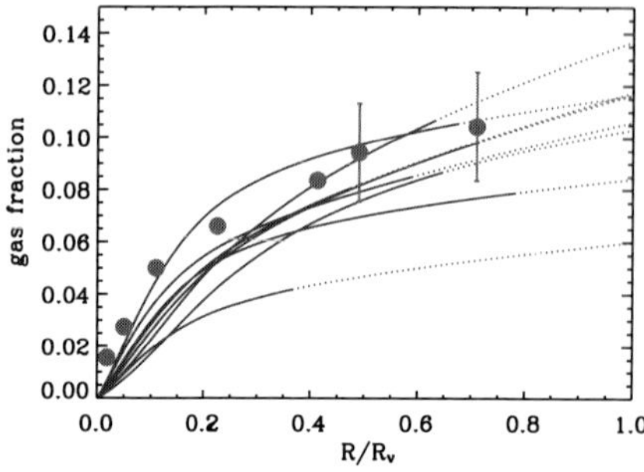

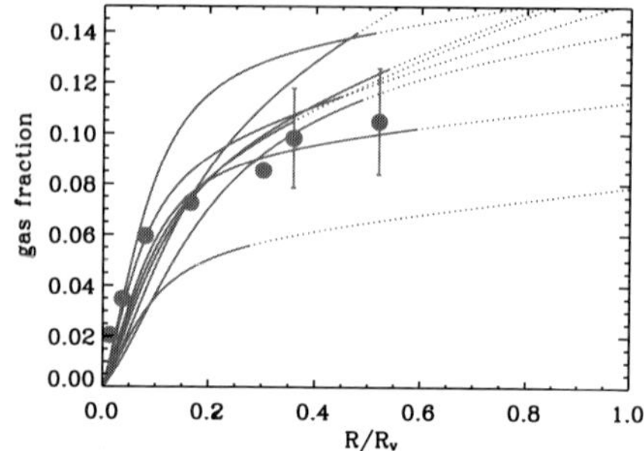

Figure 1. The scaled apparent f_{gas} profiles of high-z XMM clusters (solid lines) compared to the mean scaled profile of local clusters in EdS model (on the left) and in a ΛCDM model (right side).

3.1. $f_{gas} - Temperature$ *correlation*

Before applying the test based on the assumption that f_{gas} is universal and therefore does not vary with redshift we should keep in mind two possible sources of complexity: the gas mass fraction varies with radius (see fig. 1) and possibly with temperature. Such a trend has been observed from time to time on local samples. By examining the RSB00 sample used in Fig.1, we indeed found a clear trend of f_{gas} to increase with temperature at least up to the X-ray detection limit $R_{\rm X\,lim}$ (while this trend is much less clear on the global sample). However, this sample was not designed to offer a uniform sample for X-ray studies, therefore it has been used only as a guideline in the present study. For the following analysis, we will use the VFJ99 sample and compute f_{gas} at their two fiducial radii R_{2000}, R_{1000} as well at the virial radius. It should be noticed that distant clusters are expected to be on the bright tail of the distribution inducing a Malmquist bias and therefore a bias toward hotter temperature. If f_{gas} actually varies with temperature this bias should be corrected for. In order to compare properly with our high-z sample we need to compute f_{gas} at the same radius in unit of the virial radius. In figure 2 we have plotted f_{gas} at fixed radius (R_{2000}, R_{1000} and $R_{\rm v}$) in the local and high redshift clusters versus temperature for the two cosmological models. The local sample reveals a clear trend of the apparent f_{gas} to increase with the temperature following $f_{gas} \propto T^{0.25}$. This trend is slightly more pronounced in the inner radius. The origin of this variation is unclear. It might either be an actual variaton of f_{gas} in clusters, but it might also be due to a variation of the clumping of the gas with temperture. Whatever this variation is, it implies an evolution of the scaling. Comparing our XMM clusters with local clusters at the virial radius where the $f_{gas} - T$ correlation is less pronouced shows that the gas mass fraction in the high-z cluster sample matches fairly well the local sample in the case of an $\Omega_M = 1$ model, while in a ΛCDM model it is deviating from the gas fraction mean value of the local sample; this effect is more pronounced at R_{1000} than at R_{2000}. Note that the gas mass fraction at the virial radius implies some extrapolation of the observed profile for clusters in our sample.

4. Cosmological application

The test based on the *ansatz* that the baryon (gas) fraction does not evolve with redshift has been used as a test for cosmological parameters determination (see Ettori et al. 2003 and Allen et al. 2002 for recent applications). From Fig. 2, at the virial radius, we can see that while the agreement between low and high-z clusters is quite

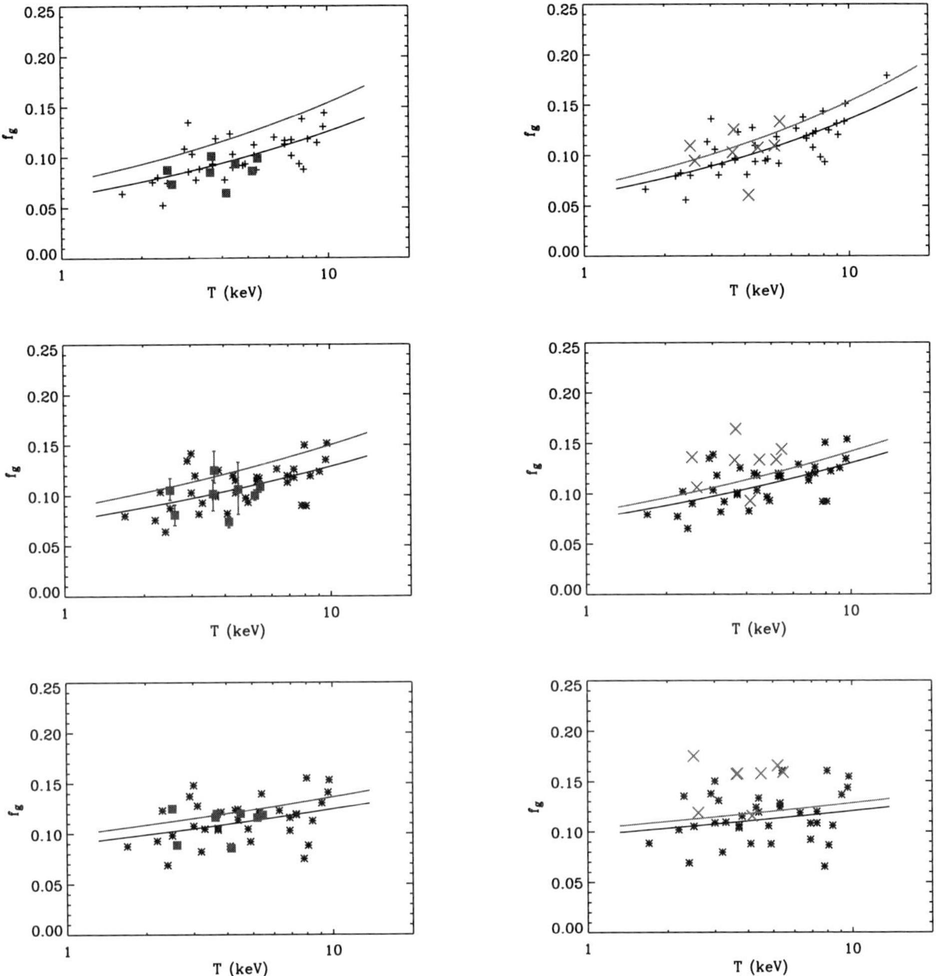

Figure 2. The apparent $f_{\rm gas}$ versus temperature measured at three different radii R_{2000} (top), R_{1000} (middle) and $r_{\rm v}$ (bottom) in the outer part of the *XMM* distant clusters compared to the same quantity evaluated at the same scaled radii in EdS model (on the left) and in a ΛCDM model (right side). The black solid line is a fit to $f_{\rm gas}$ of local clusters; we also plot as continuous colored (upper) line the predicted variation of $f_{\rm gas}$ under the scaling assumption.

good in an EdS cosmology, there is a clear offset of $f_{\rm gas}$ values computed in a ΛCDM model for high-redshift clusters compared to local ones. Several aspects however make the direct cosmological interpretation difficult. First it is to notice that the clear trend with temperature could be a source of confusion: temperatures in our sample are fainter and therefore expected $f_{\rm gas}$ are smaller. Comparing the gas fraction in our clusters with the hottest local clusters, would have led to favour the ΛCDM model. The internal stucture of the gas seems not to be strictly identical in high and low redshift clusters, declining faster in the central part of high-z clusters. This is consistent with what seems to emerge from Fig. 1. This complex internal structure could be the result of non gravitational heating of the gas which is currently advocated to explain the observed L_X–T relation. Possible systematic variations with radius, temperature and redshift, imply that the baryon fraction test should be performed with caution and as much as possible at the virial radius.

5. Conclusion

The observations of distant clusters with XMM offer a unique possibility to investigate the outer emissivity of the gas distribution in distant X-ray clusters. We have shown how the observations of the distant SHARC clusters are leading to interesting results on this issue: the shape of the apparent $f_{\rm gas}$ derived for those clusters is in rough agreement with the shape of the apparent $f_{\rm gas}$ of local clusters. This is an independent confirmation that the scaled shape of the baryon fraction in clusters is in rough agreement with the shape found in numerical simulations (Frenk et al. 1999). By comparing our distant clusters to a sequence of local clusters, we found that a clear variation with temperature and redshift of the internal structure of clusters exists which cannot be described by some simple scaling relations. We use our high-z sample to apply the cosmological test based on the assumption that $f_{\rm gas}$ should have a universal value independent of redshift. At the virial radius we have found that the data are roughly consistent with the scaling hypothesis in the Einstein-de Sitter model, but not within the standard concordance model in which the observed gas mass fraction appears higher ($4\,\sigma$) than in clusters of the local universe. The internal structure of the gas presents however some complex variations that makes difficult a fully reliable use of the gas mass fraction for the determination of the cosmological parameters.

References

Allen, S. W.; Schmidt, R. W.; Fabian, A. C. 2002 *MNRAS* **334**, L11–L15
Bartlett, J. et al. 2001, proceedings of the XXI rencontre de Moriond, astro-ph/0106098
Bryan, G.L. & Norman, M.L. 1998 *ApJ* **495**, 80–99
Ettori, S., Tozzi, & P. Rosati 2003 *MNRAS* 305, 834
Frenk, C.S, White, S.D.M et al. 1999 *ApJ* **525**, 554–582
Kaiser, N. 1986 *MNRAS* **222**, 323–345
Lumb, D. et al. 2004, astro-ph/0311344, accepted for publication in *A&A*
Mathiesen, B., Evrard, A.E. & Mohr, J.J. 1999 *ApJ* **520**, L21
Mohr, J.J., Mathiesen, B. & Evrard, A.E. 1999 *ApJ* **517**, 627–649
Navarro, J.F., Frenk, C.S., White, S.D.M. 1997 *ApJ* **490**, 493–508
Pen, Ue-Li 1997 *NewA* **2**, 309–317
Ponman, T.J., Cannon, D. B. & Navarro, J. F. 1999 *Nature* **397**, 135–137
Roussel, H., Sadat, R. & Blanchard, A. 2000 *A&A* **361**, 429–443
Sadat, R.& Blanchard, A. 2001 *A&A* **329**, 19–24
Sasaki, S. 1996 *PASJ* **48**, 119–122
Vauclair, S.C. et al. 2003 *A&AL* 412, 37–41
Vikhlinin, A., Forman, W. & Jones, C. 1999 *ApJ* **525**, 47–57

Outskirts of Galaxy Clusters: Intense Life in the Suburbs
Proceedings IAU Colloquium No. 195, 2004
A. Diaferio, ed.

DOI: 10.1017/S1743921304000365

Measuring cluster gas mass fractions with X-ray, Sunyaev-Zel'dovich effect, and weak lensing data

S. LaRoque[1], M. Bonamente[2], J. E. Carlstrom[1,3], M. Joy[4] and E. D. Reese[5]

[1]Department of Astronomy and Astrophysics, University of Chicago, Chicago, IL 60637
[2]Department of Physics, University of Alabama, Huntsville, AL 35899
[3]Kavli Institute for Cosmological Physics, Department of Physics, Enrico Fermi Institute, University of Chicago, Chicago, IL 60637
[4]Department of Space Science, NASA Marshall Space Flight Center, Huntsville, AL 35812
[5]Department of Physics, University of California, Berkeley, CA 94720

Abstract. We estimate the gas mass fractions for a sample of 39 high-z galaxy clusters using a combination of three different data sets: *Chandra* X-ray data, OVRO/BIMA centimeter-wave Sunyaev-Zel'dovich Effect (SZE) data, and published weak gravitational lensing data. The gas mass fraction f_g is calculated in three different ways. The X-ray surface brightness and SZE decrement are used independently to determine the number of baryons in the intracluster gas and subsequently the gas mass; the total mass follows assuming spherical symmetry and hydrostatic equilibrium (HSE). The third, less model-dependent approach is to calculate the projected SZE gas mass along a cylinder and divide by the total mass derived from the lensing data. The goal of this study is to compare the f_g results from the three different methods to investigate systematic uncertainties associated with clumping of the intracluster medium, cluster elongation, and the HSE assumption.

1. Introduction

Clusters of galaxies are the largest known virialized objects and can be used as cosmological probes. The mass composition on cluster scales is expected to reflect the universal mass composition; clusters are therefore a useful probe of Ω_B/Ω_M. In this preliminary analysis, we have calculated the gas mass fraction f_g for a sample of 39 clusters ($0.14 \leqslant z \leqslant 0.89$) observed by both the *Chandra* X-ray observatory and the BIMA/OVRO interferometric Sunyaev-Zel'dovich Effect (SZE) imaging experiment. The f_g calculation is performed in two parallel ways, by studying the cluster's X-ray emission and its SZE. We have also combined SZE data with weak gravitational lensing data to calculate f_g a third way for fifteen of the clusters in our sample. Due to the differing dependencies of these methods on the cluster properties, the results can be compared to better understand systematics associated with assumptions about clumping of the intracluster medium, temperature structure, cluster morphology, and hydrostatic equilibrium (HSE). A ΛCDM cosmology with $\Omega_M = 0.3$ and $\Omega_\Lambda = 0.7$ is assumed throughout and $H_0 = 70$ km s^{-1} Mpc^{-1}. All uncertainties are at the 90% level.

2. Data analysis

2.1. *X-ray data*

Clusters were observed with the *Chandra* ACIS detector. Seven clusters in the sample are seen to have "cooling flow" cores; their bright central regions were excised prior to spatial and spectral fitting. Electron temperatures are calculated by fitting the 0.7—7.0 keV spectral data to a Raymond-Smith thermal plasma model. The X-ray surface brightness S_x is fit to an isothermal spherical β-model, and the resulting density distribution is used to get the gas mass as a function of radius. Care is taken not to extrapolate the model beyond the region where X-ray emission is detected. The total gravitational mass is determined assuming HSE, and the cluster gas mass fraction is then simply $f_g = M_{\rm gas}/M_{\rm tot}$. A Markov chain Monte-Carlo (MCMC) method is used to determine the probability distributions and best-fit values of the model parameters and masses. Gas mass fractions are reported within r_{2500}, the radius at which the mean mass density of the cluster is equal to 2500 times the critical density, $\rho_c(z)$, at that epoch.

2.2. *SZE data*

SZE observations were conducted at the Berkeley Illinois Maryland Association (BIMA) and Owens Valley Radio Observatory (OVRO) interferometric arrays at 28.5 GHz (BIMA and OVRO) and at 30 GHz (OVRO). The X-ray derived T_e is assumed, and the SZE brightness profile is fit to an isothermal spherical β-model. The density distribution is then used to get the gas mass. As before, the total masses are calculated under the assumption of HSE. A MCMC is used to determine probability distributions for the model parameters and masses. The masses are best constrained at a radius of 65" (Grego *et al.* 2001) and results are reported there.

2.3. *Lensing data*

The gas mass fraction is calculated in a third way for fifteen of the clusters by comparing the SZE-derived gas masses with the total masses determined from weak lensing data. Lensing masses come from Clowe *et al.* (1998) and Dahle *et al.* (2002). Gravitational lensing and the SZE both probe a cluster's projected mass distribution within a cylinder whose long axis is the line of sight. No deprojection is therefore necessary to calculate f_g. Comparison of the results from all three methods should illuminate any trends in cluster elongation, and the T_e independence of the lensing masses should help constrain temperature structure and non-thermal pressure support. Using the SZE-derived β-model parameters, we calculate the gas mass in a cylindrical volume of radius 65" and compare it with the gravitational lensing mass at the same radius to get f_g. Since the published lensing masses are calculated assuming a flat universe with $\Lambda = 0$, the cylindrical SZE gas masses are calculated assuming the same cosmology. A rough scaling is then applied to make the SZE/lensing values consistent with the ΛCDM cosmology assumed for the SZE and X-ray gas mass fractions.

3. Results

The X-ray, SZE, and SZE/lensing gas mass fractions are shown in Figure 1. Uncertainties are statistical only. There is no evidence of evolution with redshift for this cosmology, and the gas mass fractions for individual clusters agree reasonably well within the uncertainties. There are possible trends apparent in the ensemble averages, however. We have calculated mean gas mass fractions for the three datasets in log space, where the uncertainties are nearly Gaussian (cf. Figure 1). We find $\overline{f}_g = 0.094 \pm 0.002$ for the X-ray data, $\overline{f}_g = 0.079^{+0.010}_{-0.007}$ for the SZE data, and $\overline{f}_g = 0.075^{+0.017}_{-0.011}$ for the SZE/lensing

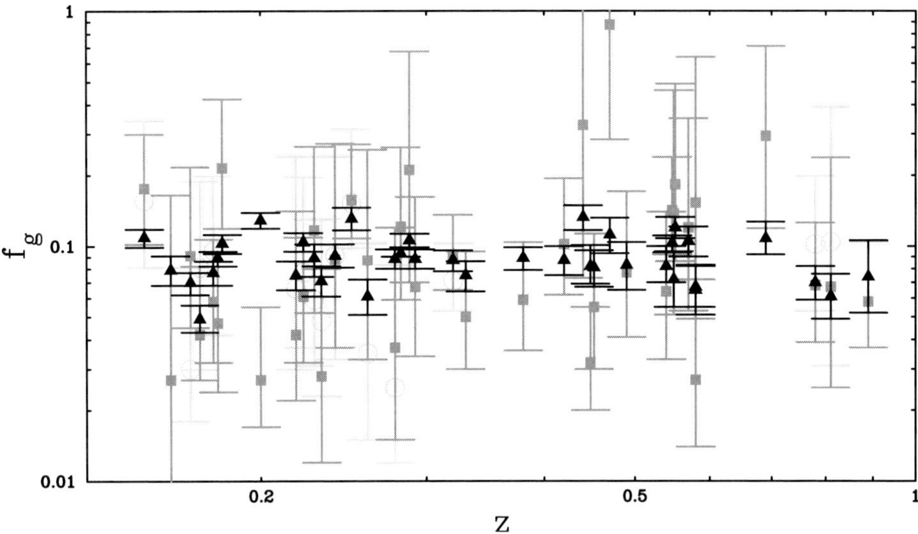

Figure 1. X-ray (black triangles), SZE (dark grey squares), and SZE/lensing (light grey circles) gas mass fractions versus redshift.

analysis. The good agreement between the SZE and SZE/lensing results is encouraging and suggests that the HSE assumption is appropriate, but the offset relative to the X-ray value needs to be explored. Clumping of the intracluster medium may bias the X-ray gas fractions high, while the isothermal assumption may bias the SZE results low, since T_e as measured from the X-ray data is an n_e^2-weighted value. Furthermore, the assumption of a flat gas mass fraction profile may cause the SZE gas mass fractions to be lower than the X-ray values, since 65″ is generally comparable to or less than r_{2500}. These are preliminary results from a work in progress, and a forthcoming study will incorporate more realistic temperature profiles, SZE gas fractions scaled with radius, and insights from a concurrent cluster distance project.

Acknowledgements

This material is based upon work supported by NSF grant No. 0096913 and by NASA LTSA grant NAG5-7986. S. L. gratefully acknowledges support from NASA GSRP Fellowship NGT5-50173.

References

Clowe, D., Luppino, G., Kaiser, N., Henry, I. & Gioia, I. 1998 *ApJ* **497**, L61–L64.

Dahle, H., Kaiser, N., Irgens, R., Lilje, P., & Maddox, S. 2002 *ApJS* **139**, 313–368.

Grego, L., Carlstrom, J., Reese, E., Holder, G., Holzapfel, W., Joy, M., Mohr, J., & Patel, S. 2001 *ApJ* **552**, 2–14

Hans Böhringer explains to Sandro Bardelli how relevant white wine is to understand cosmology *(by J.M. Colberg)*.

Rashid Sunyaev enjoys the discussion at the welcome party *(by J.M. Colberg)*.

Section B
Building clusters

Outskirts of Galaxy Clusters: Intense Life in the Suburbs
Proceedings IAU Colloquium No. 195, 2004
A. Diaferio, ed.

DOI: 10.1017/S1743921304000377

The state of wide-field surveys and their contributions to the study of cluster peripheries

Ian P. Dell'Antonio

Physics Department, Brown University, Providence, RI 02912, USA email: ian@het.brown.edu

Abstract. The last decade has seen an explosion in the number and scope of wide-field surveys. The data provided by these surveys will greatly increase our understanding of the relationships between the outskirts and the cluster centers and the field. I summarize the status of current deep-wide optical surveys and describe the plans for the next generations of surveys, with particular emphasis on the applications of these surveys to weak gravitational lensing. As a concrete example of such a survey, I will present results from the Deep Lens Survey (DLS).

---

1. Introduction

The last ten years have seen an explosion in the depth and breadth of imaging, comparable in terms of the increase in number of objects detected to the effect of the introduction of photographic film at the beginning of the last century. This huge advance has been driven largely by the availabilty of large format CCDs, and, most importantly, of CCD Mosaics (Wittman *et al.* 1998, Starr *et al.* 2000; Wolfe *et al.* 1998). The effect of this explosion has been to change our perception of what a deep imaging survey is. Just as the 2400 redshifts of the original Cfa Survey (Davis *et al.* 1982) now represent a night's work for the large multi-object spectrographs, so too the scope of the surveys today has changed. Up to now, we have surveyed nearly 1/2 of the sky to $R \sim 23$, about 250 square degrees to $R \sim 24.5$, and about 80 square degrees to $R > 26$. The next generation of surveys underway, and especially the dedicated survey missions proposed for the next decade will image the entire sky $R \sim 28$.

With the explosion of available data the types of problems that can be addressed has also expanded. For example, we can now map the distribution of galaxies and of mass (via gravitational lensing) out to angular distances of almost $30'$ from the cluster center. This corresponds to about 3.5 Mpc for a $z = 0.1$ cluster in the currently favored cosmological models (Spergel *et al.* 2003). (Throughout the paper I will assume a cosmological model with $h = 0.71$, $\Omega_m = 0.27$, $\Omega_\Lambda = 0.73$.)

2. Wide-field surveys – current status

The primary distinguishing characteristics of wide field imaging surveys are sky coverage, depth, and spectral coverage (or number of filters). Sky coverage is critical both for obtaining large enough samples of objects (for example clusters), and for overcoming cosmic variance. For large and/or nearby objects, sky coverage also ensures that objects are fully sampled. Depth is essential for source density and the detection of high redshift (and thus primarily faint) objects. Surveys seeking to study properties over a wide range of redshifts obviously require depth, and this is also critical for gravitational lensing (§ 4). Spectral coverage is essential for determining properties of the objects detected,

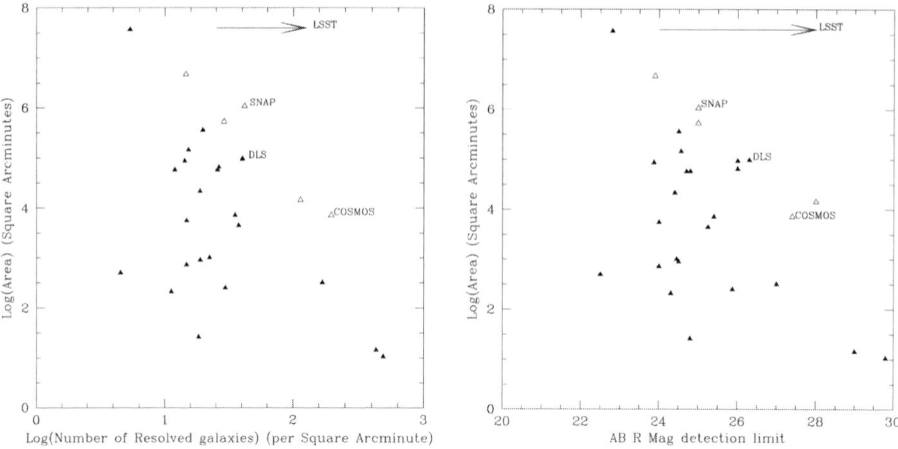

Figure 1. Left: the area coverage versus surface density of resolved objects, the relevant statistic for weak gravitational lensing studies. Right: the area coverage versus photometric depth for a selection of completed, ongoing and future survey projects (adapted from Jannuzi 1999).

and as a way of estimating redshifts for objects too faint or too numerous to pursue spectroscopically.

Because the available telescope time is (still!) not infinite, different survey projects have chosen to emphasize different aspects of this. This can be summarized by Figure 1, right panel, (adapted and expanded from Jannuzi 1999), which plots area coverage versus limiting magnitude in some photometric band for a large selection of current and future surveys, thus probing two of the three dimensions of (size, depth, spectral coverage) parameter space. Note the empty region on the top right–this corresponds to the part of parameter space that is currently inaccessible due to the finite amount of photon-collecting power available. Despite the fact that this is the standard way to describe these surveys, plotting area versus magnitude can be misleading. A better way to plot the choices made by the surveys is to plot area coverage versus surface density of objects (galaxies, resolved galaxies, QSO, etc.), where the nature of the object depends on the goal of the survey. For example, if the aim of a hypothetical survey were to study angular clustering on a variety of scales, the relevant quantities would be number density of galaxies and area. For the detection of weak lensing shear the relevant quantities are the area and the number density of resolved galaxies, which depends on the depth *and* the image quality (and on the filter). A survey studying 3-d clustering would add some measure of redshift-space resolution. In Figure 1, left panel, an estimate of the source density for weak gravitational lensing is given in place of the limiting magnitude. This folds in information about the depth, resolution and filter choice of the survey. As might be expected, space-based surveys such as SNAP (Rhodes *et al.* 2003) or COSMOS (Scoville 2004) fare comparatively better by this metric.

In general, surveys have either aimed to be wide but shallow (e.g. SDSS York *et al.* 2000), or ultra-deep (HDF N+S Williams *et al.* 1996, 1998, UDF Beckwith 2004). Another popular strategy, adopted by the CFHTLS (CFHT, 2004), attempts to combine strategies by having a nested scheme in which large areas are covered to a more shallow depth, with a smaller internal region covered to great depth. The strategy is largely driven by the scientific results desired. For example, the HDF images tell us (almost) nothing about the distribution of matter at $z < 0.1$. Similarly, the source density of resolved galaxies at $z > 2$ in the SDSS is too low to yield any interesting results. Future missions such as LSST (Tyson *et al.* 2002) and perhaps the VST will image all the visible sky multiple times, building up depth (and thus source density) incrementally.

3. Clusters science from surveys

Wide-field surveys (almost without exception) do not target areas of the sky because of the presence of clusters of galaxies. Indeed, at the lowest redshifts, most of the deeper wide-field surveys are biased against large structures– nobody wants Virgo Cluster galaxies blocking the view of more distant objects! Instead, the power of the wide-field imaging surveys comes about because the clusters extracted from the survey are a truly representative sample down close to the selection limits (whether they be in redshift, photometry or even source density) of the survey. In this situation, one of the most fundamental result from a survey is simply the number density per comoving volume of clusters. However, the problem remains of detecting the clusters in a complete and unbiased way.

3.1. *Photometric methods*

In the absence of spectroscopic redshift information, surveys have moved to photometric techniques such as the red sequence (Yee *et al.* 1999) or photometric redshifts to obtain three dimensional information about the distribution of matter. These methods have been very successful in isolating clusters of galaxies. Other surveys, such as the Las Campanas Distant Galaxy Cluster Survey (LCDGCC) (Gonzalez *et al.* 2000) have used surface brightness fluctuations to detect clusters of galaxies efficiently using less observing time than would have been required to detect the individual galaxies (Dalcanton 1996).

3.2. *Weak lensing*

Today, several surveys are underway that seek to use gravitational lensing shear to detect clusters of galaxies (for example DLS Wittman *et al.* 2002, the CFHTLS surveys, and Suprime33 Miyazaka *et al.* 2003). In addition, several other surveys (RCS Yee *et al.* 1999, NDWFS Jannuzi 1999), although not specifically designed with gravitational lensing in mind, are using gravitational lensing measurements to supplement or complement photometric searches.

Weak gravitational lensing relies on measuring the small induced ellipticities in the shapes of background galaxies due to foreground masses. The basic result (Kaiser & Squires 1993) is that the surface density Σ at any point divided by the critical surface density Σ_{crit} (denoted by κ) is derived by the weighted average of the tangential ellipticity e_t (corrected for PSF smearing) of the background galaxies, where the weight function is approximately $1/r^2$ (but deviates for small r to avoid noise divergences). Therefore, by measuring the shapes and sizes of background galaxies (and hence e_t), we can obtain an estimate of κ.

Compared to photometric techniques, weak lensing has the great advantage of not being baryon-biased. In particular, the detectability of a mass clump via weak lensing does not depend on the number of luminosity function of galaxies in the clump. This opens up the possibility of testing whether the photometric techniques are biased. Indeed, if "dark clusters" exist, as is periodically proposed (e.g. Erben *et al.* 2000), weak lensing might well be the only technique that will reveal them.

Another advantage of weak gravitational lensing is that the current generation of surveys can routinely measure shear values of about 0.005. For realistic clusters, this means detecting the mass out to 2.5-3 h^{-1} Mpc, far beyond the radius where the X-ray emission has fallen below background and the galaxy overdensity has become undetectable. Thus, weak lensing is one of the best techniques for measuring how the mass in the outskirts of clusters merges into the large-scale structure distribution.

However, for all the advantages, weak shear lensing does suffer from some serious limitations. One of the most fundamental is that because the measurement depends on shape measurements for galaxies far behind the cluster, the photometric depth (and thus the exposure time) required to detect a typical cluster is significantly greater than for photometric methods. Furthermore, because the information content of each background galaxy is small, weak gravitational lensing necessarily produces a smoothed estimate of the mass distribution. This makes small-scale substructure very hard to map out. In fact, the problem is compounded for the study of the outskirts of clusters, because the effective resolution that can be supported depends on the strength of the lensing signal. Because the shear decreases by a factor of almost 100 from the inner regions of the cluster to the outskirts, the resolution of mass mapping will suffer consequently.

Finally, as pointed out by White *et al.* (2002), weak gravitational lensing suffers significantly from superpositions. Because the window function $D_{LS}/D_L D_S$ for gravitational lensing is not a steep function of the cluster redshift for a given source redshift, galaxies of a given redshift z_s "feel" the effects of a broad range of lens redshifts z_l. This means that unrelated clusters aligned along the line of sight will be detected as if they were one cluster with $M \lesssim M_1 + M_2$. Furthermore, because the large scale structure is filamentary, there will be a contribution to the mass of clusters from filaments aligned along the line of sight, a contribution that cannot be separated from the mass of the cluster itself. Because the contrast between the filament contribution and the cluster contribution scales as the cluster mass, this is expected to be an ever-increasing problem as we work our way down the cluster mass function.

4. Mass clustering in the Deep Lens Survey

4.1. *Survey description*

The Deep Lens Survey (Wittman *et al.* 2002, see also http://dls.het.brown.edu) is a 5 year project (now in its fifth year) that is using the NOAO Mosaic I and II cameras on the 4m telescopes on Kitt Peak and Cerro Tololo to image 24 square degrees of the sky in B, V, R, z′. Total exposure times are 18000 s in R, and 12000s each in each of the three other bands. However, because of the gaps in the CCD Mosaics, and the fact that the regions we seek to image (subfields) are 40′ by 40′ compared to a single exposure footprint of 36′ by 36′, the effective depth of any part of the survey is only about 14500s in R, and about 10000s in the other bands. This results in a 1σ surface brightness limit in R of 28.7 magnitudes per square arcsecond. In terms of source density, this implies roughly 45 resolved sources per square arcminute.

The individual exposures are kept to 600 seconds (900 in R) in order to avoid too high a sky background and to provide temporal coverage (for the transient object part of the DLS survey Becker *et al.* 2003, 2004). Exposures are typically taken in a sequence of 5 dithered pointings, offset to provide complete coverage through the gaps and also some ovelap with neighboring subfields. Image-to-image differences yield fast-time transients (asteroids, NEOs, bursters and flare stars). Night-to-night, month-to-month, or year-to-year differences yield catalogs of Supernovae, variable stars and AGN. We have made strong efforts to produce subfields of even image quality–when the seeing is better than 0.9″ we observe in R, with the result that the entire R-band dataset is nearly uniform in image quality. This is very important for weak lensing because the number density of usable sources (as opposed to detected sources) depends quite strongly on the seeing in

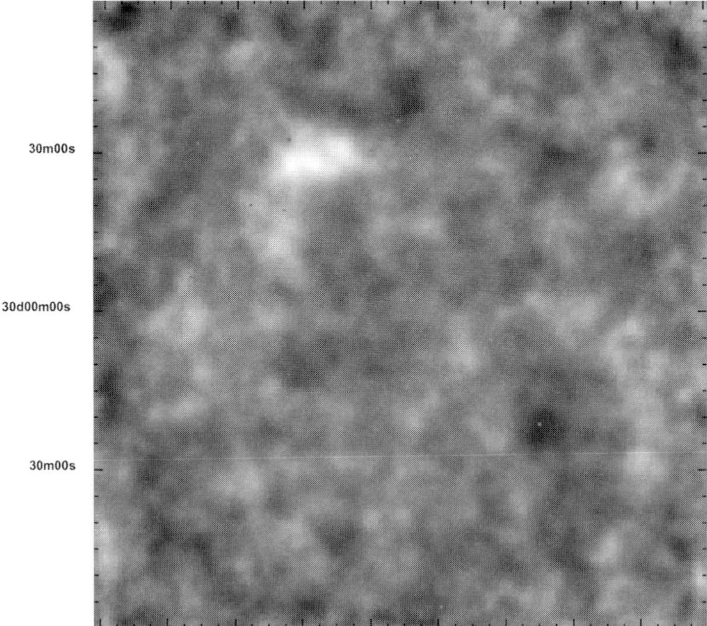

Figure 2. A map of the surface mass density κ for a 4-square degree region of the sky. Light regions represent regions of high surface mass density, dark regions are regions of low surface mass density.

this seeing/depth range. Therefore, non-uniformities in both the photometric depth and image quality result in non-uniformities in the mass sensitivity.

The shape and photometric properties of each individual exposure are calculated based on the properties of the unsaturated stars in the images. The typical density of stars in the images is 100/pointing, or roughly 300 per square degree. After the usual processing, all the images of a given subfield/filter combination are stacked, with each image being simultaneously convolved with a spatially varying circularizing kernel which makes the images of stars round.

4.2. *Lensing results*

Figure 2 shows the mass distribution in the first 2x2 degree field completed by the DLS. About 7×10^5 galaxies were used for this image, which has an effective resolution of about $2'$ by $2'$. The higest-mass concentrations in this region are being followed up spectroscopically with Gemini and through X-ray measurements. To date, we have had Chandra/XMM followup of 10 clusters.

The most obvious set of mass clumps in the image correspond to a known object, Abell 781 ($z = 0.29$). The cluster is morphologically very complex, with two other large mass subclumps in addition to the previously identified A781 core (these subclumps are also traced, albeit less strongly, by the galaxy density). Chandra imaging of the cluster (Hughes *et al.* 2004) shows the same three clumps in the X-rays. However, the M/L_x ratio differs considerably in each clump. The azimuthally averaged ellipticity induced by A781 can be traced out to $r \sim 20'$ from the cluster center, corresponding to a distance of 5 Mpc for our assumed cosmology. In addition to detailed studies of previously known clusters, the DLS has been uncovering new clusters. For example we (Wittman *et al.* 2003) recently discovered a cluster (CL1055-0348) (Figure 3) and estimated its redshift

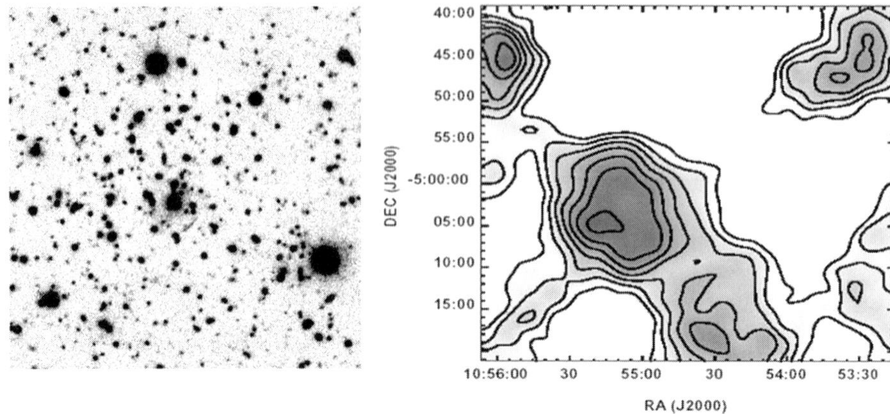

Figure 3. Left: the central region of CL1055-0503. A candidate arc can be seen below and to the right of the central galaxy. Right: The weak lensing map of that portion of the field–the higest peak corresponds to the CL1055-0503.

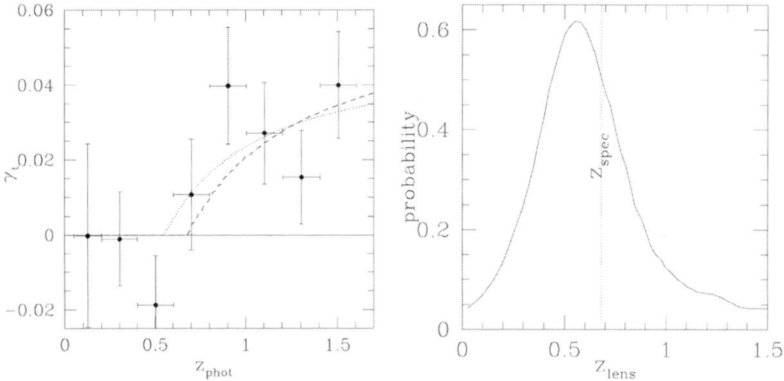

Figure 4. Left: the mean tangential shear for background galaxies as a function of their photometric redshift. Right: the probability distribution for the lens redshift as determined from the data from the left-hand plot (From Wittman *et al.* 2003).

solely from the weak lensing data combined with the photometric redshift estimates of the background galaxies (Figure 4). Spectroscopic follow-up with Keck confirmed the predicted redshift $z = 0.68$, demonstrating the power of weak lensing and photometric redshifts combined to reveal the three-dimensional distribution of mass in the Universe.

5. Conclusions and future developments

Wide-field surveys are providing unique data for the study of clusters of galaxies and their environment. As the depth to which these surveys reach has increased, the redshift range over which clusters can be detected has grown, allowing the study of the mass function of clusters and its evolution to be probed. In particular, the growth of weak gravitational shear as a way of detecting and measuring clusters is providing a clear look at the mass distribution of clusters and its evolution, one that is unbiased with respect to baryons (and, at least for the upper end of the mass-function), relatively insensitive to the large-scale structure distribution. Surveys such as the DLS and others have mapped around 0.1 percent of the sky. The next generation of instruments will provide all-sky maps at comparable (or greater) depth.

A whole host of wide-field surveys are in the different stages of development. Over the next five years we will see the large CFHTLS survey programs and the VST on Cerro Paranal, as well as VISTA in the NIR. On longer timescales, LSST, PanSTARRS and SNAP promise to provide deep information over the whole sky.

Acknowledgements

I would like to acknowledge my collaborators on the DLS– my co-PIs Tony Tyson and Dave Wittman, and also Vera Margoniner, Andy Becker, Dinesh Loomba, Gillian Wilson, Dara Norman, Jack Hughes, Jeff Kubo and Hossein Khiabanian. This work has been partially supported by NSF grant AST-0134753.

References

Becker, A.C.; DLS Collaboration, 2002, AAS 201.7802.

Becker, A.C, *et al.*, 2004, ApJ, in press.

Beckwith, S.V.W., *et al.*, 2004, in preparation.

Brown, M. J. I., *et al.*, , "Early-type Galaxy Clustering in the NOAO Deep Wide-field Survey", 2004, in Clusters of Galaxies: Probes of Cosmological Structure and Galaxy Evolution, from the Carnegie Observatories Centennial Symposia. Carnegie Observatories Astrophysics Series. Edited by J.S. Mulchaey, A. Dressler, and A. Oemler, 2004. Pasadena: Carnegie Observatories, http://www.ociw.edu/ociw/symposia/series/symposium3/proceedings3.html

CFHTLS Survey Home page: http://www.cfht.hawaii.edu/Science/CFHTLS/

Dalcanton, J.J., 1996, ApJ, 466, 92.

Erben, T., *et al.*, 2000, A&A 355, 23.

Gonzalez, A.H., Zaritsky, D., Dalcanton, J.J., Nelson, A.E., "The Las Campanas Distant Cluster Survey: First Results", in "Clustering at High Redshift", ASP Conference Series (ed. A. Mazure, O. Le Fevre, and V. Le Brun), 200, 416

Miyazaki, S., *et al.*, 2003, IAU colloquium 216, in press.

Jannuzi, B., NDWFS official webpage, http://www.noao.edu/noaodeep

Rhodes, J. *et al.*, 2003, astro-ph 0304417.

Scoville, N., COSMOS webpage, http://www.astro.caltech.edu/cosmos

Spergel, D., et al., 2003, ApJS, 148, 175.

Starr, B.M., *et al.*, 2000, Proc. SPIE, 4008, 1022.

Tyson, J.A., *et al.*, 2002, BAAS, 201.13401.

White, M., van Waerbeke, L., & Mackey, J., 2002, ApJ, 575, 640.

Williams, R.E., *et al.*, 1996, AJ, 112, 1335.

Williams, R., *et al.*, 1998, BAAS 193.7501.

Wittman, D., *et al.*, 1998, Proc. SPIE, 3355, 626.

Wittman, D., *et al.*, 2002, Proc. SPIE, 4836, 73.

Wittman, D., *et al.*, 2003, ApJ, 597, 218.

Wolfe, T., *et al.*, 1998, Proc. SPIE, 3355, 487.

Yee, H.K.C., Gladders, M.D., Lopez-Cruz, O., 1999, "The Color-Magnitude Relation of Early-Type Galaxies: A Tool for Cluster Finding and Redshift Determination", in ASP Conference Series, Vol 191, Proceedings of the OCIW/Pasadena Workshop on Photometric Redshifts, page 166.

York, D., *et al.*, 2000, AJ, 120, 1579.

Outskirts of Galaxy Clusters: Intense Life in the Suburbs
Proceedings IAU Colloquium No. 195, 2004
A. Diaferio, ed.

DOI: 10.1017/S1743921304000389

Optical/infrared observations of high-redshift galaxy clusters

H. K. C. Yee

Department of Astronomy and Astrophysics, University of Toronto, Toronto, Ontario, M5S 3H8, Canada email: hyee@astro.utoronto.ca

Abstract. Galaxy clusters at $z \sim 1$ are particularly important in the study of cluster evolution and the application of clusters to cosmological studies. We briefly discuss the difficulties in creating well-defined, large, high-redshift cluster samples, and some of the techniques used in optical/IR imaging to overcome them. We give a summary of the current state of optical/IR observations of the relatively small number of high-redshift clusters identified so far, including discussions of the galaxy luminosity function, the Fundamental Plane, the color-magnitude relation, and the Butcher-Oemler effect. The application of photometric redshift techniques is also highlighted.

1. Introduction

Galaxy clusters, being the largest mass concentrations in the Universe, have long played an important role in both the study of galaxy evolution and the determination of cosmological parameters. In the former, galaxy clusters offer a unique laboratory for investigating the relationship between galaxy formation and evolution and the environment; in the latter, the mass-to-light ratio and the evolution of the mass function of clusters allow us to measure the mass density parameter Ω_m, the amplitude of the perturbation spectrum σ_8 (e.g., Oukbir & Blanchard 1992), and the dark matter equation of state w (e.g., Levine et al. 2002, and many others). In both cases, the leverage and advantage offered by having a sample with a large redshift coverage are significant. In the case of cluster evolution, at the redshift of ~ 1 we are approaching the formation epoch of clusters, often put at $z \sim 2$ to 5, so that we would be in a much better position to observe the evolutionary effects. For the determination of cosmological parameters, a large baseline of redshift allows us to measure the evolution of the mass function, breaking the degeneracy between Ω_m and σ_8. The measurement of the mass function at $z \sim 1$ of a sufficiently large sample will also provide a chance to constrain w. However, the number of known high-redshift galaxy clusters, even just at $z > 0.8$, remains small, and these very often were found either as the tail of the distribution of a systematic survey, or by serendipity.

In this review we will examine the roles of optical/IR observations in the study of high-redshift galaxy clusters. (I will limit high-redshift to mean $z > 0.8$.) We will first address the important issue of defining a useful sample of clusters at high redshifts, specifically finding clusters using optical/IR imaging (Section 1). We then examine the still relatively small amount of current observations of clusters at $z \sim 1$. Section 3 focuses on imaging observations. This includes discussions of the cluster galaxy luminosity function, the Fundamental Plane of early-type galaxies, the color-magnitude relation, and the Butcher-Oemler effect. A brief discussion of spectroscopy and the application of photometric redshift techniques to high-redshift clusters is presented in Section 4. We end with a brief look at the future prospects of optical/IR observations of $z \sim 1$ clusters.

2. The search for high-redshift clusters

As recent as five or six years ago, the number of $z > 0.8$ clusters was on the order of a dozen. Most of these were found as the high-redshift tail of the distribution from X-ray surveys. A few were discovered by their association with $z \sim 1$ radio galaxies. The difficulty in finding high-redshift clusters is due to separate problems in the two major methods of finding clusters. For X-ray surveys, the available instruments simply do not go deep enough and cover sufficient areas. Rosati et al. (2002) provide a summary of the state of $z \sim 1$ clusters selected by X-ray observations as of a couple of years ago. For the more classical optical imaging surveys, the increased column of foreground galaxies presents a daunting problem of projection contamination which increases with redshift. Postman et al. (1996) used the so-called match-filter technique to optimize the identification of enhanced density clumps of galaxies that have the properties expected of clusters. Gonzalez et al. (2003) use an ingenious method of measuring background light fluctuations from shallow images from a small telescope as markers for possible high redshift over-density regions. However, these attempts to use optical imaging to search for high-redshift clusters suffer a typical contamination rate of $\sim$ 25 to 30%.

Gladders & Yee (2000) proposed a method of using two-filter optical imaging to search for high-redshift clusters to overcome the projection problem. This technique uses the observation that the cores of clusters are dominated by early-type galaxies which follow the well-known color-magnitude relation. Thus, by searching for galaxy density enhancements in successive slices of colors, one can effectively isolate early-type galaxies at different redshift intervals, removing the projection problem if the proper filters are chosen. The recent arrival of large mosaic CCD cameras on 4m-class telescopes also matches the depth and areal coverage required for meaningful high-redshift cluster searches.

The Red-Sequence Cluster Survey (RCS) is a 92 square degree imaging survey designed to optimize the detection of galaxy clusters with $0.5 < z < 1.2$, exploiting these recent developments. The survey, conducted with the CFHT 12k and CTIO Mosaic 8k cameras, obtained images in z' and R_C filters, which straddle the 4000Å break for early-type galaxies at $0.45 < z < 1.3$. The survey technique is very successful at identifying clusters at redshifts up to ~ 1. The redshift distributions of clusters for different richness bins from the first catalogs of $\sim$ 10 square degrees (Gladders & Yee 2004) are shown in Figure 1. Spectroscopic confirmations of over 60 clusters with redshifts from 0.2 to 1.1 have been obtained. Blindert et al. (2004, these proceedings) show velocity histograms for seventeen $0.2 < z < 0.65$ clusters, demonstrating that all but one (which has poor data) are consistent with significant clumps in velocity space. Extrapolating from the well-analysed 10 square degrees from Gladders & Yee (2004), we can expect the order of 350 clusters of richness Abell 0 and higher at redshift between 0.8 and 1.2 to be found in the RCS, greatly expanding the number of known $z \sim 1$ clusters.

The RCS demonstrates the efficiency of searching for clusters using the red-sequence as a marker of excess galaxy density. Similar techniques using a larger number of filters to search for clusters in multi-color/angular position phase space have also been developed for the lower redshift data of SDSS. With the commissioning of the MegaCam, a 36-CCD, one square degree field-of-view camera at the CFHT, we have begun the RCS2, a 1000 square degree optical cluster survey. This survey, modeled after the RCS but with somewhat shallower images in z', r, and g, will discover approximately 30,000 clusters, with 1000's at $z \sim 1$. A main scientific goal is to measure the cluster mass function evolution with sufficient statistical accuracy to constrain w, the equation of state of dark energy (see, e.g., Levine et al. 2002). The RCS2 requires $\sim$ 43 nights of telescope time, and is expected to be completed by the end of 2005.

They found that the colors of the CMR have evolved only moderately compared to that of Coma, by about a rest frame $\Delta(U - J)\sim 0.5$, and that the slope and dispersion of the CMR at $z \sim 0.8$ are similar to lower z clusters. Subsequently, various authors added a small number of higher redshift clusters (e.g., van Dokkum et al. 2001; Blakeslee et al. 2003; Lidman et al. 2004). Holden et al. (2004) extended the redshift range of Stanford et al. (1998) using six clusters with HST images with redshifts between 0.78 and 1.27 with IR and optical photometry and HST morphologies, with the primary result that the last significant episode of star formation in early-type galaxies is likely at $z > 3$.

Gladders et al. (1998) demonstrated that the slope of the CMR can provide constraints on the formation epoch of early-type galaxies in clusters. Combining a sample of $z \sim 0.1$ Abell clusters and HST imaging of a few high-redshift clusters up to $z \sim 0.85$, they compared the measured slope to model slopes from Kodama & Arimoto (1997), and concluded that, at least for the relatively rich clusters at $z \sim 0.8$, their cores were formed at $z > 2.5$. However, the small sample at the higher redshifts clearly does not preclude a scatter of slope values which may arise from cluster cores having a variety of formation epochs and durations.

It should be noted that all the high-redshift clusters studied so far are basically the same set of about a dozen clusters at $z > 0.8$, and these are in general extremely rich or X-ray luminous clusters. More recently, Gladders et al. (2004, in preparation) examined the dispersion of the $i - K$ CMR from a large sample or RCS clusters at $z \sim 1$, with the intriguing preliminary result showing that the dispersion of the CMR appears to be larger for poorer clusters. If this result is confirmed, it would suggest that less massive clusters may have a more recent formation time, compared to the massive ones.

3.4. *The Butcher-Oemler effect*

The Butcher-Oemler (BO) effect (Butcher & Oemler 1984) was amongst the first observational evidence that galaxies evolve over cosmic time. The increase in the blue galaxy fraction (f_b) with redshift has been connected to the greater amounts of star formation in galaxies in higher-redshift clusters. It is increasing clear that this higher blue fraction is primarily a reflection of the evolution of the infalling population (as opposed to the evolution of the CMR, which can be considered the "original" population more closely associated with the formation of the cluster core). In the simplest form the BO effect can be interpreted as the result of a combination of cosmologically driven changing galaxy infall rate and the truncation of star formation in the infalling galaxies (Ellingson et al. 2001), without reference to an evolving mechanism that turns galaxies red. However, despite two decades since its original discovery, both the magnitude (or existence) and the interpretation of the BO effect are still being debated vigorously.

Rakos & Schombert (1995) pushed the BO effect to $z \sim 0.9$ (with three clusters at $z > 0.8$) using photometry in the rest frame Stromgren filter system, and found that there is a general increase of f_b, from ~ 0.1–0.2 at low-z to as high as ~ 0.9 at $z \sim 0.9$. De Propris et al. (2003) argued that a strong BO effect is not observed in the De Propris et al. (1999) sample, when the cluster galaxies are selected by K-band magnitude, concluding that the BO effect mostly occurs in low-mass galaxies. However, it also appears that, based on their plots, one cannot rule out a BO effect for their sample. The f_b computed using a large subsample of RCS clusters with $0.4 < z < 1.2$ with z'-band selected galaxies inside r_{200} shows a general increase up to $f_b \sim 0.8$. The seemingly confusing and conflicting results from the many investigations in the BO effect likely arise from the very different operational definitions used to measure the blue fractions. For example, it is clear that the cluster-centric radius used to sample the galaxies greatly influences the results (e.g., see Ellingson et al. 2001). A dynamically defined radius such as

r_{200}, where one would sample galaxies in similar cluster dynamical state, would be a much more consistent way of selecting galaxies, compared to, say, a fixed metric radius (see, e.g., Yee 2003). Andreon et al. (2004) give an interesting discussion on the importance of the many aspects that impact on the measurement of the BO effect, including the magnitude limit, color-cut, cluster radius, and error estimation.

4. Spectroscopic observations

4.1. *Spectroscopy*

Spectroscopy of cluster galaxies provides two crucial sets of information: the redshifts – allowing membership information (though still not necessarily definitive) and dynamical analysis; and spectral features – providing much more detailed information on stellar populations than that from just colors. However, at $z \sim 1$, spectroscopic data with coverage and sample size similar to those at lower redshift is difficult and expensive. Currently, most $z > 0.8$ clusters have on the order of a dozen or fewer redshifts per cluster (with the exception of MS1054–03 with 30 redshifts, Wuyts et al. 2004). The EDisCS survey (Halliday et al., this conference) contains detailed spectroscopic studies of a set of $z \sim 0.7$ to 0.8 clusters. Spectroscopic data sets at $z \sim 1$ comparable to the moderate redshift ($0.2 < z < 0.6$) cluster samples of the CNOC1 (Yee et al. 1996) and the MORPH collaboration (Dressler et al. 1999) are still a few years away.

4.2. *Very low-resolution spectroscopy: photometric redshift*

The difficulties in obtaining spectroscopic data for a large sample of galaxies in high-redshift clusters make the application of photometric redshift techniques an attractive method for studying these clusters. Typical uncertainty of photometric redshifts is much larger than the velocity dispersion of a rich cluster; hence, photo-z's can only be considered as a tool to limit the foreground and background contamination. Statistical background corrections based on a blank field galaxy sample with identical photo-z cuts are still needed. The application of photo-z techniques is especially beneficial for the study of the outskirts of clusters and high-redshift clusters. In both cases the relative ratio of the numbers of field and cluster galaxies is much larger. Using a photo-z sample of cluster galaxies should improve the derivation of the cluster galaxy LF and various analyses of galaxy populations based on photometric/color data. Furthermore, for high-redshift clusters, photo-z's will greatly improve the efficiency of multi-object spectroscopy, especially in the outskirts of clusters, in that galaxies can be preselected for observation based on photo-z, greatly reducing the number of non-cluster spectra. A number of projects using photo-z techniques to select cluster members are discussed in these proceedings (e.g., Kodama et al., Nakata et al., Li et al.).

5. Future prospect and summary

The many results from different wavelength regimes indicate to us that galaxy clusters are intrinsically complex entities. They are intimately intertwined with their large scale environment, and the physical processes involved, from the outskirts to the core, are often complex and not subject to simple observations and analysis. Hence, to study the evolution of clusters, investigations using considerably larger samples (by more than an order of magnitude) than what have been done are essential to unravel the workings of these large aggregates of galaxies. The most robust preliminary conclusion about $z \sim 1$ clusters is that the core population of the rich clusters were likely formed at $z > 3$.

In the last few years there has been a general convergence of developments which will generate considerable advances and excitement in the study of high-redshift galaxy clusters. Large and well-defined cluster surveys, in optical/IR, X-ray, and SZ-effect, are poised to produce significant $z \sim 1$ (and beyond) cluster samples. Instrumentation on 8–10m-class telescopes increasingly provide powerful tools to do detailed studies on these high-redshift clusters. An example is the wide-field multi-object spectrograph IMACS on the Magellan 6.5-m telescope, which offers the capability of obtaining up to 1000 spectra in a single observation. Similar capabilities are also being made available on the VLT, and the Keck and Subaru telescopes. The RCS has provided a large sample of clusters at $z > 0.8$ for which extensive spectroscopic, multi-color optical and IR-imaging are being carried out, along with some X-ray and SZ effect imaging. This, plus many projects being carried out at other large telescopes, will no doubt bring us closer to understanding the processes affecting the evolution of galaxy clusters.

Acknowledgements

I would like to thank my RCS collaborators. The RCS project is supported in part by operating grants from NSERC, from the University of Toronto, and from the Canada Research Chair Program.

References

Andreon, S., Lobo, C., & Iovino, A. 2004 *MNRAS*, **349**, 889–898.
Aragon-Salamanca, A., Ellis, R.S., Couch, W.J., & Carter, D. 1993 *MNRAS*, **262**, 764–794.
Arimoto, N. & Yoshii, Y. 1987 *A.&A.* **173**, 23–38.
Bachall, N., et al. 2003 *Ap.J.S.*, **148**, 243–273.
Blakeslee, J.P. et al. 2004 *Ap.J.*, **599**, L143–L146.
Bower, R., Lucey, J.R. & Ellis, R.S 1993 *MNRAS* **254**, 601–613.
Butcher, H. & Oemler, A., Jr. 1984 *Ap.J.* **285**, 426–438.
De Propris, R., Stanford, S.A., Eisenhardt, P.R., & Dickinson 2003 *Ap.J.* **598**, 20–35.
De Propris, R., et al. 1999 *A.J.* **111**, 719–729.
Dressler, A., et al. 1999 *Ap.J.S.* **122**, 51–80.
Ellingson, E., Lin, H., Yee, H.K.C., & Carlberg, R.G 2001 *Ap.J.*, **547**, 609–622.
Ellis, S.C. & Jones, L.R. 2004 *MNRAS*, **348**, 165–175.
Gladders, M.D., Lopez-Cruz, O., Yee, H. K. C., & Kodama, T. 1998 *Ap.J*, **501**, 571–577.
Gladders, M.D. & Yee, H.K.C. 2000 *A.J.*, **120**, 2148–2162.
Gladders, M.D. & Yee, H.K.C. 2004 *Ap.J.S.*, in press.
Gonzalez, A.H., Zaritsky, D., Dalcanton, J.J., & Nelson, A. 2001 *Ap.J.S.* **137**, 117–138.
Holden B.P., Stanford, S.A., Eisenhardt, P.R., & Dickinson M. 2003 *A.J.* **127**, 2484–2510.
Kodama, T. & Arimoto, N. 1997 *A.&A.* **320**, 41–53.
Kodama, T. & Bower, R. 2003 *MNRAS*, **346**, 1–12.
Levine, E.S., Schulz, A.E., & White, M. 2002 *Ap.J.* **567**, 569–578.
Lidman, C. et al. 2004 *A.&A.* **416**, 829–837.
Oukbir, & Blanchard, A. 1992 *A.&A.* **262**, L21–L24.
Rakos, K.D. & Schombert, J.M. 1995 *Ap.J.* **439**, 47–59.
Rosati, P., Borgani, S., & Norman, C. 2002 *ARAA* **40**, 539–577.
Stanford, S.A., Eisenhardt, P.R., & Dickinson M. 1998 *Ap.J.* **492**, 461–479.
Postman, M., et al. 1996 *A.J.* **111**, 615–641.
Toft, S., Soucail, G., Hjorth, J. 2003 *MNRAS* **334**, 337–346.
van Dokkum, P. & Stanford, S.A. 2001 *Ap.J.* **585**, 78–89.
van Dokkum, P., Franx, M., Kelson, D.D., & Illingworth, G.D. 1998 *Ap.J.* **504**, L17–L21.
van Dokkum, P. & Stanford, S.A. 2003 *Ap.J.* **585**, 78–89.
Wuyts, S., van Dokkum, P., Kelson, D., Franx, M., & Illingworth, G. 2004 *Ap.J.* **604**, 677–688.
Yee, H.K.C., & Ellingson, E. 2003 *Ap.J.*, **585**, 215–226.
Yee, H.K.C., Ellingson, E., & Carlberg, R.G. 1996 *Ap.J.S.*, **102**, 269–287.
Yee, H.K.C. 2003, *Astrophys. & Space Science*, **285**, 269-277.

Outskirts of Galaxy Clusters: Intense Life in the Suburbs
Proceedings IAU Colloquium No. 195, 2004
A. Diaferio, ed.

DOI: 10.1017/S1743921304000390

Wide–field spectroscopy of A1689 and A1835 with VIMOS: first results

Oliver Czoske

IAEF, Universität Bonn, Auf dem Hügel 71, 53121 Bonn, Germany

Abstract. We are using VIMOS to conduct a wide-field spectroscopic survey covering fields of $0.5 \times 0.5\,\mathrm{deg}^2$ around the X-ray luminous clusters of galaxies Abell 1689 ($z = 0.185$) and Abell 1835 ($z = 0.25$). Here we describe the observations and first results on the redshift distribution of subsamples of cluster galaxies to $R \simeq 22$ for which we at present have obtained secure redshifts. These subsamples constitute $\sim 40\%$ of the total spectroscopic sample and contain 525 and 630 cluster members in Abell 1689 and Abell 1835, respectively, placing them amongst the largest redshift samples available for any cluster of galaxies.

1. Introduction

Imaging surveys of the infall regions of clusters of galaxies which use photometric redshift techniques to isolate cluster galaxies have made strong progress in recent years (e. g. Kodama et al. 2001; Gray et al. 2003). While photometric surveys have the advantage of being complete down to well below M^*, only spectroscopic surveys can add a third dimension, velocity, to the galaxy distribution in and around clusters. The largest wide–field spectroscopic samples at present exist for near-by clusters, in particular as obtained by the CAIRNS project (Rines et al. 2003). An example for a previous wide–field redshift survey at higher redshift is described by Czoske et al. (2001, 2002), who present a catalogue of redshifts for 300 cluster members with $V \lesssim 22$ in Cl0024+1654 at $z = 0.395$. In that case, it was only the redshift information extending to large cluster-centric distances which revealed the complex structure of what appeared in other observations to be a relaxed rich cluster.

The recent advent of high-multiplex spectrographs on 8–10 meter class telescopes has made it possible to obtain large numbers of high-quality spectra of galaxies and around clusters of galaxies in a short amount of time. The data described by Czoske et al. (2001) were obtained over the course of four years. Samples larger by a factor of $2 \ldots 3$ can now be obtained in ~ 10 hours of observation time. Here I present the first results from a spectroscopic survey of the two X-ray luminous clusters Abell 1689 and Abell 1835 at redshifts $z = 0.185$ and $z = 0.25$ respectively. We use the VIsible imaging Multi-Object Spectrograph (VIMOS) on VLT UT3/Melipal.

2. Observations

The field of view of VIMOS available for spectroscopy consists of four quadrants of $\sim 7 \times 7\,\mathrm{arcmin}^2$, the separation between the quadrants is $\sim 2\,\mathrm{arcmin}$. Using the LR-Blue grism, one can place $\sim 100 \ldots 150$ slits per quadrant. The resulting spectra cover the wavelength range $3700 \ldots 6700\,\text{Å}$ with a resolution $R \simeq 200$.

We use as the basis for object selection panoramic multi-colour images obtained with the CFH12k camera on CFHT (Czoske 2002), covering $40 \times 30\,\mathrm{arcmin}^2$ in BRI for A1689 and *VRI* for A1835. The input catalogue has been cleaned of stars (Bardeau et al. 2004, in preparation).

We attempted to cover the entire CFH12k field of view by using 10 VIMOS pointings for each cluster. Due to technical problems with VIMOS only 8 and 9 masks were observed in service mode for A1689 and A1835, respectively. Four of these masks did not fully meet the observational requirements although the spectra are at least partly usable.

In the observed masks, slits were placed on 3373/3542 objects in A1689 and A1835 respectively. The sample was cut at $R = 23$, where objects fainter than $R = 22$ were allowed to appear on several masks to provide higher signal-to-noise for these faint objects. For each mask three exposures were obtained for a total integration time of 54 minutes. The data were reduced using the VIPGI pipeline (Scodeggio et al., in preparation†). The system subtracts a bias frame, automatically localises the 2D spectra on the chip and determines their curvature. After wavelength calibration from HeNeAr standard lamp observations the 1D spectra were optimally extracted.

Redshift determination involves an identification step using cross correlation with a set of templates and a subsequent interactive check of the correctness of the redshift identification, and a measurement step using again a correlation with the allowed redshift range constrained to a narrow interval around the correct identification. At the time of writing secure redshifts have been determined for 1312/1427 objects in the two clusters. All further results presented in this contribution will be based on these subsets which reach $R \sim 22$ and constitute $\sim 40\%$ of the total samples.

3. Redshift space structure

3.1. *Abell 1689*

Abell 1689 is one of the richest ($R = 4$) clusters in the Abell catalogue and shows a complex structure in its inner parts, including a large number of strongly gravitationally lensed arcs and arclets. Its X-ray luminosity and temperature are $L_X = 5.2 \times 10^{44}\, h^{-2}\, \mathrm{erg\, s^{-1}}$ (Ebeling et al. 1996) and $T_X = 9.3\, \mathrm{keV}$ (Andersson & Madejski 2004), respectively. Based on 96 redshifts in the cluster centre, Girardi et al. (1997) divide the redshift distribution into three distinct subgroups.

The left panel of Fig. 1 shows the redshift distribution for galaxies near the mean cluster redshift of Abell 1689 as derived from our VIMOS data. We include redshifts from the literature and previous observations of the cluster centre using 4m class telescopes. We define cluster membership by the redshift range $0.169 < z < 0.202$ and find 525 cluster members in the subsample.

The distribution on large scales is very regular, with clearly defined "caustics" whose amplitude decreases strongly with clustercentric distance. This is also reflected in the velocity dispersion profile which decreases from $\simeq 2100\, \mathrm{km\, s^{-1}}$ in the central bins to $\simeq 1200\, \mathrm{km\, s^{-1}}$ for the outer bins. The redshift histogram in conjunction with the redshift-distance scatter plot hints at the presence of a distinct subgroup in the cluster center at a clustercentric velocity of $\sim 3000\, \mathrm{km\, s^{-1}}$ which might artificially enhance the central velocity dispersion.

In order to probe for the presence of substructure in Abell 1689 we apply the test described by Dressler & Shectman (1988). This test computes for each galaxy the statistic $\delta \sim [(\overline{v}_{\mathrm{local}} - \overline{v})^2 + (\sigma_{\mathrm{local}} - \sigma)^2]$ to compare the local velocity distribution based on the 10 nearest neighbours to the global distribution. The left panel in Fig. 2 shows the result

† `http://cosmos.mi.iasf.cnr.it/marcos/vipgi/vipgi.html`

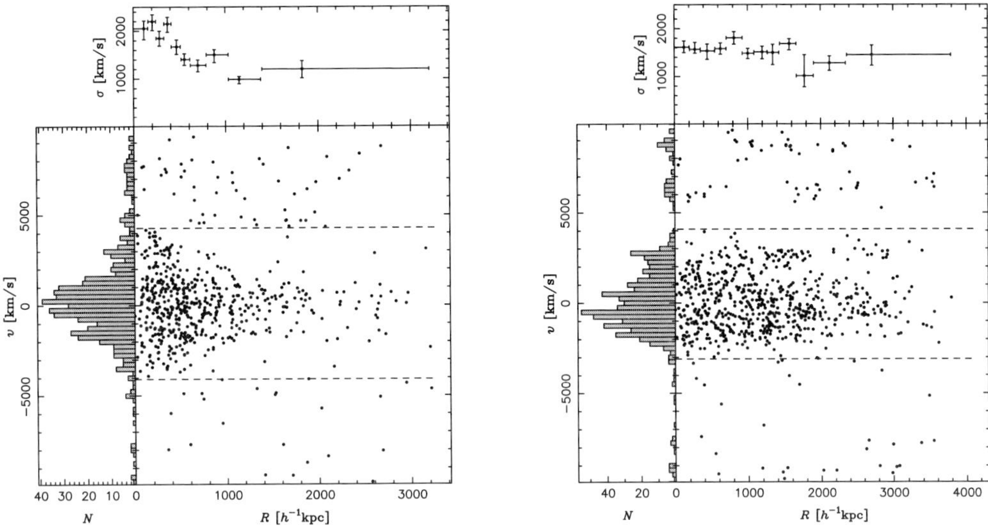

Figure 1. The redshift distribution for Abell 1689 (left) and Abell 1835 (right). Redshifts are given as cosmologically corrected velocities relative to the mean cluster redshifts. The velocity dispersion profiles in the top panels were computed in bins containing 50 galaxies each within the cluster limits marked in the central panel.

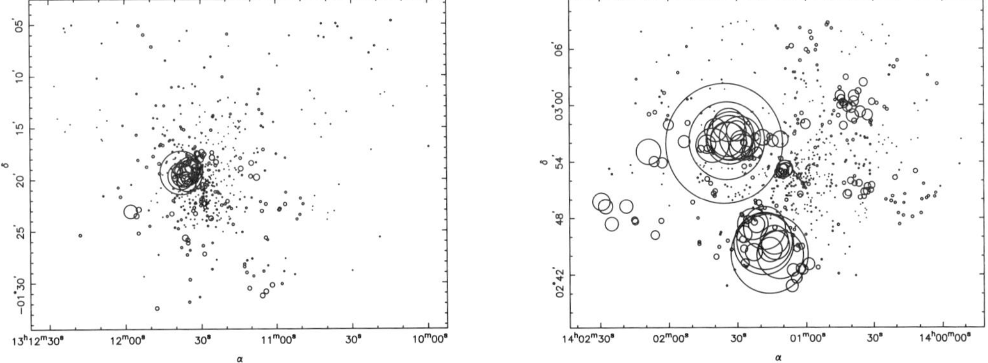

Figure 2. Dressler-Shectman plots for Abell 1689 (left) and Abell 1835 (right). The circles have radii proportional to $\exp(\delta)$, where δ is the Dressler-Shectman statistic (see text for details).

for Abell 1689. The only apparent distinct group of galaxies lies at $\sim 2'$ to the north-east of the cluster center. The location of this group, which has been seen in previous redshift samples, corresponds to a group of bright galaxies in images of the cluster. The distribution of redshifts of these galaxies is skewed towards higher redshifts: 51% of them lie at $z > 0.192$, whereas only 13% of all cluster members lie at this redshift. The distribution of all galaxies with $z > 0.192$ covers a much larger field but seems concentrated on the location identified by the Dressler–Shectman test. Andersson & Madejski (2004) use X-ray data obtained with XMM/Newton to derive a redshift map for the emitting intra-cluster gas and find evidence for a higher redshift in their eastern quadrant, roughly coinciding with our substructure.

3.2. *Abell 1835*

Abell 1835, with $9.6 \times 10^{44}\,h^{-2}\,\mathrm{erg\,s^{-1}}$, is the most luminous cluster in the ROSAT Brightest Cluster Sample (Ebeling et al. 1998). Temperature measurements with ASCA and XMM-Newton find a comparatively modest global temperature of $\sim 7.5\,\mathrm{keV}$ (Ota 2001; Majerowicz et al. 2002) with a cool core of about 4 keV. The surface brightness distribution is nearly circular to a radius of $\sim 0.8\,h^{-1}\,\mathrm{Mpc}$. From observations of the central parts of Abell 1835 it therefore appears that this is a very massive and relaxed cluster.

The redshift distribution around Abell 1835 is shown in the right hand panel of Fig. 1. Defining cluster membership by $0.24 < z < 0.27$, we find 630 cluster members. The distribution of Abell 1835 seems more extended than the distribution of Abell 1689, although a detailed investigation of the completeness of the samples is necessary to assess the validity of this statement. The velocity dispersion profile (VDP) of the cluster galaxies in Abell 1835 is flat at $\sim 1500\,\mathrm{km\,s^{-1}}$. From the histogram and the redshift–distance plot it is clear, however, that the VDP at large radii includes a contribution from a seemingly separate band of redshifts at $\sim 3000\,\mathrm{km\,s^{-1}}$ and might thus be overestimated at $R \gtrsim 1\,h^{-1}\,\mathrm{Mpc}$.

The result of the Dressler-Shectman test on Abell 1835 is shown in the right hand panel of Fig. 2. It is immediately apparent that the infall region is much more structured than the one of Abell 1689, indicating that these clusters inhabit rather different environments. The Dressler-Shectman test provides evidence for at least two distinct structures in the outskirts of Abell 1835, one $\sim 1.5\,h^{-1}\,\mathrm{Mpc}$ to the east of the cluster center, the other $\sim 1.4\,h^{-1}\,\mathrm{Mpc}$ to the south-east. The eastern structure is centred on a compact group of eight galaxies, all with $z > 0.26$, significantly higher than the mean cluster redshift. The southern structure is centred around a bright elliptical galaxy for which unfortunately no redshift is available. The mean redshift of the galaxies in this group is 0.245, somewhat lower than the cluster redshift.

4. Conclusions

The project described in this contribution is still at an early stage. So far, secure redshifts have been obtained for only $\sim 40\%$ of the observed samples. Nevertheless, the subsamples presented here include 525 and 630 members galaxies of the clusters Abell 1689 and Abell 1835, respectively. Already, these samples are the largest available for any cluster with the exception of Coma (Rines et al. 2003), and demonstrate the power of VIMOS to provide unprecedented redshift samples for clusters of galaxies at intermediate redshifts.

We have described the redshift distribution for the two clusters out to $\gtrsim 3\,h^{-1}\,\mathrm{Mpc}$ ($\gtrsim 2r_{\mathrm{vir}}$). The infall regions of the two clusters are markedly different, with Abell 1835 showing evidence for strong clustering in its outskirts whereas the outskirts of Abell 1689 seems much more homogeneous. Curiously, this is at contrast to the structure of the central parts of the clusters which is complex for Abell 1689 but seems relaxed for Abell 1835. The spatially more concentrated galaxy distribution in Abell 1689 as compared to Abell 1835 agrees with results from weak lensing analyses by Clowe & Schneider (2001, 2002), who find concentration parameters of $c = 7.9$ for Abell 1689 and $c = 2.9$ for Abell 1835.

The first goal of future work is to complete the redshift determination for the spectral samples and thus to extend the study of the cluster structures to fainter magnitude. After proper assessment of the completeness characteristics of the resulting galaxy catalogues it will be possible to provide mass profiles based on integration of the Jeans equation in the inner parts and analysis of the "caustic" structure in the outer parts, following Diaferio (1999), which can then be compared to mass profiles derived from weak lensing (Clowe & Schneider 2001, 2002; Bardeau et al. 2004).

Further work will relate the spectral types of the galaxies to cluster-centric distance, the local galaxy density and surface mass density as determined from weak gravitational lensing.

References

Andersson, K. E. & Madejski, G. M. 2004, astro-ph/0401604

Bardeau, S., Kneib, J.-P., Czoske, O., Richard, J., Soucail, G., Smail, I., Ebeling, H., & Smith, G. P. 2004, in preparation

Clowe, D. & Schneider, P. 2001, *Astron. & Astroph.* , 379, 384

—. 2002, *Astron. & Astroph.* , 395, 385

Czoske, O. 2002, PhD thesis, Université Toulouse III – Paul Sabatier

Czoske, O., Kneib, J.-P., Soucail, G., Bridges, T. J., Mellier, Y., & Cuillandre, J.-C. 2001, *Astron. & Astroph.* , 372, 391

Czoske, O., Moore, B., Kneib, J.-P., & Soucail, G. 2002, *Astron. & Astroph.* , 386, 31

Diaferio, A. 1999, *Mon. Not. Roy. Ast. Soc.* , 309, 610

Dressler, A. & Shectman, S. A. 1988, *Astron. Journ.* , 95, 985

Ebeling, H., Edge, A. C., Böhringer, H., Allen, S. W., Crawford, C. S., Fabian, A. C., Voges, W., & Huchra, J. P. 1998, *Mon. Not. Roy. Ast. Soc.* , 301, 881

Ebeling, H., Voges, W., Böhringer, H., Edge, A. C., Huchra, J. P., & Briel, U. G. 1996, *Mon. Not. Roy. Ast. Soc.* , 281, 799

Girardi, M., Fadda, D., Escalera, E., Giuricin, G., Mardirossian, F., & Mezzetti, M. 1997, *Astroph. Journ.* , 490, 56

Gray, M. E., Wolf, C., Meisenheimer, K., Taylor, A., Dye, S., Borch, A., & Kleinheinrich, M. 2003, astro-ph/0312106

Kodama, T., Smail, I., Nakata, F., Okamura, S., & Bower, R. G. 2001, *Astroph. Journ.* , 562, L9

Majerowicz, S., Neumann, D. M., & Reiprich, T. H. 2002, *Astron. & Astroph.* , 394, 77

Ota, N. 2001, PhD thesis, University of Tokyo

Rines, K., Geller, M. J., Kurtz, M. J., & Diaferio, A. 2003, *Astron. Journ.* , 126, 2152

Outskirts of Galaxy Clusters: Intense Life in the Suburbs
Proceedings IAU Colloquium No. 195, 2004
A. Diaferio, ed.

DOI: 10.1017/S1743921304000407

Optical and X-ray analysis of the galaxy cluster A3921

C. Ferrari[1,2]†, C. Benoist[2], S. Maurogordato[2], A. Cappi[3], E. Slezak[2], E. Belsole[4,5]‡, J. L. Sauvageot[4], H. Bourdin[2] and G. W. Pratt[4,6] ¶

[1] Institut für Astrophysik, Universität Innsbruck, Technikerstraße 25, A-6020 Innsbruck, Austria

[2] CERGA, UMR 6527, CNRS, Observatoire de la Cte d'Azur, BP 4229, Le Mont-Gros, 06304 Nice Cedex 4, France

[3] INAF, Osservatorio Astronomico di Bologna, via Ranzani 1, I-40127 Bologna, Italy

[4] Service d'Astrophysique, CEA Saclay, L'Orme des Merisiers Bât 709, F-91191 Gif-sur-Yvette, France

[5] H.H. Wills Physics Laboratory, University of Bristol, Tyndall Avenue, Bristol BS8 1TL, U.K.

[6] Max-Planck-Institut für extraterrestrische Physik, Postfach 1312, 85741 Garching, Germany

Abstract. We present the analysis and results of a new spectroscopic and photometric survey of the central 1.5 Mpc^2 region of the galaxy cluster A3921 (z=0.094). We detect the presence of two dominant clumps of galaxies with a mass ratio of ∼4-5: a main cluster centered on the BCG (A3921-A), and a NW sub-cluster (A3921-B) hosting the second brightest cluster object. By comparing optical results to the X-ray analysis of XMM observations (Belsole *et al.* 2004), we find that A3921-B is tangentially traversing the main cluster along a SW/NE direction. The analysis of stellar populations of more than one hundred cluster members reveals that the merger event has probably affected the kinematics and spectral properties of the active galaxies in A3921.

1. Introduction

In the standard cosmological scenario of hierarchical structure formation, galaxy clusters form by the merger of less massive systems. Combined optical and X-ray studies have been particularly successful in revealing the complex dynamics of merging clusters (e.g. Davis *et al.* 1995, Durret *et al.* 1998, Arnaud *et al.* 2000). This field is more and more active since precise spectro-imaging data in X-rays is now available with Chandra and XMM, allowing us to derive spatially resolved temperature and density maps, in which very typical signatures of merging have been detected.

In this paper we present the analysis of new spectroscopic and imaging data relative to the cluster A3921 with EFOSC2 and WFI at the ESO 3.6 and 2.2 m telescopes respectively (Ferrari *et al.* 2004). These optical results are compared with the new XMM-Newton X-ray analysis of the cluster (Belsole *et al.* 2004).

† This research was supported in part by Marie Curie individual fellowships MEIF-CT-2003-900773.
‡ This research was supported in part by PPARC.
¶ This research was supported in part by Marie Curie individual fellowship MEIF-CT-2003-500915.

2. Optical morphology

In the top panel of Fig. 1, the iso-density contours of the projected distribution of galaxies belonging to the red-sequence of A3921 up to R=18.2 (corresponding to R=R*+2) are overlaid on the R-band image of the central field of the cluster. A3921 is clearly characterized by a bimodal morphology, with two main sub-clusters (hereafter A and B), as confirmed by the application of a classical segmentation algorithm to the iso-density map of A3921. Using a threshold of 3 times the r.m.s of the map, this algorithm detect two domains. For each of them, the position of the pixel with the highest density and the two first barycentric moments are estimated, allowing us to define the center and the orientation (*i.e.* the position angle) of each clump. A visual inspection of the optical iso-density contours reveals that the double morphology of A3921 is quite distorted. In particular, the clump A3921-B is characterized by a twisting of the iso-density contours from the internal to the external regions of the sub-cluster, and by the presence of a less dense Southern tail. This analysis is quantitatively confirmed by applying our segmentation algorithm with a threshold of 4 times the r.m.s of the map. In this case (dashed lines in the top panel of Fig. 1), the general elongation of the clump B varies significantly from the 3σ clipping case. A3921-B is more elongated towards the South for the fainter density cut, following the Southern tail of the clump and confirming a twisting of the optical iso-density contours. The shape of A3921-A is much more regular, but in both cases an offset of the optical density maxima from the two BCGs' positions is observed.

3. Pre- or post-merger?

The optical morphology of A3921 suggests that this cluster consists of two interacting systems. Therefore, in the following we will try to understand if they are in a pre- or post-merger phase.

3.1. *Dynamical analysis*

Both the velocity distribution of the whole cluster and those of the two clumps separately do not show any evidence for merging, as they are not significantly different from a Gaussian distribution. Moreover, we observe a very small difference in the mean velocities of the two sub-clusters (Δv$\simeq$90 km/s, well within the velocity dispersion of each individual subcluster). Using the estimate of the velocity dispersion of each sub-cluster, we have calculated the virial masses of A3921-A and B inside the two circles of the top panel of Fig. 1 (R$\simeq$0.4 Mpc), obtaining a quite high mass ratio between these two components ($\sim$4-5).

The analysis of the velocity distribution, however, does not answer our main question about the dynamical state of the cluster. We have thus applied the two-body dynamical formalism to the sub-structures of A3921. Because parametric solutions to the equations of motion can be derived both in the bound and unbound case, this method allows us to determine the dynamical state of the cluster by studying the solutions of the equation of motions as a function of α, *i.e.* the angle between the line connecting the two systems and the plane of the sky (Beers *et al.* 1982). We have made several tests under the hypothesis that a) the two sub-clusters have never collided before, or have collided b) 0.25, c) 0.50 and d) 1.00 Gyr ago. The results of these tests are summarized in Tab. 1.

3.2. *Comparison with X-ray results and discussion*

From Tab. 1 we can see that several solutions could explain the observed dynamics of A3921. A comparison with the X-ray properties of the cluster can allow us to understand which of them we can accept.

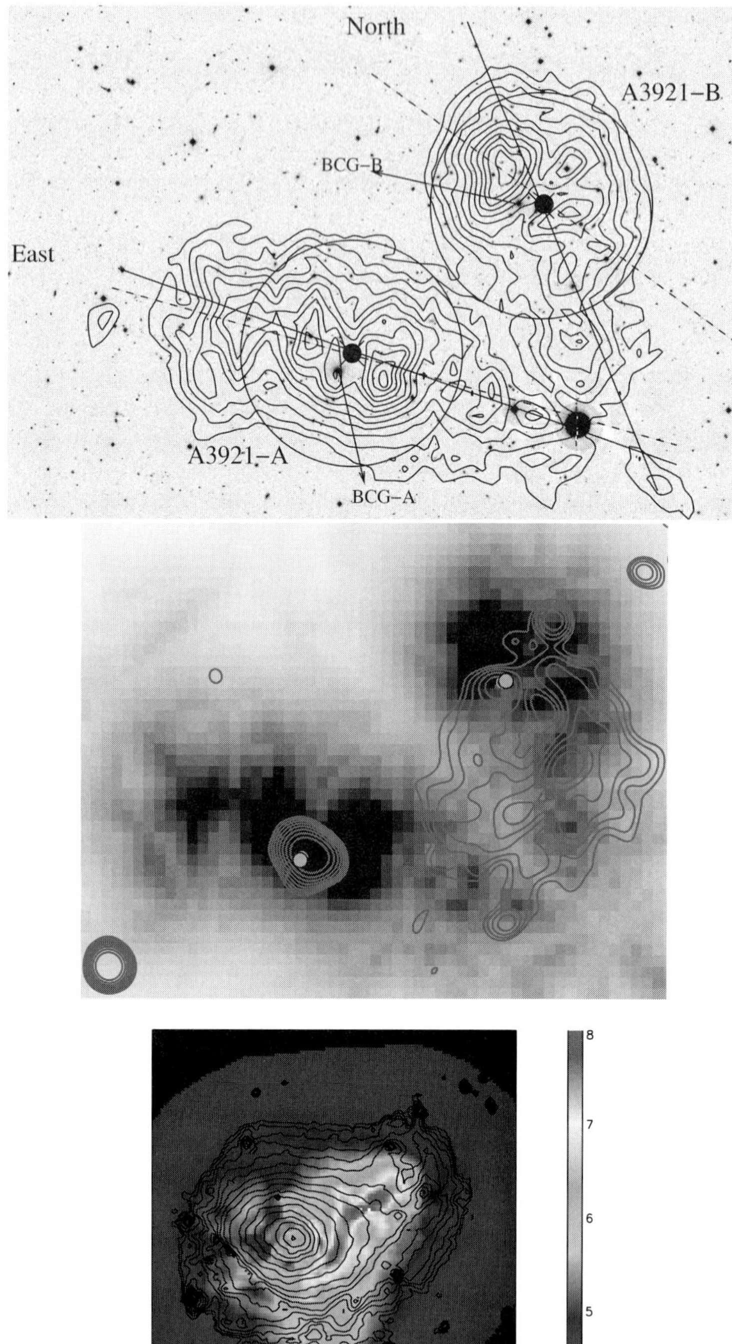

Figure 1. Top: Optical iso-density contours overlaid on the R–band image of A3921 central field (30×20 arcmin^2). The lines (solid and dashed) and the black small circles show respectively the orientation and the centre of the two subclusters (see text). The dynamical analysis of each clump has been done on the galaxies inside the two black circles (R$\simeq$ 0.4 Mpc). The positions of the brightest galaxies of the two sub-clusters are also shown – **Middle:** X-ray residuals above a 2D-β model overlaid on the red sequence galaxy iso-density map up to R=R*+2. The positions of the brightest galaxies of the two sub-clusters are indicated by small circles – **Bottom :** Temperature map of A3921 obtained by applying the wavelet algorithm described in Bourdin *et al.* 2004.

Scenario	T_0 [Gyr]	V [km/s]	R [Mpc]	α [deg]	Solutions
a1	12.60	90.43	7.31	84.19	outgoing
a2	"	90.01	6.04	82.97	incoming
a3	"	2211	0.741	3.822	incoming
b	0.25	1582	0.741	3.29	outgoing
c	0.50	198	0.83	26.99	outgoing
d1	1.00	73.45	1.28	54.67	outgoing
d2	"	116.6	1.166	50.60	incoming
d3	"	1069.3	0.7424	4.175	incoming

Table 1. Possible solutions for the two-body dynamical model. Col.1: ID of the possible scenario – Col.2: time since last interaction between the two clumps – Col.3: relative velocity between the two clumps – Col.4: spatial separation between the two systems – Col.5: angle between the plane of the sky and the line connecting the centers of the two clumps – Col.6: state of the systems for the possible solutions

The analysis of XMM-Newton observations by Belsole *et al.* 2004 reveals that A3921 is characterized by a bimodal morphology in the X-ray domain too. The X-ray emission of A3921-A can be modelled with a 2Dβ-model, leaving a distorted residual structure toward the NW, coincident with A3921-B (middle panel of Fig. 1). The main cluster detected in X-ray is centered on the BCG position, while the X-ray peak of the NW clump is offset from the brightest galaxy of A3921-B. The temperature map of the cluster shows an extended hot region oriented parallel to the line joining the centres of the two sub-clusters (bottom panel of Fig. 1). A comparison of this image with numerical simulations by Ricker & Sarazin (2001) suggests that we are observing the central phases of an off axis merger between two unequal mass objects, where clump A is the most massive component, consistent with optical results.

By combining the signatures of merging derived both from the optical iso-density map and from X-ray results, we can now reconsider the solutions of the two-body dynamical model (Tab. 1). The high angle solutions of cases a1) and a2) would imply a very large real separation for the two sub-clusters, which is very unlikely taking into account the clear signs of interaction between the two clumps. We can also exclude the c) solution, as we expect a higher value of the relative velocity ($\geqslant$1000 km/s) between the two clumps for obtaining a so evident hot bar in the temperature map. Finally, the comparison of the observed and simulated temperature maps clearly exclude an older merger (e.g. t_0=1 Gyr ago), as we should not detect such obvious structure in the temperature map. Therefore, only the solutions corresponding to the very central phases of merging seem to be possible to explain all our observational results, implying a collision axis nearly perpendicular to the line of sight consistent with the observed velocity distribution.

The superimposition of the X-ray residuals on the optical iso-density map (middle panel of Fig. 1) shows that the bulk of X-ray emission in A3921-B is offset towards SW from the main concentration of galaxies. As numerical simulations show that the non-collisional component is much less affected by the collision than the gas distribution, this offset suggests that A3921-B is tangentially traversing A3921-A along a SW/NE direction, with its galaxies in advance with respect to the gaseous component. This off-axis collision scenario is consistent with the shape of the feature in the temperature map (Belsole *et al.* 2004). From a comparison of the galaxy density maps to those obtained in numerical simulations with different impact parameters before and after the collision, the orientation of the NW group tail also suggests that it has already tangentially crossed the main cluster (Ferrari *et al.* 2004). As the structure of the group seems preserved, the collision has probably occured on the edge of the cluster, confirming the results from the X-ray analysis.

4. Star-formation properties of A3921 galaxies

We have analysed the effects of the merging event on star formation in A3921 galaxies. Following the classification scheme by Dressler *et al.* (1999), we have divided the cluster members into passive (k-type), post-starburst (k+a and a+k types) and emission-line galaxies (e type). 76% of A3921 galaxies belong to the first group, while the remaining 24% show evidence of recent or ongoing star formation. In particular we find a high percentage of emission line galaxies with respect to similar studies on other low redshift clusters (e.g. Barrena *et al.* 2002). These objects seem to be localized in the region surrounding the hotter temperature bar detected in the X-ray. Even if a wider spectroscopic coverage is necessary to assess the presence of such a spatial correlation, a 2D Kolmogorov-Smirnov test rejects the hypothesis that the positions of k and e-types galaxies are drawn from the same parent population with an extremely low significance level ($\sim$0.1%). Moreover, by analysing separately the velocity distributions of passive and active galaxies, we find that a) active objects are characterized by a significantly higher velocity dispersion and a non-Gaussian velocity distribution, and b) a K-S test rejects the hypothesis that the two velocity datasets are drawn from the same parent population with a significance level of $\sim$8%. It is therefore possible that both the kinematics and spectral properties of the active galaxies in A3921 have been influenced by the merging event.

References

Arnaud, M., Maurogordato, S., Slezak, E., & Rho, J., 2000, A&A, 355, 461

Barrena, R., Biviano, A., Ramella, M., Falco, E.E., & Seitz, S., 2002, A&A, 386, 861B

Beers, T.C., Geller, M.J.& Huchra, J.P., 1982, ApJ, 257, 23

Belsole, E. et al., 2004, in preparation

Bourdin, H., Sauvageot, J.L., Slezak, E., Bijaoui, A., & Teyssier, R., 2004, A&A, 414, 429

Davis, D.S., Bird, C.M., Mushotzky, R.F., & Odewahn, S.C., 1995, ApJ, 440, 48

Dressler, A., Smail, I., Poggianti, B.M., Butcher, H., Couch, W.J., Ellis, R.S., & Oemler, A., 1999, ApJ, 122, 51

Durret, F., Forman, W., Gerbal, D., Jones, C., & Vikhlinin, A., 1998, A&A, 335, 41

Ferrari, C. et al., 2004, in preparation

Ricker, P.M., & Sarazin, C.L., 2001, ApJ, 561, 621

Outskirts of Galaxy Clusters: Intense Life in the Suburbs
Proceedings IAU Colloquium No. 195, 2004
A. Diaferio, ed.

DOI: 10.1017/S1743921304000419

The distribution of mass and light in cluster infall regions

K. Rines

Yale Center for Astronomy and Astrophysics, Yale University, New Haven, CT 06511, USA
email: krines@astro.yale.edu

Abstract. The CAIRNS (Cluster And Infall Region Nearby Survey) project is a large spectroscopic survey of the infall regions surrounding nine nearby rich clusters of galaxies. I describe the survey and use the kinematics of galaxies in the infall regions to estimate the cluster mass profiles. At small radii, these mass profiles are consistent with independent mass estimates from X-ray observations and Jeans analysis. I demonstrate the dependence of mass-to-light ratios on environment by combining these mass profiles with Two-Micron All-Sky Survey (2MASS) photometry. Near-infrared light is more extended than mass in these clusters, suggesting that dense cluster cores are less efficient at forming galaxies and/or more efficient at disrupting them. At large radii, galaxy populations in cluster infall regions closely resemble those in the field. The mass-to-light ratio at these radii should therefore be a good probe of the global mass-to-light ratio. The mass-to-light ratio in the infall region yields a surprisingly low estimate of $\Omega_m \sim 0.1$.

---

1. Introduction

The relative distribution of matter and light in the universe is one of the outstanding problems in astrophysics. Clusters of galaxies, the largest gravitationally relaxed objects in the universe, are important probes of the distribution of mass and light. Zwicky (1933) first computed the mass-to-light ratio of the Coma cluster using the virial theorem and found that dark matter dominates the cluster mass. Recent determinations using the virial theorem yield mass-to-light ratios of $M/L_{B_j} \sim 250hM_\odot/L_\odot$ (Girardi et al. 2000 and references therein). Equating the mass-to-light ratio in clusters to the global value provides an estimate of the mass density of the universe; this estimate is subject to significant systematic errors introduced by differences in galaxy populations between cluster cores and lower density regions (Carlberg et al. 1997, Girardi et al. 2000). Indeed, some numerical simulations suggest that cluster mass-to-light ratios exceed the universal value (Diaferio 1999, Kravtsov & Klypin 1999, Bahcall et al. 2000).

Determining the global matter density from cluster mass-to-light ratios therefore requires knowledge of the dependence of mass-to-light ratios on environment. Bahcall et al. (1995) show that mass-to-light ratios increase with scale from galaxies to groups to clusters. Ellipticals have larger overall values of M/L_B than spirals, presumably a result of younger, bluer stellar populations in spirals. At the scale of cluster virial radii, mass-to-light ratios appear to reach a maximum value. Some estimates of the mass-to-light ratio on very large scales ($>10h^{-1}$Mpc) are available (Bahcall et al. 1995), but the systematic uncertainties are large.

There are few estimates of mass-to-light ratios on scales between cluster virial radii and scales of $10h^{-1}$Mpc (Rines et al. 2000, Rines et al. 2001a, Biviano & Girardi 2003, Katgert et al. 2003, Kneib et al. 2003, Rines et al. 2004, Tully 2004 and references therein). On these scales, many galaxies near clusters are bound to the cluster but not yet in equilibrium (Gunn & Gott 1972). These cluster infall regions have received relatively little scrutiny because they are mildly nonlinear, making their properties very difficult to predict analytically. However, these scales are exactly the ones in which galaxy properties change dramatically (e.g., Ellingson et al. 2001, Lewis et al. 2002, Gómez et al. 2003, Treu et al. 2003, Balogh et al. 2004, Gray et al. 2004). Variations in the mass-to-light ratio with environment could have important physical implications; they could be produced either by a varying dark matter fraction or by variations in the efficiency of star formation with environment. In blue light, however, higher star formation rates in field galaxies could produce lower mass-to-light ratios outside cluster cores resulting only from the different contributions of young and old stars to the total luminosity (Bahcall et al. 2000, Tully 2004).

Galaxies in cluster infall regions produce sharp features in redshift surveys. Early investigations of this infall pattern focused on its use as a direct indicator of the global matter density Ω_m. Unfortunately, random motions caused by galaxy-galaxy interactions and substructure within the infall region smear out this cosmological signal (Diaferio & Geller 1997, Vedel & Hartwick 1998). Instead of sharp peaks in redshift space, infall regions around real clusters typically display a well-defined envelope in redshift space which is significantly denser than the surrounding environment (Rines et al. 2003, hereafter Paper I, and references therein).

Diaferio & Geller (1997) and Diaferio (1999) analyzed the dynamics of infall regions with numerical simulations and found that in the outskirts of clusters, random motions due to substructure and non-radial motions make a substantial contribution to the amplitude of the caustics which delineate the infall regions. Diaferio & Geller (1997) showed that the amplitude of the caustics is a measure of the escape velocity from the cluster; identification of the caustics therefore allows a determination of the mass profile of the cluster on scales $\lesssim 10h^{-1}$Mpc.

Diaferio & Geller (1997) and Diaferio (1999) show that nonparametric measurements of caustics yield cluster mass profiles accurate to $\sim$50% on scales of up to 10 h^{-1} Mpc. This method assumes only that galaxies trace the velocity field. Indeed, simulations suggest that little or no velocity bias exists on linear and mildly non-linear scales (Kauffmann et al. 1999a, Kauffmann et al. 1999b). The caustic method has been applied to systems as large as the Shapley Supercluster (Reisenegger et al. 2000) and as small as the Fornax cluster (Drinkwater et al. 2001) as well as to many nearby clusters (Paper I). Biviano & Girardi (2003) applied the caustic technique to an ensemble cluster created by stacking redshifts around 43 clusters from the 2dF Galaxy Redshift Survey. Rines et al. (2000) found an enclosed mass-to-light ratio of $M/L_R \sim 300h$ within 4 h^{-1}Mpc of A576. Rines et al. (2001) used 2MASS photometry and the mass profile from Geller et al. (1999) to compute the mass-to-light profile of Coma in the K-band. They found a roughly flat profile with a possible decrease in M/L_K with radius by no more than a factor of 3. Biviano & Girardi (2003) find a decreasing ratio of mass density to total galaxy number density. For early-type galaxies only, the number density profile is consistent with a constant mass-to-light (actually mass-to-number) ratio.

Here, we calculate the infrared mass-to-light profile within the turnaround radius for the CAIRNS clusters (Paper I), a sample of nine nearby rich, X-ray luminous clusters. We use photometry from 2MASS, the Two Micron All Sky Survey (Skrutskie et al. 1997) and add several new redshifts to obtain complete or nearly complete surveys of galaxies up to 1-2 magnitudes fainter than $M^*_{K_s}$ (as determined by Cole et al. 2001 and Kochanek et al. 2001). Infrared light is a better tracer of stellar mass than optical light; it is relatively insensitive to dust extinction and recent star formation. Despite these advantages, there are very few measurements of infrared mass-to-light ratios in clusters (Tustin et al. 2001, Rines et al. 2001a, Lin et al. 2003, Rines et al. 2004).

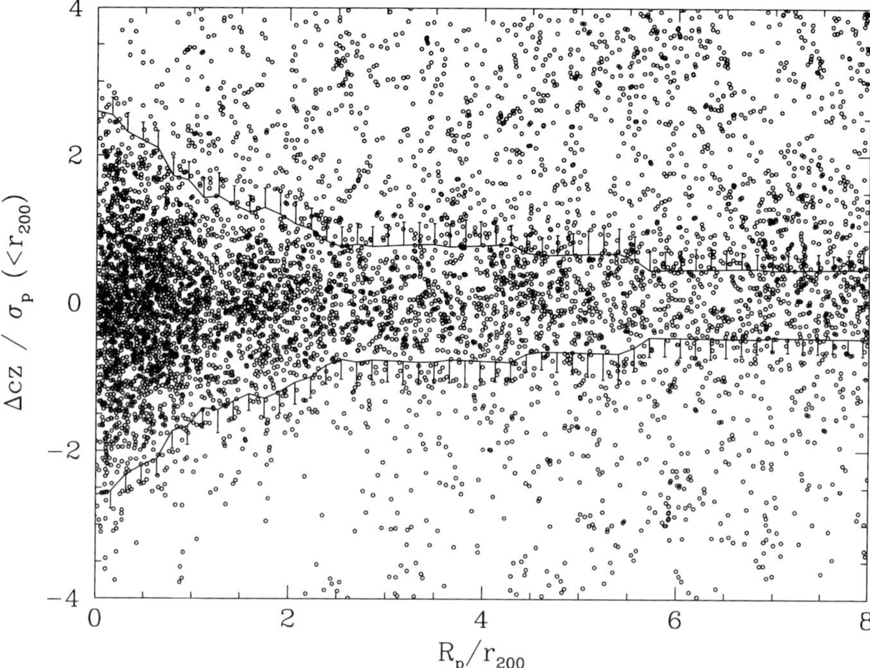

Figure 1. Redshift versus projected radius for the combined CAIRNS cluster. The solid lines indicate the caustics and the errorbars show 1-σ uncertainties.

2. Observations

The CAIRNS clusters are 8 of the 14 Abell clusters that are: nearby ($cz < 15,000$ km s^{-1}), with Abell richness class $R \geqslant 1$, X-ray luminous ($L_X > 2.5 \times 10^{43} h^{-2}$erg s^{-1}), and with declination $\delta > -15°$. Between 1997 and 2003, we collected 5607 redshifts (both new and remeasured) in large areas around nearby clusters with the FAST spectrograph on the FLWO 1.5-m telescope in Arizona. We targeted galaxies within a projected radius of $\sim 10 h^{-1}$Mpc of the clusters, selecting targets first from POSSII 103aE plates and later from 2MASS when it became available. The redshift catalogs in Rines et al. (2004) contains complete K_s band magnitude-limited samples extending 1-2 magnitudes fainter than the characteristic magnitude $M^*_{K_s}$. These samples include ~60-85% of the total light in the clusters and their infall regions.

3. Mass profiles

Figure 1 shows the redshifts of galaxies surrounding nearby clusters as a function of projected radius (normalized to r_{200}, the radius within which the average density is 200 times the critical density). The caustic pattern (a dense envelope in phase space with well-defined edges) is evident in the combined cluster as well as in each of the 9 clusters. Using the phase space distribution of galaxies in cluster infall regions, we apply the kinematic mass estimator of D99 to these clusters. The resulting mass profiles agree well with NFW (Navarro et al. 1997) and Hernquist (1990) models, but exclude an isothermal sphere.

These mass profiles agree with X-ray masses at small radius as well as virial masses (after correction for the surface pressure term) at slightly larger radii (Rines et al. 2002,

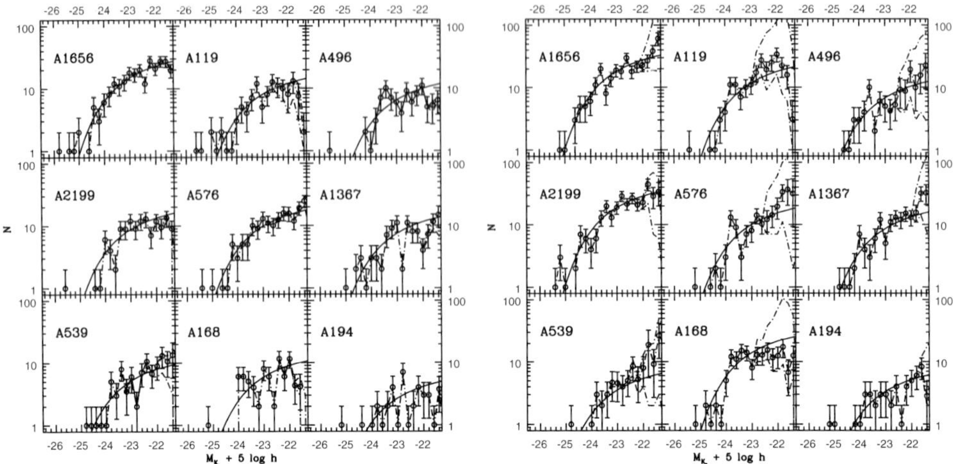

Figure 2. K_s-band luminosity functions for the CAIRNS clusters. The left panel shows galaxies in the virial regions (projected within r_{200}); the right panel shows galaxies in the infall regions (projected outside r_{200}). Solid lines show the shape of the 2dF/2MASS K_s-band LF with arbitrary normalization.

Paper I). The latter is primarily a consistency check as the caustic technique utilizes the same kinematic data as the virial theorem. This consistency can be further demonstrated with the velocity dispersion profiles (see Paper I and Rines et al. 2004).

4. Near-infrared luminosity functions

When using the mass-to-light ratio in clusters to estimate Ω_m, one must determine whether the luminosity functions of field and cluster galaxies differ significantly. Given the well-known morphology-density relation, it is possible that the two LFs differ significantly. We use 2MASS photometry to determine the near-infrared luminosity functions for the CAIRNS clusters and infall regions (Rines et al. 2004). The cluster and infall region LFs are very similar to each other (Figure 2) and to the field LF for relatively bright galaxies ($M_{K_s} \lesssim M^*_{K_s} + 2$). Because of this similarity, we can correct for the luminosity in faint galaxies using the field LF. Our redshift surveys include 60–85% of the total light.

5. Near-infrared mass-to-light profiles

Both the surface number density and surface luminosity density profiles of cluster/infall region members are more extended than the mass profiles of Paper I. Figure 3 shows the mass-to-light profiles of the CAIRNS clusters. The mass-to-light profiles are either flat or show a decreasing M/L_K with increasing radius. The mean mass-to-light ratio within r_{200} is a factor of 1.8 ± 0.3 larger than the mean value outside r_{200}.

The decreasing mass-to-light profiles could be caused by gradients in stellar populations with radius; such effects have been invoked to account for a similar result for B band mass-to-light profiles in simulations (Bahcall et al. 2000). In K band, however, changes in stellar populations with radius are expected to change the mean stellar mass-to-light ratio by $\lesssim 20\%$. Thus, the CAIRNS mass-to-light profiles provide tentative evidence for variations in the efficiency of galaxy formation and/or disruption. Environments with

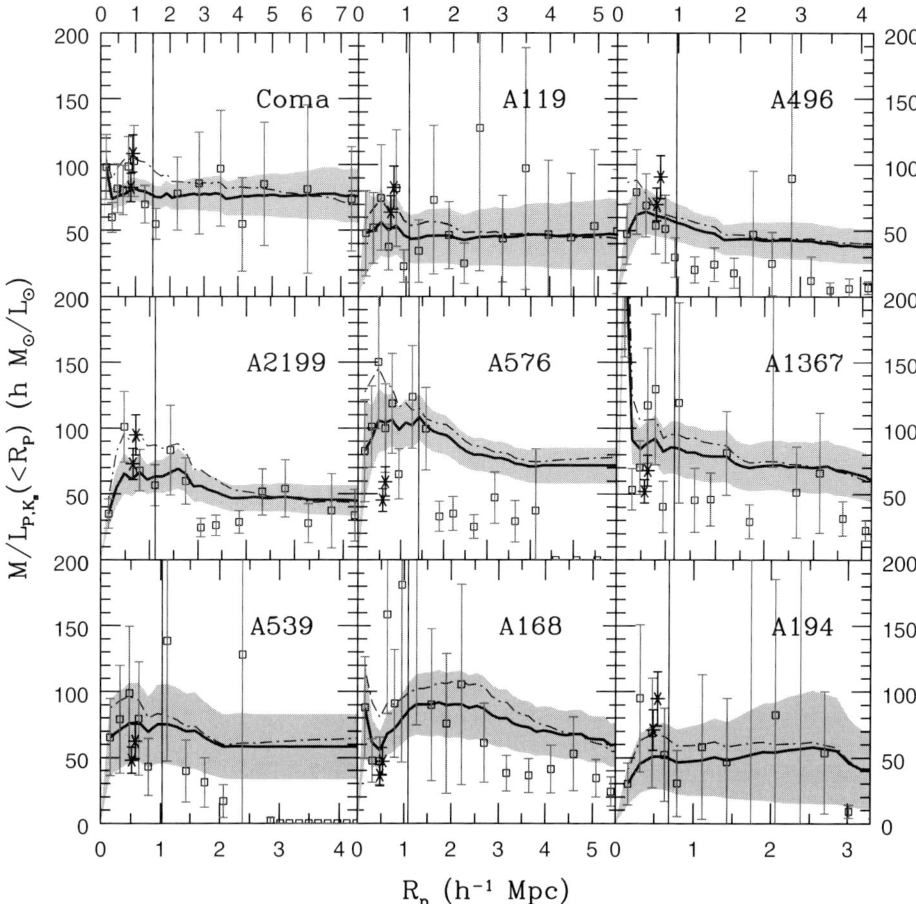

Figure 3. Mass-to-light ratio in K_s band as a function of radius for the CAIRNS clusters. The solid lines show the caustic mass profile $M(<r)$ divided by the projected luminosity profile $L_{K_s}(<R_p)$ and the shaded regions indicate the associated 1-σ uncertainties. The open squares show the mass-to-light ratio $M(r, r+dr)/L_{K_s}(R_p, R_p+dR_p)$ in radial shells. The dash-dotted line shows the projected best-fit Hernquist mass profile $M_H(<R_p)$ divided by $L_{K_s}(<R_p)$. The solid stars show the mass-to-light ratio calculated using the X-ray temperature and the mass-temperature relation to estimate the mass. The lower point shows $M_X(<r_{500})/L_{K_s}(<R_{500})$ and the upper point shows $M_X(<R_{500})/L_{K_s}(<R_{500})$ assuming $M_X(<R_{500}) = 1.3M_X(<r_{500})$ as is true for an NFW mass profile with c=5.

higher virial temperatures (like cluster centers) are more efficient at disrupting galaxies and/or less efficient at forming them.

A related trend has been noted by Lin et al. (2003, 2004), who show that K-band mass-to-light ratios within r_{500} increase with increasing cluster mass. This result indicates that more massive clusters (with larger virial temperatures) have less efficient galaxy formation and/or more efficient galaxy disruption. We confirm this trend in Rines et al. (2004). Because cluster infall regions should be composed of galaxies inhabiting less massive systems (Rines et al. 2001b, 2002) and/or regions with lower virial temperatures, the above trend predicts that cluster infall regions should have smaller mass-to-light ratios than virial regions, consistent with the CAIRNS results.

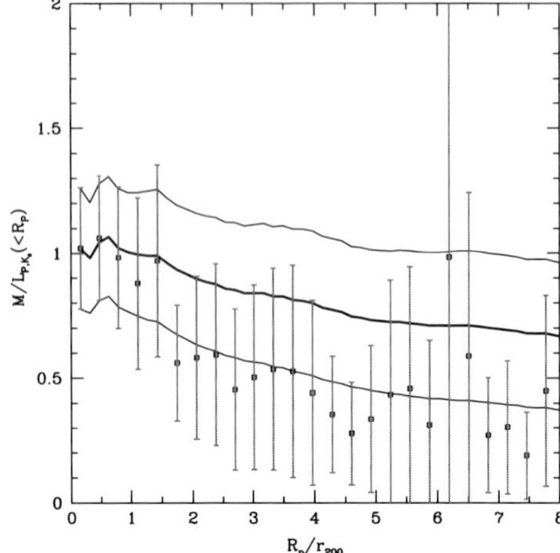

Figure 4. Mass-to-light ratio in K_s band as a function of radius for the combined CAIRNS cluster. The thick and thin solid lines show the caustic mass profile $M(< r)$ divided by the projected luminosity profile $L_{K_s}(< R_p)$ and the associated 1-σ uncertainties. The open squares show the mass-to-light ratio $M(r, r + dr)/L_{K_s}(R_p, R_p + dR_p)$ in radial shells.

Figure 4 shows the mass-to-light profile of the combined CAIRNS cluster (normalized to unity at r_{200}). The mass-to-light profile clearly decreases with radius. The combined cluster should be less susceptible to substructure than the individual clusters. Figure 4 shows that the mass-to-light ratio in shells decreases by about a factor of 2 from the virial region to the infall region, consistent with the results for individual clusters. Taking the mass-to-light ratio in cluster infall regions as an estimate of the global value and the SDSS luminosity density extrapolated to K_s band (Blanton et al. 2003), we estimate $\Omega_m = 0.10 \pm 0.02$ (statistical). This estimate is somewhat lower if we use the K_s band luminosity density of either Cole et al. (2001) or Kochanek et al. (2001). We discuss potential systematic effects in Rines et al. (2004). Most of these effects would flatten the observed mass-to-light profiles relative to the true profiles, suggesting that the observed decrease in mass-to-light ratio with radius is real.

6. Conclusions

Cluster infall regions contain more galaxies than their virial regions. If currently popular cosmological models are correct, the mass in infall regions will eventually accrete onto the parent clusters, and their final masses will increase by a factor of about 2.

Near-infrared luminosity functions depend only weakly on environment, at least at the bright end. We show that near-infrared light is more extended than mass in cluster infall regions, suggesting environmental dependence of the efficiency of galaxy formation and/or disruption. If more efficient galaxy disruption is responsible, intracluster stars might be a significant component of stellar mass in clusters. The mass-to-light ratios in infall regions suggest a low $\Omega_m \sim 0.1$. Future work is needed to determine the significance of the conflict of this result with the currently favored $\Omega_m \sim 0.3$.

Acknowledgements

I would like to thank Margaret Geller and Antonaldo Diaferio for their many contributions as my primary collaborators in this work. I would also like to thank Tom Jarrett,

Michael Kurtz, Joe Mohr, Gary Wegner, and John Huchra for their contributions to the CAIRNS project. Perry Berlind, Mike Calkins, and Susan Tokarz collected and reduced most of the spectroscopic data. I thank the Smithsonian Institution for support.

References

Bahcall, N. A., Cen, R., Davé, R., Ostriker, J. P., & Yu, Q. 2000, *Astrop. J.*, 541, 1

Bahcall, N. A., Lubin, L. M., & Dorman, V. 1995, *Astrop. J.*, 447, L81

Balogh, M. et al. 2004, *MNRAS*, 348, 1355

Biviano, A. & Girardi, M. 2003, *Astrop. J.*, 585, 205

Blanton, M. R. et al. 2003, *Astrop. J.*, 592, 819

Carlberg, R. G., Yee, H. K. C., & Ellingson, E. 1997, *Astrop. J.*, 478, 462

Cole, S. et al. 2001, *MNRAS*, 326, 255

Diaferio, A. 1999, *MNRAS*, 309, 610

Diaferio, A. & Geller, M. J. 1997, *Astrop. J.*, 481, 633

Drinkwater, M. J., Gregg, M. D., & Colless, M. 2001, *Astrop. J.*, 548, L139

Ellingson, E., Lin, H., Yee, H. K. C., & Carlberg, R. G. 2001, *Astrop. J.*, 547, 609

Geller, M. J., Diaferio, A., & Kurtz, M. J. 1999, *Astrop. J.*, 517, L23

Gómez, P. L. et al. 2003, *Astrop. J.*, 584, 210

Girardi, M., Borgani, S., Giuricin, G., Mardirossian, F., & Mezzetti, M. 2000, *Astrop. J.*, 530, 62

Gray, M. E. and Wolf, C. and Meisenheimer, K. and Taylor, A. and Dye, S. and Borch, A. and Kleinheinrich, M. 2004, *MNRAS*, 347, L73

Gunn, J. E. & Gott, J. R. I. 1972, *Astrop. J.*, 176, 1

Hernquist, L. 1990, *Astrop. J.*, 356, 359

Katgert, P., Biviano, A., & Mazure, A. 2003, *Astrop. J.*, 600, 6457

Kauffmann, G., Colberg, J. M., Diaferio, A., & White, S. D. M. 1999a, *MNRAS*, 303, 188

—. 1999b, *MNRAS*, 307, 529

Kneib, J.-P. et al. 2003, *Astrop. J.*, 598, 804

Kochanek, C. S. et al. 2001, *Astrop. J.*, 560, 566

Kravtsov, A. V. & Klypin, A. A. 1999, *Astrop. J.*, 520, 437

Lewis, I. et al. 2002, *MNRAS*, 334, 673

Lin, Y., Mohr, J. J., & Stanford, S. A. 2003, *Astrop. J.*, 591, 749

Lin, Y., Mohr, J. J., & Stanford, S. A. 2004, *Astrop. J.*, in press, astro-ph/0402308

Navarro, J. F., Frenk, C. S., & White, S. D. M. 1997, *Astrop. J.*, 490, 493

Reisenegger, A., Quintana, H., Carrasco, E. R., & Maze, J. 2000, *Astron. J.*, 120, 523

Rines, K., Geller, M. J., Kurtz, M. J., & Diaferio, A., 2003, *Astron. J.*, 126, 2152

Rines, K., Geller, M. J., Diaferio, A., Mahdavi, A., Mohr, J. J., & Wegner, G. 2002, *Astron. J.*, 124, 1266

Rines, K., Geller, M. J., Diaferio, A., Mohr, J. J., & Wegner, G. A. 2000, *Astron. J.*, 120, 2338

Rines, K., Geller, M. J., Kurtz, M. J., Diaferio, A., Jarrett, T. H., & Huchra, J. P. 2001a, *Astrop. J.*, 561, L41

Rines, K., Mahdavi, A., Geller, M. J., Diaferio, A., Mohr, J. J., & Wegner, G. 2001b, *Astrop. J.*, 555, 558

Rines, K. , Geller, M. J., Diaferio, A., Kurtz, M. J., & Jarrett, T. H., 2004, *Astron. J.*, submitted (astro-ph/0402242)

Skrutskie, M. F. et al. 1997, in ASSL Vol. 210: The Impact of Large Scale Near-IR Sky Surveys, 25

Treu, T., Ellis, R. S., Kneib, J., Dressler, A., Smail, I., Czoske, O., Oemler, A., & Natarajan, P. 2003, *Astrop. J.*, 591, 53

Tully, R. B., 2004, *Astrop. J.*, submitted, astro-ph/0312441

Tustin, A. W., Geller, M. J., Kenyon, S. J., & Diaferio, A. 2001, *Astron. J.*, 122, 1289

Vedel, H. & Hartwick, F. D. A. 1998, *Astrop. J.*, 501, 509

Zwicky, F. 1933, Helv. Phys. Acta, 6, 110

Outskirts of Galaxy Clusters: Intense Life in the Suburbs
Proceedings IAU Colloquium No. 195, 2004
A. Diaferio, ed.

DOI: 10.1017/S1743921304000420

M/L of X-ray luminous clusters from weak lensing

H. Dahle

Institute of Theoretical Astrophysics, University of Oslo,
P.O. Box 1029, Blindern, N-0315 Oslo, Norway email: hdahle@astro.uio.no

Abstract. We present results from a weak lensing survey of a sample of 53 clusters of galaxies at $0.15 < z < 0.4$, selected based on their high X-ray luminosities. For each cluster, we measure the mass-to-light ratio out to a radius $R \sim (0.5 - 1.5)h^{-1}$ Mpc. By measuring the lensing masses of a large sample of clusters in a consistent way, we quantify the intrinsic scatter in M/L and determine its average value for massive clusters. In the near future, an ongoing spectro-photometric and weak lensing survey of a supercluster region in Aquarius will extend M/L measurements to much larger radii and explore its dependence on scale and environment.

1. Introduction

Gravitational lensing measurements based on deep optical or near-IR imaging provides the most direct way of determining the mass-to-light ratios of galaxy clusters. Previous results based on weak lensing studies of clusters (see e.g. the compilation by Mellier 2002) show a wide scatter in M/L values, but it is uncertain whether this reflects a true scatter in M/L or whether it is mostly caused by differences in reduction and analysis techniques, assumptions about source galaxy distances, different passbands being used etc. On the other hand, several weak lensing studies have also revealed a few "dark cluster" candidates that have been detected based on their gravitational lensing effect and which appear to have extreme M/L values compared to normal clusters. At least one reported "dark cluster" candidate has a measured bolometric mass-to-light ratio $M/L_{\rm bol} \geqslant 1000h$ in solar units (Erben et al. 2000; Gray et al. 2001).

By analyzing weak lensing data of a large sample of clusters in a consistent way, it should be possible to accurately determine M/L at the high-mass end of the cluster mass function and also quantify the intrinsic scatter in M/L.

2. X-ray luminous cluster sample

Weak gravitational lensing measurements of 38 X-ray luminous galaxy clusters were presented by Dahle et al. (2002), and subsequent data have been obtained for 15 additional clusters with similar selection criteria. Clusters in the redshift range $0.1 < z < 0.4$ were selected from different ROSAT-based cluster surveys (Briel & Henry 1993; Ebeling et al. 1996, 1998, 2000; Böhringer et al. 2000), and almost all targets have X-ray luminosity $L_{\rm X,0.1-2.4keV} \geqslant 10^{45}{\rm erg\,s^{-1}}$. A subsample of 12 clusters were observed using the UH8K camera at the University of Hawaii 2.2m telescope (covering a field $\sim 3h^{-1}$ Mpc across at the distance of the clusters), while the rest were observed with 2k CCD cameras at the UH 2.2m telescope and at the Nordic Optical Telescope (covering smaller fields of $\sim 1h^{-1}$ Mpc). Details on the reduction and weak lensing analysis of this data set are provided by Dahle et al. (2002).

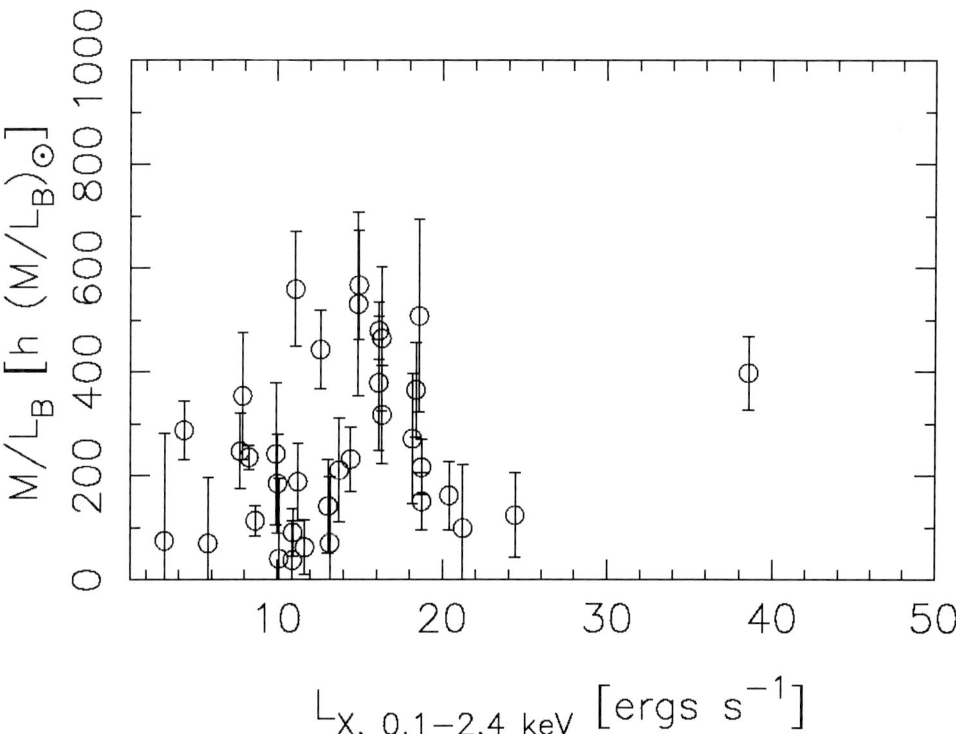

Figure 1. B-band mass-to-light ratio for 35 clusters in our X-ray luminous sample as a function of X-ray luminosity.

2.1. *Measured and predicted surface mass density*

The shear estimator introduced by Kaiser (2000) was used to measure the weak gravitational shear. The shear inversion techniques of Kaiser & Squires (1993) and Squires & Kaiser (1996) were used to reconstruct the projected dimensionless mass surface density κ_M from UH8K and 2k CCD data, respectively. We generated sets of source galaxy catalogues with randomized shear, keeping the galaxy positions as in the real catalogues, while randomly shuffling the measured galaxy shapes, such that the spread of values in the ensemble of resulting mass maps could be used to evaluate the uncertainties in the mass maps.

Assuming a mass-traces-light model for the clusters, a predicted κ_L for the projected mass density was generated based on $I-$band photometry. In an image of κ_L with a pixel size corresponding to a solid angle $d\Omega$, the contribution of a galaxy of apparent magnitude m_I and redshift z to the pixel on which the galaxy is located, is given by:

$$\kappa_L d\Omega = \frac{M}{L_B} \frac{M_\odot}{(10\mathrm{pc})^2} \frac{(1+z)^4}{\Sigma_{\mathrm{crit}}(z)} 10^{0.4[M_{B\odot} - m_I - K_{BI}(z)]}, \tag{2.1}$$

where M/L_B is the rest-frame B-band mass-to-light ratio (in solar units) used for the prediction, and $\Sigma_{\mathrm{crit}}(z)$ is the effective critical density (where the distribution of background galaxy redshifts is estimated from HDF-based photometric redshifts). The quantity $K_{BI}(z)$ combines the conventional K-correction and the rest-frame $B - I$ color. It is possible to reliably select early-type galaxies and estimate their redshifts based on their $V - I$ colors (Wilson et al. 2001), but we initially adopted a different approach, assuming that the observed galaxy flux in the fields was contributed by galaxies at the cluster redshift plus a background with constant surface flux density. A few bright galaxies that were obvious foreground objects were removed by hand.

A radially dependent correction for the contamination by cluster galaxies in our source galaxy catalogue was calculated, and the κ_L maps were corrected accordingly, to mimic the corresponding effect on the κ_M maps.

2.2. *Results for M/L*

The rest-frame B-band mass-to-light ratios in the cluster fields were calculated by taking the discrete Fourier transform of the predicted and measured surface density images and comparing the amplitudes of the Fourier modes. In Figure 1, the resulting values for $M/L_{\rm B}$ are plotted against X-ray luminosity. The M/L values were calculated for a concordance model universe with $(\Omega_\Lambda, \Omega_m) = (0.7, 0.3)$.

From a subset of 35 clusters, we find $\langle M/L_B \rangle = (219 \pm 11)h(M/L_B)_\odot$, and a median value of $M/L_B = 236h(M/L_B)_\odot$. Taken at face value, the spread of observed M/L values indicate an intrinsic scatter in M/L_B of about $\sigma(M/L_B) = 118 \pm 19$ about the mean value. However, part of this scatter could be due to incomplete removal of foreground/background contamination in the galaxy catalogues, and our underlying assumption that mass traces light may not be valid for all systems. Since our data indicate a significant intrinsic cluster-to-cluster variation in M/L, this may at least partly explain the wide range of cluster M/L values reported in the literature.

The halo-mass dependence of M/L is still not a fully resolved issue: While some dynamical studies indicate that mass increases faster than luminosity (Girardi et al. 2002), such a trend is not seen from a recent X-ray based study (Sanderson & Ponman 2003). We note that since our clusters only cover a modest range in mass, our results must be combined with other data sets probing less massive systems to be able to address this issue.

3. Survey of the Aquarius supercluster region

The highest projected density of $R \geqslant 1$ Abell/ACO clusters in the sky is located around $\alpha = 23.2^h$, $\delta = -22\,\deg$ (Abell 1961). This high density is caused by two superclusters (at $z = 0.08$ and $z = 0.11$) seen in projection, along with a filament extending outwards along the line of sight from the far end of the $z = 0.11$ supercluster.

A combination of weak lensing data with galaxy spectroscopy and photometry allows to study the relationship between mass and light in this region, in particular the scale- and type-dependence of galaxy bias. An ongoing survey combines galaxy spectroscopy with the 2dF instrument at the 3.9m AAT with a weak-lensing survey based on wide-field imaging from the WFI at the ESO 2.2m telescope. In a total of 20 WFI pointings we cover an area of 6.2 square degrees (corresponding to a transverse scale $\sim 20h^{-1}$Mpc at the supercluster redshift), containing 25 previously known galaxy clusters. We have obtained spectra of ~ 3000 galaxies down to $R = 19.5$ in this region and use this data to identify 26 previously unknown clusters and groups and estimate their redshifts and velocity dispersions along with the known clusters (Smith et al. 2004). Our results indicate that the $z = 0.08$ and $z = 0.11$ superclusters in Aquarius are both very massive systems, comparable to the Shapley supercluster or the Corona Borealis supercluster. The imaging

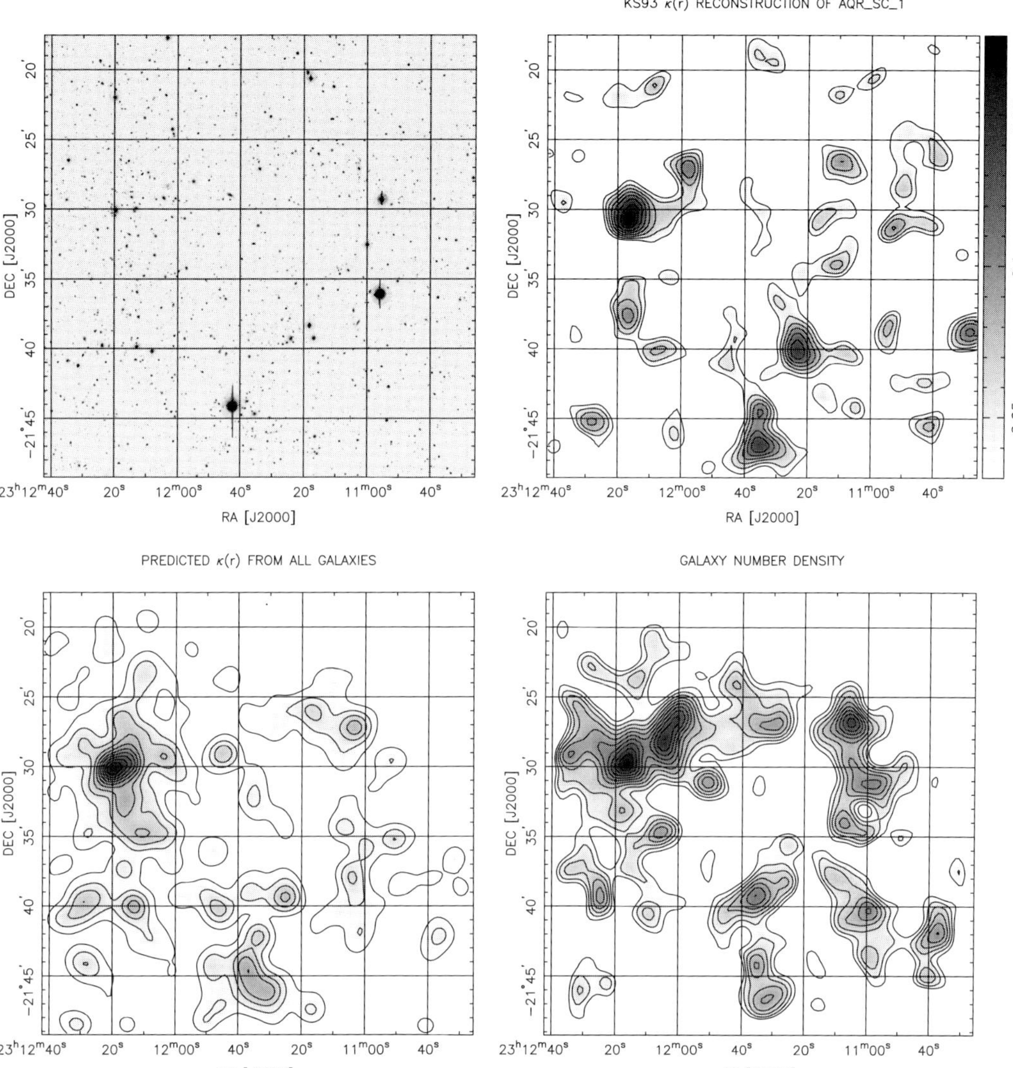

Figure 2. Preliminary results for 1 of the 20 pointings within the Aquarius supercluster region. Top left: R-band image based on ESO WFI imaging. Top right: The surface mass density κ_M, determined from weak shear measurements of background galaxies. Bottom left: Predicted surface mass density associated with the light from galaxies in the field, κ_L. To generate this plot, the galaxies were all assumed to lie at a single redshift, but in our ongoing analysis we use $B-R$ color information combined with our 2dF spectroscopic redshifts to generate a more accurate measurement of κ_L. Bottom right: The number density of all galaxies in the field. Two Abell clusters are located within this field, A2554 ($z = 0.111$; upper left) and A2550 ($z = 0.122$; bottom center), in addition to several lesser clusters and galaxy groups.

part of this survey is still ongoing, and preliminary results for one of the 20 WFI pointings is shown in Figure 2. Several peaks that correspond to previously known clusters are visible in the maps, along with some new systems.

Acknowledgements

I thank my collaborators R. E. Smith, P. B. Lilje and S. J. Maddox.

References

Abell, G. O., 1961, AJ, 66, 607

Böhringer, H. et al. 2000, ApJS, 129, 435

Briel, U., & Henry, J. P. 1993, A& A, 278, 379

Dahle, H., Kaiser, N., Irgens, R. J., Lilje, P. B., & Maddox, S. J. 2002, ApJS, 139, 313

Ebeling, H., Voges, W., Böhringer, H., Edge, A.C., Huchra, J. P. & Briel, U.G. 1996, MNRAS, 281, 799

Ebeling, H., Edge, A. C., Böhringer, H., Allen, S., Crawford, C. S., Fabian, A.C., Voges, W., & Huchra, J. P. 1998, MNRAS, 301, 881

Ebeling, H., Edge, A. C., Allen, S. W., Crawford, C. S., Fabian, A. C., & Huchra, J. P. 2000, MNRAS, 318, 333

Erben, T., van Waerbeke, L., Mellier, Y., Schneider, P., Cuillandre, J.-C., Castander, F. J. , & Dantel-Fort, M. 2000, A& A, 355, 23

Girardi, M., Manzato, P., Mezzetti, M., Giuricin, G., & Limboz, F. 2002, ApJ, 569, 720

Gray, M. E., Ellis, R. S., Lewis, J. R., McMahon, R. G., & Firth, A. E. 2001, MNRAS, 325, 111

Kaiser, N. 2000, ApJ , 537, 555

Kaiser, N. & Squires,G. 1993, ApJ, 404, 441, KS93

Mellier, Y. 2002, Space Science Reviews, 100, 73

Sanderson, A. J. R. & Ponman, T. J. 2003, MNRAS, 345, 1241

Smith, R. E., Dahle, H., Maddox, S. J., & Lilje, P. B. 2004, astro-ph/0401641

Squires, G., & Kaiser, N., 1996, ApJ, 473, 65

Wilson, G., Kaiser, N., Luppino, G. A., & Cowie, L. L. 2001, ApJ, 555, 572

Outskirts of Galaxy Clusters: Intense Life in the Suburbs
Proceedings IAU Colloquium No. 195, 2004
A. Diaferio, ed.

DOI: 10.1017/S1743921304000432

Action model of infall into the Virgo cluster

R. Brent Tully[1,2] **and Roya Mohayaee**[2]

[1]Institute for Astronomy, University of Hawaii, Honolulu, HI 96822, USA email: tully@ifa.hawaii.edu

[2]Observatoire de la Côte d'Azur, 06304 Nice, France email: roya@obs-nice.fr

Abstract. The observed infall of galaxies into the Virgo cluster puts strong constraints on the mass of the cluster. A non-parametric fully non-linear description of the infall can be made with orbit reconstructions based on Numerical Action Methods. The mass of the cluster is determined to be $1.2 \times 10^{15} M_\odot$. The mass-to-light ratio for the cluster is found to be seven times higher than the mean ratio found across the region within $V = 3000$ km/s.

1. Introduction

The general problem we will consider is the reconstruction of orbits that galaxies might have followed to arrive at their currently observed positions on the sky and in redshift. On this occasion we will focus on the particularly interesting circumstances associated with infall into the Virgo cluster. The program involves three distinct components.

First, we require as complete a map as possible of the angular positions and velocities of galaxies. It will be assumed that there is some correlation between the distribution of galaxies in the catalog and the actual distribution of mass, which is the parameter that really interests us. However it is to be anticipated from the outset that the relationship between what we see – galaxies of various types – and the distribution of mass may be complex (Tully 2003, 2004). It is also to be appreciated that large scale tides may be dynamically significant so one needs a map of the distribution of galaxies that extends well beyond the region of immediate focus. In the present instance, our interest is in the environs of the Virgo cluster at a distance of 16.8 Mpc. Our map of the distribution of galaxies is based on a sample of 3151 galaxies within ~ 40 Mpc. The sample is complete to $0.1L^\star$ within 25 Mpc at high galactic latitudes. Selection function corrections are made as a function of distance and 'fake' sources are added at low latitudes to account for missing sources in the zone of obscuration.

The second critical component is a catalog of accurate distances to galaxies. Distance measures, d, allow a separation of observed velocities, $V_{\rm obs}$, and peculiar velocities, $V_{\rm pec}$, since $V_{\rm pec} = V_{\rm obs} - {\rm H}_0 d$. Here, $\rm H_0$ is the Hubble constant which is taken to be 80 km/s/Mpc. For purposes of separating $V_{\rm pec}$ from $V_{\rm obs}$ it is only required that distances and the Hubble constant be on the same scale; that is, with averaging that takes into account large scale flows: ${\rm H}_0 =< V_{\rm obs}/d >$. If distances are, say, systematically measured too large then a self-consistent value of $\rm H_0$ will be too low. Derived values of $V_{\rm pec}$ are independent of the distance scale zero-point. Another factor with regard to distances to appreciate is that errors are a percentage of the distance, 10% to 20% depending on the methodology, with the consequence that errors in the derived $V_{\rm pec}$ grow linearly with distance. At the Virgo cluster a 10% error corresponds to ± 140 km/s in $V_{\rm pec}$. This uncertainty is tolerable for the Virgo infall problem but for more distant clusters the situation would be unsatisfactory. The distance catalog itself is a synthesis of our own observations (Pierce and Tully, in preparation) that exploit the luminosity–linewidth method (Tully and Fisher 1977) and material from the literature. That literature material provides distances from a variety of methodologies: luminosity–linewidth (Mathewson et al. 1992, Lu et al. 1993, Tully and Pierce 2000), Cepheid (Freedman et al. 2001), Tip of the Giant Branch (Karachentsev et al. 2003), Planetary Nebula Luminosity Function (Jacoby et al. 1990), and Surface Brightness Fluctuation (Tonry et al. 2001). This latter important source is distinguished separately in the ensuing discussion. In total we have distances to almost 900 galaxies from the luminosity–linewidth and other assorted methods and 292 distances from the Surface Brightness Fluctuation method.

The third ingredient is the theoretical machinery to convert information on the amplitude of the peculiar velocities of galaxies into a map of the mass distribution. In linear theory there is a direct relation between $V_{\rm pec}$ and matter density fluctuations. The relationship is more complex in the vicinity of collapsed structures, such as the Virgo cluster. We make use of Numerical Action Methods which we have referred to as 'Least Action'. The numerical techniques have been discussed by Peebles (1989,1995), Shaya et al. (1995), and Phelps (2002). The procedure allows for orbit reconstructions in highly nonlinear regimes, though ambiguity arises if the orbits are complex, especially if dynamical friction or orbital energy exchange is important. These are not serious issues for the Virgo infall problem.

2. Numerical Action Methods

The 'Action' is defined as the integral along paths of the Lagrangian, the kinetic minus the potential energy. The equations of motion for particle orbits satisfy minima of the derivatives of the action with respect to position and momentum. Orbits of particles that represent saddle points of the action and agree with boundary conditions are also *physically plausible*.

Orbits are described by the 6 elements of phase space (α_i, δ_i, d_i, V_i^α, V_i^δ, V_i^r) under the influence of the ensemble of mass elements, m_i, and a time frame. In the ensuing discussion we will assume a flat universe with the matter density as a fraction of the closure density $\Omega_m = 0.21$, the vacuum energy fractional density $\Omega_\Lambda = 0.79$, and $\mathrm{H}_0 = 80$ km/s/Mpc. This choice of parameters sets the clock. We solve a mixed boundary value problem constrained by the 3 well-known elements of phase space today (α_i, δ_i, V_i^r) and the theoretically motivated condition that initially the 3 components of velocity were negligible; i.e. taken to be zero. Then given mass assignments to the components of our galaxy catalog, orbits can be constructed.

The observed luminosities of the components of the catalog, ℓ_i, provide a guide for the mass assignments. Given $m_i = \ell_i(m_i/\ell_i)$, the uncertainties in m_i can be transformed into uncertainties in m_i/ℓ_i. Hence, with specification of mass-to-light values we can constrain plausible orbits. There would be too many degrees of freedom if all m_i/ℓ_i were allowed to be different so, to begin, we go to the other limit and assume all objects have the *same* value, M/L.

Given a choice of M/L within the context of a specified cosmology we can constrain plausible orbits, but how can discrimination be made between alternatives? It is here that we make use of distance measurements. The orbits that have been defined are constrained as boundary conditions to have specified positions on the sky and radial velocities at the current epoch. The end point distances are an output of the model. For a substantial fraction of the objects we have observed distances. These cases can be used to evaluate the quality of the model through a χ^2 estimator that compares differences between model and observed distances (the comparison is done in the logarithms using distance moduli).

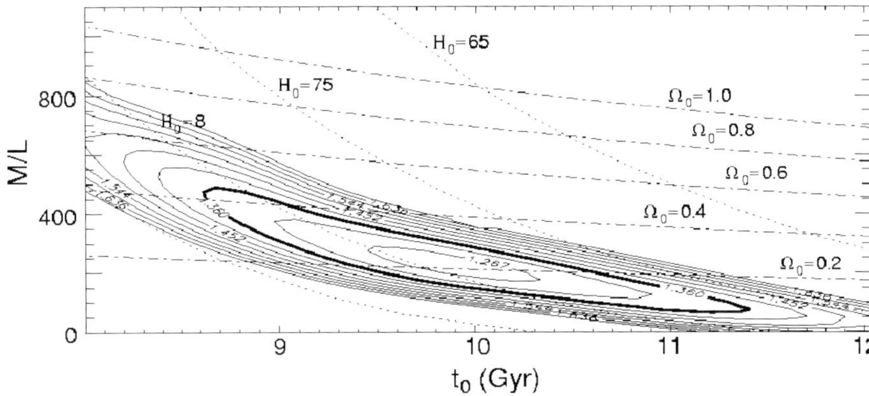

Figure 1. Contours of constant χ^2 from an Action analysis of a catalog restricted to $V < 3000$ km/s. In this case, $\Omega_\Lambda = 0$ is assumed. The heavy contour represents the 2σ confidence level.

Figure 1 provides an example of χ^2 contours in the domain of M/L vs. t_0, the age of the universe. The analysis presented here assumed $\Omega_\Lambda = 0$ and the distance scale zero point was 10% above our current preference. These differences particularly affect the time domain and, at the 10% level, the M/L values but do not affect the validity of the Ω_m estimates.

It is to be emphasized what one should not and what one should take seriously with the Action models. One should *not* take seriously the specific orbit reconstructions. The orbits that are derived are physical but not unique. The uniqueness problem becomes important for large-scale reconstruction where a different method which guarantees uniqueness, can be employed (Frisch et al. 2002). One *should* take seriously the constraints on mass. A thought experiment can make this point clear. Suppose we consider two cases of interaction between a test particle and a substantial mass where the difference between the two cases is simply the mass of the major attractor. Suppose, as in the real world, we can measure the line-of-sight velocity of the test particle with high accuracy and measure the separation between the test particle and attractor with some error. We want to estimate the mass of the attractor. Admitting that the line-of-sight velocity is only a component of the true velocity we only have a statistical test. However it should be clear that the measure of the separation between particles provides a measure of the mass. For a specified infall velocity, if the attractor has more mass then the test particle has to be farther away. The Action models are responding to this effect. If there is generally more mass, then for fixed velocities and within a specified cosmological model, objects are farther apart. Our distance observations are distinguishing between the mass possibilities.

We are careful to demand that the cosmological model be specified because, as is clear from Fig. 1, there is a trade off between mass and time. A model with more mass and less time can duplicate a model with less mass and more time. Essentially the effect of $\Omega_\Lambda > 0$ is to change the clock.

3. Virgo infall

Probably of all the problems we could consider with the Action machinery, the simplest application is in connection with infall into the Virgo cluster. Here we are giving consideration to the first approach infall of essentially 'test particles' toward a dominant mass. Near to the cluster, the radial motions toward the cluster (reaching 1500 km/s) are much larger than any transverse motions ($\sim$ 100 km/s). Since the orbits are essentially radial, on first approach, and have not yet reached the cluster, there are no serious concerns about orbital energy exchange or dynamical friction. The interactions involve only small perturbations on a two-body encounter.

We begin with Action model parameters that provide a reasonable χ^2 fit to distance measures throughout the region with $V < 3000$ km/s. Initially, a single value of M/L is taken for all components. Fig. 1 provides an example of a good solution. In the present study, we assume a flat cosmological model with $\Omega_m h^2 = 0.135$ (where $h = \mathrm{H}_0/100$) consistent with microwave background measurements (Spergel et al. 2003) but take the slightly nonstandard values $h = 0.80$ and $\Omega_m = 0.21$.

As we have reported earlier (Shaya et al. 1995), good fits to the ensemble of the data within $V = 3000$ km/s with a single M/L value assigned to all components requires a low $M/L \sim 200 M_\odot/L_\odot$. However a part of the data is poorly fit with this solution. This model implies a sufficiently modest mass for the Virgo cluster that it cannot generate the large infall velocities that are observed.

The so-called 'triple-value' region (Tonry and Davis 1981) around a cluster provides a signature of infall that depends on the mass of the cluster. Theoretical models demonstrate the properties of infall in 3-dimensions (Bertschinger 1985). The structure is simple in the volume within the zone of first collapse into the radius of a caustic which is the outer extent of orbits of objects that have passed through the cluster once and reached second turnaround. Within this zone between outer (first turnaround) and inner (second turnaround) radii the infall pattern translates to the now familiar 'triple-value' wave, whereby there are three locations that give the same velocity. In such instances, one of the locations lies outside the first turnaround – or zero velocity surface for the cluster – and the other two are within the collapse region. There are two infall locations that give the same velocity because of a line-of-sight geometrical effect. Since the motions represent infall toward the cluster, objects on the front side of the cluster have higher velocities than the cluster mean and objects on the back side of the cluster have lower velocities than the cluster mean.

There are two key points to understand in connection with the present study. The first is the obvious point that *the amplitude of the triple-value wave depends directly on the mass of the cluster.* Hence, if one is convinced that an object is within the infall region (rather than at the location outside the zero-velocity surface) then the mass of the cluster needs to be sufficient to explain the velocity of that object. In many cases in which the velocities are similar to that of the cluster the demand will not be too great. The challenge is to explain the extreme cases.

The second point to appreciate is that the Virgo cluster is near enough that we generally can unambiguously determine whether a target is within or outside the infall region. We may not be able to distinguish between the two infall options but that is of minor consequence. The essential point is that if the object is within the infall region then that object in itself puts a minimal demand on the mass interior to its position within the infall domain.

The general domain of the Virgo infall region is demonstrated in Figure 2. It is observationally determined that the outer radius of first turnaround is about 28° from the center of the cluster (a radius of 8 Mpc) while the caustic of the second turnaround is at about 6° (1.8 Mpc). In Fig. 2, all the objects that are highlighted (plus others), except for the object 102, are within the Virgo cluster zero-velocity surface. It is seen that many of these objects are to the left of the cluster in this plot. These objects lie in the 'Virgo Southern Extension' which is a cloud of galaxies impinging on the cluster at the present epoch.

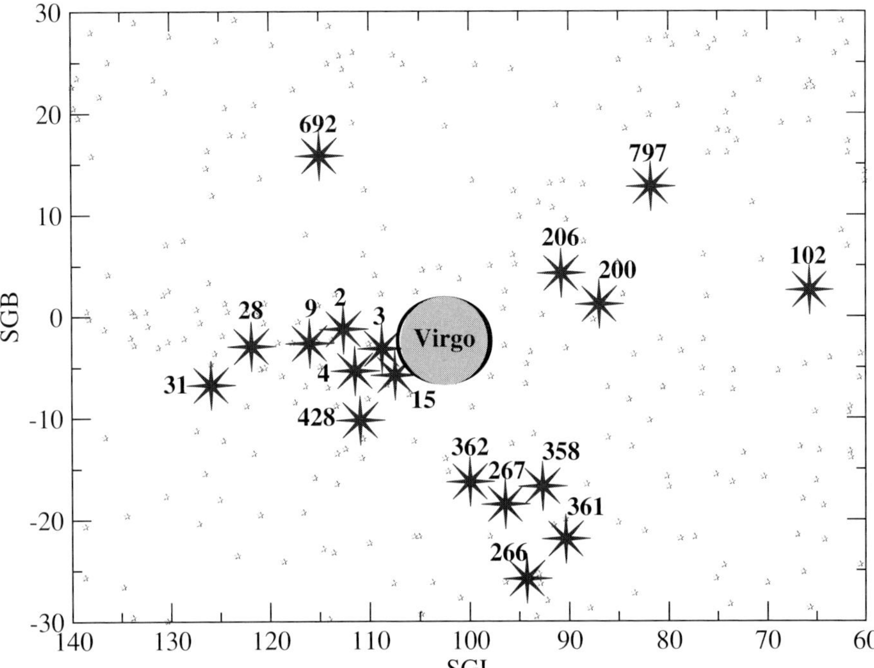

Figure 2. Projection on the sky of the Virgo cluster and surrounding region. The various symbols indicate the positions of objects projected in the vicinity of the cluster and those noted by large stars and numbered are particularly 'interesting' from the standpoint of the Virgo infall problem. The infall region extends to $\sim 28^{\circ}$ around the cluster, encompassing all but the object labelled 102 on this plot.

The objects highlighted in Fig. 2 have received special attention because they have good distance measures and many of them provide strong constraints on the cluster mass. An example is presented in Figure 3. The two panels illustrate the triple-value wave in the direction of a specific group of galaxies (object 4 in Fig. 2: the 11-4 Group in Tully 1988) with the only difference between the two cases being the Virgo cluster mass assignment. In both cases, the Action model was solved assuming $M/L = 125 M_{\odot}/L_{\odot}$ as a baseline. In the case shown on the right, however, the mass of the clusters dominated by early type systems (for present purposes that means the Virgo cluster) was augmented by 7 to $M/L = 875 M_{\odot}/L_{\odot}$. The mass of the cluster was increased from $1.6 \times 10^{14} M_{\odot}$ (left panel) to $1.2 \times 10^{15} M_{\odot}$ (right panel). The measured distance and velocity of the 11-4 Group is indicated by the point with error bars. The group of stars indicate the distance and velocity of the Virgo cluster as modelled by a series of Action trials. The 'x' indicate distances and velocities determined for the 11-4 Group with the Action trials. The points that describe waves rising from zero distance and velocity to 40–45 Mpc at 3000 km/s are the orbital end points of test particles scattered along the line-of-sight of the 11-4 Group.

There is a clear difference between the two figures in the amplitude of the triple-value wave in the vicinity of the Virgo cluster. This difference is a consequence of the different cluster mass assignments. The wave defined by the test particles shows where *objects could lie* in the context of the specified model. In the right panel of Fig. 3, the data point representing the 11-4 Group lies on the triple-value wave, hence the model accommodates this datum. In the left panel, however, the same data point lies well away from the much

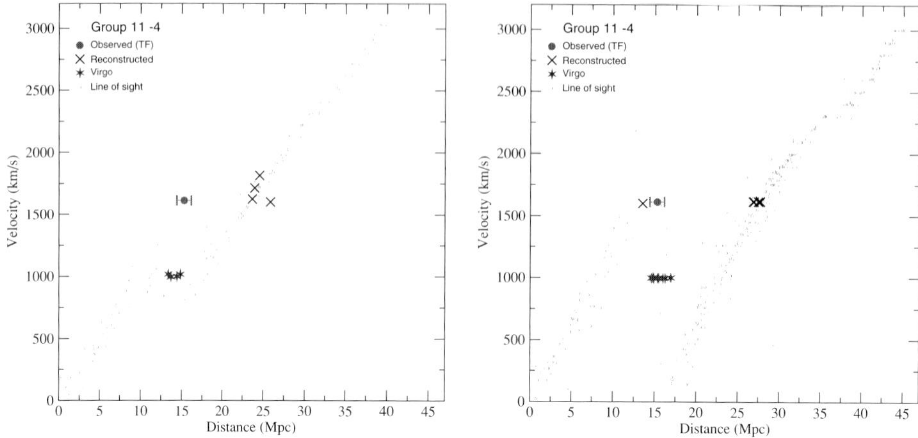

Figure 3. The left plot is the (limited) triple-value wave associated with the line-of-sight directed at the infall group 11-4 in action models with $M/L = 125M_{\odot}/L_{\odot}$ for all objects. In this case the mass of the Virgo cluster is $1.7 \times 10^{14} M_{\odot}$. The mean Virgo cluster distance and velocity given by several Action trials is indicated by the stars. The locations found for the 11-4 group in different trials with the Action models are given by the 'x'. The point with error bars indicate the *observed* position and velocity of the group. The right plot is the (much more extensive) triple-value wave along the same line of sight toward the 11-4 group generated by action models still with $M/L = 125M_{\odot}/L_{\odot}$ for most objects but now with the mass of the Virgo cluster (and other E/S0 groups in the catalog) amplified by a factor 7. The Virgo cluster is now given a mass of $1.2 \times 10^{15} M_{\odot}$. Symbols have the same meaning as in the left panel.

more modest triple-value wave. We are forced to conclude that either the data point is unreliable or the model shown in this figure is bad. The 'x' symbols that illustrate the positions found for the 11-4 Group in the Action models reinforce the point. In the left panel, the Action models put the group in the vicinity of 25 Mpc, in expansion away from the cluster, rather than at it's measured distance of 15.3 ± 0.9 Mpc. In the right panel, the Action models frequently put the group at a similar background position but on an occasion a saddle point of the action was chosen in good agreement with the observed distance. Clearly, in a computation of the χ^2 evaluator of the distance agreement this latter case is strongly favored. The important point to be made is that *in order for an infall solution to be available for the 11-4 Group the mass of the cluster has to be high enough to generate a triple-value wave with an amplitude that catches the velocity observed for the group.*

4. Conclusions

Similar analyses can be made of dozens of other lines-of-sight through the infall domain around the Virgo cluster. If there is a distance measurement that confirms that a target is inside the zero-velocity surface around the cluster then the line-of-sight component of its velocity puts a constraint on the cluster mass. In each case, the cluster mass must be at least enough to provide an Action infall solution.

We determine that a minimum but sufficient Virgo cluster mass for the models is $1.2 \times 10^{15} M_{\odot}$. This mass is 50% larger than the mass determined from application of the virial theorem (Tully and Shaya 1984). However, the virial radius for the cluster is ~ 0.8 Mpc while the mass measured by infall is on a scale $\geqslant 1.8$ Mpc. The infall analysis, which is based on very simple physics, gives an estimate of the *global* mass of the cluster. It can be anticipated that this mass will exceed the masses determined over more limited scales by gravitational lensing, X-ray, or virial studies.

Unfortunately, the analysis is not easily duplicated on other clusters, at least not so cleanly. The Virgo cluster is the only environment *near* enough that distance measurements distinguish between infall and expansion regimes, and *massive* enough to have an extensive, well populated infall domain. It provides a single good case. What we find, though, is that whereas overall the Action models find preference for $M/L \sim 125 M_\odot/L_\odot$ and $\Omega_m \sim 0.2$, in the Virgo cluster we need $M/L \sim 900 M_\odot/L_\odot$.

Acknowledgements

This research is being carried out in close collaboration with Jim Peebles, Steven Phelps, and Ed Shaya. Support for RBT is provided by JPL Contract 1243647. RM is supported by the European Union Marie Curie Fellowship HPMF-CT-2002-01532.

References

Bertschinger, E. 1985, ApJ, 58, 39
Freedman, W.L. et al. 2001, ApJ, 553, 47
Frisch U., Matarrese S., Mohayaee R. and Sobolevskii A. 2002, Nature, 417, 260
Jacoby, G.H., Ciardullo, R., Ford, H.C. 1990, ApJ, 356, 332
Karachentsev, I.D. et al. 2003, A&A, 398, 493
Lu, N.Y. et al. 1993, ApJS, 88, 383
Mathewson, D.S., Ford, V.L., Buchhorn, M. 1992, ApJS, 81, 413
Peebles, P.J.E. 1989, ApJ, 344, L53
Peebles, P.J.E. 1995, ApJ, 449, 52
Phelps, S.D. 2002, ApJ, 575, 1
Shaya, E.J., Peebles, P.J.E., Tully, R.B. 1995, ApJ, 454, 15
Spergel D.N. et al. 2003, ApJ, 148, 175
Tonry, J.L., Davis, M. 1981, ApJ, 246, 680
Tonry, J.L. et al. 2001, ApJ, 546, 681
Tully, R.B. 2003, astro-ph/0312441
Tully, R.B. 2004, astro-ph/0402516
Tully, R.B. 1988, *Nearby Galaxies Catalog*, Cambridge University Press
Tully, R.B., Fisher, J.R. 1977, A&A, 54, 661
Tully, R.B., Pierce, M.J. 2000, ApJ, 533, 744
Tully, R.B., Shaya, E.J. 1984, ApJ, 281, 31

Outskirts of Galaxy Clusters: Intense Life in the Suburbs
Proceedings IAU Colloquium No. 195, 2004
A. Diaferio, ed.

DOI: 10.1017/S1743921304000444

Infall from peculiar velocities

María Laura Ceccarelli[1], Carlos Valotto[1], Nelson Padilla[2], Diego G. Lambas[1], Riccardo Giovanelli[3] and Martha Haynes[3]

[1]IATE Group, Observatorio Astronómico, Universidad Nacional de Córdoba, Argentina
[2]Department of Physics, University of Durham, UK
[3]Department of Astronomy, Cornell University, USA

Abstract. We perform a statistical analysis to study the infall of galaxies onto groups and clusters in the nearby Universe. The study is based on the UZC/SSRS2 group catalogs and galaxy peculiar velocity data. We find a clear signature of infall of galaxies onto groups in a large range of scales 5 Mpc $\mathrm{h}^{-1} < r < 30$ Mpc h^{-1}. We obtain a significant increase of the infall amplitude with group mass derived from virial equilibrium and from the luminosity of the member galaxies. We compare the observational results with mock catalogs derived from numerical simulations of semi-analytical models where we obtain a behavior similar to observations.

---

1. Introduction

Inhomogeneities in the distribution of matter are the source of the gravitational field in the Universe. As a response to the existence of such a field, objects embedded in it develop peculiar velocities. The enveloping velocity field takes the form of a collapsing streaming motion towards the local density maximum, with a velocity amplitude which depends only on the distance to the mass overdensity (Peebles 1980). In this work we investigate the structure of the peculiar velocity field around groups of galaxies in the nearby universe. We use groups identified from the UZC/SSRS2 catalogs (Merchan et al. 2000) and a recent compilation of galaxy peculiar velocities (CPV) obtained by Giovanelli & Haynes (2004) in order to study directly the infall and the dependence on different group properties.

2. Statistical methods and results

We may describe the mean infall of galaxies onto groups and clusters using galaxy peculiar velocities which give only the line of sight projection of the three dimensional peculiar velocity of a galaxy. This is achieved by:

$$V_{pr}(r,\theta) = V_{infall}(r)\cos\theta, \tag{2.1}$$

where θ is the angular separation between the observer and the galaxy as seen from the group centre. We have neglected in this analysis group peculiar velocities whose effects are expected to cancel out in this study.

We analyse the dependence of galaxy peculiar velocities V_{pr} on $\cos\theta$, in different spherical concentric shells around the individual groups and calculate averages $\langle V_{pr}(\theta)\rangle$ for different bins in $\cos\theta$ for the total group sample.

In Figure 1 we plot $\langle V_{pr}\rangle$ vs $\cos\theta$ at different scales. As it can be seen in the figure, at small scales (panel a) the uncertainties are larger than the mean peculiar velocity. At larger scales, there is a clear infall signature (panels b, c and d). The maximum amplitude occurs between 10 Mpc h^{-1} and 20 Mpc h^{-1}. Only at the largest scales explored (panels e and f) the infall is significantly reduced.

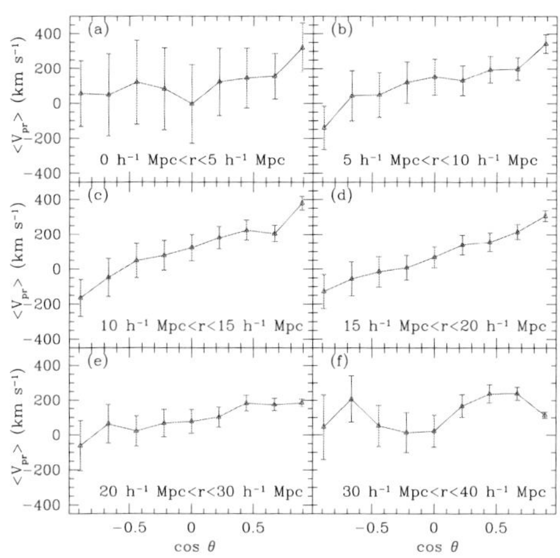

Figure 1. Mean values $\langle V_{pr} \rangle$ as a function of the angle θ for CPV galaxies and UZC groups.

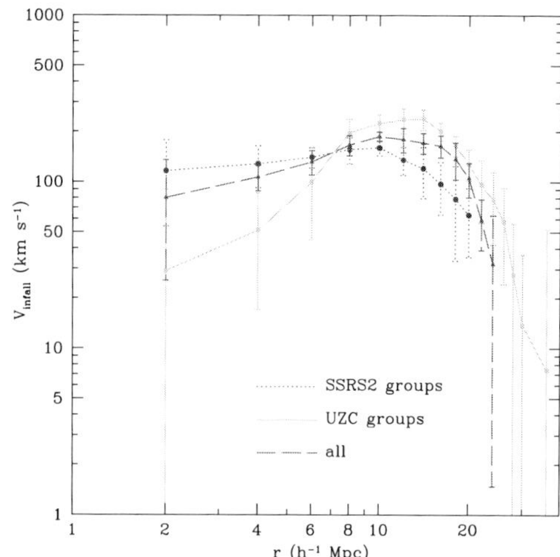

Figure 2. Amplitude of infall velocity of CPV galaxies onto UZC/SSRS2 groups.

The mean infall amplitude $V_{infall}(r)$ as a function of scale r is derived from the least square method applied to the observed $\langle V_{pr} \rangle$ vs $\cos\theta$. Figure 2 shows the resulting V_{infall} as a function of r where it can be seen that in the range of scales 5 Mpc $\mathrm{h}^{-1} < r <$ 20 Mpc h^{-1}, the amplitude of the infall velocities is in the range 100 km/s to 200 km/s.

It is expected that the amplitude of the infall velocities increases monotonically with group mass. To estimate group masses we have used either the virial mass M_V or the total group luminosity in the B band L_g. For groups with $M_V < 10^{13} M_\odot$, we obtain $V_{infall} \simeq$ 200 km/s whereas for $M_V > 10^{13} M_\odot$ a larger infall is observed, $V_{infall} \simeq$ 350 km/s. For groups with $L_g < 10^{15} L_\odot$ we obtain $V_{infall} \simeq$ 190 km/s whereas for $L_g > 10^{15} L_\odot$, $V_{infall} \simeq 490$ km/s. The results are shown in Figure 3 where it can be seen that, by dividing the sample into two roughly equal number subsamples by virial mass or luminosity, the difference in the amplitude of infall velocities is larger when dividing by luminosity than by mass.

The mock catalogues used to compare the results of the observations were extracted from a high resolution N-body simulation populated with galaxies using a semi-analytic model, by the theory group at the University of Durham. The simulation uses standard ΛCDM parameters and we have added observational constraints to construct suitable mock samples corresponding to both UZC groups and CPV galaxies.

We have tested the dependence of the amplitude of infall velocity on group virial mass and luminosity in the mock catalogues using the same procedure applied to the observational data. The mean infall velocities obtained are plotted in Figure 4, where we can see that the results from the mock samples are in qualitative agreement with those of observational data which indicates that galaxies and groups in semi-analytic models within the ΛDCM scenario provide a suitable agreement with the observations.

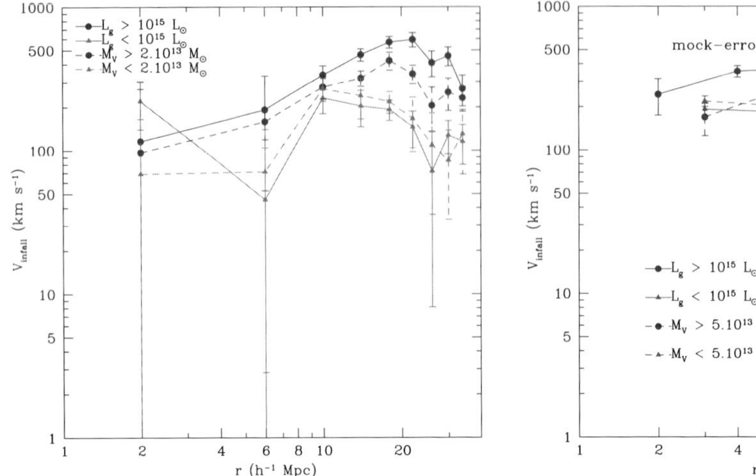

Figure 3. Mean infall velocity onto subsamples in UZC galaxy groups.

Figure 4. Mean infall velocity onto subsamples in mock catalogs.

3. Discussion

We find a clear signature of infall of galaxies onto groups in a large range of scales from galaxy peculiar velocity data. The amplitude of the infall velocities is of the order of few hundreds km/s. There is a significant dependence of the infall amplitude on group virial mass and in a similar fashion the total luminosity of the groups is an important parameter that determines the infall amplitude. We observe a larger difference in the infall amplitude for groups samples divided according to luminosity than to virial mass. We obtain a similar behaviour for the infall of galaxies onto groups in the numerical models which provides galaxy formation through semi-analytic prescriptions. The results for different virial mass and luminosity thresholds are consistent with those of the observational data, indicating that the models are suitable to reproduce the observations.

References

Peebles, J.P.E. 1980 *The Large Scale Structure of the Universe, Princeton University Press*

Merchán, M.E. et al. 2000 *ApJ.* **544**, 2

Outskirts of Galaxy Clusters: Intense Life in the Suburbs
Proceedings IAU Colloquium No. 195, 2004
A. Diaferio, ed.

DOI: 10.1017/S1743921304000456

Dynamical masses of RCS galaxy clusters

Kris Blindert[1], H. K. C. Yee[1], M. D. Gladders[2] and E. Ellingson[3]

[1]Department of Astronomy & Astrophysics, University of Toronto, 60 St. George St., Toronto, Ontario M5S 3H8, Canada email: blindert@astro.utoronto.ca, hyee@astro.utoronto.ca

[2]Carnegie Observatories, 813 Santa Barbara Street, Pasadena, California 91101, USA email: gladders@ociw.edu

[3]Center for Astrophysics and Space Astronomy, University of Colorado, 389 UCB, Boulder, Colorado 80309-0389, USA email: elling@kestrel.colorado.edu

Abstract. A multi-object spectroscopy follow-up survey of galaxy clusters selected from the Red-sequence Cluster Survey (RCS) is being completed. About forty clusters were chosen with redshifts from 0.15 to 0.6, and in a wide range of richnesses. One of the main science drivers of this survey is a study of internal dynamics of clusters. We present some preliminary results for a subset of the clusters, including the correlation of optical richness with mass, and the mass-to-light ratio as a function of cluster mass.

1. The survey

The RCS is a large imaging survey in R and z', which uses galaxy overdensity along with colour information to efficiently detect clusters out to $z>1$ (Gladders & Yee 2004). Detection of a cluster with the RCS technique also provides z_{phot}, a photometric estimate of the cluster redshift. In order to study cluster dynamics and galaxy populations, we are completing a spectroscopic follow-up survey of a subset of about 40 RCS clusters. These clusters have a wide range of optical richnesses (parametrized by the cluster-centre galaxy correlation amplitude B_{gc}; see Yee & Ellingson 2003, hereafter Y03), and are chosen with redshift estimates $z < 0.6$ so as to allow efficient follow-up. We have performed multi-object spectroscopy of galaxies in the fields of these clusters using the LDSS-2 spectrograph on the Magellan telescopes and the MOS/SIS on the CFHT. We observe two masks per field, for improved spatial sampling as well as some redundant observations (which help in estimating redshift errors). In order to obtain spatial coverage out to a few virial radii for each cluster, we observe between one and three fields depending on the redshift and richness of the cluster. Our selection criteria for mask design depends only on geometry and magnitude: we make no colour cuts, so as to ensure a fair sample for use in studying galaxy populations in these clusters. Overall, with two to six masks per cluster, roughly 30 spectra per mask with Magellan/LDSS-2 and 100 spectra per mask with CFH/MOS (thanks to a band limiting filter which we use to multi-tier our spectra), we have over 7000 spectra in total. Of course we cannot obtain redshifts from all spectra, nor are all the galaxies cluster members; ultimately we expect to have about 1750 members for the clusters in this survey.

2. Preliminary results

Here we present first results from 17 of the clusters in our sample for which we have obtained redshifts and computed weights using the CNOC technique (Yee, Ellingson, & Carlberg 1996). This subset of clusters spans our full redshift and richness ranges: $0.2 < z_{phot} < 0.6$ and $200 < B_{gc} < 1900$. We use the scheme of Zabludoff et al. (1990)

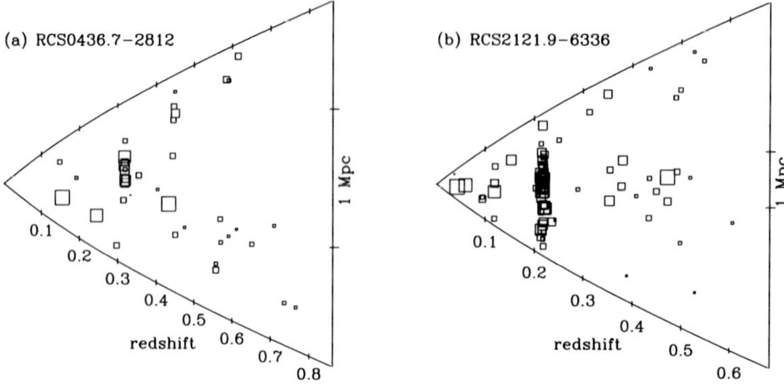

Figure 1. Pie diagrams of two example clusters; diagram (a) shows a poor cluster, while (b) is one of our richest systems. Each square is a galaxy, with size related to apparent magnitude.

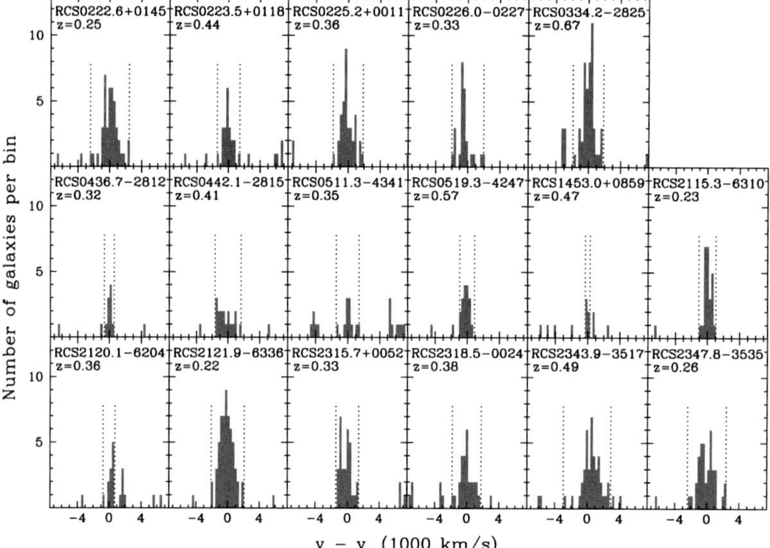

Figure 2. Velocity histograms of the clusters in 300 km/s bins. Dotted lines show the limits chosen by our interloper rejection scheme.

to reject non-members, which is quite strict (and hence may somewhat underestimate cluster properties such as velocity dispersion). Velocity dispersions are calculated using weights to account for incompleteness (we use magnitude, colour, and geometric weights). We estimate cluster masses using the formula $M = (3/G)\,\sigma_1{}^2\ r_{vir}$. The virial radius is usually calculated from pointwise galaxy separations, which can be noisy (Carlberg et al. 1996, hereafter C96); proper calculation of r_{vir} will require further investigation. In the meantime, we substitute $r_{200} = \sqrt{3}\sigma_1/10H(z)$ (Carlberg et al. 1997). K-corrected R_c-band cluster luminosities, computed within r_{200}, are also estimated using weights.

2.1. *Confirmation of the systems*

Figure 1 shows pie diagrams for two of the clusters in our subset, one poor and one rich. Velocity histograms for all 17 clusters are in Figure 2. We find many members in

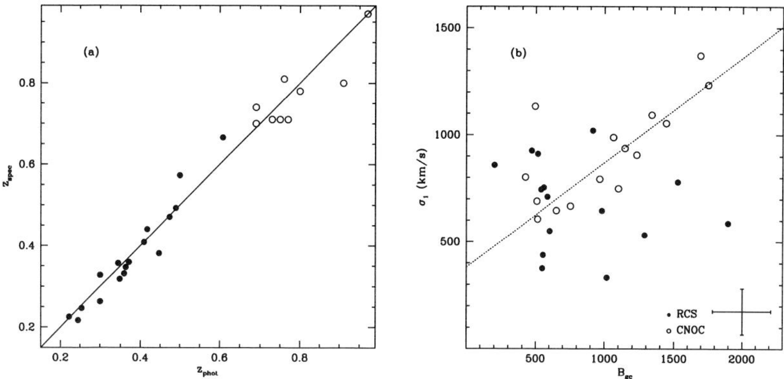

Figure 3. (a) Quality of the RCS redshift estimates. Filled circles are our subsample, and open circles are higher redshift clusters observed with Magellan LDSS-2 by F. Barrientos and M. Gladders. The solid line is not a fit to the data, but rather the $z_{phot} = z_{spec}$ line. **(b)** Optical richness versus velocity dispersion. Filled circles are our subsample, and open circles are CNOC clusters (Y03). For reference, the best fit to the CNOC clusters is shown as a dotted line. A typical error bar is shown at bottom right.

all but two systems: RCS0436.7-2812 and RCS1453.0+0859. The former is a real peak in the galaxy distribution, both on the sky and in velocity space (see Figure 1a), but is simply a very poor system: its optical richness is only B_{gc}=256. The latter system cannot be confirmed as a physical system: we see no clear peak in either velocity or on the sky. However, observing conditions were very poor for this cluster and few redshifts were obtained; in fact there does seem to be a small concentration of galaxies on the sky at this location, but redshifts couldn't be determined for any of them. Thus, though we cannot confirm this cluster candidate, neither can we dismiss it as a 'false positive'.

2.2. *Redshifts and richnesses of the clusters*

The RCS technique predicts reliable redshift estimates, even at low redshift where the R and z' filters no longer straddle the 4000Å break (see Figure 3a). The relation between optical richness and velocity dispersion, shown in Figure 3b, is not as tight. Though the B_{gc} and σ_1 values are roughly consistent with those found by Y03, the scatter is larger. In particular, we find one outlier with very low σ_1 compared to its optical richness. It is possible that this is an example for which our interloper rejection algorithm is too strict (see the velocity histogram of RCS0334.2-2825 in Figure 2). It may be that more sophisticated rejection algorithms will include these galaxies and increase the velocity dispersion of the cluster.

2.3. *Mass-to-light ratios*

The R-band mass-to-light ratios are plotted against velocity dispersion in Figure 4. We find an increase in M/L as a function of σ_1; in other words, more massive clusters have higher mass-to-light ratios. We caution the reader against over-interpreting this result, since our mass calculations are preliminary and our cluster sample is still relatively small. However, this result is in qualitative agreement with a comparison of CNOC clusters against CNOC2 groups (also shown in Figure 4), as well as with the findings of Eke et al. (2004) on groups in the 2dF galaxy survey and with K-band results for the CAIRNS rich clusters (Rines et al. 2004, also presented at this conference).

Outskirts of Galaxy Clusters: Intense Life in the Suburbs
Proceedings IAU Colloquium No. 195, 2004
A. Diaferio, ed.

DOI: 10.1017/S1743921304000468

Diffuse intracluster light in 10 clusters at $0.05 < z < 0.31$

J. E. Krick and R. A. Bernstein

Department of Astronomy, University of Michigan, Ann Arbor, MI 48109, USA
email: jkrick@umich.edu

Abstract. Surface brightness measurements of the diffuse intracluster light (ICL) in centrally concentrated, relaxed clusters have demonstrated that stars bound only to the cluster potential can contain anywhere from 5 to 50% of the total cluster flux. These stars are a by-product of the evolution of galaxies in clusters, and can be used to constrain models for the structure and evolution of clusters and their member galaxies. The ICL will be an important test of cosmological models as it is a baryonic component of clusters that is likely to evolve with redshift. To learn about these processes we are undertaking a program to measure the characteristics of the ICL (total flux, color, and substructure) as a function of cluster mass, morphology, and redshift. We have obtained data for a sample of 10 clusters at low and moderate redshifts. We discuss here our preliminary results for one cluster in that sample, A3888. With 4-5 hour integration times on-source per filter (V & gunn-r), we reach a 1 sigma surface brightness limit of 29.9 mag/arcsec^2 in the V band. We have identified an intracluster light component that is $\sim$ 10% of total cluster light, extends to 600 kpc from the center of the cluster, and is redder in color than the cluster galaxies ($V - r = 0.4$ near the center, increasing to $V - r = 0.8$ at r = 500 kpc). In addition we find 3 low surface brightness features which are blue ($V - r \simeq 0$), and in total are equivalent to one $m_V = 20.8$ mag galaxy.

1. Introduction

Diffuse intracluster light comes from stars which are bound only to the cluster potential and not to an individual galaxy. Studying the ICL can give insight into the physical processes which drive the formation, morphological evolution, and star formation histories of galaxies. In particular the ICL is interesting because it depends on the structure of the galaxies themselves, the interaction histories of the galaxies, and the details of star formation and how it is triggered inside (or outside) of galaxies. Additionally, the ICL, which is typically not included in the baryon census, will effect the baryon fraction of clusters and the change of baryon fraction as a function of time. If the ICL fraction does indeed evolve with redshift, it will have a systematic effect on the redshift dependence of the baryon fraction. Recent work by Allen, Schmidt, Fabian, & Ebeling (2003) has used a change in baryon fraction with redshift of a few percent to constrain cosmological parameters.

ICL measurements in the literature indicate that the ratio of ICL to total cluster light is $5 - 50\%$ (e.g., Vilchez-Gomez, Pello, & Sanahuja 1994; Bernstein, Nichol, Tyson, Ulmer, & Wittman 1995; Gonzalez, Zabludoff, Zaritsky, & Dalcanton 2000; Feldmeier, Mihos, Morrison, Harding, Kaib, & Dubinski 2004). Theoretical models and N-body simulations over the last 20 years have also predicted similar values for the ICL fraction (e.g., Malumuth & Richstone 1984; Moore, Katz, Lake, Dressler, & Oemler 1996; Dubinski 1998; Mihos 2004). Both the measurements and the modeling have now become sophisticated enough to make real advances in this field.

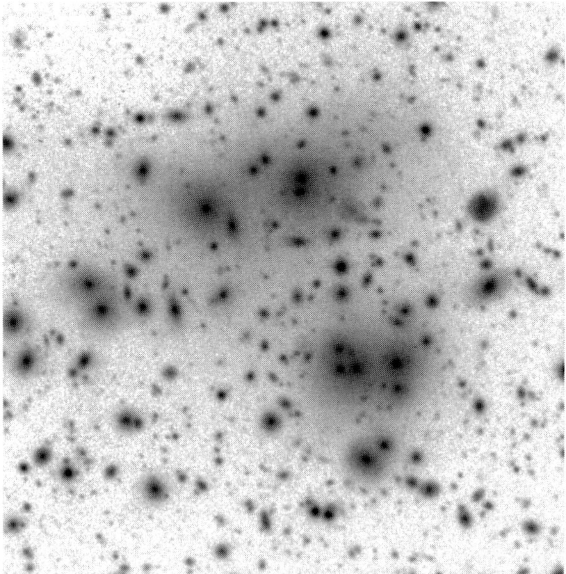

Figure 1. The center of A3888 in V. The grey–scale is linear over the range 20.4–29.5 mag arcsec^{-2}. An extended diffuse component is evident in the cluster core.

2. The program

We have observed a sample of 10 clusters as an addition to the forming statistical sample in the literature. These clusters were selected based on 3 main properties; (1) available X-ray observations giving an estimate of the masses, (2) the clusters are representative of a range in cluster mass, morphology, richness and redshift, and (3) they are at low HI column density and high galactic latitude to avoid the effects of galactic extinction and light (see Table 1). A few selected clusters have lensing mass estimates as a complimentary method of determining cluster mass.

We have fully reduced one cluster in our sample, A3888, for which we have $\sim$ 4.5 hours total integration time on source with the LCO 100" duPont telescope in both V and gunn-r. In addition 1/3 of the total exposure time over the run was devoted to blank sky images to create night sky flats. The data were processed in the standard way with special attention paid to flat fielding and sky subtraction. A large-scale flat fielding accuracy of 0.1% of the sky value is reached to achieve our goal of measuring the ICL to 29.9 mag/arcsec^2 in the V band. Figure 1 shows the fully reduced image of the central Mpc of A3888. It is evident that this cluster does not have a cD galaxy, however the three central clumps of galaxies are confirmed members. These central galaxies are not well fit by deVaucouleurs profiles or any combination of Sersic and exponential disk profiles. This cluster is dynamically active, which will be important in interpreting the flux and substructure in the ICL.

3. Substructure

Using the technique of unsharp masking, we find 3 possible tidal features all within the central 500 kpc of A3888. There are 2 small arcs (10 kpc x 3 kpc) near to the center of the cluster and 1 large (90 kpc x 13 kpc) diffuse, tail-like feature at $\sim$ 500 kpc from the center. These objects are unlikely to be gravitational arcs since they are not oriented tangentially to the cluster distribution. The diffuse nature of the large feature

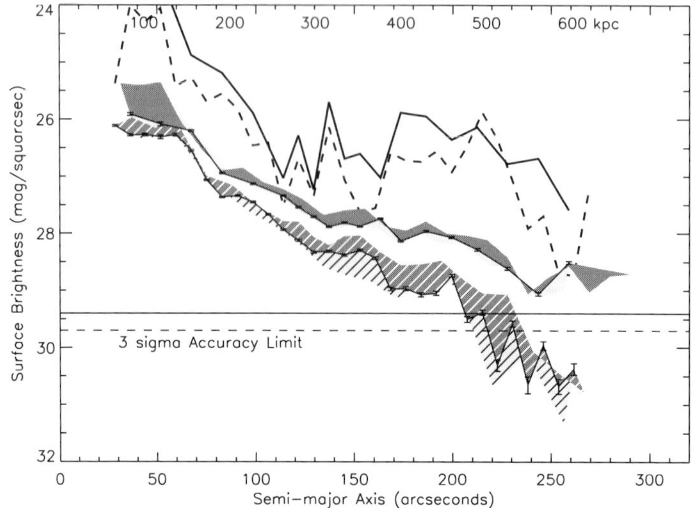

Figure 2. Radial profile of A3888 in V and gunn-r. The solid lines surrounded by shading show our measurement of the SB of the ICL as a function of radius in the r (solid shadings) and V band (hatched shadings). Error bars on the solid lines are standard deviation on the mean in each of the elliptical annuli. As mask size is a concern, we have adjusted the mask size±30% and plotted those surface brightness' as the shaded lines above and below the solid lines respectively. Total cluster flux is shown in the r (solid) and V (dashed) lines. A 3-sigma accuracy limit from the large scale flat fielding is shown for both r (solid) and V(dashed).

also suggests that it is tidal in nature. We cannot yet rule out the possibility that the 2 smaller features are low surface brightness galaxies seen edge-on.

All three features are blue in color, $V - r \simeq 0.0$, implying a young population coming from late-type galaxies. The total flux in these 3 features is equal to one r = 20.8 mag galaxy, which is 10% of an M^* galaxy. Given a cluster crossing time for this cluster of 3.5 Gyr and assuming these tidal features will dissipate in ~ 1 crossing time, over 1 Hubble time approximately $\frac{1}{2}$ of an M^* galaxy will be contributed to the ICL through the visible tidal features.

4. Surface brightness

To measure the flux in the ICL we remove all flux in the image due to detected stars and galaxies. We use SExtractor (Bertin & Arnouts 1996) to identify all sources down to $m_V = 24.8$ mag with sizes greater than 1.5" at a SNR of 5. We determine the PSF of this telescope to a radius of 120" (450 pix) by combining the profiles of super-saturated stars with non-saturated stars. The PSF is well fit by a Moffat function, which is used to subtract all stars from the image. The galaxy profiles, on the other hand, are not easy to quantify as they do not have a simple analytic form. In order to avoid errors from imperfect galaxy subtraction techniques, we mask all galaxies to 2.0 - 2.5 times their isophotal radii as defined by SExtractor. The masks are large enough that mask size has a negligible effect on the inferred ICL as described below.

After subtracting the stars and masking the galaxies, we smooth the image with a 5x5" boxcar, and then fit the resulting image with the IRAF routine ELLIPSE, (see figure 2). ELLIPSE is unable to fit isophotes in the very center where many pixels are masked, but quickly takes up a position angle similar to that in the x-ray image at larger radii, indicating that the ICL is following the underlying mass distribution of the cluster. Figure 2 also shows 3σ accuracy limits for the large scale flat fielding and background subtraction.

To identify the surface brightness profile of the resolved cluster galaxies we estimate cluster membership based on the color magnitude relation for this cluster. Those galaxies which lie within 1σ of a biweight fit to the color magnitude diagram are taken to be cluster members. In addition to this color cut we have also made a measurement of the contribution due to foreground and background galaxies by measuring the galaxy flux in a background annulus 1Mpc from the center of the cluster. More accurate measurements of cluster membership will come from a luminosity function (LF).

5. Discussion and conclusions

We have identified an intracluster component in A3888 out to 600 kpc from the center of the cluster down to ~ 30.0 mag/arcsec^2 in the V and gunn-r band. We show that this component does not follow the same light profile as the resolved sources, but rather has a steeper surface brightness profile. This ICL component is found to be aligned with the hot gas in the cluster as evidence of its correlation with the underlying mass distribution.

The ICL is redder than the galaxies, implying origins in an older population of stars. We find preliminary evidence for a color gradient in the ICL, where the ICL becomes redder with increasing radius ($V - r = 0.4 - 0.8$).

Comparing the ICL to cluster galaxy flux, we find that the ICL component in A3888 accounts for $\sim 10\%$ of the total cluster flux. A more accurate estimate of cluster flux will come from a LF, for which data has kindly been made available by Pimbblet, Smail, Kodama, Couch, Edge, Zabludoff, & O'Hely (2002). A LF will also provide us with an estimate of the flux from LSB galaxies below our detection threshold, which we are currently unable to disentangle from the ICL.

Reduction of the remaining sample of 9 clusters continues. Once complete, we will be able to examine the exciting physics locked up in the dependence of the ICL on cluster characteristics.

References

Abell, G. O., Corwin, H. G., & Olowin, R. P. 1989, *ApJS*, **70**, 1

Allen, S. W., Schmidt, R. W., Fabian, A. C., & Ebeling, H. 2003 *MNRAS* **342**, 287

Batuski, D. J., Miller, C. J., Slinglend, K. A., Balkowski, C., Maurogordato, S., Cayatte, V., Felenbok, P., & Olowin, R. 1999, *ApJ*, **520**, 491

Bernstein, G. M., Nichol, R. C., Tyson, J. A., Ulmer, M. P., & Wittman, D. 1995 *AJ* **110**, 1507

Bertin, E. & Arnouts, S. 1996 *A&AS*, 117, 393

Busarello, G., Merluzzi, P., La Barbera, F., Massarotti, M., & Capaccioli, M. 2002, *A&A*, **389**, 787

Chen, J., Huchra, J. P., McNamara, B. R., & Mader, J. 1998, *Bulletin of the American Astronomical Society*, **30**, 1307

Ciardullo, R., Ford, H., & Harms, R. 1985, *ApJ*, **293**, 69

Collins, C. A., Guzzo, L., Nichol, R. C., & Lumsden, S. L. 1995, *MNRAS*, **274**, 1071

Couch, W. J. & Newell, E. B. 1984, *ApJS*, **56**, 143

Couch, W. J. & Sharples, R. M. 1987, *MNRAS*, **229**, 423

Couch, W. J., Balogh, M. L., Bower, R. G., Smail, I., Glazebrook, K., & Taylor, M. 2001, *ApJ*, **549**, 820

den Hartog, R. 1995, Ph.D. Thesis

De Propris, R., Couch, W. J., Colless, M., Dalton, G. B., Collins, C., Baugh, C. M., Bland-Hawthorn, J., Bridges, T., Cannon, R., Cole, S., Cross, N., Deeley, K., Driver, S. P., Efstathiou, G., Ellis, R. S., Frenk, C. S., Glazebrook, K., Jackson, C., Lahav, O., Lewis, I., Lumsden, S., Maddox, S., Madgwick, D., Moody, S., Norberg, P., Peacock, J. A., Percival, W., Peterson, B. A., Sutherland, W., & Taylor, K. 2002, *MNRAS*, **329**, 87

Dubinski, J. 1998 *ApJ* **502**, 141

Ebeling, H., Voges, W., Bohringer, H., Edge, A. C., Huchra, J. P., & Briel, U. G. 1996, *MNRAS*, **281**, 799

Feldmeier, J. J., Mihos, J. C., Morrison, H. L., Harding, Kaib, P. N., & Dubinski, J. 2004 *ApJ*, in press

Girardi, M., Borgani, S., Giuricin, G., Mardirossian, F., & Mezzetti, M. 1998, *ApJ*, **506**, 45

Girardi, M. & Mezzetti, M. 2001, *ApJ*, **548**, 79

Gonzalez, A. H., Zabludoff, A. I., Zaritsky, D., & Dalcanton, J. J. 2000 *ApJ* **536**, 561

Kowalski, M. P., Ulmer, M. P., & Cruddace, R. G. 1983, *ApJ*, **268**, 540

Mazure, A., Katgert, P., den Hartog, R., Biviano, A., Dubath, P., Escalera, E., Focardi, P., Gerbal, D., Giuricin, G., Jones, B., Le Fevre, O., Moles, M., Perea, J., & Rhee, G. 1996, *A&A*, **310**, 31

Malumuth, E. M. & Richstone, D. O. 1984 *ApJ* **276**, 413

Mihos, J. C. 2004 In *Clusters of Galaxies: Probes of Cosmological Structure and Galaxy Evolution, from the Carnegie Observatories Centennial Symposia.* (ed. J.S. Mulchaey, A. Dressler, & A. Oemler). Carnegie Observatories Astrophysics Series, p. 278. Cambridge University Press.

Moore, B., Katz, N., Lake, G., Dressler, A., & Oemler, A. 1996 *Nature* **379**, 613

Pimbblet, K. A., Smail, I., Kodama, T., Couch, W. J., Edge, A. C., Zabludoff, A. I., & O'Hely, E. 2002 *MNRAS* **331**, 333

Stein, P. 1996, *A&AS*, **116**, 203

Struble, M. F. & Rood, H. J. 1999, *ApJS*, **125**, 35

Teague, P. F., Carter, D., & Gray, P. M. 1990, *ApJS*, **72**, 715

Vilchez-Gomez, R., Pello, R., & Sanahuja, B. 1994 *A&AS* **283**, 37

Wu, X., Xue, Y., & Fang, L. 1999, *ApJ*, **524**, 22

Table 1: Cluster characteristics

cluster name	ra (J2000) (hr:mm)	dec (J2000) (dd:mm)	l (degrees)	b (degrees)	z	B/M	Richness Class	L_x (0.1–2.4 keV) 10^{44}ergs/s[l]	σ_v km/s	Number of confirmed members	n(HI)[l] 10^{20}cm^{-2}	lensing measurement
A4059	23:57	-34:40	356.8	-76.06	0.048[r]	I	1	3.09	845^{+280}_{-140}[t]	45[d,p]	1.1	—
A3880	22:28	-30:34	17.99	-58.50	0.058[r]	II	0	1.86	827^{+120}_{-79}[m]	122[q,f,k]	1.1	—
A2734	00:11	-28:52	19.46	-80.98	0.062[r]	III	1	2.55	628^{+61}_{-57}[m]	127[j,p,k]	1.8	—
A2556	23:13	-21:38	41.37	-66.97	0.087[r]	II-III	1	2.47	1247^{+249}_{-249}[t]	5[e,o,b]	2.0	—
A4010	23:31	-36:30	359.0	-70.60	0.096[r]	I-II	1	5.55	625^{+127}_{-95}[m]	30[j,p]	1.4	—
A3888	22:34	-37:43	3.96	-59.40	0.151[r]	I-II	2	14.52	1102^{+137}_{-107}[n]	69[s]	1.2	—
A3984	23:15	-37:48	359.0	-67.19	0.181[r]	II-III	2	9.18	—	—[f]	1.7	—
A0141	01:06	-24:35	175.3	-85.93	0.23[r]	III	3	12.62	—	—[a]	1.8	Dahle et al. (2002)
AC118	00:14	-30:2	8.90	-81.24	0.308[r]	III	3	22.05	1947^{+292}_{-201}[h]	363[c,g,h]	1.7	Smail et al. (1991)
AC114	22:58	-34:47	8.32	-64.78	0.31[a]	II-III	2	7.0	1388^{+128}_{-71}[n]	380[g,i,h]	2.0	Smail et al. (1991)

Notes: a: Abell, Corwin, & Olowin (1989). b: Batuski et al. (1999). c: Busarello et al. (2002) d: Chen et al (1998). e: Ciardullo, Ford, & Harms, (1985). f: Collins et al. (1995). g: Couch & Newell (1984). h: Couch & Sharpless (1987). i: Couch et al (2001). j: den Hartog (1995). k: De Propris et al. (2002). l: Ebeling et al. (1996). m: Girardi et al. (1998). n: Girardi & Mezzetti (2001). o: Kowalski, Ulmer, & Cruddace, (1983). p: Mazure et al. (1996). q: Stein (1996). r: Struble & Rood (1999). s: Teague, Carter, & Gray, (1990). t: Wu, Xue, & Fang (1999).

Outskirts of Galaxy Clusters: Intense Life in the Suburbs
Proceedings IAU Colloquium No. 195, 2004
A. Diaferio, ed.

DOI: 10.1017/S174392130400047X

Intracluster light at z∼0.25 from SDSS imaging data

Stefano Zibetti and Simon D. M. White

Max-Planck-Institut für Astrophysik, Garching bei München, Germany email: zibetti@mpa-garching.mpg.de

Abstract. We investigate the broadband optical emission of diffuse intergalactic stars in galaxy clusters at $z \sim 0.25$ by means of an image stacking technique. The images of 654 clusters, selected with the max-BCG algorithm from a subsample of the SDSS-Data Release 1 (DR1), have been stacked after masking all the sources detected down to very low S/N. The resulting images in the g, r and i bands provide reliable photometric data at the level of $\gtrsim$ 30 mag/□" (r band), out to $\gtrsim$ 600 kpc from the Brightest Cluster Galaxy (BCG). Our analysis shows that: i) the IntraCluster Light (ICL) is much more concentrated than the galaxy light, contributing ∼30% of the total cluster optical emission at $R = 100$ kpc, but less than 10% at $R > 500$ kpc; ii) the ICL contributes between 15 and 20% of the total cluster optical luminosity between the optical radius ($\mu_r = 25$ mag/□") of the BCG and 500 kpc; iii) the colours of the ICL are consistent with the global colours of the cluster galaxies, with little evidence for redder g−r.

1. Introduction

Firstly proposed by Zwicky(1951), the presence of a diffuse population of intergalactic stars in galaxy clusters is now a well established observational fact (see e.g. Arnaboldi 2003 and references therein, Krick and Bernstein this conference). Such a population is commonly believed to originate from the disruption of small galaxies and from the tidal stripping of the outskirts of large galaxies in high-density environments. Studying the properties of the ICL as a function of redshift and of the richness and morphology of the parent cluster can give a significant insight in the interaction processes that drive galaxy evolution in dense environments.

Observing the intracluster light (ICL) is extremely challenging, due to the very low surface brightness involved (> 26 mag/□"), which is less than 1% of the typical surface brightness of the sky. During the last years the increasing sensitivity of CCD detectors and the development of new observational techniques has made it possible to start a number of optical surveys aiming at the study of the ICL, by means of both broadband imaging of unresolved stellar populations (e.g. Feldmeier *et al.* 2002) and of narrow band detection of intracluster planetary nebulae (e.g. Arnaboldi *et al.* 2003). Nevertheless, for a statistically representative number of clusters of different morphology and richness and located at different redshifts we still lack reliable determinations of the basic properties of the ICL, such as the fractional amount with respect to the total optical emission of the cluster, the radial distribution, the link with (possible) cD envelopes, or the colours. In fact, observations of the ICL in individual clusters are extremely demanding in telescope time and require an incredibly high control of the flat fielding and of the sources of light pollution, such as fore- and background sources and the internal reflections of the camera.

We have chosen to tackle the problem with a purely statistical approach, adopting an image stacking technique, similar to that used by Zibetti, White, & Brinkmann (2004). The Sloan Digital Sky Survey (SDSS, York *et al.* 2000) is providing a one-quarter-of-sky coverage with 5-band imaging (Fukugita *et al.* 1996, Gunn *et al.* 1998) and spectroscopy.

From a subsample of its first data release (DR1, Abazajian *et al.* 2003), we have extracted a catalogue of 654 clusters and rich groups in the redshift range 0.2-0.3, whose imaging in the g, r, and i bands was processed and stacked, as described in Sec. 2. The results are presented and discussed in Sec. 3.

2. Method: the stacking technique

The very essence of the stacking technique consists in averaging a large number of images of clusters, in order i) to increase the signal-to-noise ratio and measure surface brightnesses as faint as $\mu_r \sim 30$ mag/□", and ii) to statistically remove the effects of flat fielding inhomogeneities and the contamination from scattered light, internal reflections, and fore- and background sources. Given the typical S/N in the SDSS imaging and the limitation imposed by the present day coverage of the survey, a suitable number of images ranges from several hundreds to roughly 1,000.

2.1. *The SDSS sample*

The sample selection focused on clusters in the redshift range 0.2-0.3 for two main reasons. First, we want the spatial extension of the individual clusters not to exceed the dimensions of the frame, since an annulus of at least 1 Mpc around the cluster must be included on the same frame for a reliable background subtraction. Given the size of the SDSS frames, the chosen redshift range represents the best trade-off between this spatial constraint and the desire for minimum cosmological dimming of the apparent surface brightness and for maximum efficiency in detecting galaxies. On the other hand, at $z = 0.25$ the g−r colour conveniently maps the 4000Å-break, thus providing important information about the intracluster stellar population.

The cluster sample is drawn from ~ 1500 deg^2 of photometric data in the SDSS DR1, using the maxBCG method by Annis *et al.* (2004) (see also Bahcall *et al.* 2003 for a description of this technique). We require estimated photometric redshift $0.2 < z < 0.3$, 17 or more galaxies identified as red-sequence members within 1 Mpc projected distance† from the BCG, and at least 13 of them within 330 kpc. The clusters selected in this way have richness roughly in the same range as the Abell clusters. Clusters in vicinity of bright foreground stars or in frames contaminated by scattered light have been pruned from the sample.

2.2. *Image processing*

The SDSS imaging data are available as bias subtracted, flat-field corrected frames. For each cluster we estimate the sky background in an annulus of 1 Mpc, 100 kpc thick, centered on the BCG, after masking all the sources detected at very low S/N with SExtractor, in at least one of the three passbands. After subtracting the background, for each image two kinds of masks are built, "ICL" and "galaxy". The former are obtained from SExtractor segmentation images, adopting fixed surface brightness detection thresholds (μ_r=24.5, μ_g=25.0, μ_i=24.0) and minimum detection area of 10 contiguous pixels. Corrections for galactic extinction and cosmological dimming $(1 + z)^4$ with respect to the median redshift of the sample are applied. The masks of bright sources are grown by 30 pixels in order to avoid scattered light, while the BCG is left unmasked. The final "ICL" mask results from the OR combination of the masks obtained for the three passbands. In the "galaxy" masks only the bright sources are masked.

The images and the corresponding masks are geometrically transformed, centered on

† H_0=70 km sec^{-1} Mpc^{-1}, Ω_0=1, Ω_Λ=0.7 throughout this paper.

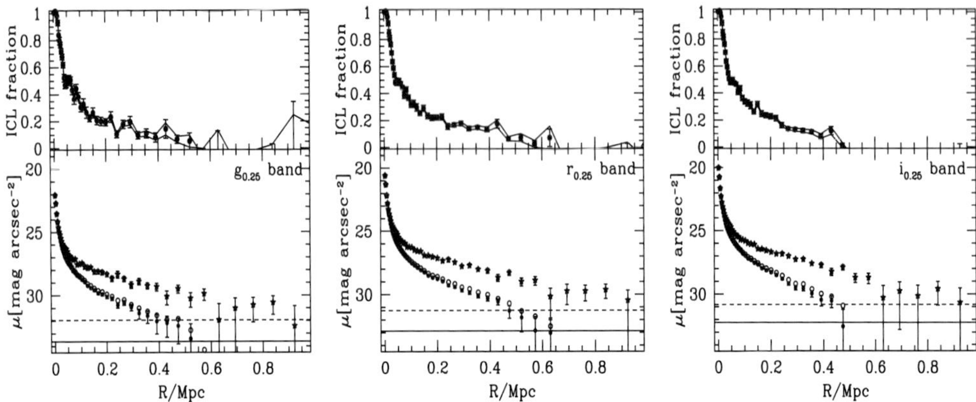

Figure 1. Lower panels: SB profiles for the total light (stars), the measured diffuse light (open circles) and the corrected ICL (filled circles). Lines display the background uncertainty for the total light (dashed line) and for the diffuse component (solid line). Errorbars are 1-σ errors, including random noise and systematic background uncertainty. Upper panels: the local ICL/total ratio. Errorbars account for random errors only, while the solid lines represent the 1-σ intervals when accounting for systematic errors in the background estimation.

the BCG, rescaled to the same physical units and randomly rotated. Finally, the images are averaged excluding the pixels masked according either to the "ICL" or to the "galaxy" masks, so that it is possible to estimate the average contribution from the diffuse light (plus the BCG) and the average total light, respectively.

3. Results

In Fig. 1 (lower panels) we show the azimuthally averaged SB profiles extracted from the stacked images, after subtracting the background evaluated at 1 Mpc, in the g, r, and i band respectively. The star points represent the total cluster light ("galaxy" masks), while the open circles are the diffuse light ("ICL" masks). The solid and the dashed lines display the level of the background uncertainty for the diffuse component and for the total light, respectively. As apparent from these plots, we are able to measure SB† as faint as $\mu_r \simeq 32$ for the diffuse component, and $\mu_r \simeq 30$ for the total light, the latter being limited by the shot noise of the background galaxy distribution.

Whether the diffuse light we measure is "true" ICL is questionable, depending on the actual definition of ICL. In principle one should consider as ICL only the light coming from stars which are unbound from individual galaxies and freely fluctuating in the cluster potential. Since this is clearly not a viable operational definition, we consider as ICL all the light coming from outside of the optical radius of galaxies (i.e. the isophotal radius at $\mu_r = 25$). This is a well defined quantity, that can be also easily tested against theoretical predictions. However, due to the magnitude limits of the SDSS (m_r=22.2), we expect significant contribution to the diffuse light from undetected cluster galaxies. We quantify this contamination by simulating a set of mock galaxy clusters, in the same observational conditions (redshift, galactic extinction, seeing) as the observed ones. Each simulated cluster includes 1,000 galaxies, whose apparent magnitude is assigned according to a montecarlo realization of a Schechter luminosity function (LF). We use the distribution of Sersic profile parameters as a function of luminosity as observed by

† All surface brightnesses are expressed in the observer's frame filter system for clusters at z=0.25.

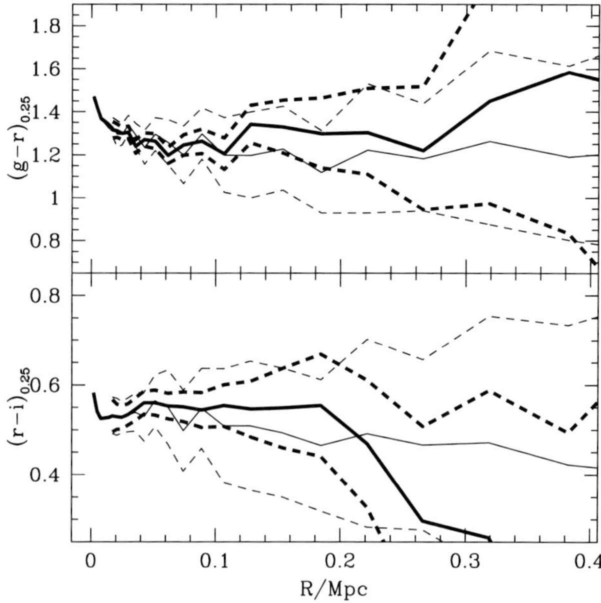

Figure 2. g−r and r−i colour profiles. Dashed lines represent the 1-σ confidence intervals, thick lines are for the ICL, thin for the total cluster light.

Blanton *et al.* (2003) to create a realistic 2D image of each galaxy, truncated at the $\mu_r = 25$ isophote. The images are then convolved with the observed PSF, background and poissonian noise are added. After processing and stacking the mock images in the same way as the observed ones, we measure the fraction of galaxy light that is missed by the masking algorithm and contributes to the diffuse component. This fraction depends on the assumed LF, and ranges from a few percent for a flat LF, up to 15% in the case of a steep faint end. Assuming the LF given by Goto *et al.* (2002) ($M_r^* = -22.21$, $\alpha = -0.85$), this fraction is 5% (r band).

The corrected SB of the ICL is represented by filled circles in the lower panels of Fig. 1. We find that the ICL is much more concentrated than the total light in the cluster, irrespective of the passband. The local fractional contribution of the ICL is shown in the upper panels of Fig. 1. Note that, since we do not mask the BCG, the fraction is 1 in the center. At 100 kpc the fraction of light contributed by the ICL is $\sim$ 30%, and drops to $\sim$ 5% at 500 kpc. The overall fraction of ICL between the average optical radius of the BCG (30 kpc) and 500 kpc is 20% in the r band, and 19% in the g and i bands, with typical uncertainties of $\pm$2%. Assuming a steeper LF can lower these figures by up to $\sim$ 5%.

Using the three observed bands, we study the g−r and r−i colour profiles of the ICL out to 400 kpc from the BCG. They are shown in Fig. 2 as the thick solid lines. The total-light colours are shown as thin solid lines. Dashed lines display the $1-\sigma$ confidence intervals. We note a strong negative colour gradient in the central 20 kpc, that is inside the BCG. The outer profiles do not display any significant gradient in the galaxy component, although marginally bluer r−i is measured at larger radii. Neglecting the r−i at $R > 0.2$ Mpc, whose value is largely undetermined due to the very low signal, we find that the ICL colours are very similar to those of the galaxy component, with perhaps weak evidence for marginally redder g−r.

This result and the strong, non-linear dependence of the ICL surface brightness on the galaxy density, are consistent with the idea of ICL being emitted by passively evolving stellar populations, stripped from the cluster galaxies in the highest density regions.

The forthcoming analysis of a larger sample of clusters (Zibetti *et al.* 2004, in preparation), differentiated according to their richness and morphology (e.g. Bautz-Morgan type), will contribute new clues about the origins of the ICL.

Acknowledgements

We thank Jim Annis for kindly providing us with the maxBCG catalog of clusters used in this work.

Funding for the creation and distribution of the SDSS Archive has been provided by the Alfred P. Sloan Foundation, the Participating Institutions, the National Aeronautics and Space Administration, the National Science Foundation, the U.S. Department of Energy, the Japanese Monbukagakusho, and the Max Planck Society. The SDSS Web site is http://www.sdss.org/. The SDSS is managed by the Astrophysical Research Consortium (ARC) for the Participating Institutions.

References

Abazajian, K. *et al.* 2003, *AJ*, **126**, 2081–2086

Annis, J. *et al.* 2004, in preparation

Arnaboldi, M. 2003, In *Recycling intergalactic and interstellar matter* (Eds. P. A. Duc, J. Braine, & E. Brinks). IAU Symposium Series, vol. 217 (*astro-ph/0310143*)

Arnaboldi, M. *et al.* 2003, *AJ*, **125**, 514–524

Bahcall, N. A. *et al.* 2003, *ApJS*, **148**, 243–274

Blanton, M. R. *et al.* 2003, *ApJ*, **594**, 186–207

Goto, T. *et al.* 2002, *PASJ*, **54**, 515–525

Feldmeier, J. J., Mihos, J. C., Morrison, H. L., Rodney, S. A., & Harding, P. 2002, *ApJ*, **575**, 779–800

Fukugita, M., Ichikawa, T., Gunn, J. E., Doi, M., Shimasaku, K., & Schneider, D. P. 1996, *AJ*, **111**, 1748–1756

Gunn, J. E., *et al.* 1998, *AJ*, **116**, 3040–3081

York, D. G. *et al.* 2000, *AJ*, **120**, 1579–1587

Zibetti, S., White, S. D. M., & Brinkmann, J. 2004, *MNRAS*, **347**, 556–568

Zibetti, S. *et al.* 2004, in preparation

Zwicky, F. 1951, *PASP*, **63**, 61

Outskirts of Galaxy Clusters: Intense Life in the Suburbs
Proceedings IAU Colloquium No. 195, 2004
A. Diaferio, ed.

DOI: 10.1017/S1743921304000481

Suprime-33: Suprime-Cam weak lensing cluster survey

Takashi Hamana

National Astronomical Observatory of Japan, Mitaka, Tokyo 181-8588, Japan

Abstract. We present the ability of weak lensing surveys to locate massive clusters. We use both analytic models of dark matter halos and mock weak lensing surveys generated from a large cosmological N-body simulation. We find that, for a weak lensing survey with a galaxy number density of $n_g = 30$ arcmin^{-2} with a mean redshift $z = 1$, the mean number of galaxy clusters which are expected to be identified is $N(\mathrm{S/N} > 4) = 7$ per 1 deg^2. Weak lensing surveys thus provide a reasonably efficient way to search for massive clusters. We also describe our project on the weak lensing cluster survey exploiting Suprime-Cam (the wide-field prime focus camera on the 8m Subaru telescope), called Suprime-33 and present current status. We also present results from its pilot survey in the Suprime-Cam 2.1 deg^2 field.

---

1. Introduction

Recently, it has become feasible to locate massive clusters directly as density enhancements using weak gravitational lensing. Miyazaki et al. (2002) indeed discovered many clusters in a weak lensing halo survey over a 2.1 square degree field, proving the ability of weak lensing to identify massive clusters. Unlike conventional optical or X-ray selected cluster catalogs, weak lensing cluster catalogs are free from a bias toward the luminous objects because the cluster finding is not based on the flux (or galaxy number density) enhancement but on the *projected mass* enhancement. This is a great advantage of gravitational lensing. However, lensing has its own disadvantages as every other cluster survey technique (optical, X-ray and Sunyaev-Zel'dovich survey) does; for example, the projection effect from unrelated structures in the same line-of-sight (Hamana, Takada & Yoshida 2004). Since clusters of galaxies are one of most important cosmological probes, it is of fundamental importance to construct large unbiased samples. Given advantages and disadvantages in each survey technique, it is desirable to combine, in a complementary manner, several techniques to construct an unbiased catalog. Such catalogs provide a valuable data set to investigate, for example, the physical state of the intra-cluster medium. To do this, it is important to correctly understand the selection function and completeness of each survey technique. For this purpose, we examined ability of a weak lensing cluster survey using both analytic models and mock numerical simulations (Hamana et al. 2004). We summaries the basic results from the research in section 3.

In section 2, we describe our project on the weak lensing cluster survey exploiting Suprime-Cam (the wide-field prime focus camera on the 8m Subaru telescope), called Suprime-33 and present current status. We also present results from its pilot survey in the Suprime-Cam 2.1 deg^2 field.

2. Suprime-33

Suprime-33 is a weak lensing survey which exploits the unique capabilities of Suprime-Cam. Its primary purpose is to construct, for the first time, a *mass selected* cluster catalog. Conventionally, cluster catalogs were generated by either an optical or X-ray observation. Galaxy clusters in such catalogs are selected based on the (either optical or X-ray) flux enhancement, therefore they must be biased toward luminous objects. Importantly, a weak lensing survey has the potential to detect massive but optical/X-ray faint clusters which conventional surveys have missed.

In our pilot weak lensing cluster search in a GTO 2 deg^2 field, we had indeed detected 5 clusters with a significant signal (S/N> 5) (Miyazaki et al. 2002), which demonstrated that the weak lensing survey is practical with Suprime-Cam. We are currently expanding the survey area to 33 deg^2 and expect our sample to contain about 100 clusters. We chose the Suprime-33 survey area where ROSAT/PSPC's deep pointings X-ray data ($T_{exp} > 25$ ksec) are available. This allows us to construct the mass selected cluster catalog having deep X-ray data.

The primary science goal of Suprime-33 is to test the structure formation paradigm with a well defined mass-limited cluster catalog. One of the most important and unambiguous tests of this paradigm concerns the abundance of halos and its evolution with redshift. These have been extensively studied using N-body simulations, and are regarded as well-established theoretically. Therefore, the comparison between the theoretical prediction and our weak lensing measurement is one of the simplest and the most direct method to test the paradigm. This test has not yet been achieved because it requires large observational resources. Also, using this first mass selected cluster catalog which covers a redshifts up to ~ 0.8, we will be able to make important contributions to the understanding of cluster physics.

3. Ability of weak lensing cluster surveys

The ability of weak lensing cluster surveys to locate massive clusters was examined in detail by Hamana et al. (2004) using numerical simulations. They showed that the expected number of clusters detected in weak lensing surveys is 7 per 1 deg^2 for a high detection signal of S/N> 4 and for the survey depth of Suprime-33 ($R_{\rm lim} = 26$). This is comparable to a deep X-ray selected cluster counts (1 (5) per 1 deg^2 for $f_{[0.5-2{\rm keV}]} > 1 \times 10^{-13}$ ($> 2 \times 10^{-14}$) erg cm^{-2} s^{-1}). The exposure time to reach $R_{\rm lim} = 26$ is only 30min for Suprime-Cam on the 8m Subaru telescope, and Suprime-Cam has 27×34 FOV. Thus a survey cost is 2 hours per 1 deg^2. Weak lensing surveys thus provide a reasonably efficient way to search for massive clusters.

Acknowledgements

T.H. acknowledges supports from Japan Society for Promotion of Science (JSPS) Research Fellowships. The N-body simulations used in this work (section 3) were carried out by the Virgo Consortium at the computer center at the Max-Planck-Institut, Garching. Numerical computations presented in this paper were partly carried out at ADAC (the Astronomical Data Analysis Center) of the National Astronomical Observatory, Japan, and at the Yukawa Institute Computer Facility.

References

Miyazaki S., et al., 2002, ApJ, **580**, L97
Hamana T., Takada M., Yoshida N., 2004, MNRAS, in press

Outskirts of Galaxy Clusters: Intense Life in the Suburbs
Proceedings IAU Colloquium No. 195, 2004
A. Diaferio, ed.

DOI: 10.1017/S1743921304000493

HIROCS – a galaxy cluster survey at high redshifts

S. Falter[1], H.-J. Röser[1], H. Hippelein[1], C. Wolf[2] and E. Bell[1]

[1]Max Planck Institute for Astronomy, Königstuhl 17, D-69117 Heidelberg, Germany
email: falter@mpia-hd.mpg.de

[2]Dep. of Physics, Denys Wilkinson Bldg., Univ. of Oxford, Keble Road, Oxford, OX1 3RH, UK

Abstract. Clusters of galaxies are the largest bound gravitational systems in the universe. They outline the large-scale structure and thus are able to test predictions of cosmological models. Detailed studies of galaxy populations in clusters at high z offer insights into galaxy evolution as a function of look-back time. Unfortunately, large homogeneous samples of galaxy clusters only exist at redshifts < 0.5. There are around 1000 candidate clusters with photometric or otherwise estimated redshifts above 0.5, but spectroscopically confirmed clusters are still scarce in this redshift regime. Some clusters with $z > 1$ have recently been identified from X-ray surveys.

1. Survey outline

HIROCS (= Heidelberg InfraRed/Optical Cluster Survey) intends to homogeneously select candidate clusters with redshifts up to 1.5. Our survey is part of a large programme (MANOS = MPIA for Astronomy Near-infrared Optical Surveys) exploring the evolution of galaxies. To reach out to redshifts of 1.5 we are observing in the infrared (H) supplemented by four optical (B, R, i, z) bands. HIROCS will cover 11 □° in four fields, three stripes with $3° \times 1°$ and one with $2° \times 1°$ (see Table 1). This ensures a sufficient coverage of large scale structure features in the foreseen redshift range. Furthermore, predictions for the number of galaxy clusters above unity from Bartelmann & White (2002) for the SDSS show that our dedicated survey area should yield a statistically significant cluster sample. They predict more than 5 clusters per □° with $z > 1$.

With one exception (CADIS-16h field) we have equatorial fields with few bright stars enabling follow-up studies from both hemispheres with the VLT and the LBT (available in the course of 2004). The right ascension of the fields is distributed such that observations are possible during the whole year. Sufficiently low and uniform galactic extinction over the field area is another requirement.

The MUNICS and CADIS fields provide extensive information for fixing the selection function of the reduced HIROCS filter set. In these fields we have multi-color data sets as well as a large number of spectra for QSOs, galaxies and stars. We chose the

Field name	RA (J2000)	DEC (J2000)	Size	E(B−V)
MUNICS-S2F1	$03^h\,06^m\,41^s.0$	$00°\,01'\,12''$	$3° \times 1°$	0.080 mag
COSMOS-10h	$10^h\,00^m\,28^s.0$	$02°\,12'\,21''$	$2° \times 1°$	0.018 mag
CADIS-16h	$16^h\,24^m\,32^s.0$	$55°\,44'\,32''$	$3° \times 1°$	0.006 mag
HIROCS-22h	$22^h\,00^m\,00^s.0$	$02°\,14'\,20''$	$3° \times 1°$	0.065 mag

Table 1. HIROCS fields – coordinates, sizes and galactic extinction from Schlegel, Finkbeiner & Davis (1998).

	$m^{\star}_{k-corr}$	E-corr	$m^{\star}$	obsLim	Δt	S/N	$\Delta t/\Box^{\circ}$
B_J	29.0	-5.975	23.1	24.1	1500	19.0	6 ksec
R_J	27.3	-2.878	24.4	25.4	5000	5.3	20 ksec
i	26.7	-2.291	24.4	25.4			~20 ksec
z	25.0	-2.480	22.5	23.5	6000	5.2	24 ksec
H	21.2	-1.348	19.9	20.9	3000	5.0	48 ksec

Table 2. Limiting magnitudes (col. obsLim) and exposure times to cover $1\Box^{\circ}$ for the HIROCS filters. For the calculation a standard E galaxy SED is scaled to produce an absolute magnitude of an $L^{\star}$-galaxy in the Gunn r-band of $M_r = -21.75$ mag (Garilli, Maccagni & Andreon 1997, Paolillo, Andreon, Longo, *et al.* 2001). This spectrum is k-corrected ($z = 1.5$) and corrected for evolutionary effects after Poggianti (1997). The observations are dithered and split into short exposures. The summed integration time per pointing is shown in the column Δt. The last column incorporates the field of view of the detector and specifies the required time to cover $1\Box^{\circ}$. We assume that one hour of exposure time yields a net science exposure time of 3 ksec. Cosmology: $H_{\circ} = 65$, $\Omega_M = 0.3$, $\Omega_{\Lambda} = 0.7$.

COSMOS-10h field because the COSMOS survey will obtain high-resolution images from HST. The HIROCS-22h field is a new field.

Our survey aims at a limiting magnitude of $L^{\star} + 1$ for an elliptical galaxy at $z = 1.5$ (5σ). Table 2 shows the limiting magnitudes and the required exposure times.

The multi-colour-classification scheme developed for the CADIS and COMBO-17 surveys (Wolf, Meisenheimer & Röser 2001) will classify all objects and assign photometric redshifts. However, the accuracy of the redshifts will not be as accurate though sufficient for cluster detections.

2. Cluster finding

We will employ a 3-dimensional cluster finding algorithm based on finding overdensities in position on the sky and in redshift space. According to its redshift z_0 derived from the multi-colour-classification the projected Abell radius R_A for each object is calculated and the number of objects N_{obj} within R_A counted. Only objects with velocity differences below a given threshold are added to N_{obj}:

$$\frac{v}{c} = \frac{(1+z_1)^2 - (1+z_0)^2}{(1+z_1)^2 + (1+z_0)^2} \leqslant \left(\frac{v}{c}\right)_{lim} . \tag{2.1}$$

Candidate galaxy clusters are found as excess numbers in N_{obj} which exceed values expected for a random object distribution. From these excess numbers we define an overdensity parameter.

A first test application to the object catalog of the Chandra Deep-Field South (CDFS) observed by COMBO-17 (field size $31' \times 30'$) reveals galaxy overdensities at $z \sim 0.5$, 0.7 and 1.1 (Figure 1).

In the past several other techniques for identifying clusters of galaxies have been developed. We intend to compare the performance of our algorithm with the Voronoi tessellation (Ramella, Boschin, Fadda & Nonino 2001), the matched-filter and the friends-of-friends algorithm. Because of the imaging in five filters we will also be able to apply the red sequence approach established for the Toronto Red-sequence Cluster Survey (Gladders & Yee 2000) for various colour combinations.

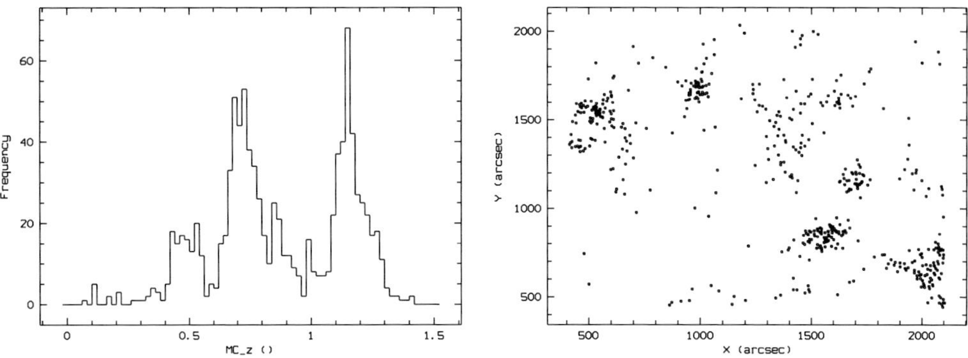

Figure 1. Left: z-histogram shows object enhancements at z ~ 0.5, 0.7, 1.1. Objects are selected by overdensity; Right: Object distribution in the CDFS. The condensations of points indicate candidate galaxy clusters (coordinates in arbitrary units).

3. Project status

Since August 2003 we are collecting the survey data at the Calar Alto 3.5 m telescope and since December 2002 with the Wide-Field-Imager (WFI) at the ESO 2.2 m telescope. A first impression of the quality of the IR data gives a pipeline-reduced mosaic of 12 H-band images in the HIROCS-22h field observed with the NIR wide-field camera OMEGA2000 on this webpage:

http://www.caha.es/newsletter/news04a/OMEGA2000_mosaic.pdf

It covers almost $3/4\,\square^\circ$; the PSF is $\sim 0.''8$. Each frame is composed of 25 single exposures with 1 minute exposure time, the limit is 19.9 mag (5σ). The holes around bright objects are due to over-corrected background, as the pipeline, whose result is shown, uses only a local sky determination.

References

Bartelmann, M. & White, S. D. M. 2002 *A&A* **388**, 732–740.
Garilli, B., Maccagni, D. & Andreon S. 1997 *A&A* **342**, 408–416.
Gladders, M. D. & Yee, H. K. C. 2000 *AJ* **120**, 2148–2162.
Paolillo, D. S. F., Andreon, S., Longo, G., et al. 2001 *A&A* **367**, 59–71.
Poggianti, B. M. 1997 *A&A* **122**, 399–407.
Ramella, M., Boschin, W., Fadda, D. & Nonino, M. 2001 *A&A* **368**, 776–786.
Schlegel, D. J., Finkbeiner, D. P. & Davis, M. 1998 *ApJ* **500**, 525S.
Wolf, C., Meisenheimer, K. & Röser, H.-J. 2001 *A&A* **365**, 660–680.

Outskirts of Galaxy Clusters: Intense Life in the Suburbs
Proceedings IAU Colloquium No. 195, 2004
A. Diaferio, ed.

DOI: 10.1017/S174392130400050X

ESO distant cluster survey: spectroscopy†

C. Halliday[1], B. Milvang-Jensen[2], S. Poirier[3], B. M. Poggianti[1], P. Jablonka[3], A. Aragón-Salamanca[4], R. Pelló[5], R. P. Saglia[2], G. De Lucia [6], L. Simard[7], D. I. Clowe[8], G. Rudnick[6] and S. D. M. White[6]

[1] Osservatorio Astronomico, vicolo dell'Osservatorio 5, 35122 Padova
[2]Max-Planck Institut für extraterrestrische Physik, Giessenbachstrasse, D-85748 Garching
[3]GEPI, CNRS-UMR8111, Observatoire de Paris, F-92195 Meudon Cedex, France
[4]School of Physics and Astronomy, University of Nottingham, NG7 2RD, UK
[5]Laboratoire d'Astrophysique, UMR 5572, Observatoire Midi-Pyrenees, 31400 Toulouse
[6]Max-Planck-Institut für Astrophysik, Karl-Schwarzschild-Str. 1, D-85741 Garching
[7]Herzberg Institute of Astrophysics, N. R. C. of Canada, Victoria, BC V9E 2E7, Canada
[8]Steward Observatory, University of Arizona, 933 North Cherry Avenue, Tucson, AZ 85721

Abstract. We present first results for spectroscopic observations of galaxies in 4 clusters at $z = 0.7 - 0.8$ and one cluster at $z = 0.5$ observed by the ESO Distant Cluster Survey (EDisCS). Our spectroscopic catalogues contain 236 cluster members of our 5 clusters, and the number of members per cluster ranges from 30 to 67. Our cluster velocity dispersions are between ~400 and over 1000 km s^{-1}. Galaxy redshift distributions are found to be non-Gaussian and we find evidence for significant substructure in two clusters, one at $z \sim 0.79$ and another at $z \sim 0.54$; both clusters have velocity dispersions exceeding 1000 km s^{-1}. These systems have clearly not yet virialised at these epochs in qualitative agreement with CDM scenarios and their cluster velocity dispersions should not be used in the measurement of cluster mass. Our clusters have a wide range of different cluster velocity dispersions, richnesses and substructuring, and our spectroscopic data set is allowing a comprehensive insight into cluster galaxy evolution as a function of redshift and environment.

1. Introduction

We present first spectroscopic results for the ESO Distant Cluster Survey (EDisCS). EDisCS (PI: White), is a photometric and spectroscopic survey of galaxies in 20 clusters at redshifts 0.4 to 0.8 selected from the Las Campanas Distant Cluster Survey (LCDCS) (Gonzalez et al. 2001). A first test of cluster reality for our EDisCS clusters was completed using imaging in two passbands (Gonzalez et al. 2002). The EDisCS project is investigating galaxy morphology, structure, masses, stellar populations and star formation histories as a function of both redshift and cluster environment, and study cluster mass and structure. Our programme involves deep optical photometry with the VLT (14 nights) (White *et al.* in preparation), near-IR photometry with the NTT (20 nights) (Aragón-Salamanca *et al.* in preparation), and multi-slit spectroscopy using the FORS2 spectrograph at the VLT (20 nights) (Halliday *et al.* A&A accepted; Milvang-Jensen *et al.* in preparation). This ground-based data set has been complemented by 80 orbits of HST ACS imaging (PI: Dalcanton) for 10 EDisCS clusters (Desai *et al.* in preparation).

† Based on observations collected at the European Southern Observatory, Chile (166.A–0162).

Cluster name (1)	No. of members (2)	$z_{cluster}$ (3)	$\sigma_{cluster} \pm \delta\sigma_{cluster}$ (4)	P_{DS} (5)
Cl 1040	30	0.7043	$418\ ^{+55}_{-46}$	0.08
Cl 1054-11	49	0.6972	$589\ ^{+78}_{-70}$	0.64
Cl 1054-12	36	0.7498	$504\ ^{+113}_{-65}$	0.14
Cl 1216	67	0.7943	$1018\ ^{+73}_{-77}$	0.05
Cl 1232	54	0.5414	$1080\ ^{+119}_{-89}$	0.01

Table 1. Cluster members and velocity dispersions are presented for each EDisCS cluster indicated in column 1; column 2 - number of cluster members; column 3 - cluster redshift; column 4 - cluster velocity dispersions ($\sigma_{cluster}$) and errors in km s^{-1}; and column 5 - measure of probability of cluster substructuring from a Dressler-Shectman analysis.

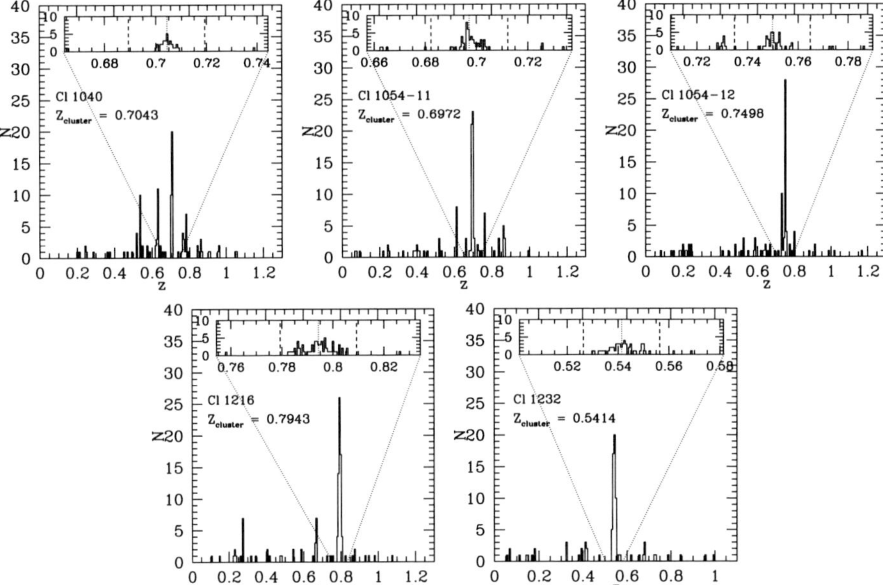

Figure 1. Redshift histograms for all 5 EDisCS clusters. Inset panel of each plot shows zoomed histogram for the redshift interval $z_{cluster} - 0.04 \leqslant z_{cluster} \leqslant z_{cluster} + 0.04$, where the cluster redshift $z_{cluster}$ is indicated by a dotted line. Dashed lines indicate the interval $z_{cluster} - 0.015 \leqslant z_{cluster} \leqslant z_{cluster} + 0.015$. Redshift bins are 0.00625 (1874 km s^{-1}) for main plot and in inset plot 0.000625 (187 km s^{-1}) (observed frame).

2. Spectroscopy of EDisCS: first results

Deep spectroscopic exposures have been obtained to date for 9 of our high-z clusters and 4 of our low-z clusters. Galaxies were selected from our I-band photometric catalogues to have I magnitude < 22 for our low-z clusters, and I < 23 for our high-z clusters. Probabilities of cluster membership and star and galaxy classification were generated using our photometric redshift codes (Pelló, Rudnick *et al.* in preparation). Our spectroscopy has defined cluster membership and we are constraining the ages and metallicities of cluster absorption-line galaxies using evolutionary population synthesis models and studying early- and late-type galaxy evolution using emission line equivalent widths, the Fundamental Plane and Tully-Fisher scaling relations and interpretation with galaxy morphologies. We present first results for the VLT spectroscopy data of 5 clusters (4 high-z and 1 low-z) from Halliday *et al.* Our redshift histograms in Figure 1 show that

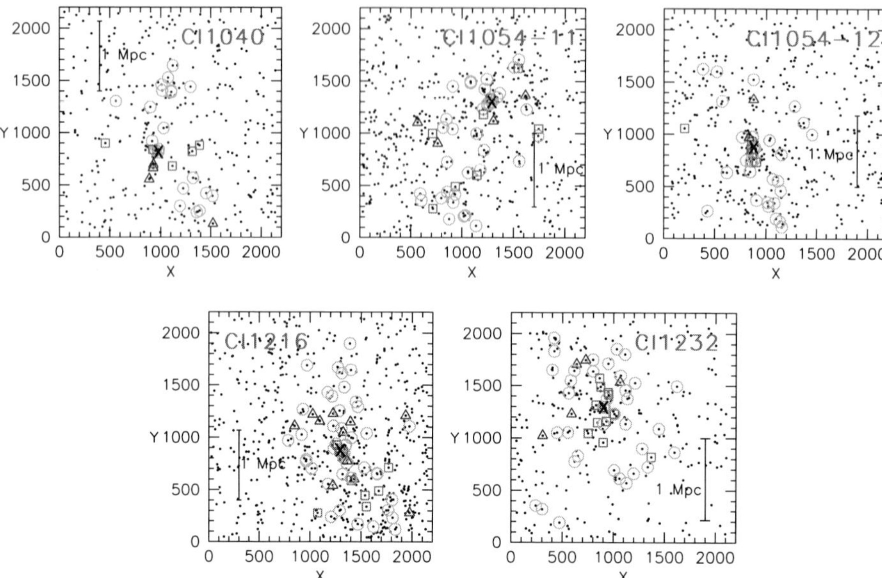

Figure 2. Plots of galaxy positions. Axis scales are in units of CCD pixels (0.2 arcsec/pixel). The cluster name is indicated at the top of each plot. Open symbols indicate the positions of cluster members: green open circles show galaxies within ± one z_σ from the cluster redshift $z_{\rm cluster}$, red open squares indicate galaxies whose redshift is more than one z_σ higher than $z_{\rm cluster}$, and blue open triangles represent galaxies whose redshift is lower than $z_{\rm cluster}$ by more than one z_σ, where z_σ is the redshift interval corresponding to the cluster velocity dispersion in the *observed* frame. Dots mark positions of galaxies selected from our I-band photometric catalogues as possible cluster members. The position of the brightest cluster galaxy (BCG) is given by a black cross symbol. A distance of 1 Mpc, calculated for a ΛCDM cosmology with $\Omega = 0.3$, $\Lambda = 0.7$ and $h = 0.7$, is provided in each plot.

the cluster velocity distributions of the 5 EDisCS clusters are non-Gaussian. The XY positions of cluster members are given in Figure 2. For two clusters (Cl1232 and Cl1216), substructure is detected with more than 95% probability by a Dressler-Shectman (DS) analysis (Dressler & Shectman 1988) (see Column 5 of Table 1). In Cl1232 the red-shifted galaxies are mainly in projection towards the cluster centre. In Cl1216 blue-shifted galaxies are offset North-East of the cluster centre. For two clusters no significant substructure is detected (Cl1040 and Cl1054-12). No evidence for substructure is found for Cl1054-11. *In conclusion, the clusters of the highest cluster velocity dispersion exceeding 1000* km s^{-1} *(Cl1216 and Cl1232) display the most significant substructuring and are probably still undergoing collapse in qualitative agreement with CDM hierarchical scenarios. Their cluster velocity dispersions should not be used in the measurement of cluster mass.*

References

Dressler, A. & Shectman, S. A. 1988 *A. J.* **95**, 985–995.
Gonzalez, A. H., Zaritsky, D., Dalcanton, J. J. & Nelson, A. 2001 *Ap. J. S.* **137**, 117–138.
Gonzalez, A. H., Zaritsky, D., Simard L., Clowe D. & White, S. D. M. 2002 *Ap. J.* **579**, 577–586.

Outskirts of Galaxy Clusters: Intense Life in the Suburbs
Proceedings IAU Colloquium No. 195, 2004
A. Diaferio, ed.

DOI: 10.1017/S1743921304000511

Spectroscopy of high-redshift EIS clusters

L. F. Olsen[1], H. E. Jørgensen[1], E. Zucca[2], S. Bardelli[2], A. Biviano[3], C. Benoist[4], L. da Costa[5], M. Ramella[3] and M. Scodeggio[6]

[1]Copenhagen University Observatory, Juliane Maries Vej 3, DK-2100 Copenhagen, Denmark
email: lisbeth@astro.ku.dk

[2]INAF - Osservatorio Astronomico di Bologna, via Ranzani 1, I-40127 Bologna, Italy

[3]INAF - Osservatorio Astronomico di Trieste, Via G.B. Tiepolo 11, I-34131 Trieste, Italy

[4]Observatoire de la Côte d'Azur, CERGA, BP 229 06304 Nice, Cedex 4, France

[5]European Southern Observatory, Karl-Schwartzschild-Str. 2, D-85748 Garching b. München, Germany

[6]Istituto di Fisica Cosmica - CNR, Milano, Italy

Abstract. We present multi-object spectroscopic observations from FORS1 mounted at the VLT for 5 clusters detected in the ESO Imaging Survey. Together with the results for three other clusters presented previously, we confirm a total of 6 out of 8 clusters with estimated redshift $z_{MF} \geqslant 0.6$. The spectroscopic redshifts are in good agreement with the estimated ones.

1. Introduction

Based on the EIS I band survey Olsen (2000) presented a catalogue of 310 cluster candidates using a matched filter technique. The estimated redshifts z_{MF} range from 0.2 up to 1.3. Hansen et al. (2002) and Olsen et al. (2003) have presented spectroscopic confirmation of 14 cluster candidates at z_{MF} = 0.2-0.3 and Ramella et al. (2000) have confirmed 4 clusters at $z_{MF} \leqslant 0.7$. Some preliminary results for three EIS cluster candidates in the range $0.8 \leqslant z \leqslant 1.3$ were given by Benoist et al. (2002). Here we present results for further five EIS cluster candidates at $0.6 \leqslant z_{MF} \leqslant 1.0$.

2. Observations

We selected the target clusters $z_{MF} \geqslant 0.6$ on the basis of optical/infrared imaging data, that were used to detect simultaneous overdensities in color and space. For all the targeted clusters we detected a promising overdensity of galaxies with similar $I - K_s$ and $J - K_s$-colors (Olsen 2000). The target galaxies were selected from their optical-infrared colour properties.

The observations were carried out using FORS1 mounted at VLT-ANTU telescope. In the multi-object spectroscopy (MOS) mode FORS1 provides 21 slits. We used grism 150I+17 with the order separation filter OG590 covering the wavelength range 6000-11000Å. This grism combined with a slitwidth of 1.4 arcsecs gave a resolution of 29Å.

3. Results

To identify groups in redshift space we use the "gap"-method by Katgert et al. (1996). We use a gap-size of $0.005(1 + z)$ to separate individual groups. This separation corresponds to a rest-frame velocity of 1500 km/s. Additionally we require that there are at least 3 galaxies in a group to consider it for further analysis. To assess the significance of

Table 1. The identified groups with significance $\sigma \geqslant 99\%$.

Field	# galaxies	z	σ_v km/s	σ
EISJ0046-2951	17	0.616	1340	99.9
EISJ0046-2951	**10**	**0.671**	**890**	**99.7**
EISJ0048-2942	7	0.402	1040	99.4
EISJ0048-2942	**33**	**0.638**	**1110**	**>99.9**
EISJ0050-2941	12	0.559	1390	99.9
EISJ0050-2941	8	0.617	1060	99.1
EISJ2236-4017	12	0.509	940	99.9
EISJ2249-3958	**8**	**0.710**	**400**	**99.4**

the systems we determine the probability of detecting a group with the same number of members at the same redshift in a pure field galaxy sample. The details of the methods are given in Benoist et al. (2002).

EISJ0046-2951. We find 2 significant groups in this field as listed in Table 1. The two groups have redshifts $z = 0.616$ and $z = 0.671$. In order to decide which of the groups is the one most likely to have caused the matched filter detection we inspect the positions of the galaxies in each group. The group at $z = 0.616$ forms an elongated structure to the north from the EIS position, and thus cannot have caused the matched filter detection. The other group with $z = 0.671$ has five galaxies at the matched filter position and five members further away. Therefore, we consider this the confirmation of the EIS cluster. The group has a velocity dispersion of $\sigma_v = 890$ km/s, which makes it a fairly massive system.

EISJ0048-2942. We find two significant groups in this field with redshifts $z = 0.402$ and $z = 0.638$. Given the large number of members in the latter group there is no doubt that the cluster at $z = 0.638$ is responsible for the matched filter detection. The cluster is rich and completely dominating the field. The velocity dispersion of this system is $\sigma_v = 1110$ km/s making it a massive system.

EISJ0050-2941. We find two significant groups in this field with redshifts $z = 0.559$ and $z = 0.617$. Both groups are spread over the field and it is unclear which one corresponds to the matched filter detection. Considering the originally estimated redshift of EISJ0050-2941 of $z_{MF} = 1.0$ compared with the much lower redshifts of the groups detected here it is interesting to note that there is a group of faint galaxies at the matched filter position that has not been observed spectroscopically. From inspecting the luminosity function of excess galaxies at the matched filter position we find that the excess galaxies have magnitudes fainter than $I \geqslant 21$, which corresponds well to the magnitudes expected for the bright cluster galaxies at $z \sim 1$. Therefore, we consider this case inconclusive.

EISJ2236-4017. In this field only one group is found to be significant. This group has a redshift of $z = 0.509$ and a velocity dispersion of $\sigma_v = 940$ km/s. This group is spread over a large fraction of the field, and is thus surprising to have caused the matched filter detection. Inspecting the image of the field we find a concentration of fainter more concentrated galaxies that could be responsible for the matched filter detection. These galaxies were unfortunately not covered by our spectroscopic observations. It should be noted that we had to abandon one of our masks for this field, which could potentially have included a number of interesting galaxies.

EISJ2249-3958. Also in this field we find only one significant group. It has a redshift of $z = 0.710$ and a velocity dispersion of $\sigma_v = 400$ km/s corresponding to a rather poor system. The structure is rather elongated and one may therefore speculate that it could be a non-virialized system.

4. Conclusions

An important question for the potential of the EIS cluster candidate catalog for populating the high-redshift domain of known clusters is related to the reliability of the matched filter estimated redshifts, z_{MF}. For the high-z cluster sample of 8 systems (5 reported in this work and 3 reported by Jørgensen et al. 2004) we compare the spectroscopic redshifts with the original z_{MF}. Altogether we have 6 confirmed systems which we use to investigate the redshift reliability of the EIS cluster candidates. We use the difference, $\Delta z = z_{spec} - z_{MF}$, between the spectroscopic and matched filter estimated redshift as a measure of the redshift reliability. For the five systems we find that the spectroscopic redshifts ranges from $\Delta z = -0.23$ to $\Delta z = 0.21$. The average deviation is $\Delta z = -0.02$ with a standard deviation of 0.16. Altogether the matched filter estimated redshifts are on average consistent with the spectroscopic ones, even though deviations can be up to $\Delta z \sim 0.23$. This result is valid all the way to the highest redshifts found in the catalog and thus makes the EIS cluster candidate catalog a good source for populating the high-z universe.

Acknowledgements

LFO thanks the Carlsberg Foundation for financial support.

References

Benoist, C., da Costa, L., Jørgensen, H.E., et al., 2002, A&A, 394, 1

Hansen, L., Olsen, L.F., Jørgensen, H.E., 2002, A&A, 388, 1

Jørgensen, H.E., et al., 2004, in preparation.

Katgert, P., Mazure, A.,Perea, J., et al., 1996, A&A, 310, 8

Kinney, A.L., Calzetti, D., Bohlin, R.C., et al., 1996, ApJ 467, 38

Olsen, L.F., 2000, Ph.D.-dissertation, Copenhagen University.

Olsen, L.F., Hansen, L., Jørgensen, H.E., et al., 2003, A&A, 409, 439

Ramella, M., Biviano, A., Boschin, W., et al., 2000, A&A, 360, 861

Outskirts of Galaxy Clusters: Intense Life in the Suburbs
Proceedings IAU Colloquium No. 195, 2004
A. Diaferio, ed.

DOI: 10.1017/S1743921304000523

Scaling relations for galaxy clusters

C. Marmo[1], G. Fasano[2], E. Pignatelli[2], B. Poggianti[2], D. Bettoni[2], C. Halliday[2], J. Varela[3], M. Moles[4], P. Kjærgaard[5], W. Couch[6] and A. Dressler[7]

[1]Astronomy Department, University of Padova, Italy email: marmo@pd.astro.it
[2]INAF, Osservatorio Astronomico di Padova, Italy
[3]IMAFF, CSIC, Madrid, Spain
[4]Instituto de Astrofísica de Andalucía, Granada, Spain
[5]Copenhagen University Observatory, Copenhagen, Denmark
[6]School of Physics, University of New South Wales, Sydney, Australia
[7]Observatories of the Carnegie Institution of Washington, Pasadena, USA

Abstract. We use preliminary results of the WINGS survey (Fasano *et al.*) to obtain determinations of optical scaling relations for galaxy clusters. Passing from one- to two- parameter scaling relations we pay particular attention to the Kormendy relation (KR) and to the Fundamental Plane (FP) of galaxy clusters, comparing them with scaling relations of elliptical galaxies.

1. Cluster profiles

Spherical symmetry has been assumed in determining the cluster surface brightness profile. Then, galaxies have been counted in a series of a concentric rings and integrated magnitudes have been obtained multiplying each galaxy by its own light in the V band. We correct the V magnitudes for galactic extinction, according to Burstein & Heiles (1982). The correction for back and foreground is made statistically, referring to the galaxy counts of Cabanac et al. (2000). Due to the dimension and shape of the wide–field detectors we had to apply an area correction, in order to circularize the rectangular limits of the field. We counted galaxies in each ring down to an apparent V magnitude of 20.5 mag. Then, we corrected for luminosity function incompletness. This correction turned out to be of the order of 1% of the total luminosity. Finally we fit the growth curve of the luminosity profile with a de Vaucouleurs law.

2. One-parameter scaling relations

While King and, in general, cored profiles reproduce the galactic luminosity profile of clusters better than the de Vaucouleurs one, we use the luminosity extrapolated from the de Vaucouleurs profile to make a comparison with ellipticals. Moreover, different definitions of luminosity can strongly affect the determination of some scaling relations.

In Figure 1 the optical luminosities of our clusters are plotted versus X–Ray luminosities and velocity dispersions from the literature. In particular, the top panels refer to the total luminosities derived from de Vaucouleurs fits of the profiles, respectively. Instead, the bottom panels report the luminosities obtained from the galaxy counts in our catalogues, just correcting for background contamination and for incompleteness of the area (L_{Acorr}). The solid lines in the plots represent our linear, weighted best fits to the data. It is evident that the slopes and *r.m.s.* of our fits decrease from the top to the bottom panels. On the other hand, the quantity L_{Acorr}, rather than of the total cluster light,

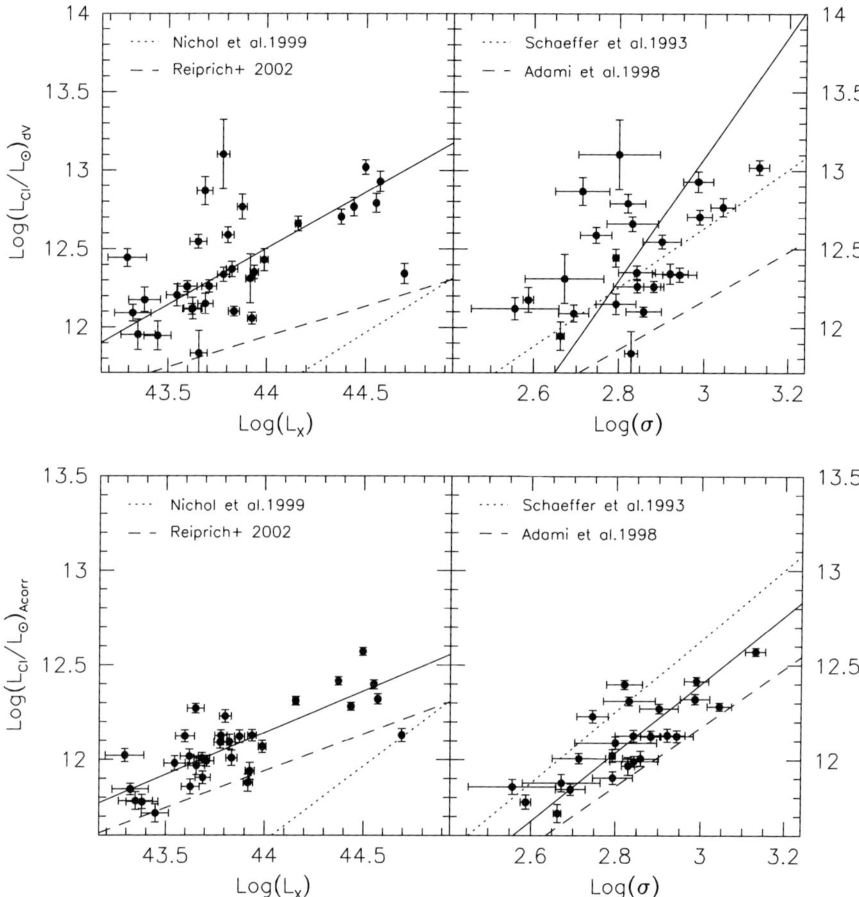

Figure 1. Left panels: total V luminosity versus X–ray luminosity: Nichol et al. (1999) (dotted lines) and Reiprich & Böringer (2002) (dashed lines). Right panels: total V luminosity versus velocity dispersion: dotted lines represent $L_V - \sigma$ fit from Schaeffer et al. (1993), dashed lines that from Adami et al. (1998).

is likely indicative of the light coming from some inner cluster region, presumably (and roughly) corresponding to the dynamically/structurally relaxed core. This would explain why it shows better correlations with dynamical and mass-density indicators, like σ and L_X.

3. The Kormendy relation for galaxy clusters

In spite of the fact that stars (in elliptical galaxies) and galaxies (in clusters) experience quite different dynamical regimes, the impressive similarity among KR relative to systems having so much different scales (see Figure 2) is likely to reflect a unified density–radius relation for luminous matter. The large scatter shown by galaxies on the $< \mu_e > -R_e$ plane could possibly reflect a more complex dynamical and structural evolution of galaxies with respect to clusters. These aspects of the extended Kormendy relation will require a careful analysis and an exaustive discussion.

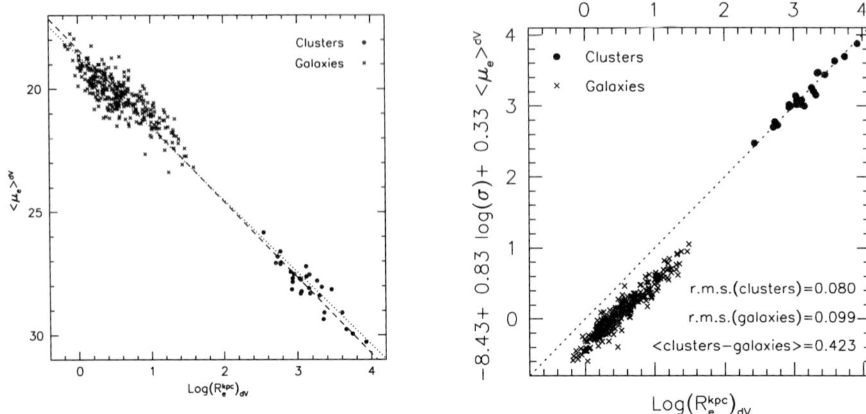

Figure 2. The Kormendy relation (left) and the Fundamental Plane (right) for elliptical galaxies and for clusters. Data for elliptical galaxies are taken from Bettoni et al. (2001).

4. The Fundamental Plane for galaxy clusters

By analogy with the 'extended Kormendy relation', we use our sample of clusters to illustrate in Figure 2 an 'extended FP' of gravitationally bound systems. This was first suggested by Schaeffer et al. (1993) with the aim to combine, in the same parameter space, systems spanning very different scales of mass and radii. Again, the slope of the FP of clusters is very close to that of galaxies, but in this case the zero points are very different, due to the fact that the mass-to-light ratio of clusters is about two orders of magnitudes greater than that of galaxies.

References

Adami, C., Mazure, A., Biviano, A., Katgert, P., Rhee, G. A&A 1998, 331, 493.

Bettoni, D., Falomo, R., Fasano, G., Govoni, F., Salvo, M., Scarpa, R. A&A 2001, 380, 471.

Burstein, D., Heiles, C. AJ 1982, 87, 1165.

Cabanac, R. A., de Lapparent, V., Hicson, P. A& A 2000, 364, 349.

Fasano, G., Bettoni, D., Marmo, C., Pignatelli, E., Poggianti, B. M., Moles, M., Kjærgaard, P. ASP Conf. 2002, 268, 361.

Nichol, R. C., Melott, A., Miller, C. J. AAS 1999, 195, 1001.

Reiprich, T. H., Böringer, H., ApJ 2002, 567, 716.

Schaeffer, R., Maurogordato, S., Cappi, A., Bernardeau, F. MNRAS 1993, 301, 21L.

Outskirts of Galaxy Clusters: Intense Life in the Suburbs
Proceedings IAU Colloquium No. 195, 2004
A. Diaferio, ed.

DOI: 10.1017/S1743921304000535

Application of the Voronoi tessellation technique for galaxy cluster search in the Münster Red Sky Survey

E. Panko[1] and P. Flin[2]

[1]Astronomical Observatory of the Nikolaev State University, Nikolaev, Ukraine
[2]Institute of Physics, Pedagogical University, Kielce, Poland

Abstract. We present the preliminary result of our project, consisting in studying the properties of a large sample of galaxy clusters. The Münster Red Sky Survey, which is a large galaxy catalogue covering an area of about 5000 square degrees in the southern hemisphere, serves as our observational basis. It is complete up to $r_F = 18^m.3$. Creation of a cluster catalogue is the first step of our investigation. We propose to use the 2D Voronoi tessellation technique for identifying galaxy clusters in this 2D catalogue. Points with high values of the inverse Voronoi tessel area will be regarded as galaxy cluster centroids. We show that this approach works correctly.

1. Introduction

In this paper we present the first element of our project. It consists in studying the properties of a large sample of galaxy clusters. In order to perform such studies, we need a sample of clusters extracted in a uniform manner from a homogeneous set of data. Therefore, we chose the Münster Red Sky Survey as our observational basis. The first step of investigation is to create the catalogue of galaxy clusters. There are three basic cluster detection algorithms: the matched filter algorithm (Postman et al. 1996), the adaptive matched filter algorithm (Kepner et al. 1999) and the Voronoi tessellation technique (Icke & van de Weygaert 1987, Zaninetti 1989, Ramella et al. 1999, 2001). Kim et al. (2002) made a comparison of these cluster-finding algorithms, using a Monte Carlo experiment with simulated clusters. We decided to apply the Voronoi tessellation technique for cluster detection. The Voronoi tessellation technique is completely non-parametric, and therefore sensitive to both symmetric and elongated clusters, allowing correct studies of non-spherically symmetric structures. For a distribution of seeds, the Voronoi tessellation creates polygonal cells containing one seed each and enclosing the whole area closest to its seed. This is the definition of a Voronoi cell in 2D. This natural partitioning of space by the Voronoi tessellation has been used to model the large-scale distribution of galaxies.

2. Observational data

The Münster Red Sky Survey constitutes the observational basis of our work. It contains scans of 217 adjoining plates of the ESO Southern Sky Atlas R covering more than 5000 degrees squared. Plates were scanned using two PDS 2020 GM^{plus} microdensitometers of the Astronomical Institute in Münster. The star/galaxy classification was performed by an automatic procedure using effective radius, central intensity and

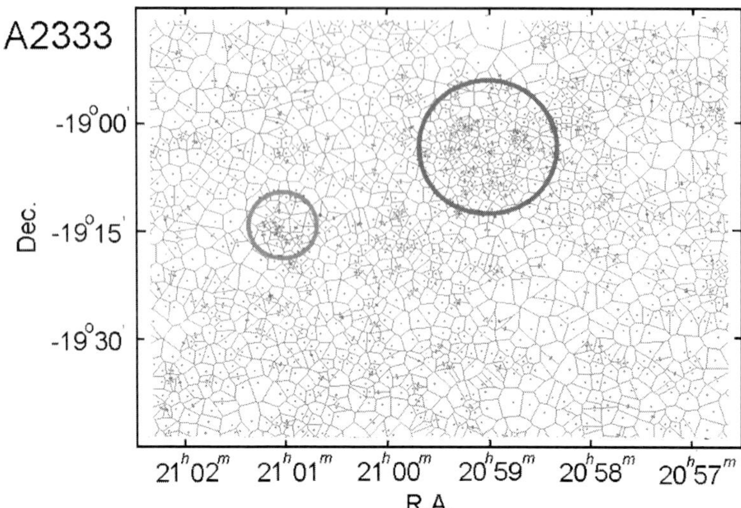

Figure 1. The region containing the A2333 after Voronoi tessellation. Each Voronoi cell encloses one galaxy. At least one more dense region is clearly seen.

surface brightness profiles. After that, visual inspection of over 2,775,000 objects was performed, allowing their re-classification. Uniformity of the whole catalogue was obtained through careful corrections of magnitudes across each plate. The overlapping regions of neighbouring plates allowed us to obtain a uniform instrumental system over the whole investigated region. The CCD photometry for 1037 galaxies and 1085 stars in 92 fields allowed one to establish the r_F magnitudes for each galaxy in the survey. The catalogue includes 5.5 million galaxies and is complete up to $18^m.3$ (Ungruhe 1999).

3. Analysis and result

We selected a fragment of the ESO/SERC 598 field with Abell cluster 2333. The A2333 cluster has 75 members, while in the field from $20^h56^m.4$ to $21^h02^m.4$ (R.A.) and from $-18°45'$ to $-19°45'$ (Dec.) there are 1781 galaxies. In the first stage of our investigation, we performed an analysis taking into account galaxy positions only. The galaxy positions provide input as seeds for the 2D Voronoi tessellation, and the Voronoi cell around each galaxy is interpreted as the effective area that each galaxy occupies in the projection plane. The result is shown in Fig. 1. Taking the inverse of these areas will be the next step of our procedure. In such a manner, the 2D local density of galaxies is determined at each point containing a galaxy. This information is then used to select galaxy members that live in highly overdense regions, which we identify as clusters. We do so by calculating the density contrast at each galaxy position as: $\Delta = (\rho - \langle\rho\rangle)/\langle\rho\rangle = (\langle\sigma\rangle - \sigma)/\sigma$, where σ is the area of the Voronoi cells, and $\langle\sigma\rangle$ is the mean area of all the cells.

4. Discussion and summary

We show that the Voronoi tessellation allows us to find the overdense regions. The main point of a further step in the search of galaxy clusters is the determination of the contrast level. This is crucial for finding real clusters. This can be done using various methods. One of them consists in using the number overdensity with respect to the average background. Another possibility is to use the colour-magnitude relation combining the data from the

Münster Red Sky Survey with those from the APM catalogues. The determination of photometric redshifts can be considered as well. At present, we are able to conclude that the applied analysis allows us to find correctly the overdense regions in two-dimensional data.

Acknowledgements

We thank Dr R. Ungruhe for sharing his data before their publication. For this research we have used NASA's Astrophysics Data System. PF has been partially supported by the Pedagogical University grant BS 052.

References

Icke, V., & van de Weygaert, R. 1987, A&A, 184, 16

Kepner J., Fan X., Bahcall N., Gunn J., Lupton R., Xu G. 1999, ApJ, 517, 78

Kim R.S.J., Kepner J.V., Postman M., Strauss, M.A., Bahcall, N.A. et al. 2002, AJ, 123, 20

Postman M., Lubin, L., Gunn, J. E., Oke, J. B., Hoessel, J. G., Schneider, D. P., & Christensen, J. A. 1996, AJ, 111, 615

Ramella, M., Nonino, M., Boschin, W., & Fadda, D. 1999, in ASP Conf. Ser. 176, Observational Cosmology: The Development of Galaxy Systems, ed. G. Giuricin, M. Mezzetti, & P. Salucci (San Francisco: ASP), 108

Ramella M., Boschin W., Fadda D., Nonino M. 2001, A&A 368, 776

Ungruhe 1999, Ph.D. thesis, Astron. Inst. Univ. Münster, Germany

Zaninetti, L. 1989, A&A, 224, 345

Outskirts of Galaxy Clusters: Intense Life in the Suburbs
Proceedings IAU Colloquium No. 195, 2004
A. Diaferio, ed.

DOI: 10.1017/S1743921304000547

On the ellipticity of galaxy clusters

P. Flin, J. Krywult, M. Biernacka
Institute of Physics, Pedagogical University, Kielce, Poland

Abstract. We analysed 246 scanned ACO clusters and we do not find any dependence of galaxy cluster ellipticity on redshift.

1. Introduction

The recent substantial evolution in the ellipticity of rich galaxy clusters was noted by Mellot, Chambers & Miller (2001). This result has been supported by Plionis (2002). In particular, he argues that the ellipticity of ACO clusters is correlated with their redshifts. Using, among others, 309 ACO clusters taken from APM, for which about 180 have determined redshifts, he claimed the existence of a significant ellipticity - redshift dependence.

2. Data

Our sample consists of 246 ACO clusters. All data come from 48" Schmidt telescopes. Original plates were scanned at the Rome and Edinburgh Astronomical Observatories. The majority of galaxy catalogues were obtained from DSS using FOCAS packages. For this sample, visual verification of object classification was made.

3. Method of analysis

It is well known that ellipticity depends also on the distance from cluster centre (e.g. Carter & Metcalfe 1980, Flin 1984, Trevese et al. 1992, Struble & Ftaclas 1994). Therefore, we decided to check the investigated relation for various distances from cluster centres. The radius changes from 0.5 to 1.5 Mpc ($h = 0.75, q_0 = 0.5$) with a step of 0.25 Mpc from the cluster centre. Thus, we have measured ellipticities at 5 distances from the cluster centre for each of the 246 clusters. The ellipticities have been determined in a standard way, this is from the covariance matrix. The redshifts are taken from the literature.

4. Results

For each cluster we have two parameters, the calculated ellipticity of the cluster and cluster redshift. A straight line was fitted to each set of data. The results of the fitting procedure for the line $e = A + B \cdot z$ are given in Table 1. The parameter R in Mpc is the radius of the circle in which the ellipse has been calculated, A and B are the fit parameters and r is the correlation coefficient. We also calculated the mean value of the five ellipticities for various R radii. The dependence of such mean ellipticities on redshift is given in Fig.1. The last line of Table 1 gives the relevant parameters.

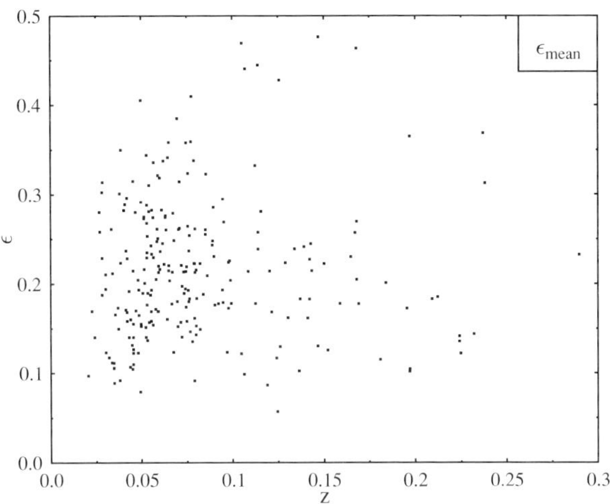

Figure 1. The relation between ellipticity and redshift for 246 investigated ACO clusters

Table 1. The result of the correlation analysis of the investigated samples

N	R	A	B	r
1	0.50	0.298	0.218	0.05
2	0.75	0.215	0.118	0.05
3	1.00	0.207	-0.097	0.05
4	1.25	0.186	-0.115	0.06
5	1.50	0.176	-0.061	0.04
6	Mean	0.216	0.013	0.01

5. Conclusions

It is clear that we do not observe any relation between ellipticities and redshifts. This differs greatly from the Plionis (2002) result. However, the claimed dependence of ellipticity on redshift reported by him is very weak. Our result is affected by such factors as e.g. the adopted cluster centre, the way of smoothing the distribution and the applied method of ellipticity determination. These factors are being carefully studied now. The current result is consistent with Mulchaey & Zabludoff (1998) and Zabludoff & Mulchaey (1998) observation reporting the lack of recent significant dynamical evolution for poor groups and cD clusters. Moreover, the lack of the eccentricity evolution is in agreement with the latest hydrodynamic simulations of clusters (Floor et al. 2003a, 2003b).

Acknowledgements

This paper was financially supported by Pedagogical University grants: BS 052, BW 116 and BW 236.

References

Carter, D., Metcalfe, N. 1980 MNRAS, 191, 325
Flin, P. 1984 in Clusters and Groups of Galaxies, eds. Mardirossian, F., et al., Dordrecht, p. 163
Floor, S.N., Melott, A.L., Miller, C.J., Bryan, G.L. 2003a, ApJ, 591, 741
Floor, S.N., Melott, A.L., Motl, P.M. 2003b, astro-ph/0307539
Melott, A., Chambers, S. W., Miller, C. J. 2001, ApJ, 559, L75
Mulchaey, J. S., Zabludoff, A. I. 1998, ApJ, 496, 73
Plionis, M. 2002, ApJL, 572, 67
Struble, M. C., Ftaclas, F. 1994, AJ, 108, 1
Trevese, D., Flin, P., Migliori, L., Hickson, P., Pittella, G. 1992, A&AS, 94, 237
Zabludoff, A. I., Mulchaey, J. S. 1998, ApJ, 496, 39

Section C
Theoreticians at work

Outskirts of Galaxy Clusters: Intense Life in the Suburbs
Proceedings IAU Colloquium No. 195, 2004
A. Diaferio, ed.

DOI: 10.1017/S1743921304000559

Intracluster thermodynamics

G. Mark Voit

Department of Physics and Astronomy, Michigan State University, East Lansing, MI 48824, USA email: voit@pa.msu.edu

Abstract. Most of the baryons in the universe remain undetected. They are thought to be floating in intergalactic space, and their thermal state is highly uncertain. The only places where we have reasonably complete observations of the overall baryon population are in clusters of galaxies, where the baryons reveal themselves through their X-ray emission. There we find that only about 10% of the baryons have turned into stars, probably because feedback processes intervene to prevent more baryons from condensing. These galaxy formation processes seem to have left an imprint on the thermal state of the intracluster medium, but in order to interpret that imprint, one needs to pay close attention to how entropy develops in the intergalactic gas. Many of the global properties of clusters turn out to be determined primarily by the entropy threshold for cooling within a Hubble time. However, *XMM-Newton* observation are revealing a sort of altered similarity among the entropy profiles of clusters seeming to indicate that galactic winds may have smoothed the local intergalactic medium before it accreted onto clusters.

1. Introduction

What is the thermodynamical state of gas as it enters a cluster of galaxies? The question is difficult to answer by direct observation because the X-ray surface brightness of clusters and groups of galaxies is so low in the vicinity of the virial radius. One would like to know whether the accreting gas has been preheated because early energy input by supernova-driven galactic winds and energy injection by active galactic nuclei strongly affect the global properties of clusters. Furthermore, an understanding of how galaxy formation alters the properties of clusters is necessary in order to derive precise values for cosmological parameters from the number density and evolution rate of the overall cluster population.

Because it is so difficult to observe the intergalactic gas outside of clusters, one of the best ways to gather information about that gas is to observe its properties after it has fallen into a cluster. Once that happens, the thermodyamical state of the gas can be determined from its X-ray emission. However, proceeding from those observations back to what must have happened to the gas to produce its current state requires paying close attention to the processes that determine intracluster entropy.

2. Why entropy matters

Entropy is of fundamental importance for two reasons: it determines the structure of the intracluster medium and it records the thermodynamic history of the cluster's gas. Entropy determines structure because high-entropy gas floats and low-entropy gas sinks. A cluster's intergalactic gas therefore convects until its isentropic surfaces coincide with the equipotential surfaces of the dark-matter potential. Thus, the entropy distribution of a cluster's gas and the shape of the dark-matter potential well in which that gas sits completely determine the large-scale X-ray properties of a relaxed cluster of galaxies. The gas density profile $\rho_g(r)$ and temperature profile $T(r)$ of the intracluster medium in this state of convective and hydrostatic equilibrium are simply manifestations of its entropy distribution.

Here we will adopt the approach of other work in this field and define "entropy" to be $K \equiv kT/(\mu m_p \rho_g^{2/3})$. The quantity K is the constant of proportionality in the equation of state $P = K\rho_g^{5/3}$ for an adiabatic monatomic gas, and is directly related to the standard thermodynamic entropy per particle, $s = k \ln K^{3/2} + s_0$, where s_0 is a constant that depends only on fundamental constants and the mixture of particle masses. A cluster achieves convective equilibrium when $dK/dr \geqslant 0$ everywhere, and the entropy distribution that determines the gas configuration in this state can be expressed as $K(M_g)$, where the inverse relation $M_g(K)$ is the mass of gas with entropy $< K$.

Comparisons between the entropy distributions of clusters that differ in mass can be simplified by casting those distributions into dimensionless form. One can define the mass of a cluster to be the mass M_{200} inside the radius r_{200} within which the mean mass density is 200 times the critical density $\rho_{\rm cr}$. Combining the scale radius r_{200}, the global baryon fraction $f_b = \Omega_{\rm b}/\Omega_{\rm M}$, and the characteristic halo temperature $kT_{200} = GM_{200}\mu m_p/2r_{200}$ then gives the characteristic entropy scale

$$K_{200} = \frac{kT_{200}}{\mu m_p (200 f_b \rho_{\rm cr})^{2/3}} = \frac{1}{2}\left[\frac{2\pi}{15}\frac{G^2 M_{200}}{f_b H(z)}\right]^{2/3} . \tag{2.1}$$

For $f_b = 0.022h^{-2}$, this entropy scale corresponds to

$$kTn_e^{-2/3} = 362\, kT_{\rm lum}\,{\rm cm}^2 \left(\frac{T_{200}}{T_{\rm lum}}\right)\left[\frac{H(z)}{H_0}\right]^{-4/3}\left(\frac{\Omega_{\rm M}}{0.3}\right)^{-4/3} . \tag{2.2}$$

Writing the entropy scale in this way makes explicit the fact that the observed temperature of a cluster is not necessarily a reliable guide to the characteristic entropy K_{200} of its halo. If the intracluster medium of a real cluster is either hotter or cooler than T_{200}, then one must apply the correction factor $T_{200}/T_{\rm lum}$ when computing the cluster's value of K_{200}, where $T_{\rm lum}$ is the emissivity-weighted mean temperature of the intracluster medium.

Hierarchical structure formation without radiative cooling or non-gravitational heating produces entropy purely through the shock heating associated with structure formation. Because there is no particular entropy scale associated with these shocks, other than the scale K_{200} set by the halo mass and the redshift, the entropy profiles of clusters simulated in the absence of non-gravitational processes are self-similar, with nearly identical entropy profiles (e.g., Voit et al. 2003).

3. Broken similarity

Observations have shown that real clusters are not like the self-similar objects that emerge from simulations without galaxy formation. The X-ray luminosity-temperature relation of clusters simulated without cooling obeys the self-similar expectation $L_{\rm X} \propto T_{\rm lum}^2$ (e.g., Navarro et al. 1995). This scaling law has been known for over a decade to deviate from the observed luminosity-temperature scaling: $L_{\rm X} \propto T_{\rm lum}^{\alpha}$, with $\alpha \approx 2.5 - 3$ (e.g., Edge & Stewart 1991). In the last several years, the observed mass-temperature relation of clusters has been shown to deviate from that of clusters simulated without cooling, in that observed clusters of a given temperature seem to have masses up to 40% lower than expected (Horner et al. 1999; Nevalainen et al. 2000; Finoguenov et al. 2001). These deviations need to be understood because the $L_{\rm X}$-$T_{\rm lum}$ and M-$T_{\rm lum}$ relations are fundamental tools for measuring cosmological parameters using cluster surveys.

Early approaches to the problem of similarity breaking in clusters postulated that some sort of heating process imposed a universal minimum entropy—an "entropy floor"—on the intergalactic gas before it collected into clusters (Evrard & Henry 1991; Kaiser 1991). Imposing a global entropy floor helps to bring the theoretical L_X-$T_{\rm lum}$ relation into better agreement with observations because this extra entropy makes the gas harder to compress in cluster cores, where entropy is smallest, particularly in the shallower potential wells of low-temperature clusters. This resistance to compression breaks cluster similarity by lowering the core density, and therefore the X-ray emissivity, in low-T clusters more than in high-T clusters, thereby steepening the L_X-$T_{\rm lum}$ relation.

According to this preheating picture, the core entropy level and scaling relations of clusters should reflect the global entropy floor produced at early times. Initial measurements of entropy at the core radius $r_{0.1}$ demonstrated that low-temperature clusters had greater amounts of entropy than expected from self-similarity and suggested that the level of the entropy floor was $\sim 135\,{\rm keV\,cm^2}$ (Ponman et al. 1999). This result matched well with numerical simulations of cluster formation in which preheating levels of $50-100\,{\rm keV\,cm^2}$ produced clusters with approximately the right L_X-$T_{\rm lum}$ relation (Bialek et al. 2001).

However, simple preheating now appears to be too crude an explanation for similarity breaking. In the preheating picture, low-temperature clusters should have large isentropic cores, but this prediction disagrees with the observations showing that the shapes of cluster entropy profiles do not depend significantly on temperature (Pratt & Arnaud 2003). In addition, the abundant evidence for intergalactic gas at $\lesssim 10^5$ K from quasar absorption line studies clearly shows that preheating cannot be global at $z \gtrsim 2$, and the preheating models themselves do not explain why the level of the entropy floor should be $\sim 135\,{\rm keV\,cm^2}$.

In contrast, the observed entropy scale of similarity breaking emerges naturally from the process of radiative cooling. Cooling that radiates an energy Δq per particle reduces the entropy by $\Delta \ln K^{3/2} = \Delta q/kT$. The entropy threshold $K_c(T)$ below which gas at temperature T cools within the universe's lifetime turns out to be quite close to the entropy floor inferred from cluster observations (Voit & Bryan 2001). Voit & Ponman (2003) further quantify this point. Figure 1 shows how entropy measurements at $0.1r_{200}$ in a large sample of clusters compare with the cooling threshold $K_c(T)$ for gas with heavy-element abundances equal to 30% of their solar values relative to hydrogen. Both the measured core entropies and the entropy threshold for cooling scale as $T^{2/3}$, and they are approximately equal, although the scatter in the data is quite significant.

Cooling therefore appears to set the entropy scale for similarity breaking, but it cannot act alone. The cooling threshold in low-temperature clusters at the present time is ~20% of the characteristic entropy K_{200} and greater than that if emission-line cooling from heavy elements is included. At earlier times, the dimensionless cooling threshold is even higher, meaning that a large proportion of the baryons belonging to the progenitor objects, that ultimately assembled into present-day clusters, would have condensed into stars or cold gas clouds if there were no feedback. This is one of the manifestations of the classic overcooling problem of hierarchical galaxy formation. Because the observed mass ratio of stars to hot gas in clusters is only $\lesssim 10\%$ (e.g., Balogh et al. 2001), wholesale baryon condensation doesn't seem to have happened.

Recognition of this overcooling problem led Voit & Bryan (2001) to propose a way for radiative cooling to determine the entropy scale of similarity breaking without acting alone. The basic idea is that gas with entropy less than $K_c(T)$ cannot persist indefinitely. It must either cool and condense or be heated until its entropy exceeds $K_c(T)$. At any given time, feedback is triggered by condensing gas parcels with entropy less than the cooling threshold and acts until those parcels are eliminated by either cooling, heating,

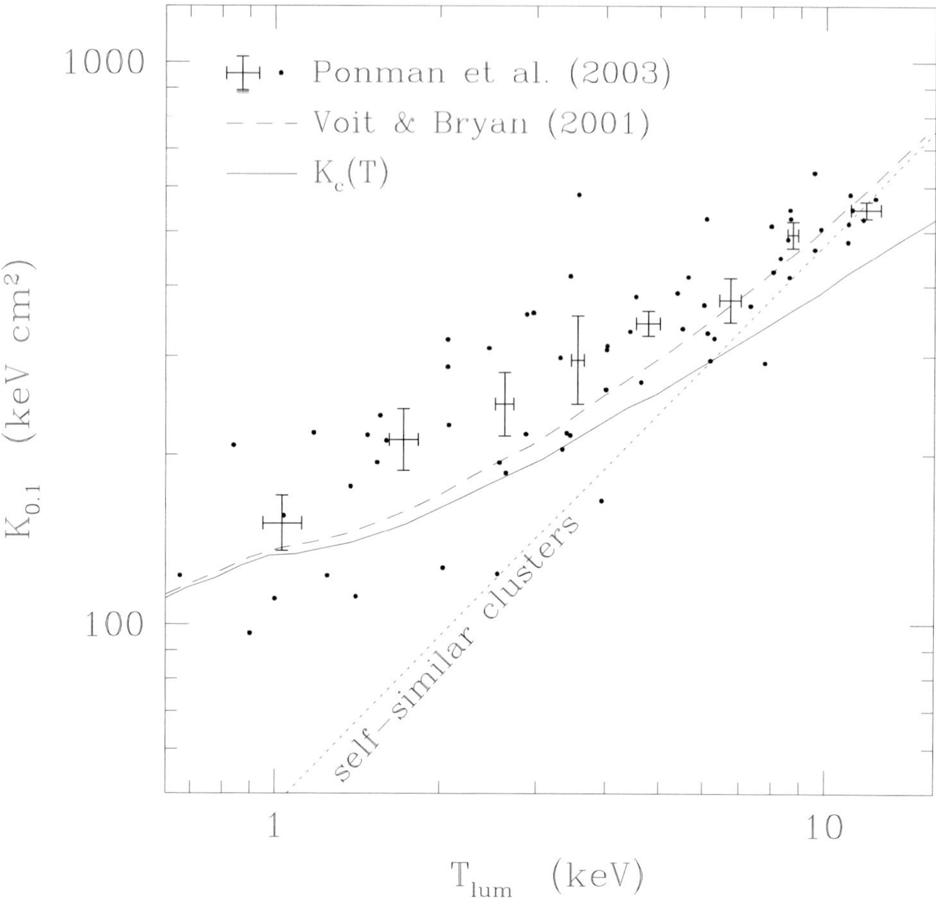

Figure 1. Comparison between entropy measured at $0.1r_{200}$ and the cooling threshold in a large sample of clusters. Small points show the entropy $K_{0.1}$ measured at $0.1r_{200}$ in a sample of 64 clusters, and points with error bars show the mean entropy measurement in temperature bins of eight clusters each. The dotted line gives the mean entropy predicted by simulations of clusters without radiative cooling or feedback. The solid line shows the value of the cooling threshold $K_c(T)$ computed for heavy-element abundances 0.3 times their solar values and $t_0 = 14$ Gyr. The dashed line shows the entropy predicted at $0.1r_{200}$ by the simple analytical model of Voit & Bryan (2001).

or some combination of the two. Thus, the joint action of cooling and feedback imprint an entropy scale roughly corresponding to the cooling threshold, regardless of how strong the feedback is.

While it might seem paradoxical, allowing the intracluster medium to radiate thermal energy actually causes its luminosity-weighted temperature to rise. The reason for this behavior is that cooling selectively removes low-entropy gas from the intracluster medium, raising the mean entropy of what remains. In non-radiative cluster simulations, the entropy of gas in the vicinity of the cluster core is below the cooling threshold K_c. This aspect of non-radiative models is unphysical, because gas with entropy less than K_c would radiate an amount of energy greater than its total thermal energy content over the course of the simulations. When cooling is allowed to occur, this low-entropy core gas condenses out of the intracluster medium and is replaced by higher entropy core gas having a higher temperature, a lower density, and therefore a lower luminosity.

4. Entropy modification

A simple analytical model for entropy modification illustrates the effect of the cooling threshold on the L_X-$T_{\rm lum}$ and M_{200}-$T_{\rm lum}$ relations (Voit & Bryan 2001; Voit et al. 2002; Wu & Xue 2002). The model assumes that the intracluster entropy distribution in the absence of galaxy formation would lead to a gas density profile similar to that of the dark matter, which can be approximated with the NFW fitting formula $\rho(r) \propto r^{-1}(1 + c_{200}r/r_{200})^{-2}$, where c_{200} is the concentration parameter. Assuming that this gas is also in hydrostatic equilibrium yields a baseline entropy distribution for the no-cooling, no-feedback case. Because condensation and feedback both act to eliminate gas below the cooling threshold, one can approximate the effects of the cooling threshold by simply truncating the baseline entropy distribution at $K_c(T_{200})$ and discarding all the gas with lower entropy. One can interpret this gas removal either as condensation or as extreme feedback that heats the sub-threshold gas to a much higher entropy level. This cooling and feedback need not occur at the center of the cluster. In a hierarchical cosmology, much of the low-entropy gas cools, condenses into galaxies, and produces feedback long before the cluster is finally assembled.

Computing the hydrostatic configuration of the modified entropy distribution in the original dark-matter potential gives L_X and $T_{\rm lum}$ as a function of the mass M_{200} and concentration c_{200} of the dark-matter halo. The resulting L_X-$T_{\rm lum}$ and M_{200}-$T_{\rm lum}$ relations agree well with observations but may slightly overpredict L_X for objects cooler than $\sim 2\,{\rm keV}$. Notice that there are no free parameters in this model, other than the cosmological parameters, because the M_{200}-c_{200} relation and the age of the universe used to compute K_c depend only on cosmology, and the heavy-element abundance used to compute the cooling threshold is taken from observations.

5. Avoiding overcooling

So what happens to the gas that cools? Supernovae are the most obvious candidate for supplying the feedback that suppresses condensation, but it is not clear that supernova heating and the galactic winds it drives can provide enough entropy to keep the fraction of condensed baryons below about 10%. Heavy-element abundances in clusters imply that the total amount of supernova energy released during a cluster's history amounts to $\sim 1\,{\rm keV}$ per gas particle in the intracluster medium. The amount of energy input needed to explain the mass-observable relations while avoiding overcooling is also ~ 1 keV, but the transfer of supernova energy to the intracluster medium would have to be highly efficient, which seems unlikely (e.g., Kravtsov & Yepes 2001). Supernova energy would have to be converted almost entirely to thermal energy with very little radiated away.

In order to avoid radiative losses, supernova heating must raise the entropy of the gas it heats to at least $100\,{\rm keV\,cm^2}$. An evenly distributed thermal energy input of order $1\,{\rm keV}$ would have to go into gas significantly less dense than $10^{-3}\,{\rm cm^{-3}}$ to avoid such losses. Gas near the centers of present-day clusters, not to mention the galaxies where supernovae occur, is denser than that, particularly at earlier times when most of the star formation happened. Simulations that spread supernova feedback evenly therefore produce too many condensed baryons in clusters (e.g., Borgani et al. 2002). Artificial algorithms that target supernova feedback at gas parcels that would otherwise cool are more successful at preventing overcooling (Kay et al. 2003). However, efforts to implement a more realistic version of targeted feedback in the form of galactic winds are still not entirely successful at preventing overcooling (Borgani et al. 2004).

If supernovae cannot prevent overcooling, then perhaps supermassive black holes in the nuclei of galaxies are what stops it. The omnipresence of supermassive black holes at the centers of galaxies and the excellent correlation of their masses with the bulge and halo properties of the host galaxy strongly suggest that the growth of black holes in the nuclei of galaxies goes hand-in-hand with galaxy formation. Furthermore, the centers of many clusters with low-entropy gas whose cooling time is less than the age of the universe also contain active galactic nuclei that are ejecting streams of relativistic plasma into the intracluster medium. It is therefore plausible that supermassive black holes at the centers of clusters provide feedback that suppresses further cooling whenever condensing intracluster gas accretes onto the central black hole.

Such a feedback loop is attractive and consistent with the circumstantial evidence, but the precise mechanism of heating remains unclear. The bubbles of relativistic plasma being inflated by the active galactic nuclei in clusters appear not to be expanding fast enough to shock heat the intracluster medium because the rims of the bubbles are no hotter than their surroundings. Also, if active galactic nuclei simply injected heat energy into the center of a cluster, then one would expect to see a flat or reversed entropy gradient in clusters with strong nuclear activity, indicating that convection is carrying heat outward. Instead, the entropy gradients in these cluster cores increase monotonically outward (David et al. 2001; Horner et al. 2004). One possibility is that heating is episodic (Kaiser & Binney 2003) and that we have not yet found a cluster in the midst of an intense heating episode.

Unfortunately, none of these heating mechanisms have yet been tested in the context of cosmological structure formation, so we do not know their overall impact on either baryon condensation or the global entropy profiles of clusters. Also, many aspects of the relationship between cosmology and nuclear activity in galaxies are highly uncertain. A major role for quasar feedback is plausible. However, the connection between the growth of central black holes in galaxies and galaxy formation itself is not well understood, and the efficiency with which black holes convert accretion energy into outflows is unknown.

6. Presmoothing by galactic winds?

Observations of entropy in the *outer* parts of clusters suggest that galactic winds driven by supernovae or active galaxies may indeed have a pronounced impact on the intergalactic medium, at least in the vicinity of clusters. The entropy profiles of clusters and groups can now be measured out to a significant fraction of the virial radius r_{200}, within which the mean mass density is 200 times the critical density. Surprisingly, these measurements are hinting that the $K(r/r_{200}) \propto T^{2/3}$ scaling relation that applies at $0.1r_{200}$ also applies to the entire entropy profile. Deep *XMM-Newton* observations of two clusters whose temperatures differ by a factor $\sim$3.5 show that the scaled profile $T_{\rm lum}^{-2/3}K(r/r_{200})$ is independent of cluster temperature (Pratt & Arnaud 2003). Likewise, an analysis of lower-quality data on a larger number of clusters also suggests that $K(r/r_{200}) \propto T_{\rm lum}^{2/3}$ at the scale radius r_{500}, within which the mean matter density is 500 times the critical density (Ponman et al. 2003).

Voit et al. (2003) and Ponman et al. (2003) have proposed that entropy input from galactic winds preceding the accretion of gas onto clusters could lead to a form of entropy amplification that would explain the observations. If galactic winds are strong enough to significantly smooth out the lumpiness of the intergalactic gas in their vicinity, then the mode of accretion of this gas onto clusters will be closer to smooth accretion than to hierarchical accretion, boosting the entropy generated through accretion shocks without changing the entropy profile's characteristic shape. This effect is a plausible explanation for the characteristics of the observed entropy profiles (Pratt & Arnaud 2003), but it has not yet been thoroughly tested in simulations. Intriguing results by Kay (2004) show that an extremely targeted feedback model, in which feedback triggered by cooling heats the local gas to 1000 keV cm^2, successfully reproduces both the normalization and shape of the observed entropy profiles. However, it remain to bee seen whether this mechanism works in detail.

References

Balogh, M. L., Pearce, F. R., Bower, R. G., & Kay, S. T. 2001, *MNRAS* **326**, 1228.
Bialek, J. J., Evrard, A. E., & Mohr, J. J. 2001, *Ap.J.* **555**, 597.
Borgani, S. et al. 2002, *MNRAS* **336**, 409.
Borgani, S. et al. 2004, *MNRAS* **348**, 1078.
David, L. P.. et al. 2001, *Ap.J.* **557**, 546.
Edge, A. C., & Stewart, G. C. 1991, *MNRAS* **252**, 414.
Evrard, A. E., & Henry, J. P. 1991, *Ap.J.* **383**, 95.
Finoguenov, A., Reiprich, T. H., & Böhringer 2001, *A&A* **368**, 749.
Horner, D. J., Mushotzky, R. F., & Scharf, C. A. 1999, *Ap.J.* **520**, 78.
Horner, D. J., Donahue, M., & Voit, G. M. 2004, in preparation.
Kaiser, C, & Binney, J. 2003, *MNRAS* **338**, 837.
Kaiser, N. 1991, *Ap.J.* **383**, 104.
Kay, S. T. 2004, *MNRAS* **347**, 13.
Kay, S. T., Thomas, P. A., & Theuns, T. 2003, *MNRAS* **343**, 608.
Kravtsov, A., V., & Yepes, G. 2000, *MNRAS* **318**, 227.
Navarro, J. F., Frenk, C. S., & White, S. D. M. 1995, *MNRAS* **275**, 720.
Nevalainen, J., Markevitch, M., & Forman, W. 2000, *Ap.J.* **536**, 73.
Ponman, T. J., Cannon, D. B., & Navarro, J. F. 1999, *Nature* **397**, 135.
Ponman, T. J., Sanderson, A. J. R., & Finoguenov, A. 2003, *MNRAS* **343**, 331.
Pratt, G. W., & Arnaud, M. 2003, *A&A* **408**, 1.
Voit, G. M., Balogh, M. L., Bower, R. G., Lacey, C. G., & Bryan, G. L. 2002, *Ap.J.* **576**, 601.
Voit, G. M., & Bryan, G. L. 2001, *Nature* **414**, 425.
Voit, G. M., Bryan, G. L., Balogh, M. L., & Bower, R. G. 2002, *Ap.J.* **576**, 601.
Voit, G. M., & Ponman, T. J. 2003, *Ap.J.* **594**, L75.
Wu, X., & Xue, Y. 2002, *Ap.J.* **572**, L19.

Outskirts of Galaxy Clusters: Intense Life in the Suburbs
Proceedings IAU Colloquium No. 195, 2004
A. Diaferio, ed.

DOI: 10.1017/S1743921304000560

Studying clusters of galaxies with hydrodynamical simulations

Volker Springel

Max-Planck-Institute for Astrophysics, Karl-Schwarzschild-Str. 1, 85740 Garching, Germany
email: volker@mpa-garching.mpg.de

Abstract. I review recent results of cosmological hydrodynamic simulations for clusters of galaxies. Such simulations have developed into an increasingly powerful tool to make theoretical predictions for clusters, and to confront them with observations. I focus, in particular, on the continuing challenge to reproduce the observed cluster X-ray scaling relations in CDM simulations. Self-consistent simulations that include radiative cooling, star formation and supernova feedback generally overpredict the X-ray luminosity on group scales and exhibit overly strong cooling flows. Solving these problems may require the consideration of more physics in cluster simulations, for example thermal conduction or heating by a central AGN.

---

1. Introduction

Beginning with the pioneering study by Evrard (1990), hydrodynamical simulations have developed into an indispensable tool for studying the formation of clusters of galaxies, and to explore the physics of the intra-cluster medium (ICM). Direct simulation is the only available technique to capture the dynamics of clusters in full generality, and hence to make detailed theoretical predictions for cluster properties expected in CDM cosmologies. Unsurprisingly, the list of topics investigated with hydrodynamic simulations has therefore become very long. It includes basic cluster properties like the run of gas density and temperature with radius, the X-ray emission of clusters, or their radiative cooling properties. Recently, this list has been augmented with studies of the thermal and kinetic Sunyaev-Zeldovich effects, AGN heating processes by jets and radio bubbles, cold fronts in the ICM, dissipative heating of the ICM by sound waves, metal enrichment, thermal conduction, magnetic field amplification during cluster formation, and gravitational lensing using simulated cluster mass models.

A fundamental topic of particular interest in numerical simulations of structure formation are the basic scaling relations of clusters, which have been addressed as a primary goal in numerous simulation studies. As I will discuss in more detail in Section 2, the theory here still faces substantial challenges in accurately matching the observed scaling relations.

Another serious challenge for cluster simulations lies in the observed absence of strong cooling flows in most clusters of galaxies, despite their copious X-ray emission. Apparently, a not yet conclusively identified heating source offsets the cooling losses. It appears likely that this problem is intertwined with that of the scaling relations, i.e. a successful solution may be able to resolve both at once. In Section 3, I will discuss some recent lines of work to study thermal conduction as a potential heating source for the centers of clusters, followed by a brief summary and a discussion of future directions in Section 4.

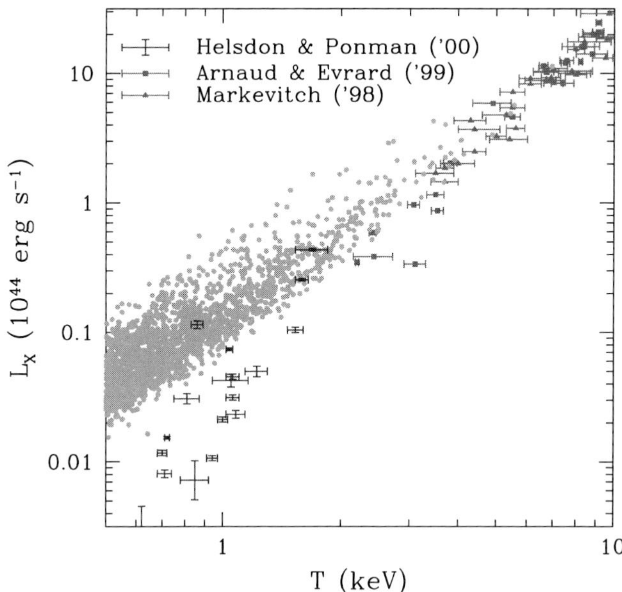

Figure 1. Relation between X-ray luminosity and cluster temperature measured for clusters formed in a large cosmological hydrodynamical simulation. For comparison, symbols with error bars give different observational data sets.

2. Cluster scaling relations

Assuming that clusters of galaxies are approximately self-similar, one expects scale-free scaling relations between basic cluster properties, for example $L_X \propto T^2(1+z)^{3/2}$ for the relation between X-ray luminosity and temperature, or $M \propto T^{3/2}/(1+z)^{3/2}$ for the mass-temperature relation. Already Navarro, Frenk & White (1995) were able to show that these relations are also recovered in CDM simulations where the gas is treated in a non-radiative, 'adiabatic' approximation. This result has been confirmed in numerous subsequent studies with higher resolution, although some small systematic deviations from the ideal self-similar scaling relations exist even here, depending on how cluster properties are measured in detail (see the poster by Sijacki & Tormen).

However, observed cluster-scaling relations differ strongly from self-similar models. In particular, the X-ray luminosity-temperature relation has a considerably steeper slope than the self-similar value of 2, meaning that poor clusters emit comparatively little compared to rich systems, much less than would be expected based on a self-similar scaling. Of course, it is to be expected that the scales imprinted by the physics of radiative cooling invalidate the assumption of self-similarity of clusters, so it is not really unexpected that the self-similar scalings do not hold for real clusters. However, perhaps somewhat more surprising is the fact that two seemingly opposite explanations have been put forward to explain the deficit of luminosity on the group scale.

One idea is to allude to the effect of radiative cooling alone (Bryan 2000; Voit, Bryan, Balogh & Bower 2002). Indeed, if one compares the ICM cooling times with the Hubble time and assumes that all the gas that can cool according to this criterion has dropped out of the ICM, one finds a substantial steepening of the scaling relations, simply because poorer clusters cool more efficiently, i.e. a comparatively larger amount of gas is removed

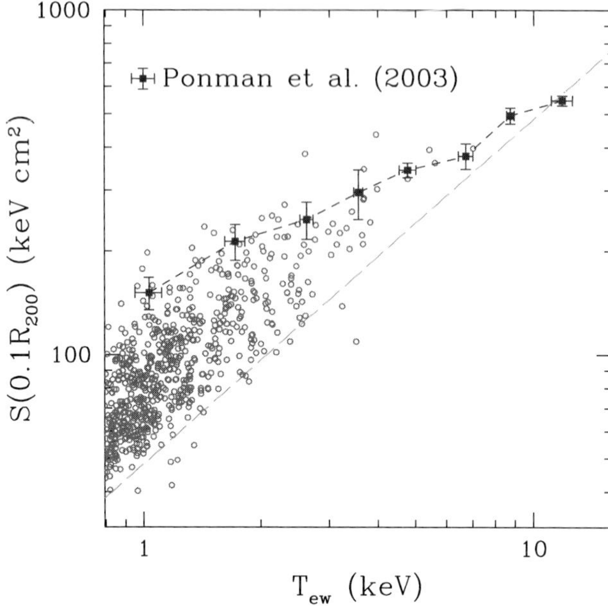

Figure 2. Entropy of clusters as a function of cluster temperature.

from the diffuse gaseous phase. The gas in the center that cools out will then be replaced by inflowing gas of higher entropy, an effect that can then also account for the observed 'entropy excess' on the group scale. This scenario has also been explored in simulations (e.g. Davé, Katz & Weinberg 2002).

The central problem with the 'cooling-only' model is that the amount of gas that is postulated to have cooled out appears to be uncomfortably large, reaching up to $\sim 40\%$ of all the baryons in the cluster. This is at least a factor 3-4 larger than the mass of baryons locked up in long-lived stars (e.g. Balogh, Pearce, Bower & Kay 2001), and much larger than the mass one can hide in supermassive black holes, cold gas, or dark stellar remnants.

The alternative explanation therefore suggests that the high entropy of the intracluster gas in poor clusters is caused by non-gravitational heating of gas, either before it collapsed into the cluster in an epoch of 'preheating', or in situ, for example by AGN or star-formation activity. This lowers the gas densities and X-ray luminosities without necessarily removing the gas from the ICM. These heating processes occur concurrently with cooling, such that the observed cluster scaling relations arise due to a non-linear coupling of heating and cooling, in ways that are actively studied with detailed hydrodynamical simulations (e.g. Muanwong et al. 2002, Tornatore et al. 2003, Kay et al. 2003) that use ad-hoc prescriptions for the heating. A self-consistent, fully successful model has still to be found, however.

Nevertheless, steady progress in hydrodynamical simulations has been made. In Figure 1, results from a recent study by Borgani et al. (2004) are shown, based on a very large cosmological SPH simulation with a particle number of 2×480^3, gas mass resolution of $6.9 \times 10^8\, h^{-1}\mathrm{M}_\odot$, and gravitational softening length of $7.5\, h^{-1}$kpc. The simulation followed radiative cooling, star formation and feedback by supernova and weak galactic winds, using the sub-resolution multiphase model for the interstellar medium proposed

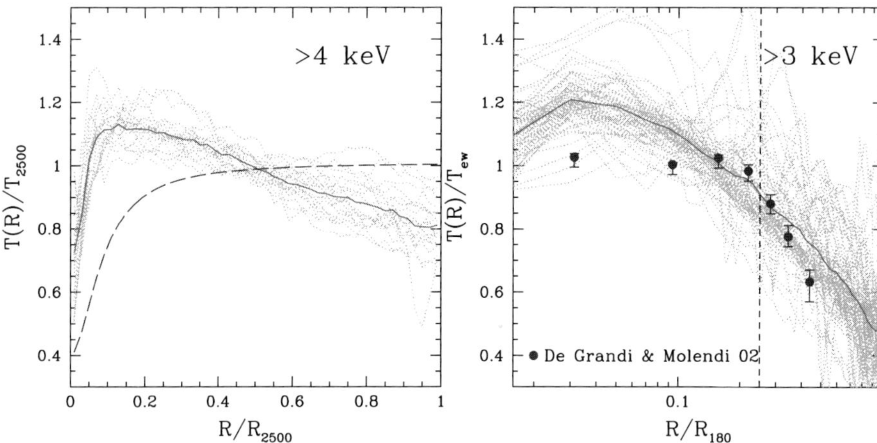

Figure 3. Projected temperature profiles of simulated clusters. The left panel compares with the best-fitting universal temperature profile of Allen, Schmidt & Fabian (2001), while the right panel compares with data from De Grandi & Molendi (2002).

by Springel & Hernquist (2003a,b). The simulation is in fact one of the largest computations carried out thus far with this amount of physics. The large simulation box size of $192\,h^{-1}$Mpc allowed the identification of a sizable cluster sample, including a few rich systems. Figure 1 shows the X-ray temperature relation measured for the clusters in this simulation, and compares it with observational data points. While there is good agreement for the most massive clusters, poorer systems on the group scale are clearly too luminous, or in other words, the simulated $L_X - T$ relation is not steep enough.

Figure 2 shows the relation between entropy and cluster temperature for the same simulation, compared both to the self-similar expectation, and to observed data, which exhibits a progressively larger 'entropy excess' towards lower mass systems. While the simulation results show clear signs of non-gravitational entropy injection due to feedback on the group scale, the effect is too weak to reconcile them with the observations. We should note however that the simulation shows a gratifying consistency with the observed abundance of clusters as a function of temperature. Also, the mass-temperature relation is in reasonable agreement with observations, highlighting that the bulk properties of clusters are quite well reproduced.

A problematic point concerns the radial temperature profiles of simulated clusters, which tend to be in significant conflict with the observed shape, as illustrated in Figure 3. Instead of showing a largely flat run of temperature with radius, and a shallow decline towards the center, the temperature rises in the inner parts of the simulated clusters and turns into a sharp drop only at the very center where the gas begins to cool rapidly. This is a manifestation of the strong cooling flows that are present in the simulated clusters, which are not observed in this form in reality. This problem appears to be quite universal in simulations with radiative cooling and has been found in similar form by a number of authors (e.g. Lewis et al. 2000; Pearce et al. 2000; Valdarnini 2002).

3. Thermal conduction

The observed absence of strong cooling flows in clusters, despite their copious X-ray emission, is a considerable challenge for hydrodynamical simulations. A number of possible physical heating sources have been proposed that may offset central cooling losses in the cluster. Suggestions include heating by buoyant bubbles inflated by central AGN activity, viscous dissipation of sound waves in the ICM, or supernova explosions in the

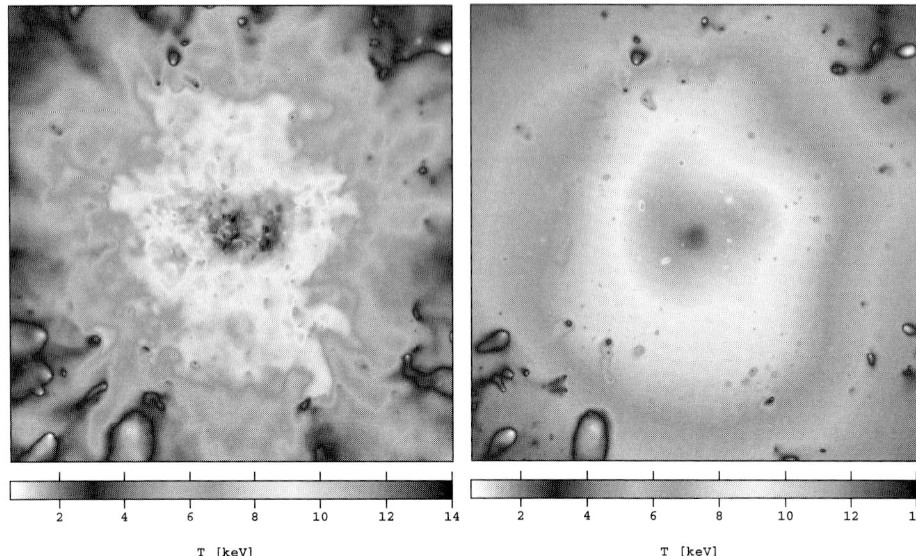

Figure 4. Thermal conduction in a massive cluster of galaxies. The panel on the left shows a temperature map in a simulation with cooling and star formation only, while the panel on the right includes thermal conduction as well.

ICM itself. Recently, Narayan & Medvedev (2001) proposed that thermal conduction may be a viable candidate for such a heating source. Due to the observed declining temperature towards the inner parts of clusters, the resulting temperature gradient should give rise to a heat current into the central parts of the cluster, provided the conductivity of the ICM is large enough. Note that the conductivity of an ionized plasma can be reduced substantially by magnetic fields, so that a major uncertainty in this conjecture lies in the unknown value of the remaining effective conductivity of the ICM.

Zakamska & Narayan (2003) have shown that for conductivities in the range 0.1-0.3 of the Spitzer value, the temperature profiles of a number of observed clusters can be explained well using a hydrostatic model where local radiative losses are assumed to be compensated by conductive heating. These required conductivities may be plausible even in the presence of magnetic fields, provided the field is sufficiently turbulent such that electrons can effectively 'diffuse' from field line to field line (Narayan & Medvedev 2001). A weakly tangled magnetic field may however reduce the conductivity by a factor > 100 compared with the unmagnetized Spitzer rate.

Depending on the magnetic field topology, there may hence be room for conduction in clusters. Note that due to the strong temperature dependence of the conductivity, $\kappa \propto T^{5/2}$, it can however be expected that it will affect the richest clusters most strongly, while becoming rapidly less important for lower mass systems.

Leaving the uncertainties about the real effective thermal conductivity of the ICM aside, it is then interesting to ask whether temperature profiles that support conductive heating of the inner parts of clusters could naturally arise in self-consistent cosmological simulations of cluster formation. Due to the non-linear interplay of cooling and conduction, with their different temperature dependences, this question is difficult to answer without hydrodynamical simulations from cosmological initial conditions.

Jubelgas, Springel & Dolag (2004) have introduced a new formulation of conduction in the SPH formalism, and implemented it in the cosmological simulation code GADGET-2.

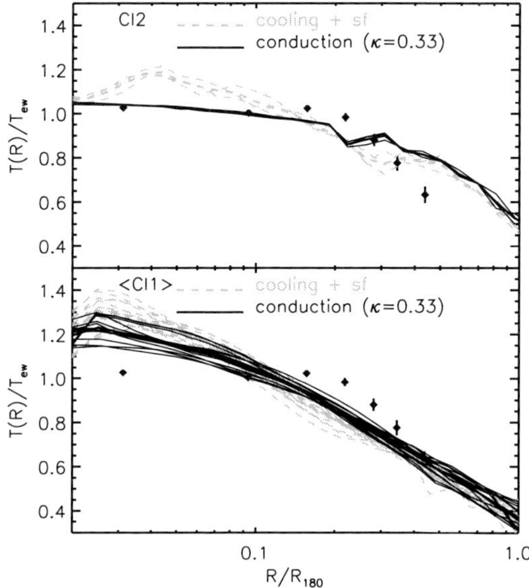

Figure 5. Radial temperature profiles in simulated clusters, for a very massive cluster (top) and several poor systems (bottom). In each case, simulations with conduction (solid lines) are compared with simulations that just included radiative cooling and star formation (grey dashed lines).

Dolag et al. (2004) used the method to study the formation of poor and rich clusters, assuming an effective isotropic conductivity equal to 1/3 of the Spitzer value $\kappa_{\rm Sp}$.

Figure 4 shows temperature maps of a massive cluster which was simulated twice (Dolag et al. 2004), once with cooling and star formation only, and a second time by also including conduction with an effective isotropic conductivity of $\kappa = 1/3\,\kappa_{\rm Sp}$. It is seen that the temperature structure is drastically modified by conduction. Small-scale temperature fluctuations are largely wiped out, except for the cold gas still bound to the centers of cluster galaxies. Also, the outer cluster regions are heated relatively to the simulation without conduction. We note that the heat loss to the outer IGM is however bounded by saturation effects, which are included in the simulation and become important for very low gas densities.

4. Discussion

Cosmological hydrodynamical simulations have proved instrumental to advance the theoretical understanding of cluster formation. While the bulk thermodynamic properties of clusters can be reproduced reasonably well with the present generation of high-resolution simulations, reproducing the detailed thermodynamic state of the ICM gas is still an elusive goal. Early hopes that inclusion of radiative cooling and some form of (comparatively weak) feedback associated with star formation would resolve the problems have not been fulfilled.

Based on the first dynamical calculations with thermal conduction, it also appears increasingly unlikely that conduction may solve the cooling flow riddle. It may still play a role in rich systems, although the strong temperature dependence of conduction may cause new problems in the scaling relations.

Future simulations of cluster formation may therefore need to include additional physics, with the most promising candidate arguably being AGN feedback. Using idealized simulations of buoyant bubbles, Churazov et al. (2001) and Quilis, Bower & Balogh (2001) have shown that heated bubbles may rise in the cluster and uplift cool gas. Combing this with radiative cooling, Dalla Vecchia et al. (2004) explored a model with periodic bubble heating, showing that the cooling losses in the central cluster region can be completely compensated for by periodic AGN activity. The required energy and duty cycle appear plausible, showing that the AGN certainly has enough power to provide the necessary heating. One key problem is how much of the energy released by the accreting black hole is coupled into the ambient gas. A number of detailed simulations of the interaction of the AGN-driven jet with the diffuse gas have already been carried out (e.g. Brüggen & Kaiser 2002; Reynolds, Heinz & Begelman 2002), and will continue to shed more light on this issue in the future. What is missing at the moment are cosmological simulations that follow cluster formation *and* AGN activity from high-redshift to the present. Such self-consistent simulations will be needed to show whether AGN feedback will bring us yet a step closer to simulation models that will eventually capture the most important physics of real clusters of galaxies.

Acknowledgements

I would like to thank all my collaborators who contributed to the work that is discussed here, and the organizers of this conference for a very stimulating and productive meeting.

References

Allen, S., Schmidt, R. W., Fabian & A. C. 2001 *MNRAS* **328**, 37.
Balogh, M. L., Pearce, F. R., Bower, R. G. & Kay, S. T. 2001 *MNRAS* **326**, 1228.
Borgani, S., Murante, G., Springel, V., Diaferio, A., Dolag, K., Moscardini, L., Tormen, G., Tornatore, L. & Tozzi, P. 2004 *MNRAS* **348**, 1078.
Brüggen, M. & Kaiser C. R. 2002 *Nature* **418**, 301.
Bryan, G. 2000 *ApJ* **540**, 39.
Churazov, E., Brüggen, M., Kaiser, C. R., Bhringer, H. & Forman, W. 2001 *MNRAS* **323**, 93.
Dalla Vecchia, C., Bower, R., Theuns, T., Balogh, M., Mazzotta, P. & Frenk, C. S. 2004 *MNRAS* submitted, astro-ph/0402441.
Davé, R., Katz, N. & Weinberg, D. 2002 *ApJ* **579**, 23.
De Grandi, S. & Molendi, S. 2002 *ApJ* **567**, 163.
Dolag, K., Jubelgas, M., Springel, V., Borgani, S. & Rasia, E. 2004 *ApJ* **606**, 97.
Evrard A. 1990 *ApJ* **363**, 349.
Jubelgas M., Springel, V. & Dolag K. 2004 *MNRAS* **351**, 423.
Kay, S. T., Thomas, P. A. & Theuns, T. 2003 *MNRAS* **343**, 608.
Lewis, G. F., Babul, A., Katz, N., Quinn, T., Hernquist, L. & Weinberg, D. H. 2000 *ApJ* **536**, 623.
Muanwong, O., Thomas, P. A., Kay, S. T. & Pearce, F. R. 2002 *MNRAS* **336**, 527.
Narayan, R.& Medvedev, M. V. 2001 *ApJ* **562**, 129.
Navarro, J., Frenk, C. & White, S. D. M. 1995 *MNRAS* **275**, 720.
Pearce, F. R., Thomas, P. A., Couchman, H. M. P. & Edge, A. C. 2000 *MNRAS* **317**, 1029.
Quilis, V., Bower, R., Balogh, M. 2001 *MNRAS* **328**, 1091.
Reynolds, Heinz & Begelman 2002 *MNRAS* **332**, 271.
Springel, V. & Hernquist, L. 2003a *MNRAS* **339**, 289.
Springel, V. & Hernquist, L. 2003b *MNRAS* **339**, 312.
Tornatore, L., Borgani, S., Springel, V., Matteucci, F., Menci, N. & Murante, G. 2003 *MNRAS* **342**, 1025.
Valdarnini, R. 2002 *ApJ* **567**, 741.
Voit, G. M., Bryan, G. L., Balogh, M. L. & Bower, R. G. 2002 *ApJ* **576**, 601.
Zakamska, N. L. & Narayan, R. 2003 *ApJ* **582**, 162.

Outskirts of Galaxy Clusters: Intense Life in the Suburbs
Proceedings IAU Colloquium No. 195, 2004
A. Diaferio, ed.

DOI: 10.1017/S1743921304000572

Dynamical evolution of the ICM

Giuseppe Tormen

Dipartimento di Astronomia, Università di Padova, Italy email: tormen@pd.astro.it

Abstract. We use numerical simulations of galaxy clusters to study the dynamical and thermal evolution of the ICM from high redshift to the present time. We measure the properties of the satellites accreting on the cluster main progenitor and then measure their self-bound mass fraction as a function of time after the merging. We also calculate the mean properties of their orbits and investigate the time evolution of their internal velocity dispersion and gas temperature. We measure and model the mean pericentric and apocentric times, distances and velocities. Finally we show how the properties of the ICM at redshift zero can be understood by decomposing them as a function of the cluster merging history.

---

1. Properties of satellites at merging time

Tormen (1997) made an analysis of the properties of dark matter satellites measured at the time $t_{mer}$when they first are accreted onto the main progenitor of a galaxy cluster. The main results of this research are the following:

• the infall of satellites onto the main cluster progenitor is very anisotropic. As shown on the left panel of Figure 1, the mean pericenter - measured from the orbits at merging time - is of order 30 percent of the virial radius of the cluster at that time. Orbits are neither too circular, nor very elongated: their circularity $\epsilon = J/J_c$ (ratio of the satellite's angular momentum to the angular momentum of a circular orbit with the same energy; $\epsilon = 0$ for purely radial orbits, $\epsilon = 1$ for purely circular ones) is peaked around $\epsilon = 0.5$, and roughly symmetric.

• Massive satellites tend to approach the main progenitor along slightly more radial orbits compared to less massive ones. This is consistent with the fact that most of the mass accreted onto a cluster comes along large scale filaments, where massive satellites are more likely to live.

• The infall pattern of satellites is largely responsible for the final shape and orientation of the cluster mass and velocity distribution, as shown on the right panels of Figure 1. This means that, if we model the shape of the infall pattern with an ellipsoid, this tends to be aligned (typically within 30 degrees) with the mass and velocity ellipsoids.

2. Properties of merged satellites

The results presented in this section are throughly discussed in Tormen, Moscardini and Yoshida (2004), where the reader is addressed to for full details of the simulation data and method. After satellites accrete onto the main cluster progenitor, their properties evolve quite rapidly, by the different interaction with the gravitational field and the ICM of the host. Simulations (e.g. Tormen, Diaferio and Syer 1998; Tormen, Moscardini and Yoshida 2004) show that massive sub haloes penetrate deeper into the main cluster, and unbind their dark matter more quickly than less massive ones. The ICM is stripped immediately from small satellites, while the gravitational potential well of massive ones is deep enough to prevent stripping of the hot gas until the whole satellite is destroyed.

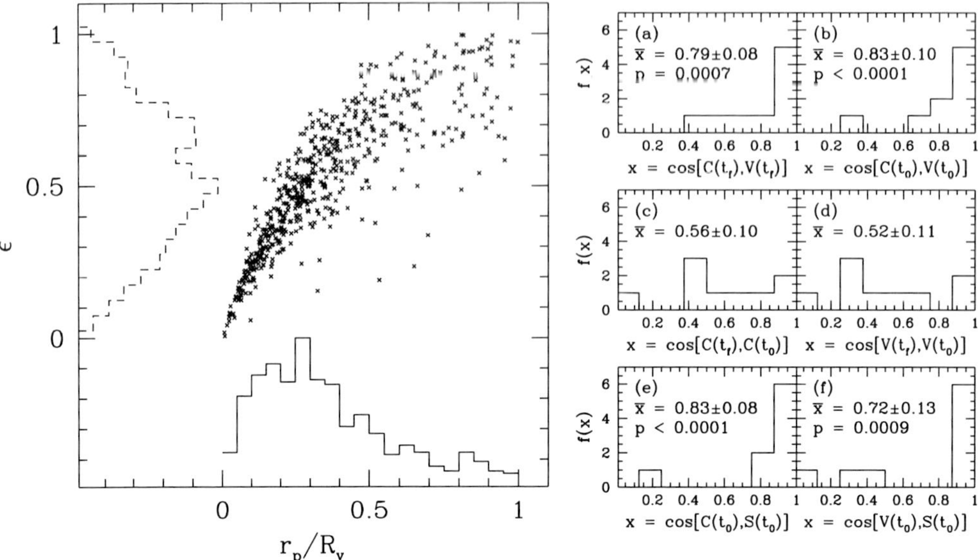

Figure 1. Left: Scatter plot of the orbital circularity ϵ versus the pericenter r_p/R_v in units of the virial radius of the main halo progenitor at merging time. The histograms show the projected distributions. Right: Correlation between the major axis of the cluster mass ellipsoid (C), the velocity ellipsoid (V) and the infall pattern of satellites (S). The histograms show distributions of the cosine of misalignment angles between the different ellipsoids. Where relevant, p gives the probability that such a correlation arises by chance, measured by a KS test (from Tormen 1997, reprinted with permission).

In the host halo, thermalization of the dark matter and hot gas accreted through the satellites happens on different timescales for the two components. The left panels of Figure 2 show that the ICM temperature of the particles initially belonging to the satellites increases faster during the 2 Gyr immediately following the merging, and stabilizes or increases more slowly thereafter. Globally the new thermal equilibrium is approached in roughly 2 Gyr, almost regardless of the satellite mass (solid bands). On the other hand, the internal velocity dispersion of the satellite dark matter takes longer to increase, and it does so more quickly for massive satellites (hatched bands).

The evolution of the same quantities, for particles which remain bound to the substructure, is quite different, as shown in the right panel of Figure 2. We remind that these particles constitute the cores of satellites, and so are those which would be actually identified as substructure in a simulation or in an observation. The figure shows that the dark matter velocity dispersion of the satellite cores never increases, but actually decreases: satellite cores become colder, the effect being more dramatic for the most massive satellites (bottom right panel). The ICM temperature displays a similar behavior, except for the smallest satellites (upper left), where the potential well is too shallow to shield the satellite ICM from the global heating. This results suggests that one should place particular attention when trying to estimate the mass of substructure from the temperature of its diffuse gas or from the velocity dispersion of its galaxies.

A summary of the orbital properties of satellites at their first pericenter and at the following apocenter are shown in Figure 3. The upper panels show the time in Gyr needed for a satellite to reach its first pericenter after crossing the virial radius of the main cluster (left), the cluster-centric distance at pericenter in units of the cluster virial

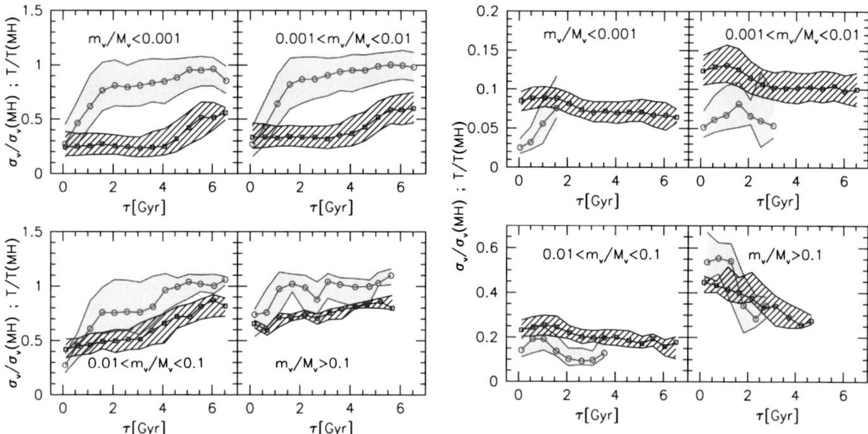

Figure 2. Time evolution of the dark matter velocity dispersion (hatched bands) and of the gas temperature (solid bands). The left panel shows evolution for all particles initially belonging to satellites, while the right panel shows evolution of the particles which at each time τ are still bound to the satellite. Different sub-panels refer to satellites of different mass, in units of the mass M_v of the host cluster measured at merging time. Satellite temperatures and velocity dispersions are normalized by the corresponding value measured for the main cluster at each time τ after merging (from Tormen, Moscardini & Yoshida 2004, reprinted with permission).

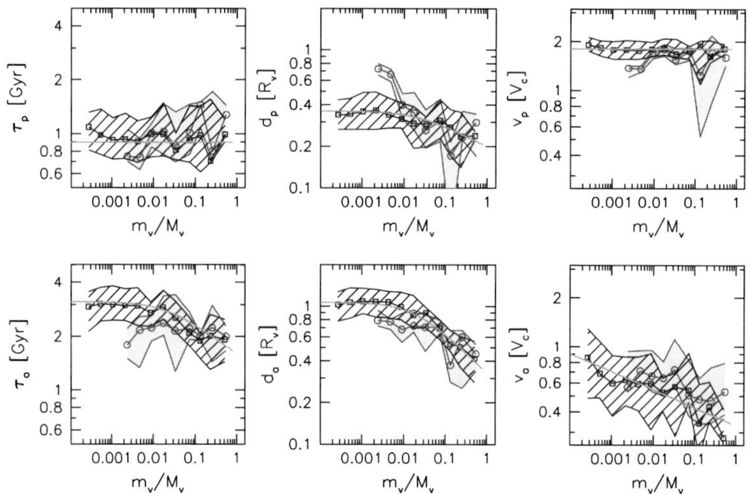

Figure 3. Orbital quantities at first pericentric passage (upper panels) and at the following apocenter (lower panels), shown as a function of the satellite mass with respect to the mass of the main cluster progenitor, m_v/M_v. Solid bands refer to gas particles, hatched bands refer to dark matter. Times are expressed in Gyr, distances are in units of the virial radius of the main cluster at that time, while velocities are in units of the circular velocity of the main cluster at the same time (from Tormen, Moscardini and Yoshida 2004, reprinted with permission).

radius (center) and the corresponding orbital velocity in units of the circular velocity of the main halo (right). In each panel, these quantities are plotted versus the satellite mass in units of the mass of the main cluster. The lower panels show the same quantities at first apocenter.

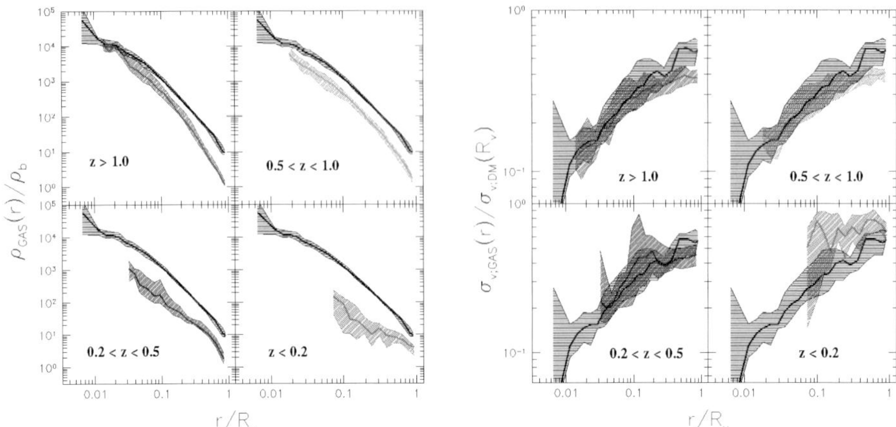

Figure 4. ICM decomposition in redshift: gas density (left plot) and gas velocity dispersion (right plot). In each plot the hatched (black) bands show global mean profiles over our SPH cluster sample (plus or minus one sigma of the mean), which are the same for all four panels in each plot. Colored bands show the mean profile over the cluster sample, but considering only the indicated subset of particles. For example, the upper left panels in each plot contrasts the global profiles with those of particles accreted at redshift larger than one.

We can see that pericentric times are reached in roughly 1 Gyr, regardless of the satellite mass; pericentric distances are smaller for massive satellites, and in rough agreement with those predicted by the initial orbits (Figure 1). Pericentric velocities are also mass-independent, with values around twice the circular velocity of the main halo. On the other hand, massive satellites reach their apocenter more quickly (lower left); they stay closer to the cluster center (lower center), and are significantly slower than small satellites (lower right). These results are consistent with the expectation that massive satellites slow down their motion quickly after their first encounter with the cluster, and merge with the cluster itself after at most a couple of pericentric passages.

3. Decomposition of the final ICM

A point of view orthogonal to that taken in the previous section is that of decomposing the ICM of the cluster at redshift zero in subsamples, by selecting the particles that were accreted by the main cluster progenitor at different redshifts. Yet another alternative is that of separating the ICM particles depending on mass of the satellite that carried them at merging time.

A simple study then is to analyze the ICM radial profiles: here we looked at the gas density, temperature, velocity dispersion and entropy. The interest of this experiment is then to be able to link the profiles to the merging history of the cluster, understanding the possible differences in the light of the different dynamical and thermal histories of the different particle subsets.

3.1. *Decomposition in redshift*

Figure 4 shows the mean gas density (left) and velocity dispersion (right) profiles averaged over our cluster sample at redshift $z = 0$. The black band is the mean profile ($\pm 1\sigma$ of the mean) obtained using all the gas particles of each cluster. The colored profiles correspond instead to subsets of particles selected according to the redshift when they

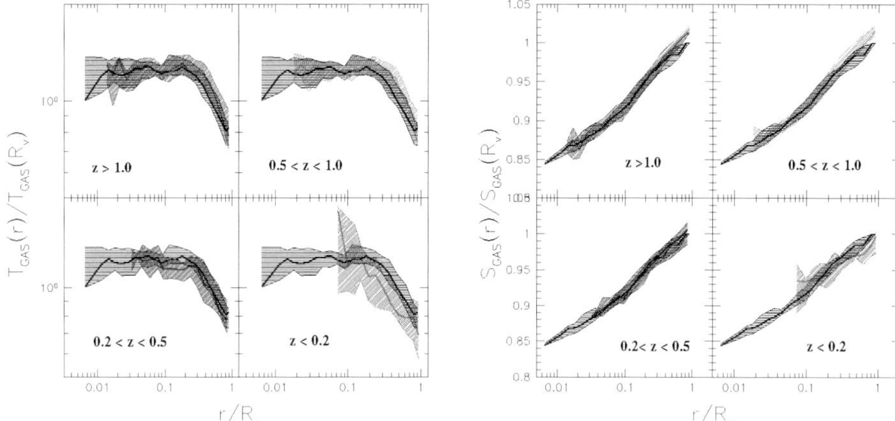

Figure 5. ICM decomposition in redshift: gas temperature (left plot) and entropy (right plot). Here also the hatched black bands correspond to global mean profiles, and the colored ones to profiles made using a subset of particles.

first were accreted by the main cluster progenitor. We immediately see a segregation in mass in the density plot: while particles accreted at redshift $z > 1$ (upper left panel) end up closer to the center in the final cluster, and in fact account for most of the gas mass in the inner regions, moving to lower and lower accretion redshifts the gas distribution becomes shallower and shallower, while more and more gas is confined to the external regions of the clusters. In particular, the lower right panel of the left plot indicates that most of the gas found around the virial radius at redshift $z = 0$ has been accreted very recently, at $z < 0.2$. This trend is effectively a secondary infall signature, as gas accreted at later times is, on average, stopped at larger and larger radii by the outgoing front of gas already accreted.

The gas velocity dispersion, measured with respect to the mean cluster velocity, is an indication of both the net gas bulk motion and of small-scale (but still macroscopic) random motions. In general the velocity dispersion decreases towards the cluster center, as shown by the black band in the right plot of Figure 4. However, this global behavior is again the superposition of different populations of particles with contrasting properties, as shown by the colored bands. While following the general trend of decreasing towards the center, the gas accreted at lower and lower redshift has larger and larger net velocities at large radii, and in the case of the most recent accretion ($z < 0.2$) the velocity dispersion is flat at all radii reached by the particles. This is consistent with what we have shown above for the gas thermalization times: the gas accreted more recently is still converting its bulk motions into internal energy. If so, it should also have an average temperature which is lower than the mean at large radii. This is indeed the case, as illustrated in the lower left panel of the left plot of Figure 5. Moreover, the same figure shows that the slower-than-average gas present around the virial radius and accreted at higher redshift is also slightly hotter than the mean, in agreement with the expectation. Finally, the right plots of Figure 5 gives the behavior of gas entropy $S \propto T/\rho^{2/3}$, which again confirms this trend, by showing that at large radii the entropy of gas accreted at lower redshift (which is slightly colder and denser) is lower than average, while the opposite is true for gas accreted at higher redshift, which at large radii is lower in density and higher in temperature.

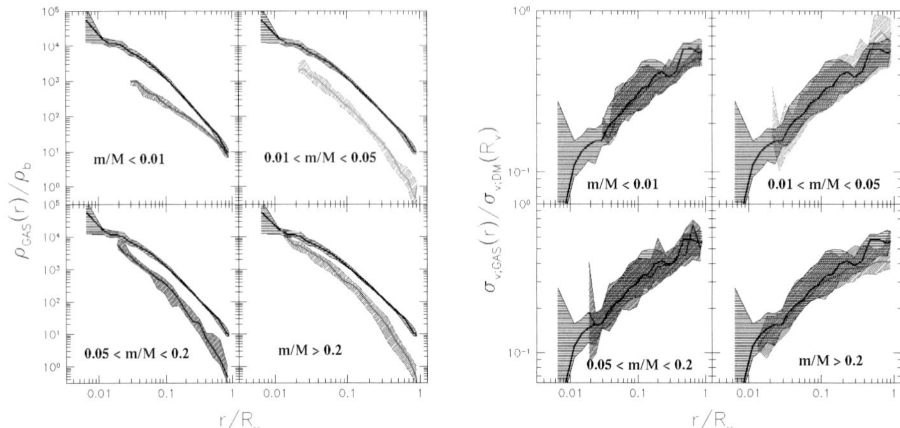

Figure 6. ICM decomposition in satellite mass: gas density (left plot) and gas velocity dispersion (right plot). Colors are as in Figure 4. Here the decomposition is done depending on the mass m_v of the satellite carrying the particles into the main cluster of mass M_v at merging time. Each panel in each plot refer to a different bin in mass ratio m_v/M_v. For example, the lower right panel in the left plot compares the mean (over the cluster sample) gas density profile made with all particles (black band) with the mean (over the cluster sample) gas profile made only with the particles which accreted in satellites with mass larger than 20 percent of the main progenitor mass, that is, accreted in major mergers.

3.2. *Decomposition in satellite mass*

If we play a similar game of decomposing the gas profiles according to the mass of the satellites that carried the particles, we obtain the profiles of Figures 6 and 7, which are the analogous of Figures 4 and 5 respectively. The gas density profiles in the left plot of Figure 6 clearly show a segregation in gas by this mass selection criterion: gas accreted in small satellites mainly sits in the outer cluster regions, while inner regions are mainly populated by the gas brought in by major mergers. The right plot of the same figure suggests a hint of trend in the outer regions, in the directions of a velocity dispersion lower than the mean both for very small and for very large satellites. In the case of small satellites this could simply reflect the fact that the gas in small satellites is stopped more easily as it enters the main cluster. For major mergers the interpretation is less obvious.

Going to the temperature profiles, in the left plot of Figure 7, we observe a more complex trend. Particles from small satellites end up with a temperature slightly higher then the mean: this is what we would expect if shocks are more efficient in heating the gas up as soon as the satellites enter the main cluster. Satellites with intermediate masses deposit significantly cooler gas in the outer regions: in this case the gas is shielded more efficiently from the stripping and heating. However, going to even larger satellites (lower right panel) we do not observe a continuity in this trend; the resulting temperature profile is well in agreement with the global one. Perhaps this simply reflects the fact that most of the cluster gas is indeed carried in by few massive mergers; in this case the mean cluster temperature would be defined by the temperature of the particles from the most massive satellites.

The right plot in Figure 8 shows the entropy profiles. Now particles from small satellites have density and temperature both higher than the average in the cluster outskirts, therefore the effect on the entropy is almost canceled. Intermediate mass satellites have lower entropy at large radii due to their lower temperature, while massive satellites donate

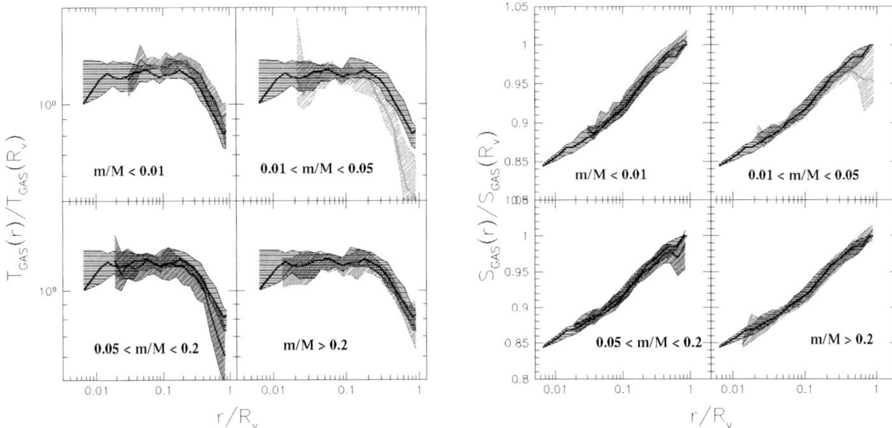

Figure 7. ICM decomposition in satellite mass: gas temperature (left plot) and entropy (right plot). Colors are as in Figure 4.

particles which end up with average entropy. Although it is not easy to understand all these behaviors in detail, we think that this decomposition game can lead to a better understanding of the global properties of clusters observed at any given redshift.

4. Conclusions

We have seen that a number of properties of galaxy clusters can be understood by looking at their merging history. The high anisotropy of the matter infall leaves an imprint on the final cluster shape, both in mass and velocity dispersion.

The thermalization of substructure in the main cluster proceeds differently for the outer and inner parts of satellites: while the former are quickly heated to temperatures of the order of the virial temperature of the main cluster, the cores of satellites tend to cool, probably due to the stripping of hot particles (both in dark matter and gas). The decomposition of matter in the final cluster, as a function of either the merging redshift or the satellite mass at merging time, brings further insight into the cluster internal structure, and suggests that at redshift zero different regions of the clusters are unevenly populated by matter accreted in different ways.

Acknowledgements

I would like to acknowledge Antonaldo Diaferio for organizing so lively a meeting, and for inviting me to give this talk. Partial support was provided by the Italian MIUR (Grant 2001, prot. 2001028932, "Clusters and groups of galaxies: the interplay of dark and baryonic matter") and by ASI. Blackwell Publishing Ltd. is acknowledged for granting the permission to reprint Figures 1, 2 and 3 from the original publications.

References

Tormen, G. 1997 *MNRAS* **290**, 411–421
Tormen, G., Moscardini, L., and Yoshida, N. 2004 *MNRAS* **350**, 1397–1408

Outskirts of Galaxy Clusters: Intense Life in the Suburbs
Proceedings IAU Colloquium No. 195, 2004
A. Diaferio, ed.

DOI: 10.1017/S1743921304000584

The structure of the ICM from high-resolution SPH simulations

G. Yepes[1], Y. Ascasibar[2], R. Sevilla[1], S. Gottlöber[3] and V. Müller[3]

[1]Grupo de Astrofísica, Universidad Autónoma de Madrid, Madrid E-280049, Spain
[2]Theoretical Physics, University of Oxford, 1 Keble Road, Oxford OX1 3NP, United Kingdom
[3]Astrophysikalisches Institut Potsdam, An der Sternwarte 16, Potsdam D-14482, Germany

Abstract. We present results from a set of high (512^3 effective resolution), and ultra-high (1024^3) SPH adiabatic cosmological simulations of cluster formation aimed at studying the internal structure of the intracluster medium (ICM). We discuss the radial structure and scaling relations expected from purely gravitational collapse, and show that the choice of a particular halo model can have important consequences on the interpretation of observational data. The validity of the approximations of hydrostatic equilibrium and a polytropic equation of state are checked against results of our simulations. We also show the first results from an unprecedented large-scale simulation of 500 h^{-1} Mpc and 2×512^3 gas and dark matter particles. This experiment will make possible a detailed study of the large-scale distribution of clusters as a function of their X-ray properties.

---

1. Introduction

Galaxy clusters are a unique laboratory to test the hierarchical paradigm of structure formation. They are the best probes of the large scale structure of the Universe and have often been used as a diagnostic of the cosmological parameters. The intrinsic non-linear nature of gravitational collapse and gas dynamics makes numerical simulations the most useful tool to study in detail the process of cluster formation and evolution.

Simple analytical models for the structure of the ICM can be derived from the hypotheses of hydrostatic equilibrium and polytropic equation of state, $P \propto \rho_g^\gamma$. But real clusters might not be well described by these two hypotheses. For instance, kinetic energy makes a significant contribution to the energy budget of merging systems, and therefore thermally-supported hydrostatic equilibrium ceases to be a valid approximation. Even in relaxed systems, this assumption is not very accurate in the outermost parts, where gas motions become more important. Departures from spherical symmetry can also play a role in the final structure of the ICM (Lee & Suto 2003) and, last but not least, there is no obvious physical reason for the gas to follow a polytropic relation. From our numerical experiments, we showed (see Ascasibar et al. 2003) that hydrostatic equilibrium is fulfilled within $\sim$ 20 per cent accuracy by all simulated clusters, as long as they are not heavily disturbed. A polytropic equation of state seems to be a good approximation as well, although its reliability near the centre is still a matter of debate (see e.g. Rasia et al. 2003). From our data we derive a polytropic index of $\gamma \sim 1.18$.

We compared four different analytical halo models with our simulations. The first two models assume that haloes are well described by Navarro et al. (1997) and Moore et al. (1999) fitting formulae, while the other two assume that the gas follows a β-model (Cavaliere & Fusco-Femiano 1976). We consider a 'canonical' version of the β-model, in which the gas is isothermal ($\gamma = 1$) and $\beta = 2/3$, and a polytropic version with $\gamma = 1.18$ and $\beta = 1$. The same value of the polytropic index has been used for the first two models

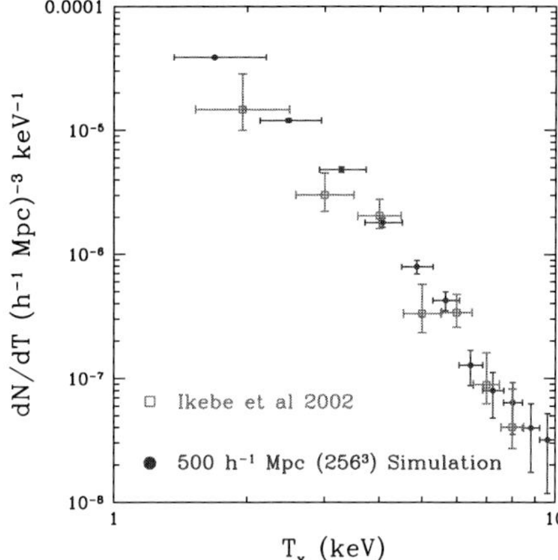

Figure 1. Comparison of the simulated X-ray temperature function with observational estimates.

as well. Hereafter we will use the abbreviations NFW, MQGSL, BM and PBM to refer to these models. For a detailed description, the reader is referred to Ascasibar et al. (2003).

2. Numerical experiments

We have carried out a series of high-resolution gasdynamical simulations of cluster formation in a flat ΛCDM universe ($\Omega_{\rm m} = 0.3$; $\Omega_\Lambda = 0.7$; $h = 0.7$; $\sigma_8 = 0.9$; $\Omega_{\rm b} = 0.02\,h^{-2}$). THe imulations have been done with the parallel GADGET code (Springel et al. 2001), with a novel version of SPH in which the entropy is explicitly conserved (Springel & Hernquist 2002). In a cubic volume of 80 h^{-1} Mpc on a side, an unconstrained realization of the power spectrum of density fluctuations corresponding to the ΛCDM model was generated for a total of 1024^3 Fourier modes. The density field was then resampled to a grid of 128^3 particles, which were displaced from their Lagrangian positions according to the Zeldovich approximation up to $z = 49$. Their evolution until the present epoch is traced by means of a pure N-body simulation with 128^3 dark matter particles. A sample of clusters selected from this preliminary low-resolution experiment were re-simulated with higher resolution by means of the multiple mass technique (see Klypin et al. 2001, for details). Mass resolution is then increased by using smaller masses in the Lagrangian volume depicted by these particles, including the additional small-scale waves from the ΛCDM power spectrum in the new initial conditions. We use 3 levels of mass refinement, reaching an effective resolution of 512^3 CDM particles (2.96×10^8 h^{-1} $\mathrm{M}_\odot$). Gas was added in the highest resolved area only. The total number of particles (dark+SPH) in this area is greater than $1 - 2 \times 10^6$ for all clusters. The gravitational softening length was set to $\epsilon = 2 - 5$ h^{-1} kpc, depending on the number of particles within the virial radius (Power et al. 2003). The SPH minimum smoothing length was fixed to the same value as ϵ.

In order to study the effects of resolution in the determination of X-ray properties of our clusters, we have resimulated one of the objects with 8 times more mass resolution, reaching an effective resolution of 1024^3 particles (i.e. $m_{\rm dark} \sim 3 \times 10^7$ $h^{-1} M_\odot$; $m_{\rm sph} \sim 5 \times 10^6 M_\odot$). The total number of particles within the virial radius was $11,106,465$.

The list of our simulated clusters extracted from the 80 h^{-1} Mpc volume span a relative small range in X-ray emission temperature (from 0.6 to 3 keV). In order to extend our numerical sample of clusters to a wider temperature (mass) range, we have simulated a considerable much bigger volume (500 h^{-1} Mpc) in which a random realization of the ΛCDM power spectrum was generated with 2048^3 particles. In this way, we will have a similar resolution for our clusters than in the previous experiments. We have resimulated the whole 500 h^{-1} Mpc box with different mass resolutions: 2×128^3, 2×256^3, and 2×512^3 dark and SPH particles. We identified all halos in the lowest resolution run (128^3). Using the same technique as before, we selected clusters in this run and resimulated them at full resolution (2048^3), using 5 different species of dark particles and SPH with $m_{\rm sph} \sim 1.6 \times 10^8 h^{-1} M_\odot$. The clusters selected cover the mass range $2 \times 10^{14} - 2 \times 10^{15}$ $M_\odot$. In this regard, we could extend the temperature range of our cluster sample up to 11 keV. The total number of particles within the virial radius in these clusters is comparable to the number of particles of the previous simulations ($\sim 10^6$).

We have recently finished the run with 2×512^3 gas and dark particles that, to our knowledge, is one of the largest adiabatic SPH simulations of large scale structure done so far. The mass resolution is $m_{\rm dark} \sim 6 \times 10^{10} h^{-1}$ $M_\odot$, which means that we can resolve from galactic halos (100+100 particles) to the biggest galaxy clusters (4×10^6 particles). We identified a total of 4×10^5 dark matter halos, with 10 or more particles, in this simulation. Well resolved halos (7000 dark particles or more) correspond to clusters with emission weighted temperatures > 3 keV. If we go down to 1000 particles, the halos resolved have temperatures > 1 keV, and temperatures go down to ~ 0.6 keV for halos with 500 or more particles. Due to the large simulated volume, we have a statistical significant sample of clusters and groups. The total number of objects with $T_X > 0.6$ keV, excluding substructure, is $\sim 30,000$. A total of 126 hot clusters have been found with $T_X > 5$ keV. The X-ray temperature function for clusters in this simulation is shown in Figure 1, together with recent observational estimates. As can be seen, the number density of clusters for the highest temperatures (> 6 keV) found in simulations is compatible with observations (see Borgani et al. 2004 for results from non-adiabatic simulations).

The resulting $M - T_X$ and $L_X - T_X$ relations are depicted in Figure 2, in which we also show 6 of the hottest clusters resimulated at high resolution. They show a convergence of results at least for clusters with $T_X > 3$ keV. The fit to the $M - T_X$ relation for clusters with $T_X > 1$ keV is $(M_{vir}/M_0) = (T_x/\text{keV})^\alpha$, with $\alpha = 1.56 \pm 0.05$ and $M_0 = 4.4 \times 10^{13} h^{-1} M_\odot$. The slope is compatible with simulations which include non adiabatic effects, although the zero point, M_0, is a factor of 2 higher. Note however that we are using the virial mass instead of the mass within an overdensity of 500. This implies that the $M - T$ relation is rather insensitive to the energy transfers due to non adiabatic processes associated to star-gas interactions.

3. Radial structure of the ICM

We compare the universal halo models described earlier, with results from our simulations. For NFW and MQGSL, we obtain the characteristic density and radius of each cluster by fitting the dark matter distribution. Gas density and temperature profiles are genuine predictions of these models. For BM and PBM, we fit the gas distribution and predict the gas temperature and the dark matter density. In Figure 3 we plot the average dark matter and gas density profiles. Not surprisingly, both NFW and MQGSL provide good fits to the dark matter density. These models are able to accurately predict the gas density, but they are too steep at large radii due to a systematic departure from hydrostatic equilibrium.

On the other hand, β-models (BM and PBM) show a core in the dark matter density that is not seen in the numerical data. Moreover, they do not give an accurate description

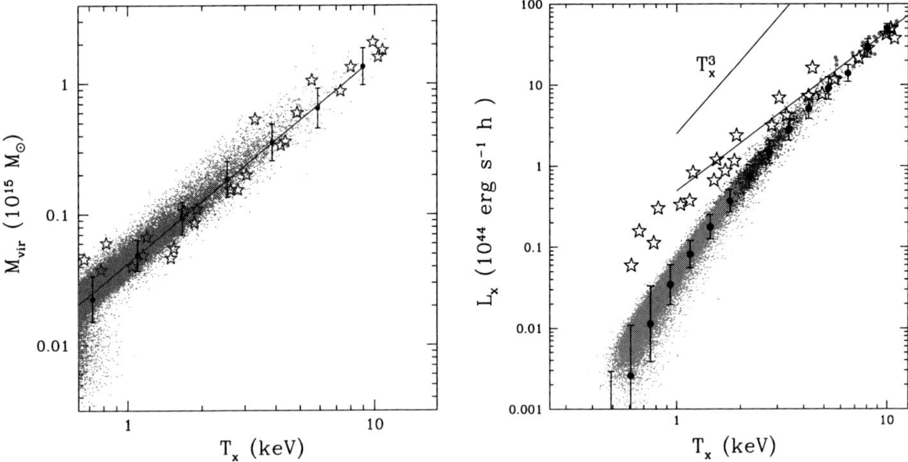

Figure 2. Results from a 500^3 h^{-3} Mpc^3 simulated volume. Left, the $M_{vir} - T_x$ relation. Right, the $L_X - T_X$ relation. Stars represent the clusters resimulated with high resolution. The dotted line is a fit to the hottest halos ($>$ 6 keV). The slope is $\sim$ 2, as in the self-similar scaling behaviour.

of the gas density profile. The inner regions are better described with low values of β, while the outer parts require higher values for this parameter.

One of the most striking results from our simulations is that they favour the hypothesis of a 'universal' temperature profile. As can be seen in Figure 3, the ICM is not isothermal, but the temperature decreases by a factor of three or four from the centre to the virial radius. The projected X-ray emission-weighted temperature profile is plotted in the bottom-left panel of Figure 3. We compare our simulations with previous work by Loken et al. (2002) based on Eulerian simulations. These authors propose a 'universal' form $T_{\rm p}(r) = T_0(1 + r/a_{\rm x})^{-\delta}$. Our results are well described by this relation. Although real clusters seem to be consistent with a polytropic equation of state (Markevitch et al. 1998), recent observations indicate the presence of a large isothermal core (De Grandi & Molendi 2002). Apart from this feature, which is not observed in our objects, adiabatic simulations are in good agreement with observational data beyond $0.2R_{200}$.

Entropy profiles are plotted in the bottom-right panel of Figure 3. Contrary to the common view, neither the analytical models nor the simulation data yield a pure power-law profile, despite the fact that purely adiabatic gasdynamics has been considered. As shown in Ascasibar et al. (2003), the standard implementation of the SPH algorithm can lead to misleading results in the inner regions due to artificial entropy losses (see e.g. Springel & Hernquist 2002). We find that the shape of the entropy profile does not depend on the cluster mass or temperature, in agreement with recent observations (Ponman et al. 2003). NFW or MQGSL models provide a better estimate of the entropy profile than the β-model, but the low gas densities predicted at large radii yield a very steep slope at $r \sim R_{200}$.

4. Convergence of results

To check for convergence of results in terms of numerical resolution, we compare, in Figure 4, radial profiles for one of our clusters from the ultra-high resolution simulation described in the previous section. As it can be appreciated, the structure of the ICM is quite similar, at least from the virial radius down to 1% of that. The features at the most

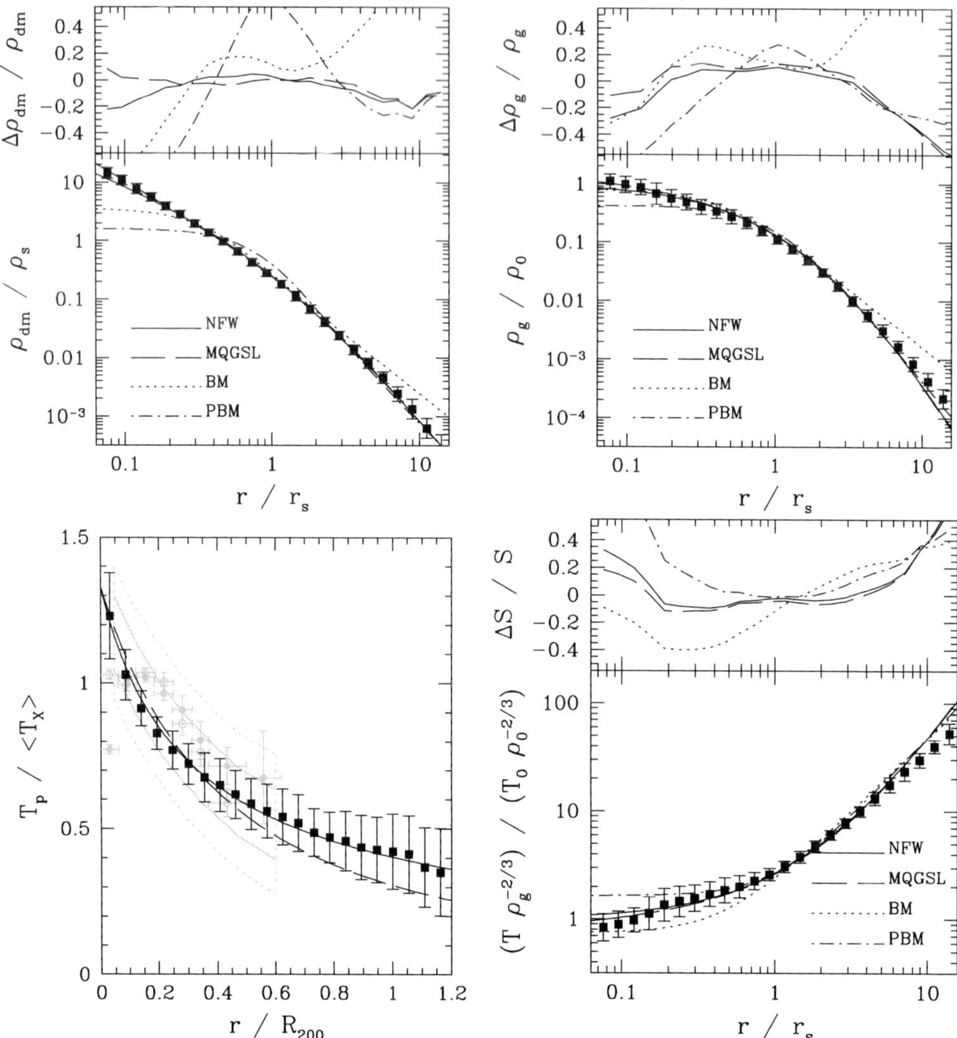

Figure 3. Universal radial profiles. Upper panels: dark matter and gas density. Lower panels: projected temperature and gas entropy. Black squares represent the numerical data, averaged over all clusters except major mergers. Error bars denote one-sigma scatter. Lines are used to plot the analytical models, and top panels quantify deviations from the simulated profile. Dashed lines in the projected emission-weighted temperature profile, shows the 'universal' profile proposed by Loken et al. (2002), while solid line represents the best fit to our data. Observations by De Grandi & Molendi (2002, circles) and Markevitch et al. (1998, boxed region) are plotted for comparison.

internal parts of the cluster are related to the different dynamical stage of the cluster. The positive results of this comparison is that the overall properties of the halos are well described with the resolution adopted in our simulations in which 500,000-1,000,000 SPH particles are used.

Acknowledgements

This work has been partially supported by the MCyT (Spain, AYA-0973), by the *Acciones Integradas Hispano-Alemanas* HA2000-0026 and by DAAD (Germany). We

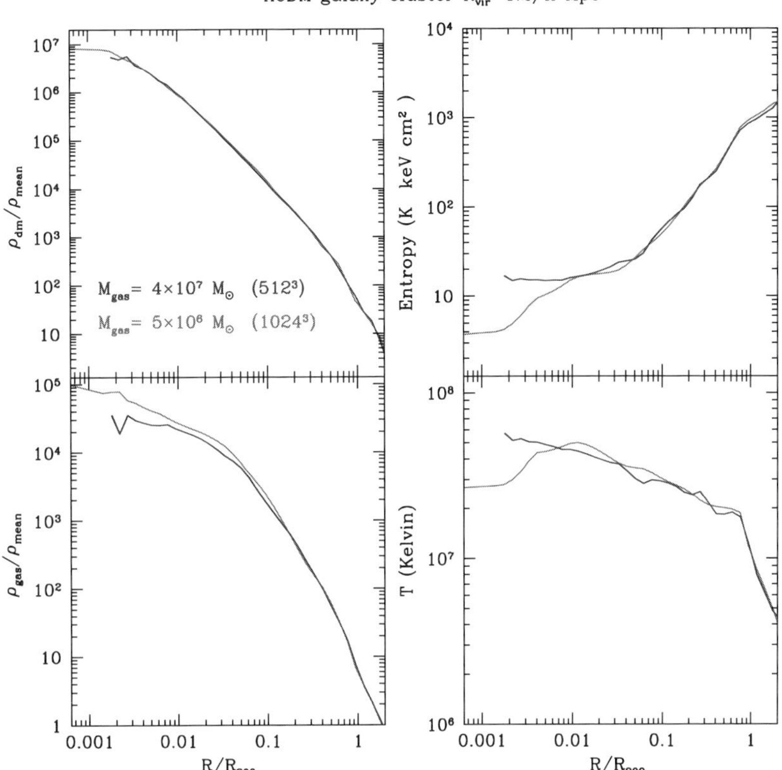

Figure 4. Radial profiles for a cluster run with two different resolutions.

thank the Forschungszentrum Jülich for allowing us to use the IBM p690+ supercomputer.

References

Ascasibar, Y., Yepes, G., Müller, V., & Gottlöber, S. 2003, MNRAS, 346, 731
Borgani, S., Murante, G., Springel, V., et al. 2004, MNRAS, 348, 1078
Cavaliere, A. & Fusco-Femiano, R. 1976, A&A, 49, 137
De Grandi, S. & Molendi, S. 2002, ApJ, 567, 163
Klypin, A., Kravtsov, A. V., Bullock, J. S., & Primack, J. R. 2001, ApJ, 554, 903
Lee, J. & Suto, Y. 2003, ApJ, 585, 151
Loken, C., Norman, M. L., Nelson, E., et al. 2002, ApJ, 579, 571
Markevitch, M., Forman, W. R., Sarazin, C. L., & Vikhlinin, A. 1998, ApJ, 503, 77
Moore, B., Quinn, T., Governato, F., Stadel, J., & Lake, G. 1999, MNRAS, 310, 1147
Navarro, J. F., Frenk, C. S., & White, S. D. M. 1997, ApJ, 490, 493
Ponman, T. J., Sanderson, A. J. R., & Finoguenov, A. 2003, MNRAS, 343, 331
Power, C., Navarro, J. F., Jenkins, A., et al. 2003, MNRAS, 338, 14
Rasia, E., Tormen, G., & Moscardini, L. 2003, astroph/0309405
Springel, V. & Hernquist, L. 2002, MNRAS, 333, 649
Springel, V., Yoshida, N., & White, S. D. M. 2001, New Astronomy, 6, 79

Outskirts of Galaxy Clusters: Intense Life in the Suburbs
Proceedings IAU Colloquium No. 195, 2004
A. Diaferio, ed.

DOI: 10.1017/S1743921304000596

The dynamics of cluster substructure

Stuart P. D. Gill, Alexander Knebe and Brad K. Gibson

Centre for Astrophysics & Supercomputing, Swinburne University, Mail #31, P.O. Box 218, Hawthorn, VIC 3122, Australia email: sgill@astro.swin.edu.au

Abstract. To understand cluster formation and evolution it is important to understand the evolution of cluster substructure. The work described here focuses upon a set of fully self-consistent numerical simulations within a ΛCDM cosmology. Within the simulations we follow in detail the temporal and spatial properties of individually tracked satellite galaxies. The host galaxy clusters were chosen to sample a variety of formation histories, ages, and triaxialities. Despite their obvious differences, we find striking similarities within the associated substructure populations. Namely, the satellite galaxy orbital distributions are indistinguishable between the galaxy clusters.

1. Introduction

There has been growing convergence to a "concordance model" for cosmological structure formation, with a "standard" ΛCDM universe comprised of 28% dark matter, 68% dark energy, and luminous baryonic matter (i.e. galaxies, stars, gas, and dust) at a mere 4% (cf. Spergel et al. 2003). In such a universe structures form bottom-up with small objects forming first that subsequently merge to form progressively larger objects (e.g. White & Reese 1978; Davis et al. 1985). Hence, galaxies and galaxy clusters are constantly fed by smaller dark halos that begin to orbit within the encompassing dark matter host. If this model is correct, it is important to understand the orbital evolution of these objects and investigate how they accrete onto the host halo. The work presented here is based upon a series of high-resolution numerical simulations within said concordance model. The analysis focuses on the detailed orbital parameters of satellite galaxies and internal properties.

The first fully self-consistent simulations targeting the subject were performed by Tormen et al. (1997, 1998). Both studies were excellent efforts. However they lacked the temporal, spatial, and mass resolution required to explore a wide range of environmental effects. Unable to follow the satellite distribution within the host's virial radius, satellites were instead tracked only up to and including the point of "accretion". This allowed an analysis of the infall pattern, rather than the orbital evolution of the satellites. Ghinga et al. (1998) also investigated the dynamics of satellite galaxies in live dark matter host halos. Although greatly increasing the mass and spatial resolution, they still lacked the temporal resolution to explicitly track the satellite orbits, rather they approximated the orbits using a spherical static potential again. In these proceedings we will present a series of simulations with the required resolution to follow the satellites even within the very central regions of the host potential ($\geqslant$5–10% of the virial radius) and the time resolution to resolve the satellite dynamics with unprecedented accuracy ($\Delta t \approx$170 Myrs).

2. Simulations and halo identification

The following section outlines our set of eight high-resolution galaxy clusters each consisting of order more than a million dark matter particles and the methods used to

Halo	$R_{\rm vir}$	$M_{\rm vir}$	$z_{\rm form}$	age	$N_{\rm sat}(<R_{\rm vir})$
# 1	1.34	2.87	1.16	8.30	158
# 2	1.06	1.42	0.96	7.55	63
# 3	1.08	1.48	0.87	7.16	87
# 4	0.98	1.10	0.85	7.07	57
# 5	1.35	2.91	0.65	6.01	175
# 6	1.05	1.37	0.65	6.01	85
# 7	1.01	1.21	0.43	4.52	59
# 8	1.38	3.08	0.30	3.42	251

Table 1. Properties of the eight dark matter host halos. Distances are measured in h^{-1} Mpc, velocities in km s^{-1}, masses in $10^{14}h^{-1}$ M$_\odot$, and the age in Gyr. We applied a mass-cut of $M > 2 \times 10^{10}h^{-1}$ M$_\odot$ (100 particles) which explains the rather 'low' number for $N_{\rm sat}(<R_{\rm vir})$.

analyse them. These clusters formed in dissipationless N-body simulations characterised by a ΛCDM cosmology ($\Omega_0 = 0.3, \Omega_\lambda = 0.7, \Omega_b h^2 = 0.04, h = 0.7, \sigma_8 = 0.9$). The runs have a mass resolution of $m_p = 1.6 \times 10^8 h^{-1}$ M$_\odot$ and achieved a force resolution of $\approx 2h^{-1}$ kpc allowing us to resolve the host halos down to about the central 0.25% of their virial radii $R_{\rm vir}$. High temporal information was required to determine the orbital development of the satellites. We therefore stored 17 outputs from $z = 2.5$ to $z = 0.5$ equally spaced with $\Delta t \approx 0.35$ Gyr. From $z = 0.5$ to $z = 0$ we have 30 outputs spaced $\Delta t \approx 0.17$ Gyr. A summary of the eight host halos, is presented in Table 1.

These simulations were analysed using two new methods for identifying gravitationally bound objects, `MHF` (`MLAPM` Halo Finder) and `MHT` (`MLAPM` Halo Tracker). A more elaborate description of this technique can be found elsewhere (Gill, Knebe & Gibson 2004). The finder (`MHF`) utilizes the adaptive meshes of the open source N-body code `MLAPM`† (Knebe, Green & Binney 2001). The adaptive meshes of `MLAPM` are recursive: refined regions can also be refined. This creates a hierarchy of refinement meshes of different resolutions covering regions of interest. The refinements are created cell by cell they are not constrained to have a rectangular (or any other) shape. Thus, the grid structure naturally surrounds the (satellite) galaxies as they manifest themselves as over-densities in the underlying background field. The tracker (`MHT`) takes an initial output of `MHF` and then tracks the particles that were a member of that satellite at later times.

3. The orbital parameters of cluster substructure

In this section we investigate the orbital parameters of the cluster substructure. Figure 1 on the LHS plots the satellite eccentricity against pericentre normalised by the virial radius of the host. The crosses represent the satellites that survived until $z = 0$ while the diamonds represent the disrupted satellites. The orbital eccentricity of the substructure is defined to be:

$$e = 1 - \frac{p}{a}\,, \tag{3.1}$$

where the pericentre p and apocentre a distances are the last measured in the satellite's orbit.

The interesting feature of this figure is that there seems to be a distinct population of satellites with an extremely tight anti-correlation between eccentricities $e \approx 0.6 - 0.9$ and pericentres $p \approx 0.3 - 0.8$. Upon detailed investigation these satellites were identified to have just completed one orbit and hence entering the halo for the first time.

† `MLAPM` can be downloaded from the web page `http://astronomy.swin.edu.au/MLAPM`

Acknowledgements

The simulations presented in this paper were carried out on the Beowulf cluster at the Centre for Astrophysics & Supercomputing, Swinburne University.

References

Bullock J., Kolatt T.S., Sigad Y., Somerville R.S., Kravtsov A.V., Klypin A.A., Primack J.R., Dekel A., 2001 MNRAS **559**, 321

Davis M., Efstathiou G., Frenk C. S., & White S. D. M., 1985 ApJ **292**, 371

Ghigna, S., Moore, B., Governato, F., Lake, G., Quinn, T., Stadel, J. 1998 MNRAS **300**, 146

Gill S.P.D., Knebe A., Gibson B.K., 2004, MNRAS in press

Knebe A., Green A., Binney J., 2001 MNRAS **325**, 845

Navarro J., Frenk C.S., White S.D.M., 1997 ApJ **490**, 493

Spergel D. N., Verde L., Peiris H. V., Komatsu E., Nolta M. R., Bennett C. L., Halpern M., Hinshaw G., Jarosik N., Kogut A., Limon M., Meyer S. S., Page L., Tucker G. S., Weiland J. L., Wollack E., Wright E. L., 2003 ApJ Suppl. **148**, 175

Tormen G., Bouchet F., White S. D. M., 1997 MNRAS **286**, 865

Tormen G., Diaferio A., Syer D., 1998 MNRAS **299**, 728

White S. D. M., & Rees M., 1978 MNRAS **183**, 341

Outskirts of Galaxy Clusters: Intense Life in the Suburbs
Proceedings IAU Colloquium No. 195, 2004
A. Diaferio, ed.

DOI: 10.1017/S1743921304000602

Enrichment of the intracluster medium from infalling galaxies

Eelco van Kampen[1], Wilfried Domainko[1], Wolfgang Kapferer[1], Sabine Schindler[1], Stefan Kimeswenger[1], Otonyo Mangete[2] and Max Ruffert[2]

[1]Institute for Astrophysics, University of Innsbruck
Technikerstraße 25, A-6020 Innsbruck, Austria
Eelco.v.Kampen@uibk.ac.at
`http://astro.uibk.ac.at/astroneu/hydroskiteam/index.htm`

[2]Department for Mathematics and Statistics, University of Edinburgh
JCMB, Mayfield Road, Edinburgh, EH9 3JZ, UK

Abstract. We present first results of numerical simulations of the dynamics of all three components of a galaxy cluster: the dark matter, the intracluster gas, and the member galaxies. Our first aim is to understand the enrichment of the gas component by studying various interaction processes between cluster galaxies and the intracluster gas, notably ram-pressure stripping and galactic winds. Our second aim is to find out how fast metals originating from such interaction processes diffuse throughout the cluster, in order to understand observed metallicity maps of galaxy clusters obtained from X-ray observations.

---

1. Introduction

From X-ray spectra it is evident that the Intra-Cluster Medium (ICM) is abundant in metals. As heavy elements are only produced in stars the processed material must have been ejected by cluster galaxies into the ICM. Several different mechanisms for interaction between the cluster galaxies and the ICM are possible but their efficiency of metal enrichment of the cluster and their time evolution are still not known. Suggested processes which remove enriched material from the galaxies are ram-pressure stripping, galactic winds, superwinds driven by starburst activity and jets from active galaxies. First results of high resolution hydrodynamic simulations on cluster scales are presented to investigate the effect of these different pollution mechanisms.

2. Numerical method

We devised a combined N-body, hydrodynamic, and phenomenological approach to simulate the interaction between the ICM and the cluster galaxies. Large-scale structure formation is simulated using an N-body tree code with an additional semi-numerical model for galaxy formation (van Kampen et al. 1999). From these simulations halo merger trees are extracted, which describes where and when galactic dark haloes form. A simple description of gas dynamics and star formation provides a means to calculate the amount of stars forming in these haloes. Stellar population synthesis models then provide the spectral evolution, i.e. luminosities and colours, of these galaxies.

The large-scale dark matter potential and the trajectories and evolving properties of the galaxies are stored, to be used as input for a hydrodynamical simulation of the ICM. These simulations are performed on cluster scale (20 h^{-1} Mpc), and employ a shock capturing grid based Piecewise Parabolic Method (PPM; Colella & Woodward 1984) on four

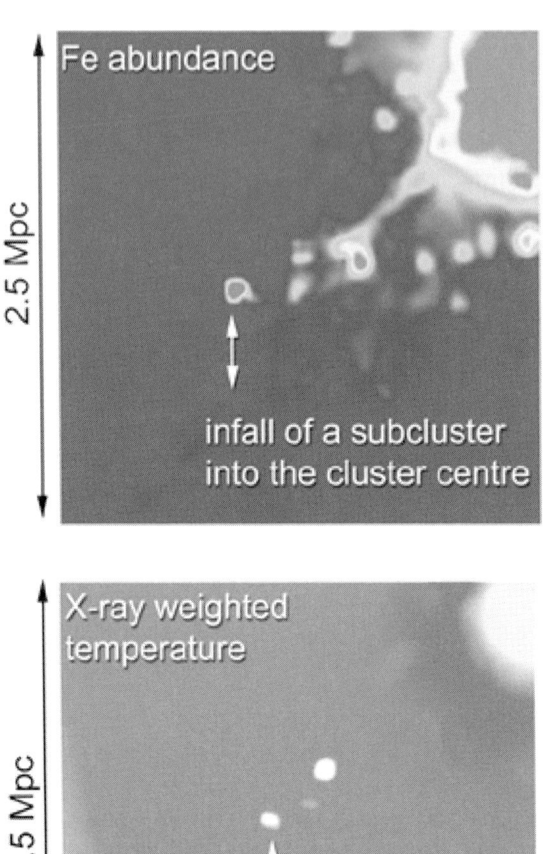

Figure 1. As galaxies and subclusters fall onto the centre of a cluster they feel more and more pressure of the intra-cluster gas. At some point this can lead to ram-pressure stripping. The galaxies lose their cool, metal-enriched gas to the intra-cluster gas. Here we show the distribution of the iron abundance and the temperature in a simulated cluster. The centre of the cluster is in the upper right corner. Top: regions of high iron abundance are visible, where subclusters and galaxies have been stripped. Bottom: the cool gas stripped off galaxies is visible for a while in the temperature map. It is surrounded by the hotter intra-cluster medium.

nested grids (Ruffert 1992), which allows us to cover the cluster center, where most of the stripping is expected to happen, with high enough resolution. The simulations include the effect of ram-pressure stripping using the criterion of Gunn and Gott (1972). Metallicity is used to trace the stripped material. For more information on the simulation method, as well as images and movies, see `astro.uibk.ac.at/astroneu/hydroskiteam/index.htm`.

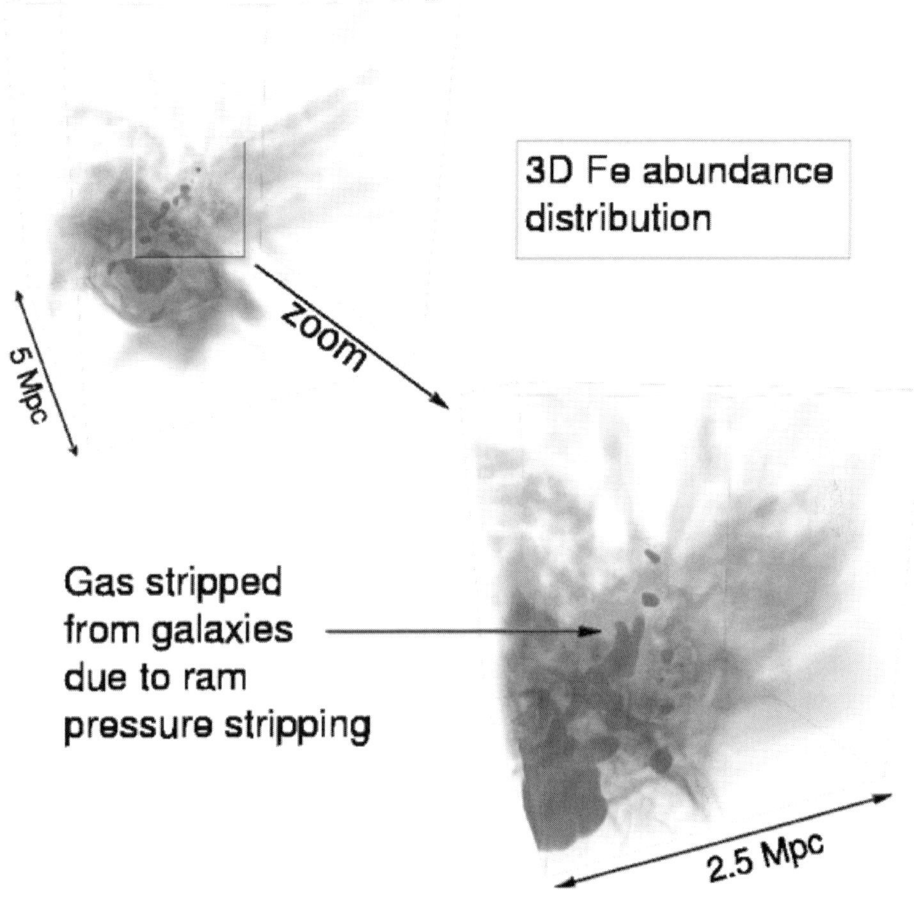

Figure 2. 3D distribution of the iron abundance in the ICM after the ISM has been stripped off the galaxies. Two different scales (5 Mpc and 2.5 Mpc) are shown.

3. Interaction between galaxies and the ICM

The metal content of the intra-cluster gas is on average roughly one third of the solar value. As the total amount of metals in the intra-cluster gas is about the same as in all galaxies together, a lot of gas must have been transported from the galaxies into the ICM. Several transport processes have been suggested: ram pressure stripping (Gunn & Gott 1972), galactic winds (De Young 1978), galaxy-galaxy interaction, jets from AGN and others. A closer look at the different metal enrichment processes is therefore very important for the understanding of cluster formation and galaxy evolution.

So far numerical simulations including only some of the processes gave quite discordant results on the efficiency of the different processes (e.g. Cen & Ostriker 1999; Aguirre et al. 2001; Metzler & Evrard 1994, 1997). To improve on this we have started a comprehensive project, using the hydrodynamic simulations described above, and including all possible enrichment processes. We calculate simulated metallicity maps for direct comparison with observations.

First results of the simulations are shown in Figs. 1 and 2. The simulations shown here include so far only ram-pressure stripping. While many other simulations have concentrated on the effect of the stripping on the galaxies (Abadi et al. 1999; Mori & Burkert 2000; Quilis et al. 2000; Vollmer et al. 2001; Schulz & Struck 2001; Toniazzo & Schindler 2001; Otmianowska-Mazur & Vollmer 2003), we concentrate here on the effects of the stripping process on the intra-cluster gas.

The gas that is stripped off the galaxies is visible not only in the metallicity maps of the intra-cluster gas, but also in the temperature maps, because this gas is cooler than the surrounding intra-cluster medium. It takes a while for the gas from the galaxies to mix with the intra-cluster gas. Sometimes one can see even shock waves passing over these cool regions and the cool gas is not heated up immediately. Also the metallicity distribution is not expected to be homogeneous shortly after the stripping, because the gas is not mixed immediately. This expectation is indeed supported be the metallicity variations observed in the metallicity maps mentioned above. In the simulations we see the highest stripping rate and hence the highest metallicities in regions where subclusters fall onto the main cluster. It also appears that shocks originating from subcluster mergers provide the means to spread stripped metals all over the cluster, but the timescale for this process has not been established yet.

Acknowledgements

The authors acknowledge the support of the Austrian Science Foundation (FWF) through grant P15868.

References

Abadi, M.G., Moore, B., Bower, R.G., 1999, MNRAS, 308, 947

Aguirre, A., Hernquist, L., Schaye, J., Katz, N., Weinberg, D.H., et al. 2001, ApJ, 561,521

Castillo-Morales, A., Schindler S., 2003, A&A, 403, 433

Cen, R., Ostriker, J.P., 1999, ApJ, 519, L109

Colafrancesco, S., Dar, A., De Rujula, A., 2003, submitted to A&A, astro-ph/0304444

Colella P., Woodward P.R., 1984, JCP, 54, 174

De Filippis, E., Schindler, S., Castillo-Morales, A., 2003, A&A, 404, 63

De Young, D.S., 1978, ApJ, 223, 47

Dolag, K., Schindler, S., 2000, A&A 364, 491

Ettori, S., Tozzi, P., Rosati, P., 2003, A&A 398, 879

Evrard, A.E., Metzler, C.A., Navarro, J.N. 1996, ApJ, 469, 494

Furusho T., 2002, Proceedings of the Japan-German workshop on "Galaxies and Clusters of Galaxies", Shuzenji, Japan, http://www-x.phys.metro-u.ac.jp/docs/workshop/cluster-2002-10-29/proceedings/

Furuzawa A., Morishita T., Hudaverdi M., Tanaka T., Takahashi S., et al., 2002, Proceedings of the Japan-German workshop on "Galaxies and Clusters of Galaxies", Shuzenji, Japan, http://www-x.phys.metro-u.ac.jp/docs/workshop/cluster-2002-10-29/proceedings/

Giovannini G., Feretti, L., 2002, in "Merging Processes in Galaxy Clusters", Astrophysics and Space Science Library, Feretti L., Gioia I.M., Giovannini G. (eds.), Kluwer, p.197

Gil-Merino R., Schindler, S., 2003, A&A, in press

Gunn, J.E., Gott, J.R.III, 1972, ApJ, 176, 1

Hudaverdi M., Akimoto F., Furuzawa A., Shibata R., Tanaka T., et al., 2002, Proceedings of the Japan-German workshop on "Galaxies and Clusters of Galaxies", Shuzenji, Japan, http://www-x.phys.metro-u.ac.jp/docs/workshop/cluster-2002-10-29/proceedings/

Johnstone, R.M., Allen S.W., Fabian A.C., Sanders, J.S., 2002, MNRAS 336, 299

Lenzen, F., Schindler, S., Scherzer, O., 2003, submitted to A&A

Metzler, C.A., Evrard, A.E., 1994, ApJ, 437, 564

Metzler, C.A., Evrard, A.E., 1997, astro-ph/9710324

Mori, M., Burkert, A., 2000, ApJ, 538, 559

Otmianowska-Mazur, K., Vollmer, B., 2003, A&A, 402, 879

Piffaretti, R., Jetzer, Ph., Schindler, S., 2003, A&A 398, 41
Quilis, V., Moore, B., Bower, R., 2000, Science, 288, 1617
Reiprich, T., Böhringer, H., 2002, ApJ, 567, 716
Rosati, P., Borgani, S., Norman, C., 2002, ARA&A 40, 539
Ruffert M., 1992, A&A 265, 82
Schindler, S. 1996, A&A, 305, 756
Schmidt, R. W., Fabian, A. C., Sanders, J. S., 2002, MNRAS 337, 71
Schulz, S., Struck, C., 2001, MNRAS, 328, 185
Toniazzo, T., Schindler, S., 2001, MNRAS, 325, 509
van Kampen, E., Jimenez, R., Peacock, J. A., 1999, MNRAS, 310, 43
Vollmer, B., Cayatte, V., Balkowski, C., Duschl, W. J., 2001, ApJ, 561, 708
Tanaka T., Hudaverdi M., Akimoto F., Furuzawa A., Tawara Y., Yamashita K., 2002, Proceedings of the Japan-German workshop on "Galaxies and Clusters of Galaxies", Shuzenji, Japan, http://www-x.phys.metro-u.ac.jp/docs/workshop/cluster-2002-10-29/proceedings/
Yamasaki N.Y., Furusho T., Ohashi T., 2002, Proceedings of the Japan-German workshop on "Galaxies and Clusters of Galaxies", Shuzenji, Japan, http://www-x.phys.metro-u.ac.jp/docs/workshop/cluster-2002-10-29/proceedings/

Outskirts of Galaxy Clusters: Intense Life in the Suburbs
Proceedings IAU Colloquium No. 195, 2004
A. Diaferio, ed.

DOI: 10.1017/S1743921304000614

Chemical enrichment of ICM in a hierarchical galaxy formation model including SNe Ia

Masahiro Nagashima[1,2], Cedric G. Lacey[1], Carlton M. Baugh[1], Carlos S. Frenk[1] and Shaun Cole[1]

[1]Department of Physics, University of Durham, South Road, Durham DH1 3LE, U.K.

[2]Department of Physics, Kyoto University, Sakyo-ku, Kyoto 606-8502, Japan email: masa@scphys.kyoto-u.ac.jp

Abstract. Metal enrichment of intracluster medium (ICM) is reanalysed in the hierarchical galaxy formation scenario. In order to investigate metal abundances of α and iron-peak elements, we incorporate metal production processes by type Ia supernovae (SNe Ia) as well as type II supernovae (SNe II) into the GALFORM model of semi-analytic galaxy formation. We find that a top-heavy initial mass function (IMF) of stars with a power index $x \simeq 0.35$ for starburst is required under the constraint that the IMF for quiescent disc star formation is the Kennicutt IMF. Our model shows an only mild evolution of [Fe/H] as suggested by a recent observation.

Type Ia supernovae (SNe Ia) give us useful information on the formation and evolution of galaxies because of the delayed explosion. Furthermore, since the abundance patterns are different between SNe Ia and type II SNe (SNe II), to investigate metal abundances of various elements is quite important to understand galaxy formation. To do this, we incorporate the metal enrichment process due to SNe Ia in addition to that due to SNe II with the GALFORM model of semi-analytic galaxy formation (Cole et al. 2000).

The initial mass function (IMF) of stars for solar-neighbourhood stars has been considered to be a Salpeter-like bottom-heavy one, which has been confirmed by both theories and observations, including an analysis using a semi-analytic model (Nagashima & Okamoto 2003). However, some studies on intracluster medium (ICM) based both on traditional monolithic collapse models and numerical simulations have suggested that the IMF should be rather top-heavy. Such a top-heavy IMF is recently suggested for the starburst galaxy M82 by some observations. Then, assuming the Kennicutt IMF for quiescent disc star formation and a top-heavy IMF ($x = 0.35$) for starburst here, we examine the metal abundances of ICM in a framework of the hierarchical galaxy formation. The life-time and chemical yields of SNe Ia are computed in a consistent manner with the adopted IMFs and stellar evolution.

In Figure 1, we show metal abundances of O, Fe, Mg and Si of ICM. The dots and crosses denote models with and without the top-heavy IMF for starburst. These models take into account the superwind, which expels gas and metals from dark haloes, and subsequent recaptures. Both models have the Kennicutt IMF for quiescent star formation. The open symbols indicate observations corrected by taking into account metallicity gradients. As clearly shown, the top-heavy IMF is required to be consistent with the observed metal abundances. We have also confirmed that another model with thermal conduction to stop gas cooling in massive haloes without superwind is also in agreement with the observation when the top-heavy IMF is adopted.

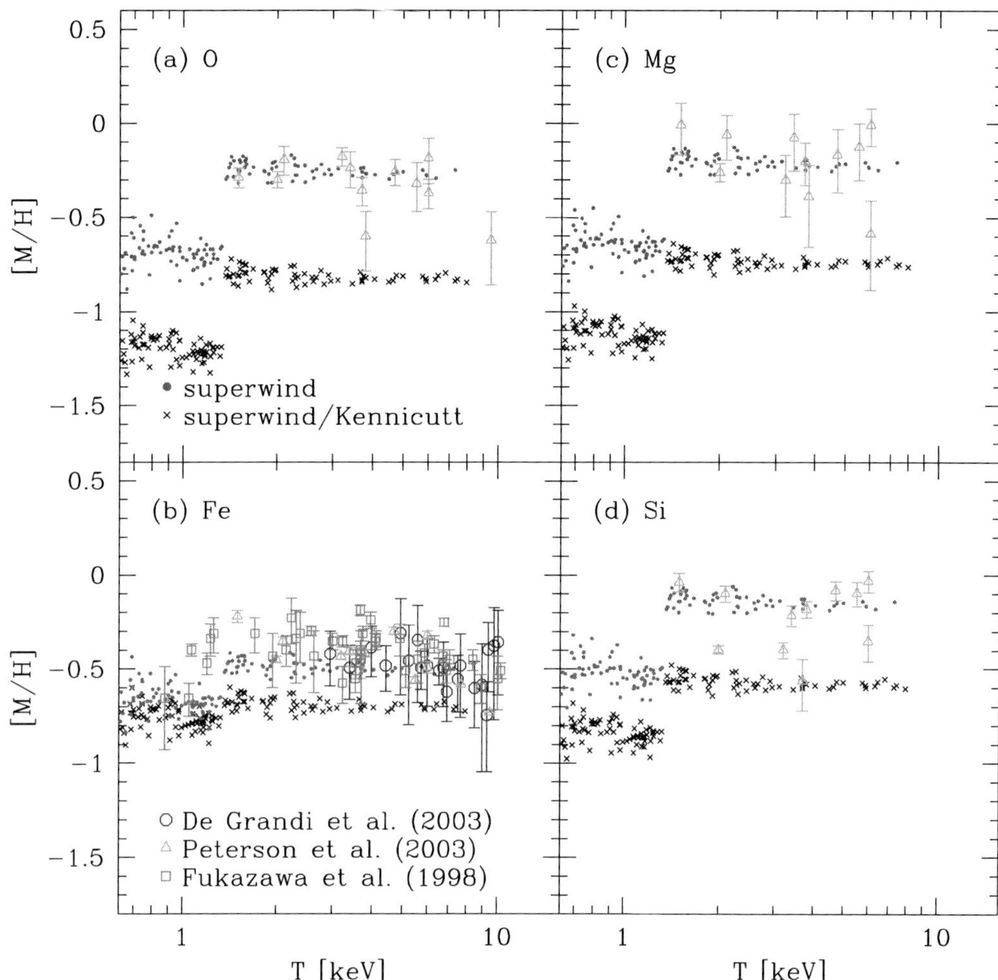

Figure 1. Metal abundances of ICM. (a) [O/H]. (b) [Fe/H]. (c) [Mg/H]. (d) [Si/H]. The dots and crosses indicate the superwind models with the top-heavy IMF and with the Kennicutt IMF for starburst. The open triangles in all panels denote the observed data by *XMM-Newton* (Peterson et al. 2003). The open circles and squares in panel (b) are data by *BeppoSAX* (De Grandi et al. 2003) and by *ASCA* (Fukazawa et al. 1998), respectively.

Figure 2 shows redshift evolution of [Fe/H]. The iron metallicity gradually decreases toward higher redshift. The difference between $z = 0$ and 1 is only ~ 0.2 dex. While in the observations low temperature clusters show systematically higher [Fe/H] than high temperature ones, the model clusters do not show such a tendency. This apparent discrepancy would be solved by increasing observational data and by systematic studies of evolution of metallicity gradients.

The model with the superwind shows slightly lower [Fe/H] than the model with thermal conduction, presumably because of its strong feedback. The difference between the two models hardly changes even at high redshift.

All of the model clusters show almost the same metallicities, metallicity ratios, and redshift evolution of metallicities, in contrast with the observations. The diversity of observed metallicities may be caused by the variety of metallicity gradients, which is presumably induced by the diversity of formation and assembly histories of the clusters.

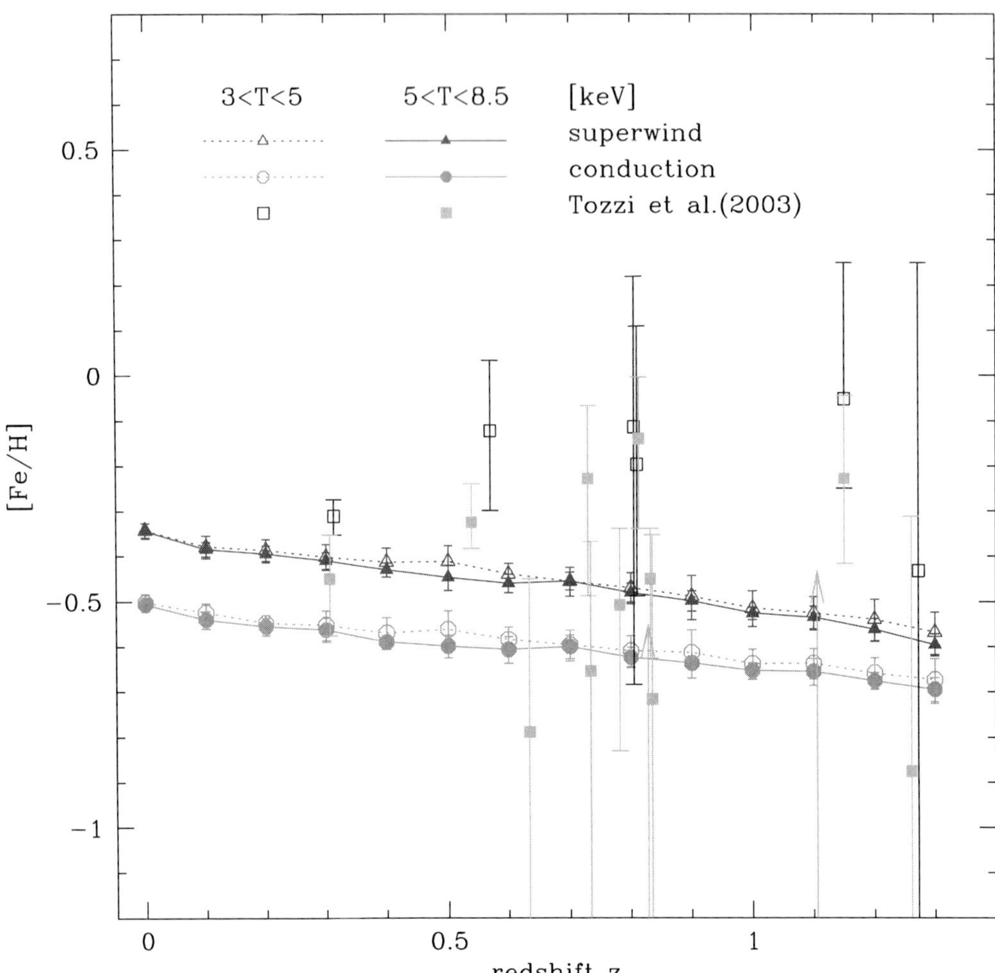

Figure 2. Redshift evolution of [Fe/H] of ICM. The triangles and circles indicate the superwind model and the conduction model, respectively. The squares are observed data by Tozzi et al. (2003). Samples are divided into two populations, high (filled symbols) and low (open symbols) temperature clusters. Note that, for two clusters, only the upper limits of [Fe/H], indicated by arrows, are measured.

It will provide important information to measure metallicities in the outer envelopes of clusters of galaxies.

References

Cole, S., Lacey, C. G., Baugh, C. M. & Frenk, C. S., 2000, MNRAS, **319**, 168

De Grandi, S., Ettori, S., Longhetti, M. & Molendi S., preprint (astro-ph/0310828)

Fukazawa, Y., Makishima, K., Tamura, T., Ezawa, H., Xu, H., Ikebe, Y., Kikuchi, K. & Ohashi T., 1998, PASJ, **50**, 187

Nagashima, M. & Okamoto, T., 2003, submitted

Peterson, J.R., Kahn, S.M., Paerels, F.B.S., Kaastra, J.S., Tamura, T., Bleeker, J.A.M., Ferrigno, C. & Jernigan, J.G., 2003, ApJ, **590**, 207

Tozzi, P., Rosati, P., Ettori, S., Borgani, S., Mainieri, V. & Norman, C., 2003, ApJ, **593**, 705

Outskirts of Galaxy Clusters: Intense Life in the Suburbs
Proceedings IAU Colloquium No. 195, 2004
A. Diaferio, ed.

DOI: 10.1017/S1743921304000626

Masclet: a new multidimensional AMR cosmological code

Vicent Quilis

Departament d'Astronomia i Astrofísica, Universitat de València, 46100-Burjassot (Valencia), Spain

Abstract. A new cosmological multidimensional hydrodynamic and N-body code based on an Adaptive Mesh Refinement scheme is described and tested. The hydro part is based on modern *high-resolution shock-capturing* techniques, whereas the N-body approach is based on a Particle Mesh method. The code has been specifically designed for cosmological applications.

---

1. Introduction

Numerical simulations are crucial tools in theoretical cosmology. The gaseous component – baryons – is also a crucial ingredient in cosmological structures. Therefore, realistic simulations aimed to understand structure formation require hydrodynamical simulations able to describe the evolution of baryons. Two general approaches can be mentioned in order to broadly classify the numerical techniques used to evolve the gaseous component: Lagrangian (SPH) and Eulerian. Among the Eulerian schemes, techniques based on Riemann solvers have been specially successful (Ryu et al. 1993; Quilis, Ibáñez & Sáez 1994; Bryan, Norman, Stone, Cen & Ostriker 1995; Gheller, Pantano & Moscardini 1998 and others). Due to their Eulerian character these techniques are constrained by a poor spatial resolution. We present a new numerical code which combines the best of both, Lagrangian and Eulerian, approaches. The basic idea is to improve the numerical resolution by implementing a scheme known as Adaptive Mesh Refinement (AMR) described originally by Berger & Oliger (1984) and Berger & Colella (1989). In order to do this, we use an Eulerian approach as the ones described above but being able to gain resolution – spatial and temporal – by refining the original grid. The AMR schemes have shown as extremely powerful in many fields related with fluid dynamics. In cosmology, in the recent past, AMR implementations have been designed by Bryan & Norman (1997), Kravtsov, Klypin & Hoffman (2002) and Teyssier (2002).

2. Basic numerical procedure

Our basic hydro solver is based on a particular implementation of the HRSC (high resolution shock-capturing) methods. Its main ingredients are the following: i) It is written in *conservation form*, ii) *the reconstruction procedure* allows the method to gain resolution by reconstructing through interpolations the distribution of the quantities within the numerical cell, iii) a *Riemann solver* to compute the numerical fluxes across cell interfaces, and iv) a third order Runge-Kutta which preserves the TVD properties of the scheme. This implementation is able to handle shocks, discontinuities and strong gradients in the integrated quantities, and it has excellent conservation properties.

The dark matter is treated as a collisionless system of particles. In our code we solved these equations by using a Lax-Wendroff scheme which is second order. In order to recover the continuous density field for dark matter component, we use a triangular shaped cloud (TSC) scheme at each time step.

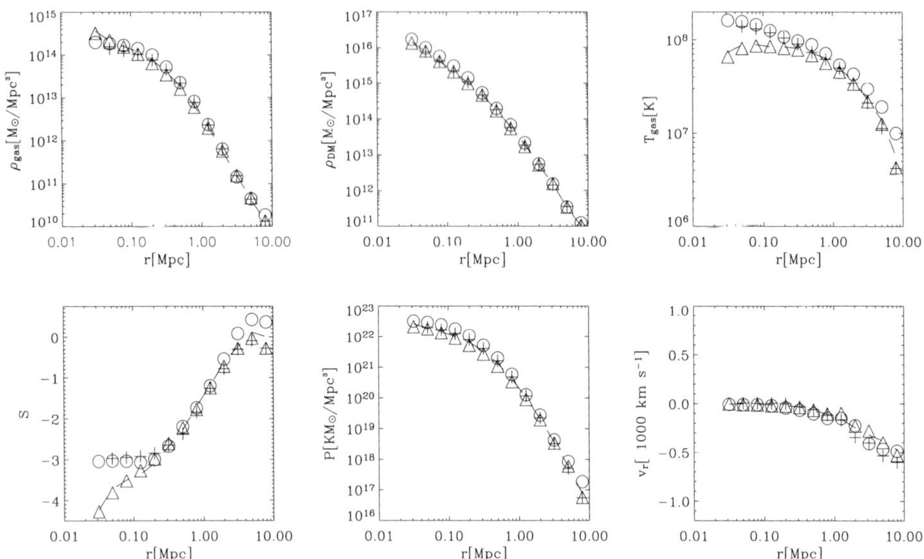

Figure 1. Radial profiles of the Santa Barbara cluster. Dashed lines stands for the average results of all the simulations displayed in the original paper by Frenk et al. (1999). Triangles and crosses show the results of simulations carried out by Jenkins (SPH) and Bryan (SAMR), respectively. Open circles display our results.

3. The Adaptive Mesh Refinement strategy

The fundamental idea behind the AMR strategy is to overcome the lack of resolution associated with the Eulerian description on a fix grid. The basic idea is simple. Regions in the original computational domain, are picked up according to some criteria. These new computational domains which we call *child grids* or *patches*, are remapped with a higher number of cells and therefore with better resolution. The value of the different quantities defined on the child grids are obtained by interpolating from the *parent grid*. Once the child grids are built, they can be evolved as an independent computational domain by using the same methods used for a fix grid. Although conceptually simple, there are severe technical complications related to the communication among the different patches and the boundary problems at different levels. Our implementation of the AMR schemes follows the one described in Berger & Colella (1989).

4. Cosmological simulation: the Santa Barbara cluster

The best test we can think of for a cosmological code in a realistic setting is to compare with the results presented in Frenk et al. (1999). In this reference, authors performed adiabatic simulations of a galaxy cluster with several codes. The initial conditions were generated in order to obtain a galaxy cluster $\sim 10^{15} M_{\odot}$ located at the center of a computational box of 64 Mpc side length. The coarse grid (l=0) has 64^3 cells. The first refinement (l=1) is fixed covering the central 32 Mpc. Higher refinements are carried out according to a local density criterion. In figure 1 radial profiles for gas density, dark matter density, temperature, entropy ($S = \ln[T\rho_{gas}^{-(\gamma-1)}]$), pressure, and gas radial velocity are shown, respectively. The profiles have been computed by averaging the quantities in spherical shells of 0.2 dex logarithmic width. The inner radius of the first shells is 10 kpc and the outer radius of the last one is 10 Mpc. Every panel displays the average results of all simulations (dashed line), results from Jenkins and Pearce simulation using Hydra SPH (triangles), and the simulation of Bryan and Norman using the SAMR Eulerian code (crosses). Open circles stand for the results of our simulation. Our results show very good agreement with the general trends shown by the average results of all simulations. There is a special good match with the profiles from Bryan' SAMR code.

Results using AMR simulations – with resolution comparable to those of SPH – showed substantial differences in quantities like temperature and entropy with respect to SPH results. This is an important conclusion – confirmed by our simulations – which points out crucial differences between the SPH and the Eulerian methods when describing the physics of structure formation.

The AMR like codes as the one presented in this paper will be crucial tools in the future of Numerical Cosmology. They combine successfully the best properties of previous traditional approaches: i.e. excellent spatial resolution as the SPH techniques, and an accurate description of the hydrodynamics of the problem (shock and discontinuities handling, void regions, accurate thermodynamics, turbulence, and some others) inherited from the Eulerian codes based on Riemann solvers.

Acknowledgements

VQ is a Ramón y Cajal Fellow of the Spanish Ministry of Science and Technology. This works has been partially supported by grant AYA2003-08739-C02-02 (partially financed with FEDER funds) and by the Generalitat Valenciana ACyT project GRUPOS03/170. Simulations were carried out in the Servei d'Informática of the Universitat de València.

References

Berger M.J. & Colella P. 1989 *J. Comp. Phys.* **82**, 64

Berger M.J. & Oliger J. 1984 *J. Comp. Phys.* **53**, 484

Bryan G.L., Norman M.L., Stone J.M.m Cen R., Ostriker J.P. 1995, *Comput. Phys. Comm.* **89**, 149

Bryan G.L. & Norman M.L. 1997 in *Computational Astrophysics; 12th Kingston Meeting on Theoretical Astrophysics* (ed. D. A. Clarke & M. J. West). ASP Conf. Ser. vol. 123, 363

Frenk C.S. et al. 1999 *ApJ* **525**, 554

Gheller C., Pantano O., Moscardini L. 1998, *MNRAS* **295**, 519

Kravtsov A.V., Klypin A.A., Hoffman Y. 2002 *ApJ* **571**, 563

Quilis V., Ibáñez J.Mª. Sáez D. 1994, *A&A* **286**, 1

Ruy D., Ostriker J.P., Kang H., Cen R. 1993 *ApJ* **414**, 1

Teyssier R. 2002, *A&A* **385**, 337

Outskirts of Galaxy Clusters: Intense Life in the Suburbs
Proceedings IAU Colloquium No. 195, 2004
A. Diaferio, ed.

DOI: 10.1017/S1743921304000638

Simulating the formation of galaxy clusters

Daisuke Nagai and Andrey V. Kravtsov

Department of Astronomy and Astrophysics, Center for Cosmological Physics, The University of Chicago, Chicago, IL 60637 USA email: daisuke,andrey@oddjob.uchicago.edu

Abstract. We study the effects of radiative cooling, star formation and stellar feedback on the properties and evolution of galaxy clusters using high-resolution Adaptive Mesh Refinement N-body+gasdynamics simulations of clusters forming in the ΛCDM universe. Cooling leads to the condensation of gas in the inner regions of clusters, which in turn leads to steepening of the dark matter profile. The cooling gas is replaced by the higher-entropy gas from the outer regions, which raises the entropy and temperature of gas in the cluster core. The magnitude of these effects is likely overestimated in the current simulations because they suffer from the overcooling problem: a much larger fraction of baryons is in the form of cold gas and stars than is observed. We find that the thermal stellar feedback alone does not remedy this problem. Additional ad-hoc preheating can lower the amount of cold gas but a simple uniform preheating results in incorrect star formation history, as it delays the bulk of star formation until $z < 1$. Our analysis shows that the overcooling in a cluster as a whole is really the overcooling in the central galaxy and its progenitors at high redshifts. This indicates that an additional heating mechanism that can continuously heat the gas in the cluster core is required to reproduce the observed cluster properties. Energy injection by the Active Galactic Nuclei, which may provide such heating, may thus be an important missing ingredient in the current theoretical models of cluster formation.

---

1. Effects of cooling, star formation and feedback

Below, we highlight some of the important effects of gas cooling and star formation on the properties of galaxy clusters:

- Gas cooling induces condensation of gas in the cluster core region ($\lesssim$ 100kpc), and the baryon mass in the cluster core at $z = 0$ is dominated by the stars associated with the central galaxy.
- The condensation of baryons causes contraction of dark matter and steepens the dark matter profile in the central region from $r^{-1.5}$ in the adiabatic runs to $r^{-2.0}$ in the simulations with cooling.
- Gas cooling induces inflow of high-entropy gas from the outside and raises the level of entropy and gas temperature in the cluster core region.
- Stellar feedback and/or additional preheating suppress the effects described above by regulating gas cooling. The magnitude of the effects on the distribution of dark matter, entropy and temperature of gas is, therefore, sensitive to the total fraction of baryons which cooled to form stars and cold gas clouds.

2. The overcooling problem

The cluster simulations with cooling and star formation revealed a significant problem: 35-55% of baryons within the virial radius are in the form of cold gas and stars at $z = 0$, which is significantly larger than the observed value of $\lesssim$10-20% (*e.g.*, Balogh 2001; Lin, Mohr & Stanford 2003). This problem was previously reported in the simulations using

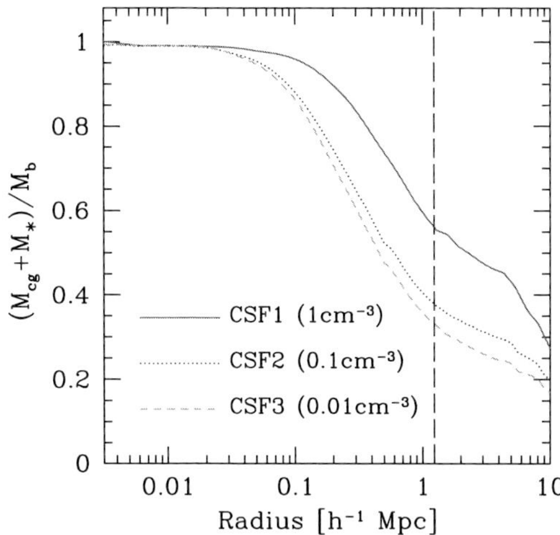

Figure 1. The cold gas + stellar mass fraction of the simulated cluster that includes cooling, star formation and stellar feedback. Three lines indicate the CSF runs with varying star formation: the star formation is allowed to proceed only in the region where the local gas density is larger than 1.0 (solid), 0.1 (dotted) and 0.01cm^{-3} (dashed). The figure shows that 35-55% of baryons within the virial radius (indicated by the dashed line) are in the form of cold gas and stars at $z = 0$, which is significantly larger than the observed value of 10-20%.

the Smoothed Particle Hydrodynamics (SPH) simulations (*e.g.*, Suginohara & Ostriker 1998; Lewis et.al. 2000; Tornatore et.al. 2003). We find that the same problem exists in our higher-resolution Eulerian simulations that include metallicity-dependent radiative cooling. Moreover, we find that the cold and stellar mass fractions depend on the details of star formation implementaton. However, we were not able to solve the overcooling problem by varying parameters of thermal stellar feedback within reasonable limits, as shown in Figure 1.

3. Effect of preheating

Additional preheating can help resolve some of the discrepancies between cluster simulations and observations. Such preheating, presumably associated with the epoch of vigorous star formation, is often advocated as the solution to the overcooling problem and the discrepancies between the predicted and observed cluster scaling relations (*e.g.*, Bialek et.al. 2001; Muanwong et.al. 2002, da Silva et.al. 2004). Our results, however, indicate that the additional preheating results in star formation history significantly different from observations, as shown in Figure 2. If the preheating occurs relatively early ($z \sim 3$) the gas manages to cool back by $z \sim 1$, and the resulting star formation rate remains high ($\sim 1000 M_\odot \text{yr}^{-1}$) at $z \lesssim 0.5$. This results in an incorrect star formation history: most of the cluster stars form in the last 7-10 billion years in stark contrast with observations, which indicate that less than 30% of cluster stars form during this epoch. If preheating occurs late, the amount of the required energy is unrealistic and the expected effect on the intergalactic medium is in contradiction to observations of the low-redshift Lyman-α forest. These results provide a strong argument against the simple uniform "preheating" scenario.

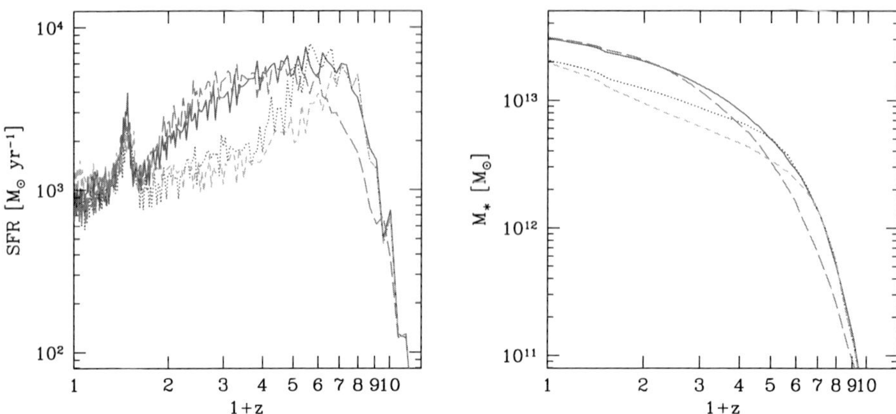

Figure 2. The star formation rate (left) and the stellar mass (right) as a function of redshift. These star formation history was constructed using stellar particles in the Lagrangian volume enclosed within the virial radius of the simulated cluster at $z = 0$. Lines indicate the simulations that include cooling, star formation and stellar feedback (solid), and those with additional preheating of 0.3 keV/particle at $z = 4.5$ (dotted), $z = 6$ (dashed) and $z = 9$ (long-dashed).

4. Discussions and conclusions

The overcooling problem is a serious challenge for the cluster formation simulations. Our analysis shows that the overcooling in a cluster as a whole is really the overcooling in its core. In other words, the central cluster galaxies formed in simulations are too massive and bright compared to observations. Analysis of cluster evolution indicates that the problem sets in early at high redshifts, and is not limited to a single object or region. This indicates that an additional heating mechanism is necessary for reproducing the observed cluster properties. These results lend support to the idea that feedback from Active Galactic Nuclei is an important missing ingredient in the current simulations of cluster formation.

References

Balogh, M. L., Pearce, F. R., Bower, R. G., & Kay, S. T. 2001, MNRAS, 326, 1228

Bialek, J. J., Evrard, A. E., & Mohr, J. J. 2001, ApJ, 555, 597

da Silva, A. C., Kay, S. T., Liddle, A. R., & Thomas, P. A. 2004, MNRAS, 348, 1401

Katz, N. & White, S. D. M. 1993, ApJ, 412, 455

Lewis, G. F., Babul, A., Katz, N., Quinn, T., Hernquist, L., & Weinberg, D. H. 2000, ApJ, 536, 623

Lin, Y. T., Mohr, J. J., & Stanford, S. 2004, ApJ submitted (astro-ph/0402308)

Muanwong, O., Thomas, P. A., Kay, S. T., & Pearce, F. R. 2002, MNRAS, 336, 527

Suginohara, T. & Ostriker, J. P. 1998, ApJ, 507, 16

Tornatore, L., Borgani, S., Springel, V., Matteucci, F., Menci, N., & Murante, G. 2003, MNRAS, 342, 1025

Outskirts of Galaxy Clusters: Intense Life in the Suburbs
Proceedings IAU Colloquium No. 195, 2004
A. Diaferio, ed.

DOI: 10.1017/S174392130400064X

Intracluster entropy from quasar activity

A. Lapi and A. Cavaliere

Università "Tor Vergata", Via Ricerca Scientifica 1, I-00133 Roma, Italy

Abstract. We discuss how the energy feedback from active galactic nuclei and quasars is essential in explaining the observed X-ray luminosities and entropy levels of the intracluster and intragroup medium.

---

1. Introduction

Clusters, groups, and many single galaxies shine in X-rays because they contain large masses of hot plasma that radiate strong luminosities by thermal bremsstrahlung and line emission. Simple conditions are found to prevail in rich clusters; they emit copious X-ray powers $L_X \propto n^2 R^3 \sqrt{T} \sim 10^{45}$ erg s^{-1} integrated over sizes $R \sim$ Mpc. The temperatures $kT \gtrsim 4$ keV are close to the virial values $kT_v \propto GM/R$ in the gravitational potential wells provided by dark matter (DM) masses $M \sim 10^{15} M_{\odot}$. The gas densities $n \sim 10^{-3}$ cm^{-3} are consistent with baryon to DM fractions $m/M \sim 1/5$, the universal value.

However, surprises arise in moving to groups and galaxies. Similarly simple conditions would imply the "gravitational" scaling $L_X \propto T^2$ (Kaiser 1986), or $L_X \propto T$ when lines become important. Instead, the luminosities, recently observed or revised, are *lower* by factors $10^{-1} - 10^{-2}$, see Fig. 1. This also shows how the emissions from small groups and galaxies *scatter* widely and often downward, a feature of non-Gaussian, largely intrinsic nature (Mushotzky 2004).

Thus in the smaller structures the plasma is clearly underdense for its observed temperatures. The issue is reinforced on considering that the specific entropy s of the plasma (or rather the adiabat $K = kT/n^{2/3} \propto e^{2\,s/3\,k}$, see Bower 1997) is enhanced relative to the scaling $K \propto T$ that is expected if the gas had been gravitationally heated by shocks in the cluster outskirts as it was accreted, see Fig. 2. This implies that extra energy *inputs* occurred during the formation of groups and clusters.

2. External preheating from AGNs

We focus on the substantial energies fed back to the gaseous baryons, when part of them condense into massive stars then exploding as type II supernovae (SNe), or accrete on to central supermassive black holes energizing active galactic nuclei (AGNs; see Valageas & Silk 1999; Nath & Roychowdhury 2002).

During the hierarchical buildup such energy inputs *preheat* the gas before it falls into new groups and clusters, so hindering its flow into the latter. In smaller systems the accretion shocks are increasingly *modulated*, thus inducing lower baryon densities and higher entropy levels relative to a pure gravitationally-driven infall (Lapi, Cavaliere & Menci 2004).

But SNe provide only a mild preheating, around 1/4 keV per particle; the resulting luminosity depressions or entropy enhancements are shown by the light strips in Figs. 1 and 2 to be limited both in amplitude and in scatter (see Cavaliere, Lapi & Menci 2002).

Including the additional contribution of AGNs raises the preheating to 3/4 keV per particle, if their energy outputs are coupled to the ambient baryons at fractional levels

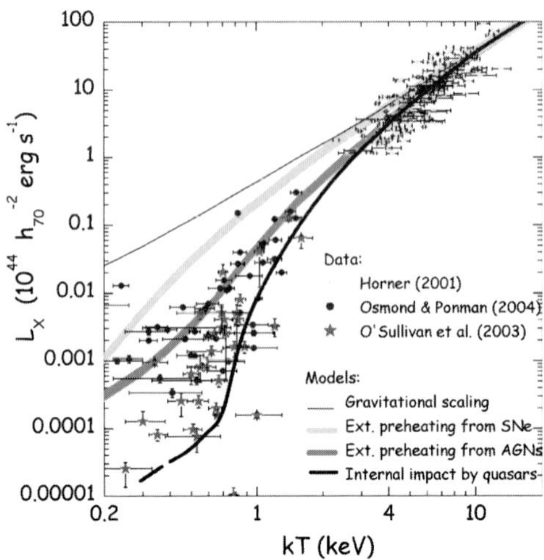

Figure 1. Integrated X-ray luminosity L_X vs. X-ray temperature T.

$f \sim 5\%$, as independently gauged by observations with CHANDRA and XMM-Newton (see Chartas, Brandt & Gallagher 2003; Pounds et al. 2003). Such a preheating value provides a sizeable step toward the locus of the data, as shown in Figs. 1 and 2 by the heavy strips.

However, the observations in groups and galaxies are only marginally fitted, while the wide scatter in the $L_X - T$ correlation is still unaccounted for.

3. Internal impact of quasars

But right in groups and galaxies the impulsive inputs by powerful quasars have an additional impact from inside. This occurs when the energy $\Delta E \sim 2\times 10^{62}\, f\,(M_\bullet/10^9\, M_\odot)\,(1+z)^{-3/2}$ ergs injected over 10^8 yr for an accreted mass $M_\bullet$ competes with the value $E \sim 2\times 10^{61}\,(kT/\mathrm{keV})^{5/2}\,(1+z)^{-3/2}$ ergs for the gas in equilibrium. The ratio $\Delta E/E \sim 0.5\,(f/5\%)\,(M_\bullet/10^9 M_\odot)\,\,(kT/\mathrm{keV})^{-5/2}$ approaches unity in poor groups or galaxies. Then over 10^2 pc a quasar wind can act as an efficient piston to drive through the gas a blastwave terminated by a shock (Granato et al. 2004).

We derive (Lapi, Cavaliere & De Zotti 2003) self-similar solutions for the blast flow perturbing an equilibrium plasma density $n(r) \propto r^{-2}$ (or steeper) under the push of the energy addition $\Delta E(t) \propto t$ (or flatter). These blasts also include the restraints set by a finite initial pressure and by DM gravity; still, the ratio of the kinetic to the thermal energy ranges from 10^{-1} up to 2.

Thus within a dynamical time the blast *ejects* beyond the cluster outskirts a considerable gas mass $\Delta m/m \simeq \Delta E/2\,E$ (for $\Delta E/E \lesssim 2$), and cause an average temperature raise $\Delta T/T \simeq \Delta E/10\,E$ of the residual medium. In the equilibrium recovered after ejection, the X-ray luminosity $L_X \propto (1-\Delta m/m)^2\, T^2$ is *depressed* and the entropy $K \propto (1-\Delta m/m)^{-2/3}\, T$ *enhanced* relative to the initial values (see solid lines in Figs. 1 and 2). We find that the current X-ray data require the diffuse baryons to absorb a fraction $f \sim 5\%$ of the full quasar output. Furthermore, in terms of increasingly strong, sporadic quasar impacts we can understand why the intrinsic scatter seen in the data *widens* toward poor groups.

On approaching the galactic scales no further density depletion takes place (knee in the solid line of Fig. 1), because the internal feedback from quasars is *self-regulated* (Silk

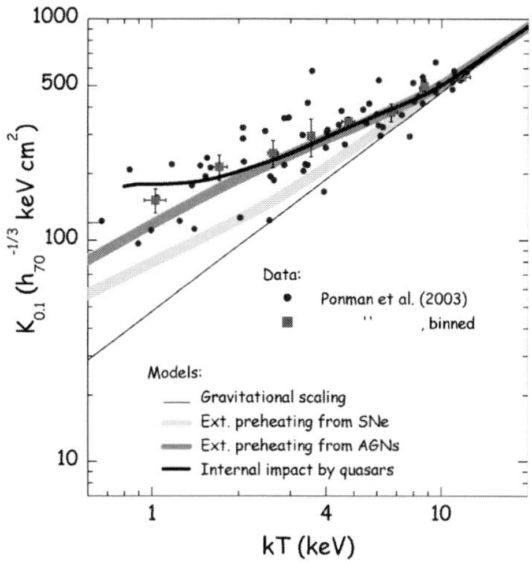

Figure 2. Central entropy K vs. X-ray temperature T.

& Rees 1998; King 2003). In other words, $\Delta E/E$ cannot exceed a few, lest the energizing accretion is stopped (Cavaliere et al. 2002). Pleasingly, with the same value $f \sim 5\%$ indicated by the X-ray data, such a limiting condition yields the *steep* correlation $M_{\bullet} \approx 5 \times 10^8\, M_{\odot}\, (f/5\%)^{-1}\, (\sigma/300\mathrm{kms}^{-1})^5 \propto \sigma_{\star}^4$ observed in the optical band (Ferrarese 2002) between the black hole masses and the galaxian velocity dispersions. In the microwave and submm bands (Lapi et al. 2003) *real-time* evidence of quasar impact in action may be caught from Sunyaev-Zel'dovich signals enhanced by overpressure in running blastwaves.

Acknowledgements

We are grateful to N. Menci for stimulating comments, and to the organizers for an interesting Colloquium.

References

Bower, R.G. 1997 *MNRAS* **288**, 355.
Cavaliere, A., Lapi, A., & Menci, N. 2002 *ApJ* **581**, L1.
Chartas, G., Brandt, W.N., & Gallagher, S.C. 2003 *ApJ* **595**, 85.
Ferrarese, L. 2002 *ApJ* **578**, 90.
Granato, G.L., De Zotti, G., Silva, L., Bressan, A., & Danese, L. 2004 *ApJ* **600**, 580.
Horner, D.J. 2001 Ph.D. Thesis, Univ. of Maryland.
Kaiser, N. 1986 *MNRAS* **222**, 323.
King, A.R. 2003 *ApJ* **596**, L27.
Lapi, A., Cavaliere, A., & De Zotti, G. 2003 *ApJ* **597**, L93.
Lapi, A., Cavaliere, A., & Menci, N. 2004 *ApJ*, submitted.
Mushotzky, R.F. 2004 *Carnegie Observatories Astrophysics Series* **Vol. 3** (Cambridge: Cambridge Univ. Press).
Nath, B.B., & Roychowdhury, S. 2002 *MNRAS* **333**, 145.
Osmond, J.P.F., & Ponman, T.J. 2004 *MNRAS*, submitted (preprint astro-ph/0402439).
O'Sullivan, S., Ponman, T.J., & Collins, R.S. 2003 *MNRAS* **340**, 1375.
Ponman, T.J., Sanderson, A.J.R., & Finoguenov, A. 2003 *MNRAS* **343**, 331.
Pounds, K.A., King, A.R., Page, K.L., & O'Brien, P.T. 2003 *MNRAS* **346**, 1025.
Silk, J., & Rees, M.J. 1998 *A&A* **331**, L1.
Valageas, P., & Silk, S. 1999 *A&A* **350**, 725.

Outskirts of Galaxy Clusters: Intense Life in the Suburbs
Proceedings IAU Colloquium No. 195, 2004
A. Diaferio, ed.

DOI: 10.1017/S1743921304000651

Reproducing the entropy structure of clusters and groups

G. González-Casado[1], J. M. Solanes[2], A. Manrique[2] and E. Salvador-Solé[2]

[1] Dep. Matemática Aplicada II, Univ. Politècnica de Cataluña, Pau Gargallo 5, Edificio U, E-08028 Barcelona, Spain

[2] Dep. Astronomia i Meteorologia, Universidad de Barcelona, Martí Franquès 1, E-08028 Barcelona, Spain

Abstract. We describe a semi-analytic model for the X-ray emitting gas in clusters and groups in which the gas is preheated before halo formation. The model relies on physically sound prescriptions for the formation and evolution of halo structure. For a gas temperature gradient corresponding to a polytropic index of 1.2 and an energy injection of 0.6 keV per gas particle, this model successfully reproduces the observed correlations between gas luminosity, gas temperature and halo mass. We use the model to investigate the detailed entropy structure of groups and clusters of galaxies and compare the results with the observational data.

1. Dark-matter model

We assume that the formation of halos, as well as their internal structure and dynamics following virialization, is dominated everywhere by the dark-matter (DM) component. The evolution of DM halos is computed according to the model developed by Salvador-Solé *et al.* (1998) and Raig *et al.* (1998, 2001). This model self-consistently includes a distinction between minor and major mergers in the usual extended Press-Schechter formalism by means of a phenomenological frontier Δ_{mer} for the fractional mass increase separating two basic mass aggregation regimes for halos: gentle mass accretion, where the halo structure is assumed to evolve undisturbed inside-out through the continuous aggregation of small clumps, and major merger events, where participant halos are disrupted giving rise to the formation of a completely new system. The epoch of formation of a given halo can be naturally defined as the redshift at which the halo experiences its last major merger. From this definition one can derive a probability distribution of formation times as well as a typical mass-accretion evolution for halos of any given mass, both in agreement with the predictions of N-body simulations (Raig *et al.* 2001). In this work we adopt the median of the formation time distribution as the typical halo formation time t_f, while the corresponding upper and lower quartiles will measure the dispersion associated to t_f. As shown by Manrique *et al.* (2003), taking the appropriate value of Δ_{mer} the model is able to reproduce the universal density profile and the mass-density correlation of halos predicted by N-body simulations.

In the present study, we consider the following cosmological context: a flat universe with $\Omega_M = 0.3$ and $\Lambda = 0.7$, a cold dark matter power spectrum of density fluctuations with normalization $\sigma_8 = 1$, a baryon content $\Omega_b = 0.04$ and a normalized Hubble constant $h = H_0/(100 \text{ km s}^{-1} \text{ Mpc}^{-1}) = 2/3$. For our DM model we adopt the empirical correlation between halo mass and halo concentration proposed by Bullock *et al.* (2001) to fit their N-body simulations (this requires a value of $\Delta_{mer} = 0.2$). In this way, the DM evolution is determined without free parameters.

2. Gas model

Instead of considering a detailed model for the shock-heating of gas accreted by halos we adopt a set of physically motivated and simplified hypotheses with the aim of describing the expected post-shock conditions of virialized gas into halos even if there is an extra energy injection previous to halo virialization (which we refer as preheating). In the present model preheating is assumed to be universal, i.e., independent of halo mass. However, we do not need to assume that preheating occurs at any concrete redshift or even at a single redshift since we simply consider that the gas has been preheated before the time of halo formation (or last major merger). On the other hand, it must be noticed that our halo model predicts redshifts of formation which are not greater than 1.5 for halo masses greater than 10^{12} solar masses.

After halo formation and relaxation we consider that the hot diffuse gas is in thermal pressure-supported hydrostatic equilibrium within the gravitational potential well of the halo created by the DM distribution (we neglect gas and galaxy contributions to the gravitational halo mass in a first approximation). The distributions of gas and DM are taken to have the same total radius at any time. The gas mass fraction at the virial radius of present day halos is taken to be equal to the cosmological baryon fraction Ω_b/Ω_M. It is assumed that the gas is in a single phase, behaves as a monoatomic ideal gas, has a null metallicity gradient, and its density, ρ_g, and temperature T_g, are related through a polytropic equation $\rho_g \propto T_g^{1/(\gamma-1)}$. The polytropic index, γ, is left as a free parameter within the range $1 \leqslant \gamma \leqslant 5/3$. Finally, we take into account the balance between the specific energies of gas and DM that must hold for a virialized halo (in absence of mass losses during formation and any subsequent loss of energy). According to this energy balance, the specific energy of gas should be equal to the specific energy of DM plus any form of specific energy increment caused by preheating, $\Delta E/\bar{\mu}_{\rm ph}$, where $\bar{\mu}_{\rm ph}$ is the mean mass per gas particle when the preheating takes place (we assume a gas with zero metallicity at that time) and ΔE is the energy per gas particle due to preheating which is left as a free parameter.

3. Comparison with observational results

Since we have fixed the cosmological context of our study, the model has only two free parameters, the polytropic index, γ, and the energy injected to the system by preheating, ΔE. We derive the model luminosities using the Sutherland & Dopita (1993) cooling function table for a gas with one third solar metallicity, whereas model temperatures are emission weighted within the halo virial radius. Then, by comparing our model predictions with the X-ray luminosity-temperature correlation observed for real systems with temperatures in the range from 0.5 keV to 10 keV, we find that the best fit to the real data is achieved for $\gamma = 1.2$ and $\Delta E = 0.6$ keV per particle. These values also fit quiet well the observed mass-temperature correlation. In Figure 1 we show the comparison between the model results from the previous best-fit parameters and the observational measurements of the gas entropy at various distances from the system center. One can see that our model again reproduces quite well the real data and, in particular, it predicts a minimum entropy close to the value observed for systems with gas temperature around 1 keV (Lloyd-Davies *et al.* 2000). However, for temperatures below 1 keV the inner entropy increases when the gas temperature decreases. More extended data samples are required in order to determine wether this trend is certainly present in real systems (see, e.g. , the recent work of Osmond & Ponman 2004). In the case that the observations of real systems with temperatures below 1 keV confirm that the entropy significantly decreases

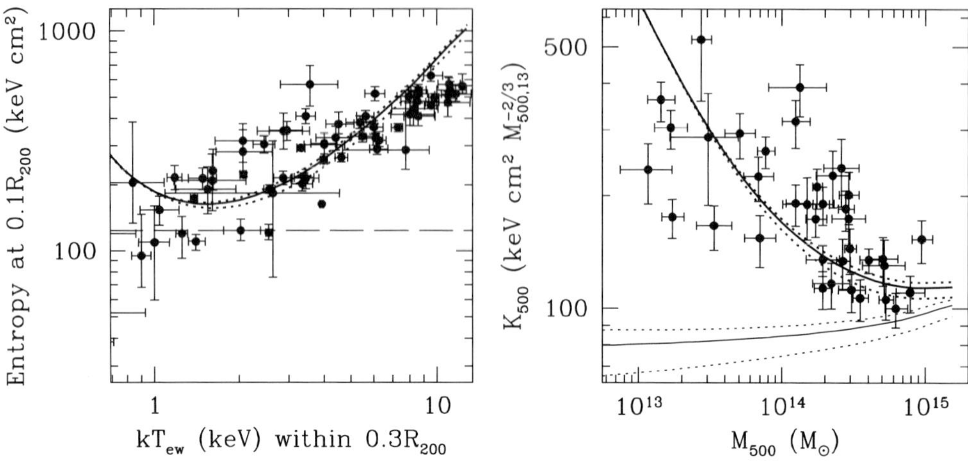

Figure 1. Comparison of the model predictions for the best-fit values of γ and ΔE in §3 with the measured gas entropy in the inner and outer regions of systems with different gas temperatures. Left panel: entropy at $0.1R_{200}$ vs emission-weighted temperature, where R_{200} is the radius enclosing a halo overdensity equal to 200. The points with error bars are taken from Ponman *et al.* (2003). The horizontal long-dashed line is the observed entropy floor from Lloyd-Davies *et al.* (2000). The thick solid line is the model prediction when halos are formed at t_f while the dotted lines indicate the dispersion corresponding to the halo formation time distribution function (see §1). Right panel: entropy at the radius enclosing a halo overdensity of 500 vs mass within such radius ($M_{500,13}$ is the latter mass in units of 10^{13} $M_{\odot}$). Points with error bars are from the observational study of Finoguenov *et al.* (2002). Again thick solid an dotted lines are the model predictions. To illustrate the effects of preheating we also plot the model results for $\Delta E = 0$ and $\gamma = 1.2$ in thin solid and dotted lines.

with gas temperature, then our model results would imply that a universal preheating of gas is inconsistent with observations.

References

Bullock, J.S., Kolatt, T.S., Sigad, Y. *et al.* 2001 *MNRAS* **321**, 559
Finoguenov, A., Jones, C., Böhringer, H. & Ponman, T.J. 2002 *ApJ* **578**, 74
Lloyd-Davies, E.J., Ponman, T.J. & Cannon, D.B. 2000 *MNRAS* **315**, 689
Manrique, A., Raig, A., Salvador-Solé, E. *et al.* 2003 *ApJ* **593**, 26
Osmond, J.P.F. & Ponman, T.J. 2004 *MNRAS* in press, astro-ph/0402439
Ponman, T.J., Sanderson, A.J.R. & Finoguenov, A. 2003 *MNRAS* **343**, 331
Raig, A., González-Casado, G. & Salvador-Solé, E. 1998 *ApJ* **508**, L129
Raig, A., González-Casado, G. & Salvador-Solé, E. 2001 *MNRAS* **327**, 939
Salvador-Solé, E., Solanes, J.M. & Manrique, A. 1998 *ApJ* **499**, 542
Sutherland, R.S. & Dopita, M.A. 1993 *ApJSup* **88**, 253

Outskirts of Galaxy Clusters: Intense Life in the Suburbs
Proceedings IAU Colloquium No. 195, 2004
A. Diaferio, ed.

DOI: 10.1017/S1743921304000663

Scaling laws of simulated galaxy clusters

Debora Sijacki[1,2] and Giuseppe Tormen[2]

[1]MPA, Garching, Germany email: deboras@mpa-garching.mpg.de

[2]Dipartimento di Astronomia, Università di Padova, Italy email: tormen@pd.astro.it

1. Introduction

Using high-resolution non radiative SPH simulations we study the scaling laws of galaxy clusters. In particular, we analyze the relations existing between total mass, gas temperature, X-ray luminosity and dark matter velocity dispersion. Our aim is to understand how these scaling laws depend on numerical, statistical and physical parameters, in order to model and thus reduce potential systematical uncertainties. The simulations were performed by the *GADGET* code in a ΛCDM cosmology, applying resimulation techniques; their mass resolution ranges between 2×10^9 and $6.8 \times 10^9 h^{-1} M_\odot$ for dark matter particles and between 3×10^8 and $7 \times 10^8 h^{-1} M_\odot$ for gas particles.

2. Method

We evaluated the scaling relations on a regular grid of radii (from $0.05R_{vir}$ to $5R_{vir}$ with step of $0.02R_{vir}$) and overdensities (from $2 \cdot 10^4 \rho_c$ to $30\rho_c$ with an adaptive step)†. Scaling laws were determined at each radius (or overdensity) using cumulative physical quantities corresponding to that radius (or overdensity) and were compared with the isothermal model (see Figs. 2 and 3). The redshift evolution of scaling relations was evaluated by taking the five snapshots closest to $z = 0$, 0.25, 0.5, 0.75 and 1. We investigated how the zero point and slope of the scaling laws depend on the number of dark matter particles used to describe a halo (see Fig. 1). We conclude that the value N_{lim} is a good compromise between numerical effects (which influence more strongly systems with less than 1000-2000 DM particles), convergence criteria (scaling laws stabilize conservatively above N_{lim}) and dynamical evolution (more massive systems are more likely to be less dynamically relaxed and thus may introduce a bias in the evaluation of scaling laws).

3. Properties at redshift zero

The slope of the $M(T)$ relation in the inner regions ($< 0.2R_{vir}$) is consistent with the self-similar prediction, while moving to larger radii it increases. This happens because at the large radii the energy budget has a non negligible contribution from the gas bulk motions (consistent with the β parameter of ~ 1.17 at R_{vir}), causing a temperature decrease. Variation with radius in the $L_X(T)$ and $L_X(M)$ relations are less evident. These two relations also require a higher number of objects to be accurately defined and to reduce the scatter. The slope of the $L_X(T)$ relation is to very good approximation consistent with the theoretically predicted value at all radii, while the slope of the $L_X(M)$ relation is somewhat underestimated. This results may reflect the fact that the incomplete gas thermalization affects the evaluation of L_X. The slope of the $M(\sigma_{DM})$ relation is close

† Especially considering the redshift evolution of scaling laws these two approaches can lead to different results, due to their correspondence to different regions of clusters.

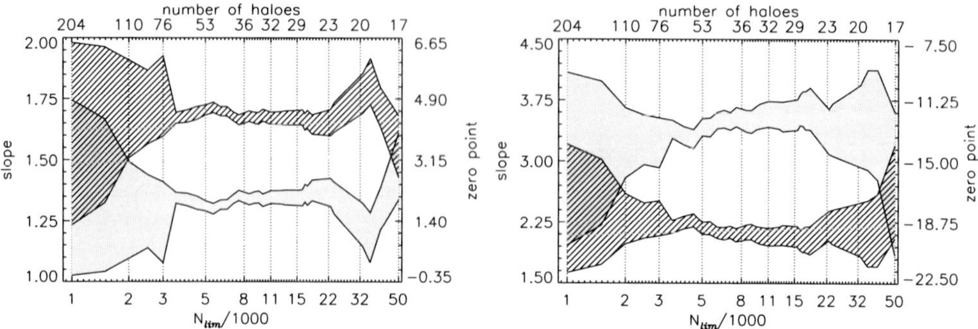

Figure 1. Left panel: $M_{vir}(T)$ scaling law, right panel: $L_X(T)$ relation. Hatched bands correspond to the slope of the relation (scaling law + its inverse) and the gray bands to the zero point. The bottom x-axis indicates the threshold in the number of dark matter particles on a logarithmic scale and the upper one shows the number of galaxy clusters having at least that threshold number.

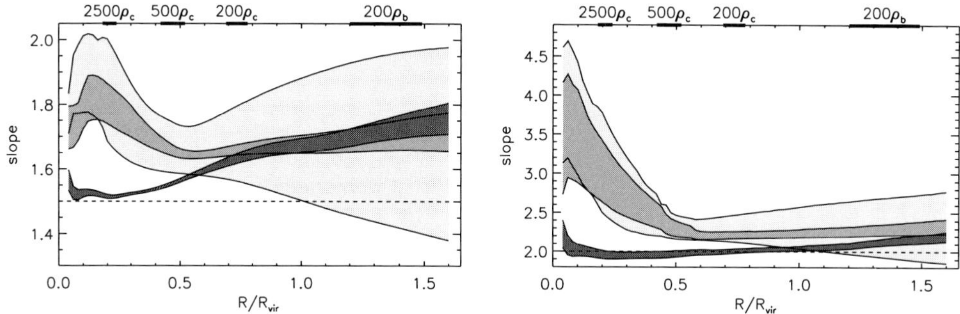

Figure 2. Slope of the $M(T)$ and $L_X(T)$ scaling laws evaluated on a regular grid of radii for 3 choices of N_{lim}=2000, 4000 and 8000 (from light to dark gray respectively), for which the scatter is substantially reduced. The continuous line gives the theoretically predicted value.

to the theoretical prediction at large radii, while in the inner regions it is systematically lower. This is due to the mass dependence of the surface pressure term present in the generalized virial theorem equation. The $\sigma_{DM}(T)$ scaling law shows in a direct way effects of non complete thermalization, which are more prominent in outer regions and in massive clusters.

4. Redshift evolution of scaling laws

The different scaling laws exhibit a redshift dependence that can be expressed by power laws of $H(z)$ and Δ_c. If we factor this out, no evolution should be observed. To first approximation, this is true also for the relations followed by our simulated clusters. Nevertheless, some systematical trends remain (see Fig. 4). In the case of the $M(T)$ and $M(\sigma_{DM})$ scaling laws, in the inner clusters regions, for a given T (or σ_{DM}) the total mass inside the fixed radius decreases with increasing redshift. This can be understood as follows: the mass profile in the inner regions is more affected by the value of the concentration parameter, and the mean concentration of our cluster sample at $z = 1$ is ~ 4.9, while at $z = 0$ it increases to ~ 8.1. The $\sigma_{DM}(T)$ relation follows the self-similar evolutionary scenario in the inner regions, while at larger radii there is evidence that the temperature systematically increases with time. This can be also seen for the

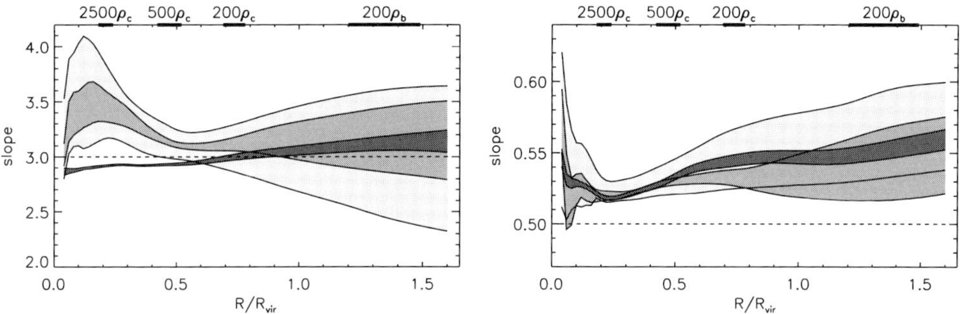

Figure 3. Slope of the $M(\sigma_{DM})$ and $\sigma_{DM}(T)$ relations. Same description as in the previous figure.

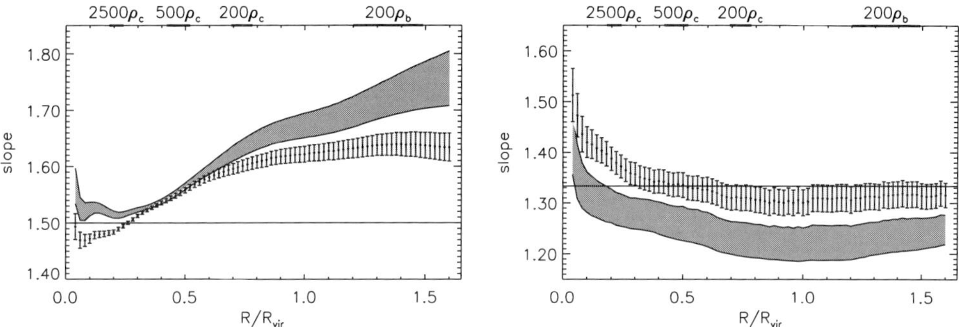

Figure 4. Slope of the $M(T)$ and $L_X(M)$ scaling laws evaluated on a regular grid of radii at redshift zero (gray band) and at 5 redshifts taken together (band with error bars).

other scaling laws involving T, and the same can be concluded regarding L_X. In order to verify if these trends can be explained by the redshift evolution of the concentration, we modeled it (as predicted by Eke et al. 1998) obtaining substantial improvement and better agreement with the self-similar evolutionary picture.

5. Conclusions

Generally speaking, substructures affect cluster properties in a more prominent way at larger radii, causing a bigger scatter in the estimated quantities. On the other hand, the mass enclosed in a given radius depends on the concentration parameter mostly in the inner regions, and this leads to systematic biases in the scaling laws compared to the isothermal prediction. Moreover, the redshift evolution of halo concentrations leads to further disagreement with respect to the self-similar picture. Once clusters are resolved with sufficient numerical accuracy, the scaling laws are more tightly defined and are less influenced by the cluster dynamical state in the inner regions. Finally, it is interesting to note that deviations from self-similarity appear in the scaling relations also in adiabatic hydrodynamics: this may suggest that a more accurate physical model of scaling laws, that models in a realistic manner profiles of various physical quantities in question, should be used.

References

Borgani, S., Murante, G., Springel, V., Diaferio, A., Dolag, K., Moscardini, L., Tormen, G., Tornatore, L. and Tozzi, P. 2004 *MNRAS* **348**, 1078–1096.

Bryan, G. L. and Norman, M. L. 1998 *ApJ* **495**, 80

Eke, V. R., Navarro, J. F., and Frenk, C. S. 1998 *ApJ* **503**, 569

Evrard, A. E., Metzler, C. A., and Navarro, J. F. 1996 *ApJ* **469**, 494

Kuhlen, M., Strigari, L. E., Zentner, A. R., Bullock, J. S., and Primack, J. R. 2004 *draft version (preprint astro-ph/0402210)*

Navarro, J. F., Frenk, C. S., and White, S. D. M. 1995 *MNRAS* **275**, 720–740.

Rasia, E., Tormen, G. and Moscardini, L. 2004 *MNRAS, in press (preprint astro-ph/0309405)*

Tormen, G., Moscardini, L., and Yoshida, N. 2003 *MNRAS, submitted (preprint astro-ph/0304375)*

Outskirts of Galaxy Clusters: Intense Life in the Suburbs
Proceedings IAU Colloquium No. 195, 2004
A. Diaferio, ed.

DOI: 10.1017/S1743921304000675

Modeling DM and ICM in simulated galaxy clusters

E. Rasia[1], G. Tormen[1] and L. Moscardini[2]

[1]Dipartimento di Astronomia, Università di Padova, vicolo dell'Osservatorio 2, I-3512, Padova, Italy email: rasia@pd.astro.it

[2]Dipartimento di Astronomia, Università di Bologna, via Ranzani 1, I-40127, Bologna, Italy

Abstract. Using the results of an extended set of high-resolution non-radiative hydrodynamic simulations of galaxy clusters we obtain simple analytic formulae for the dark matter and hot gas distribution, in the spherical approximation.

---

1. Introduction

Galaxy clusters are the largest virialized systems we observe in the universe today. Any comparison with observations in the optical or via gravitational lensing requires modeling at least the dark matter (DM) component of galaxy clusters, while X-ray or millimetric observations (e.g. the Sunyaev-Zel'dovich effect) require knowledge of the internal structure of the Intra Cluster Medium (ICM). Both aspects can in principle be studied by means of numerical simulations.

In this paper we summarize some of our results: an extended discussion can be found in Rasia et al. (2004).

Numerical Simulations. The sample includes 17 high-resolution clusters, simulated with the hydro-N-body code GADGET (Springel et al. 2001). Their virial masses range between $3.1 \times 10^{14} h^{-1} M_{\odot}$ and $1.7 \times 10^{15} h^{-1} M_{\odot}$ and virial radii, $R_{\rm v}$, between 1.4 and $2.5 h^{-1}$ Mpc. Our analysis is restricted to the $z = 0$ outputs. Thanks to the random criterion used to select the re-simulated haloes, the modeling we are doing will be representative of an average cluster, in an average environment and dynamical configuration.

2. Modeling dark matter profiles

The underlying idea of this work is to build a self-consistent model for the radial profiles describing the dynamical properties of galaxy clusters. For each cluster we compute the radial profile of the interested quantity, then we average the profiles and fit this using at most double power-law analytic fits. We decided to consider all the profiles presented in this work only at radii larger than $\log x = -1.85$, where $x \equiv r/R_{\rm v}$. Within this radius our clusters contain on average ≈ 800 dark matter particles. In all plots the solid line is the average profile, the shaded region represents the error on the mean and the vertical dashed line is the inner-radius limit of our model.

Density structure. A model for the dark matter density profile is straightforward to obtain by combining our coarse-grained phase-space density model, $\overline{f}(x) \equiv \rho(x)/\sigma_r^3(x) \propto x^{-1.875}$, with the radial velocity dispersion profile $\tilde{\sigma}_r(x) = 0.67 x^{0.3}(x + x_p)^{-0.48}$, with $x_p = 10^{-1.16}$ and solving for the density. We easily get the following approximate fit for

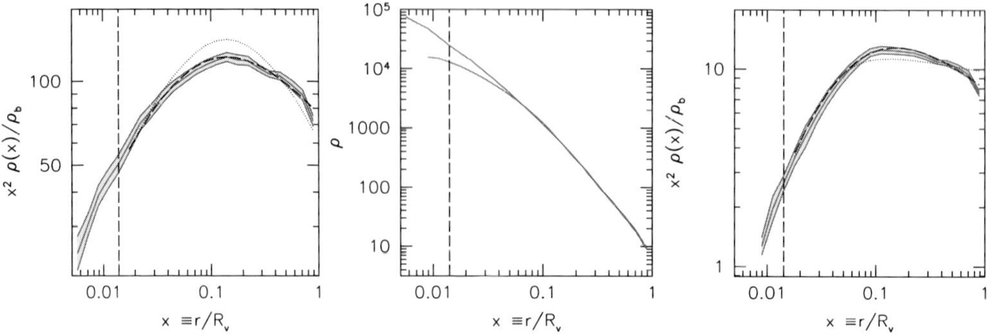

Figure 1. Density profiles. Left (Right) panel: average dark matter (gas) profile $\rho(x)x^2$. The dashed line is our fit, while the dotted lines is the NFW model (β-model). Central panel: gas (solid) and dark matter (dashed) density profile normalized so that they match at the virial radius.

the dark matter spatial density distribution:

$$\tilde{\rho}(x) = \frac{\rho_0}{x(x+x_p)^{1.5}}; \rho_0 = \frac{(1-f_{\rm b})\Delta_{\rm v}}{6\left[(1+2x_p)/(1+x_p)^{1/2} - 2x_p^{1/2}\right]}, \tag{2.1}$$

where $\Delta_{\rm v}$ is the virial over-density specified by the cosmological model and $f_{\rm b} = 0.097$ is the average baryonic fraction.

The mean density profile, averaged over the whole sample of simulated clusters, is shown in Fig. 1, our model (eq. 2.1) is represented by the dashed line, while with the dotted line we represent the NFW model (Navarro, Frenk & White 1997), with a concentration parameter equal to $c = 1/x_s = 6.3$.

Mass estimates. Given the previous relations, the cluster mass inside any given radius can be obtained in different ways: by integrating our density profile (eq. 2.1), or by integrating the NFW profile, or through the Jeans equation. We found that the Jeans equation is a proper description of the mass, being the error always smaller than 10 per cent and that the mass derived by integrating our density profile is in very good agreement with the actual mass.

3. Modeling gas profiles

Density structure. In analogy to what has been done in the previous section for the dark matter profiles, in this section we will apply a similar analysis to the radial profiles of the hot gas component. Let us start from the density. In the central panel of Fig. 1 we compare the profiles of gas and dark matter, normalized so that they match at the virial radius.

The profiles of gas and dark matter are roughly self-similar for $x > 0.06$, while the gas profile becomes flatter in the internal region. The self-similarity of the DM and gas profiles in the outer part of the cluster suggested us to fit the gas density profile with the same functional form used for the dark matter, with a different concentration value, relaxing the constraint on the inner slope and assuming of course a different normalization

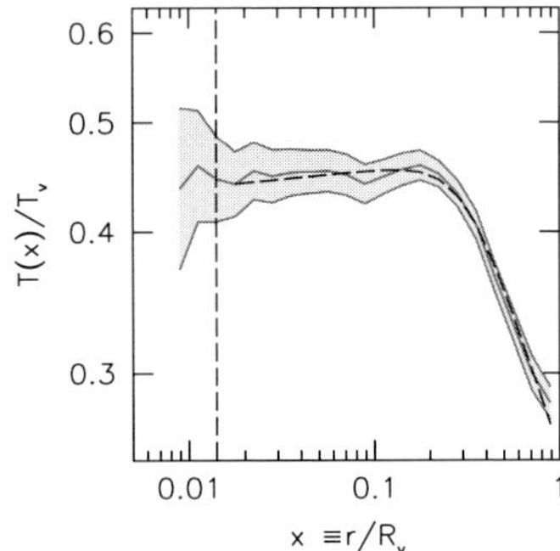

Figure 2. Average gas temperature profile for our model (dashed curve).

based on the average baryon fraction:

$$\tilde{\rho}(x) \equiv \frac{\rho(x)}{\rho_\mathrm{b}} = \frac{\rho_0}{(x+x_p)^{2.5}}, \rho_0 = \frac{f_\mathrm{b}\Delta_\mathrm{v}}{3} \times \left[\frac{16x_p^3/3 + 40x_p^2/3 + 10x_p + 2}{(x_p+1)^{2.5}} - \frac{16x_p^{1/2}}{3}\right]^{-1} \tag{3.1}$$

where $x_p = 0.04$;

In the right panel of Fig. 1 we compare the average profile for the gas density obtained from our simulated clusters (solid curve) with the proposed fit (dashed curve) and with the β-model (Cavaliere & Fusco-Femiano 1976, 1978)(dotted line).

Temperature structure. Observations suggest the presence of a central isothermal region followed by a smooth decline towards the outskirts and there are strong indications that the value of the temperature at the virial radius is approximately a factor of 2 smaller than at the center.

In Fig. 2 we show the average temperature profile normalized by the estimate of the virial temperature, $T_\mathrm{v} \equiv (G\mu m_p M_\mathrm{v})/(k_\mathrm{b} R_\mathrm{v})$. In this relation G is the gravity constant, $\mu = 0.59$ is the mean molecular weight, m_p is the proton mass and k_b is the Boltzmann constant.

The sudden temperature drop at $x \approx 0.2$ cannot be fit by a double power-law as those used so far. Instead, we model this sharp transition by using the following expression: $T(x)/T_\mathrm{v} = (0.255\, x^{0.016})/[(x^4 + x_p^4)^{0.13}]$, where $x_p = 10^{-0.51}$.

Mass estimates. A typical mass estimator using ICM as a tracer and assuming a spherical and static gravitational potential is the equation of the hydrostatic equilibrium. However, the hypotheses underlying this equation are too restrictive for at least two reasons. First, our simulations show that the mean velocity anisotropy $\beta(x)$ is not zero, second the residual bulk motions of the gas must be taken into account.

A more proper mass estimate, including the radial velocity dispersion and the velocity anisotropy, is given by:

$$M^E(<x) = -xM_v\tilde{T}(x)\left[\frac{d\ln\tilde{\rho}(x)}{d\ln x} + \frac{d\ln\tilde{T}(x)}{d\ln x}\right] + xM_v\tilde{\sigma}_r^2(x)\left[\frac{d\ln\tilde{\rho}(x)}{d\ln x} + \frac{d\ln\tilde{\sigma}_r^2(x)}{d\ln x} + 2\tilde{\beta}(x)\right]. \tag{3.2}$$

We found that the β-model underestimates the true cluster mass at most radii, whereas using our density profile improves the estimate at all radii and reduces the error at the virial radius to 15 per cent. The complete model is the most accurate at all radii, with an error of 5 to 10 per cent at $x > 0.2$.

References

Cavaliere A., Fusco-Femiano R. 1976, *A&A* **49**, 137
Cavaliere A., Fusco-Femiano R. 1978, *A&A* **70**, 677
Navarro J.F., Frenk C.S., White S.D.M. 1997 *ApJ* **490** 493
Rasia E., Tormen G., Moscardini L. 2004, *MNRAS*, **351**, 237
Springel V., Yoshida N., White S.D.M. 2001 *NewA* **6** 79

Outskirts of Galaxy Clusters: Intense Life in the Suburbs
Proceedings IAU Colloquium No. 195, 2004
A. Diaferio, ed.

DOI: 10.1017/S1743921304000687

Simulating Chandra observations of galaxy clusters

E. Rasia[1], A. Gardini[2], P. Mazzotta[3], G. Tormen[1], S. De Grandi[4] and L. Moscardini[5]

[1]Dipartimento di Astronomia, Università di Padova, vicolo dell'Osservatorio 2, I-3512, Padova, Italy email: rasia@pd.astro.it

[2]Departement of Astronomy, University of Illinois, 1002 W. Green Street, Urbana, IL 61801, USA

[3]Dipartimento di Fisica, Università di Roma, via della Ricerca Scientifica, 1, Roma, Italy

[4]INAF-Osservatorio Astronomico di Brera, via Bianchi 46, I-23807, Merate (LC), Italy

[5]Dipartimento di Astronomia, Università di Bologna, via Ranzani 1, I-40127, Bologna, Italy

Abstract. The direct comparison of observations to numerical hydro-N-body simulations, although simple in principle, is not always trivial because of possible artificial effects produced by the instrument response and by instrumental and sky background. To overcome this problem we build the software package X-MAS (X-ray MAp Simulator) devoted to simulate X-ray observations of galaxy clusters obtained from hydro-N-body simulations.

---

1. Introduction

Theoretical studies of the dynamical processes underlying the formation and evolution of galaxy clusters are now mainly based on the results of numerical hydro-N-body simulations. Their results clearly show that clusters of galaxies experience violent events that release an enormous amount of energy in the intra-cluster medium (ICM). These events induce strong, but transient, variation of both ICM density and temperature, so that their distribution is far to be smooth and spherically symmetric. From the observational side, thanks to the superb angular and spectral resolution of the latest generation X-ray satellites, *Chandra* and XMM-Newton, we now know that many galaxy clusters, including the ones previously identified as relaxed, actually present a great deal of spatial features and have a rather complex thermal structure. Due to their complex nature and the impossibility of using simple de-projection techniques, most of these observed features can be quantitatively studied only through a direct comparison to numerical simulations. Ideally, to make these comparisons straightforward one needs to re-process the simulations themselves through a sort of virtual observatory that takes into account the instrumental response and the sky and the instrumental background.

2. X-ray MAp Simulator: the method

In this section we describe how the X-MAS package works. The package can be divided into two main units. The first unit is quite general and does not depend on the specific characteristics of the X-ray telescope. The second unit simulates the data relevant to an observation with a specific X-ray telescope and detector for a defined amount of time. At present our software package simulates *Chandra* observations in ACIS-S3 configuration only. The application of our simulation package to simulate *Chandra* in the ACIS-I mode and XMM-Newton observations requires an adaptation of this second unit only. In the following we summarize the steps of the two units, while a complete description is in Gardini et al. (2004).

First Unit: generating differential flux maps: The input of this unit is the output of a hydro-N-body simulation. For each gas particle in the simulation, we compute the emissivity and distribute it over the corresponding volume. After selecting a line of sight for the simulated observation, we compute the projected spectrum for each pixel and then the differential flux for each angular coordinate in bins of energy. The final step is adding Galactic absorption.

Second Unit: simulating Chandra observations: We estimate the expected number counts and iteratively subdivide the tile region until the counts become smaller than a given threshold; we then use the command FAKEIT in the utility XSPEC to convolve the spectral model of each subregion with the response of the ccd and to add the sky background; at the end we generate the final photon event file.

3. ACIS-S3 observation of a simulated cluster

We generate a 300 ks ACIS-S3 *Chandra* observation of a high-resolution hydro-N-body simulation of a galaxy cluster presenting a major merger. It is selected from a sample of 17 clusters presented elsewhere (Rasia, Tormen & Moscardini 2004; Tormen,Moscardini & Yoshida 2004). Since this particular simulation does not provide information on the cluster metallicity, we fixed its value to $Z = 0.3\,Z_\odot$ and we assumed a galactic equivalent column density of $N_H = 5 \times 10^{20}$ cm^{-2}.

The cluster flux map, obtained with the first unit of our X-MAS, is presented in Fig. 1 (left panel). The figure shows the flux in the [0.1,10.0] keV energy interval. The superimposed iso-contours are obtained from the image after a Gaussian smoothing with $\sigma = 8''$. In the right panel of Fig. 1 we show the photon image of the 300 ks *Chandra* ACIS-S3 observation of the simulated cluster, as obtained by applying the second unit of our software X-MAS. The image, extracted from the event file in the [0.3,9.0] keV energy band, is background-subtracted, vignetting-corrected, and binned to 1 arcsec. We notice that, as expected, after being observed with *Chandra*, the spatial features present in the simulation, but fainter than the instrument background, are no longer detected (for example, the three faint sub-clumps on the North, North-West and South-West, clearly visible). Conversely all the brighter features appear to be well reproduced by our simulator. We find that this is true even on scales as small as a few arcseconds.

4. Spectral analysis: temperature maps

In literature we usually find that the projected spectroscopic temperature T_{spec} (obtained from the spectral analysis of the data), is directly compared with the so-called emission-weighted projected temperature T_{em} of the simulation. Unfortunately, in some cases T_{spec} and T_{em} do not identify the same quantity. In fact, if the cluster has a complex thermal structure, the spectroscopic temperature is *de-facto* the result of the fit a "wrong" model (a single temperature model) to a multi temperature observed spectrum.

Using X-MAS we produced a *Chandra* observation of the simulated cluster and by performing a standard spectral analysis, we derived the spectroscopic temperature. This quantity is shown in the left panel of Fig. 2, while in the right panel we represent the emission-weighted temperature as derived directly from the simulation and computed using the same resolution of the spectroscopic temperature. If we now compare the two maps in Fig. 2, we notice that, although qualitatively similar, they show a number of important differences. Among others, we point our attention on the fact that: i) the

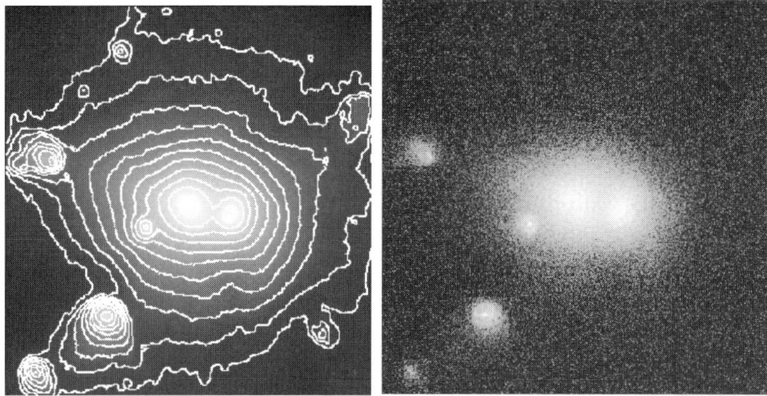

Figure 1. Left panel: Flux map of the simulated galaxy cluster, as obtained from the first unit. Right panel: photon image of the 300 ks ACIS-S3 observation.

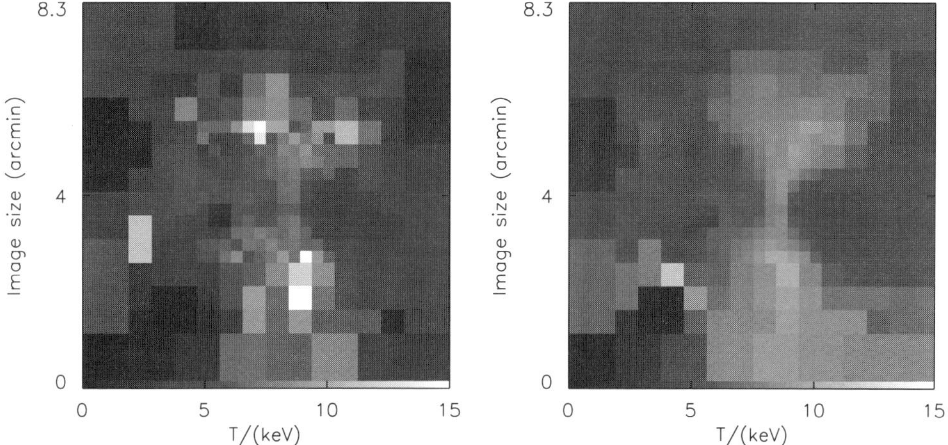

Figure 2. Temperature maps for the galaxy cluster. Left panel: map of the projected temperature, as derived from the spectroscopic analysis of the *Chandra* observation. Right panel: map of the emission-weighted temperature.

central merging blobs appear to be colder in the observed spectroscopic temperature map than in the T_{em}; ii) the shock front produced by the motion of the innermost subclump in the lower-left corner, clearly visible in the emission-weighted map, is no longer detected in the spectroscopic temperature map.

To conclude we stress the fact that a proper comparison simulations-observations can only be done through the actual comparison of the spectral properties of both. This means that, in order to perform direct comparisons simulations-observations the simulated clusters must also be properly "observed" using X-ray observatory simulators like, for example, our X-MAS.

References

Gardini A., Rasia E., Mazzotta P., Tormen G., De Grandi S., Moscardini L. 2003, *MNRAS in press*, astro-ph/0310844

Rasia E., Tormen G., Moscardini L. 2004, *MNRAS*, **351**, 237

Tormen G., Moscardini L., Yoshida N. 2004,*MNRAS*, **350**, 1397

Outskirts of Galaxy Clusters: Intense Life in the Suburbs
Proceedings IAU Colloquium No. 195, 2004
A. Diaferio, ed.

DOI: 10.1017/S1743921304000699

Velocity field in cluster outskirts in ΛCDM models

Maximiliano C. Pivato[1], N. D. Padilla[2] and D. G. Lambas[1]

[1]Grupo IATE, Observatorio Astronómico de Córdoba, UNC, Argentina email: maxip@oac.uncor.edu; dgl@oac.uncor.edu

[2]Dept. Physics, University of Durham, South Road, Durham, DH1 3LE, UK email: n.d.padilla@durham.ac.uk

Abstract. We analyse statistically the infall of mass onto clusters in Λ-CDM numerical simulations. Also, we consider numerical simulations which include galaxy formation through semi-analytical models to compare the expected infall of mass and galaxies. We find the most probable values at $V \sim 300-400$ km/s and a significant tail of large infall velocities ($V > 1000$ km/s). The flow onto cluster is more pronounced for high local density regions. Mass and galaxies behave similarly although with a slightly lower amplitude.

1. Introduction

The dynamical behaviour of objects near regions of high density contrast can be described by the spherical infall model with a collapsing streaming motion with amplitude depending only on the distance to the local density maximum (Diaferio & Geller 1997). Assuming linear theory, $v_p(r) = -1/3\Omega_0 H_0 r\delta(r)$; in the non-linear regime, Rëgos & Geller (1989) find an approximate solution given by $v_p(r) = -1/3\Omega_0 H_0 r\delta(r)/[\delta(r)+1]^{0.25}$ which slightly improve the agreement with simulations (Croft, Dalton & Efstathiou 1999). Here, we explore the characteristics of streaming motions in cluster vicinity using large high-resolution N-body simulations (VLS) carried out by the Virgo Supercomputing Consortium. VLS simulations consider a Λ-CDM primordial spectrum ($\Omega_\Lambda = 0.7$, $\Omega_m = 0.3$, $h = 0.7$) in a $479h^{-1}$Mpc box size with resolution $6.8 \times 10^{10} M_\odot$.

2. Analysis

In figure 1, we show the results of the mass infall onto the most massive clusters in the simulation box. The arrows indicate the velocity flow where it can be clearly appreciated the systematic infall around the most prominent clusters.

The histograms of figure 2 show the distribution of the cluster-centric component of the peculiar velocities of particles at different distances from the most massive clusters ($M > 10^{14} M_\odot$) in the simulation. It can be seen the large spread of velocities and the shift of the mean values toward negative velocites by an amount of $v \sim 200$ km/s.

The distributions corresponding to the net infall at a given cluster-centric distance are displayed in figure 3. It can be appreciated the non gaussianity of the distribution of infall velocities regardless of the distance to the cluster. The distributions show the most probable values $V \sim 200 - 400$ km/s and the tails of large infalling velocities, $V > 1000$ km/s.

We expect the amplitude of the infall velocities at a given distance from the group center to increase monotonically with cluster mass. Therefore, we have explored the mean infall at different distances to clusters with a range of masses $M < 1.1 \times 10^{13} M_\odot$

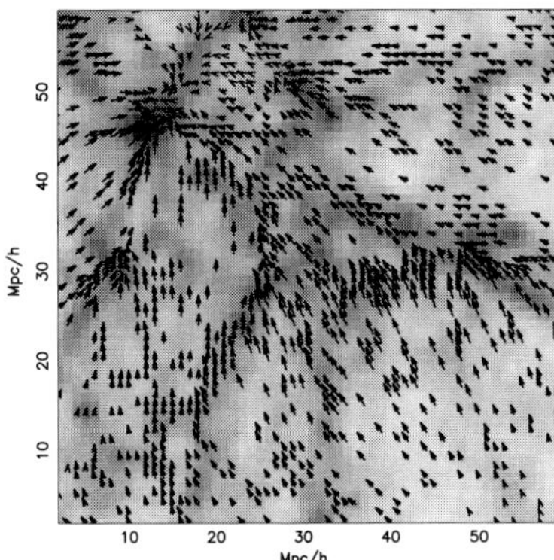

Figure 1. Infall pattern in a typical region of the simulation box.

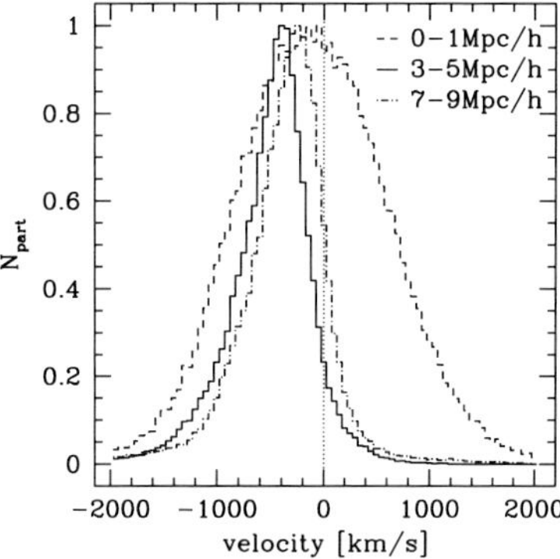

Figure 2. Distribution of particle velocities in the direction to cluster centers, normalized to the maximum values.

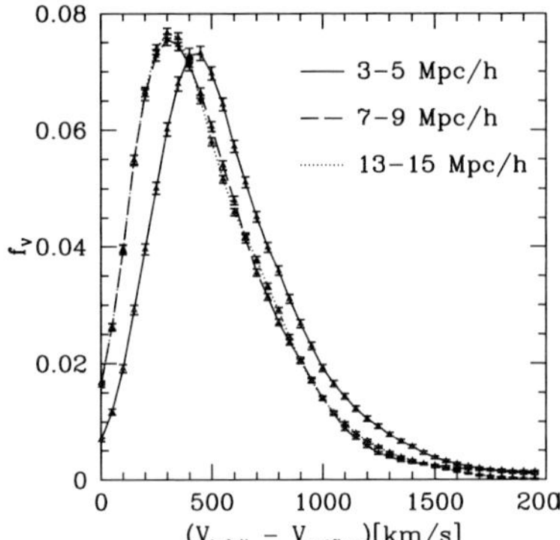

Figure 3. Distribution of the net infall velocities.

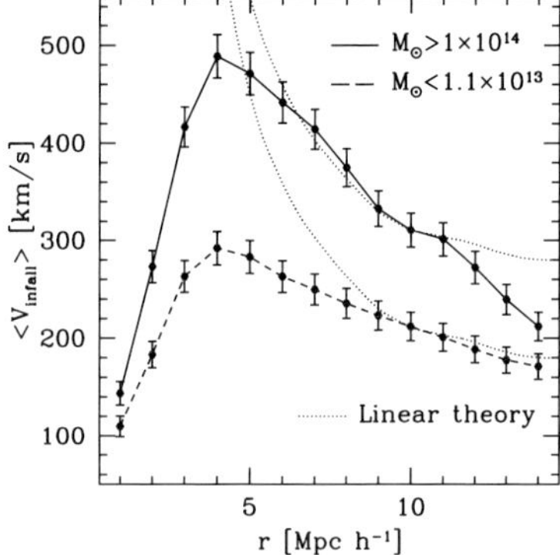

Figure 4. Mean infall velocities as a function of distance for two ranges of cluster masses.

and $M > 10^{14}M_{\odot}$. The results displayed in figure 4 show a significant increase of the maximum infall amplitude onto the most massive clusters on scales $r \sim 4h^{-1}$Mpc.

We also explored the variation of the infall pattern according to the local particle density, smoothed on $3h^{-1}$Mpc scale, appropiate to consider the collapsing streaming motions onto clusters from different regions such as filaments or voids. In figure 5 are shown the results, where it can be appreciated that particles in high density regions flow most rapidly onto large mass concentrations than those in underdense environments.

Finally, we have considered a high resolution N-body simulation populated with galaxies using the semi-analytic model, GALFORM (Cole et al. 2000; Benson et al. 2002). The simulation uses standard ΛCDM parameters, with normalization given by $\sigma_8 = 0.80$ and a primordial spectral index of $n = 0.97$. The N-body simulation box is $250h^{-1}$Mpc on a side and contains 1.25×10^8 dark matter particles of mass $1.04 \times 10^{10}h^{-1}M_{\odot}$.

Figure 6 shows the mean infall velocity of DM particles and semi-analytic galaxies in the simulation cube, as a function of distance to cluster centres. We show results for the full sample of semi-analytic galaxies and for those with absolute magnitudes $B_j < -18$. This figure shows the results for subsamples of haloes defined using different lower mass limits. As can be seen, galaxies show a behaviour similar to DM particles even when restricting to brighter galaxies, although infall velocities seem to be marginaly lower for galaxies than for the dark matter.

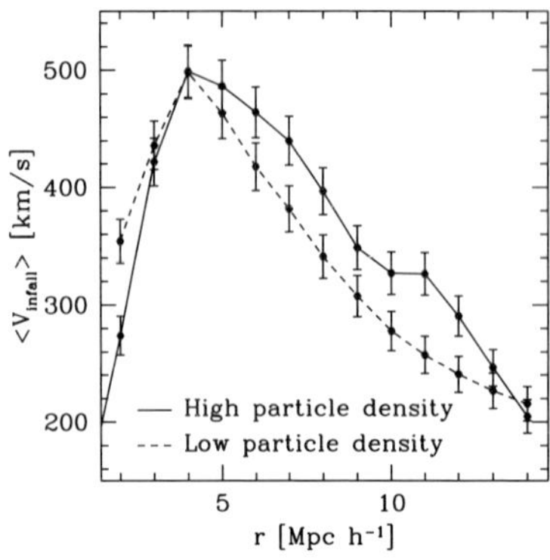

Figure 5. Distribution of net infall velocities for particles in high (solid line) and low (dashed line) density regions.

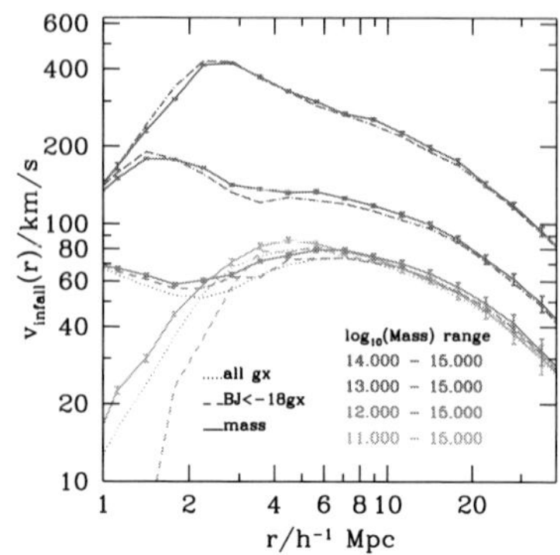

Figure 6. Mean infall for galaxies and dark matter particles in semianalytical models.

Acknowledgements

MP and DL acknowledge kind support from Dr. Antonaldo Diaferio to attend the IAU Colloquium 195.

References

Benson A.J., Lacey C.G., Baugh C.M., Cole S. & Frenk C.S. 2002 *Mon. Not. R. Astr. Soc.* **333**, 156–176.

Cole S., Lacey C.G., Baugh C.M., & Frenk C.S. 2000 *Mon. Not. R. Astr. Soc.* **319**, 168–204.

Croft R.A.C., Dalton G.B. & Efstathiou G. 1999 *Mon. Not. R. Astr. Soc.* **305**, 547–562.

Diaferio A. & Geller M.J. 1997 *Astrophys. J.*, **481**, 633–643.

Rëgos E. & Geller M.J. 1989 *Astr. J.*, **98**, 755–765.

Outskirts of Galaxy Clusters: Intense Life in the Suburbs
Proceedings IAU Colloquium No. 195, 2004
A. Diaferio, ed.

DOI: 10.1017/S1743921304000705

Constraints to cosmological parameters through cluster evolution

A. Del Popolo and N. Ercan

Boğaziçi University, Physics Department, 80815 Bebek, Istanbul, Turkey

Abstract. In this paper, we revisit the constraints obtained by several authors (Reichart et al. 1999; Eke et al. 1998; Henry 2000) on the estimated values of $\Omega_{\rm m}$, n and σ_8 in the light of recent theoretical developments: 1) new theoretical mass functions (Sheth & Tormen 1999, Sheth, Mo & Tormen 2001, Del Popolo 2002b); 2) a more accurate mass-temperature relation, also determined for arbitrary $\Omega_{\rm m}$ and Ω_Λ (Del Popolo 2002a).

Firstly, using the quoted improvements, we re-derive an expression for the X-ray Luminosity Function (XLF), similarly to Reichart et al. (1999), and then we get some constraints to $\Omega_{\rm m}$ and n, by using the *ROSAT* BCS and EMSS samples and maximum-likelihood analysis. Then we re-derive the X-ray Temperature Function (XTF), similarly to Henry (2000), re-obtaining the constraints on $\Omega_{\rm m}$, n, σ_8. In the case of both the XLF and the XTF, the changes in the mass function and M-T relation produces an increase in $\Omega_{\rm m}$ of $\simeq 20\%$ and similar results in σ_8 and n.

1. Introduction

It is well known that clusters are strong X-ray emitters whose study can put constraints on fundamental cosmological parameters. There are different methods to trace the evolution of the cluster number density: a) The X-ray temperature function (XTF) has been presented for local (e.g. Henry & Arnaud 1991) and distant clusters (Eke et al. 1998; Henry 2000). b) The evolution of the X-ray luminosity function (XLF).

The results obtained for $\Omega_{\rm m}$ and other cosmological parameters are in many cases discrepant the one with the other. Several studies in literature, show that the parameters values span the entire range of acceptable solutions: $0.2 \leqslant \Omega_{\rm m} \leqslant 1$ (see Reichart et al. 1999). The reasons leading to the quoted discrepancies has been studied in several papers (Eke et al. 1998; Borgani et al. 2001): 1) The inadequate approximation given by the PS (e.g., Bryan & Norman 1997). 2) Inadequacy in the structure formation as described by the spherical model leading to changes in the threshold parameter $\delta_{\rm c}$ (e.g., Governato et al. 1998). 3) Inadequacy in the M-T relation obtained from the virial theorem (see Del Popolo 2002a). 4) Effects of cooling flows. 5) Determination of the X-ray cluster catalog's selection function. 6) Evolution of the L-T relation. 7) Optimization methods used in the analysis. These reasons lead us to re-calculate the constraints on $\Omega_{\rm m}$, n and σ_8, using the XLF and XTF.

2. Constraints to $\Omega_{\rm m}$ and n from the XLF

Similarly to Reichart et al. (1999), we re-derived an expression for the XLF, using an improved version of the mass function and $M-T$ relation, obtained in Del Popolo (2000a,b), taking account of the effects of asphericity and tidal interaction with neighbors. Then we got some constraints to $\Omega_{\rm m}$ and n, by using the *ROSAT* BCS and EMSS samples.

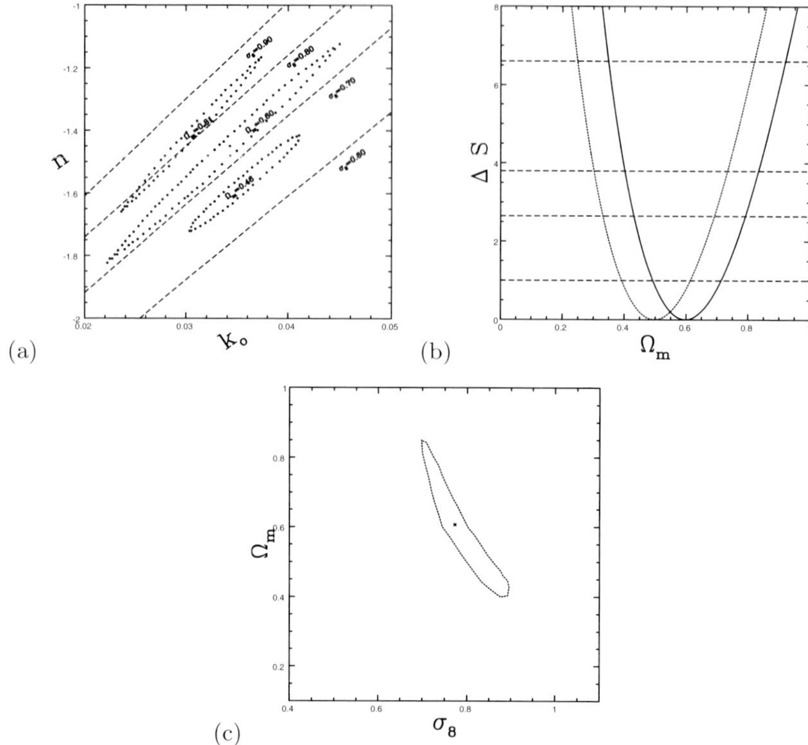

Figure 1. (a) The 68% confidence contours for the parameters n, $k_{\rm o}$ and $\Omega_{\rm m}$ for the open model. The dashed lines are lines of constant σ_8. (b) Δ(likelihood) for the parameter $\Omega_{\rm m}$. The solid line is obtained from the model of this paper while the dotted line is that calculated by Henry (2000). The dashed lines represent various confidence levels (65%, 90%, 95%, 99%). (c) The 68% confidence contours for the parameters σ_8, and $\Omega_{\rm m}$ for the open model (see also Henry 2000, Fig. 9).

As described in Del Popolo (2000b), the mass function can be approximated by:

$$n(m,z) \simeq 1.21 \frac{\overline{\rho}}{m^2} \frac{d\log(\nu)}{d\log m} \left(1 + \frac{0.06}{(a\nu)^{0.585}}\right) \sqrt{\frac{a\nu}{2\pi}} \exp\{-a\nu \left[1 + \frac{0.57}{(a\nu)^{0.585}}\right]^2 /2\} \quad (2.1)$$

where $a = 0.707$. Eq. (2.1) can be converted from a mass function to a luminosity function using an $L-T$ relation (we use that of Mathiesen & Evrard 1998), and a $T-M$ relation. This last is the one obtained in Del Popolo (2000a), and is based on the merging-halo formalism of Lacey & Cole (1993), accounting for the fact that massive clusters accrete matter quasi-continuously, and again take account of angular momentum acquisition by protostructures:

$$kT \simeq 8\mathrm{keV} \left(\frac{M^{\frac{2}{3}}}{10^{15} h^{-1} M_{\odot}}\right) \frac{\left[\frac{1}{m_1} + \left(\frac{t_\Omega}{t}\right)^{\frac{2}{3}} + \frac{K(m_1,x)}{M^{8/3}}\right]}{\left[\frac{1}{m_1} + \left(\frac{t_\Omega}{t_0}\right)^{\frac{2}{3}} + \frac{K_0(m_1,x)}{M_0^{8/3}}\right]} \quad (2.2)$$

(see Del Popolo 2000a for a derivation of the previous equation and the definition of the terms in it). The luminosity function is obtained as in Reichart et al. (1999), using the mass function and the $M-T$ relation previously introduced (see Del Popolo 2003 for a detailed analysis). A Bayesian inference analysis used to constrain the model parameters shows that: $\Omega_m = 1.15^{+0.40}_{-0.33}$ and $n = -1.55^{+0.42}_{-0.41}$. The previous result shows that the change in the mass function and M-T relation gives rise to an increase of Ω_m and n of $\simeq 20\%$ with respect to Reichart's results. The lesson from the previous calculation is that taking account of non-sphericity in collapse and the fact that massive clusters accrete matter quasi-continuously gives rise to a noteworthy change in the prediction of cosmological parameters, as Ω_m. In order to check the previous trend, we have also estimated the value of Ω_m following Borgani et al. (2001). Analyzing the ROSAT Deep Cluster Survey (RDCS) and using the XLF to obtain constraints on cosmological parameters, Borgani et al. (2001) found that $\Omega_m = 0.35^{+0.13}_{-0.10}$. Using their method and data, but our mass function and M-T relation, one obtains larger values of Ω_m ($\Omega_m \simeq 0.4 \pm 0.1$) that, differently from the previous analysis (Reichart et al. 1999), exclude an Einstein-de Sitter model.

3. Constraints to Ω_m, n, and σ_8 from the XTF

As previously reported, the mass function (MF) is a critical ingredient in putting strong constraints on cosmological parameters (e.g. Ω_m). In the following, we will re-calculate the constraints obtained by Henry (2000), by using the mass function and the M-T relation modified as described in the previous section eq. (2.1) and eq. (2.2) (see Del Popolo 2003). we use a maximum likelihood fit to the unbinned data in order to determine various model parameters. The method is described in Marshall et al (1983). The likelihood function is given by their eq. (2), adapted to our present situation. At this point, we can fit the data described in Section. 2 of Henry (2000) to the theory previously described using the quoted maximum likelihood method. The most general description of the results requires the three parameters of the fit (Ω_m, σ_8 and n). These values shows that the correction introduced by the new form of the mass function and M-T relation gives rise to higher values of Ω_m ($\Omega_m = 0.6 \pm 0.13$, while it is $\Omega_m = 0.49 \pm 0.12$ for Henry 2000) and $n = -1.5 \pm 0.32$ ($n = -1.72 \pm 0.34$ in Henry 2000). Constraints are relatively tight when considering this single parameter. We find that $\Omega_m = 0.6^{+0.12}_{-0.11}$ at the 68% confidence level and $\Omega_m = 0.6^{+0.23}_{-0.2}$ at the 95% confidence level for the open model.

Concluding, our analysis shows that improvements in the mass function and M-T relation increases the value of Ω_m and that even small correction in the physics of the collapse can induce noteworthy effects on the constraints obtained.

References

Borgani S., et al. 2001, ApJ 561, 13
Bryan G.L., Norman M.L., 1997, ApJ 495, 80
Del Popolo A., Gambera M., 1998, A&A 337, 96
Del Popolo A., 2002a, MNRAS 336, 81
Del Popolo A., 2002b, MNRAS 337, 529
Del Popolo A., 2003,ApJ 599, 723
Eke V. R., Cole S, Frenk C. S., Henry J.P., 1998, MNRAS 298, 1145
Governato F., et al., 1999, MNRAS 307, 949
Henry J.P., 2000, ApJ 534, 565
Lacey C., Cole S., 1993, MNRAS 262, 627
Marshall H. L., et al., 1983, ApJ 269, 35
Reichart D. E., et al., 1999, ApJ 518, 521
Sheth R. K., Tormen G., 1999, MNRAS 308, 119
Sheth R. K., Mo H. J., Tormen G., 2001, MNRAS 323, 1 (SMT)

Roman Juszkiewicz reminds when the low-density universe was born *(by J.M. Colberg)*.

Brent Tully, Volker Müller and Gustavo Yepes at Palazzo Barolo *(by J.M. Colberg)*.

Giuseppe Tormen and Tommaso Treu at the conference dinner *(by J.M. Colberg)*.

Rien van de Weygaert toasts to the success of both the Colloquium and the Barolo wine *(by J.M. Colberg)*.

Part 3

FORMATION AND TRANSFORMATION OF GALAXIES

Section A

Nurture and genetics of galaxies

Outskirts of Galaxy Clusters: Intense Life in the Suburbs
Proceedings IAU Colloquium No. 195, 2004
A. Diaferio, ed.

DOI: 10.1017/S1743921304000717

The properties of galaxies on the outskirts of clusters

Erica Ellingson

Center for Astrophysics and Space Astronomy, University of Colorado, Boulder, CO 80309, USA email: elling@casa.colorado.edu

Abstract. The outskirts of clusters are dynamically complex regions populated by galaxies falling into the cluster potential for the first time as well as some which have already visited the cluster core. Addressing the properties of galaxies in these transition regions requires the integration of cluster dynamics, an understanding of environmental influences on galaxies at both high and low densities and hierarchical scenarios of cluster formation and growth. In this review I describe some recent observational results on galaxies in the outskirts of clusters, and the evolution of galaxies in clusters as a function of local density and radius.

1. Introduction

The regions between the central virialized cluster cores and the general field hold special interest for both galaxy evolution and cosmological investigations. While the over densities measured on large scales in this region are small, these regions provide constraints on environmental mechanisms which might occur in lower-density environments prior to infall into clusters. The infall regions at the edges of clusters also form a direct link between the cluster and the rest of the universe, sampling the filamentary structures that eventually flow into clusters.

The standard portrait of massive galaxy clusters is on the surface quite simple. The cores are dominated by red, early-type galaxies containing very little new star formation. Observations at both low and high redshifts indicate that the stars in many of these galaxies formed at very high redshifts ($z > 2$; Bower, Lucey & Ellis 1993, Stanford et al. 1998, van Dokkum et al. 1998 and others) and have been evolving passively since then. Later type galaxies are found more commonly at larger radii (e.g., Dressler 1980, Whitmore et al. 1993), with spatial and dynamical properties consistent with being relatively new arrivals into the cluster potential well. This phenomenon was shown exceptionally well by Biviano & Katgert 2004, in these proceedings, who show that both red and blue populations are in dynamical quasi-equilibrium with the cluster potential, but that the red galaxy population is a much better spatial tracer of the underlying mass distribution. Blue galaxies show a broader spatial extent, but are in equilibrium with the same gravitational potential. These blue galaxies are expected over time to transform into a less-active population due to some environmental process associated with the cluster environment which curtails star formation. The ensuing gradients in cluster populations, from red cores to bluer outskirts, reflect the balance between the infall of star-forming galaxies and the efficiency with which star formation is squelched.

The physical mechanisms which may be responsible for the lower fraction of star-forming galaxies in dense environments include attacks on both the physical structure and gas content of the galaxies. Individually or in combination, they appear to be sufficient to transform normal field spirals into the observed cluster core populations. Possible mechanisms include galaxy-galaxy mergers, "harassment", ram-pressure stripping of gas from infalling galaxies by a dense intra-cluster medium, the disruption of halo gas reservoirs and tidal disruption from the cluster potential (e.g., Lavery & Henry 1986, Dressler & Gunn 1983, Moore et al. 1996, Fujita 1998 and others). Despite this litany of violence, the mechanisms which dominate galaxy evolution in and near clusters are not well constrained. Studies of galaxies in the regions intermediate between cluster and field provide more specific clues as to which mechanisms dominate in which environments.

2. Properties of galaxies in the outskirts of clusters

Until recently, comprehensive studies of the outskirts of galaxy clusters have been difficult. At lower redshifts, the large field sizes required have been a poor match to previous instrumentation, and at higher redshifts (or fainter magnitudes at lower redshift), contamination from unrelated galaxies overwhelms the low-density cluster counts. Wide field detectors, photometric redshift techniques and very large spectroscopic surveys have made extraordinary advances in the past few years, producing detailed inventories of cluster outskirts.

Results from the 2dF Survey (Lewis et al. 2002) provided a first spectroscopic measure of star formation in cluster galaxies to very large radii (up to 10 times the virial radius). They found strong gradients in the star formation rates as measured via Hα emission, with cluster cores containing very little star formation, and a smooth gradient leading to field galaxy levels. Somewhat unexpectedly, star formation in galaxies was measurably suppressed well beyond the cluster virial region, where the most powerful environmental processes are expected to occur. Gradients in galaxy star formation rates via [OII] were also measured for intermediate redshift clusters, with qualitatively similar findings, to about twice the virial radius (Balogh et al. 1997).

Gradients in galaxy properties have also been measured by several investigations based on the Sloan Digital Sky Survey (SDSS) database. Gomez et al. (2003) found that star formation rates in clusters at $3-4$ virial radii were suppressed relative to the field population. Goto et al. (2003) traced gradients in the morphologies of galaxies in clusters using SDSS data, finding that elliptical galaxies dominate the cores, but that intermediate-type morphologies (generally S0 galaxies according to their classification scheme) are seen at $0.2-2$ virial radii. This scenario appears to be consistent with the paradigm where field spirals are transformed to S0 galaxies via some environmental mechanism.

It has long been known that cluster spirals are deficient in their HI gas content (Giovanelli & Haynes 1985; see also Solanes et al. 2001). As with star formation rates and galaxy morphologies, radial gradients show a smooth transition from gas-poor cores to the properties of normal field galaxies, possibly also with an extension beyond the virial radius (Sanchis et al. 2004, Mamon et al. 2004).

3. Suburban lifestyles near the big city

These detailed studies suggest, unsurprisingly, that cluster environments are hazardous for gas-rich, star-forming and disk galaxies, and that one or more of the mechanisms suggested for affecting galaxy properties is at work in and near the clusters. The large radial range of affected galaxies, well beyond the virial region, suggests that these effects may begin to occur well before galaxies are incorporated into the cluster potential, thus diminishing the roles of ram-pressure stripping and other mechanisms which require a dense intra-cluster medium or a deep gravitational potential and promoting galaxy interactions and other mechanisms which dominate in small group environments.

However, the transition regions between the cluster virialized region and the general field are dynamically complex and the galaxies in these regions are a mixture of both infalling objects and objects which have passed once through the inner regions and have scattered out again. Timescales for travel in and through the clusters are generally shorter than the stellar evolution timescales which govern the wholesale spectral and morphological changes seen from field to cluster core. This will inevitably produce a blurring in the galaxy populations near the virial radius. Within the suburban analogy, galaxies in cluster outskirts correspond to residents of small communities near the big cities. Local teenagers have certainly visited the city at some point and returned home (at least for awhile). While their present observed environment may seem tame and suburban, it is not unsurprising that many of them will have purple hair, tattoos and pierced tongues.

How many of the galaxies in the infall regions have actually fallen through the central core regions and are outside the virial radius as "backsplash?" Balogh, Navarro & Morris (2000) suggested roughly 50% of galaxies between 1 and 2 virial radii were previously inside the virial radius (see also Moore et al., in these proceedings). A detailed dynamical simulation of the Virgo cluster by Mamon et al. (2004) also showed some scatter beyond the virial radius, but very little outside of 2 virial radii. Clearly, some of the decrease in star formation in cluster galaxies near the virial radius may still be due to deep-potential mechanisms.

Galaxies which are truly infalling for the first time probably dominate at radii larger than 2 virial radii. However, these galaxies are also not expected to be identical to an untouched field population. The well-established morphology-density effect (Dressler 1980) shows a continuous decline in the spiral population throughout several orders of magnitude in projected galaxy density. While density is degenerate with radius for massive, regular clusters, studies of irregular clusters and groups suggest that local density is a primary driver in galaxy evolution. In hierarchical scenarios, clusters grow via the merging of these smaller structures, implying that a significant amount of pre-processing may occur in the galaxy population while they live in group-sized structures and before they are incorporated into a massive system. Thus the outskirts of clusters should be composed of a dynamic mixture of both infalling and back-splashing galaxies, each of which will have stellar populations and morphologies which are different from a pristine field population, but which may reflect the effects of very different environmental mechanisms.

The apparent strength of the influence of local density on galaxy properties begs the question: do cluster-scale environments matter at all? If the infall and merging process allows galaxies in general to retain their local density, then the observed cluster radial gradients could simply be a spatial restructuring of the morphology-density effect with no additional environmental influence required from the cluster itself. Several recent studies suggest that there are small but detectable differences in the morphology-density relation depending on whether the galaxies are near massive clusters. Balogh et al. (2002) found that high X-ray luminosity (presumably more massive) clusters had a marginally smaller fraction of star forming galaxies at a given local galaxy density than did clusters with lower X-ray luminosities. Subsequent studies of SDSS clusters (Balogh et al. 2004) suggested that star formation rates are primarily correlated with local density, but that galaxies in the densest regions in and near clusters show lower star formation rates than more isolated systems of similar density. These studies suggest that cluster-scale structures do indeed have an additional effect on galaxy star formation rates.

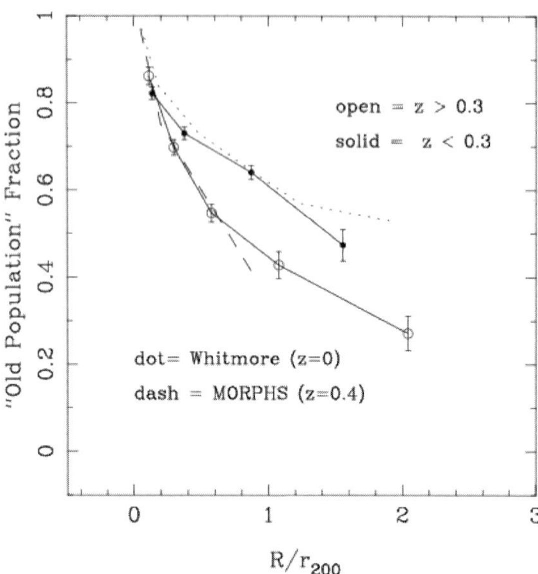

Figure 1. Fraction of cluster galaxies with old (early-type) spectral features, as a function of radius for $\sim$ 900 galaxies in 16 clusters from the CNOC sample at $0.2 < z < 0.55$ (Ellingson et al. 2001). The cores of the clusters are consistently dominated by old, red galaxies, while the outskirts have a higher fraction of star forming and intermediate age galaxies at higher redshifts. Data from $z = 0$ (Whitmore et al. 1993) and regular clusters from $z = 0.4$ (MORPHS; Dressler et al. 1997) are overplotted, assuming that their morphological elliptical and S0 types correspond to old spectral types. Note that for this composite cluster radius and density are essentially degenerate.

4. Evolution in cluster population gradients

These results from the low density outskirts of cluster reinforce the concept that clusters are not closed boxes, and that understanding even the core cluster properties requires an understanding of the structures from which the cluster was created. Studies of clusters at intermediate and high redshift suggest significant evolution in cluster properties since $z \sim 1$, which may be related to changing infall rates and the evolving field and infalling population more than evolution of the physical cluster environment.

The Butcher-Oemler effect is a classic measure of cluster evolution; the fraction of blue galaxies within clusters is seen to increase from a few percent at $z = 0$ to a substantial fraction of the cluster population at $z \sim 0.5$ and higher (e.g., Butcher & Oemler 1978, 1984; and many others). The origin of this effect can be constrained by examining the radial gradients in these clusters as a function of redshift. Figure 1 shows the fraction of spectroscopically-identified old (early-type spectra) population galaxies as a function of radius for clusters in the CNOC cluster sample (Canadian Network for Observational Cosmology; Yee, Ellingson & Carlberg 1996). The Butcher-Oemler effect is confirmed explicitly in the sample photometry; here the evolution is seen to be dominated by changes in the cluster outskirts. While the core regions appear to evolve very little, the gradients between 0.3 and 1.5 r_{200} are steeper at higher redshift. This suggests that cluster evolution even within the virial radius is most likely related to changes within the infalling population rather than physical properties of the core region.

Radial distributions of the galaxies in this sample show that the old population has a remarkably constant profile, while the remaining galaxies (star-forming and intermediate age populations) form a spatially broader profile that declines over time. This pattern

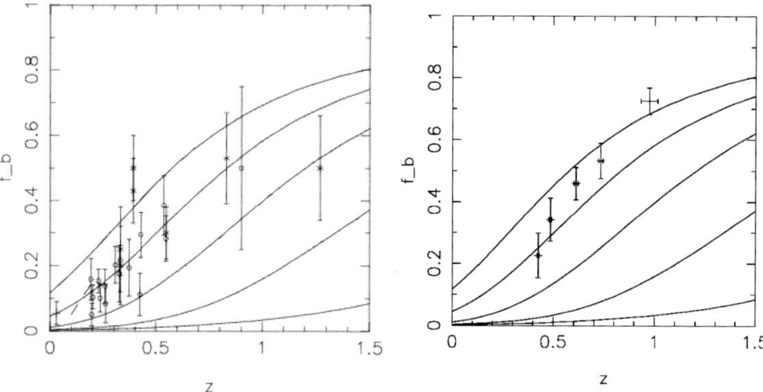

Figure 2. (a, left) The Butcher-Oemler effect from CNOC clusters and clusters gathered from the literature. Measures of galaxy properties, counting radii and magnitude limits are roughly similar for different investigations, but please note that the data are still somewhat inhomogeneous. Model curves are for expected infall rates and transformation timescales of 0.5, 1.0, 1.5, 2.0 and 2.5 Gyr, from lower to upper (see text for details). (b, right) Preliminary results from 85 clusters from the RCS survey with richness > Abell class 0. Average blue fractions in the highest redshift bin are corrected for the slight selection effect expected from a red-sequence-selected cluster sample; corrections at lower redshift are expected to be minimal.

mirrors that from dynamical studies of clusters at lower redshift (e.g. Biviano & Katgert 2004, in these proceedings), which identify spatially broad, bluer galaxy populations as newly infalling galaxies. These results suggest that the Butcher-Oemler effect is driven primarily by a decline in the cluster infall population over time.

Such a decline in the overall infall rates is expected for currently accepted cosmological models. Figure 2a shows the fraction of blue galaxies within the virial radius as a function of redshift for a number of clusters. Model curves show the expectation from extended Press-Schechter (Lacy & Cole 1993) infall rates through the cluster virial radii for a Λ–CDM cosmology. The infalling galaxy population is modeled after an evolving field population. After infall, galaxies are transformed from "blue" to "red" on a given exponential timescale. These models are very simple, but illustrate that the Butcher-Oemler effect is generally consistent with expected infall rates and a constant transformation timescale of about 2 Gyr. This timescale seems short; even with an abrupt halt to star formation at the virial radius, stellar evolution timescales would still require at least $1 - 1.5$ Gyr. Significant pre-processing of the infalling galaxies could (and probably does) lengthen this timescale to allow for one or two crossing-times to pass before star formation is completely quenched.

Figure 2b shows a similar plot using 85 clusters in the optically-selected cluster sample from the Red Sequence-Cluster Survey (Gladders & Yee 200). These results suggest an even steeper Butcher-Oemler effect at $z > 0.5$. While these results are very preliminary, they can be used to frame the pertinent questions. A steeper relation could mean that inner-cluster mechanisms are less efficient in quenching star formation in the infalling galaxies. This would correspond to a longer transformation timescale due to either a changing physical environment or more resilient galaxies with deeper gas reservoirs at higher redshifts. The other possibility is that less pre-processing is in effect at higher redshift.

Evolution in the morphology-density relation at lower densities indeed supports a scenario where at least part of the explanation is due to less pre-preprocessing. Morphological studies of intermediate redshift clusters (Dressler et al. 1997, Treu et al. 2003) show a significant morphology-density relation which, like the low-redshift relation, appears to be broadly consistent for a given local projected density, regardless of whether the galaxy is in the throes of the cluster core environment or in its outskirts. The intermediate redshift relation is offset towards a lower fraction of early-type galaxies at a given density relative to the $z = 0$ relation, particularly at the group/poor-cluster densities which form the likely infalling population.

However, linking this evolution specifically to the evolution of populations within clusters requires a more detailed understanding of the structures falling into clusters as a function of redshift. Tracing the low-density filamentary structures around clusters is a very difficult observational task, particularly at higher redshifts where foreground/background contamination dominates the galaxy counts. Photometric redshift techniques show significant promise in tracing this structure beyond the cluster core. Kodama et al. (2001, in these proceedings; see also Gray et al., and Li et al., in these proceedings) have traced groups and substructures around clusters at intermediate redshifts, confirming a correlation of galaxy colors with galaxy surface density. Such studies provide critical observational tests of hierarchical cluster formation.

5. Summary

Like their suburban counterparts, the outskirts of clusters are complicated regions where inner-city and rural populations blend. New wide-field surveys of low-redshift clusters have documented an extension of environmental effects on galaxies to very low density regions around clusters. These results underscore the need for detailed dynamical modeling of the clusters to several virial radii in order to disentangle galaxies falling into the cluster for the first time from those who have already visited the denser regions. The strong influence of local density on infalling galaxies provides a basis for evaluating the amount of pre-preprocessing in groups and lower-density structures, and the additional effects of mechanisms at work within the deep cluster potential. At this point, evidence for additional damage by the cluster is becoming well-established, but may be subordinate to the effects of local environment before and during the infall phase. If this is the case, then substructure must be retained to a significant extent during and after infall.

The evolution of cluster properties, particularly in the outskirts, appears to be strongly dependent on the infalling population, both in terms of infall rates and the properties of the infalling galaxies. As at low redshift, understanding the correlations of galaxy properties and local density at low densities is necessary for assessing the relative importance of environmental mechanisms inside and outside of the cluster virial regions. Currently our picture of cluster evolution is incomplete, with significant uncertainties in the infall rates, the amount of pre-processing, and the efficiency of the cluster core at diminishing star formation. New clusters samples at high redshift and ongoing work on measuring cluster properties to higher redshifts currently hold promise for providing the necessary context for tracing cluster evolution in a fully cosmological context.

References

Balogh, M. L., et al., 1997, ApJ, 527, 54
Balogh, M. L., Navarro. J., Morris, S. L., 2000, ApJ, 540, 113
Balogh, M. L., et al. 2002, MNRAS, 335, 10
Balogh, M. L., 2004, MNRAS, 348, 1355
Bower, R. G., Lucey, J. R., Ellis, R. S. 1992, MNRAS, 254, 601
Butcher, H., & Oemler, A., Jr. 1978, ApJ, 226, 559

Butcher, H., & Oemler, A., Jr. 1984, ApJ, 285, 426
Dressler, A. 1980, ApJ, 236, 351
Dressler, A., et al. 1997, ApJ, 490, 577
Ellingson, E., Lin, H., Yee, H. K. C., Carlberg, R. G., 2001, ApJ, 547, 609
Fujita, Y. 1998, ApJ, 509, 587
Giovanelli, M.P. & Haynes, M., 1985, ApJ, 292, 404
Gladders, M. D., Lopez-Cruz, O. Yee, H. K. C., Kodama, T., 1998, ApJ, 501, 571
Gladders M. D. & Yee, H. K. C., 2000, AJ, 120, 21
Gomez, P., et al., 2003, ApJ, 584, 210
Goto, T., et al, 2004, MNRAS, 348, 515
Kodama, T., et al. 2001, ApJ, 562, 9
Lacey, C. & Cole, S., 1993, MNRAS, 262, 627
Lavery, R. J., & Henry, J. P. 1988, ApJ, 330, 596
Lewis, I. et al., 2002, MNRAS, 334, 673
Mamon, G., 2004, A&A, 414, 445
Moore, B., et al., 1996, Nature, 379, 613
Sanchis, T., et al., 2004, A&A, 418, 393
Solanes, J. M., 2001, ApJ, 548, 97
Stanford, S. A., Eisenhardt, P. R., Dickinson, M., 1998, ApJ, 492, 461
Treu, T., et al., 2003, ApJ, 591, 53
van Dokkum, P. G., et al., 1998, ApJ, 504, L17
Whitmore, B. C., Gilmore, D. M., Jones, C. 1993, ApJ, 407, 489

Outskirts of Galaxy Clusters: Intense Life in the Suburbs
Proceedings IAU Colloquium No. 195, 2004
A. Diaferio, ed.

DOI: 10.1017/S1743921304000729

Galaxy clusters at low redshift

Michael L. Balogh

Department of Physics, University of Durham, South Road, Durham UK, DH1 3LE email: M.L.Balogh@Durham.ac.uk

Abstract. We present a new analysis of galaxy colours and star formation rates as a function of environment, from the Sloan Digital Sky Survey and the 2dF Galaxy Redshift Survey. The galaxy distribution is well characterised by only two distinct populations: one that is blue and star–forming, and one that is red and quiescent. The colour and Hα distribution of each of these populations is approximately independent of environment, while their relative abundances depend strongly on local galaxy density. Even in the lowest-density environments, a substantial ($\sim$ 30 per cent) population of galaxies have red colours and exhibt no detectable Hα emission. We compare these data with theoretical predictions of galaxy formation, properly accounting for observational selection biasses and uncertainties. We conclude that galaxy transformations must either be rapid (with a timescale $\tau \lesssim 0.5$ Gyr), or must occur preferentially at high redshift.

1. Introduction

It has been known for a long time that many (perhaps most) galaxy properties in the nearby Universe correlate strongly with their environment. Early work by Dressler (1980) and Postman & Geller (1984) showed that the distribution of galaxy morphologies depends strongly on local galaxy density, in the sense that early-type galaxies (E/S0) dominate in locally dense environments, while late-types (spiral and irregular galaxies) become increasingly important toward lower densities. An important result of this work was that the correlation is strongest with estimates of the *local* density, with little or no further dependence on the cluster type (i.e. degree of central concentration, or velocity dispersion of the system). In particular, Postman & Geller (1984) showed that the trend exists in small galaxy groups, as well as in clusters. Although there may be additional effects at work in the very cores of clusters (e.g. Whitmore et al. 1993; Domínguez et al. 2002), it is clear that *most* of the correlation is driven by the local density of galaxies (or, perhaps, dark matter: Gray et al. 2004, and these proceedings).

It remains an open question to what extent the different galaxy populations in clusters are due to the direct influence of their environment, and how much of it is due to factors related to their mass, or formation time. Part of the reason for this uncertainty is that even in the lowest density environments studied by Dressler (1980), the distribution of galaxy morphologies still differs markedly from the global average; studies of cluster outskirts are therefore required to connect the two environments, where one might expect galaxy transformations may be taking place (Treu et al. 2003, and these proceedings). The advent of large redshift surveys like the Sloan Digital Sky Survey (SDSS, York et al. 2000) and 2dF Galaxy Redshift Survey (2dFGRS, Colless et al. 2003) make it possible to do this at low redshift for the first time.

Unfortunately, galaxy morphology is notoriously difficult to quantify in an objective manner; in particular, no automated method has proven effective at distinguishing elliptical galaxies from S0 galaxies. Nor is it clear that morphology is in any way the fundamental parameter of interest, as it correlates strongly with luminosity (Marzke et al. 1994; Cuesta-Bolao & Serna 2003), colour and SFR (Kennicutt 1992), among other quantities. Since the latter properties are both easier to measure and easier to relate to theoretical predictions, more studies have focussed on the environmental dependence of these properties, most notably galaxy colours (e.g. Butcher & Oemler 1984; Kodama & Bower 2001) star formation rates (SFRs, e.g. Balogh et al. 1997) and HI content (Solanes et al. 2001b; van Gorkom 2004, see also contributions from M. Verheijen and J. Solanes, in this volume). These studies and many more have shown that galaxies in dense environments tend to be redder, with lower current star formation rates and low amounts of HI gas for their mass. Although these trends are partly consistent with the morphology–density relation, there is evidence that galaxies of a fixed morphological type have lower star formation rates (Balogh et al. 1998; Poggianti et al. 1999) and less neutral hydrogen (Solanes et al. 2001a) in clusters. In general these trends vary continuously with local density, with little or no dependence on larger scale structures. For example, there is little or no trend in the global fraction of blue galaxies with cluster velocity dispersion (De Propris et al. 2004; Margoniner et al. 2001).

These trends suggest that galaxies in dense environments are starved of the raw materials needed to make new stars (i.e. cold gas). Many mechanisms have been proposed to cause this, including ram-pressure stripping (Gunn & Gott 1972), galaxy harassment (Moore et al. 1999) and strangulation (the stripping of a hot gas halo, Larson et al. 1980; Balogh et al. 2000). In particular, the strong evolution seen in the cluster population suggests such processes may have occurred in the recent past (Dressler et al. 1997, and these proceedings). Further clues lie in the exceptions to the general trend of low activity in dense environments. For example, Moss & Whittle (1993, 2000) find that early–type spiral galaxies in clusters actually show stronger Hα emission than galaxies of the same morphological type in the field. Far-infrared observations suggest that some clusters may contain a population of dusty starburst galaxies in excess of that expected in the field (Duc et al. 2002, and contribution to these proceedings), and the presence of k+a galaxies in the Coma cluster are also indicative of recent starbursts that may have been induced through interactions with neighbouring galaxies or the cluster gas (Poggianti et al. 2004, and contribution to these proceedings). These observations suggest that interactions may temporarily enhance star formation, leading to rapid consumption of cold gas. Ultimately (over ~ 1 Gyr timescales), the net result is a lack of fuel and a low star formation rate.

Until recently, most comprehensive cluster studies have been at moderate redshift, where the smaller angular size of the virialised cluster region facilitates full coverage. The 2dFGRS and SDSS now make it possible to study clusters over arbitrarily wide fields at low redshift, and we have taken advantage of this to trace the trend in Hα emission line strength (directly related to SFR) over the full dynamic range of galaxy densities in the local Universe (Lewis et al. 2002; Gomez et al. 2003). Here, the local density Σ_5 is based on the projected distance to the fifth–nearest neighbour, within ± 1000km s^{-1} and brighter than $M_r \sim -20$. The average emission line strength was observed to increase strongly with decreasing Σ_5, in a way holds both within clusters as well as outside them. The trend shows evidence for a critical density at $\Sigma_5 \sim 1$ Mpc^{-2}, below which the correlation with environment is much weaker. This density is typical of cluster infall regions, and the observed break suggests that there may be additional physics at work in these environments. Here, we report on a continuation of this work, based on an analysis of colour and SFR distributions in the 2dFGRS and SDSS, and make some initial comparisons with simple models of galaxy formation to offer an interpretation of the results.

2. The colour distribution

Galaxy colours are a simple, useful diagnostic of the galaxy population that can be easily traced to higher redshifts (e.g. Butcher & Oemler 1984, see also contributions from T. Kodama and F. Nakata in these proceedings). In particular, the fraction of blue galaxies, as defined by a simple, magnitude–dependent colour cut, is known to increase strongly with decreasing galaxy density (e.g. Margoniner & de Carvalho 2000; De Propris et al. 2004). Using data from the SDSS, Baldry et al. (2004) showed that the galaxy colour distribution is well parametrised by a simple, two-component model where each population is a Gaussian. This allows us to divide the population in an objective way that is insensitive to systematic and statistical uncertainties in the data. The mean and dispersion of each distribution is a strong function of luminosity; in particular, brighter galaxies are redder, an effect that may be partly due to trends in metallicity and dust with stellar mass (Brinchmann et al. 2003).

Recently, we have investigated how these model parameters depend on environment (Balogh et al. 2004a), using the first data release of the SDSS (Abazajian et al. 2003). We select a magnitude–limited sample of 24346 galaxies with $z < 0.08$, but we still base our measurement of Σ_5 on the number of galaxies (within a ± 1000 km s^{-1} redshift slice) brighter than $M_r = -20$ to ensure a uniform definition over the full redshift range.

The dominant trend with environment in the colour distribution is the change in the relative number of galaxies in the blue and red populations. The fraction of red galaxies at fixed luminosity increases from 10–30 per cent to ~ 70 per cent toward denser environments. Furthermore, this increase is seen for galaxies at all luminosities, which implies that whatever mechanism is responsible for establishing the correlation must affect bright and faint galaxies in a similar way. We also use the C4 cluster catalogue of Nichol, Miller et al. (in preparation) to test whether the global environment of a surrounding cluster has an additional influence on the colour distribution. Considering all galaxies within the virial radius, the fraction of red galaxies (at fixed luminosity) appears to be independent of velocity dispersion, although the uncertainties are large, due to the relatively small sample of clusters. We conclude, as others have before us (e.g. Dressler 1980; Fairley et al. 2002; De Propris et al. 2004), that the population composition depends primarily on the local environment, and is much less sensitive to the large scale structure.

In contrast, the mean and dispersion of each Gaussian distribution are approximately independent of environment, which is surprising given the large change in their relative abundance. For example, if the increased abundance of red galaxies were due to interactions (e.g. mergers and harassment) that increase in frequency with local galaxy density, we might have expected the blue distribution to become increasingly non-Gaussian with increasing density, and to gradually blend into the red distribution. The fact that this is not observed means that any such transformation must either be sufficiently rapid, or sufficiently rare, to keep the overall colour distribution unchanged. This has the important implication that galaxy clusters do not inhibit star formation in *all* member galaxies, and we conclude that the properties of star-forming galaxies (dust, metallicity, recent SFR history) are determined primarily by the internal structure of the galaxy, rather than by their environment.

To quantify a limit to the rate of this transformation process, we investigate a simple evolutionary scenario, using the Bruzual & Charlot (2003) stellar population models, to reproduce the colour distribution of galaxies in dense regions ($\Sigma_5 > 1$ Mpc^{-2}), assuming that the red population results from quenching of star formation in bluer field galaxies of the same luminosity, at a uniform rate over a Hubble time. If the star formation is terminated abruptly (the "truncation" scenario) the simple 2–Gaussian model is hardly affected. This is because the colour evolution is rapid, and the number of galaxies caught in transformation at the present day is small (~ 1 per cent). Therefore, the success of the simple 2–Gaussian model at most luminosities and environments does not rule out the possibility that galaxy transformations occur on short timescales. On the other hand, if star formation is allowed to decay with an exponential timescale $0.5 < \tau < 4$ Gyr (as expected in the strangulation scenario Larson et al. 1980; Balogh et al. 2000), the number of galaxies with intermediate colours at the present day would increase by a factor ~ 2. This would clearly distort the observed distribution in a way that is inconsistent with the observations.

We therefore conclude that short-timescale transformations could play a role at all densities and luminosities, without disrupting the Gaussian model fit. The small fraction of galaxies predicted to be in the transition phase is comparable to the fraction of spectroscopically identified post-starburst galaxies (Goto et al. 2003a; Quintero et al. 2004) which may be the signature of such transformations (e.g. Dressler & Gunn 1992, see also contributions from A. Dressler, B. Poggianti and K.-V. Tran, in this volume). Another possible candidate is the class of anemic (passively evolving) spirals (van den Bergh 1976; Goto et al. 2003b) which might arise because morphology changes on a longer timescale than the instantaneous SFR. However, we caution that our constraint on the timescale is sensitive to the assumption that the transformations occur uniformly in time. If, instead, transitions were more common in the past, the data would accomodate a slower rate of SFR decay.

3. Star formation rates

With uncertainties $\lesssim 1$Å on line equivalent widths, the distribution of $W_0(H\alpha)$ also exhibits a bimodality (Gomez et al. 2003). Although the decomposition is not as simply characterised as for the colours, we can approximately divide the sample into star–forming galaxies with $W_0(H\alpha)>$ 4Å, and the non–star–forming population with lower equivalent widths. We recently used the 2dFGRS and SDSS to investigate the environmental dependence of this distribution (Balogh et al. 2004b) and found, as expected, that the fraction of galaxies with emission lines declines strongly with increasing density. This correlation persists at *all* densities, and never asymptotes to a value that is independent of density. Therefore there is no sharp transition, although at densities greater than $\Sigma_5 = 1$ Mpc^{-2} the emission–line fraction does begin to decrease more strongly. Interestingly, however, we found that the observed distribution of $W_0(H\alpha)$ above 4Å is *independent* of local density. Therefore, relative SFRs are not low in clusters, among the star–forming population. This is consistent with the colour analysis, above, and with the fact that both the scatter in the Tully–Fisher relation (e.g. Giuricin et al. 1986; Biviano et al. 1990) and the shape of the UV luminosity function (Cortese et al. 2003) are independent of environment. The result is still surprising, given that these blue galaxies are actively growing and might be thought to be most susceptible to external influences.

Even at the lowest densities, the fraction of emission line (or blue) galaxies is not 100 per cent. We selected a sample of isolated galaxies, based on three criteria. First, we measure the 3-D galaxy density using two smoothing Gaussian kernels with standard deviations of 1.1 and 5.5 Mpc, respectively, and select galaxies in the lowest few per cent of *both* densities. Furthermore, we require that these isolated galaxies do not belong to any group in the comprehensive catalogue of Eke et al. (2004). Even among the brightest of these isolated galaxies, ~ 30 per cent show $W_0(H\alpha)<$ 4 Å, and red colours. Therefore, present-day local environment may not be the only driver of galaxy evolution, since star formation has shut off in these isolated galaxies. One possibility is that they are fossil groups (Ponman et al. 1994; Mulchaey & Zabludoff 1999; Romer et al. 2000), which have completely merged to form a single galaxy.

4. Model predictions

The results above could arise in one of at least two ways. One possibility is that the stellar mass of the galaxy determines its properties, and that the stellar mass function varies with environment (as suggested by differences in the K−band luminosity functions of cluster and field galaxies, e.g. Balogh et al. 2001; De Propris et al. 2003). The second possibility is that the age distribution at fixed stellar mass depends on environment, implying an environmentally–induced transformation. Kauffmann et al. (2004) have derived stellar masses from the SDSS galaxy sample, and concluded that this is indeed what happens.

Although we do not have stellar masses for the 2dFGRS sample, we can interpret our results by comparison with mock catalogues generated from the updated model predictions of Cole et al. (2000, GALFORM), incorporating the survey selection function and observational biasses. The models we consider are similar to those in Cole et al. (2000) in most respects. The most significant differences are that we use a lower baryon fraction as required by other observations (e.g. Burles et al. 2001) and include the effects of superwind feedback (Benson et al. 2003).

The main success of the model is that it reproduces the observed trend of emission–line fraction on local density extremely well, over two orders of magnitude in density (Figure 1a). Although the agreement in normalisation results partly from the choice of parameters in the dust model, the trend with density is a robust prediction and the good match to the data is remarkable. In Figure 1b we show the $W_0(H\alpha)$ distribution, for $W_0(H\alpha)> 4$Å, in three environments. Below $\Sigma_5 \sim 1$ Mpc^{-2} (the critical density identified by Lewis et al. 2002), the predicted $W_0(H\alpha)$ distribution of star-forming galaxies does not depend on local density, in agreement with the observations in Balogh et al. (2004b). At the highest densities, however, the entire $W_0(H\alpha)$ distribution shifts to lower values, in contrast with the observational result.

These predicted trends are not due to a change in the stellar mass function with environment. The superwind feedback limits the number of very massive galaxies by expelling gas from haloes at early times, which results in a mass function that is nearly independent of environment. Therefore the successful prediction of the environmental trend is due to the only transformation mechanism present in these models: strangulation. However, the change in predicted $W_0(H\alpha)$ distribution at the highest densities implies that the timescale for this transformation may be shorter than predicted.

5. Conclusions

Broadly speaking, galaxy populations are simply characterised by two distinct classes: one that is blue and star–forming, and one that is red and quiescent. The most surprising result of this study is that the colour and Hα distribution of each of these populations is approximately independent of environment in the local Universe. This suggests that SFRs, dust and metallicity distributions are determined primarily by individual galaxy types (e.g. stellar mass) rather than by their surroundings. The fraction of red galaxies increases strongly with increasing density, which is likely the result of transformations that are either rapid, or occur preferentially at high redshift.

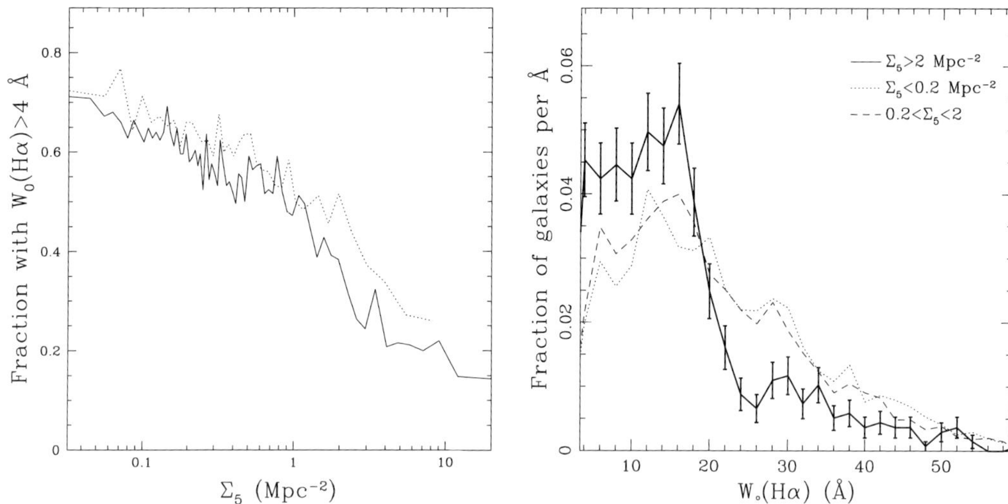

Figure 1. a) Left: The fraction of galaxies with $W_0(H\alpha)>$ 4Å, as a function of local, projected density Σ_5. The *solid line* shows the GALFORM model predictions, and the *dashed line* represents the 2dFGRS data, from Balogh et al. (2004b). There are 250 galaxies in each bin, so the Poisson-distributed uncertainties are typically $\sim$ 0.05. **b) Right:** The predicted $W_0(H\alpha)$ distribution for star-forming galaxies in the GALFORM model, with $W_0(H\alpha)>$ 4Å. The *solid*, *dashed* and *dotted* lines represent galaxies in low, intermediate, and high density environments, respectively. We show Poisson-distributed error bars on the high-density subsample, which is the smallest of the three. The change in the shape at highest densities is not observed in the 2dFGRS or SDSS data (Balogh et al. 2004b).

Acknowledgements

I would like to thank all of my collaborators whose work, both published and unpublished, is represented here. In particular, R. Bower, R. Nichol, I. Baldry, C. Miller, K. Glazebrook, V. Eke, I. Lewis, A. Gray, C. Baugh, S. Cole and C. Frenk and the 2dFGRS team have all contributed to these results.

References

Abazajian, K. et al. 2003, AJ, 126, 2081

Baldry, I. K., Glazebrook, K., Brinkmann, J., Ivezić, Z., Lupton, R. H., Nichol, R. C., & Szalay, A. S. 2004, ApJ, 600, 681

Balogh, M. L., Baldry, I. K., Nichol, R. C., Miller, C., Bower, R. G., & Glazebrook, K. 2004a, ApJL, submitted

Balogh, M. L., Christlein, D., Zabludoff, A. I., & Zaritsky, D. 2001, ApJ, 557, 117

Balogh, M. L., Morris, S. L., Yee, H. K. C., Carlberg, R. G., & Ellingson, E. 1997, ApJL, 488, 75

Balogh, M. L., Navarro, J. F., & Morris, S. L. 2000, ApJ, 540, 113

Balogh, M. L., Schade, D., Morris, S. L., Yee, H. K. C., Carlberg, R. G., & Ellingson, E. 1998, ApJL, 504, L75

Balogh, M. L. et al. 2004b, MNRAS, 348, 1355

Benson, A. J., Bower, R. G., Frenk, C. S., Lacey, C. G., Baugh, C. M., & Cole, S. 2003, ApJ, 599, 38

Biviano, A., Giuricin, G., Mardirossian, F., & Mezzetti, M. 1990, ApJS, 74, 325

Brinchmann, J., Charlot, S., et al. 2003, MNRAS, in press, astro-ph/0311060

Bruzual, G. & Charlot, S. 2003, MNRAS, 344, 1000

Burles, S., Nollett, K. M., & Turner, M. S. 2001, ApJL, 552, L1

Butcher, H. & Oemler, A. 1984, ApJ, 285, 426
Cole, S., Lacey, C. G., Baugh, C. M., & Frenk, C. S. 2000, MNRAS, 319, 168
Colless, M. et al. 2003, astro-ph/0306581
Cortese, L., Gavazzi, G., Boselli, A., Iglesias-Paramo, J., Donas, J., & Milliard, B. 2003, A&A, 410, L25
Cuesta-Bolao, M. J. & Serna, A. 2003, A&A, 405, 917
De Propris, R. et al. 2003, MNRAS, 342, 725
—. 2004, MNRAS, in press, astro-ph/0402642
Domínguez, M. J., Zandivarez, A. A., Martínez, H. J., Merchán, M. E., Muriel, H., & Lambas, D. G. 2002, MNRAS, 335, 825
Dressler, A. 1980, ApJ, 236, 351
Dressler, A. & Gunn, J. E. 1992, ApJS, 78, 1
Dressler, A., Oemler, A. J., Couch, W. J., Smail, I., Ellis, R. S., Barger, A., Butcher, H., Poggianti, B. M., & Sharples, R. M. 1997, ApJ, 490, 577
Duc, P.-A., Poggianti, B. M., et al. 2002, A&A, 382, 60
Eke, V. R. et al. 2004, MNRAS, 348, 866
Fairley, B. W. et al. 2002, MNRAS, 330, 755
Giuricin, G., Mezzetti, M., & Mardirossian, F. 1986, A&A, 157, 400
Gomez, P. L., Nichol, R. C., et al. 2003, ApJ, 584, 210
Goto, T., Nichol, R. C., et al. 2003a, PASJ, 55, 771
Goto, T., Okamura, S., et al. 2003b, PASJ, 55, 757
Gray, M. E., Wolf, C., et al. 2004, MNRAS, 347, L73
Gunn, J. E. & Gott, J. R. 1972, ApJ, 176, 1
Kauffmann, G., White, S. D. M., Heckman, T. M., Menard, B., Brinchmann, J., Charlot, S., Tremonti, C., & Brinkmann, J. 2004, MNRAS, submitted, astro-ph/0402030
Kennicutt, R. C. 1992, ApJ, 388, 310
Kodama, T. & Bower, R. G. 2001, MNRAS, 321, 18
Larson, R. B., Tinsley, B. M., & Caldwell, C. N. 1980, ApJ, 237, 692
Lewis, I., Balogh, M., et al. 2002, MNRAS, 334, 673
Margoniner, V. E. & de Carvalho, R. R. 2000, AJ, 119, 1562
Margoniner, V. E., de Carvalho, R. R., Gal, R. R., & Djorgovski, S. G. 2001, ApJL, 548, L143
Marzke, R. O., Geller, M. J., Huchra, J. P., & Corwin, H. G. 1994, AJ, 108, 437
Moore, B., Lake, G., Quinn, T., & Stadel, J. 1999, MNRAS, 304, 465
Moss, C. & Whittle, M. 1993, ApJL, 407, L17
—. 2000, MNRAS, 317, 667
Mulchaey, J. S. & Zabludoff, A. I. 1999, ApJ, 514, 133
Poggianti, B. M., Bridges, T. J., Komiyama, Y., Yagi, M., Carter, D., Mobasher, B., Okamura, S., & Kashikawa, N. 2004, ApJ, 601, 197
Poggianti, B. M., Smail, I., Dressler, A., Couch, W. J., Barger, A. J., Butcher, H., Ellis, R. S., & Oemler, A. J. 1999, ApJ, 518, 576
Ponman, T. J., Allan, D. J., et al. 1994, Nature, 369, 462
Postman, M. & Geller, M. J. 1984, ApJ, 281, 95
Quintero, A. D., Hogg, D. W., et al. 2004, ApJ, 602, 190
Romer, A. K., Nichol, R. C., et al. 2000, ApJS, 126, 209
Solanes, J., Manrique, A., García-Gómez, C., González-Casado, G., Giovanelli, R., & Haynes, M. P. 2001a, ApJ, 548, 97
Solanes, J., Manrique, A., García-Gómez, C., et al. 2001b, ApJ, 548, 97
Treu, T., Ellis, R. S., et al. 2003, ApJ, 591, 53
van den Bergh, S. 1976, ApJ, 206, 883
van Gorkom, J. H. 2004, in Clusters of Galaxies: Probes of Cosmological Structure and Galaxy Evolution, from the Carnegie Observatories Centennial Symposia. Published by Cambridge University Press, as part of the Carnegie Observatories Astrophysics Series. Edited by J.S. Mulchaey, A. Dressler, and A. Oemler, 2004, p. 306
Whitmore, B. C., Gilmore, D. M., & Jones, C. 1993, ApJ, 407, 489
York, D. G., Adelman, J., et al. 2000, AJ, 120, 1579

Outskirts of Galaxy Clusters: Intense Life in the Suburbs
Proceedings IAU Colloquium No. 195, 2004
A. Diaferio, ed.

DOI: 10.1017/S1743921304000730

Infall and starbursts in z $\sim$ 0.5 clusters

Alan Dressler

Carnegie Observatories, 813 Santa Barbara Street, Pasadena, CA 91101

Abstract. In this paper I address three topics. The first regards the use of composite spectra as a tool to study the spectral evolution of galaxies in intermediate-redshift clusters. This technique is used to demonstrate the robustness of measurements of strong Hδ absorption, indicative of mild-to-moderate starbursts, in a significant fraction of galaxies in these clusters. Second, I present some spectroscopic data on Abell 851 ($z = 0.41$) that suggests that the starbursts in this cluster have been triggered by tidal encounters in infalling groups in the cluster. Third, I briefly describe the IMACS imaging-spectrograph, which has been built for the Magellan-Baade telescope, and, following the idea that the outskirts of clusters are "where the action is," describe a large spectroscopic program to study "cluster building" using IMACS.

The results reported here are the work of many people, and are described in detail in Dressler *et al.* (2004) and Oemler *et al.*(2004). The author would appreciate citations to these papers, with the addition, if appropriate, of "as reported in Dressler (2004)" – this conference.

1. Introduction

Butcher & Oemler's (1978) report of a significant number of star forming galaxies in intermediate-redshift clusters, in contrast to the largely dormant cluster galaxies of the present era, was quite a surprise given the relatively small lookback time of only 4-5 Gyr. They speculated, correctly it now seems, that the star formation they were seeing was associated with disk galaxies, rather than ellipticals, which they postulated were already long residents of these clusters. Further, they suggested that these disk galaxies were likely infalling spirals whose fate was to add to the large S0 population. The mechanism most frequently (almost exclusively) credited with this conversion was ram-pressure stripping Gunn & Gott (1972), a process that would connect strongly to the very core of the cluster — a global process, as opposed to a local one involving interactions between galaxies. Perhaps equally surprising then, was the discovery by Gunn and myself (Dressler & Gunn 1983) that the spectra of these star forming galaxies were typically those of starbursts and post-starbursts, rather than spectra typical of spirals undergoing continuous, relatively constant star formation. Of course, the spectra of a truncated star formation history is qualitatively similar to a post-starburst spectrum, but quantitative estimates of the strength of the bursts suggested that these were significant enhancements in the star formation rates and that most (all?) of the infalling, starforming galaxies seem to be involved.

The Morphs group greatly expanded the spectroscopy of galaxies in a collection of 10 clusters at $0.35 < z < 0.55$, and found, in support of the early work, a $\sim 20-30\%$ fraction of cluster galaxies that have strong Hδ indicative of subsiding or recent starbursts (Dressler *et al.* 1997). However, particularly puzzling was the absence of bright ongoing starbursts that could account for the later-stage examples. A solution to this problem, proposed by Poggianti *et al.* (1999), is that one class of these galaxies, those with strong Hδ and some [O II], are in fact dusty starbursts whose true burst strength is hidden. A variety of evidence now supports this view, including data from the ISO satellite discussed

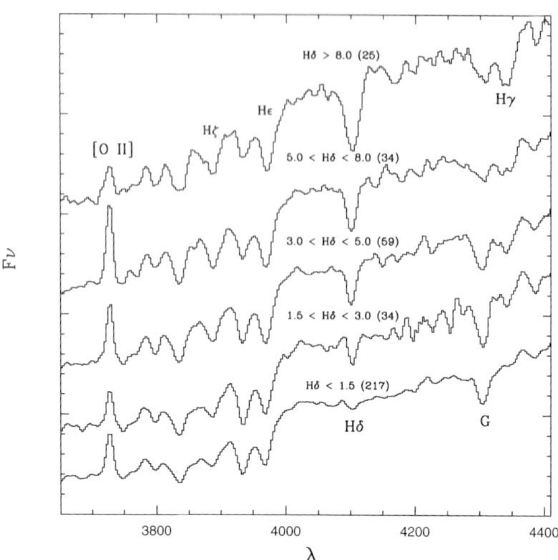

Figure 1. Composite spectra for Morphs sample galaxies with Hδ absorption recorded in 5 ranges of strength. The clear trend of increasing Hδ strength and corresponding other members of the Balmer series (note, in particular, Hγ, Hϵ, and Hζ marked on top spectrum) shows the reality of the Hδ measurement, even in the presence of substantial noise in the continuum in individual spectra.

at this conference and the detection of these many sources in the radio continuum (e.g., Glenn Morrison's PhD thesis).

2. Using composite spectra to compare stellar populations

Also in 1999, however, came a challenge to the Morphs claim of a large fraction of starbursts from Balogh *et al.* (1999) whose analysis of the CNOC cluster sample failed to show a large fraction of galaxies with strong Hδ. From the perspective of several years on, it is easy to see a number of reasons for this difference, which are discussed in Dressler *et al.* (2004, hereinafter D04). Regardless, a consequence of the attempts to reconcile these two studies was a realization by the Morphs that the relatively low S/N of spectra in their and the CNOC study, typically ∼10 per resolution element, makes the comparison of results problematical, especially when different studies adopt different schemes of spectral classification, as was the case here. As a remedy, we decided to form composite spectra of each of our clusters, and to compare them with composite spectra of a present-epoch cluster sample assembled by Shectman and myself in the mid-1980s. As described in D04, we coadded the spectra after normalizing them and weighting them by luminosity (by their photometric magnitudes), so as to similulate as closely as possible a composite stellar population for the whole cluster.

As expected, the S/N of these spectra is vastly improved over individual spectra. This allowed us to answer one question raised by Balogh *et al.*, whether a significant number of the detections of Hδ in the Morphs study might be the result of the high noise level. As shown in Figure 1, by forming composites of the different spectra as a function of Hδ strength, the resulting spectra show a clear sequence of increasing contribution from A stars. In particular, not just Hδ, but the other Balmer absorption lines, are evident in the spectra. The Morphs measurements of Hδ, though often noisy, are robust.

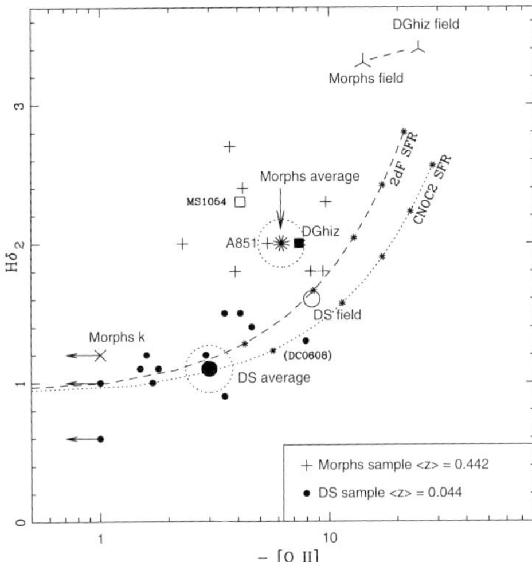

Figure 2. Measurements of EW([O II]) and EW(Hδ) for field and cluster samples There is a clear separation of the "mean" values of Hδ for the Morphs and the DS (low-redshift) samples, but there is also no overlap in the individual measurements for the clusters studied. As discussed in Dressler *et al.* (2004), the two lines represent different mixes of passive galaxies and continuously star–forming galaxies with [O II] distributions from the 2dF Survey (dashed line) and the CNOC survey (dotted line). Although such mixes adequately describe the low–redshift DS clusters and field, they fail to match any of the higher–redshift cluster or field samples, demonstrating the importance of starbursts in these galaxy populations.

With such composite spectra, cluster-by-cluster, we compare the stellar populations, as characterized by [O II] and Hδ, in Figure 2. The intermediate–redshift clusters all average Hδ strengths of ∼2.0Å and have substantial [O II] emission varying between -3Å and -10Å equivalent width; both are indicative of substantial ongoing star formation. In contrast, each of the low–redshift clusters shows much weaker EW(Hδ) ∼ 1.3Å— this is not just 35% weaker, but typical of a population with no significant star formation for several Gyr — and much weaker EW([O II]) = -2Å to -5Å. This difference in spectral characteristics appears to be primarily dependent on epoch rather than a function of cluster properties, but of course, the sample is still too small to be sure.

Figure 2 also plots two curves representing the locus of galaxies with histories of continuous star formation, added in different amounts to passively-evolving galaxies, formed as explained in D04. These two curves, which represent a stronger (CNOC "field") and weaker (2dF "group") star formation rate, describe the $z = 0$ local sample of field and cluster galaxies very well, but fail to account the location of points representing the $z > 0.35$ clusters and field. We take this as strong evidence that these intermediate-redshift populations contain a significant fraction of starburst galaxies not common today (among luminous galaxies), and that this represents an important departure in the mode of star formation in the earlier universe. We note that higher gas fractions in the star forming galaxies, and greater interactions in an earlier epoch of a lambda-dominated universe, may contribute to this difference.

The starburst galaxies observed in the Morphs sample have the morphology of disk galaxies, albeit disturbed disk galaxies, in 80-90% of the cases. It is reasonable, then, to imagine that the starburst is involved in the conversion of a spiral galaxy to an S0, although whether a starburst can be strong enough to accomplish this by itself is unknown at this time. What does seem clear, though, is that a simple truncation of star formation, the picture favored by the CNOC collaboration in their modeling of [O II] strengths as a function of radius from the cluster center (Ellingson *et al.* 2001) doesn't tell the whole story with respect to the Morphs observations. This highlights the value of collecting high S/N spectra with sufficient resolution to measure absorption as well as emission lines.

3. Spectroscopic results for Abell 851: early evidence that the triggering of starbursts happens in the cluster outskirts

These results from the Morphs studies, for the most part, come from the inner parts of these clusters, roughly out to $R \sim 1$ Mpc. To identify the mechanism(s) responsible for the starbursts it is sensible to push such observations as far out into the cluster environment as possible, due to the different environmental dependences of these proposed processes (Treu *et al.* 2003). Abell 851 is a very rich clusteri with several distinct centers in both optical and X-ray images, evidence which points to a propitious moment in its history when subclumps are merging to form a much larger, relaxed cluster. Also, as shown by Kodama *et al.* (2001), and much exhibited at this meeting, A851 has an obvious structure of infalling groups along what appears to be a rather filametary structure.

With new spectroscopic observations obtained with COSMIC on the 200-inch Hale telescope over the last few years, we have pushed into the outskirts of this cluster. In Figure 3 we abstract those results by showing the distribution of different spectral classes reaching out to $R \sim 3$ Mpc from the cluster core. For the region covered by our mosaic of 3×3 HST-WFPC2 frames, we find a complete mix of the different spectral types, which is to say, there is no decline in the percentage of starburst galaxies as we move further from the strongest central concentration (and principal X-ray center) of the cluster. Given the chaotic state of this cluster, and the substantial distances from the cluster center where these starbursts are found, it becomes increasingly less likely that global processes — associated with the cluster's dark matter or gas distribution, for example (i.e., ram-pressure stripping or harassment) — have triggered these starbursts.

The smoking gun, so to speak, of this argument, is the nature of the furthest cluster members found out in filaments to the north-west, which lie almost 3 Mpc from the cluster center, seen clearly in the Kodama *et al.* map. As can be seen from the coding in Figure 3, 7 out of 9 members of the top filament are starburst galaxies, a remarkable result. (Just below is another apparent structure with 4/6 starbursts.) Note that most of these are e(a) galaxies, the kind that the Morphs identify as the active, highly obscured starbursts *in progress*, which would be exactly as expected if the triggering is happening here, far from the cluster core. Our tentative conclusion, based on this limited but significant example, is that tidal interactions on these galaxies, from their neighbors and from "in situ" cluster galaxies, are triggering these gas-rich spirals to undergo substantial bursts of star formation.

The details of these observations for A851, and our interpretation, are discussed in Oemler *et al.* (2004).

4. ICBS: using IMACS to study the outskirts of distant galaxy clusters

These results on A851 are encouraging in regard to finally identifying a major cause of galaxy transformation due to environment. Gus Oemler, Mike Gladders, Bianca Poggianti, and myself, are collaborating to make much more extensive observations (in radial

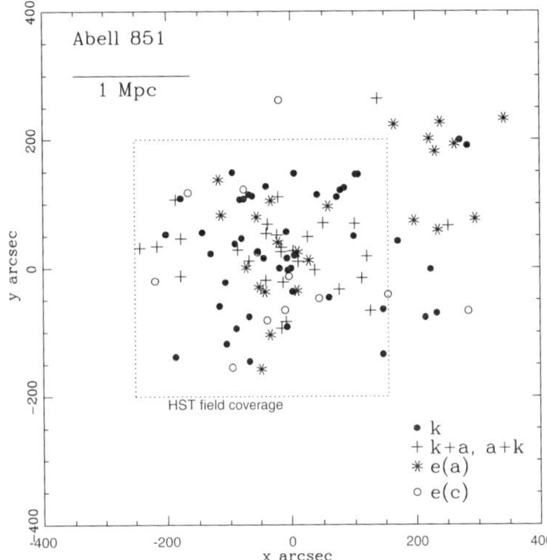

Figure 3. Distribution of the spectral types of 113 cluster members in the field of Abell 851. Within the area covered by the HST WFPC2 mosaic, shown by the dotted line, passive galaxies – k, continuously star-forming galaxies – e(c), and active- and post-starbursts – e(a) & k+a, a+k, are well mixed. There is no decline in the fraction of starbursts with cluster radius. For the filamentary structures to the northwest (upper right), active starbursts are the most common spectral type that has been detected, suggesting that the triggering mechanism of starbursts has occurred at least this far from the center of Abell 851.

distance and in numbers of cluster galaxies recovered) for a statistically significant number of intermediate-redshift clusters.

We are fortunate to have an excellent tool for this job. The Inamori-Magellan Areal Camera and Spectrograph has recently gone into routine operation on the Magellan-Baade 6.5-m telescope. IMACS is a re-imaging camera/spectrograph with a huge field, by big-telescope standards, of almost 1/2-degree in diameter, and the capacity to place multislits on hundreds of targets per exposure. This makes feasible an ambitious program of gathering a thousand or more cluster members in fields of these clusters, which we call the IMACS Cluster Building Survey.

Our strategy for this survey is, of course, to push as far out in the clusters as possible: the IMACS field covers typically out to $R \sim 5$ Mpc for $0.35 < R < 1.2$, which will be our target redshift range. However, we want to do more than study the behavior of individual clusters, but also follow the process of galaxy accretion and possible interactions over the principal epochs of cluster growth, eventually out to (at least) $z \sim 1.2$. Cluster simulations by Gladders, using the Hydra Consortium code, suggest that the galaxies that end up well-concentrated in clusters at $z = 0.3$ cover the full volume $R < 5$ Mpc at $z = 1.2$. To follow the cluster-building process, moreover, we must ensure that we are making an "apples to apples" comparison, that is, we must choose clusters at $z = 1.2$ that are progenitors of the clusters we will study at $z \sim 0.8$ and $z \sim 0.4$. We do this by selecting from equal volumes of the universe and selecting the most massive clusters at each epoch, making sure that the volume is not too large such that one-of-a-kind clusters are likely to creep into the sample.

The clusters will be chosen from the Red Sequence Cluster Survey (Gladders & Yee 2004) and Sloan Digital Sky Survey. Our initial goal is 5 clusters at $0.35 < z < 0.50$ and 5 clusters at $0.75 < z < 0.85$. With the intensive effort that we think will be necessary

in order to reach firmer conclusions. In this sense, SPITZER can provide excellent data for this purpose.

References

Altieri, B., Metcalfe, L., Kneib, J-P. et al. 1999, A&A 343, 65

Barvainis., R., Antonucci R., Helou, G. 1999, ApJ 118, 645

Biviano, A., Metcalfe, L., McBreen, B. et al. 2004, accepted for publication in A&A, astro-ph/0406391 (Paper II)

Coia, D., McBreen, B., Metcalfe, L. et al. 2004, submitted to A&A (Paper III)

Czoske, O., Moore, B., Kneib, J.-P., & Soucail, G. 2002, A&A, 386, 31

Duc, P.-A., Poggianti, B.M., Fadda, D. et al. 2002, A&A 382, 60

Elbaz, D., Cesarsky, C.J., Chanial, P. et al. 2002, A&A 384, 848

Granato, G. L., Silva, L., Bressan, A., Lacey, C. G., Baugh, C. M., Cole, S., & Frenk, C. S. 2001, ApSSS, 277, 79

Kennicutt, R. C., 1998, ARA&A, 36, 189

Kessler, M.F., Steinz, J.A., Anderegg, M. et al. 1996, A&A 315, L27

Lémonon, L., Pierre, M., Cesarsky, C.J. et al. 1998, A&A 334, L21

Metcalfe, L., Kneib, J.-P.; McBreen, B.; et al. 2003, A&A, 407, 791 (Paper I)

Silva, L., Granato, G. L., Bressan, A., & Danese, L. 1998, ApJ, 509, 103

Smail, I., Kuntschner, H., Kodama, T. et al. 2001, MNRAS 323, 839

Treu, T., Ellis, R. S., Kneib, J., Dressler, A., Smail, I., Czoske, O., Oemler, A., & Natarajan, P. 2003, ApJ, 591, 53

Outskirts of Galaxy Clusters: Intense Life in the Suburbs
Proceedings IAU Colloquium No. 195, 2004
A. Diaferio, ed.

DOI: 10.1017/S1743921304000766

A submm survey of high-redshift galaxy clusters: a submm Butcher-Oemler effect?

Tracy Webb[1], H. K. C. Yee[2], Henk Hoekstra[2,3] and Mike Gladders[4]

[1] Leiden Observatory, Niels Bohrweg 2, 2333CA Leiden, The Netherlands email: webb@strw.leidenuniv.nl

[2]Department of Astronomy & Astrophysics, University of Toronto, 60 St. George St., Toronto, M5S 3H8, Canada

[3] Canadian Institute for Theoretical Astrophysics, University of Toronto, 60 St. George St., Toronto, M5S 3H8, Canada

[4] Carnegie Observatories, 813 Santa Barbara St., Pasadena, CA, 91101, USA

Abstract. We present the initial results of a deep submillimeter survey of ten $z \sim 1$ galaxy clusters with the SCUBA bolometer array on the JCMT. Recently, an excess number of dust enshrouded starbursts in high-redshift galaxy cluster fields was reported (Best 2002) and has been interpreted as increased star-formation within young clusters. An alternative explanation is that these starbursts are background objects behind clusters with enhanced lensing cross-sections, over that predicted from simple lensing models, so-called "super-lenses" seen in recent optical cluster surveys. To investigate this issue, we are studying a set of clusters which show strong optical arcs (super-lenses) and a set of clusters of equal redshift and richness but which are not super-lenses. The preliminary results, based on 50% of the sample, tentatively suggest that, while the dusty starbursts are limited to the super-lensing cluster fields, they are not background objects but lie within the clusters themselves. Cluster-cluster merging offers a means of increasing both the lensing cross-section, thereby accounting for the multiple strong optical arcs, and increasing the star formation within the clusters through galaxy interactions.

---

1. Introduction

Submillimeter (submm) observations of the high redshift universe provide a unique and crucial window on the formation and evolution of galaxies (Eales et al. 2000, Webb et al. 2003). They trace regions of extreme luminosity and star-formation and, in particular, are sensitive to dusty systems which may be completely obscured at optical wavelengths. Furthermore, the unique negative K-correction produces an approximately constant observed 850μm flux with redshift; therefore submm-luminous systems above a given intrinsic luminosity can be detected to very high redshifts of $z \sim 6$. Thus, submm observations offer a potentially powerful tool with which to study the formation and evolution of galaxies, and the formation of the massive ellipticals in clusters, which likely formed in intense and dusty starbursts at high redshifts.

Submm astronomy is still in its infancy and only a handful of submm facilities exist which are capable of carrying out useful submm observations. The program presented here has used the Submm Common User Bolometer Array (SCUBA) on the James Clerk Maxwell Telescope (JCMT) in Hawaii, operating at 850μm. Above a redshift of $z \sim 1$ the galaxies detected with SCUBA have far-infrared (FIR) luminosities in excess of $10^{12} L_{\odot}$, and are similar to the ultraluminous infrared galaxies (ULIRGs) in the local universe. Like present day ULIRGs, they harbour a powerful energy source of either intense star formation and/or an active galactic nucleus, which is so deeply enshrouded by dust that the bulk of their energy is absorbed by the dust and re-emitted at FIR wavelengths. The flux limit of SCUBA is set by the confusion limit of the telescope at 850μm, $S_{850\mu m} \sim$ 2mJy, below which individual sources cannot be resolved. However, a number of groups have employed galaxy clusters as gravitational lenses to stretch the source plane and push the confusion limit to deeper levels (Smail et al. 2002). At low redshift ($z < 0.5$) where clusters are dominated by old, early-type galaxies with little or no dust and star formation, clusters do not contribute to the submm source counts and this is a relatively clean method of sampling the background population.

The recent observations of Best (2002) however show a possible excess of submm sources toward $z \sim 1$ cluster fields over that expected from simple isothermal sphere lensing models. Since lensing appears insufficient to explain the counts, this has been interpreted as an increase in dusty star formation in high redshift clusters, that is, a submm Butcher-Oemler effect. An increase in star formation in higher redshift galaxy clusters is not a new idea (Butcher & Oemler 1984) but the star formation rates implied by the 850μm luminosities are extreme (> 100 M$_\odot$ yr^{-1}) and must be caused by previously unstudied processes. In general the processes which act to enhance or stifle star formation in cluster and infalling galaxies are poorly understood and the existence of such intense star bursts add a new dimension to the problem. However, the number of clusters studied by Best (2002) is very small, and the increase in dusty starburst in high-redshift clusters is by no means an established global effect but must be verified through larger samples.

The above cluster-member interpretation rests on the fact that there are more submm-luminous systems than expected from simple lensing considerations, but if the lensing is stronger than the simple models predict, these systems could be highly magnified background objects. In fact, there is mounting evidence that some clusters at $z \sim 1$ do have enhanced lensing cross-sections. In the 90 square degree Red-sequence Cluster Survey (RCS, Gladders & Yee 2000), Gladders et al. (2003) found a fraction of multiple optical arcs for clusters at $z > 0.6$ much larger than expected from a ΛCDM cosmology. This, along with similar results from other high/medium redshift cluster samples (e.g. Bartlemann et al. 1998), leads to the interpretation that at higher redshift there is a subset of rich clusters which have much higher lensing cross-sections than that estimated from a standard isothermal mass model. These "super-lenses" could be due to substructure, orientation effects, or a high central mass surface density, and could account for the excess submm sources seen toward some clusters.

2. The RCS submm survey: observations and preliminary results

We have begun an observational program with SCUBA at 850μm to verify the results of Best (2002) and, if a true excess exists, to determine if the objects are primarily background lensed systems or objects at the same redshift as the cluster, either proper cluster members or infalling galaxies. We are imaging a sample of ten $0.6 < z < 1.1$ clusters selected from the RCS, which has produced a large and systematically selected sample of $z \sim 1$ clusters (see also Blindert et al. and Yee et al., these proceedings). In order to address the question of whether this effect is limited to strong lensing clusters we have selected 5 strong-lensing clusters (i.e., those that show strong optical arcs) and 5 clusters of equal richness and redshift but which do not show enhanced strong lensing.

We reach a 3σ depth of $\sim$ 3 mJy which requires $\sim$10 hours of integration per cluster. The field of view of SCUBA is $\sim 2.5{\times}2.5$ arcmin^2 and is well matched to the inner regions of the cluster fields (that is, within the virial radius). This is the approximate flux density that Arp 220 (a typical ULIRG in the local universe) would have if placed at the cluster redshifts and thus we are sensitive to extreme dusty star bursts. At such a depth we have essentially reached the confusion limit at 850μm and therefore little is gained by integrating deeper. The maps are reduced using the standard SURF pipeline and the noise is estimated and sources extracted following the technique of Eales et al. (2000). We have completed five clusters to date and the maps are shown in Figure 1 with signal-to-noise contours highlighting the significant objects. There are three clusters with strong optical arcs (super-lenses), RCS0224, RCS2319, and RCS1419, and two non-super-lenses, RCS1122 and RCS1326.

3. Discussion: cluster-cluster mergers?

All five clusters show a single detection near the center of the cluster (roughly (0,0) in Figure 1). These central sources could be detections of the central galaxies (Edge et al. 1999) or background lensed galaxies. The lensing is strongest within the inner ~ 20 arcsec and it is also within this region that the optical arcs are found. Unfortunately, the relatively large beam size of the JCMT at 850μm (15 arcsec FWHM) introduces an uncertainty in the submm position of up to 8 arcseconds. This, and the fact that the counterparts are dusty and likely optically faint means that it is not possible to unambiguously identify the source of the submm emission using submm and shallow optical data alone, but requires deep optical, near-infrared and radio imaging.

For the two non-super lenses, RCS1122 and RCS1326, the single central submm source is the only one detected with significance (the objects visible at the very edges of these images are too close to the edge to be considered secure detections). However, the three super-lensing clusters contain additional objects further out from the center which are significantly brighter than the central object (4 mJy$-$9 mJy). All of these systems are 1.0$-$1.5 arcminutes from the center of the cluster where, given simple lensing considerations, the magnification is expected to be very small (< 1.2) and it therefore seems unlikely that these systems are lensed unless significant substructure exists, and there is currently no evidence for this in the present data.

A more likely scenario, based on the current limited evidence, is that these systems are cluster members which are undergoing very intense starbursts due to the effects of the cluster environment and interactions with other galaxies. However, the fact that the submm luminous systems are only seen in the super-lenses implies that something is still unique about these particular clusters, and suggests a common cause for the enhanced lensing cross-section (as indicated by the multiple optical lensed arcs) and the excess submm sources in the outskirts. We note that we use the term outskirts loosely as the submm sources, though far from the central region of strong lensing, are still within the virial radii of the clusters, though we have no information regarding the extent of their distribution beyond the limits of the SCUBA maps.

Cluster-cluster merging offers a possible explanation for both of these phenomena. In the standard paradigm of structure formation clusters of galaxies form hierarchically through major mergers and the accretion of smaller clusters and groups. Clusters in the local universe are, for the most part, relaxed and virialized, but at higher redshift, where clusters are younger and still forming, these processes will become important. Cluster merging and accretion must also have a profound effect on the galaxies within clusters and may be one of the dominant processes at high redshift which drives their evolution.

Torri et al. (2004) have developed a model which shows that during a major cluster merger the lensing cross section is significantly, though temporarily, increased. This lensing enhancement can bring the number of multiple optical arcs seen in the RCS and other cluster surveys into agreement with the predictions of ΛCDM models. The effect on the cluster galaxies is not well understood though a cluster-cluster merger likely leads to

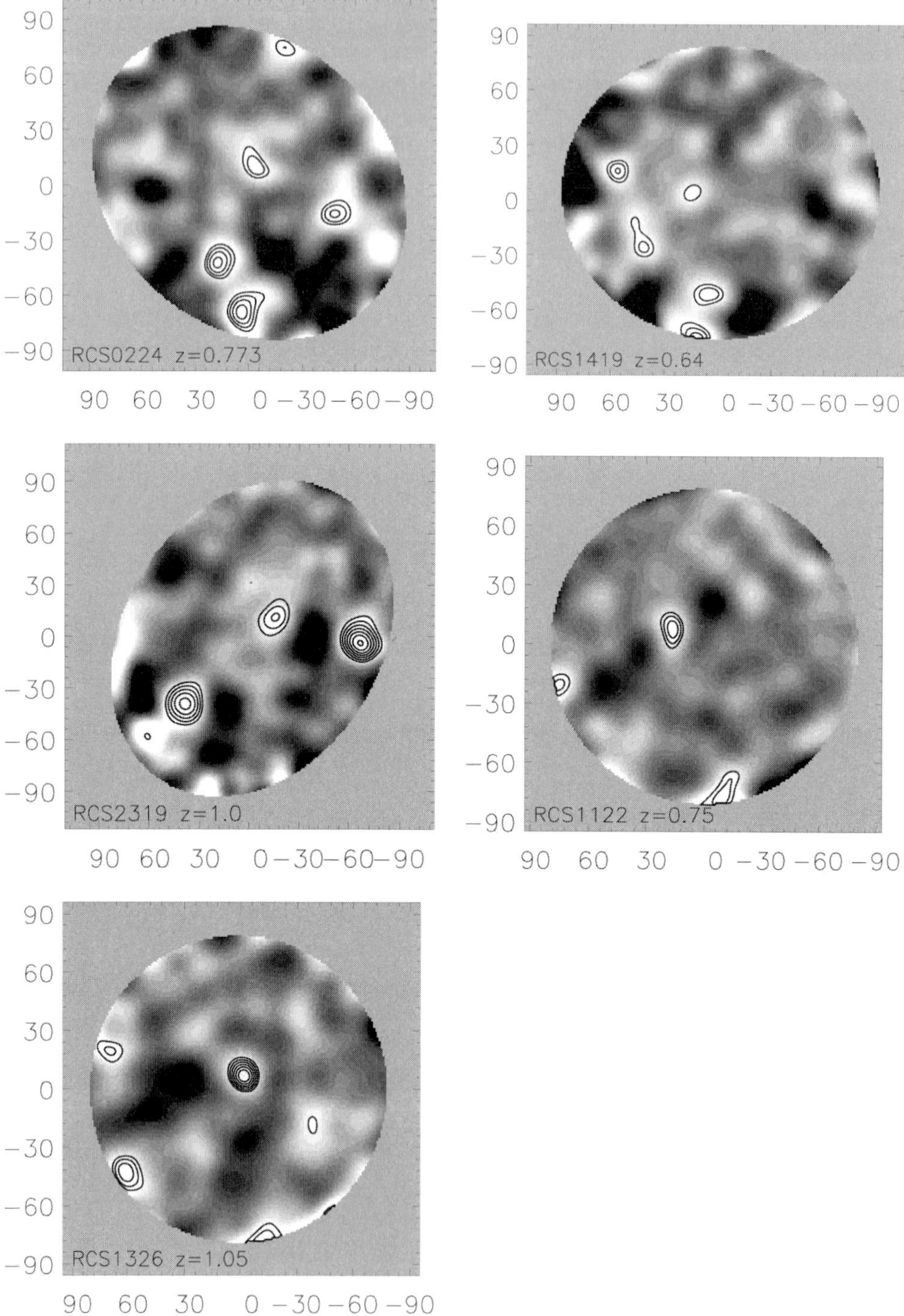

Figure 1. The 850μm maps of five RCS clusters with complete SCUBA imaging. The grey scale images correspond to the 850μm flux, and overlaid are signal-to-noise beginning at 2.5σ and increasing in 0.5σ steps for all clusters expect for RCS2319, which has 1σ contour steps. The images have been smoothed with a 14 arcsec Gaussian. The strong lensing clusters are RCS0224, RCS2319, and RCS1419. Note the large number of detections in these three clusters relative to the two non-strong-lensing clusters. This will be verified with larger number statistics upon completion of the survey.

an increased interaction rate between galaxies (see Mihos 2004 for discussion). Though major galaxy mergers are negligibly rare within cluster environments due to the high velocities of the interactions, this may not be the case in the outskirts of clusters during a merging event, where the galaxy-galaxy merger rate may increase and possibly lead to ULIRG-like star-bursts as seen in the field. Indeed, extreme starbursts have been seen in the mid-infrared studies of $z \sim 0.5$ clusters with ISO (e.g., Duc et al. these proceedings), though the trigger of these bursts is not clear.

If this scenario is correct we are observing the ongoing formation of clusters and in particular have caught some of the cluster galaxies in the process of forming the bulk of their stars. As we push cluster observations to higher redshift this effect will become increasingly important as the clusters will be younger, merging and accretion more common, and the galaxies themselves younger and possibly more gas rich. While imaging a large sample of galaxies over a broad redshift range is currently limited by the mapping speed of SCUBA (and the dearth of clusters at $z > 1$), the commissioning of the SCUBA2 array on the JCMT in 2006 will allow wide field and deep submm observations of large samples of high-redshift clusters in very reasonable amounts of time.

4. Ongoing and future work

We have completed the submm imaging for 50% of the cluster sample, or five clusters, and have proposed to the JCMT to complete the remaining five clusters by January 2005. With a sample of 10 clusters we will be able to determine statistically if an excess of submm galaxies is found in high-redshift cluster fields and in strong lensing $z \sim 1$ clusters in particular. The success of this program however depends crucially on extensive follow-up data. As previously mentioned, the large beam of the JCMT at 850μm produces an uncertainty in the submm position and deep optical, near-infrared and radio imaging are required to identify the counterparts. Without this information we cannot study the nature and cause of the starburst or place the object behind or within the cluster through a secure redshift. Deep multi-wavelength imaging is also necessary to investigate our suggestion that the super-lensing clusters are undergoing major cluster mergers, while the non-super-lenses are not. To this end we have begun follow-up imaging on a number of clusters with the VLA, CFHT, HST, VLT and Chandra, and will continue to pursue further observations in the future for the entire cluster sample.

References

Best, P. 2002 *MNRAS* **336**, 1293.

Bartelmann, M., Andreas, H., Colberg, J.M., Jenkins, A., & Pearce, F.R. 1998 *A&A* **330**, 1.

Butcher, H. & Oemler, A. 1984, *ApJ*, **285**, 426.

Eales, S., Lilly, S.J., Webb, T., Dunne, L., Gear, W., Clements, D., & Yun, M. 2000, *ApJ* **120**, 2244.

Edge, A.C., Ivison, R.J., Smail, I., Blain, A.W., & Kneib, J.P. 1999, *MNRAS* **306**, 599.

Gladders, M. & Yee, H.K.C. 2000 *AJ*, **120**, 2148.

Gladders, M.D., Hoekstra, H., Yee, H.K.C., Hall, P., & Barrientos, L.F. 2003, *ApJ* **593**, 48.

Mihos, C. 2004, In *Clusters of Galaxies: Probes of Cosmological Structure and Galaxy Evolution*, (ed. J.S. Mulchaey, A. Dressler, and A.Oemler), p. 278.

Smail, I., Ivison, R.J., Blain, A.W., & Kneib. J.-P. 2002 *MNRAS* **331**, 495.

Torri, E., Meneghetti, M., Bartelmann, M., Moscardini, L., Rasia, E., & Tormen, G. 2004 *MNRAS* **submitted**.

Webb, T., et al. 2003, *ApJ* **587**, 41.

Outskirts of Galaxy Clusters: Intense Life in the Suburbs
Proceedings IAU Colloquium No. 195, 2004
A. Diaferio, ed.

DOI: 10.1017/S1743921304000778

Newcomers meet the intracluster medium in the Coma cluster

Bianca M. Poggianti[1], T. J. Bridges[2], M. Yagi[3], Y. Komiyama[4], D. Carter[5], B. Mobasher[6], S. Okamura[7] and N. Kashikawa[3]

[1] INAF-Osservatorio Astronomico di Padova, Italy
[2] Anglo-Australian Observatory, Australia
[3] National Astronomical Observatory, Mitaka, Tokyo, Japan
[4] Subaru Telescope, Hilo, HI, USA
[5] Liverpool John Moores University, Birkenhead, Wirral, UK
[6] Space Telescope Science Institute, Baltimore, USA
[7] Department of Astronomy, University of Tokyo, Japan

Abstract. A main topic at this meeting is how galaxies are affected when they *enter* for the first time the cluster environment from the outskirts. Most of the times we are forced to infer the environmental effects indirectly, relying on systematic variations of galaxy properties with environment, but there aren't many examples of direct observations able to unveil ongoing transformations, and the corresponding mechanism producing it. We present a case in which it is possible to identify the cluster environment, and in particular the intracluster medium and the recent infall history of galaxies onto the cluster, as the cause for a recent, abrupt change in the evolutionary history of galaxies.

1. Post-starburst galaxies and substructure in Coma

This study is based on a photometric and spectroscopic survey of galaxies in the Coma cluster (Mobasher et al. 2001), which is distinctive from other surveys in three ways: for the large galaxy magnitude range covered (almost 7 mag, down to $M_B \sim -14$), for the large area surveyed (two regions, towards the cluster center and to the South-West, $\sim 1 \times 1.5$ Mpc each), and for being a simply magnitude limited sample, with no morphological selection criteria adopted for spectroscopy. In this survey we have found that a significant fraction (~ 10 %) of the cluster dwarf galaxy population at $M_V > -18.5$ has post-starburst/post-starforming spectra (Poggianti et al. 2004, hereafter P04). This type of spectrum ("k+a", or "E+A") indicates a galaxy with no current star formation activity which was forming stars at a vigorous rate in the recent past (last 1.5 Gyr). In the $B-R$ color-magnitude diagram, a group of blue and a group of red k+a galaxies can be easily distinguished in Coma. The average EW(Hδ) of the blue group is significantly stronger than that of the red group. The blue, strong k+a's most likely correspond to "young" k+a's (observed soon after the termination of star formation, < 300 Myr) and the red, weaker k+a's are "old" ones (observed at a later stage of the evolution, 0.5-1.5 Gyr).

A suggestive clue about the possible physical mechanism responsible for the k+a spectra comes from the recent X-ray mosaic observations of Coma obtained with *XMM-Newton.* Coma has two central dominant galaxies, NGC 4874 (a cD galaxy) and NGC4889 (a very bright elliptical), and another cD galaxy, NGC4839, that dominates a substructure South-West of the center (Fig. 1). Neumann et al. (2003) have identified and

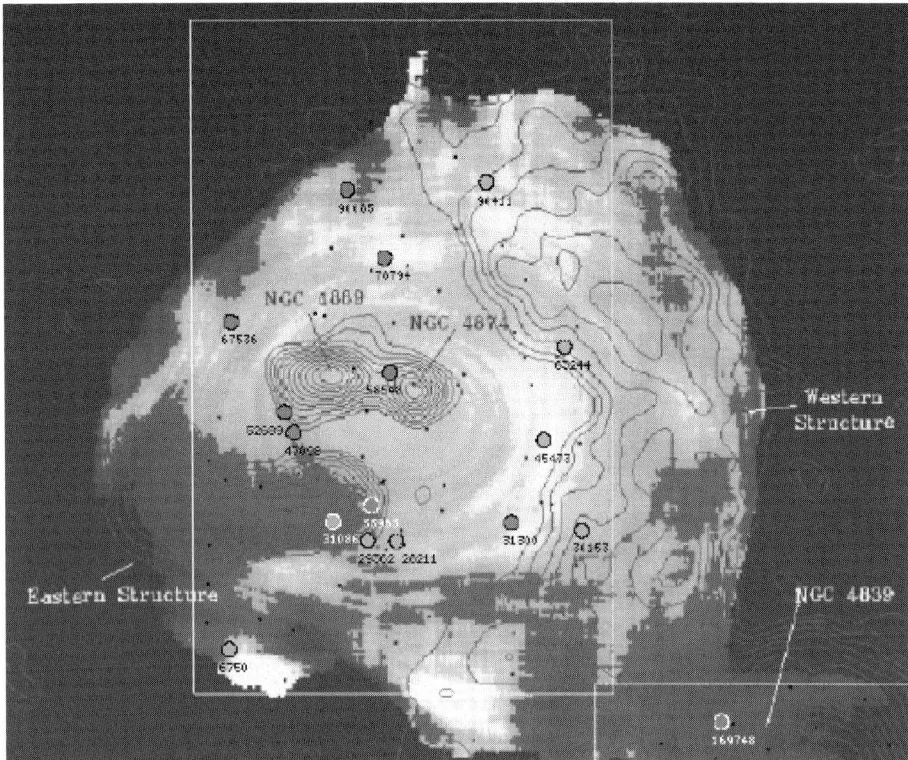

Figure 1. *N.B. This plot is best viewed in color, see Poggianti et al. 2004.* Position of k+a galaxies with respect to X-ray substructure and X-ray temperature map. Only the central field of Coma is shown here, see P04 for a full map. Strong-lined k+a's with EW(Hδ) > 5 Å are shown as light-coloured large dots, while weaker k+a's are plotted as darker large dots. Tiny black dots are dwarf Coma members with velocities > 7200 km s^{-1}. X-ray residuals from Neumann et al. (2003) are plotted as contours and clearly identify two substructures (Western and Eastern substructures), in addition to the NGC4839 peak in the South-West and the excess of emission towards the two central galaxies (NGC4874 and NGC4889). The lowest contour and the step width between two contours are each 5 σ. The hardness ratio image (2-5 keV/0.5-2keV, Neumann et al. 2003) is also shown. Darker regions correspond to temperatures below 8 keV, intermediate-grey regions to kT > 8 keV and light-coloured regions to kT > 10 keV. The rectangles show the limits of the two fields of our photometric and spectroscopic survey (Coma1 towards the cluster center and Coma3 in the South-West). Each rectangle is about 1 by 1.5 Mpc.

discussed X-ray substructure by fitting a smooth profile and subtracting it from the data. The residuals reveal several structures, that are shown as contours in Fig. 1: besides the well known NGC4839 South-West group, Neumann et al. find a large residual to the West of the cluster centre ("Western structure" in Fig. 1) elongated along the North-South direction, and a filament-like structure South-East of the centre ("Eastern structure" in Fig. 1), elongated along the East-West direction. The temperature map shown in Fig. 1 sheds further light on the accretion history of Coma. Neumann et al. conclude that the region of high temperature observed between the Western structure and the Coma center is caused by the infall of this structure, either via compression or via shock waves. These authors consider the two maxima in the western structure to be likely the result of the disruption of a galaxy group during its infall, instead of two galaxy groups falling at the same time. In contrast, the South-Eastern structure is cooler than the mean cluster temperature and is associated with a low-mass galaxy group dominated by two large galaxies, NGC4911 and NGC4921. Based on the filamentary form of this structure, the same authors conclude it is observed during the infall process while being affected by ram pressure stripping close to the cluster centre.

The coincidence of the position of the strongest k+a galaxies and the X-ray structures is striking. Four k+a's with EW(Hδ) > 5 Å (light large dots in Fig. 1) trace the edge of the Western structure towards the Coma centre. Another three are associated with the Eastern structure, all at its western boundary. Thus, young post-starbursts are distributed close to the edge of infalling substructures. In the case of the Western substructure this edge is the infalling front, while for the Eastern substructure it is unclear whether the group is moving to the West, as suggested by the appearance of the X-ray residuals, or to the East, as suggested by the positions of NGC4911 and NGC4921 (Neumann et al. 2003). Overall, this strongly suggests that the k+a spectra, i.e. the truncation of the star formation activity in these galaxies and possibly the previous starburst, could be the result of an interaction with the hot intracluster medium (ICM).

In contrast, the location of the red k+a galaxies in Fig. 1 does not appear to be correlated with the X-ray residuals. The red k+a phase has a timescale that is comparable to the core crossing time in a cluster like Coma, and any signature of the link between the truncation of star formation and the location within a substructure is thus erased in the older k+a's, while it is still detectable in the youngest subsample of blue k+a's.

It is instructive to note that looking for a spatial segregation in the location of galaxies on the sky would not allow to establish a correlation between the star formation history of the k+a galaxies and the substructure: the link with the dynamical history of Coma appears evident only once a detailed X-ray map reveals the complicated structure in the hot intracluster gas.

The blue k+a's do show, however, a radial velocity distribution that is significantly different from that of the red k+a's and the global Coma dwarf population. Their mean radial velocity is 8120±709 km s^{-1}, with all but one at $v > 7200$ km s^{-1}. In contrast, both the red k+a's and all faint galaxies with passive spectra have much lower mean velocities: 6992±761 and 6854±244 km s^{-1}, respectively.

2. Implications

Two main conclusions can be drawn from the analysis of the Coma k+a galaxies.

a) *The relation between k+a galaxies and substructure within the Coma cluster.* The position of faint post-starburst galaxies in Coma relative to X-ray substructure in the cluster strongly suggests that the interruption of the star formation activity in these galaxies is a cluster-related phenomenon, most likely due to the impact with the intracluster medium of recently in-fallen galaxy groups. Moreover, the physical process that caused the halting of the star formation must have acted on a *short timescale*: such timescale must be significantly shorter than 1 Gyr, the timescale for the visibility of the k+a signature in the spectra, for a k+a spectrum to be produced.

b) *Comparing the properties of k+a galaxies in Coma and in clusters at z=0.4-0.5: the downsizing effect.*

Numerous spectroscopic surveys of galaxies in distant clusters have found significant populations of luminous k+a galaxies (see for example Couch & Sharples 1987, Abraham et al. 1996, Dressler & Gunn 1992, Fisher et al. 1998, Dressler et al. 1999, Tran et al. 2003, but see also Balogh et al. 1999). However, k+a galaxies as luminous as those in distant clusters ($M_V \leqslant -20$) are absent in Coma, where k+a's are detected at magnitudes typically fainter than $M_V \sim -18$ (P04).

Post-starburst galaxies in Coma are thus much fainter than those observed in distant clusters. K+a spectra in clusters appear to be a luminous phenomenon at $z = 0.5$ and a faint one at $z = 0$. In Coma we are observing late-type star-forming galaxies becoming dwarf spheroidals, while the descendants of k+a's at high redshift will be among the most massive early-type galaxies today (see also Tran et al. 2003, and these proceedings). The observed evolution of the maximum luminosity of k+a galaxies most probably reflects a change in the galaxy populations falling into clusters, and provides further evidence of a "downsizing effect": going to lower redshift, the maximum luminosity/mass of galaxies with significant star formation activity progressively decreases, in *all environments*, and active star formation in low mass galaxies seems to be more protracted on average than in massive galaxies.

3. Related (mostly open) questions

1. Based on the strength of the lines, most of the blue k+a galaxies in Coma are inequivocally post-starburst systems, in which a starburst preceded the halting of the star formation. *Was the starburst triggered by the impact with the intracluster medium?* Based on timescale arguments, it is possible that the interaction with the ICM produces both the starburst and the subsequent halting of the star formation. Low–mass star–forming galaxies, however, often have a burst-like, non–regular star formation history, and thus the possibility that the starburst could be an intrinsic non–environmentally related phenomenon cannot be excluded.

2. *Could the termination of the star formation activity in these galaxies be due to "strangulation"?* The Hδ strength of the blue k+a galaxies implies that star formation was truncated in these galaxies on a short timescale, i.e. short compared to the k+a timescale of 1-1.5 Gyr. A slowly declining star formation activity such as that envisaged if galaxies simply lost their gas halo reservoir when becoming part of a group ("strangulation", e.g. Bower & Balogh 2003) is not able to produce such strong Balmer lines.

3. *Do k+a galaxies in distant clusters originate from the same physical process (the interaction with the ICM) that produced k+a galaxies in Coma?* This is yet unknown, and needs to be uncovered by studies of distant clusters (e.g. see Dressler et al. these proceedings).

4. *Are there k+a dwarf galaxies in distant clusters?* So far, spectroscopy of distant cluster galaxies has only been obtained for relatively bright galaxies, thus it is unknown how many faint k+a's exist in clusters at high-z, and whether the luminous ones that have been observed are only the "tip of the iceberg".

5. *Are field k+a galaxies at low redshift dwarf galaxies too?* Giant, massive k+a galaxies are known in the field in the local Universe (e.g. Yang et al. 2004). Most likely field k+a's originate from a different mechanism (such as galaxy-galaxy mergers, Zabludoff et al. 1996) that has the same final effect of the dense environment: terminate the star formation activity.

6. In hierarchical models, small structures form first, and as a consequence low mass galaxies can be expected to have stellar populations on average *older* than more massive galaxies. *Is the result presented here thus inconsistent with the predictions of semi-analytic modeling within a hierarchical scenario?* This depends on the prescriptions adopted for the star formation recipe. Some recent models (e.g. De Lucia et al. in preparation) assume a star formation efficiency that increases with the halo mass, and thus obtain a trend of increasing age with galaxy mass for elliptical galaxies.

7. And finally, *what is the origin of the downsizing effect, i.e. of the fact that star formation in low mass galaxies appears to be more protracted on average than in massive galaxies?* Most likely, since this phenomenon is known both in clusters (e.g. Smail et al. 1998, Kodama & Bower 2001, De Lucia et al. 2004) and in the field (e.g. Cowie et al. 1996, Kauffmann et al. 2003), the mass dependence of the star formation history we observe in clusters simply reflects the evolution of the galaxy populations falling in from the field and it is intrinsic to the galaxy itself (e.g. related to its mass and due to the combination of gravity and self–regulating star formation, see Chiosi & Carraro 2002), and not an environmental effect.

References

Bower, R.G. & Balogh, M.L. 2003 *in Carnegie Observatories Astrophysics Series, Vol.3: Clusters of Galaxies: Probes of Cosmological Structure and Galaxy Evolution* eds. J.S. Mulchaey, A. Dressler and A. Oemler (Cambridge: Cambridge University Press), http://www.ociw.edu/ociw/symposia/series/symposium3/proceedings.html.

Chiosi, C., Carraro, G. 2002 *MNRAS* **335** 335.

Cowie, L.L., Songaila, A., Hu, E.M., Cohen, J.G. 1996 *AnJ* **112** 839.

De Lucia, G., and the EDisCS collaboration 2004 *submitted* (see also De Lucia et al. these proceedings).

Kauffmann, G., Heckman, T.M., White, S.D.M. et al. 2003 *MNRAS* **341** 54.

Kodama, T., Bower, R.G. 2001 *MNRAS* **321** 18.

Mobasher, B., Bridges, T. J., Carter, D., Poggianti, B. M., et al. 2001 *ApJS* **137** 279.

Poggianti, B.M., Bridges, T.J., Komiyama, Y., Yagi, M., Carter, D., Mobasher, B., Okamura, S., Kashikawa, N. 2004 *ApJ* **601** 197.

Smail, I., Edge, A.C., Ellis, R. S., Blandford, R.D. 1998 *MNRAS* **293** 124.

Tran, K.-V.H., Franx, M., Illingworth, G., Kelson, D.D., van Dokkum, P. 2003 *ApJ* **599** 865.

Yang, Y., Zabludoff, A.I., Zaritski, D., Lauer, T.R., Mihos, J.C. 2004 *ApJ, in press* (astro-ph/0402062).

Zabludoff, A. I., Zaritsky, D., Lin, H., Tucker, D., Hashimoto, Y., Shectman, S.A., Oemler, A., Kirshner, R.P. 1996 *ApJ* **466** 104.

Outskirts of Galaxy Clusters: Intense Life in the Suburbs
Proceedings IAU Colloquium No. 195, 2004
A. Diaferio, ed.

DOI: 10.1017/S174392130400078X

Galaxy evolution in clusters from z=1 to z=0

S. Andreon[1], J. Willis[2] †, H. Quintana[2], I. Valtchanov[3], M. Pierre[4] and F. Pacaud

[1]INAF–Osservatorio Astronomico di Brera, Milano, Italy
[2]Dep. de Astronomía y Astrofísica, Pontificia Universidad Católica de Chile, Santiago, Chile
[3] Imperial College, London, UK
[4]CEA/DSM/DAPNIA, Service d'Astrophysique, Gif-sur-Yvette, France

Abstract. The XMM–LSS project is detecting distant clusters of low mass, quite comparable in mass to the ones in the local universe. This allows a direct comparison of galaxy properties at different redshifts in "similar" clusters. We present here first results on the evolution of the reddest galaxies in 25 clusters/groups at $0.3 \lesssim z \lesssim 1.0$ and for the whole galaxy population in the same clusters. The emerging picture from the current study is that the counterparts of present day clusters tend to show two or more distinct populations: a relatively old ($z_f > 2-5$) population evolving passively, together with a younger population, ostensibly responsible for the apparent brightening of the characteristic magnitudes, m^*.

1. Introduction

The nature and the time scale of the processes that shape galaxies properties in clusters and groups is still unclear, in spite of significant progress in the past years.

The window opened by the redshift dependence of the galaxies properties has been used to give constraints on the time scales of the processes (e.g. Butcher & Oemler 1984; Dressler et al. 1997; Stanford, Eisenhardt & Dickinson, 1998; Treu et al. 2003). However, many of the clusters compared at different redshifts have different masses (or X-ray luminosities), in such a way that in some occasions "we are comparing unripe apples to ripe oranges in understanding how fruit ripens" (Andreon & Ettori 1999). Furthermore, many of intermediate redshift clusters are optically selected, with a risk of biasing the optical properties of galaxies (see, for example, Andreon, Lobo & Iovino, 2004). Finally, rich clusters are rare environments where evolution is thought to be accelerated (Kauffman 1996).

There is therefore a compelling need to study galaxy properties of low mass (much below Coma) clusters at intermediate redshift ($0.3 \lesssim z \lesssim 1$). They have masses similar to common clusters in the local universe. Selecting them independently of the optical band will limit the risk of bias. This is one of the aims of the XMM–Large Scale Structure collaboration¶ (Pierre et al. 2004, Andreon, Pierre et al. 2003) that takes advantage of the multi-wavelength observations in progress on XMM, CTIO, CFHT Legacy Survey, UKIDSS, Spitzer, Galex and other facilities. Figure 1 shows individual X–ray luminosities of more than 1300 clusters of galaxies (black points) drawn from the literature (and listed in the BAX database‖). Their median X-ray luminosity (spline) increases with redshift, as shown in Andreon & Ettori (1999) for the Butcher & Oemler (1984) sample. There is, instead, a good match in X–ray luminosity between XMM–LSS clusters (the region at

† Present Address: Dep. of Physics and Astronomy, Univ. of Victoria, Victoria, Canada
¶ see http://vela.astro.ulg.ac.be/themes/spatial/xmm/LSS/
‖ http://bax.ast.obs-mip.fr/

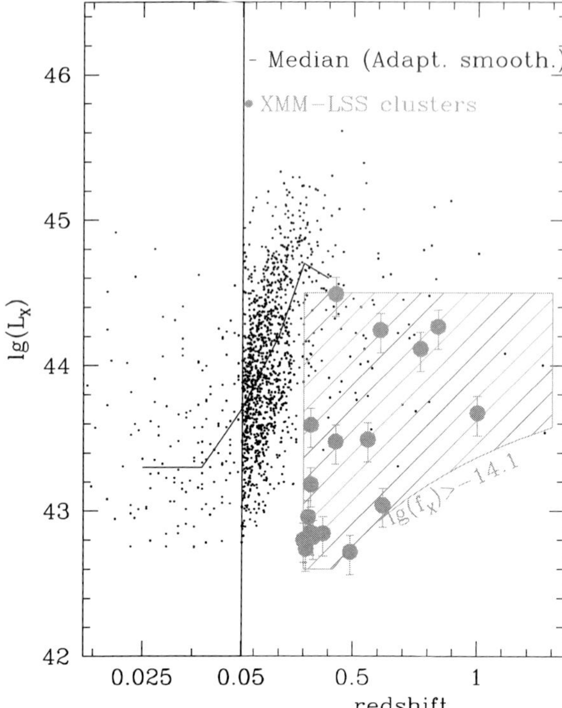

Figure 1. X-ray luminosity of about 1300 literature clusters (small points), their median average (spline) and the region accessible to XMM–LSS, given the 0.8×10^{-14} erg s^{-1} cm^{-2} flux limit in the [0.5-2] keV band. Green points are clusters confirmed during the 2002 spectroscopic campaign.

$z \gtrsim 0.3$) and the clusters in the local Universe (considered, say, in the morphology–density paper by Dressler 1980 or in the Butcher–Oemler paper by Butcher & Oemler 1984).

In a single ~ 5 deg^2 area XMM–LSS has currently spectroscopically observed some 70 clusters at $z \gtrsim 0.3$, the most distant being at $z = 1.05$ for the time being††. A high redshift ($z > 1$) cluster sample is also being assembled, characterized by extended X–ray emission associated with faint or absent optical galaxies and significant galaxy overdensity in deep NIR images. The sample is currently composed by at least six $z > 1.0$ clusters.

Here we report results for $\sim$ 20 clusters at $z \lesssim 1$. Most of them are X–ray selected, with intermediate–low mass (as measured by their X–ray luminosity) and low richness $R \lesssim 0$, as measured on the Abell (1958) scale. In addition, we have included a few, more massive clusters drawn from the literature, and a few clusters below the X-ray detection limit but color detected with our own version of the Gladders & Yee (2000) method (see Andreon 2003 for an early application to SDSS data) in the same XMM-LSS sky area. This analysis allows us to check the impact of the X–ray selection. More details can be found in Andreon et al. (2004).

2. The colour–magnitude relation

Figure 2 shows the colour evolution of the red envelope of the red sequence, defined as the median colour of the three brightest galaxies on the red sequence. Several model predictions are indicated††. The top (long dashed) curve neglects ageing of the stellar population. The additional models are more physically motivated. Passive stellar ageing

†† A true color image with X-ray contours of the latter can be inspected at the URL http://www.brera.mi.astro.it/~andreon/g04_06trico.png

†† Throughout this paper we assume $\Omega_M = 0.3$, $\Omega_\Lambda = 0.7$ and $H_0 = 70$ km s^{-1} Mpc^{-1}.

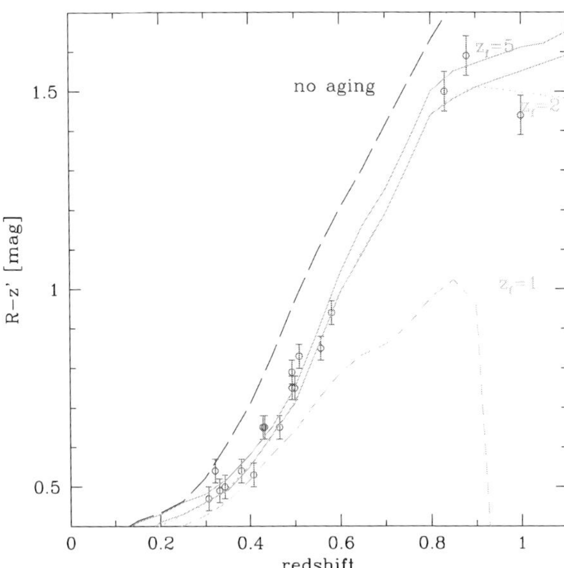

Figure 2. Observed $R - z'$ colour of the red envelope of the red sequence observed in each cluster as a function of redshift. Three galaxy evolution models are considered: a non–evolving early–type galaxy of present–day age at all redshifts, and two evolving early–type galaxy models, each characterised by a different formation redshift and mass.

and chemical evolution are described employing the model of Kodama & Arimoto (1997) and assume a formation redshift, z_f, and a total stellar mass. The two continuous green curves indicate $z_f = 5$ and a total stellar mass of $\sim 1.7 \times 10^{11} M_\odot$ and $\sim 6.4 \times 10^{10} M_\odot$. The expectation for a mass of $\sim 6.4 \times 10^{10} M_\odot$ and two lower formation redshifts ($z_f = 2$ and $z_f = 1$) are plotted as dotted and short dashed curves, respectively. The colour of the envelope of the red sequence is reproduced well by models where the stars in the oldest galaxies in the clusters form at $2 \lesssim z \lesssim 5$ in good agreement with the findings of Stanford et al. (1998) and Kodama et al. (1998), based on a set of richer clusters located within a comparable redshift range, and with Andreon et al. (2003) for a sample of low redshift clusters of low richness. Differently from previous work, our observations are quite uniform (virtually all photometric data were obtained at the same telescope+instrument in one single observing run). Furthermore, the cluster sample is mainly selected independently of the optical properties and avoids a potentially circular analysis between optically selected clusters and the properties of the red sequence. Finally, many of the clusters have the right low mass where current predictions regarding the assembly of bright, red galaxies (e.g. Kauffmann 1996 and Eggen, Lynden-Bell & Sandage 1962) display the greatest divergence.

Therefore, in these clusters, including the low mass ones, the color magnitude relation zero–point evolves passively, i.e. these galaxies must have formed the bulk of their stellar mass at $z > 2 - 5$.

3. The luminosity function

But what about the whole galaxy population, including blue galaxies? To this end, we measured the luminosity function in the standard way, and we fit a Schechter (1978) function of fixed slope to the data, hence determining m^* and its error. Figure 3 displays the redshift dependence of 47 m^* values at $z \gtrsim 0.3$ (two $z = 0.15$ calibrating points are also plotted, each one being the average of 21 X–ray selected clusters). To highlight the possible effects of active luminosity evolution upon m^*, the distance modulus and passive luminosity evolution terms were removed employing a m^* redshift evolution term,

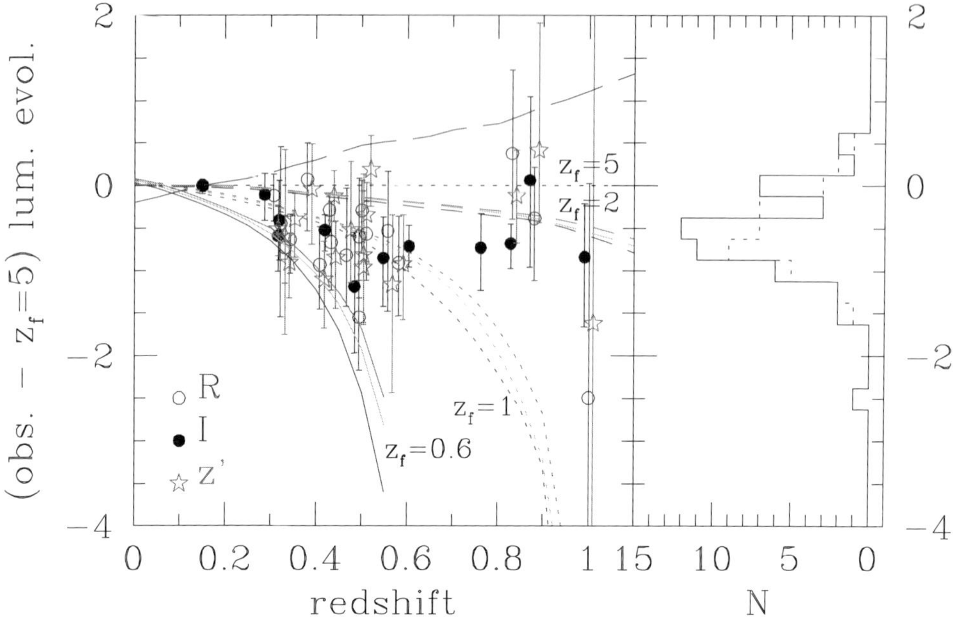

Figure 3. *Left panel:* Characteristic LF magnitude, m^*, evolution as a function of redshift having removed the constribution from passive ($z_f = 5$) stellar evolution. The labeled curves are the predictions for different formation redshifts. Points and curves of the same colour refer to the same filter, as indicated within the Figure. The long–dashed curve is the expectation neglecting R–band stellar evolution. Points are slightly offset in redshift in order to limit crawding. *Right panel:* frequency distribution of the points in the left panel (solid histogram) and of the corresponding values derived without any colour selection and excluding problematic clusters.

computed for a passively evolving stellar population formed at $z_f = 5$ using the model of Kodama & Arimoto (1997), previously employed to compute the colour of the red sequence. The model predictions are normalized to the observed m^* at $z = 0.15$. The m^* data points are systematically brighter than a model based upon an old, passively evolving stellar population (horizontal line). A non–evolving model is also strongly ruled out (rising line). The clusters presented in Figure 3 require a secondary star formation episode at $z_f < 5$ in order to generate an m^* values brighter than the passive evolution model. The same conclusion can be drawn by considering the right panel of Figure 3 (solid line) which displays the histogram of m^* values marginalized over redshift. The resulting m^* distribution is approximately 1 magnitude wide and is offset from zero toward bright magnitudes

The additional curves in Figure 3 indicate the expected m^* evolution for stellar populations formed at successively lower redshifts. In order to account for the bright m^* values observed at redshifts $z \sim 0.3$ a formation redshift as low as $z_f = 0.6$ would be required, although, by adopting such a low formation redshift, the predicted m^* value at a redshift $z = 0.15$ would be 0.2 magnitudes brighter than that reported by GMA99. The LF data points are not well described by any formation model based upon a single episode of star formation and a more complex scenario should be considered. The last (in cosmic time) star formation event should brighten the average m^* values by up to 1 mag (right panel of Figure 3). Such secondary star formation activity may be related to the Butcher–Oemler effect (Butcher and Oemler 1984), although the evidence for the latter is not compelling

(Andreon & Ettori 1999; Andreon, Lobo, Iovino 2004). A Butcher–Oemler analysis of the present sample of clusters is presently in progress.

4. Conclusions

We are at last observing the high redshift counterparts of the well studied low redshift clusters. The emerging picture from the current study is that even low mass clusters at intermediate redshift are composed by two or more distinct galaxy populations: a relatively old population formed at $2 \lesssim z_f \lesssim 5$ and evolving passively together with a younger population, ostensibly responsible for the apparent brightening of the characteristic LF magnitudes. The determination of the nature of this secondary activity (e.g. the time scale and the relationship with the cluster properties) is within the reach of the XMM–LSS project, since the $z < 1.3$ redshift regime is to be ultimately sampled by several hundreds of X–ray or color selected clusters, with supporting multi-color and spectroscopic observations.

Acknowledgements

SA thanks the whole XMM-LSS collaboration for the numerous discussions. SA wormly thanks A. Dressler for useful suggestions. This work has received support from MURST-COFIN n. 2003020150-005.

References

Abell, G. O. 1958, ApJS, 3, 211
Andreon, S. 2003, A&A, 409, 37
Andreon, S. & Ettori, S. 1999, ApJ, 516, 647
Andreon, S., Lobo, C., Iovino A., 2004, MNRAS, 349, 889
Andreon, S., Pierre, M., & the XMM-LSS collaboration 2003, Società Astronomica Italiana Memorie Supplement, 3, 188
Andreon, S., Willis, J., Quintana, H., Valtchanov, I., Pierre, M., 2004, MNRAS, submitted
Butcher, H. & Oemler, A. 1984, ApJ, 285, 426
Dressler, A., et al. 1997, ApJ, 490, 577
Gladders, M. D. & Yee, H. K. C. 2000, AJ, 120, 2148
Eggen, O. J., Lynden-Bell, D., & Sandage, A. R. 1962, ApJ, 136, 748
Kauffmann, G. 1996, MNRAS, 281, 487
Kodama, T. & Arimoto, N. 1997, A&A, 320, 41
Pierre, M. et al. 2004, PASP, submitted (astro-ph/0305191)
Schechter, P. 1976, ApJ, 203, 297
Stanford, S. A., Eisenhardt, P. R., & Dickinson, M. 1998, ApJ, 492, 461
Treu, T., Ellis, R. S., Kneib, J., Dressler, A., Smail, I., Czoske, O., Oemler, A., & Natarajan, P. 2003, ApJ, 591, 53

Outskirts of Galaxy Clusters: Intense Life in the Suburbs
Proceedings IAU Colloquium No. 195, 2004
A. Diaferio, ed.

DOI: 10.1017/S1743921304000791

Internal kinematics of spiral galaxies in distant rich galaxy clusters

K. Jäger[1], A. Böhm[1], B. L. Ziegler[1], J. Heidt[2] and C. Möllenhoff[2]

[1]Universitäts-Sternwarte Göttingen, Germany, email: jaeger@uni-sw.gwdg.de

[2]Landessternwarte Heidelberg, Germany

Abstract. We present our project on galaxy evolution in the environment of distant rich clusters aiming at disentangling the importance of specific interaction and galaxy transformation processes from the hierarchical evolution of field galaxies. Spatially resolved MOS spectra were gained with VLT/FORS to analyze the internal kinematics of disk galaxies. First results are shown for the clusters MS 1008.1−1224 ($z = 0.30$), Cl 0303+1706 ($z = 0.42$), and Cl 0413−6559 ($z = 0.51$). Out of 35 late-type cluster members, 13 galaxies exhibit a rotation curve of the universal form rising in the inner region and passing over into a flat part. The other members have peculiar kinematics. The 13 cluster galaxies for which a maximum rotation velocity could be derived are distributed in the Tully-Fisher diagram very similarly to field galaxies from the FORS Deep Field with corresponding redshifts. The same is true for seven non-cluster galaxies observed in the cluster fields. The TF-cluster spirals do not show any significant luminosity evolution as might be expected from certain phenomena specific to clusters. Contrary to that, the disturbed kinematics of the non–TF cluster spirals indicate ongoing or recent interaction processes.

1. Introduction

Galaxies in clusters reside in a special environment which is characterized by high galaxy densities, the presence of a hot intracluster medium (ICM) and vast amounts of dark matter. This environment strongly influences the evolution of the cluster members superposed on the field galaxy evolution that arises from the hierarchical growth of objects and the declining star formation rates over cosmic epochs. Besides tidal interactions (including merging) as known from the field, cluster galaxies are affected by phenomena, specific to clusters, related to the structure of the cluster (like harassment) or to the ICM (like ram-pressure stripping). These mechanisms can cause substantial distortions of the internal kinematics of disk galaxies leading to "rotation curves" (RCs) that no longer follow the universal form (Persic et al. 1996). Rubin et al. (1999), for example, classified half of their sample galaxies observed in the Virgo cluster as kinematically disturbed (asymmetric or even truncated curves). On the other hand, many cluster galaxies follow a tight Tully–Fisher relation (TFR) (e.g. Giovanelli et al. 1997).

An important idea presently discussed came from the finding that local clusters are dominated by elliptical and lenticular galaxies while the distant ones show a high fraction of spiral and irregular galaxies (Dressler et al. 1997): field spirals might fall into clusters and experience a morphological transformation into S0 galaxies. Related to this scenario it is not yet clear whether the halo of dark matter and, therefore, the total mass and the RC of a galaxy can also be affected by certain interaction phenomena.

First results on the kinematics of galaxies in distant clusters were presented by Metevier (2004) with 10 TF–galaxies in the cluster Cl 0024+1654 ($z = 0.40$) and by Milvang–Jensen et al. (2003) with 8 spirals in MS 1054.4–0321 ($z = 0.83$). While the first study sees a larger scatter of their sample galaxies in the TF diagram without evidence for an

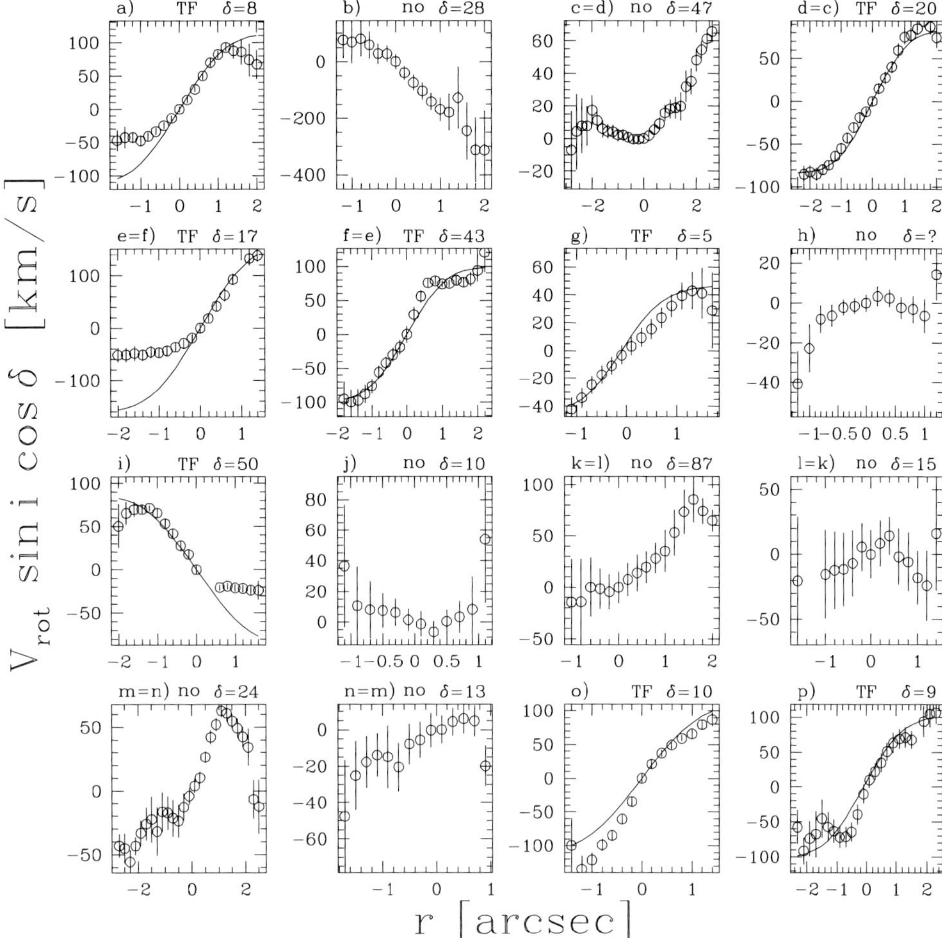

Figure 1. Spatially resolved velocity profiles for 12 disk galaxies in the cluster MS 1008–12 at $z = 0.30$. Note, panels c&d, e&f, k&l and m&n represent the same galaxy, rsp., but observed with different slit positions (the rotation angles of the two MOS masks differed by 67°). δ gives the angle between slit direction and apparent major axis. Panels a–p are ordered according to the distance of the galaxies from the cluster center (upper left: smallest proj. radius). Values of the maximum rotation velocity could be determined for cases a, d, e, f, g, i, o & p indicated by the label "TF". These galaxies enter the TFR shown in Fig. 2. In all other cases, the kinematics are too disturbed. In one case (m&n), we observe a prominent bar that causes a highly asymmetric rotation curve.

evolution of the zero point, the second one finds a trend towards brighter luminosities (0.5–1 B mag) with respect to field spirals of comparable redshifts. It is argued that either truncation of star formation or starbursts may cause an increased or decreased mass–to–light ratio, respectively.

2. Project and observations

During 5 nights at the ESO–VLT we gained spectra of galaxies within 7 clusters in the redshift range $0.3 \leqslant z < 0.6$ with FORS 1&2. One of our goals was the derivation of spatially resolved RCs to analyse the internal kinematics of the galaxies and to construct

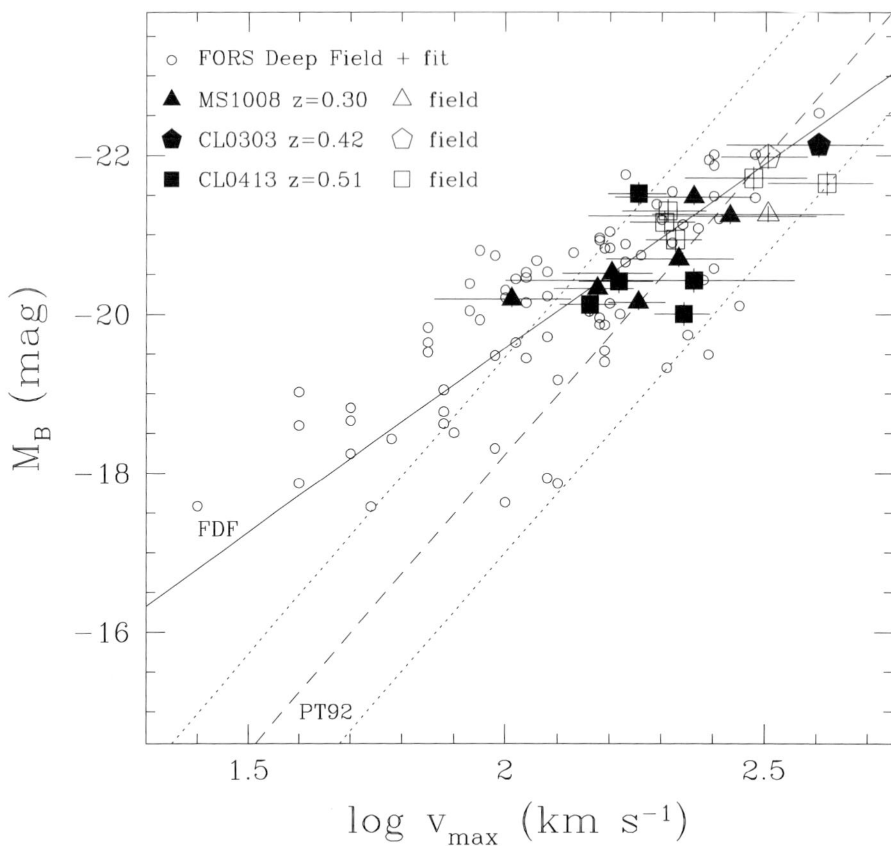

Figure 2. Tully–Fisher diagram of cluster spirals in MS 1008–12 ($z = 0.30$), Cl 0413–65 ($z = 0.51$) and Cl 0303+17 ($z = 0.42$). Also shown are 7 field objects (open symbols) that were also observed in the clusters' field–of–view. In comparison to our FORS Deep Field sample of 77 field galaxies with a mean redshift of 0.5 (small open circles), the cluster galaxies are similarly distributed and do not deviate significantly from the linear fit to the FDF sample (solid line). The cluster members follow the same trend with respect to the local TFR (the fit $\pm 3\sigma$ to the Pierce & Tully 1992 sample is given) as the distant field galaxies. Since we observed mostly bright cluster galaxies, their luminosities are not significantly increased in accordance with an undisturbed evolution. (Restframe B magnitudes were calculated for a flat $\Omega_\lambda = 0.7$ cosmology with $H_0 = 70\,\mathrm{km\,s^{-1}\,Mpc^{-1}}$.)

the TFR (Tully and Fisher 1977) of distant clusters. HST/WFPC2 images of the core regions were retrieved from the archive to complement our own ground–based imaging. Here we present results of three of the clusters, Cl 0413–65 ($z = 0.51$) and MS 1008–12 ($z = 0.30$), Cl 0303+17 ($z = 0.42$). Observations were made in 1999 and 2000 with FORS mounted at the Cassegrain focus of the VLT. We used grism 600R and get a spectral resolution of $R \approx 1200$ at a slitwidth of 1 arcsec. The spatial scale was 0.2 arcsec/pixel with a dispersion of ~1.08 Å/pixel. The total integration time was set to ≈2 hrs to meet our signal–to–noise requirements. Seeing conditions ranged between 0.4 and 1.3 arcsec FWHM. Two MOS setups have been observed for each cluster, yielding 116 objects with apparent magnitudes in the range $18.0 < R < 23$.

3. Data reduction and rotation curve modelling

A detailed description of the sample selection and data reduction including finding charts and data tables can be found in Jäger et al. (2004). For the derivation of position velocity diagrams we followed the way described in detail in Böhm et al. (2004) and Ziegler et al. (2003). Here we only give some short remarks on this topic.

All RCs were determined using either the [O II]3727, Hβ or [O III]5007 emission line. Gaussian fits have been applied stepping along the spatial axis with a median filter window of typically 0.6 arcsecs to enhance the S/N. Since the apparent disk sizes of spirals at intermediate redshifts are only slightly larger than the slit width (1 arcsec), the spectroscopy covers a substantial fraction of the two–dimensional velocity field. Thus, the maximum rotation velocity $v_{\rm max}$ (the constant rotation in the outer part of a galaxy due to the Dark Matter halo) cannot be determined "straightforwardly" from the observed rotation. To tackle this problem, we simulated the spectroscopy of each galaxies' velocity field with the respective inclination and position angle, also taking seeing and luminosity weighting into account. The simulated rotation curves which best reproduced the observed ones (solid lines in Fig. 1) yielded the values of $v_{\rm max}$.

4. Results and discussion

Redshifts and spectral types could be determined for 89 galaxies. From the 50 cluster members, 35 turned out to be late type galaxies. Only 13 cluster galaxies exhibit a rotation curve of the universal form rising in the inner region and passing over into a flat part. The other members have peculiar kinematics or too low S/N. For 7 non–cluster galaxies v_{max} could also be measured.

Only those rotation curves which show no strong perturbations are eligible for a determination of v_{max} as needed for the TF diagram which is shown in Fig. 2. v_{max} could be measured for 7(MS 1008–12), 5(Cl 0413–65), and 2(Cl 0303+17) cluster members, respectively. Luminosities were derived from total magnitudes (measured with SExtractor, see Bertin & Arnouts 1996) of V (MS 1008–12), I (Cl 0413–65), and R (Cl 0303+17) FORS images corrected for Galactic and intrinsic extinction, transformed to restframe Johnson B according to SED type, and calculated for a flat $\Omega_\lambda = 0.7$ cosmology ($H_0 = 70\,{\rm km\,s^{-1}\,Mpc^{-1}}$).

Within our TF diagram the distant cluster spirals are distributed similar to the FORS Deep Field (FDF) spirals (Ziegler et al. 2002, Böhm et al. 2004) which have been observed with exactly the same instrument configuration and which represent a similar cosmic epoch ($\langle z_{\rm FDF}\rangle = 0.5$). No significant deviation from the distant field TFR is visible and the cluster sample has not an increased scatter, but the low number of cluster members prohibits any quantitative statistical analysis. We can conclude that the mass–to–light ratios of the observed distant cluster spirals cover the same range as the distant field population indicating that no phenomenon specific to clusters dramatically changed the stellar populations. In particular, there was no starburst in the recent past of the examined cluster galaxies that would have significantly risen their luminosities. Since we mostly selected bright galaxies, the cluster members occupy a region in the TF diagram where no significant luminosity evolution is visible with respect to the local TFR.

On the other hand, this conclusion is true only for those objects that enter the TF diagram and is not valid for the whole cluster sample. More than half of our cluster galaxies can not be used for a TF analysis due to their peculiar kinematics not following the "rise–turnover–flat" RC shape of large, isolated spirals. This is in contrast to the smaller fraction of peculiar curves in our sample of the FDF spirals. The objects with peculiar RCs may actually be subject to ongoing or recent interactions. Indeed, a morphological classification of the cluster spirals in terms of their asymmetry indices (Abraham et al. 1996) reveals that they have a slightly higher degree of asymmetry on the mean than the kinematically less disturbed ones. A correlation between morphological and kinematical disturbance hints to a common origin for both in "strong" interactions like close encounters with tidal effects, accretion events or even mergers. Such processes most probably also influence the stellar populations of a galaxy changing its integrated luminosity as well.

No trend is visible of the RC form with (projected) clustercentric distance (cf. Fig.1). But since all observed galaxies are located within the virial radius, this is in compliance with dynamical models in which the galaxy population of a cluster is well-mixed within that region. In particular, we most probably do not have any new arrivals from the field in our sample.

Acknowledgements

Based on observations with the ESO–VLT (64.O–0158 & 64.O–0152). This work has been supported by the Volkswagen Foundation (I/76 520) and the Deutsche Forschungsgemeinschaft (Fr 325/46–1 and SFB 439).

References

Abraham et al. 1996 *MNRAS* **279**, L47.

Bertin E., & Arnouts, S. 1996 *A&AS* **117**,393.

Böhm, A., Ziegler, B. L., Saglia, R.P., et al. 2004 *A&A* in press (astro-ph/0309263)

Dressler, A. et al. 1997 *ApJ* **490**, 577

Giovanelli, R., Haynes, M. P., Herter, et al. 1997 *AJ* **113**, 22

Jäger, K., Ziegler, B.L., Böhm, A., et al. 2004 *A&A* submitted

Kodama, T., & Bower, R. G. 2001 *MNRAS* **321**, 18

Metevier, A. J. 2004. In Clusters of Galaxies: Probes of Cosmological Structure and Galaxy Evolution, eds. Mulchaey, J.S, Dressler, A., Oemler, A. *Carnegie Obs. Astrophysics Series* **Vol.3**

Milvang-Jensen, B., Aragón-Salamanca, A., Hau, G. K. T., et al. 2003 *MNRAS* **339**, L1

Persic, M., Salucci, P., & Stel, F. 1996 *MNRAS* **281**, 27

Pierce, M. J. & Tully, R. B. 1992 *ApJ* **387**, 47

Rubin, V. C., Waterman, A. H., Kenney, J. D. P. 1999 *AJ* **118**, 236

Tully, R.B., & Fisher, J.R. 1977 *A&A* **54**, 661

Ziegler, B.L., Böhm, A., Fricke, K.J., et al. 2002 *ApJL* **564**, L69

Ziegler, B.L., Böhm, A., Jäger, K., et al. 2003 *ApJL* **598**, L87

Outskirts of Galaxy Clusters: Intense Life in the Suburbs
Proceedings IAU Colloquium No. 195, 2004
A. Diaferio, ed.

DOI: 10.1017/S1743921304000808

Wide field imaging of distant clusters

Tommaso Treu

Division of Astronomy & Astrophysics University of California, Los Angeles, CA 90095-1562, USA email: ttreu@astro.ucla.edu

Abstract. Wide field imaging is key to understanding the build-up of distant clusters and their galaxy population. By focusing on the so far unexplored outskirts of clusters, where infalling galaxies first hit the cluster potential and the hot intracluster medium, we can help separate cosmological field galaxy evolution from that driven by environment. I present a selection of recent advancements in this area, with particular emphasis on Hubble Space Telescope wide field imaging, for its superior capability to deliver galaxy morphologies and precise shear maps of distant clusters.

1. Introduction

In the current standard cosmological paradigm (see, e.g., talks by Benson, Moore, Springel, Tormen), clusters of galaxies form in correspondence with the first overdensities in the early universe and grow by mergers and by accreting material from the surrounding regions (which I will generally refer to as the *field*).

In this scenario, the infall regions of clusters are believed to be very important – as testified by the enthusiastic participants of this meeting. As galaxies, gas, and dark matter are accreted onto the cluster, they are subject to interactions with the steep gravitational potential, along with the galaxies and hot plasma already present in the cluster. Under appropriate conditions, the outcome of these interactions can be quite dramatic, e.g., altering the morphology and star formation properties of infalling galaxies (e.g. Mastropietro's talk).

In this contribution I will focus on two particular aspects of the physics of the infall regions: i) the environmental effects on the morphological properties of galaxies and ii) the distribution of dark matter. As suggested by Antonaldo and the Scientific Organizing Committee, I will focus on wide field (> 1 Mpc radius) imaging of distant clusters as a diagnostic tool to study these processes (for other diagnostic tools see talks by, e.g., Balogh, Dressler, Ellingson, Moran, Poggianti). Studying distant clusters ($z > 0.1$) provides two important pieces of information. One the one hand, distant clusters are efficient gravitational lenses and therefore weak and strong lensing can be used to characterize their mass distribution (see also Schneider's talk). On the other hand, with current technology it is possible to reach look-back times of several Gyrs, i.e comparable to the evolutionary time-scales for these systems. Therefore, by comparing local clusters to their "progenitors"†, we can aim at empirically mapping the cosmic evolution of the infall regions. Due to space limitations, I will restrict my contribution to a personal selection of observational results. Most of them are obtained with the *Hubble Space Telescope*, for its ability to determine galaxy morphology and measure lensing signals.

The Hubble constant, the matter density, and the cosmological constant are assumed to be $H_0 = 65$ km s^{-1} Mpc^{-1}, $\Omega_m = 0.3$, and $\Omega_\Lambda = 0.7$, throughout this paper.

† With all due cautionary notes on this crucial step.

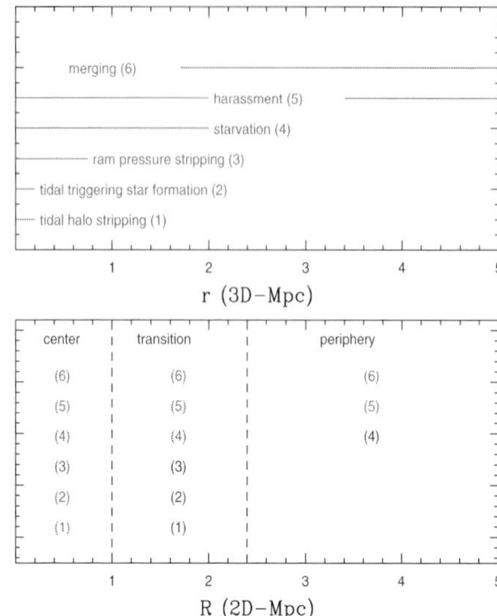

Figure 1. Summary of spheres of influence for key physical mechanisms in the galaxy cluster CL0024 ($z = 0.39$). Top panel: the horizontal lines indicate the radial region where the mechanisms are most effective (in 3-D space; note that the harassment is effective in the entire range). Lower panel: for each projected annulus we identify the mechanisms that can have affected the galaxy in the region (red). The blue numbers indicate processes that are marginally at work. See T03 for additional discussion (reproduced with permission).

2. Environmental processes and physical scales

It is empirically known that the mix of morphological types is a function of local galaxy density (Dressler 1980) and redshift (Dressler et al. 1997; Treu et al. 2003, hereafter T03; Smith et al. 2004). This is generally interpreted as the combination of two mechanisms. Firstly, high density regions are very special places in the Universe and therefore galaxies observed there might have formed under special initial conditions. Secondly, interactions between galaxies, dark matter and the intra–cluster medium (i.e. environmental processes) are likely to transform in–falling field galaxies from gas–rich spirals to gas–poor lenticular galaxies. In some sense, understanding the balance between the two processes can be thought of as a nature vs. nurture problem, although certainly the distinction is somewhat arbitrary (the environment a galaxy is formed in is, in a broad sense, part of the initial conditions). Therefore, I will use the following simplified formulation of the problem. Let us consider a general population of *field* galaxies and let them fall into a cluster. When (and if) they arrive at the cluster center, will they have changed their distribution of morphologies, as a result of interactions, so as to match that of high density regions? If the answer is no, then we have to invoke different initial conditions (e.g. mass function). If the answer is yes, then we would like to know exactly which physical mechanism is doing what.

Assuming the answer to the former question is "yes" or "in part", in order to address the latter issue (what physical mechanism is doing what) it is useful to rank the proposed mechanisms by estimating the relevant time and length scales. In the upper panel of Figure 1, I show spheres of influence for various physical mechanisms (computed as an example for CL0024 at $z = 0.39$, see T03 for definitions and details). The lower panel shows two–dimensional spheres of influence once projection effects, duration of phenomena, and travel times are taken into account (see discussion of the "back-splash"-

effect in Ellingson's and Moore's talks). It is clear that the center of the cluster is a very busy environment, where many processes are simultaneously at work. By contrast, the outer regions – the *transition* region around the virial radius and the *periphery* beyond that – provide a much cleaner environment, and therefore the opportunity to separate the effects of different mechanisms.

3. Wide field imaging

3.1. *Ground based studies*

One of the key difficulties in interpreting wide field imaging data in terms of galaxy population is identifying cluster members at large radii, where the contrast over the background/foreground is low. Abraham et al. (1996) made an important step forward in this respect by obtaining unprecedented spectroscopic follow-up – including hundreds of members – of the $z = 0.228$ cluster A2390. They obtained the first observational constraints on the infall scenario, finding that the properties of galaxies – such as colors and spectral features – changed smoothly with radius, consistent with an idea of gradual infall without significant mixing (see also Morris et al. 1998). The recent development of wide field imagers and spectrographs on big telescopes – with the ability to obtain hundreds of redshift at a time – seems to suggest that we will see more studies of this kind in the next years, with even larger redshift catalogs (Czoske at this meeting presented an ongoing project collecting *thousands* of redshifts per cluster).

To tackle the same problem, Kodama et al. (2001) followed a different approach. They used photometric redshifts based on high quality Subaru images to identify galaxies in a relatively thin redshift slice around the cluster A851 ($z = 0.39$). Although some residual contamination is present, the use of photometric redshifts allowed them to study an even larger sample of galaxies to fainter intrinsic luminosities. Based on this technique, they were able to convincingly show the filamentary structure of the accreting material. They also detected a rapid transition in galaxy colors at a characteristic density, possibly the signature of environmental processes influencing infalling galaxies (see also Gray's talk).

3.2. *HST-WFPC2 studies*

The Wide Field and Planetary Camera 2 (WFPC2) revolutionized the subject, by delivering images of galaxies in† distant clusters with sufficient resolution to accurately measure their shapes and morphologies (e.g. Couch et al. 1994; Dressler et al. 1994; Ellis et al. 1997). Unfortunately, the 4 800x800 pixel^2 CCDs provide a field of view of only ~ 2.5 arcmin on a side, corresponding to less than a Mpc at $z \sim 0.5$. Hence, performing wide field studies of distant clusters requires patient and expensive mosaicing. Using contiguous pointings, Couch et al. (1998) and van Dokkum et al. (1998) constructed the morphology-density relation and color-magnitude relation reaching the *transition* region of two clusters at $z \sim 0.3$. They found that the morphology density relations starts to flatten out beyond $\sim$ 1Mpc and that the color-magnitude relation of lenticular galaxies is less tight than in the cluster core (but see Andreon 1998 for an explanation of this latter finding in terms of morphological classification uncertainties). WFPC2 mosaics of comparable size have been obtained now for a few clusters (e.g. talks by Dressler and Tran).

In order to reach significantly beyond 1 Mpc in a finite number of HST orbits, T03 adopted a sparse sampling strategy. With 39 WFPC2 pointings they covered a circle 10 Mpc in diameter, with fractional area coverage 20-40% in the *transition* and *periphery*

† Or behind, for weak lensing purposes.

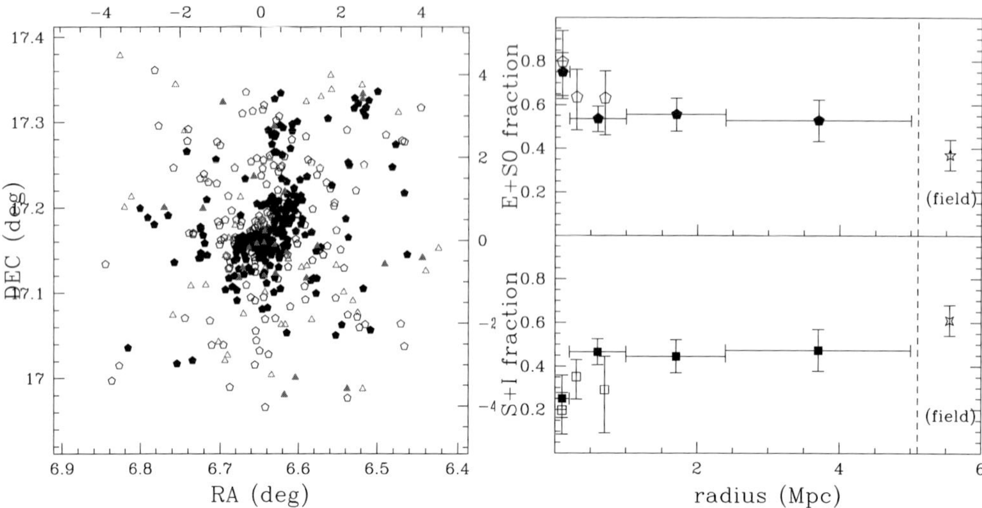

Figure 2. Left: Projected distribution of CL0024 member galaxies. Solid symbols indicate objects with available HST-WFPC2 images. The top and right scale is in Mpc. Right: Fraction of morphological types ($I < 21.1$) in CL0024+16 as a function of cluster radius: (upper panel) fraction of E+S0 galaxies; (lower panels) fraction of spirals. Fractions determined from the entire WFPC2 sample – removing the background statistically – are shown as large empty points, while fractions determined from the spectroscopic WFPC2-z catalog are shown as solid points. Points beyond the dashed line at 5.1 Mpc are field fractions. From T03 (reproduced with permission).

regions of cluster CL0024 at $z \sim 0.39$. More than 1300 redshifts, including those of 472 members, are available to identify cluster members at the periphery of the cluster.

The left panel of Figure 2 shows an updated map of the cluster (superceding that in T03) members. By combining HST data and redshift, the morphology-radius relation can be measured for a single system out to almost 5 Mpc (Figure 2, right panel; from T03, updated to include the complete redshift catalog). Beyond 1 Mpc the fraction of early-type galaxies declines gently with radius toward the lower field value. This gradient cannot be caused by mechanisms such as ram-pressure stripping or tidal interaction with the main cluster potential.

The distribution of galaxies is clearly not circularly symmetric, as show in the left panels of Figures 2 and 3. Therefore, we can learn more about the relationship between morphology and environmental effects by studying the morphology-density relation as a function of location inside the cluster (right panel of Figure 3). Remarkably, the morphology density relation does not depend – within the uncertainties – on the location inside the cluster. At a given density, the morphological mix appears to be independent of the cluster radius. A direct implication of this finding is that galaxy concentrations outside the central peak are physically associated and therefore galaxies are not accreted as individuals but rather as groups, with their own hierarchy of morphologies (and dark matter halos as I discuss later). At least part of the morphological transformations could happen at the group level where galaxies are "pre-processed" before entering the cluster. At the group level, different physical mechanisms are at work (e.g. merging is more frequent, ram pressure stripping is less effective), and therefore it is possible that different mechanisms dominate at subsequent stages.

The sparse sampled mosaic of CL0024 can also be used to study the dark matter distribution in the infall regions. A joint weak and strong lensing analysis (Kneib et

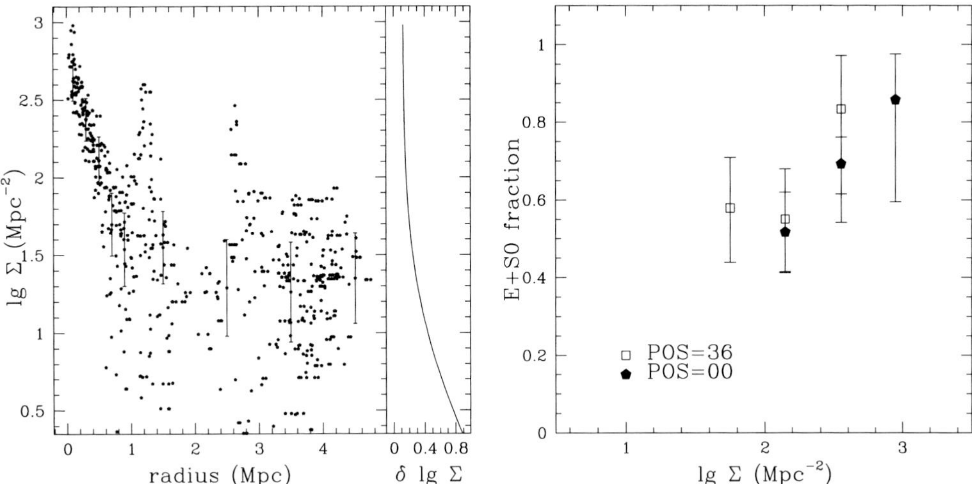

Figure 3. Left: Local surface density vs cluster radius in CL0024 revealing the large range at a given radius and the spike in local density at ~ 1 Mpc. The points with error bars indicate the average Σ within radial bins, and the scatter expected from measurement errors. The expected uncertainty as a function of Σ is shown in the right panel. Right: Fraction of E+S0 galaxies in CL0024+16 as a function of local projected density for galaxies in the central (solid pentagons) and in the NW overdensity region (open squares; POS36) corresponding to the spike in local density at ~ 1 Mpc in the left panel. Note the close match in the regions of overlapping density. From T03 (reproduced with permission).

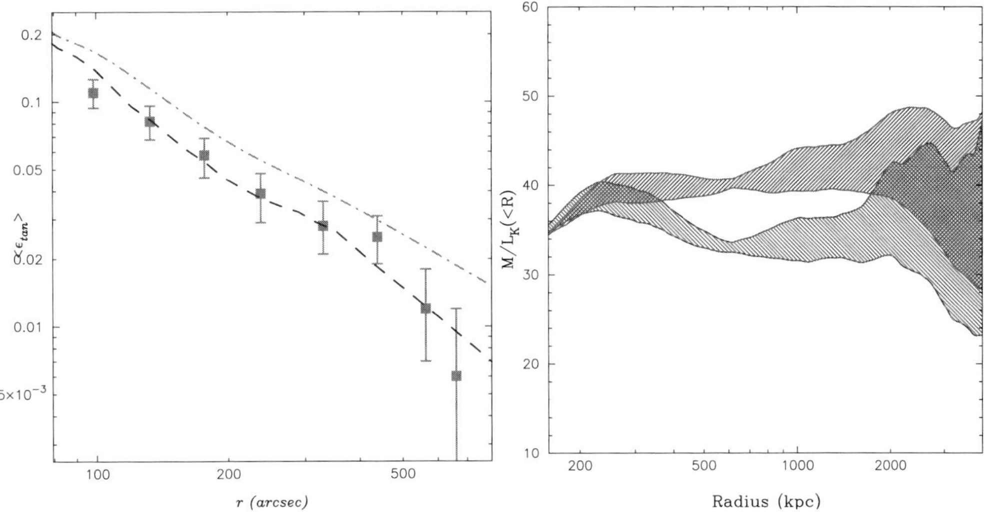

Figure 4. Left: Reduced tangential shear profile of CL0024 for the combined WFPC2 and STIS data (magenta points with error bars). The dashed line is the visual representation of the reduced tangential shear of the 2 clump NFW model that best fits both the strong and weak lensing constraints. The dot-dashed line corresponds to the reduced tangential shear of the 2 clump SIS model that best fits the strong lensing constraints but fails to fit the weak lensing measurements. Right: M/L_K ratio (rest frame solar units) of CL0024 . The (blue) lower hatched region corresponds to the M/L derived for the enclosed field-subtracted K-band sample and the (red) upper hatched region that for the color-selected sample. From K03 (reproduced with permission).

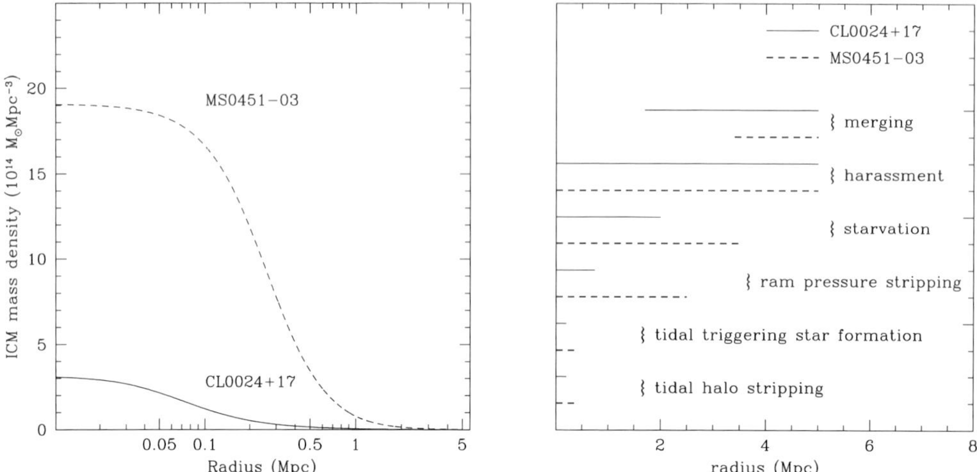

Figure 5. Left: comparison of the mass density profile of the intracluster medium (ICM) in CL0024+17 and MS0451-03. Right: Comparison of the regions of influence of key physical mechanisms for CL0024+17 and MS0451-03. Note the dramatic enhancement in the size of the sphere of influence of the ICM – ram pressure stripping and starvation – in MS0451-03.

al. 2003; hereafter K03) reveals the presence of a secondary clump of mass, coincident with the overdensity of galaxies 1 Mpc NW of the cluster core identified in Figures 2 and 3, with mass-to-light ratio consistent with that of the main clump. After appropriate azimuthal averaging, the lensing analysis can be extended to the periphery, resulting in the mass shear profile shown in the left panel of Figure 4. The joint weak and strong lensing constraints are consistent with a Navarro, Frenk & White (1997) mass profile outside the central region, and inconsistent with a singular isothermal sphere (c.f. talks by Biviano and Mamon). The right panel of Figure 4 shows the rest frame K-band mass-to-light ratio of the cluster, which is approximately constant all the way out to the periphery, indicating that the infalling material has approximately the same fraction of mass in stars as the central regions. A similar result for local clusters – based on a completely independent "caustic" technique (Diaferio & Geller 1997; Diaferio 1999) – was shown as this meeting by Rines. Remarkably, the M/L_K of CL0024 is consistent with that of local clusters, once a correction corresponding to passive evolution of an old stellar population is applied.

3.3. *HST-ACS studies*

The successful installation of the Advanced Camera for Surveys (ACS) on board HST has improved its ability to carry out wide field imaging of distant clusters by a factor of 10 (considering area and depth). Several wide field studies are currently ongoing, including for example the ACS GTO cluster program, and the follow-up of MS1054 ($z = 0.83$) by Kelson et al. This improvement makes it possible to obtain fully sampled mosaic of distant clusters over diameters corresponding to several virial radii. It is of particular interest to study and contrast the properties of galaxies in clusters with different intracluster medium (ICM) properties. For example, MS0451 ($z = 0.54$; covered by GO-9836, PI Ellis) is a rich cluster, much more luminous in the X-ray than CL0024. Figure 5 illustrates the different ICM content of the two clusters and compares the spheres of influences of the various processes in the two clusters. ICM related mechanisms, such as ram-pressure stripping and starvation, have larger spheres of influence in MS0451 than in CL0024.

With so many exciting projects on their way, and the warm hospitality of Torino, I hope that the organizing committee will reconvene another meeting on the subject in the near future.

Acknowledgements

It is a pleasure to acknowledge the numerous and valuable contributions of my collaborators – Oliver Czoske, Alan Dressler, Patrick Hudelot, Jean-Paul Kneib, Phil Marshall, Sean Moran, Priya Natarajan, Gus Oemler, Graham Smith, Ian Smail – to the work presented here. I am especially grateful to Richard Ellis for his invaluable scientific contributions and much appreciated advice and support. Finally, I would like to thank the organizers of this meeting for the exciting scientific program, and for inviting me. Financial support from IAU and from NASA through Hubble Fellowship grant HF-01167.01 is gratefully acknowledged.

References

Abraham, R. et al. 1996 *ApJ* **471**, 694–719.

Andreon, S. 1998 *ApJ* **501**, 533–538.

Couch, W.J., Barger, A. J., Smail, I., Ellis, R. S., Sharples, R. M. 1994 *ApJ* **430**, 121–138.

Diaferio, A. 1999, *MNRAS* **309**, 610–622.

Diaferio, A. & Geller, M. J. 1997, *ApJ* **481**, 633–643.

Dressler, A. 1980 *ApJ* **236**, 351–365.

Dressler, A., Oemler, A. Jr., Butcher, H. R., Gunn, J. E. 1994 *ApJ* **435**, L23–L26.

Dressler, A. et al. 1997 *ApJ* **490**, 577–591.

Ellis, R. S., Smail, I., Dressler, A., Couch, W. J., Oemler, A. Jr., Butcher, H., Sharples, R. M. 1997 *ApJ* **483**, 582–596.

Kneib, J.-P.. et al. 2003 *ApJ* **598**, 804–817.

Kodama, T., Smail, I., Nakata, F., Okamura, S., Bower, R. G. 2001 *ApJ* **562**, L9–L12

Morris, S. L., Hutchings, J. B., Carlberg, R. G., Yee, H. K. C., Ellingson, E., Balogh, M. L., Abraham, R. G., Smecker-Hane, T. A. 1998 *ApJ* **507**, 84–101.

Navarro, J., Frenk, C. & White S. 1997 *ApJ* **490**, 493–508.

Smith, G.P., Treu, T., Ellis, R.S. Moran, S.M., Dressler A. 2004 *ApJ, submitted, astro-ph/0403455*

Treu, T., Ellis, R. S., Kneib, J.-P., Dressler, A., Smail, I., Czoske, O., Oemler, A., Natarajan, P. 2003 *ApJ* **591**, 53–78.

van Dokkum, P. G., Franx, M., Kelson, D. D., Illingworth, G. D., Fisher, D., Fabricant, D. 1998 *ApJ* **500**, 714–737.

Outskirts of Galaxy Clusters: Intense Life in the Suburbs
Proceedings IAU Colloquium No. 195, 2004
A. Diaferio, ed.

DOI: 10.1017/S174392130400081X

Spectroscopic signatures of galaxy evolution in Cl0024+16 at z∼0.4

Sean M. Moran[1], **Richard S. Ellis**[1] **and Tommaso Treu**[2]

[1]California Institute of Technology, Department of Astronomy, Mail Code 105-24, Pasadena, CA 91125, USA email: smm@astro.caltech.edu, rse@astro.caltech.edu

[2]University of California at Los Angeles, Department of Physics & Astronomy, Los Angeles, CA 90095, USA email: ttreu@astro.ucla.edu

Abstract. We report the first results from a panoramic spectroscopic survey of galaxies in the rich cluster Cl0024+1654 ($z \simeq 0.4$). Using HST imaging we examine the properties of early-types as a function of cluster radius. At all cluster radii, our sample lies on a self-consistent Fundamental Plane whose zero point implies evolution since $z = 0$ corresponding to $\Delta[\log(M/L_V)] = 0.14 \pm 0.06$, an overall trend consistent with previous work. Using diagnostic [O II] emission and Balmer absorption lines, we locate a population of intrinsically faint galaxies at 1–2.4 Mpc radius undergoing a period of star formation. The luminosity-dependent radial trends are suggestive of the gradual quenching of star formation for infalling galaxies. We discuss physical mechanisms that may be responsible for this environmental evolution.

---

1. Introduction

Environmental processes have played a significant role in shaping the morphological evolution of galaxies in clusters. Galaxy clusters at intermediate redshift provide excellent laboratories to study these environmental effects, at a time when they were perhaps most active. At $z \sim 0.4$, clusters were still actively forming, with a much higher fraction of blue, star-forming galaxies than is seen today (Butcher & Oemler 1984). Yet the morphology–density relation was already in place (e.g. Treu et al. 2003), suggesting that environmental processes were already working to curtail star-formation within infalling galaxies. Compared to the local universe, there seems to be a deficit of S0 galaxies that nearly matches an excess of spirals (Dressler et al. 1997). As a result, it is thought that environmental processes may work to transform spiral galaxies into S0s, though the physical mechanisms responsible are still poorly understood.

Spectroscopy of cluster members at various stages of infall into the cluster can provide a key to the dominant environmental processes. Depending on the process at work, different physical mechanisms will leave different spectral and dynamical signatures in the affected galaxies. While many physical processes occur simultaneously in the central 0.5–1 Mpc of a cluster, the various mechanisms can be disentangled by contrasting differences observed across the entire cluster, from the periphery where galaxies are just beginning to feel the cluster potential, through the transition region near the virial radius, and into the virialized core.

We present here the first results from a wide-field spectroscopic study of Cl0024+1654 at $z \sim 0.4$, which complements the *Hubble Space Telescope (HST)* imaging on this cluster presented in Treu et al. (2003). Making use of the morphological classifications of Treu et al. (2003), we analyze spectral features and kinematics of confirmed cluster members.

By dividing our galaxies according to morphology, we can better trace how the cluster environment affects each class of galaxies, independently of spectral classification. Previous spectroscopic studies of moderate redshift clusters have either focused on the cluster core, where morphologies are available (e.g., Poggianti et al. 1999), or have relied on spectral types or colors to trace the effects of environment out to the cluster periphery (e.g. Fisher et al. 1998). With our more extended sample, we can trace the detailed spectral properties and kinematics of both early and late type galaxies across the cluster, providing a more complete picture of how infalling galaxies are affected by the cluster environment.

We focus here on cluster members classified as E+S0 by the HST imaging survey. A future paper will address the properties of spiral galaxies in Cl0024.

2. Observations

Treu et al. (2003) presented the results of a wide-field HST imaging survey of Cl0024+16. The survey includes 39 sparsely-sampled WFPC2 images taken in the F814W filter ($\sim I$ band), providing good coverage of the cluster field out to radius > 5 Mpc. Treu et al. (2003) reports morphological classifications down to $I = 22.5$. Classifications to a limiting magnitude of $I = 21.1$ were found to be very reliable. For galaxies between $I = 21.1$ and $I = 22.5$, broad classification as early or late type is still reliable, though differentiating between E and S0 galaxies, for example, is less certain.

In October 2002, and again in October 2003, we used the DEIMOS spectrograph on *Keck II* to obtain deep (2.5 hrs) spectra of $\sim$ 1000 galaxies in the field of Cl0024. Selection priority was given to known cluster members with *HST* images, followed by unknown galaxies in the *HST* survey with $I < 21.0$; galaxies without *HST* images, or those with magnitudes $21.0 < I \leqslant 22.5$ filled the slit mask gaps. Spectra were reduced using the DEEP2 DEIMOS data pipeline. In total, we have obtained spectra of over 200 cluster members to $I = 22.5$. We present initial results on 74 E+S0 galaxies, with 54 of these having high quality spectra.

In our sample, no galaxy dimmer than $I \sim 21$ yielded a high quality spectrum. This roughly matches the magnitude limit of Treu et al. (2003) for precise morphological classification. We therefore divide our E+S0 galaxies into a sample that is brighter than $I = 21.1$ and one that includes all observed E+S0 members down to $I = 22.5$. An M_V^* galaxy in Cl0024 corresponds to $I = 19.5$ for ΛCDM cosmology with $h = 0.72$ (Smail et al. 1997; Treu et al. 2003), such that our two samples represent galaxies brighter than $M_V^* + 1.6$ and $M_V^* + 3.0$, respectively.

3. Results

3.1. *Fundamental Plane of Cl0024*

For our brighter sample, we are able to study galaxy kinematics via velocity dispersions. With the addition of surface photometry from the *HST* images, we can derive the parameters of the fundamental plane (FP) in Cl0024: effective radius R_e, mean surface brightness SB_e, and central velocity dispersion σ_0 (Djorgovski & Davis 1987; Dressler et al. 1987). We study the mass-to-light ratio of cluster galaxies as a function of cluster environment and galaxy sub-class (elliptical or S0).

Previous studies have traced a mild shift in the intercept of the FP with redshift (e.g. Wuyts et al. 2004; Kelson et al. 2000c). This seems to be consistent with passive luminosity evolution of stellar populations with a high redshift of formation (Wuyts et al. 2004), though biases due to morphological evolution are difficult to quantify. All such studies, however, have focused on the central regions of the clusters studied, or on a small field sample. With our broader spatial coverage, we can uncover any significant difference

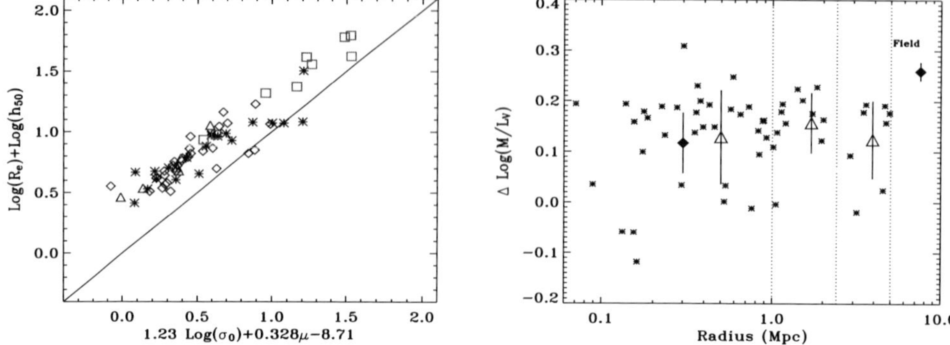

Figure 1. Left: FP of Cl0024+16, compared to Coma cluster (solid line). Asterisks are galaxies classified as E, open diamonds are S0, and open triangles are E/S0. Open squares are from van Dokkum & Franx (1996). **Right:** Change in M/L_V ratio for galaxies in Cl0024, with respect to $\langle M/L_V \rangle$ for Coma. Asterisks represent individual galaxies. Open triangles: averages for the three radial zones indicated. Error bars reflect the rms scatter. Filled diamonds: points from van Dokkum & Franx (1996) and Treu et al. (2001) for cluster core and field, respectively.

in the mean M/L_V of early types as a function of radius. Our sample also extends to fainter magnitudes than previous studies at $z \sim 0.4$, allowing us to probe M/L_V for smaller early-types that perhaps formed later than the most massive cluster ellipticals.

In order to determine velocity dispersions for early-types with high quality spectra, we fit them to a grid of stellar templates degraded to the instrumental resolution and smoothed to various velocity dispersions (Van der Marel 1994). We apply a 5% correction to match a 3.4″ aperture at the distance of Coma (Jørgensen, Franx, & Kjærgaard 1996). In addition, *GALFIT* (Peng et al. 2002) was used to derive effective radii and surface brightnesses for all galaxies with velocity dispersions.

Figure 1 presents the FP of Cl0024 compared to that of the Coma cluster. We find that the average offset from the Coma FP implies a change in the M/L_V ratio between $z = 0.4$ and $z = 0.02$ of $\langle \Delta \log (M/L_V) \rangle = 0.139 \pm 0.06$. This is a slightly smaller evolution than found in the field at $z \sim 0.4$ by Treu et al. (2001), though it is in excellent agreement with the offset determined by Kelson et al. (2000c) for a cluster at $z = 0.33$.

Red squares in Figure 1 indicate points from van Dokkum & Franx (1996), who also measured the FP in Cl0024. The two sets of data fall on the same plane, and our inferred $\langle \Delta \log (M/L_V) \rangle$ is consistent with their work. Combining our two datasets, we find that, within the errors, the tilt (Ciotti, Lanzoni & Renzini 1996) of the FP of Cl0024 matches that of the Coma cluster FP. This suggests that the tilt of the FP cannot be explained by a simple relation between galaxy age and mass (Treu et al. 2001).

When dividing our sample into three subclasses (E, S0, and E/S0), we find no evidence of significant variation between the populations, either in scatter, intercept, or tilt of the FP. Furthermore, the mean M/L_V does not vary with cluster radius, as shown on the right-hand side of Figure 1. However, the uncertainties in these preliminary results preclude us from identifying any subtle differences or trends that may exist.

3.2. *Radial trends in spectral features*

To probe for star formation in cluster early-types, we examine diagnostic spectrum lines as a function of cluster radius, both taking ensembles and galaxy-by-galaxy. This is done for the entire sample to $I = 22.5$. [OII] emission is sensitive to ongoing star formation

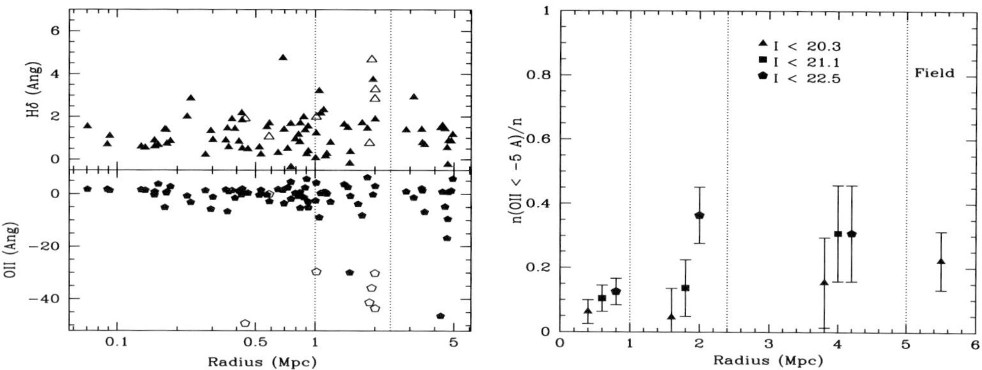

Figure 2. Left: Hδ and [OII] EWs for all E+S0 members as function of radius. Dotted lines delineate the core, transition region, and periphery. Open symbols: $21.1 < I < 22.5$. Filled symbols: $I < 21.1$. **Right:** Fraction of E+S0 galaxies with EW([OII]) < -5Å, as a function of radius. Triangles represent fraction for galaxies with $I < 20.3$. The field value is from Treu et al. (2002). Squares and pentagons are the same measure, but to $I = 21.1$ and $I = 22.5$, respectively.

and indicates the presence of gas, whereas the Balmer absorption line $H\delta$ is indicative of recently-completed activity.

Figure 2 shows these two diagnostics, measured as equivalent widths, as a function of radii, both on an individual galaxy basis and as means. On the left is a split plot with EW(Hδ) on top and EW([OII] 3727Å) below it. On the right, we plot the fraction of all E+S0 galaxies with EW([OII])< -5Å, averaged across each of the radial zones indicated. Figure 3 shows the coadded spectra of all early-types binned by radial zone; top to bottom, they trace the mean early-type spectrum from the periphery to the core. All galaxies with EW(OII) < -15Å were coadded separately; see Figure 3, right panel.

Examining individual measurements of EW([OII]), we find a population of galaxies with strong emission, which are concentrated in a narrow range in radius, close to the virial radius at 1.7 Mpc. The spectra of these galaxies are dominated by emission lines, including [OIII] and several Balmer lines (see Figure 3, right panel). These emission line galaxies are preferentially dim; as denoted by the un-filled pentagons in Figure 2, most are in the magnitude range $21.1 < I \leqslant 22.5$. Previous studies have not been sensitive to early-type cluster galaxies at these magnitudes and in this radius range. If this emission is due to star formation, these galaxies could be undergoing mergers or some interaction with the ICM.

In addition to the emission line galaxies, we see a gradual trend of increasing [OII] emission and, to a lesser extent, Balmer absorption with increasing radius. This suggests an encounter with the cluster environment which serves to suppress star formation gradually during infall. Co-added spectra of 'normal' E+S0 members in Figure 3 show a clear increase in an average galaxy's OII emission with radius, but little difference in the average galaxy's $H\delta$ line strength.

On closer examination, there is a luminosity trend in the fraction of E+S0 galaxies with EW([OII]) < -5. (Figure 2). For the brightest galaxies, $I < 20.3$, the fraction of [OII]-emitters decreases monotonically, from the field measurement of Treu et al. (2002) to the cluster core. However, when we include slightly dimmer galaxies, to $I = 21.1$, we see an enhanced fraction of emitters in the cluster periphery, though [OII] emission is still rare within 2.4 Mpc. Now including our full sample to $I = 22.5$, the fraction of star-forming galaxies in both the transition region and periphery are higher than expected. In the transition region, the jump represents the addition of the dim but strongly emitting

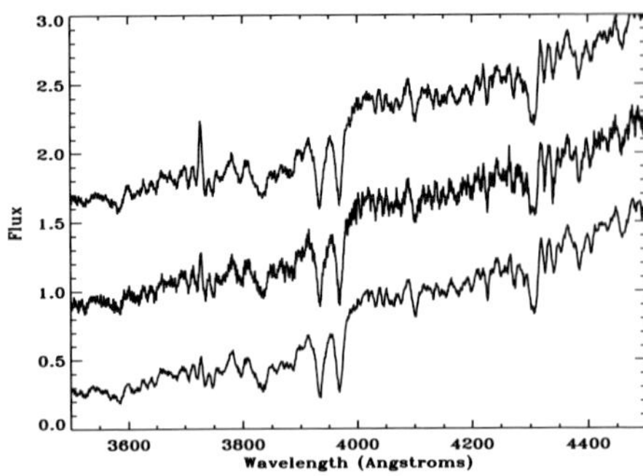

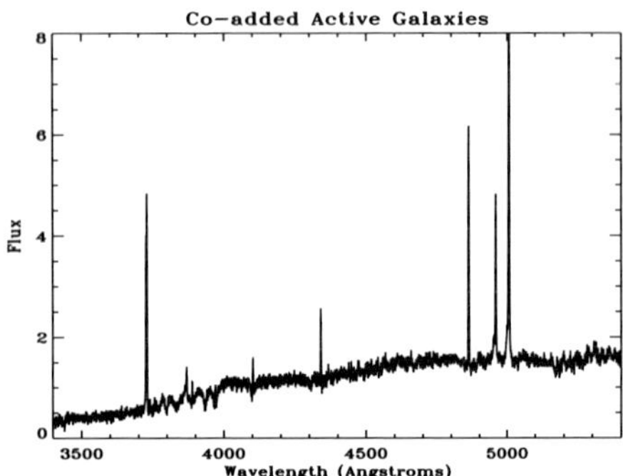

Figure 3. Left: Coadded spectra for all E+S0 members, divided by distance from cluster core. From top to bottom: periphery, transition region, and cluster core. **Right:** Coadded spectrum of all E+S0 cluster members with EW(OII) < -15Å. Wavelengths in rest frame.

galaxies discussed above. The overall trend suggests that the star formation observed in the outskirts of this cluster is suppressed earliest in the largest and brightest E+S0s.

4. Discussion

The brightest galaxies in our sample ($I < 20.3$) follow the FP and show few signs of star formation. Similarly, galaxies to $I = 21.1$ lie on the FP and exhibit star formation only in the periphery. Perhaps star formation is quenched early during infall of these galaxies, with passive luminosity evolution from that point. Because the dimmest galaxies seem to undergo big bursts of star formation (though AGN activity has yet to be ruled out) near the virial radius, it could be that they are simply swallowing small satellite galaxies. In the dimmest hosts, the addition of a small satellite with ongoing star-formation will be the most obvious. This could also happen with the slightly brighter star-forming galaxies in the periphery, but the observable effect of swallowing a satellite is smaller.

We have presented evidence for a population of actively star-forming early type galaxies, concentrated in distance from the cluster center. In addition, moderate star formation is seen across a range of radii. The radius at which this star formation disappears seems to be correlated with galaxy luminosity, possibly due to galaxy cannibalism of small satellites. We intend to expand this analysis to the spiral cluster members, in the hopes of obtaining a more complete picture of how the cluster affects infalling galaxies.

References

Butcher, H., Oemler, J., Jr., 1984, ApJ, 285, 426
Ciotti, L., Lanzoni, B. & Renzini, A., 1996, MNRAS, 282, 1
Djorgovski, S. & Davis, M., ApJ, 313, 59
van Dokkum, P. & Franx, M., 1996, MNRAS, 281, 985
Dressler, A., Oemler, A., Jr., Couch, W. J., et al., 1997, ApJ, 490, 577
Dressler, A., Lynden-Bell, D., Burstein, D., et al., 1987, ApJ, 313, 42
Fisher, D., Fabricant, D., Franx, M., & van Dokkum, P., 1998, ApJ, 498, 195
Jørgensen, I., Franx, M., & Kjærgaard, P., 1996, MNRAS, 280, 167
Kelson, D. D., Illingworth, G. D., van Dokkum, P. G., & Franx, M., 2000, ApJ, 531, 184
Peng, C. Y., Ho, L. C., Impey, C. D., & Rix, H., 2002, AJ, 124, 266
Poggianti, B. M., Smail, I., Dressler, A., et al., 1999, ApJ, 518, 576
Smail, I., Dressler, A., Couch, W. J., et al., 1997, ApJS, 110, 213
Treu, T., Ellis, R. S., Kneib, J. et al., 2003, ApJ, 591, 53

Treu, T., Stiavelli, M. Casertano, S., et al., 2002, ApJ, 564, L13
Treu, T., Stiavelli, M., Bertin, G., et al., 2001, MNRAS, 326, 237
van der Marel, R., 1994, MNRAS, 270, 271
Wuyts, S., van Dokkum, P. G., Kelson, D. D., et al., 2004, ApJ, 605, 677

Outskirts of Galaxy Clusters: Intense Life in the Suburbs
Proceedings IAU Colloquium No. 195, 2004
A. Diaferio, ed.

DOI: 10.1017/S1743921304000821

Linking star formation and environment in supercluster galaxies

M.E. Gray[1], C. Wolf[2], K. Meisenheimer[3], A. Taylor[4], S. Dye[5], A. Borch[3] and M. Kleinheinrich[3]

[1]School of Physics and Astronomy, University of Nottingham, Nottingham NG7 1GW, UK
email: Meghan.Gray@nottingham.ac.uk

[2]Department of Physics, University of Oxford, Keble Road, Oxford OX1 3RH, UK
[3]Max-Planck-Institut für Astronomie, Königstuhl 17, D-69117, Heidelberg, Germany
[4]Institute for Astronomy, University of Edinburgh, Blackford Hill, Edinburgh EH9 3HJ
[5]Astrophysics Group, Blackett Lab, Imperial College, Prince Consort Road, London SW7 2BW

Abstract. We present a unique multiwavelength dataset of the Abell 901/902 supercluster at $z = 0.16$. The environment of the structure is well characterized by weak lensing analysis, galaxy dynamics, and X-ray imaging, and we examine how galaxy properties correlate with these measures of environment. We examine the star-formation–density relation and find a surface mass density threshold above which star-formation in supercluster galaxies is truncated, and that this transformation affects primarily the faint end of the star-forming galaxy population (down to $M^* + 6$). In addition, we present a preliminary dynamical analysis of the spectroscopic sample ($n \sim 300$ galaxies), revealing a bimodal distribution of galaxy velocities and significant substructure.

---

1. The dark matter environment

The influence of environment on the structure and star-formation properties of galaxies is of key importance in the quest to understand the physical process responsible for galaxy evolution. Star-formation–density relations have been observed in the SDSS and 2dF surveys (Lewis et al. 2002, Gómez et al. 2004, Balogh et al. 2004) and in the infall region around a $z = 0.4$ cluster (Kodama et al. 2001). In each of these studies, 'environment' has been characterized by local galaxy density. By contrast, here we use gravitational lensing to examine the variation of star-formation activity directly with the *dark matter environment* in a low-redshift supercluster system.

The A901/902 supercluster was one of 4 fields targeted by the COMBO-17 band survey (Wolf et al. 2003, 2004). The structure, comprised of 3 clusters and associated filametary structure all at $z = 0.16$ is fully contained in the $0.5° \times 0.5°$ field-of-view of the ESO Wide-Field Imager (WFI). The 5 broad bands and 12 medium bands ensure accurate photometric redshifts and spectral classifications in a complete sample down to $R = 24$. Furthermore, high-quality R-band imaging with sub-arcsecond seeing provides the basis for a 2D weak lensing analysis of the region (Gray et al. 2002) as well as the first attempt at a 3D lensing inversion (Taylor et al. 2004).

Having characterized the distribution of dark matter in the field with weak lensing, we investigate correlations between the location of galaxies in dense environments and their degree of star-formation activity. Using the photometric redshift catalogue we are able to precisely isolate galaxies from a 4×4 h^{-1} Mpc region around the supercluster in the range $0.15 < z < 0.18$. We measure rest-frame photometric quantities from PEGASE template fits to the 17-band photometric data, and divide supercluster galaxies into two

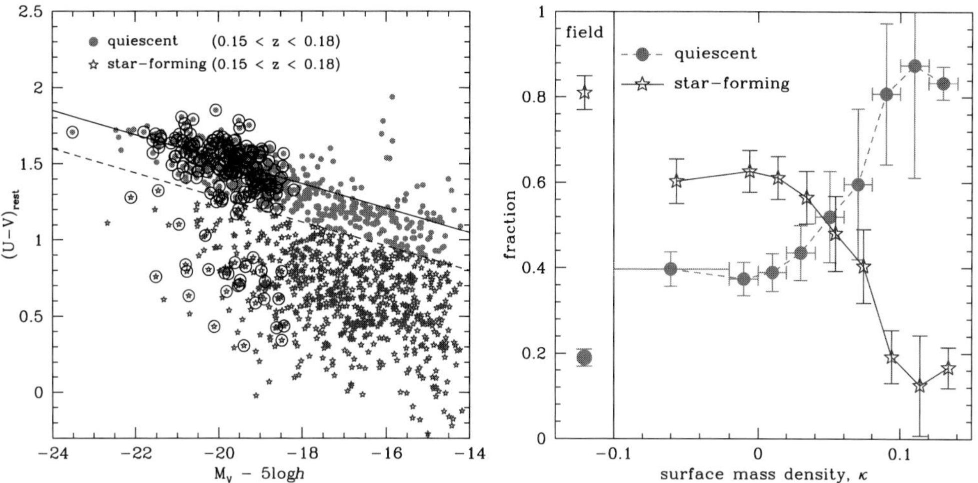

Figure 1. *Left:* The rest-frame $U-V$ vs M_V colour-magnitude diagram for supercluster galaxies with $0.15 < z_{\rm phot} < 0.18$. Open circles indicate galaxies for which 2dF spectra have been obtained. *Right:* The SF-κ relation for the A901 supercluster. A dramatic change in relative abundance of each galaxy type at a given density is seen as the proportion of star-forming galaxies declines and quiescent galaxies become more dominant in dense regions. The symbols at left show the corresponding fractions reached in the field, obtained from an identical redshift slice within two blank COMBO-17 fields (Gray et al. 2004).

classes ('quiescent' and 'star-forming') based on deviation blueward of the $(U-V)_{\rm rest}$ vs. M_V colour-magnitude sequence (Fig. 1a).

We sample the weak gravitational lensing mass-map of Gray et al. (2002) at the location of each galaxy in order to use mass as a direct measurement of environment for the first time. We observe (Gray et al. 2004):

- clear evidence of segregation by type, with the highest density regions populated almost exclusively by galaxies classified according to their rest-frame $U-V$ colours as quiescent;
- a threshold surface mass density from lensing, $\kappa = 0.05$, above which star-formation activity is rapidly suppressed (Fig. 1b);
- that this abrupt transformation affects primarily the faint end of the star-forming galaxy population (down to $M^* + 6$) and occurs at a local surface number density that is an order-of-magnitude above the density of the field;
- that when only galaxies brighter than $M_V^* + 1$ are considered the trends with environment remain, but are more gradual and extend beyond $2h^{-1}$ Mpc radius.

2. The spectroscopic sample

The A901/902 field was observed with the 2dF spectrograph on the AAT over two observing campaigns: one night in March 2002 (using the 1200B grating) and one night in March 2003 (using the 600V grating). Targets having $R < 20$ were selected in the first instance by COMBO-17 redshift indicating likely cluster membership. Additional fibres were allocated to other objects in the field where possible. Even with multiple fibre configurations (one per night) the limitations placed on fibre allocation by the crowded nature of the field meant that the sampling was incomplete and that several of the brightest cluster members were missed.

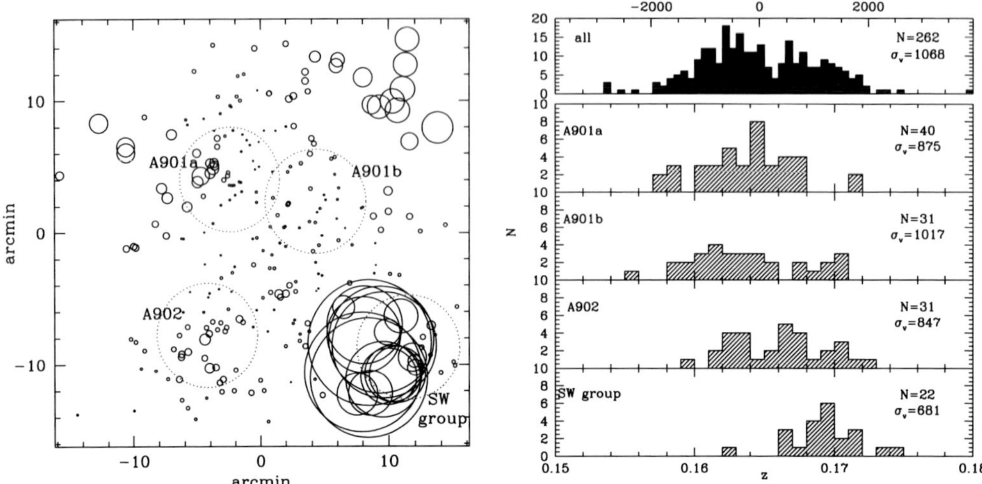

Figure 2. *Left:* The results of a Dressler-Shectman test on supercluster velocities, revealing significant substructure. The large number of large circles in the SW corner indicate a group that is distinct relative to the three main clusters both spatially and kinematically. *Right:* Redshift histograms and velocity dispersions for the full sample (top) and the three clusters and one group selected in regions indicated by the large dashed circles in the left-hand image. The full sample hints at some degree of bimodality, but the spatially-selected structures do not segregate in redshift space in a straightforward manner (with the exception of the SW group centred around $z = 0.17$).

Galaxy redshifts were determined and cross-checked according to two methods: (a) fitting of the prominent H and K absorption features by eye, and (b) cross-correlation with known templates. In total 373 galaxy redshifts were obtained, 263 of which were in the range of interest $0.15 < z < 0.18$. Of that sample, 43 were observed during both campaigns. A Dressler-Shectman (1988) test shows significant substructure in the field, particularly a group to the SW of the main structure (Fig. 2a). Monte Carlo simulations using randomized velocities show that the probability of producing this level of substructure by chance is $P < 0.0001$. Redshift histograms for the full field and spatially selected galaxies around the known mass peaks and galaxy overdensities reveal distinct components, particularly a group at $z = 0.16$ southwest of the three main clusters.

3. Further work

In addition to the aforementioned multicolour and spectroscopic dataset, we are also in the process of analyzing a deep 90ks XMM of the field. A rudimentary first-look analysis shows extensive diffuse emission from all three clusters as well as the SW group, and a large number of point sources which appear to trace the same structure. The relative prominence of the extended emission around each cluster appears contrary to naive predictions from either the weak lensing map or the galaxy distributions. Interestingly, the largest extended peak in the X-ray map corresponds to A901b: in this cluster the galaxy distribution is highly elongated and is turn misaligned with the mass peak. Thus we find that as we dissect the supercluster layer by layer, each measure of the 'environment' yields a different picture of this complex structure.

In summary, our goal is to employ the full dataset to accurately characterize the environmental influences that may be acting on the infalling galaxies, whether by the influence of the cluster dark matter, hot X-ray gas, or high galaxy number density. We hope to disentangle which interactions (whether between the galaxy and the cluster gravitational field, the galaxy and the intracluster medium, or galaxy-galaxy interactions) are responsible for the observed transformation of photometric and morphological properties.

References

M. Balogh et al. 2004 MNRAS **348**, 1355

A. Dressler & S. Shectman 1988 AJ **95**, 985

M. Gray, A. Taylor, K. Meisenheimer, S. Dye, C. Wolf, E. Thommes 2002 ApJ **568**

M. Gray, C. Wolf, K. Meisenheimer, A. Taylor, S. Dye, A. Borch, M. Kleinheinrich 2004 MNRAS **437**, 73L

P. Gómez et al. 2003 ApJ 584, 210

T. Kodama, I. Smail, F. Nakata, S. Okamura, R.G. Bower 2001 ApJ **562**, 9

I. Lewis et al. 2002 MNRAS **334**, 673

A.N. Taylor, D.J. Bacon, M.E. Gray, C. Wolf, K. Meisenheimer, S. Dye, A. Borch, M. Kleinheinrich, Z. Kovacs, L. Wisotzki 2004, MNRAS, submitted (astro-ph/0402095)

C. Wolf, K. Meisenheimer, H.-W. Rix, A. Borch, S. Dye, M. Kleinheinrich 2003 A&A **401**, 73

C. Wolf, K. Meisenheimer, M. Kleinheinrich, A. Borch, S. Dye, M. Gray, L. Wisotzki, E. F. Bell, H.-W. Rix, A. Cimatti, G. Hasinger, G. Szokoly 2004 A&A, submitted (astro-ph/0403666)

Outskirts of Galaxy Clusters: Intense Life in the Suburbs
Proceedings IAU Colloquium No. 195, 2004
A. Diaferio, ed.

DOI: 10.1017/S1743921304000833

Galaxy evolution in dense environments: a concise HI perspective

Marc A. W. Verheijen

Kapteyn Astronomical Institute, Postbus 800, 9700 AV Groningen, The Netherlands
email: verheyen@astro.rug.nl

Abstract. Observing the neutral hydrogen in galaxy clusters provides crucial insights in the physical processes that influence the evolution of gas-rich galaxies as they migrate from the lower-density filaments through the cluster outskirts into to the higher-density central regions. The morphology-density relation, the Butcher-Oemler effect, and the observed HI deficiencies in the central regions of galaxy clusters suggest that infalling galaxies experience a strong transformation of their morphologies, star formation rates, and gas content. The physical mechanisms that trigger and govern these transformations may depend strongly on environment. This contribution aims to illustrate that the morphological and kinematic characteristics of the cold gas provide a sensitive tool to determine which mechanisms dominate in which environments.

1. Motivation for observing the cold gas

Hydrogen is the most abundant primordial baryonic matter in the Universe out of which the visible galaxies and galaxy clusters have formed. Observations of the distribution and kinematics of cold neutral hydrogen at the epoch of galaxy formation, and during their subsequent evolution, provide one of the most direct windows on the physical processes involved. Numerous studies indicate that the pace of a galaxy's evolution, like its growth by mass accretion, the rate at which it forms stars, the frequency of tidal interactions and the occurrence of minor or major mergers, depends significantly on its local and global environment. In particular the environment encountered in the outskirts of galaxy clusters is very favorable to rapid evolution and transformation of galaxies as they fall in along the lower-density filaments into the higher-density cluster cores. Such infalling galaxies encounter an increased local galaxy density, which results in more frequent tidal interactions, and they may be confronted with a hot Intra-Cluster Medium (ICM) as they approach the cluster core. This change of environment can have a dramatic impact on a galaxy's morphology, its gas content and its star formation rate. This is also implied by the well-known morphology-density relation combined with the notion that clusters continue to accrete galaxies from the surrounding large scale structure in which they are embedded. Consequently, infalling galaxies must experience a significant transformation on a relatively short time scale, which is also suggested by the Butcher-Oemler effect and the interpretation that the blue galaxies in the outskirts of galaxy clusters experience a period of vigorous star formation which can end abruptly once all the cold gas is consumed or removed. These galaxies eventually find themselves in the cluster core as gas-poor systems with a passively evolving stellar population.

The presence of cold gas in galaxies is a prerequisite for star formation, and the amount and distribution of cold gas regulates the duration and intensity of star formation periods. The removal of this cold gas reservoir from infalling galaxies is a crucial element in the process that transforms these systems. The key question is which physical processes trigger and govern the enhanced star formation, and the removal of the cold gas. What is the role and ultimate fate of the cold neutral hydrogen in infalling galaxies?

Table 1. Galaxy clusters imaged in neutral hydrogen, compiled from the literature and from work by the author and his collaborators.

Cluster	z	σ km/s	N_z	T_{B-M}	C	R	L_X 10^{44}erg/s	f_b	Array	Mode	# Flds	N_{det}
(1)	(2)	(3)	(4)	(5)	(6)	(7)	(8)	(9)	(10)	(11)	(12)	(13)
Ursa Major	0.003	149	96		26				WSRT	pointed	65	70
									VLA	blind	56	
Virgo	0.004						1.00	0.04	VLA	pointed	23	23
Hydra	0.013	676	256	III	50	1	0.47		VLA	vol. lim.	13	49
Abell 3627	0.014	897	112	I	59	1	2.2		ATCA	pointed	3	2
Abell 262	0.016	415	>38	III	40	0	0.55	0.02	WSRT	pointed	5	11
Coma	0.023	880	$\sim 10^3$	II	106	2	7.21	0.03	VLA	pointed	12	39
									WSRT	blind	mosaic	
Abell 496	0.033	657	238	I:	50	1	3.54		VLA	vol. lim.	10	46
Hercules	0.037	887	>42	III	87	2	0.88	0.14	VLA	blind	4	61
Abell 754	0.054	1048	408	I-II:	92	2	8.01		VLA	vol. lim.	2	4
Abell 85	0.055	1443	275	I	59	1	8.38		VLA	vol. lim.	8	23
Abell 2670	0.076	908	219	I-II	142	3	2.55	0.04	VLA	vol. lim.	3	49
Abell 2218	0.176			II:	214	4	8.99	0.11	WSRT	vol. lim.	1	1
Abell 2192	0.188	706	36	II-III:	62	1			VLA	vol. lim.	1	1
									GMRT	vol. lim.	1	0
Abell 1689	0.183	1253	525	II-III:	228	4	20.74	0.09	VLA	vol. lim.	1	0
Abell 963	0.206		~200	I-II	134	3	10.23	0.19	WSRT	vol. lim.	1	0

Ursa Major: Verheijen & Sancisi (2001), Verheijen *et al.* (2001). Virgo: Cayatte *et al.* (1990). Hydra: McMahon (1993). A3627: Vollmer *et al.* (2001). A262: Bravo-Alfaro *et al.* (1997). Coma: Bravo-Alfaro *et al.* (2000), Beijersbergen (2003). A496: van Gorkom *et al.* (in progress). Hercules: Dickey (1997). A754: van Gorkom *et al.* (in progress). A85: van Gorkom *et al.* (in progress). A2670: Poggianti & van Gorkom (2001). A2218: Zwaan *et al.* (2001). A2192: Verheijen *et al.* (in progress). A1689: Verheijen *et al.* (in progress). A963: van Gorkom *et al.* (in progress).

2. Physical processes

Observing the cold gas in transforming galaxies inherently provides important clues about the physical processes that lead to its removal. Many mechanisms have been proposed that could result in the depletion of the cold gas. Only some are mentioned here.

The fragile extended gas disks of spiral galaxies are effective tracers of tidal interactions. Tidal interactions between galaxies, either by a few slow but effective encounters or by many 'harassing' fast encounters (e.g. Moore *et al.* 1996), may result in the removal of cold gas from the tenuous outer regions, leading to truncated HI disks. Once an infalling galaxy encounters the hot ICM, the cold gas may be removed by ram-pressure, and turbulent and viscous stripping (e.g. Schulz & Struck 2001; Abadi *et al.* 1999), provided the ICM is sufficiently dense and the galaxy moves fast enough. Ram-pressure may also compress the cold gas clouds, triggering enhanced star formation. The cold gas could also be removed by more gentle and slower processes like thermal evaporation of the ISM by the hot ICM, provided there is sufficient thermal conduction (e.g Cowie & Songaila 1977). The slow removal of the cold gas, regardless of the physical process, was labeled 'starvation' by Treu *et al.* (2003). Besides the physical removal of the cold gas from the gravitational potential, a triggered burst of star formation may quickly consume all the cold gas and lock it up in a new generation of stars. This exhaustion of the cold gas reservoir can be achieved within a few 10^7 years during a dusty and possibly obscured star burst. Luckily, the radio continuum emission, a by-product of HI synthesis imaging, can

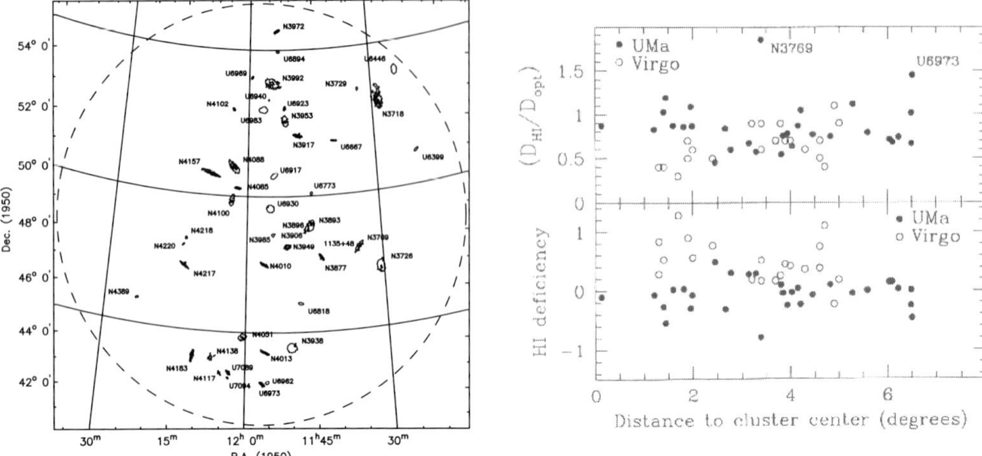

Figure 1. Left: Composition of integrated maps of atomic hydrogen in the Ursa Major cluster. Individual galaxies are 4× enlarged. **Right:** A comparison of HI properties of galaxies in the Ursa Major and Virgo clusters as a function of projected distance to the cluster center. Above: ratio of HI to optical diameters. Below: overall HI deficiency as a function projected distance.

be used to derive star formation rates which are unaffected by dust extinction, although contamination by AGN activity may be significant.

Apart from providing insights in the physical processes mentioned above, neutral hydrogen observations may also help to determine the amount of substructure, the mass accretion rate, and the dynamical state of galaxy clusters. A volume limited survey reveals the location of all the gas-rich galaxies, and helps to map the large scale structure in which these galaxy clusters are embedded. For instance, if a subclump of galaxies is observed to be gas-rich, it is quite unlikely it has crossed the high density cluster core on its infall trajectory. Furthermore, in case the cold gas is in the process of being removed from the plane of an infalling galaxy, its offset may indicate the direction in which a galaxy is moving and thus help to constrain its orbit.

Finally, observed metal abundances of the ICM suggest that at least some of it finds its origin in the Inter-Stellar Medium (ISM) and it would be interesting to know how much the effect of ram-pressure stripping contributes to this.

3. Galaxy clusters imaged in HI

In order to investigate the role that the cold gas plays in the transformation of infalling galaxies, and the influence of the environment on the various physical processes, a significant sample of galaxy clusters has been and is being imaged in the 21cm line of neutral hydrogen. Table 1 provides an overview of these clusters, complemented by existing studies from the literature. The clusters span the full range of redshifts practically accessible by present day telescopes, up to a redshift of 0.2 where the nearest Butcher-Oemler clusters are found. The clusters also span a range in richness, X-ray luminosity and dynamical state. Some clusters display significant substructure while others, like Abell 963, seem to be fully relaxed. These clusters and their surroundings encompass a large variety of environments, ranging from the most dense cluster cores to areas of significant underdensity. One example of a low density diffuse cluster is presented in the left panel of Figure 1: Ursa Major.

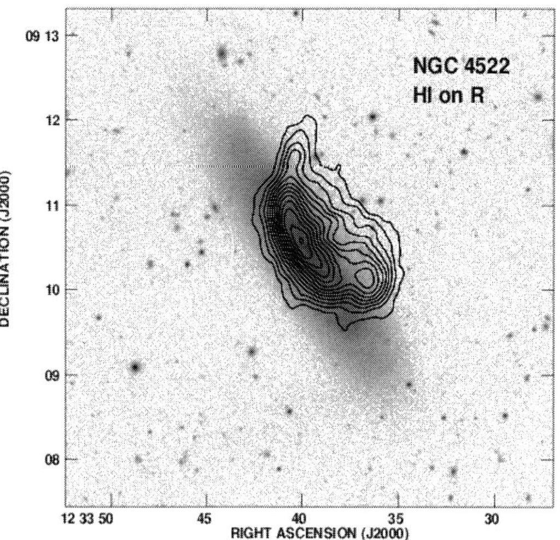

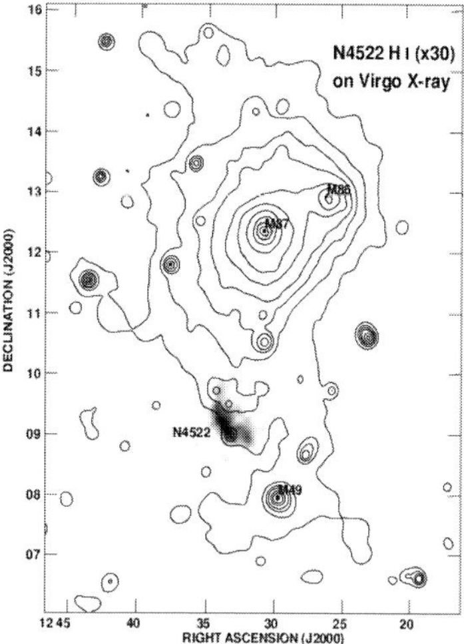

Figure 2. Left: NGC 4522 in the Virgo cluster. The HI contours are displaced from the optical disk and clearly show the sweeping effect of ram-pressure stripping. **Right:** The location of NGC 4522 with respect to the distribution of X-ray emitting gas in the Virgo cluster. Figures taken from Kenney, van Gorkom & Vollmer (2004).

4. Some phenomena observed in HI

Several intriguing and revealing phenomena have been observed in neutral hydrogen, and some of these will be illustrated here. First of all, from spatially unresolved single dish studies, there were strong hints already in the seventies, that galaxies in cluster cores are relatively poor in HI compared to their counterparts in the lower density field. The meaning of this HI deficiency became more clear with the advent of HI imaging data from the Very Large Array of galaxies in the Virgo cluster, as illustrated in the right panel of Figure 1. The gas disks of galaxies within 2 degrees from M87 are smaller than those of galaxies further away from the cluster center. At 4 degrees from the cluster center, the gas disks in Virgo seem to be of comparable size to those in the lower density Ursa Major environment (upper panel), but the Virgo galaxies are still more deficient in HI (lower panel); it suggests that the overall column densities of the gas disks in Virgo are lower compared to the Ursa Major galaxies. This illustrates that HI imaging gives crucial physical meaning to the notion of HI deficiency.

Early HI imaging data of the Coma and Virgo clusters showed that the gas disks of galaxies in the outskirts of galaxy clusters can be asymmetric, both morphologically and kinematically, and significantly offset from the stellar disk of a galaxy. The spectacular example of N4522 in the Virgo cluster has been recently published by Kenney, van Gorkom & Vollmer (2004) and is reproduced in Figure 2. The small gas disk is clearly offset from the stellar disk and seems to be bend and swept away from the plane, interpreted to be the result of ram-pressure stripping. The right panel of Figure 2 indicates the position of this galaxy with respect to the X-ray contours. Clearly, the relation between the location and orbit of this galaxy with respect to the ICM is not straightforward.

Figure 3 illustrates that ram-pressure stripping is not the only physical process to remove cold gas from a galaxy. The galaxies N4026 and N4111 are among the brightest lenticulars in the Ursa Major cluster. This dynamically young cluster has a low velocity dispersion, lacks a central concentration, shows no X-ray emission from a hot ICM, contains maybe one elliptical galaxy and a dozen lenticulars. It is an entirely different environment than the Virgo cluster. The three brightest lenticular galaxies are located

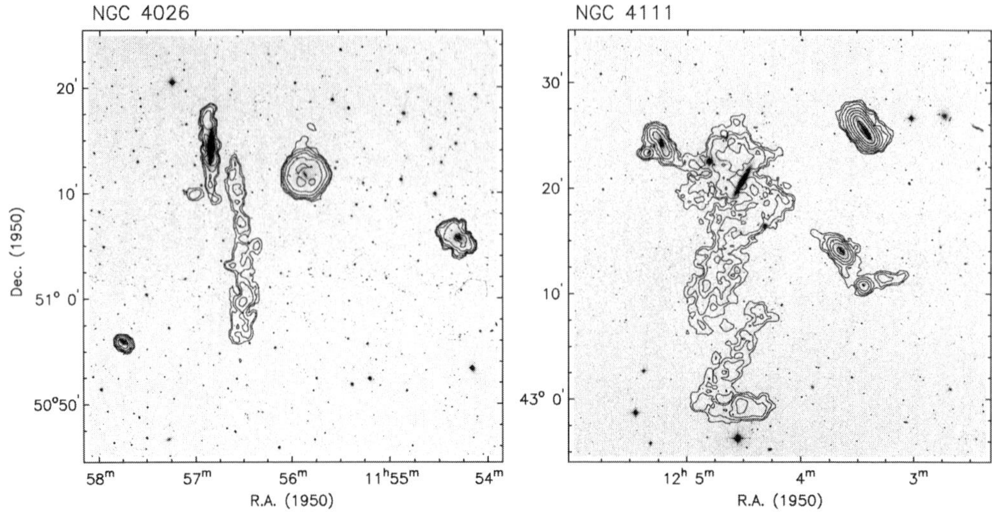

Figure 3. HI filaments observed near the brightest lenticular galaxies in the Ursa Major cluster. These lenticulars are located in small sub-groups within the Ursa Major volume.

in small subclumps within the Ursa Major volume, and all three (including N3998) show extended HI filaments. The orientation and kinematics of these filaments suggest that they were tidally stripped from the outer disks by their nearby companions. Do we witness here the transformation of a late-type spiral galaxy into a lenticular galaxy through tidal stripping, or the accretion of cold gas from the Inter-Galactic Medium (IGM)? Detailed modelling is required to answer this question. However, these lenticular galaxies support the notion that significant pre-processing of the cold gas component of galaxies may already occur in the lower density filaments before infalling galaxy are confronted with the higher density environment of the cluster outskirts.

Another intriguing observation, not illustrated here, is that the HI Mass Function (HIMF) may depend significantly on the environment. A most striking result is that the slope of the HIMF as observed from the HI Parkes All Sky Survey (HIPASS, Zwaan *et al.* 2003) is much steeper (~ -1.3) than what is observed both in the Local Group (Zwaan & Briggs 2000) and in the Ursa Major cluster where a flat HIMF slope (~ -1.0) is observed (Verheijen *et al.* 2001). Furthermore, the HIPASS survey suggests there is a morphological dependence of the HIMF, which is not too surprising.

5. Abell 2192: pushing the HI redshift frontier

In an effort to include a rich galaxy cluster with a clearly observed Butcher-Oemler effect, several massive clusters near a redshift of 0.2 were selected for HI synthesis observations (see Table 1). This is the highest redshift accessible, requiring long integration times, and currently, only pilot studies have been performed to characterize feasibility and telescope performance. To be able to address issues related to selection effects, it was decided to observe a less massive and more diffuse cluster around $z = 0.2$ as well; Abell 2192. Extensive wide-field optical imaging and multi-object spectroscopy has been obtained for this cluster. As a feasibility study, A2192 has been observed with the VLA for ~80 hours, and the resulting data cube contained at least one HI detection as illustrated in Figure 4. It concerns a regular inclined barred spiral galaxy at a redshift of 0.1887, located some $15'$ or a projected 2.8 Mpc from the cluster center. This galaxy is

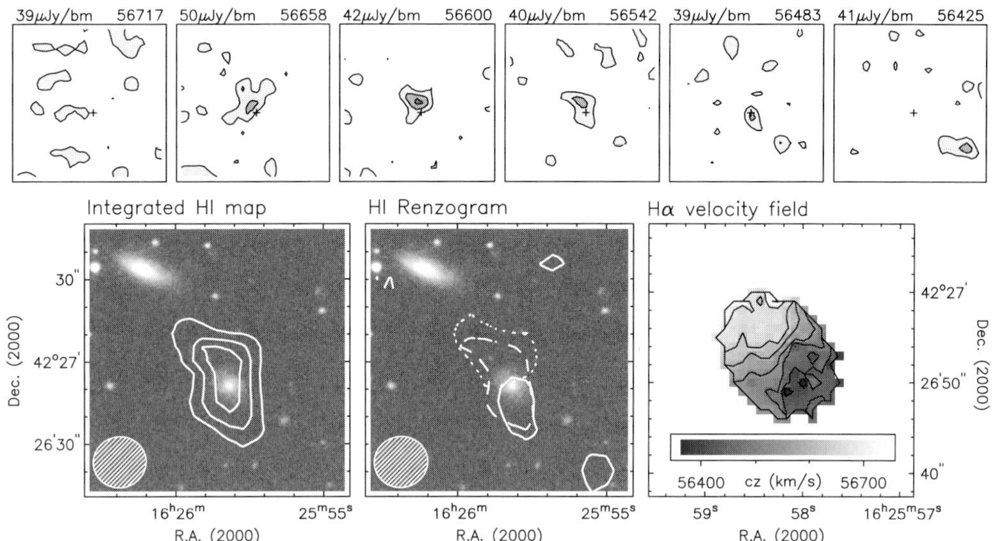

Figure 4. Atomic hydrogen detected in Abell 2192 at $z = 0.1887$. **Upper panels:** individual channel maps from the VLA datacube. The rms noise and heliocentric velocity are noted above each panel. **Lower left:** Total HI map. The contours coincide with the position of an inclined barred late-type galaxy. $M_{HI} = 7\times10^9 M_\odot$, SFR<12 $M_\odot$/yr. **Lower middle:** contours from the three channel maps plotted on top of each other; solid: $cz = 56483$ km/s, dashed: $cz = 56542$ km/s, dotted: $cz = 56600$ km/s. **Lower right:** Hα velocity field of the optical counterpart obtained with the PMAS IFU spectrograph at the 3.5m telescope on Calar Alto. There is excellent correspondce between the HI and Hα recession velocities, confirming the HI detection.

part of the large scale structure around A2192. The HI data seem to suggest a possible tidal interaction with a bright and somewhat lopsided nearby galaxy of yet unknown redshift, just to the northeast. This highest redshift HI emission detected to date demonstrates that HI imaging of Butcher-Oemler clusters at $z = 0.2$ is feasible with present day facilities.

6. Conclusions and future plans

Spatially resolved HI synthesis imaging of galaxy clusters provides crucial insights into the physical processes that trigger and govern the evolution and transformation of galaxies in the outskirts of clusters. Spatially resolved data have given physical meaning to HI deficiencies. The HI morphologies and kinematics of infalling galaxies help to distinguish ICM-ISM interactions from gravitational disturbances which are otherwise undetectable. The cold gas reservoir is a measure of the potential for star formation in a galaxy, and a close relation is expected between a galaxy's gas content and the evolutionary state of its stellar population, i.e. starburst, e(a), post-starburst, or passive. The radio continuum data provide dust-free measures of star formation rates. Furthermore, volume limited HI surveys of galaxy clusters help to determine the dynamical state of a cluster, and illustrates the dynamics of cluster mass accretion.

Feasibility studies for future HI imaging of Butcher-Oemler clusters at $z \sim 0.2$ have yielded the highest redshift HI emission detected to date at a redshift of 0.1887 in A2192.

Future plans entail the homogenization of the HI data sets for the various clusters listed in Table 1 in terms of volume and sensitivity; the continuation of collecting optical wide-field imaging and spectroscopy necessary to measure the blue galaxy fractions and to determine the evolutionary states of stellar populations in cluster galaxies; a new volume limited survey of the Hercules supercluster is planned in conjunction with wide-field OmegaCam imaging and other supporting observations.

Finally, the planned Square Kilometer Array (SKA) will allow us to measure the content, morphology, and kinematics of neutral hydrogen in galaxies beyond a redshift of 1. This will revolutionize our understanding of the physical processes that shape the galaxies and determine their evolution in every imaginable environment.

Acknowledgements

The National Radio Astronomy Observatory is a facility of the National Science Foundation, operated under cooperative agreement by Associated Universities, Inc. The Westerbork Synthesis Radio Telescope is operated by the ASTRON (Netherlands Foundation for Research in Astronomy) with support from the Netherlands Foundation for Scientific Research (NWO). PPak is developed within the ULTROS project under the German Verbundforschung grant 05AE2BAA/4. The author wishes to thank the Scientific Organizing Committee for their invitation and travel support.

References

Abadi, M.G., Moore, B. & Bower, R.G. 1999 *M.N.R.A.S.* **308**, 947–954

Beijersbergen, M. 2003 PhD thesis, Rijksuniversiteit Groningen, Groningen

Bravo-Alfaro, H., Szomoru, A., Cayatte, V., Balkowski, C. & Sancisi, R. 1997 *A.& A.S.* **127**, 537–546

Bravo-Alfaro, H., Cayatte, V., van Gorkom, J.H. & Balkowski, C. 2000 *A.J.* **119**, 580–592

Cayatte, V., van Gorkom, J.H., Balkowski, C. & Kotanyi, C. 1990 *A.J.* **100**, 604–634

Cowie, L.L. & Songaili, A. 1977 *Nature* **266**, 501–503

Dickey, J.M. 1997 *A.J.* **113**, 1939–1972

Kenney, J.D.P., van Gorkom, J.H. & Vollmer, B. 2004 *A.J.* **127**, 3361–3374

McMahon, P.M. 1993 PhD thesis, University of Columbia, New York

Moore, B., Katz, N., Lake, G., Dressler, A. & Oemler, A., Jr. 1996 *Nature* **379**, 613–616

Poggianti, B.M. & van Gorkom, J.H. 2001 In *Gas and Galaxy Evolution* (ed. J.E. Hibbard, M. Rupen & J.H. van Gorkom). ASP Conference Proceedings, vol. 240, pp. 599–606. San Francisco: Astronomical Society of the Pacific

Schulz, S. & Struck, C. 2001 *M.N.R.A.S.* **328**, 185–202

Treu, T., Ellis, R.S., Kneib, J.-P., Dressler, A., Smail, I., Czoske, O., Oemler, A. & Natarajan, P. 2003 *Ap.J.* **591**, 53–78

Verheijen, M.A.W., Trentham, N., Tully, R.B. & Zwaan, M.A. 2001 In *Gas and Galaxy Evolution* (ed. J.E. Hibbard, M. Rupen & J.H. van Gorkom). ASP Conference Proceedings, vol. 240, pp. 507–514. San Francisco: Astronomical Society of the Pacific

Verheijen, M.A.W. & Sancisi, R. 2001 *A.& A.* **370**, 765–867

Vollmer, B., Cayatte, V., van Driel, W., Henning, P.A., Kraan-Korteweg, R.C., Balkowski, C., Woudt, P.A. & Duschl, W.J. 2001 *A.& A.* **369**, 432–440

Zwaan, M.A.& Briggs, F.H. 1997 *Ap.J.* **530**, L61-L64

Zwaan, M.A., van Dokkum, P.G. & Verheijen, M.A.W. 2001 *Science* **293**, 1800-1803

Zwaan, M.A., Staveley-Smith, L., Koribalski, B.S., Henning, P., Kilborn, V.A. *et al.* 2003 *A.J.* **125**, 2842–2858

Outskirts of Galaxy Clusters: Intense Life in the Suburbs
Proceedings IAU Colloquium No. 195, 2004
A. Diaferio, ed.

DOI: 10.1017/S1743921304000845

H I view of the Virgo cluster outskirts: implications on galaxy evolution

José M. Solanes[1], Teresa Sanchis[1], Eduard Salvador-Solé[1] and Gary A. Mamon[2]

[1]Departament d'Astronomia i Meteorologia, CER d'Astrofísica, Física de Partícules i Cosmologia, Universitat de Barcelona, Martí i Franquès 1, E-08028 Barcelona, Spain
email: jm.solanes@ub.edu; tsanchis@am.ub.es; eduard@am.ub.es

[2]Institut d'Astrophysique de Paris (CNRS UMR 7095), 98 bis Bd Arago, F-75014 Paris, France email: gam@iap.fr

Abstract. The H I deficiency pattern of the spiral population in the Virgo cluster region reveals a significant number of galaxies at very large clustercentric distances with gaseous deficiencies comparable to those measured in the cluster centers. We have used the output of cosmological N-body simulations to investigate whether the gas-deficient galaxies on the outskirts of the Virgo cluster may have previously passed through its core. We find that the maximum radius reached by infalling galaxies as they bounce out of a Virgo-like cluster must be less than 2.5 virial radii, which results in a predicted velocity-distance diagram noticeably different from the one drawn by the data. The latter is fairly well reproduced, however, after including in the simulations distance errors at the 20% relative rms level. Yet, for several objects apparently over 5 Mpc in front or behind the Virgo center the assumption that they are instead close enough to the cluster to have passed already through its core strains the bounds of plausibility. Hence, unless these outlying H I-deficient Virgo's spirals have grossly incorrect distances, they cannot have lost their gas by interactions with the intracluster medium. This suggests that the evolution of spirals might begin already on the suburbs of clusters.

1. Introduction

The characterization of the large-scale 3D distribution of the neutral gas (H I) deficiency around the Virgo I Cluster region by Solanes *et al.* (2002) shows a significant number of galaxies with a dearth of atomic hydrogen at large Virgocentric distances. Apart from a central enhancement of the H I deficiency, which is essentially coincident with the cluster core, important neutral gas deficiency is found associated with several nearby galaxies at line-of-sight (LOS) distances ~10–15 Mpc, preferentially located to the north of M87 and moving away from the cluster with large relative velocities, and in a tentative background group of galaxies at LOS distances ~25–30 Mpc, mostly with systemic velocities close to the cluster mean and lying in the region dominated by the southern edge of the classical M49 subcluster as well as by clouds W′ and W. These peripheral gas-deficient objects show gaseous deficiencies deviating more than $3\,\sigma$ from normalcy — equivalent to a factor of 5.25 decrease in $M_{\rm H\,I}$ (Solanes *et al.* 1996) —, which are as strong as those measured in the centers of Virgo and other rich galaxy clusters.

One of the mechanisms that can most naturally account for the observed reduction in the interstellar gas content of cluster galaxies is the ram pressure ablation caused by the rapid motion of galaxies through the dense intracluster medium (ICM). However, the majority of the outlying H I-deficient spirals in the Virgo region appear too far from the cluster (> 5 Mpc) to have already passed through its core. Here we use the structure of simulated dark matter (DM) halos in space phase to demostrate that the maximum

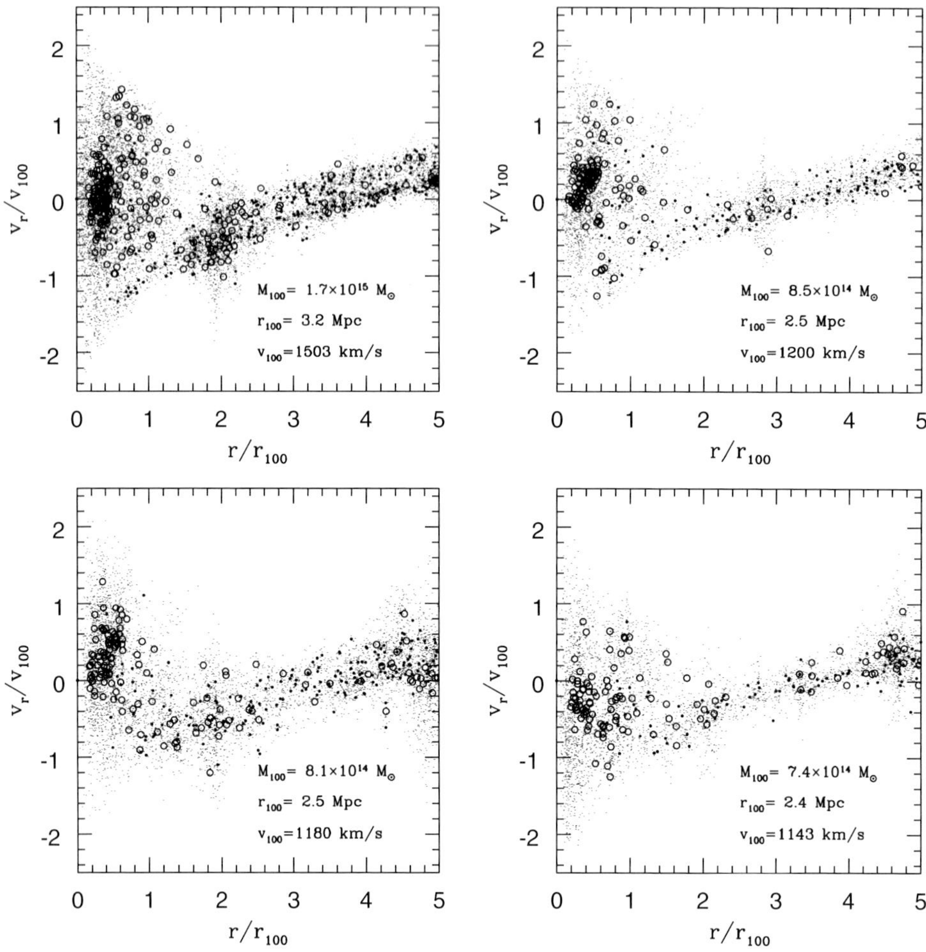

Figure 1. 3D radial phase space plots of DM particles in a ΛCDM cosmological N-body simulation for 4 massive halos at redshift $z = 0$. The *open circles* and *big dots* are the identified halos in the cosmological simulation respectively without and with galaxies within them.

rebound radius reached by galaxies falling into a cluster and bouncing out cannot explain the observed positions of these peripheral objects in velocity-distance plots. Among the alternative explanations for the presence of H I-deficient galaxies on the outskirts of the Virgo cluster that can be suggested, we pay special attention to the errors in the distance.

2. Maximum rebound radius for a Virgo-like cluster

For a better understanding of the kinematics of galaxies in the direction of the Virgo cluster, we have simulated an observation of the velocity-distance relation using the DM particles in the GalICS cosmological simulation (Hatton *et al.* 2003). This simulation is run for a flat ΛCDM universe with cosmological parameters $\Omega_0 = 1/3$, $h = 2/3$, and $\sigma_8 = 0.88$. DM halos are detected with a 'Friends-of-Friends' (FoF) algorithm with a variable linking length such that the minimum mass of the FoF groups is $1.65 \times 10^{11} M_\odot$

(20 particles) at any time step. Over 2×10^4 halos are found at the final timestep, corresponding to the present-day ($z = 0$) universe.

Figure 1 shows the radial phase space diagrams, i.e. radial velocity vs. radial distance, both relative to the halo center, for four isolated massive halos. The halo centers are provided by the simulation output and correspond to the barycenter of the FoF groups of particles. Radial distances are normalized to the virial radius, r_{100}, corresponding to the radius where the mean density is 100 times the critical density of the universe. As illustrated in this figure, particles beyond the virialized core and outside the infalling/expanding zone of phase space cannot reach any further than $\sim 2.5\, r_{100}$, in consistency with the analysis of the orbital evolution of particles in the cosmological simulations of Fukushige & Makino (2001). Moreover, particles come in groups that appear to be tidally shredded in space phase. In fact, we have verified that particle condensations associated with galaxies avoid the fairly sparse regions of phase space where the outermost outgoing particles of the simulation are seen. In summary, given our estimate of the virial radius of the Virgo cluster (see next section), *galaxies that cross through the core of Virgo cannot bounce out beyond ~4–5 Mpc from the cluster center.*

3. The velocity-distance relation of the Virgo cluster region

Among the 12 halos that `GalICS` produces at $z = 0$ with $M_{\rm FoF} > 10^{14}\, M_\odot$, we have chosen the one that resembles the Virgo cluster most, both in terms of mass and environment (see Sanchis *et al.* 2004 for details). Figure 2a shows the velocity-distance plot of the DM particles (*small points*) of this halo, conveniently rescaled to Virgo units, and from which particles statistically fainter than the typical magnitude limit of the observed catalogs (namely $B < 18$) have been removed. Any of the isolated massive halos in the `GalICS` simulation has a phase space diagram similar to that shown in Fig. 2a, once scaled to the Virgo cluster virial radius and circular velocity derived by Mamon *et al.* (2004) from X-ray observations: $r_{\rm Vir} = 1.65\, h_{2/3}^{-1}$ Mpc and $v_{\rm Vir} = 780$ km s^{-1}.

Superposed to the DM particles in Figure 2a are all the spiral galaxies (*triangles*) from Sanchis *et al.* (2002) and the early-type galaxies (*circles*) from Sanchis *et al.* (2004) lying within a cone of half-opening 9° aligned with the axis connecting the observer to the center of the halo. This figure shows that the velocity field drawn by the galaxies is noticeably different from the one followed by the DM particles of the simulation. Nevertheless, both the distance to the halo and the rescaling factors appear to be well chosen, so the differences between the Virgo cluster and our Virgo-like halo should be attributable to other factors, such as the non-negligible errors affecting distance estimates. In contrast, because the errors in velocity are negligible, it appears unlikely that the galaxies that, for a given distance, lie within more than 1000 km s^{-1} from the locus of DM particles belong to the low- or high-velocity tails of large velocity dispersion groups near the cluster, since in our simulations there are no such high velocity dispersion groups outside the main mass concentration.

As expected, the observed velocity-distance diagram is fairly well reproduced when we incorporate to the simulation the mean relative distance error for the spiral galaxy sample calculated in the study by Solanes *et al.* (2002), who found a dispersion of the Tully-Fisher relation of roughly 0.4 mag, corresponding to an uncertainty of $\sim 20\%$ in relative distance (see Fig. 2b). Yet, it is possible to identify up to 13 highly H I-deficient galaxies with apparently reliable distance estimates that 1) appear to be located well in the foreground or background of the Virgo cluster (sometimes even outside the general locus of the DM particles) or 2) lie just outside the projected virial radius of the cluster,

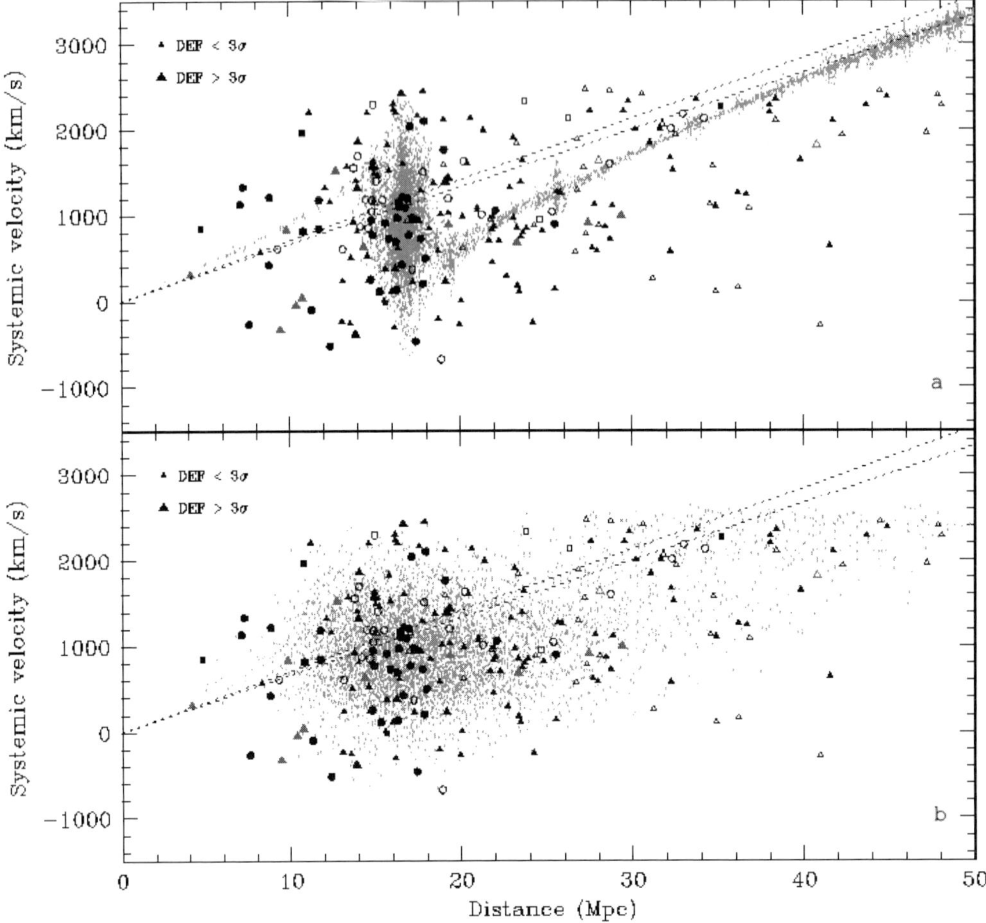

Figure 2. (a) Simulated and observed velocity-distance diagrams. *Dots* represent the velocity field traced by the DM particles. *Circles* and *triangles* represent early- and late-type galaxies, respectively. H I-deficient galaxies are shown as *filled triangles*, with the most probable cluster outliers (see Table 1) shown as *gray filled triangles*. *Squares* represent spirals with no H I deficiency data. *Open symbols* indicate galaxies with uncertain distances. *Dashed* lines show the unperturbed Hubble flow with $H_0 = 66.7$ and 70 km/s/Mpc, respectively (going upwards). (b) Same as (a) but incorporating Gaussian 20% relative distance errors for the DM particles.

for which their gas deficiency appears to be inconsistent with galaxy-ICM interactions (see the large gray filled triangles of Fig. 2b, as well as Table 1).

4. Implications of the outlying H I-deficient S's on galaxy evolution

Several alternatives for the origin of the H I deficiency of the 13 objects listed in Table 1 have been thoroughly assessed by Sanchis *et al.* (2004). Grossly incorrect distances, tidal interactions with close neighbors, and biases in the estimated H I-deficiency (e.g. S0/a's misclassified as Sa's) can provide valid explanations for almost every case of gas deficiency for these peripheral objects. Still the most plausible explanation for the gas-deficient galaxies lying behind the Virgo cluster at ∼25–30 Mpc from us is that they are in a background concentration of newcomers. The fact that 15 (60% of the total) of the

Galaxy VCC	NGC	RA (J2000)	dec	T	B_T^c	D_{25}	θ	v ($\mathrm{km\,s^{-1}}$)	D (Mpc)
—	4064	$12^h04^m11\overset{s}{.}8$	$18°26'33''$	1	11.73	$4\overset{'}{.}0$	$8\overset{\circ}{.}8$	837	9.8
522	4305	$12^h22^m03\overset{s}{.}5$	$12°44'27''$	1	12.87	$2\overset{'}{.}0$	$2\overset{\circ}{.}2$	1814	40.7
524	4307	$12^h22^m06\overset{s}{.}3$	$09°02'27''$	3	11.84	$3\overset{'}{.}5$	$4\overset{\circ}{.}0$	913	27.4
559	4312	$12^h22^m32\overset{s}{.}0$	$15°32'20''$	2	11.76	$4\overset{'}{.}7$	$3\overset{\circ}{.}7$	47	10.8
713	4356	$12^h24^m15\overset{s}{.}9$	$08°32'10''$	6	13.02	$2\overset{'}{.}6$	$4\overset{\circ}{.}2$	998	29.4
979	4424	$12^h27^m13\overset{s}{.}3$	$09°25'13''$	1	11.99	$3\overset{'}{.}4$	$3\overset{\circ}{.}1$	314	4.1
1043	4438	$12^h27^m46\overset{s}{.}3$	$13°00'30''$	1	10.55	$8\overset{'}{.}7$	$1\overset{\circ}{.}0$	–45	10.4
1330	4492	$12^h30^m58\overset{s}{.}9$	$08°04'41''$	1	13.04	$1\overset{'}{.}9$	$4\overset{\circ}{.}3$	1638	28.1
1569	—	$12^h34^m31\overset{s}{.}4$	$13°30'23''$	5	14.62	$0\overset{'}{.}8$	$1\overset{\circ}{.}4$	687	23.3
1690	4569	$12^h36^m50\overset{s}{.}5$	$13°09'54''$	2	9.63	$10\overset{'}{.}4$	$1\overset{\circ}{.}7$	–328	9.4
1730	4580	$12^h37^m49\overset{s}{.}5$	$05°22'09''$	2	12.39	$2\overset{'}{.}0$	$7\overset{\circ}{.}2$	893	19.3
1760	4586	$12^h38^m28\overset{s}{.}1$	$04°19'09''$	1	12.06	$3\overset{'}{.}9$	$8\overset{\circ}{.}3$	639	14.4
1859	4606	$12^h40^m57\overset{s}{.}7$	$11°54'46''$	1	12.28	$2\overset{'}{.}9$	$2\overset{\circ}{.}5$	1528	12.7

Table 1. H I-deficient galaxies possibly outside the Virgo cluster.

galaxies listed in this range of projected distances by Solanes *et al.* (2002) share similar positions on the sky ($12^h15^m \leqslant \mathrm{RA} \leqslant 12^h30^m$ and $+6° \leqslant \mathrm{dec} \leqslant +10°$), while 12 of them have LOS distances between 27 and 30 Mpc, and 11 have systemic velocities between ∼600–1300 $\mathrm{km\,s^{-1}}$, reinforces the impression that the presence of highly neutral gas deficient objects on Virgo's far side is real. Thus, in contrast to what it is commonly assumed, not all the H I-poor objects in the Virgo region reside in the neighborhood of the cluster core.

If our interpretation of the situation is correct, then we might have unveiled the first evidence of a group of severely H I-deficient galaxies on the suburbs of a cluster. Provided that severe cold-gas deficiency in S's deprived of their hot gas reservoirs inevitably triggers changes, first in the star formation rate and later on Hubble type, our finding suggests that the morphological evolution of disk galaxies is not restricted to the cluster environment but is also important in groups.

Certainly there is still room for an improved analysis. In particular, accurate distances obtained with Cepheids are necessary to do better on group membership assignments by reducing errors in the distance modulus below 0.2 mag (less than 10% fractional errors in distance). Similarly, VLA maps for both the highly H I-deficient galaxies lying on the outskirts of the Virgo cluster and their closest neighbors, would be most beneficial to identify the gas removal mechanism.

References

Fukushige, T. & Makino, J. 2001, *ApJ*, **557**, 533.

Hatton, S., Devriendt, J., Ninin, S., et al. 2003, *MNRAS*, **343**, 75.

Kitayama, T. & Suto, Y. 1996, *ApJ*, **469**, 480.

Mamon, G.A., Sanchis, T., Salvador-Solé, E., & Solanes, J.M. 2004, *A&A*, **414**, 445.

Sanchis, T., Solanes, J.M., Salvador-Solé, E., Fouqué, P., & Manrique, A. 2002, *ApJ*, **580**, 164.

Sanchis, T., Mamon, G.A., Salvador-Solé, E., & Solanes, J.M. 2004, *A&A*, in press.

Solanes, J.M., Giovanelli, R., & Haynes, M.P. 1996, *ApJ*, **461**, 609.

Solanes, J.M., Sanchis, T., Salvador-Solé, E., Giovanelli, R., & Haynes, M.P. 2002, *AJ*, **124**, 2440.

Outskirts of Galaxy Clusters: Intense Life in the Suburbs
Proceedings IAU Colloquium No. 195, 2004
A. Diaferio, ed.

DOI: 10.1017/S1743921304000857

Galaxies in present-day clusters: evolutionary constraints from their distributions and kinematics

Andrea Biviano[1] **and Peter Katgert**[2]

[1]INAF/Osservatorio Astronomico di Trieste, via G. B. Tiepolo 11, I-34131 Trieste, Italy
[2]Sterrewacht Leiden, Postbus 9513, Niels Bohrweg 2, 2300 RA Leiden, The Netherlands

Abstract. We discuss evidence in local, present-day clusters of galaxies (from the ENACS survey) about the way in which those clusters have evolved and about the evolutionary relationships between the galaxies of different morphological types in them. This evidence is complementary to that obtained from the study of clusters at intermediate and high redshifts. We argue that the spatial distribution and the kinematics of the various types of galaxies in and outside substructures support the following picture.

The *elliptical and S0 galaxies* have been around for a long time and have obtained an isotropic velocity distribution. The spatial distribution and kinematics of the *early spirals* are consistent with the idea that many of their kind have transformed into an S0, but that they have survived, most likely because of their velocities. The distribution and kinematics of the *late spirals* are consistent with a picture in which they have been accreted fairly recently. They have mildly radial orbits and hardly populate the central regions, most likely because they suffer tidal disruption. Finally, the distribution and kinematics of the *galaxies in substructures,* when taken at face value, imply tangential velocity anisotropy for these galaxies, but this result may be (partly) due to the procedure by which these galaxies are selected. A first attempt to take the effects of selection into account shows that isotropic (or even mildly radial) orbits of subcluster galaxies cannot be excluded.

1. Introduction and summary of the data sample

By observing the precursors of present-day clusters, it has become possible to study in a direct manner the evolution of the clusters themselves, and the evolutionary relationships between the various types of galaxies in them. Images of rich clusters at redshifts $z \approx 1$ show a high merger fraction, which is taken to be direct evidence for the formation of elliptical galaxies by merging. Yet, it is unlikely that *all* early-type galaxies formed around or before that epoch (van Dokkum & Franx 2001). In particular, the variation with redshift of the composition of the galaxy population in clusters suggests that S0 galaxies have come into being fairly recently, say, since $z \sim 0.5$, apparently at the expense of the spirals (Dressler et al. 1997; Fasano et al. 2000). The processes by which spirals transform into S0 galaxies have been studied numerically (e.g. Moore et al. 1998; Abadi et al. 1999). It appears that impulsive encounters can indeed transform a spiral into an S0 galaxy, by stripping a small fraction of the stellar disk, and by heating up the disk.

We have used the data from the ENACS to investigate if the effects of these processes are still detectable in present-day clusters. Our analysis is based on an 'ensemble' cluster of ~ 3000 galaxies built by combining 59 nearby ($0.035 \leqslant z \leqslant 0.098$) clusters from the ENACS data-set (Katgert et al. 1996, 1998), after suitable scaling of the galaxies projected clustercentric distances and velocities (Biviano et al. 2002). For most of these galaxies also spectral-type or morphological data are available (Thomas 2002). We consider five populations of cluster galaxies, selected on the basis of their different projected phase-space distributions (Biviano et al. 2002). These populations are (i) galaxies in subclusters (herafter referred to as the 'Subs' class), and, among the galaxies outside subclusters, (ii) the brightest ellipticals (with $M_R \leqslant -22 + 5 \log h$, the 'Bright' class), (iii) the other ellipticals together with the E/S0 and the S0 galaxies (the 'Early' class), (iv) the early spirals (Sa–Sb; the 'S-early' class), and (v) the late spirals and irregulars (Sbc–Ir) together with the emission-line galaxies (except those with early morphology), hereafter globally referred to as the 'S-late' class galaxies.

The substructure distinction is based on a refined version of the Dressler & Shectman (1988) criterion. Galaxies in cold and/or moving substructures were identified by the velocity dispersion and/or average velocity of their surroundings, which are smaller than the global value, or different from the cluster mean, respectively. With the significance threshold that we used, about 25% of all galaxies are in substructures. The fraction of galaxies in substructures shows a significant apparent decrease towards the cluster center. This is probably a real effect, but there is likely to be a systematic contribution from the selection algorithm (Biviano et al. 2002).

2. Clues from the spatial distributions

The projected distributions of the 5 classes are shown in the left-hand panel of Fig. 1, in the form of the smooth estimate of $I(R)$ (using the *LOWESS* technique, see Gebhardt & Fisher 1995; Katgert et al. 2004; Biviano & Katgert 2004). Note that our cluster sample has a median value $r_{200} \approx 1.2\,h^{-1}$ Mpc. The brightest ellipticals (Bright-class galaxies) are most centrally concentrated. The Early-class shows a flatter slope, however with a clear steepening beyond $\approx 0.3\,r_{200}$. The S-early- and S-late- classes both avoid the central region within $\approx 0.1\,r_{200}$, but in addition the S-late-class shows an appreciable flattening of $I(R)$ below $\approx 0.3\,r_{200}$. The Abel deprojection of $I(R)$ which yields the 3-D number density profiles shows that both the S-early- and S-late- classes are absent from the central region. Apparently, galaxies of the S-early- and S-late- classes get destroyed or transformed within $\approx 0.3\,r_{200}$, as it is unlikely that their phase-space distributions would prohibit them from entering the central regions. Galaxies in substructures do not avoid the central regions, but are less centrally concentrated than galaxies of the Early-class.

Information on possible transformation relations between the galaxies in the different classes may be gleaned from the morphology-radius (MR) and, in particular, the morphology-density (MD) relation (e.g. Dressler 1980). For the ENACS clusters those have been discussed by Thomas & Katgert (2004), who considered ellipticals (including the brightest) and S0 galaxies separately, in addition to the S-early- and S-late-classes. The MR-relations show the dependence on radius of the composition in terms of morphology, and are fully consistent with the $I(R)$'s in the upper panel of Fig. 1. The MD-relation gives the distribution of local projected density around galaxies of the different classes. Thomas & Katgert (2004) first used the original definition of projected density, i.e. that based on the 10 nearest neigbours, or Σ_{10}. However, Σ_{10} contains a very noticeable print-through of the average relation between projected density and radius. As a result, the Σ_{10} MD-relation is highly correlated with the MR-relation (and thus, with the $I(R)'s$). The ellipticals prefer the highest densities, and the remaining three classes have indistinguishable Σ_{10} distributions shifted towards lower densities (left-hand panel of Fig. 2). From this perspective, the S0 galaxies and ellipticals behave differently; part of this difference is due to the brightest ellipticals being located at the clusters centres.

Thomas & Katgert (2004) also used Σ_1, i.e. the projected density based on the distance to the nearest neighbour.As expected, Σ_1 is much noisier than Σ_{10}, but at the same time

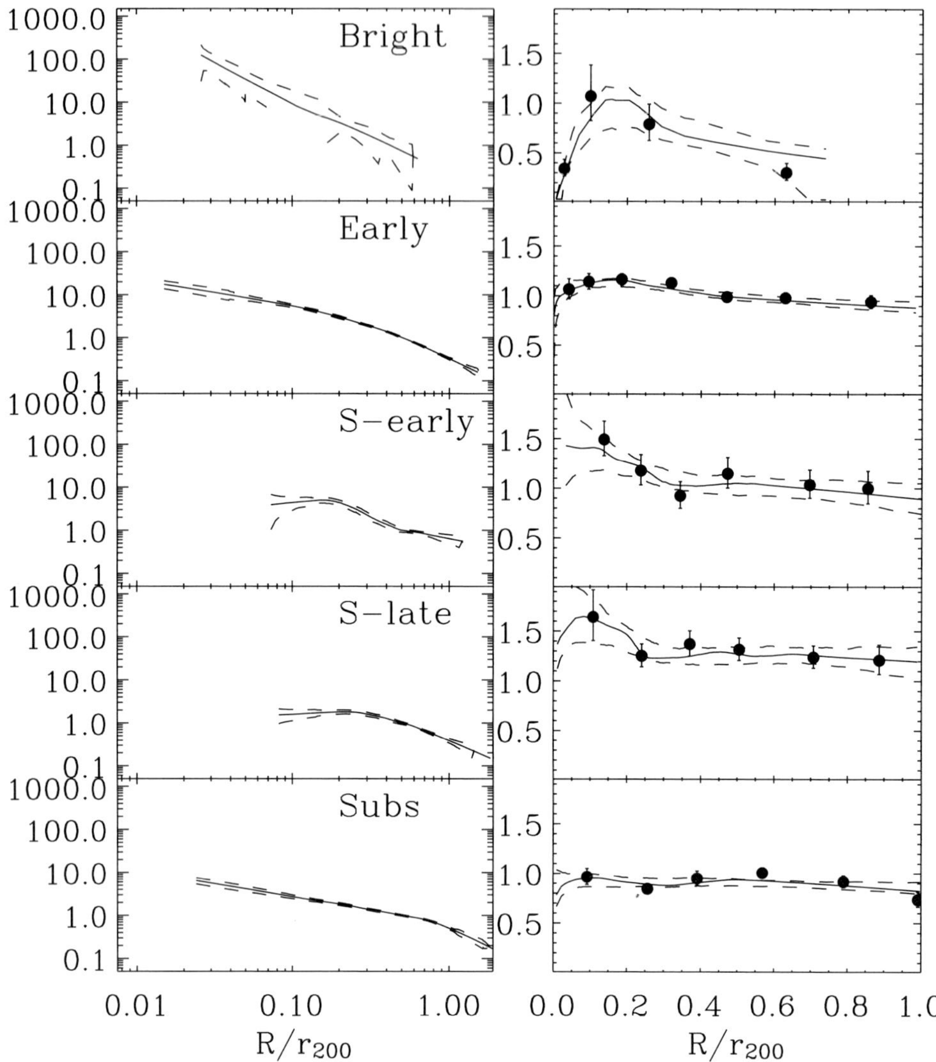

Figure 1. The projected distributions (left) and the velocity dispersion profiles (right) of the five galaxy classes. Dashed lines indicate 1-σ confidence levels. Dots (with 1-σ error bars) in the right-hand panel indicate binned estimates of the velocity dispersion profiles. Units along the y-axis in the left-hand panel are arbitrary. The y-axis in the right-hand panel is in units of the cluster global line-of-sight velocity dispersion (on average, ≈ 700 km s^{-1} for the 59 clusters that compose the ensemble cluster).

it is much less correlated with projected radius, and therefore a less biased measure of local projected density than Σ_{10}. That is probably the reason why the Σ_1-distributions of S0 and S-early galaxies are significantly different, with S0 galaxies preferring higher densities than the S-early galaxies (right-hand panel of Fig. 2).

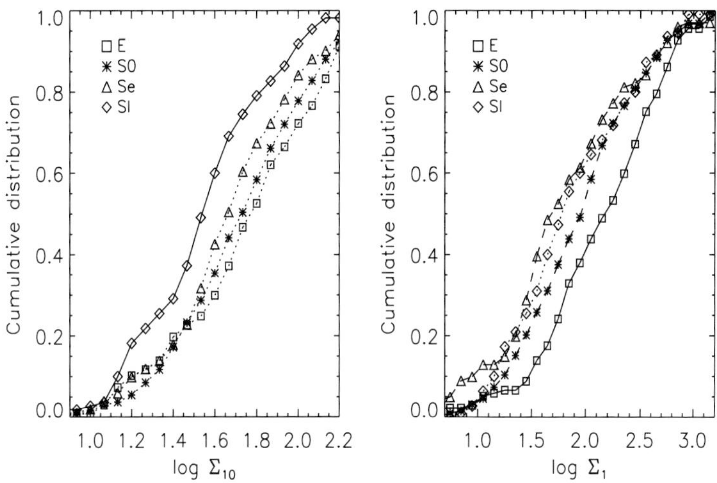

Figure 2. The morphology-density relations for Σ_{10} (left) and Σ_1 (right).

3. Clues from the kinematics

3.1. *Isotropic orbit distributions?*

For a study of the kinematics of the various galaxy classes, we have used the observed projected distributions $I(R)$ and velocity dispersion profiles (vdp's, hereafter). The vdp's of the five different classes are quite different (see right-hand panel of Fig. 1). The Bright-class vdp is quite 'cold', not only in the cluster centre. The Early-class vdp is nearly flat. The vdp's of the S-early and the S-late classes are rather similar, showing a steep decline from the centre out to $\sim 0.3\,r_{200}$, followed by a nearly flat vdp. However, the S-late vdp is 'hotter' than the S-early vdp at all radii. Finally, the vdp of the Subs-class is cold and flat, even flatter than the Early-class vdp.

In order to determine the kinematics of the different galaxy classes we first need to determine the gravitational potential in which the galaxies move. We have therefore first derived the mass-profile $M(<r)$, using the data for the Early-class galaxies, assuming that those have an isotropic distribution of orbits. That assumption is justified by the shape of their line-of-sight velocity distribution, from which Katgert et al. (2004) concluded that, for the Early-class $0.8 \leqslant \beta' \leqslant 1.05$, with

$$\beta' \equiv (\overline{v_r^2}/\overline{v_t^2})^{1/2}, \tag{3.1}$$

where $\overline{v_r^2}$, $\overline{v_t^2}$ are the mean squared components of the radial and tangential velocity, respectively (i.e. isotropic orbits correspond to $\beta'(r) = 1$). Therefore the cluster mass profile, $M(<r)$, follows from the *isotropic* Jeans equation (Binney & Tremaine 1987) applied to the Early-class. The resulting $M(<r)$ is rather similar to a NFW (Navarro et al. 1997) profile with a concentration parameter $c = 4.0^{+2.7}_{-1.5}$ (Katgert et al. 2004).

Next, this cluster $M(<r)$ was used to search for solutions of dynamical equilibrium for the other four classes (see § 1). Given the cluster $M(<r)$ and the observed number density profile of a given class of cluster galaxies, one can solve the inverse Jeans equation (van der Marel 1994) for a given $\beta'(r)$. The resulting vdp is then compared, after the usual Abel projection (Binney & Tremaine 1987), with the observed line-of-sight vdp, in a

χ^2 sense. If the two vdp's are statistically indistinguishable, the $\beta'(r)$ profile is considered acceptable. Instead of considering complicated $\beta'(r)$ profiles, we tested whether the data were consistent with an isotropic orbit distribution for the other galaxy classes, i.e. if $\beta'(r) = 1$ is acceptable. This appears to be the case for galaxies of the S-early class, but not for those of the Bright, S-late, and Subs classes.

3.2. *The orbital anisotropy profiles*

Since the assumption of (nearly) isotropic orbits is not valid for all except the Early and, possibly, the S-early classes we have investigated what the data tell us about the orbital anisotropy. Instead of trying constant anisotropy (van der Marel et al. 2000; Łokas & Mamon 2003), or anisotropy models (Carlberg et al. 1997), we determine $\beta'(r)$ directly from the data, using the procedure that Solanes & Salvador-Solé (1990) developed from that of Binney & Mamon (1982). The $\beta'(r)$ solutions are checked by inverting the Jeans equation again (van der Marel 1994), thereby deriving a vdp that can be compared with the observed one. The derived vdp's always agree very well with the observed vdp's. We also applied this check to the sample of Early-class galaxies, and indeed recovered $\beta'(r) = 1$ to within $\pm 15\%$.

Uncertainties on the β'-profiles were estimated by dividing each galaxy sample into four subsamples, each containing a different half of all the galaxies in the original sample. We then applied the procedure of Solanes & Salvador-Solé (1990) to each subsample. The rms of the four resulting β'-profiles is an approximate, but probably conservative, estimate of the uncertainty in the full sample $\beta'(r)$.

Physically acceptable solutions are found for the Early and three of the four other populations of cluster galaxies. No acceptable solution is found for the Bright class. This is most likely the consequence of a failure of the collisionless Jeans equation in describing the dynamics of these galaxies, which is probably affected by dissipative processes, such as mergers and dynamical friction (see Fusco-Femiano & Menci 1998 for a *collisional* solution).

The β'-profile obtained for the S-early class increases rapidly from $\beta' \simeq 1$ near the centre, to $\beta' \simeq 1.8$ at $r/r_{200} \simeq 0.45$, and then smoothly decreases to $\simeq 1$ again at larger radii (see Fig. 3, top panel). Taken at face value, this solution implies mild radial anisotropy just outside the cluster centre. The partial radial anisotropy at $r \simeq 0.45\, r_{200}$ could result from a natural selection effect. I.e., among S-early galaxies, those reaching the high-density inner cluster region can avoid transforming into S0's only if they move sufficiently fast. We should not overinterpret the data, though, since the isotropic solution is acceptable for galaxies of the S-early class, and the β'-profile is in fact consistent with $\beta'(r) = 1$.

The $\beta'(r)$ profile of the S-late galaxies is not different from unity out to $\simeq 0.7\, r_{200}$, but then increases linearly with radius reaching a radial anisotropy $\beta' \simeq 1.8$ at $r \simeq 1.5\, r_{200}$ (see Fig. 3, mid panel). This β'-profile is remarkably similar to those obtained for dark matter particles in numerical simulations (Ghigna et al. 1998; Diaferio 1999). It is therefore tempting to conclude, by analogy, that S-late galaxies, like dark matter particles in numerical simulations, still retain memory of the process of (mostly radial) gravitational infall along the filaments connecting to the cluster, and are therefore recent arrivals into the cluster.

Finally, the data for the galaxies of the Subs class imply orbits with substantial tangential anisotropy at all radii (see Fig. 3, solid line in the bottom panel). However, two systematic effect must be accounted for. The first one is the internal velocity dispersion of subclusters. If we subtract an assumed internal velocity dispersion of 250 km s^{-1} from the observed vdp of Subs galaxies, the orbital tangential anisotropy becomes even

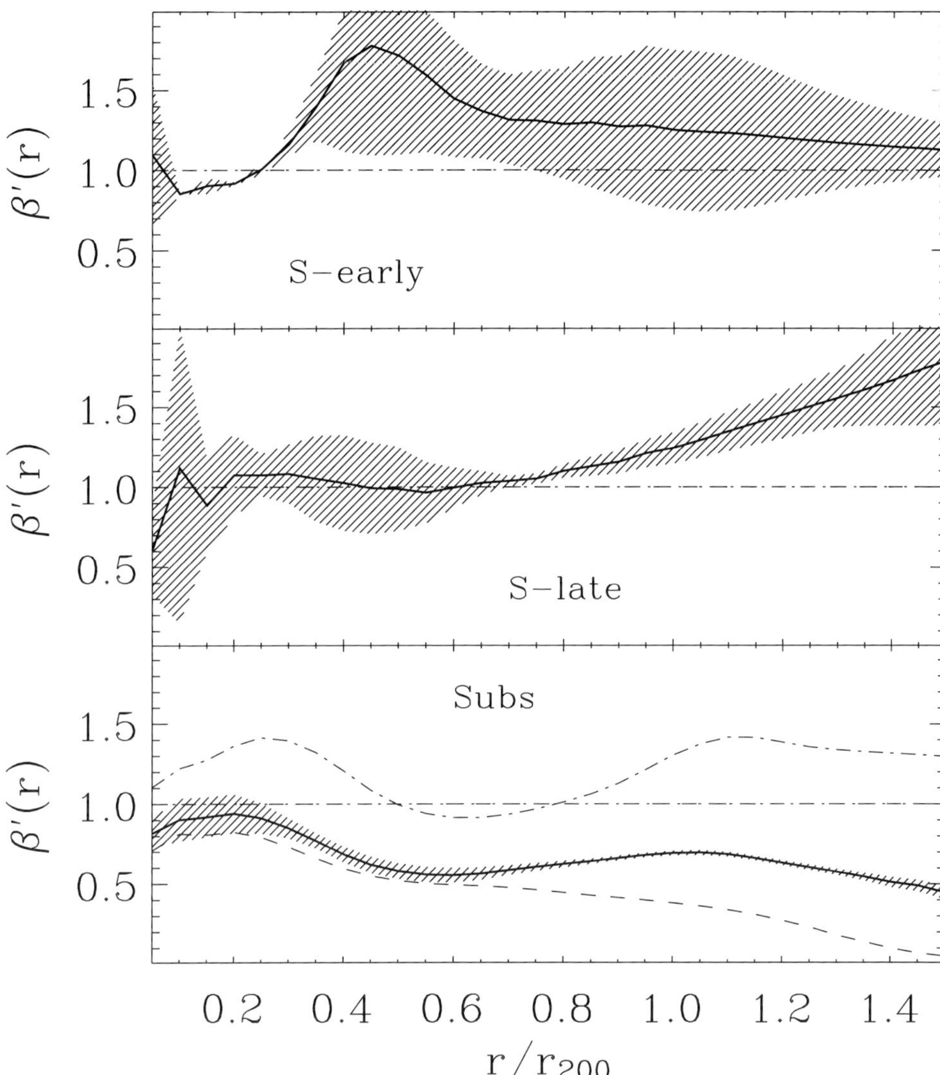

Figure 3. Top panel: The velocity anisotropy profile, $\beta'(r) \equiv (\overline{v_r^2}(r)/\overline{v_t^2}(r))^{1/2}$, of the S-early class. The hatched region represents the approximate 1-σ confidence interval. Mid panel: same as top panel, but for the S-late class. Bottom panel: same as top panel, but for the Subs class. The dashed line is the β'-profile obtained after subtracting in quadrature the internal velocity dispersion of the subclusters from the observed vdp. The dash-dotted line is the β'-profile obtained after correcting the observed number density profile and vdp of the Subs class for the systematic selection effects, according to a plausible, but by no means unique, model.

stronger (see Fig. 3, dashed line in the bottom panel). The second effect is the radial variation of the efficiency of the detection of Subs galaxies. Very simple modeling shows that the efficiency of the selection method decreases towards the cluster center. As a result, both the number density profile and the vdp of the Subs class are affected. A simple first-order correction of these profiles leads to a β'-profile which no longer implies tangential orbits (see Fig. 3, dash-dotted line in the bottom panel). A better understanding of the orbital characteristics of subclusters thus requires either an improved, radially-unbiased, selection method of galaxies in subclusters, or at least an improved modeling of the biases.

4. Summary

The Early-class galaxies (ellipticals and S0 galaxies) in clusters have a distribution of line-of-sight velocities that is consistent with an isotropic orbit distribution. These galaxies were therefore used to derive the mass profile of the ENACS ensemble cluster. Using this mass profile, we constrained the orbits of cluster galaxies of other classes. From the distribution and velocity dispersion profile of the S-early-class we conclude that these may also have isotropic orbits, although they show an apparent radial anisotropy around $\approx 0.45\, r_{200}$. The latter may be a result of the process by which the majority of the S-early-galaxies are thought to have transformed into S0 galaxies through impulsive encounters. The fact that the local density around S-early-galaxies is smaller than that around S0 galaxies provides further support for this interpretation. The S-late-class has mildly radial orbits outside $\approx 0.7\, r_{200}$, and are probably still falling in. Their absence from the central region most likely indicates that they are tidally disrupted on their crossing of the cluster core. Finally, the galaxies in substructures apparently are on tangential orbits, although isotropic (or mildly radial) orbits cannot be excluded at present.

Acknowledgements

We wish to thank the Local Organizing Committee for a very enjoyable and interesting meeting. We acknowledge useful discussions with Alain Mazure and Tom Thomas.

References

Abadi, M. G., Moore, B., & Bower, R., G. 1999, *MNRAS,* **308,** 947.
Binney, J., & Mamon, G. 1982, *MNRAS,* **200,** 361.
Binney, J. & Tremaine, S. 1987, *"Galactic Dynamics"* (Princeton Univ. Press).
Biviano, A., Katgert, P., Thomas, T., & Adami, C. 2002, *A&A,* **387,** 8.
Biviano, A., & Katgert, P. 2004, *in preparation.*
Carlberg, R.G., Yee, H.K.C., & Ellingson, E. 1997, *ApJ,* **478,** 462.
Diaferio, A. 1999, *MNRAS,* **309,** 610.
Dressler, A. 1980, *ApJ,* **236,** 351.
Dressler, A. & Shectman, S.A. 1988, *AJ,* **95,** 985.
Dressler, A., Oemler, A. Jr., Couch, W. J., et al. 1997, *ApJ,* **490,** 577.
Fasano, G., Poggianti, B. M., Couch, W. J., et al. 2000, *ApJ,* **542,** 673.
Fusco-Femiano, R. & Menci, N. 1998, *ApJ,* **498,** 95.
Gebhardt, K. & Fischer, P. 1995, *AJ,* **109,** 209
Ghigna, S., Moore, B., Governato, F., et al. 1998, *MNRAS,* **300,** 146.
Katgert, P., Mazure, A., Perea, J., et al. 1996, *A&A,* **310,** 8.
Katgert, P., Mazure, A., den Hartog, R., et al. 1998, *A&AS,* **129,** 399.
Katgert, P., Biviano, A., & Mazure, A. 2004, *ApJ,* **600,** 657.
Łokas, E. & Mamon, G. 2003, *MNRAS,* **343,** 401.
Moore, B., Lake, G., & Katz, N. 1998, *ApJ,* **495,** 139.
Navarro, J.F., Frenk, C.S. & White, S.D.M. 1997, *ApJ,* **490,** 493.
Solanes, J.M., & Salvador-Solé 1990, *A&A,* **234,** 93.
Thomas, T. 2002, *PhD thesis, Leiden Observatory.*
Thomas, T., & Katgert, P. 2004, *in preparation.*
van der Marel, R. 1994, *MNRAS,* **270,** 271.
van der Marel, R., Magorrian, J., Carlberg, R., et al. 2000, *AJ,* **119,** 2038.
van Dokkum, P. G., & Franx, M. 2001, *ApJ,* **553,** 90.

Outskirts of Galaxy Clusters: Intense Life in the Suburbs
Proceedings IAU Colloquium No. 195, 2004
A. Diaferio, ed.

DOI: 10.1017/S1743921304000869

Breaking the mass/anisotropy degeneracy in the Coma cluster

Gary A. Mamon[1], Ewa L. Łokas[2] and Teresa Sanchis[3]

[1]IAP, Paris, France email: gam@iap.fr

[2]Copernicus Center, Warsaw, Poland email: lokas@camk.edu.pl

[3]Dep. di Astronomia i Meteorologia, Univ. de Barcelona, Spain email: tsanchis@am.ub.es

Abstract. We provide the first direct lifting of the mass/anisotropy degeneracy for a cluster of galaxies, by jointly fitting the line of sight velocity dispersion and kurtosis profiles of the Coma cluster, assuming an NFW tracer density profile, a generalized-NFW dark matter profile and a constant anisotropy profile. We find that the orbits in Coma must be quasi-isotropic, and find a mass consistent with previous analyses, but a concentration parameter 50% higher than expected in cosmological N-body simulations. We then test the accuracy of our method on realistic non-spherical systems with substructure and streaming motions, by applying it to the ten most massive structures in a cosmological N-body simulation. We find that our method yields fairly accurate results on average (within 20%), although with a wide variation (factor 1.7 at $1\,\sigma$) for the concentration parameter, with decreased accuracy and efficiency when the projected mean velocity is not constant with radius.

1. Introduction

The kinematical ("Jeans") analyses of near-spherical structures using the spherical Jeans equation for stationary systems:

$$\frac{\mathrm{d}}{\mathrm{d}r}(\nu\sigma_r^2) + \frac{2\beta}{r}\nu\sigma_r^2 = -\nu\frac{GM}{r^2} ,$$

where ν is the 3D density distribution of the tracer population, suffer from the fact that this one equation involves two unknowns: the mass distribution $M(r)$ and the velocity anisotropy $\beta = 1 - \sigma_\theta^2/\sigma_r^2$ (we make use of the symmetry of spherical systems yielding $\sigma_\phi = \sigma_\theta$). In some cases of simple anisotropy profiles $\beta(r)$, the Jeans equation can be inverted to yield the radial dispersion as a single integral over radii of νM times some kernel, and one can insert this solution into the equation that links the line-of-sight velocity dispersion to the radial one, to obtain after some algebra, the line-of-sight dispersion as a single integral over νM and some other kernel.

In what follows, we attempt to lift the mass/anisotropy degeneracy by adding a second equation, namely the 4th order Jeans equation for constant anisotropy (Łokas 2002)

$$\frac{\mathrm{d}}{\mathrm{d}r}(\nu\overline{v_r^4}) + \frac{2\beta}{r}\nu\overline{v_r^4} + 3\nu\sigma_r^2\frac{\mathrm{d}\Phi}{\mathrm{d}r} = 0 ,$$

where we express the line of sight *kurtosis* as a double integral over νM and M and a third kernel (the details are given in Łokas & Mamon 2003).

2. Kinematic analysis of the Coma cluster

We apply our formalism to the nearby Coma cluster, for which we've extracted from NED the positions and line-of-sight velocities of 967 galaxies, among which 355 early-type

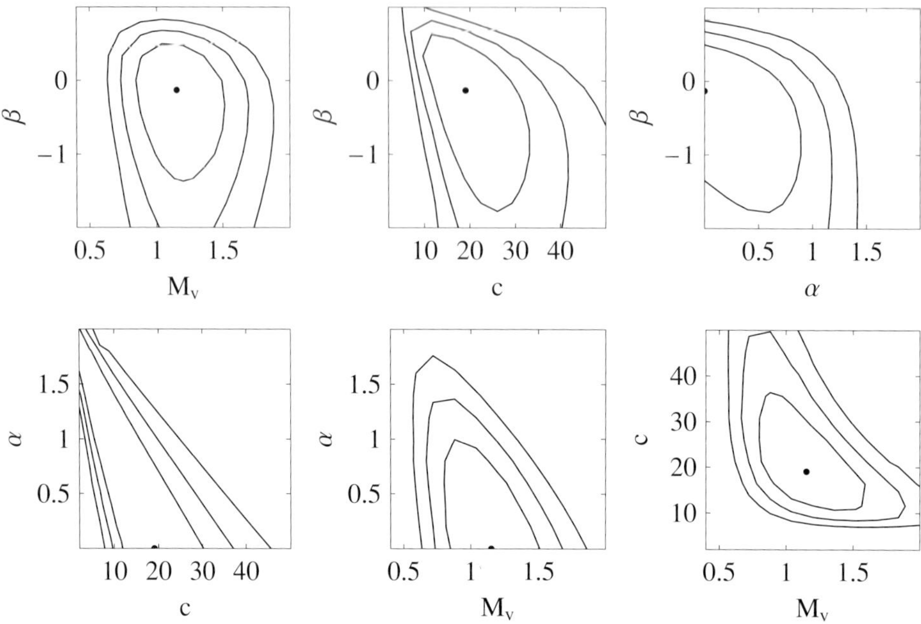

Figure 1. Probability contours (1, 2, and 3σ) for joint fits of $\sigma_{\rm los}$ and $(\log \kappa_{\rm los})^{1/10}$ (the two are found to be independent). *Circles* indicate the best-fitting parameters. The mass is in units of $10^{15} M_\odot$.

galaxies whose kinematics we analyze here (we omit spiral galaxies, which are expected to have a non negligible mean radial velocity caused by their infall and subsequent transformation at cluster pericenter). Our mass model includes 1) a stellar component, which we find is well fit by an NFW (Navarro et al. 1996) model; 2) a dark matter component, which we assume to be a generalized NFW model with arbitrary inner slope:

$$\rho(r) \propto (r/a)^{-\alpha} (1 + r/a)^{\alpha-3} , \tag{2.1}$$

where a is the scale radius of the dark matter (in general different from that of the luminous distribution), and $\alpha = 1$ for the NFW model. The profiles differ by their inner slope $-\alpha$ but have a common outer limiting behavior of r^{-3}; and 3) an isothermal gas component, taken from the ROSAT X-ray observations by Briel et al. (1992).

Using 7 radial bins at $R < 80'$, we jointly fit our four parameters, mass M_v within the virial radius r_v, velocity anisotropy β, inner slope α of the dark matter density profile, and dark matter concentration $c = r_v/a$. Contrary to the case when only velocity dispersion is studied, the minimization procedure now converges (except for a concentration / inner slope degeneracy). The best fits are shown in Figure 1. The best fit parameters are $M_v = 1.2 \times 10^{15} M_\odot$ (corresponding to dark matter virial radius $r_v = 92' = 2.7$ Mpc), $\beta = -0.13$, $\alpha = 0$ and $c = 19$ with $\chi^2/N = 6.1/10$. In other words, our best fit models are isotropic and the generalized dark matter model of equation (2.1) provides excellent fits to the data.

Our best fit radial profiles for each half-integer value of α are shown in Figure 2. The best fit density profiles are indistinguishable beyond $0.03\, r_v$, and the outer mass profiles are very close to the infall estimates of Geller, Diaferio & Kurtz (1999). The density profiles are 50% more concentrated than in structures of comparable mass within

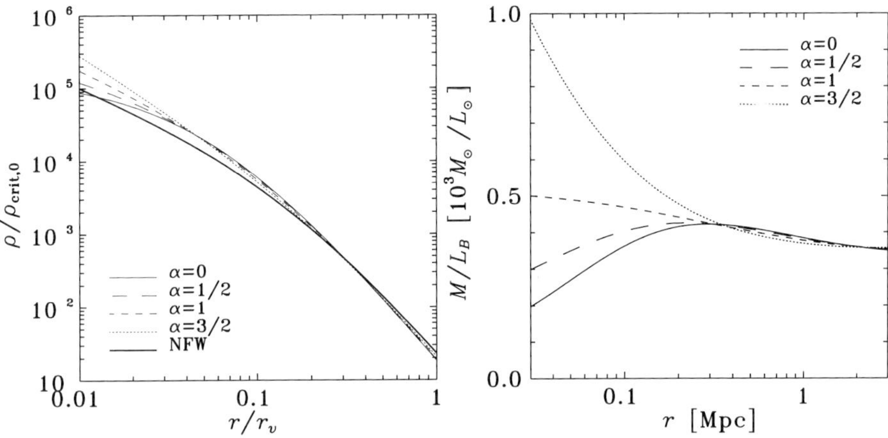

Figure 2. Best fitting dark matter profiles with different inner slopes (*left panel*) and their associated mass-to-light ratios (*right panel*), for different α increasing upwards from 0 to 1.5 (with c = 19, 14, 9 and 5, respectively). The dark matter virial radii are r_v = 2.7 Mpc for all models shown. The *heavy solid curve* in the left plot indicates the best NFW model (c = 6) from cosmological N-body simulations, which is still (marginally) consistent with the data.

cosmological N-body simulations in a ΛCDM Universe (Bullock et al. 2001). Interestingly, the M/L profiles have nearly constant slope for the NFW case ($\alpha = 1$).

Our analysis differs from previous kinematical analyses (e.g. Merritt 1987; den Hartog & Katgert 1996; Carlberg et al. 1997; van der Marel et al. 2000; Biviano & Girardi 2003; Katgert et al. 2004), as 1) we have, for a single cluster, a larger sample of galaxies, which, given their early morphological type, should be in dynamical equilibrium in the cluster potential; 2) we remove pairs from the computation of the velocity moments; 3) we include kurtosis in the analysis; 4) we model the dark matter distribution using a generalized formula inspired by the results of cosmological N-body simulations; 5) we include hot gas.

In comparison to studies based upon stacking of many clusters, our analysis of the Coma cluster benefits from not having to introduce errors in any stacking procedure, and from a cleaner removal of interlopers. On the other hand, the analyses of stacked clusters have the advantage of averaging out particular inhomogeneities of individual clusters such as Coma, expected in hierarchical scenarios of structure formation and observed in Coma, both in projected space (Fitchett & Webster 1987; Mellier et al. 1988; Briel et al. 1992) and velocity space (Colless & Dunn 1996; Biviano et al. 1996).

3. Kinematic analysis of simulated clusters

In our Jeans analysis of the Coma cluster, we assumed that the cluster was spherical, with no substructure nor streaming motions, and that projection effects were not serious. We test (details in Sanchis, Łokas & Mamon 2004) these assumptions using a ΛCDM cosmological N-body (tree-code) simulation, run by Ninin (1999), with 256^3 particles, and with $\Omega_m = 1/3$, $\Omega_\lambda = 2/3$, $h = 2/3$, and $\sigma_8 = 0.88$ (details in Hatton et al. 2003).

We analyze each of the 10 most massive halos along 3 axes: the major axis, an orthogonal one and an intermediate one. Figure 3 shows the radial profiles of line-of-sight velocity moments (mean, dispersion, skewness and kurtosis) along the principal axis. These profiles display statistically significant radial variations caused by substructures

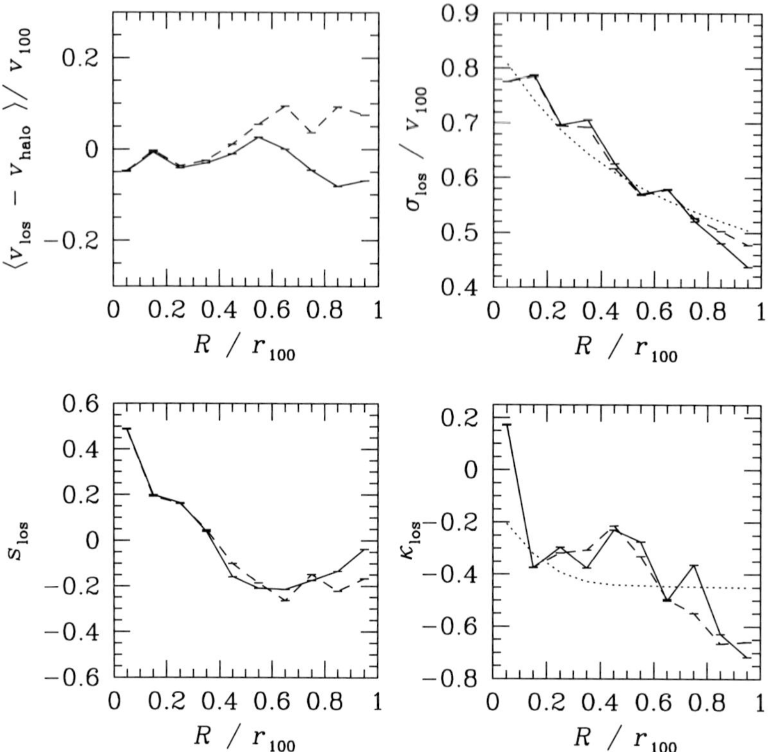

Figure 3. Projected velocity moments of the dark matter particles in halo 1 measured along the principal axis. The *upper left panel* shows the mean line-of-sight velocity with respect to the velocity of the center of the halo in units of the circular velocity at r_{100}. The *upper right panel* gives the line-of-sight velocity dispersion in the same units. The two *lower panels* give the skewness (*left*) and kurtosis (*right*). In each panel the *solid line* shows results for particles lying inside the sphere of radius r_{100}, while the *dashed line* is for all particles. The *dotted curve* shows the fits obtained from the Jeans equations.

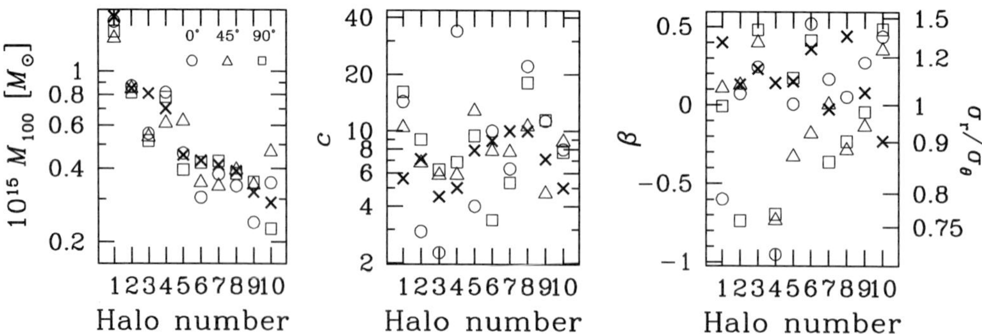

Figure 4. Fitted (on projected data) values (using 40 particles per bin) of virial mass M_{100}, concentration parameter c and anisotropy β of the ten halos for the three directions of observation with respect to the major axis of each halo: 0° (*circles*), 45° (*triangles*) and 90° (*squares*). The values measured on 3D data are shown with *crosses*.

Particles	$\Delta \log M_{100}$		$\Delta \log c$		$\Delta\beta$		$\Delta \log(\sigma_r/\sigma_\theta)$	
per bin	mean	σ	mean	σ	mean	σ	mean	σ
All	−0.03	0.09	0.20	0.18	−0.78	1.04	−0.12	0.12
40	−0.07	0.10	0.08	0.24	−0.20	0.48	−0.04	0.11

Table 1. Results of the fitting procedure

with radial streaming motions. Moreover, the moments vary considerably from halo to halo and for the 3 different projections of a given halo. Note also that the projected mean velocity differs substantially between all the particles in the projected cylinder and the subset within the virial sphere.

Figure 4 shows the results of our fits from our Jeans analysis (*open symbols*) in comparison with our fits to the 3D data (*crosses*) and Table 1 summarizes the statistical accuracy of our fitting procedure. The virial mass is typically underestimated by $15 \pm 26\%$, while the concentration parameter is overestimated by $20 \pm 74\%$ and the ratio σ_r/σ_θ underestimated by $9\% \pm 29\%$. Hence, our kinematical analysis yields good results *on average*, but with a fairly large dispersion in accuracies (with the most inaccurate estimates for halos with non zero mean line of sight velocities at some radii). Halo 4, which departs the most from the 3D measures of concentration and anisotropy, happens to have important variations in its projected mean velocity profile, whereas nearly all other halos have normal mean velocity profiles. This suggests that the joint dispersion/kurtosis Jeans analysis is adequate for structures with fairly constant projected mean velocity profiles. Given the large dispersion on the concentration parameter, our high value of c for Coma is now consistent with the 1.5 times lower value expected from cosmological N-body simulations.

References

Biviano, A. et al. 1996 *A&A* **311**, 95–112
Biviano, A. & Girardi, M. 2003 *ApJ* **585**, 205–214
Briel, U., Henry, J. P. & Böhringer, H. 1992 *A&A* **259**, L31–L34
Bullock, J. S. et al. 2001 *MNRAS* **321**, 559–575
Carlberg, R. G. et al. 1997 *ApJ* **485**, L13–L16
Colless, M. & Dunn, A. M. 1996 *ApJ* **458**, 435–454
den Hartog, R. & Katgert, P. 1996 *MNRAS* **279**, 349–388
Fitchett, M. & Webster, R. 1987 *ApJ* **317**, 653–667
Geller, M. J., Diaferio, A. & Kurtz, M. J. 1999 *ApJ* **517**, L23–L26
Hatton, S. et al. 2003 *MNRAS* **343**, 75–106
Katgert, P., Biviano, A. & Mazure, A. 2004 *ApJ* **600**, 657–669
Łokas, E. L. 2002 *MNRAS* **333**, 697–708
Łokas, E. L. & Mamon, G. A. 2003 *MNRAS* **343**, 401–412
Mellier et al. 1988 *A&A* **199**, 67–72
Merritt 1987 *ApJ* **313**, 121–135
Navarro, J. F., Frenk, C. S. & White, S. D. M. 1996 *ApJ* **462**, 563–575
Ninin, S. 1999 PhD thesis, Univ. of Paris 11
Sanchis, T., Łokas, E. L. & Mamon, G. A. 2004 *MNRAS* **347**, 1198–1207
van der Marel, R. P. et al. 2000 *AJ* **119**, 2038–2052

Outskirts of Galaxy Clusters: Intense Life in the Suburbs
Proceedings IAU Colloquium No. 195, 2004
A. Diaferio, ed.

DOI: 10.1017/S1743921304000870

Velocity of galaxies with different luminosity

Y. P. Jing[1] **and G. Börner**[2]

[1]Shanghai Astronomical Observatory, the Partner Group of MPI für Astrophysik, Nandan Road 80, Shanghai 200030, China email: ypjing@shao.ac.cn

[2]Max-Planck-Institut für Astrophysik, Karl-Schwarzschild-Strasse 1, 85748 Garching, Germany email: grb@mpa-garching.mpg.de

Abstract. We present the first determination of the pairwise velocity dispersion of galaxies at different luminosity with the final release of the Two-Degree Field Galaxy Redshift Survey (2dFGRS). Our result surprisingly shows that the random velocities of the faint galaxies are very high, around $700\,\mathrm{km\,s^{-1}}$, reaching similar values as the brightest galaxies. At intermediate luminosities slightly brighter than the characteristic luminosity $M_\star$, the velocities exhibit a well defined steep minimum near $400\,\mathrm{km\,s^{-1}}$. The result challenges the current halo model of galaxies of Yang et al. that was obtained by matching the clustering and luminosity function of 2dFGRS, and can be an important constraint in general on theories of galaxy formation, e.g., the semi-analytical model. Combining the observed luminosity dependence of clustering, our result implies that quite a fraction of faint galaxies are in massive halos of galaxy clusters as the brightest ones, but most of the $M_\star$ galaxies are in galactic halos.

1. Introduction

The clustering of galaxies in the Universe is characterized by their spatial positions, and their peculiar velocities which lead to deviations of their motion from the pure Hubble flow. The big redshift surveys assembled in recent years by the diligent work of many astronomers give angular positions and redshifts for large numbers of galaxies. A rough 3D map can be obtained by placing the galaxies at distances along the line of sight derived via Hubble's law from their redshifts. The peculiar velocity, however, also contributes to the redshift, and this leads to a misplacement of the galaxy away from its true location. The local gravitational field is the cause of the peculiar motion and thus the redshift distortion in the galaxy maps can give information on the underlying matter distribution.

The amplitude of the distortions can be estimated from the pair distribution of galaxies. For pairs of galaxies at distances much larger than their separation, one can use a plane-parallel approximation, and write for the power spectrum in redshift space,

$$P^S(k,\mu) = P(k)\frac{(1+\beta\mu^2)^2}{1+\frac{1}{2}[\sigma_v(k)k\mu]^2}\,. \tag{1.1}$$

Here μ is the cosine of the angle between the wave vector and the line of sight. The linear redshift distortion parameter β is related to the linear growth factor $f(\Omega_M) \simeq \Omega^{0.6}$ (Ω_M is the matter density), and the normalization σ_8 of the density fluctuation spectrum via $\beta = \Omega_M^{0.6}\sigma_8$. $P(k)$ is the power spectrum in real space. The pairwise velocity dispersion (PVD) $\sigma_v(k)$ describes the virial motion of galaxies in dense systems, i.e. the Finger-of-God effect. As shown by Jing & Börner (2001), the redshift power spectrum of dark matter and its biased tracers can be accurately described by Eq.(1.1). Applying this model to an observation of $P^S(k,\mu)$ may therefore yield a determination of the three quantities $P(k)$, β, and $\sigma_v(k)$, all of which are useful observables for testing galaxy formation models.

Here we use the final release of the 2dFGRS (`http://www.mso.anu.edu.au/2dFGRS`) to measure $P^S(k,\mu)$ and the PVD $\sigma_v(k)$ of galaxies. Since we are interested in the PVD at small scale, we fix β at a reasonable value of 0.45 (Peacock et al. 2001). The 2dF catalog has already been analyzed statistically with respect to the PVD (Hawkins et al 2003), but here we use a novel method to estimate the PVD. Furthermore, we take advantage of the large number of galaxies in the 2dF catalog to bin the galaxies in different luminosity intervals, and to make the first study of the luminosity dependence of the PVD. This was not possible up to now, and we shall see that remarkable tests of the galaxy formation models become possible with the luminosity dependence of the PVD. The luminosity dependence of the clustering (the two-point correlation function) of galaxies in the 2dFGRS has been investigated (Norberg et al. 2002a), but not of the PVD.

2. Statistical analysis

In order to study the luminosity dependence of the PVD, we divide the galaxies into 11 subsamples according to their absolute luminosity. The subsamples are successively brightened by 0.5 magnitude from the faintest sample $M_b = -17.0 + 5\log h$ to $M_b = -22.0 + 5\log h$, with successive subsamples overlapping by 0.5 magnitude. Here h is the Hubble constant in units of $100\,\mathrm{km\,s^{-1}Mpc^{-1}}$. For computing the absolute magnitude, we have used the k-correction and luminosity evolution model of Norberg (2002b; k + e model);, i.e., the absolute magnitude is in the rest frame b_j band at $z = 0$. We assume a cosmological model with the density parameter $\Omega_0 = 0.3$ and the cosmological constant $\lambda_0 = 0.7$ throughout this paper.

We measure the redshift two-point correlation functions $\xi_z(\mathbf{s})$ following the method of Jing, Mo, & Börner(1998). The random samples for the clustering analysis are generated in the same way as described in Jing & Börner (2004a). Each random sample for a northern or southern luminosity subsample contains 100,000 random points. We convert the $\xi_z(\mathbf{s})$ to $P^S(k,\mu)$ by the Fourier transformation. More details about our statistical method can be found in our journal paper (Jing & Börner 2004b).

3. Results

In Fig. 1 the basic measurement of the power spectrum in redshift space $P(k,\mu)$ is shown. The four panels in this figure correspond to four different luminosity intervals from faint to bright galaxies. The values of k range from $0.16\,h\mathrm{Mpc}^{-1}$ at the top to $2.5\,h\mathrm{Mpc}^{-1}$ at the bottom with an interval $\Delta\log_{10} k = 0.2$. The south and north samples agree quite well with the full survey indicating that cosmic variance is not a major problem. The power spectrum for the larger k-values decreases quite strongly with μ, more than a factor of 10 between $\mu = 0$ and $\mu = 1$. The solid lines are the best fits obtained by applying equation (1.1) to the data of the whole survey, showing that the model (1.1) can accurately describe the observed redshift power spectrum of galaxies.

The luminosity dependence of the PVD is shown most clearly in Fig. 2, where we have plotted σ_v at $k = 1\,h\mathrm{Mpc}^{-1}$ for the 11 overlapping samples. The surprising result is a strong dependence on luminosity with the bright and the faint galaxies reaching high values of $\simeq 700\,\mathrm{km\,s^{-1}}$ or more, and a well defined minimum of $\simeq 400\,\mathrm{km\,s^{-1}}$ for the galaxies of magnitude $M_b - 5\log_{10} h = -20.5$. The bright and the faint galaxies apparently have high random motions, as expected for objects in massive halos or in clusters. The $M_\star$ like galaxies are rather moderate in their PVD, and probably reside in galaxy size halos.

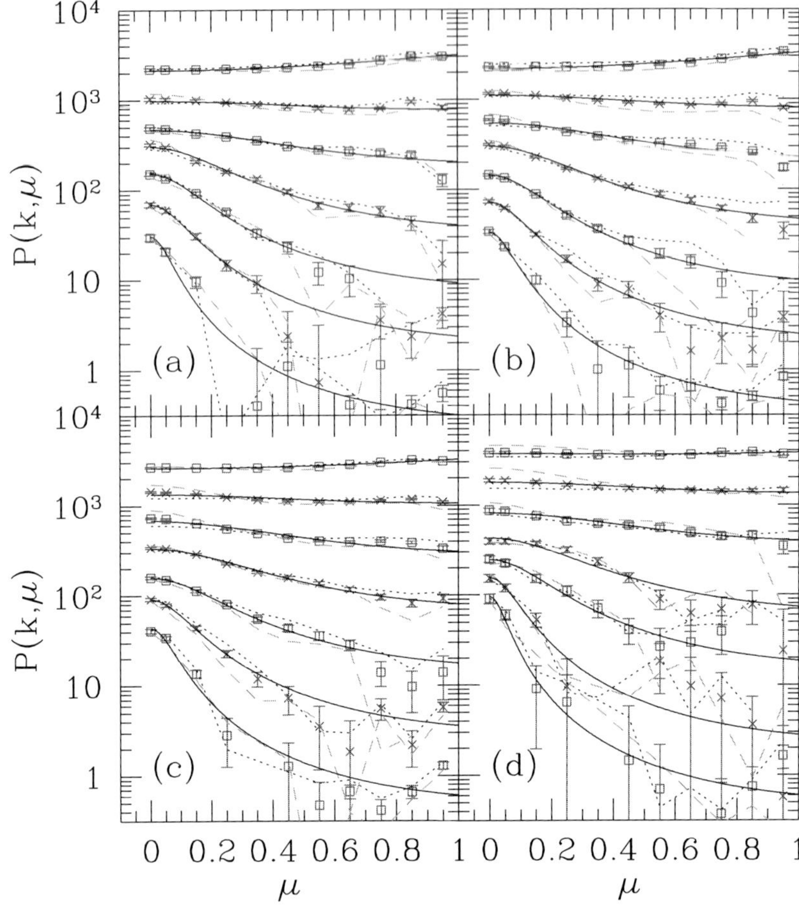

Figure 1. The redshift space power spectrum $P(k, \mu)$ measured in 2dFGRS. The symbols are for the whole survey, the dotted lines for the south subsample, and the dashed lines for the north subsample. The errors are plotted only for the whole survey, estimated with the bootstrap method. The smooth solid lines are the best fits of Eq.(1.1) to data of the whole sample. (a) for $-17.5 < M_b - 5\log h < -18.5$; (b) for $-18.5 < M_b - 5\log h < -19.5$; (c) for $-19.5 < M_b - 5\log h < -20.5$; (d) for $-20.5 < M_b - 5\log h < -21.5$. The k values in each panel are from $0.16\,h\mathrm{Mpc}^{-1}$ (top) to $2.5\,h\mathrm{Mpc}^{-1}$ (bottom).

The thick solid line represents the prediction based on the up-to-date halo model of Yang et al. (2003). For the halo model, we adopt the cosmological model that is a currently popular flat low-density model with the density parameter $\Omega_0 = 0.3$ and the cosmological constant $\lambda_0 = 0.7$ (LCDM). The shape parameter $\Gamma = \Omega_0 h$ and the amplitude σ_8 of the linear density power spectrum are 0.2 and 0.9 respectively. We use two sets of simulations, with boxsizes $L = 100\,h^{-1}$Mpc and $L = 300\,h^{-1}$Mpc, that were generated with our vectorized-parallel P^{3}M code (Jing & Suto 2002) for this model. Both simulations use 512^3 particles, so the particle mass m_p is $6.2 \times 10^8 M_\odot$ and $1.7 \times 10^{10} M_\odot$ respectively in these two cases. We populate the halos with galaxies in a similar way to that of Yang et al. (2004), but we used our own code and adopted the model parameters of Model M1 in Yang et al. (2003). The halo model, which matched nicely the luminosity function and the luminosity dependence of clustering of galaxies in 2dFGRS (confirmed by our analysis), clearly does not match our observation of the PVD. The failure of the

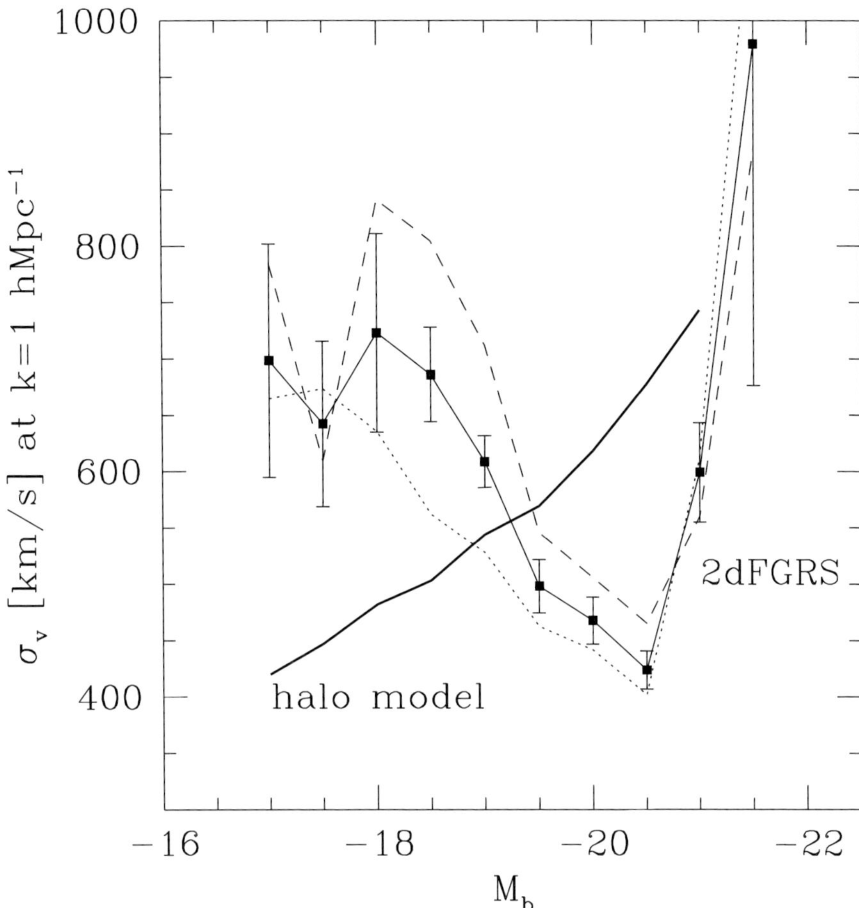

Figure 2. The pairwise velocity dispersion measured at $k = 1\,h\mathrm{Mpc}^{-1}$ in the 2dFGRS (symbols for the whole sample, dotted line for the south, and dashed line for the north), compared with the prediction based on the halo model (thick solid line). The error bars of the observed results are given by the mock samples.

halo model indicates that the prescription of how to populate halos must be adapted better.

4. Discussion

The analysis of the velocity fields of the galaxies in the 2dFGRS has led to a surprising discovery: the random velocities of the faint galaxies are very high, around $700\,\mathrm{km\,s^{-1}}$, reaching similar values as the bright galaxies. At intermediate luminosities the velocities exhibit a well defined steep minimum near $400\,\mathrm{km\,s^{-1}}$.

It seems that the galaxies in different luminosity intervals appear as different populations in their own right, defined by objective statistics. A look at figure 2 shows convincingly that this is actually the case. For this figure we have sorted the galaxies in 11 luminosity bins, each one magnitude wide, from magnitude −16.5 to −22 and plotted the value of σ_v at a wave number of $k \simeq 1\,h\mathrm{Mpc}^{-1}$. Such a finely resolved binning of galaxies in samples of different luminosities is possible for the 2dFGRS, because it is big enough to contain sufficiently many galaxies in each luminosity class.

The PVD is an indicator of the depth of the local gravitational potential. Therefore we find the interesting result that the bright and the faint galaxies move in the strongest gravitational field. They are in clusters, while the galaxies around magnitude -20, the M_* galaxies in the Schechter luminosity function, populate the field.

The bimodal nature of the correlation between the PVD and luminosity may be used as a stringent test of galaxy formation models. We have investigated the halo occupation model (Yang et al. 2003) which has been optimally fitted to reproduce the luminosity function, and the two-point correlation function of the 2dFGRS. If we adapt this model to the PVD value of the M_* galaxies, we see that it cannot give the high values found for the fainter galaxies. The PVD values of the model actually run opposite to the data and the model assigns smaller values to the fainter galaxies. This must mean that the assignment of galaxies to the dark matter halos must be done in a more intricate way as up to now. The number of faint galaxies in clusters must be increased substantially to at least recover the high PVD found for them. Also, the low value of $400\,\mathrm{km\,s^{-1}}$ found for the galaxies with magnitude -20.5 must mean that these galaxies reside in dark matter halos of galactic size. The halo population model must be fitted with a much more complex scheme of assigning galaxies to halos, if the results shown in figure 2 are to be reproduced.

Another way, widely used, to connect dark matter to galaxies is the semi-analytic modeling, where the dark matter distributions obtained from N-body simulations are supplemented with some of the physical processes important in galaxy formation using semi-analytic techniques. A test of the PVD vs luminosity for this type of models will be the aim of a subsequent paper.

Acknowledgements

JYP would like to thank the Max-Planck Institute für Astrophysik for its warm hospitality, IAU for its partial travel support, and the SOC for adjusting the talk time. The work is supported in part by NKBRSF (G19990754), by NSFC (No.10125314), and by the CAS-MPG exchange program.

References

Hawkins, E., et al. 2003, MNRAS, 346, 78

Jing, Y. P. & Börner, G. 2001, ApJ, 547, 545

Jing, Y. P. & Börner, G. 2004a, ApJ (in press) astro-ph/0311585

Jing, Y. P. & Börner, G. 2004b, in preparation

Jing, Y. P., Mo, H. J., & Börner, G. 1998, ApJ, 494, 1

Jing, Y. P. & Suto, Y. 2002, ApJ, 574, 538

Norberg, P. et al. 2002a, MNRAS, 332, 827

Norberg, P. et al. 2002b, MNRAS, 336, 907

Yang, X. et al. 2004, MNRAS (in press) astro-ph/0303524

Yang, X., Mo, H. J., & van den Bosch, F. C. 2003, MNRAS, 339, 1057

Peacock, J. A., et al. 2001, Nature, 410, 169

Outskirts of Galaxy Clusters: Intense Life in the Suburbs
Proceedings IAU Colloquium No. 195, 2004
A. Diaferio, ed.

DOI: 10.1017/S1743921304000882

Radio imaging of head-tailed galaxies from an in-falling filament of the Horologium-Reticulum supercluster

M. Johnston-Hollitt[1], M. Fleenor[2], J. Rose[2], W. Christiansen[2] and R. W. Hunstead[3]

[1]Leiden Observatory, PO Box 9513, 2300 RA Leiden, The Netherlands
email:johnston@strw.leidenuniv.nl

[2]Department of Physics and Astronomy, University of North Carolina, Chapel Hill, North Carolina 27599-3255, USA

[3] School of Physics, University of Sydney, NSW 2006, Australia

Abstract. We present a multi-wavelength study of a pair of head-tailed radio galaxies in an in-falling filament of the A3125/A3128 cluster complex which is located at the centre of the massive Horologium Reticulum Supercluster (HRS). We compare the large-scale structure evident from extensive optical observations of the region with the dynamical information gleaned from the morphology of the radio jets. We argue that in addition to the acceleration produced by in-fall toward the centre of A3125, the closest galaxy cluster, the radio jets show some evidence of bulk motion along the major filament of the HRS.

---

1. Introduction

The Horologium Reticulum Supercluster (HRS) is one of the most massive cluster concentrations observed in the nearby universe. It is second only to the Shapley Supercluster in the local 300 Mpc and provides a valuable opportunity to study the effect of the cluster environment on its constituent galaxies. Recently optical data have revealed a wealth of sub-structure in the A3125/A3128 cluster complex at the heart of the HRS, showing that hierarchical merging is taking place (Rose et al. 2002). In addition examinations of the inter-cluster regions have demonstrated a bi-modality to the entire structure (Fleenor et al. 2004) suggesting a complex dynamical evolution.

With such a wealth of activity evident in the kinematic data it is unsurprising to see evidence of dynamical activity associated with individual galaxies. In this paper we present results of radio imaging of a pair or bent-tailed sources found on the outskirts of the A3125/A3128 cluster complex. These galaxies are associated with a known in-falling filament close to A3125.

2. In-falling filament

Rose et al. (2002) have used spectroscopic data to identify several in-falling groups and clusters in the central regions of the HRS. Radio images of the region containing the in-falling filament and group structures "F2"and "G2", which have a combined total of around 25 members, were taken with the Australia Telescope Compact Array (ATCA) at 1.4 and 4.8 GHz. The 4.8 GHz data were obtained in 2×12 hours in the 6A configuration. The final image has a synthesised beam of $9''.9\times7''.9$ at P.A. 3.2° and a continuum image rms noise of 12 μJy beam^{-1}. The 1.4 GHz image was obtained from a larger 40 point mosaic of the region in the 1.5, 6A and 6D configurations with total integration time of

39 hours. The final mosaic image has a synthesised beam of $11''.7 \times 7''.5$ at P.A. 72.2° and a continuum image rms noise of 60 μJy beam^{-1}. In order to detect the low surface brightness extensions, the 1.4 GHz data were also tapered to give a lower resolution image with synthesised beam $23''.2 \times 21''.4$ and rms noise of 7.9 mJy beam^{-1}.

Figure 1 shows the radio contours going from lowest to highest resolution overlaid on the DSS image; blue contours are the tapered 1.4 GHz data, purple the untapered 1.4 GHz data and red the 4.8 GHz data, in addition known members of the filament have been marked with boxes.

Three radio sources are identified in the data; these are labelled as source A, B and C in Figure 1. Source A and B have been identified with members of the filament, whereas source C has no optical counterpart visible in the DSS and is thus likely a chance projection of a background radio galaxy.

3. Radio images

Objects A and B have been identified as PKS J0327−532 and SUMSS J032752-532613 respectively. The 1.4 GHz data shows that the jets of both galaxies have been bent and the data for PKS J0327−532 are particularly striking with multiple turns and bends present. The projected lengths have been measure at 7.75' and 5.75' for the Northern and Southern jets in PKS J0327−532 and at 5.25' for both jets in SUMSS J032752-532613. At the redshift of the host galaxies this corresponds to tail lengths of 480 and 356 kpc for PKS J0327-532 and 325 kpc for SUMSS J032752-532613. From the 1.4 GHz data it can be seen that the jet in PKS J0327-53 is at its peak brightness some distance from the optical host galaxy which is consistent with emission from a Fanaroff-Riley Class II source (Fanaroff & Riley 1974). By comparison the emission in SUMSS J032752-53261 is brightest at the location of the optical host galaxies which is consistent with a Fanaroff-Riley Class I source (Fanaroff & Riley 1974).

In the case of PKS J0327-532 the surface brightness drops dramatically about half way along the total jet length trailing away into lower surface brightness extensions which are not visible in the higher frequency data. The brighter of these low surface brightness features is associated with the southern tail and is also clearly detected in a Molonglo Observatory Synthesis Telescope image at 843 MHz (not shown). In both cases the jets in PKS J0327-532 maintain a collimated appearance the entire length of the jet. In contrast the jets in SUMSS J032752-532613 terminate in flairs of low surface brightness which are roughly circular about the jet. This is particularly evident for the southern most jet.

In the highest resolution images at 4.8 GHz the extended tails are too faint to observe and so the images only reveal the brightest parts of each source. In the case of PKS J0327-532 the 4.8 GHz image conclusively gives source APMUKS(BJ) B032601.62-533543.5 as the optical counterpart while APMUKS(BJ) B032629.42-533630.0 is determined as the optical identification for SUMSS J032752-532613 in both the 1.4 and 4.8 GHz data. This shows definitively that these sources are members of the group.

4. Relation to large-scale structure

The angular separation of the optical host galaxies is too great to consider these as a binary pair, thus we can not look to an orbital motion to explain the direction of bending of the radio tails (for comparison see Klamer et al. 2004). It is striking that both galaxies show some signs of bending toward the northwest which is roughly 90° away from the direction of in-fall to the closest cluster, A3125, but toward the main axis of the HRS. On the other hand the fainter tail of PKS J0327-532 shows bending along the line of

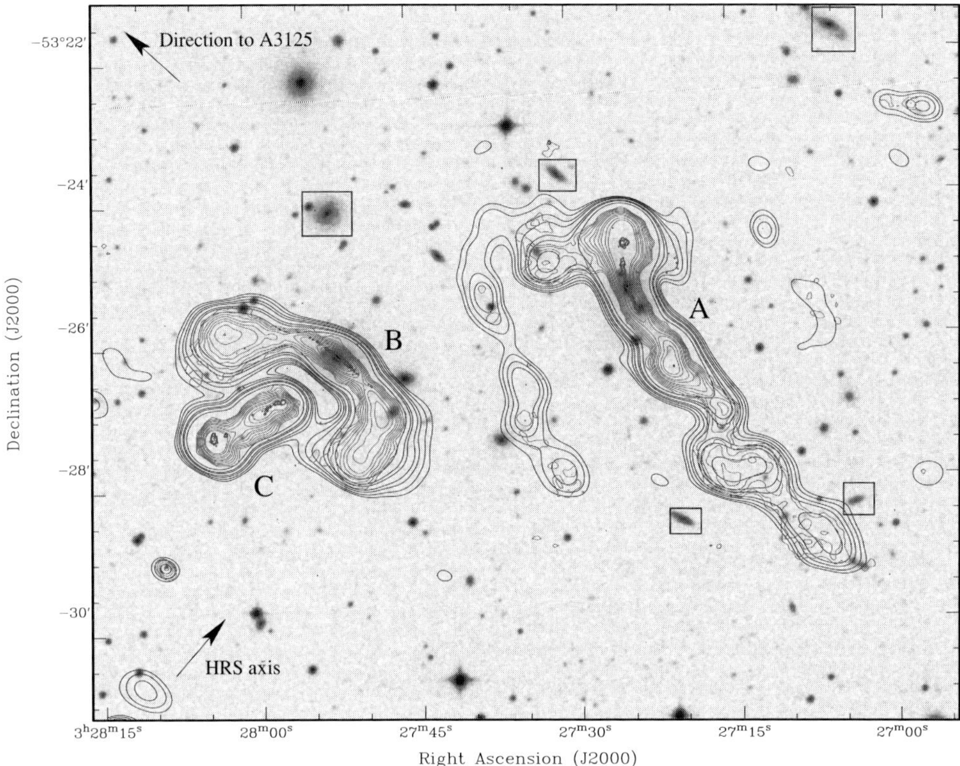

Figure 1. Image of tailed galaxies in an in-falling group: radio contours of increasing resolution are shown over an optical image from the DSS. Galaxies with boxes and those marked A and B are some of the members of the in-falling objects from Rose et al (2002). Arrows show the direction to both the nearest galaxy cluster, A3125 and the HRS main axis.

in-fall to A3125. Thus, we see evidence for motion along both the line of in-fall to the closest galaxy cluster and the main axis of the HRS which could be an indication of bulk gas flow along the main filament of the supercluster.

Acknowledgments

MJ.-H. acknowledges the support of the Leids Kerkhoven-Bosscha Fonds for travel to Torino. The Australia Telescope Compact Array is part of the Australia Telescope which is funded by the Commonwealth of Australia for operation as a National Facility managed by CSIRO.

References

Fanaroff, B. L. & Riley, J. M. 1974 *MNRAS* **167**, 31P–36P.

Fleenor, M.C., Rose, J.A., Christiansen, W.A., Hunstead, R.W., Johnston-Hollitt, M., & Drinkwater, M.J. 2004 *this volume*.

Klamer, I. J., Subrahmanyan, R. & Hunstead, R. W. 2004 *MNRAS* in press, astro-ph/0402574

Rose, J. A., Gaba, A. E, Christiansen, W. A., Davis, D. S., Caldwell, N., Hunstead, R. W., & Johnston-Hollitt, M. 2002 *AJ* **123**, 1216–1246.

Outskirts of Galaxy Clusters: Intense Life in the Suburbs
Proceedings IAU Colloquium No. 195, 2004
A. Diaferio, ed.

DOI: 10.1017/S1743921304000894

Populations of galaxy groups in clusters from wide-field multicolor imaging

I. H. Li[1], H. K. C. Yee[1], E. Ellingson[2] and G. Laurent[2]

[1]Department of Astronomy and Astrophysics, University of Toronto, 60 St. George Street, Toronto, ON, M5S 3H8, Canada email: tornado@astro.utoronto.ca, hyee@astro.utoronto.ca
[2]Center for Astrophysics and Space Astronomy, CB389, University of Colorado, Boulder, CO 80309, USA email: elling@casa.colorado.edu, Glenn.Laurent@Colorado.EDU

Abstract. We present the preliminary results of *BVRI* imaging of six clusters chosen from a follow-up project of the *CNOC* cluster survey. In an attempt to ascertain the evolution of galaxies in the cluster environment, we focus on the infalling galaxy groups, especially on their characteristics reflected by their color properties. To investigate the role of cluster environment in the evolution of galaxy groups, the color properties of galaxy groups are explored as a function of cluster radius and the local density of cluster galaxies. Both the radius and the local density affect the properties of groups, but evidence suggests that density may be more important in the relationship between cluster environment and galaxy characteristics.

1. Introduction

Clusters grow via the accretion of the field galaxies or galaxy groups. A general picture of clusters is that star-forming blue galaxies are in the suburbs while the central regions are occupied by the red passive galaxies. The infalling field galaxies and groups evolve and change their stellar population under the influence of the cluster environment.

This work focuses on the infalling groups rather than individual galaxies. The gravitational potential well possessed by groups could affect the evolution of grouped galaxies. To change the distribution of infalling galaxy populations to what we observe in clusters, the infalling groups must evolve to turn off their star forming activities while or before being accereted by the clusters. This change can be reflected in the color properties. This report presents the preliminary results of *BVRI* four-color imaging of six galaxy clusters at $0.17 \lesssim z \lesssim 0.38$. We adopt $\Omega_m = 0.3, \Omega_\Lambda = 0.7$, and $H_0 = 75$ km/s/Mpc.

2. Data and analysis

Sample and data: We have observed six clusters in *BVRI* from *CNOC* cluster survey (Yee et al. 1996) using CFH12k mosaic camera with a $28' \times 42'$ field of view. We formed photometric-redshift samples of galaxies for each cluster with an absolute magnitude limit of $M_R = -19.0$.

Photometric redshift method: To select the cluster galaxies, we use the empirical four-color fitting photometric redshift method based on that developed by Connolly et al. (1995). The CNOC2 field galaxy survey (Yee et al. 2000) patch *0920+37* is used as the training set. Both color and magnitude are used in the fitting. Different photometric redshift cutoffs based on the dispersion in $z_{spec} - z_{phot}$ as a function of galaxy color are used to select galaxies in the cluster redshift space. The 1σ redshift cutoffs range from 0.03 to 0.05.

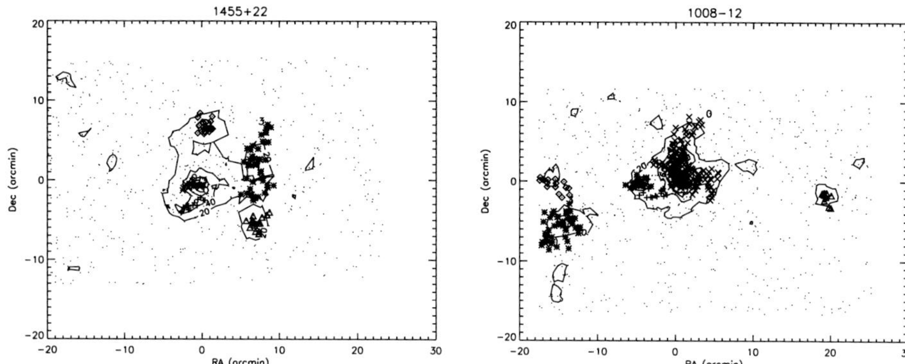

Figure 1. Map of (*a*) MS1455+22 (*b*) MS1008-12. Contours are based on the local galaxy number density without background and foreground subtraction. Different groups, as found by a friends-of-friends algorithm, are marked by different symbols.

Background and foreground determination: The background and foreground contamination are determined by using the same CNOC2 patch (with an effective area of 657 arcmin^2) with the same magnitude cutoffs as individual clusters.

Group finding: To see the local environmental effect, the local surface galaxy number density (Σ) is calculated based on the distance to the 10$^{\rm th}$ nearest neighbor, r_{10}. A friends-of-friends algorithm is used to find galaxy groups. To differentiate galaxy groups in different cluster global density regions, we use a dynamic linking length. Examples of cluster maps and groups are shown in Fig. 1. The dynamic linking length, a function of cluster galaxy density and cluster-centric radius, can uncover groups even in the high density regions close to the cluster cores.

3. Discussion

Cluster environmental effects: As an example, we plotted $B-R$ color as function of cluster-centric radius (Fig. 2a) and local galaxy density (Fig. 2b) for MS1455+22. The color gradient becomes nearly constant after passing R_{200}, suggesting that the color properties strongly correlate with cluster-centric radius within the virial radius. We also see a drop in the color trend at $1.2 \leqslant \log\Sigma \leqslant 2.0$ galaxies/Mpc^2. This color transition may indicate a threshold effect in transforming galaxy properties (Kodama et al. 2001; Gomez et al. 2003).

Blue galaxy fraction: Figure 3 shows the cluster blue fractions (with different magnitude cutoffs) as a function of cluster-centric radius (normalized by R_{200}). The blue fractions calculated using fainter magnitude cutoffs are higher, reflecting that faint galaxies are usually blue. The blue fractions at the virial radius are similar for different clusters, regardless of cluster richness. The blue fraction also tends to increase with decreasing local density (Fig. 4) for galaxies in both groups and clusters, indicating that local environment is more important than the average cluster environment. In the same density region, high-z clusters have larger fractions of blue galaxies. Similarly, groups in high-z clusters tend to contain more blue galaxies, suggesting that the Butcher-Oemler effect may also be seen in groups.

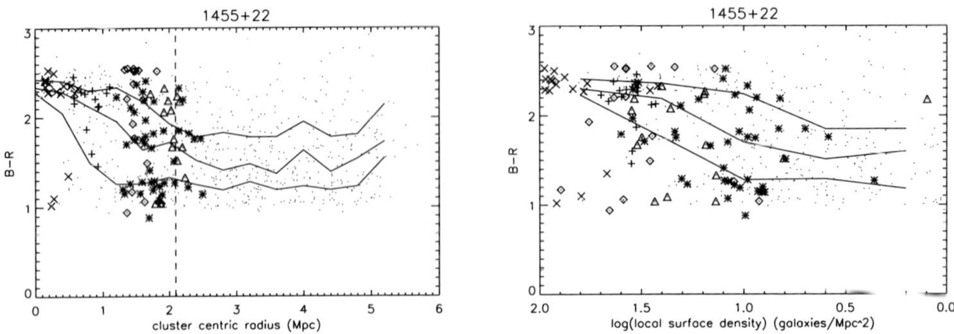

Figure 2. $B-R$ color as a function of (a) cluster-centric radius (b) local galaxy density for MS1455+22. Overlayed are 75, 50, and 25 percentile of galaxy color. Galaxies in groups are marked by different symbols. The vertical dashed line in (a) indicates R_{200}.

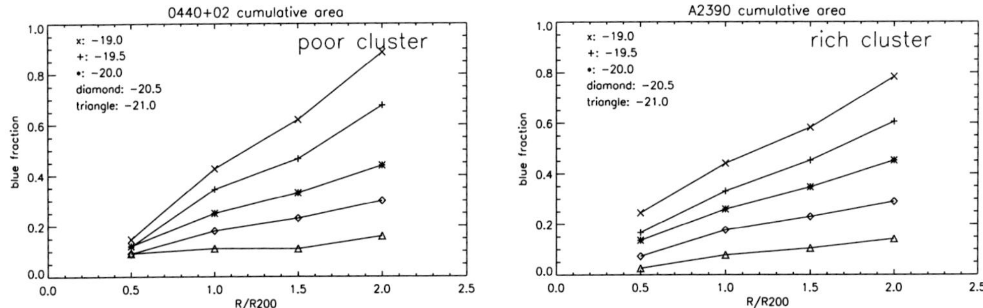

Figure 3. Blue fraction as a function of cluster-centric radius (a) MS0440+02 (b) Abell 2390. Several magnitude cutoffs are used as indicated in the upper-left in the plots. The blue fractions are similiar at $1R_{200}$ regardless of cluster richness.

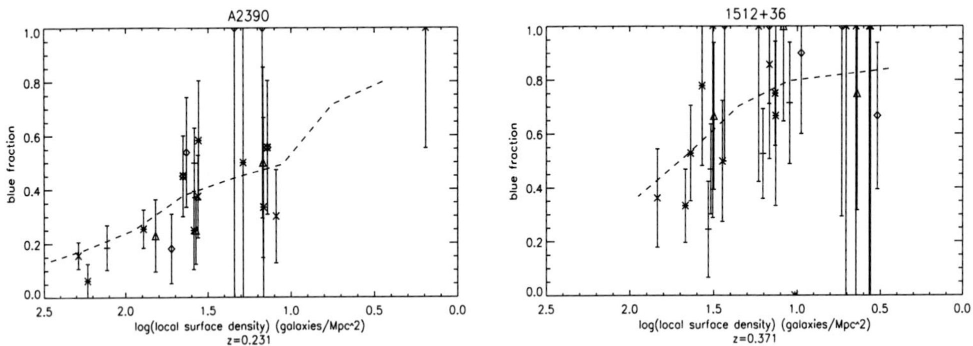

Figure 4. Blue galaxy fraction (dashed line) for the whole clusters as a function of local galaxy density for (a) Abell 2390 (b) MS1512+36. Blue fractions in different groups are indicated by different symbols. The dependence of blue fraction on local galaxy density for groups is similar to that for the whole cluster. Clusters and groups at high redshift have larger blue fractions.

References

Connolly, A. J., Csabai, I., Szalay, A.S., Koo, D.C., Kron, R.G. & Munn, J.A. 1995 *AJ*, 110–2655
Gómez et al. 2003 *ApJ*, 584–210
Kodama, T., Smail, I., Nakata, F., Okamura, S. & Bower, R.G. 2001 *ApJ*, 562, L9
Yee, H.K.C., Ellingson, E. & Carlberg, R.G. 1996 *ApJS*, 102–269
Yee, H.K.C., et al. 2000 *ApJS*, 475–492

Outskirts of Galaxy Clusters: Intense Life in the Suburbs
Proceedings IAU Colloquium No. 195, 2004
A. Diaferio, ed.

DOI: 10.1017/S1743921304000900

Photometric properties of dwarf galaxies in the Coma cluster: radial dependence

Yutaka Komiyama[1], Sadanori Okamura[2], Masafumi Yagi[3], Nobunari Kashikawa[3], Bianca M. Poggianti[4], Terry J. Bridges[5], Bahram Mobasher[6] and Dave Carter[7]

[1]Subaru Telescope, Hilo, HI, USA email: komiyama@subaru.naoj.org
[2]Department of Astronomy, University of Tokyo, Bunkyo-ku, Tokyo, Japan
[3]National Astronomical Observatory of Japan, Mitaka, Tokyo, Japan
[4]Osservatorio Astronomico di Padova-INAF, Padova, Italy
[5]Department of Physics, Queen's University, Kingston, Canada
[6]Space Telescope Science Institute, Baltimore, MD, USA
[7]Liverpool John Moores University, Birkenhead, Wirral, UK

Abstract. Photometric properties (effective surface brightness, effective radius, radial profile index, axis ratio, color, color gradient) of 328 galaxies in the Coma cluster fainter than $R = 15$ mag are examined as a function of the distance from the cluster center. No significant gradient is found for the effective surface brightness, effective radius and radial profile index. The distribution of axis ratios shows a concentration of round galaxies at the cluster center in the magnitude range $16.5 < R < 18$; most of these are found to be old and to have intermediate metal abundance, suggesting that they are nucleated dwarf ellipticals. On the other hand, we find a significant gradient in color, in the sense that galaxy colours become bluer with increasing distance from the cluster center. We conclude that this color gradient represents a metallicity gradient.

1. The data

We carried out a wide-field imaging and spectroscopic survey in the Coma cluster using the Mosaic CCD Camera II and WYFFOS multi-fibre spectrograph on the William Herschel Telescope, obtaining homogeneous images for a 2.22 deg^2 area to $R \simeq 23$ mag, and medium resolution spectra for about 500 galaxies to $R = 20$ (Komiyama et al. 2002; Mobasher et al. 2001; Poggianti et al. 2001). In this study, we analyze the photometric properties of 328 spectroscopically confirmed cluster galaxies fainter than $R = 15$ mag (121, 124 and 83 in magnitude ranges of $15 < R < 16.5$, $16.5 < R < 18$ and $18 < R$, respectively).

2. Radial dependence

Photometric parameters are examined as a function of distance from the cluster center. We fit the distribution of a given parameter P with $P = C \log d + D$, where d is the distance from the cluster center, to see if there is a radial gradient. We assume that the radial gradient is significant when C is larger than $2\sigma_C$, where σ_C is the fitting error of C.

2.1. *Luminosity structure parameters: μ_e, r_e, N, axis ratio*

We first examine the luminosity structure parameters and find no significant radial gradient for the effective surface brightness (μ_e), effective radius (r_e) and radial profile index (N) fitted to the Sersic law $I(r) = I_0 \exp[-(r/r_0)^N]$, in any magnitude range.

But we find a significant gradient of axis ratio for galaxies in the magnitude range $16.5 < R < 18$ (3.6σ significance), in the sense that round galaxies preferentially reside in the cluster center. The spectral analysis shows that most of these are old (> 9Gyr) and have a typical metal abundance ($-1 < Z < -0.15$) for galaxies in this magnitude range. Ferguson & Sandage (1989) showed that nucleated dwarf ellipticals in the Virgo and Fornax clusters tend to have round shapes and reside preferentially in the central region of the cluster. Therefore, round galaxies found in the center of Coma could be nucleated dwarf ellipticals. Their absolute magnitudes (roughly $-18 < M_R < -16.5$) also support this idea. However, we cannot verify this because the nuclei are buried in the seeing disk in our images.

2.2. *Stellar population parameter: $B_K - R_K$*

We find that the $B_K - R_K$ color shows a significant gradient in all magnitude ranges (2.5, 2.8 and 4.5 σ significance for $15 < R < 16.5$, $16.5 < R < 18$ and $18 < R$, respectively), in the sense that galaxies which are located at the outskirts of the cluster have bluer colors as seen in the left panels of Fig. 1. It is also noted that the slope of the gradient becomes steeper as galaxies become fainter.

Carter et al. (2002) reported that there is clear evidence of a metallicity gradient for the galaxies in the Coma cluster, based on the same data used in this study. However, their analysis included all galaxies, regardless of magnitude. We hence examine the metallicity distribution of galaxies in each magnitude range, in the same manner as we did for color. As seen in the right panels of Fig. 1, we find gradients in the metallicity distribution, although the significance levels are not as high as those for color (2.4, 0.9 and 2.1σ for $15 < R < 16.5$, $16.5 < R < 18$ and $18 < R$, respectively). Since no significant gradient in the age distribution is found in all magnitude ranges, we conclude that the gradient in color represents a metallicity gradient, not an age gradient. We also note that the slope of the metallicity gradient is steeper for fainter galaxies, which is also the case for color.

We suggest that the metallicity gradient can be explained by pressure confinement by the hot intracluster medium (Babul & Rees 1992). According to this scenario, a galaxy located in the inner regions of the cluster, where the pressure of the intracluster medium is high, would have a longer star-forming period and a higher metallicity compared to a galaxy in the outskirts of the cluster. In addition, since the time scale of gas confinement is partly determined by the binding energy (i.e. mass), massive galaxies would be more metal-rich than less massive galaxies for the same external pressure. This may account for the trend that the slope of the metallicity gradient becomes steeper for fainter galaxies.

2.3. *Color gradients within cluster galaxies: $d(B_K - R_K)/dr$*

The color gradients *within* galaxies show a significant variation across the cluster, for the faintest galaxies ($18 < R$). For this faint subsample, galaxies near the cluster center tend to have positive color gradients (Δ(B-R)/Δlog(r) > 0; i.e. colors become redder with increasing galactocentric distance), while galaxies near the cluster outskirts tend to have negative color gradients (Δ(B-R)/Δlog(r) < 0; i.e. colors become bluer with increasing galactocentric distance). This correlation is significant at a level of 3.3σ; no correlation is seen for brighter galaxies. The observed trend could reflect environmental effects for the evolution of dwarf galaxies, although the scatter is large. Further theoretical and observational studies are required to explain the nature of this trend.

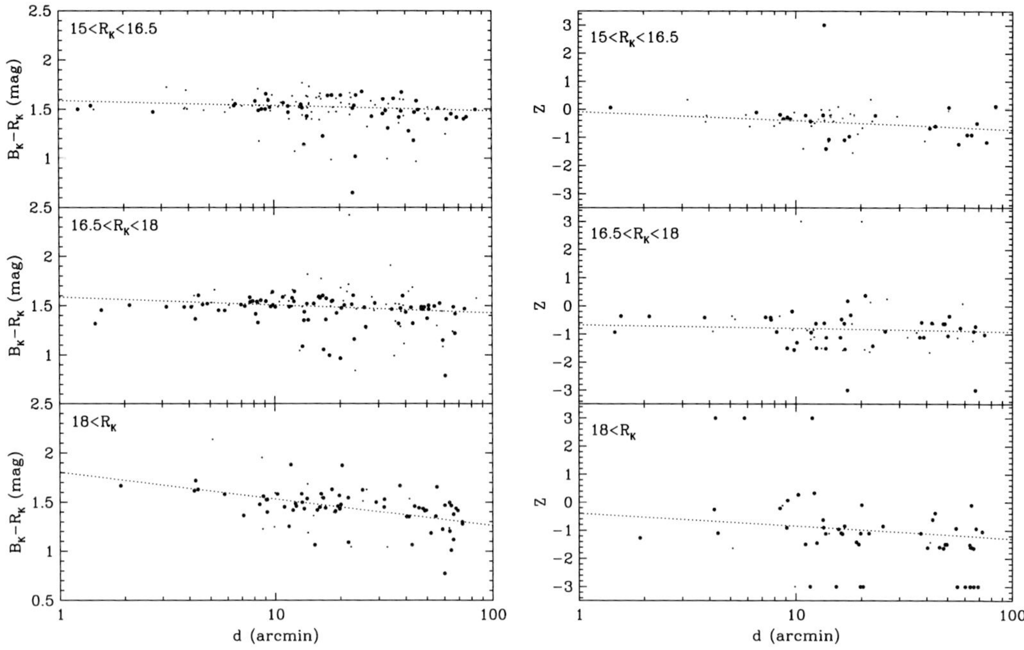

Figure 1. Left: $B_K - R_K$ color plotted as a function of the distance from the center of the Coma cluster. Isolated galaxies are plotted as large dots, while blended galaxies are plotted as small dots. The dotted lines show the best-fit lines for the data (overlying points are excluded from the fit). Right: Metallicity Z plotted as a function of the distance from the cluster center. For galaxies lying outside the age-metal model grid, an arbitrary metallicity (+3 or −3 dex) is assigned, but these galaxies are excluded from the fitting procedure.

References

Babul, A. & Rees, M. J. 1992 *MNRAS* **255**, 346

Carter D., Mobasher B., Bridges, T. J. et al. 2002 *ApJ* **567**, 772

Ferguson, H. C. & Sandage, A. 1989 *ApJ* **346**, L53

Komiyama Y., Sekiguchi, M., Kashikawa, N. et al. 2002 *ApJS* **138**, 265

Mobasher B., Bridges, T. J., Carter, D. et al. 2001 *ApJS* **137**, 279

Poggianti B. M., Bridges, T. J., Mobasher, B. et al. 2001 *ApJ* **562**, 689

Outskirts of Galaxy Clusters: Intense Life in the Suburbs
Proceedings IAU Colloquium No. 195, 2004
A. Diaferio, ed.

DOI: 10.1017/S1743921304000912

Environmental effects in the structural parameters of Coma galaxies

J. A. L. Aguerri[1], J. Iglesias-Páramo[2], J. M. Vílchez[3] and C. Muñoz-Tuñón[1]

[1]Instituto de Astrofísica de Canarias
[2]Laboratoire d'Astrophysique de Marseille
[3]Instituto de Astrofísica de Andalucía, CSIC

Abstract. We have studied the quantitative morphology of the galaxies of the Coma cluster brighter than $m_r = 17$. The surface brightness profiles of all galaxies were decomposed into the structural components: bulge and disc. We found a correlation between the scale length of the discs and the location in the cluster. Galaxies located at small projected distances show smaller discs than those located in the outermost regions of the cluster. We have also investigated the correlation betwenn the B−r color of the galaxies and their environment. For late-type galaxies there is a correlation between the color and the position in the cluster. The bluest galaxies are located in the outermost regions of the cluster and show the largest discs. These results can be explained in terms of galaxy harassment.

---

1. Introduction

Galaxy clusters are ideal to study the influence of environment on the evolution of galaxies. Galaxies in clusters can quickly evolve due to many physical processes, such as: harassment or pre-processing.

The harassment scenario (Moore et al. 1996) proposes that galaxies in clusters evolve, on short time-scales, due to the tidal interactions between them and with the global gravitational potential of the cluster. This mechanism is very efficient at transforming late-type galaxies in S0s, and it will also reduce the size of the discs. Similar to harassment, but more efficient in rich galaxy clusters, is the pre-processing, which consists of recent arrivals in the cluster being pre-processed in group or subcluster environments (Kodama & Smail 2001). The study of the quantitative morphology of galaxies can provide useful constrains to these theories.†

2. Structural parameters and color distribution

We have fitted the isophotal surface brightness profiles of the galaxies with two components: bulge and disc. The surface brightness profiles of the bulges were modeled by a Sersic law, and for the discs we used an exponential profile. The fitting algorithm was developed by Trujillo et al. (2001), and uses a Levenberg-Marquard nonlinear fitting routine. Monte Carlo simulations were carried out in order to determine the uncertainties of the recovered structural parameters. The galaxies were classified according to the bulge-to-total luminosity ratio (B/T) (see Aguerri et al. 2004). We found that there is no trend

† The results presented here correspond to the Coma galaxy cluster. We have observed 1 square degree of Coma cluster. The images were obtained using the Sloan-Gunn r filter of the WFC (see Iglesias-Páramo et al. 2002). The B band magnitudes of the galaxies were obtained from the Godwin et al. (1983) catalogue of Coma cluster galaxies. We have assumed that Coma members are those with 4000 km s^{-1} $< cz <$ 10000. The radial velocities were taken from NED.

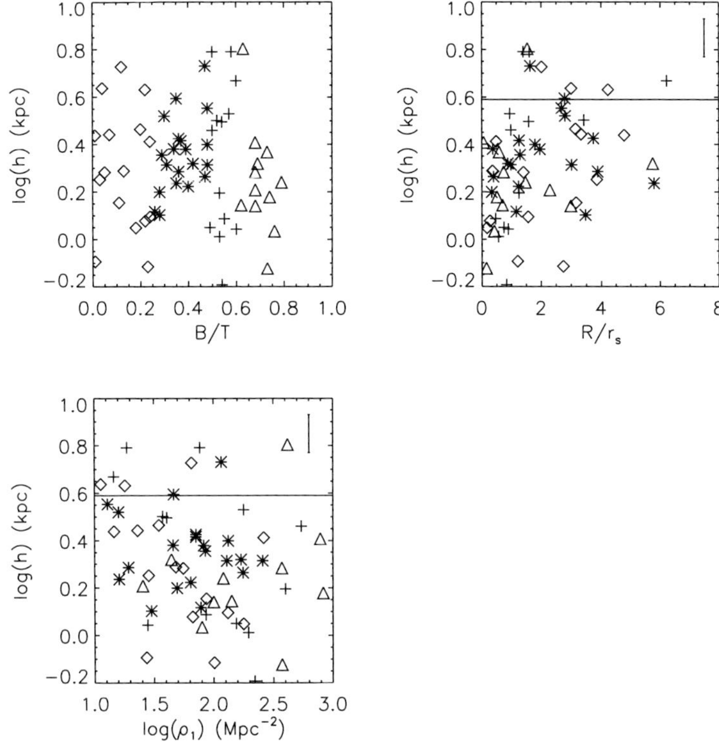

Figure 1. Scale length of the discs vs B/T (top-left), projected distance to the cluster center (top-right), and projected density (bottom-left). The horizontal dashed line shows the scale of the discs of field galaxies. The symbols represent: E with disc (triangles), S0 (crosses), early-type spirals (asterisks) and late-type spirals (diamonds).

between the scale of the discs and the morphological type of the galaxies, similarly to what is found for local field samples. The mean value of the scale of the discs for Coma galaxies is $\log h = 0.33$, which is smaller than the value for field galaxies with similar luminosities ($\log h = 0.59$; Graham 2001). There is also a correlation between the scale of the discs and the position in the cluster. No large discs are located at the center of the cluster (see Fig. 1).

We have investigated the B−r color distribution of the galaxies in Coma. The fraction of blue galaxies (as defined by Butcher-Oemler 1984) is 0.07, taking into account all galaxies brighter than $m_r = 17$. These blue objects have bigger velocity dispersions than the red ones (Colless & Dunn 1996). The color of E and S0 galaxies is constant independently of their position in the cluster. For spirals, the bluest galaxies are located in the outermost regions of the cluster. This trend is more important for late-type galaxies (see Fig. 2).

3. Discussion and conclusions

Galaxies in clusters are affected by their environment, which can be responsible for the morphological transformations between different types. Galaxy harassment is the tidally induced evolution brought about by multiple high speed encounters with other galaxies and with the cluster gravitational potential. This mechanism and the pre-processing can explain the small scale h observed in the stellar discs of the galaxies in Coma. The tidal interactions induced by the harassment can also enhance the star formation rate (SFR)

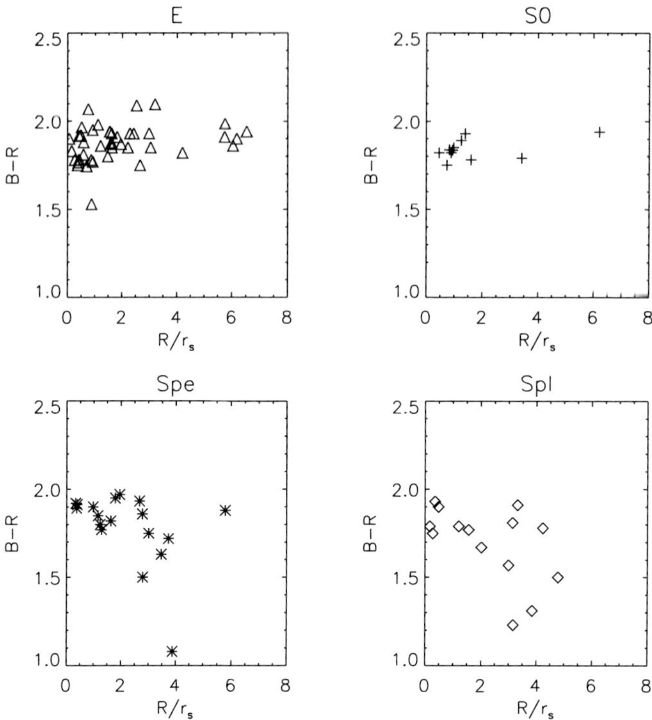

Figure 2. B−r color of the different type of galaxies vs the projected distance to the center of the cluster.

of the galaxies. If the SFR is enhanced by the tidal interactions with the cluster potential, then the bluest galaxies should be located in the center of the cluster. If the SFR is due to galaxy-galaxy encounters then the bluest galaxies are located in the outermost region of the cluster (Fujita 1998). This can explain the distribution of colors observed in the Coma spiral galaxies. The bluest galaxies are those which are falling into the cluster and they are suffering the harassment for the first time.

Acknowledgements

This work has been partially financed by the project AYA2001-3939-C03 of the Spanish Programa Nacional de AStronomía y Astrofísica of the MCyT.

References

Aguerri, J. A. L., Iglesias-Paramo, J., Vilchez, J. M., & Muñoz-Tuñón, C. 2004 *AJ* **127**, 1344–1359.

Butcher, H. & Oemler, A. 1984 *ApJ* **285**, 426–438.

Colles, M. & Dunn, A. M. 1996 *ApJ* **458**, 435–454.

Fujita, Y. 1998 *ApJ* **509**, 587–594.

Godwin, J. G., Metcalfe, N., & Peach, J. V. 1983 *MNRAS* **202**, 113–124.

Graham, A. W. 2001 *AJ* **121**, 820–840.

Iglesias-Páramo, J., Boselli, A., Cortese, L., Vílchez, J. M., & Gavazzi, G. 2002 *A&A* **384**, 383–392.

Kodama, T. & Smail, I. 2001 *MNRAS* **326**, 637–642.

Moore, B., Katz, N., Lake, G., Dressler, A., & Oemler, A. 1996 *Nature* **379**, 613–616.

Trujillo, I., Aguerri, J. A. L., Gutiérrez, C. M., & Cepa, J. 2001 *AJ* **122**, 38–54.

Outskirts of Galaxy Clusters: Intense Life in the Suburbs
Proceedings IAU Colloquium No. 195, 2004
A. Diaferio, ed.

DOI: 10.1017/S1743921304000924

The K-band luminosity functions of galaxies in high-redshift clusters

Simon Ellis[1] **and Laurence Jones**[2]

[1]Anglo-Australian Observatory, PO Box 196, Epping, NSW, Australia email: sce@aaoepp.aao.gov.au

[2]University of Birmingham, UK email: lrj@star.sr.bham.ac.uk

Abstract. K band luminosity functions (LFs) of three, massive, high redshift clusters of galaxies are presented. The evolution of K^*, the characteristic magnitude of the LF, is consistent with purely passive evolution, and a redshift of formation $z_f \approx 1.5$–2.

1. Introduction

Near-infrared photometry of galaxies in high redshift clusters provides an excellent method of studying evolution in the baryonic masses of cluster galaxies. The K band is rather insensitive to the enhanced blue emission associated with young stellar populations. This means that observations made in the K band are measuring emission from the old stellar populations, thus the near-infrared (NIR) luminosity is much more closely correlated with the stellar mass of a galaxy than optical luminosities which may be significantly inflated by the presence of only a small, but young, stellar population. The K band has the added advantages that it has relatively small k-corrections, and these are very similar for galaxies of different spectral-type, and the extinction due to dust is also small. Therefore by measuring the LFs of galaxies in high redshift clusters, and comparing to well established LFs of nearby systems, it is possible to determine how the distribution of stellar mass throughout the cluster population has evolved.

2. Sample and observations

We have obtained K band photometry of the central regions of three of the most massive ($\sim 10^{14} - 10^{15} M_\odot$), high redshift ($z = 0.83 - 1.03$) clusters of galaxies known. All three were discovered in the WARPS serendipitous X-ray survey (Scharf et al. 1997). Observations were made with UFTI on the UKIRT telescope on Mauna Kea, Hawaii. The seeing was typically half an arcsecond in K. The observations typically covered 3 arcminutes square, corresponding to 30%–40% of the virial radii. Further details of the observations and analysis can be found in Ellis & Jones (2004).

3. Modelling passive evolution

In the traditional monolithic collapse picture of galaxy formation (Eggen et al. 1962) all the stars in a galaxy are formed in an inital burst, and thereafter the galaxy evolves only in a passive, quiescent manner as its stars make their journey along the main sequence. This results in a gradual dimming of the stars, and consequently the galaxy. Any further episodes in the evolution of galaxies, such as secondary bursts of (or gradual) star-formation activity, mergers, and gravitional effects such as harrassment, stripping, starvation, etc. must take place against this background of the passive evolution of the pre-existing stars.

Thus, in order to establish whether any of these 'extra-passive' processes are important in the evolution of cluster galaxies (or more strictly, if they are important in the evolution of the K band luminosities of cluster galaxies), we adopt the approach of modelling a passively evolving system, and searching for irregularities between the predictions of the model and observations. Unless the K band is insensitive to these extra-passive events they should show up as differences. Thus it may be expected that the following analysis is more senstive to issues regarding the mass-assembly of the galaxies, than it is to the star-formation history of galaxies.

We have modelled passive evolution using the synthetic stellar population libraries of Bruzual & Charlot (2003). Stellar populations with a Salpeter initial mass function and solar metallicity were assumed. The evolution was modelled for instantaneous bursts at z_f =1.5, 2 and 5, along with predictions of no evolution.

4. Results

LFs were constructed individually for each cluster after applying a statistical subtraction of foreground/background galaxies based on number counts of galaxies from nearby offset fields observed at the same time as the clusters, and on spectroscopic information for the brighter cluster members, where available. The LFs for ClJ0152 (top left), ClJ1226 (top right) and ClJ1415 (bottom left) are shown in figure 1.

Schechter (1976) functions were fit to the LFs of each cluster individually, with the faint end slope fixed at $\alpha = -0.9$ (the local value for the Coma cluster, De Propris et al. 1998, and the field, Gardner et al. 1997), since the data were not deep enough to constrain both α and K^* precisely. The evolution of K^* with redshift is shown in the lower right panel of figure 1, along with lower redshift points from De Propris et al. (1999). The models (described in the previous section) are normalised at low redshift to K^* of the Coma cluster (De Propris et al. 1998). The evolution is seen to be consistent with passive evolution models, with $z_{\mathrm{f}} \approx 1.5$ if the models are normalised to Coma, or a higher redshift of formation if the models are normalised to the low redshift points from De Propris et al. (1999).

Purely passive evolution of early-type galaxies is consistent with several other studies including the evolution of the K band luminosity function (De Propris et al. 1999, Toft et al. 2003), evolution of the fundamental plane in terms of mass-to-light ratios (van Dokkum et al. 1998), and studies of the scatter of the colour-magnitude relation (see e.g. Ellis et al. 1997, Stanford et al. 1998).

Whilst the observed evolution is fully consistent with passive evolution, it may be reconciled with the framework of hierarchical formation of structure, so long as any merging between luminous galaxies in massive clusters takes place before $z \approx 1$. As the systems under consideration here are unusually massive, they will be associated with regions of large over-density, and thus it is perhaps expected that objects will assemble at early epochs in such an environment. Furthermore, the large velocity dispersions associated with massive clusters makes the likelihood of mergers small (see e.g. Diaferio et al. 2001). For a more thorough discussion of this work see Ellis & Jones (2004).

Acknowledgements

The authors would like to thank warmly both Ben Maughan and Harald Ebeling. UKIRT staff have been very efficient and helpful; service time observations were

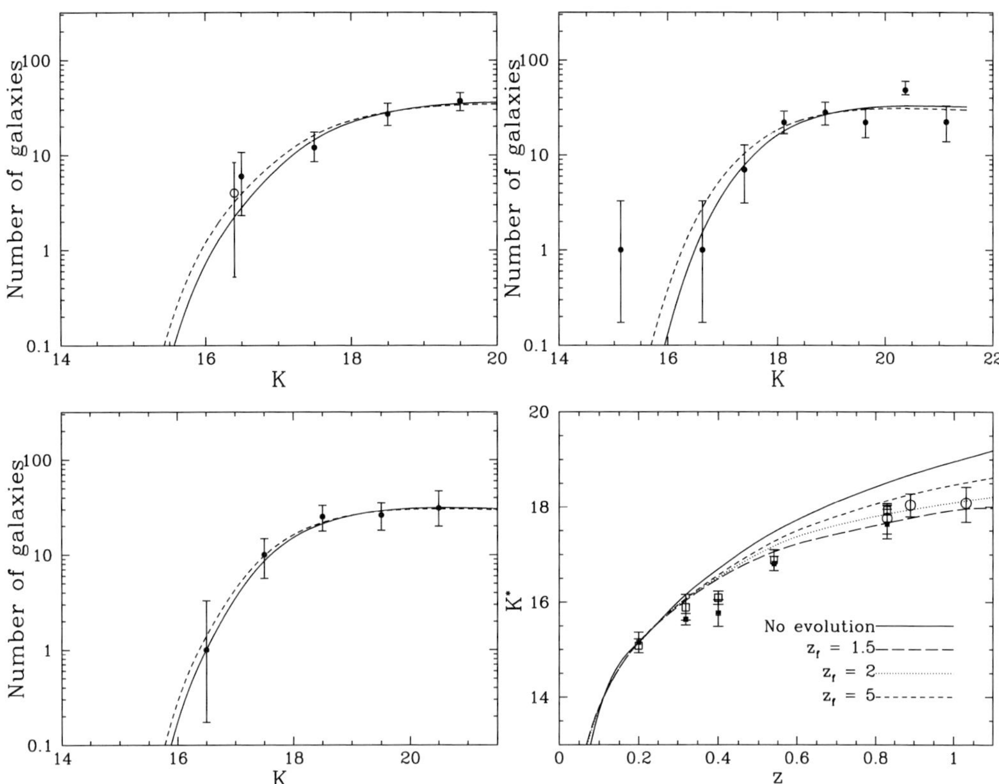

Figure 1. The evolution of the K band LF. The first three panels show the LFs of the clusters of galaxies individually: top-left, ClJ0152; top-right, ClJ1226; bottom-left, ClJ1415. The lower right-hand quadrant shows the evolution of K^*. The circles are data from this paper. The squares are from De Propris et al. (1999), open symbols being low L_X systems and closed symbols being high L_X systems.

performed by some of them. The authors would also like to thank Stefano Andreon and Antonaldo Diaferio for their help.

References

Bruzual G., Charlot S., 2003, MNRAS, 344, 1000

Butcher H., Oemler A., 1984, ApJ, 285, 426

De Propris R., Eisenhardt P. A., Stanford S. A., Dickinson M., 1998, ApJ, 503, L45

De Propris R., Stanford S. A., Eisenhardt P. A., Dickinson M., Elston R., 1999, AJ, 118, 719

Diaferio A., Kauffmann G., Balogh M. L., White S. D. M., Schade D., Ellingson E., 2001, MNRAS, 323, 999

Eggen O. J., Lynden-Bell D., Sandage A. R., 1962, ApJ, 136, 748

Ellis R. S., Smail I., Dressler A., Couch W. J., Oemler A. J., Butcher H., Sharples R. M., 1997, ApJ, 483, 582

Ellis S. C., Jones L. R., 2004, MNRAS, 348, 165

Gardner J. P., Sharples R. M., Frenk C. S., Carrasco B. E., 1997, ApJ, 480, L99

Scharf C., Jones L. R., Ebeling H., Perlman E., Malkan M., Wegner G., 1997, ApJ, 477, 79

Schechter P., 1976, ApJ, 203, 297

Stanford S. A., Eisenhardt P. R., Dickinson M., 1998, ApJ, 492, 461

Toft S., Soucail G., Hjorth J., 2003, MNRAS, 344, 337

van Dokkum P. G., Franx M., Kelson D. D., Illingworth G. D., 1998, ApJ, 504, L17

Outskirts of Galaxy Clusters: Intense Life in the Suburbs
Proceedings IAU Colloquium No. 195, 2004
A. Diaferio, ed.

DOI: 10.1017/S1743921304000936

The luminosity function of galaxies in the Hercules cluster

R. Sánchez-Janssen[1], J. Iglesias-Páramo[2], C. Muñoz-Tuñón[1], J. A. L. Aguerri[1] and J. M. Vílchez[3]

[1]Instituto de Astrofísica de Canarias, Spain email: ruben@ll.iac.es

[2]Laboratoire d'Astrophysique de Marseille, France

[3]Instituto de Astrofísica de Andalucía, CSIC, Spain

Abstract. We have imaged ~ 1 deg^2 in the V-band in the direction of the Hercules cluster (Abell 2151). The data are used to compute, for the first time, the luminosity function (LF) of galaxies in the cluster down to the dwarf regime ($M_{lim} \sim -13.85$). The global LF is well described by a Schechter function (Schechter 1976) with best-fit parameters $\alpha = -1.30 \pm 0.06$ and $M_V^* = -21.25 \pm 0.25$. The radial dependence of the LF has also been studied, finding that it turns out to be almost constant within the errors even further away than the virial radius. Given the presence of significant substructure within the cluster, we have analized the LFs in different regions. While the LFs of the two subclusters present are consistent with each other and with the global one, the southernmost one exhibits a somewhat steeper faint-end slope.

1. Introduction

The luminosity function of galaxies (LF), that is the probability density of galaxies in a certain population having a given luminosity, is one of the key observable quantities for galaxy evolution theories. The comparison between the LF of galaxies in clusters and in the field should clarify the role played by the environment in regulating galaxy evolution. Recent results by large surveys (De Propris et al. 2003) point towards the composite LF of cluster galaxies having a steeper faint-end slope than field galaxies, revealing that this influence indeed exists. A2151 is a nearby ($z = 0.0367$), irregular and spiral-rich cluster ($\sim 50\%$). There is strong evidence (from optical and X-ray studies) suggesting that the cluster is still in the process of collapse: the lack of hydrogen deficiency in the spiral population (Giovanelli & Haynes 1985), the bumpy distribution of the hot intracluster gas and its low X-ray flux (Huang & Sarazin 1996), and the presence of at least three distinct subclusters (Bird, Davis & Beers 1995, BDB hereafter) point towards A2151 being a young and relatively unevolved cluster, thus making it an excellent target for the study of the LF and stablishing comparisons with more evolved systems.

2. Observations

The observations were carried out with the WFC at the 2.5m INT. We have imaged a mosaic of five overlapping pointings covering an area of about $1^\circ \times 1^\circ$ across the cluster (Fig. 1, left). The calibration was achieved using aperture magnitudes of bright galaxies taken from Takamiya, Kron & Kron (1995). We have also observed the nearby SA107 control field for purposes of background galaxy counts decontamination. The V magnitudes were corrected for Galactic extinction using the Schlegel, Finkbeiner & Davis (1998) dust map and the Cardelli, Clayton & Mathis (1989) extinction curve.

3. Source extraction and photometry

The identification and extraction of sources was carried out using SExtractor (Bertin & Arnouts 1996). The separation between stars and galaxies was performed based on SExtractor's stellarity index (S/G), which assigns a numerical value close to 1 if the object is a star and values closer to 0 if it is a galaxy-like object. Pitifully, the S/G classifier breaks down at the lower magnitude end, complicating the identification. In order to test the reliability of the index we simulated artificial images of isolated gaussian-PSF stars and recovered them with the same selection criteria as used for the Hercules fields. FWHMs ranged the values of our frames. We obtained values of S/G $\geqslant 0.85$ for star-like objects and with accurate photometry down to V $\sim$ 22.

4. Sample completeness

One of the key problems in the study of LFs is the knowledge of detection limits of the data. We have carried out Monte Carlo simulations of artificial galaxy images, and, as done with stars, recovered them using SExtractor. We generated galaxies with four different morphologies: three exponential profiles with $r_{eff} = 0.75$, 1.5 and 3 kpc and one Sérsic profile with $r_{eff} = 0.5$ kpc, each in the magnitude range $17 < V < 23$ ($-18.85 < M_V < -12.85$ at the distance of A2151 and assuming a value of $H_0 = 75$ km s^{-1} Mpc^{-1}). Those are typical parameters of dE and dSph galaxies in nearby clusters (Binggeli & Jerjen 1998) which are supposed to dominate the faint end of the LF. Results of our simulations show that, as expected, galaxies with higher effective radii are hardly detected, while recovered magnitude differences are always below 0.3 mag, which do not play an important role in our subsequent analysis, since for the LF we are using bins of 1 mag width. If we make the simple hypothesis that all these galaxies occur with the same probability in the cluster, our sample is 80% complete down to V = 22.

5. The luminosity functions

Once we had the number counts and effective areas for both the cluster and control field samples, we computed the Hercules cluster LF as the statistical difference between these counts. In Fig. 1 (right) we plot the cluster LF down to V = 22 ($M_V \sim -13.85$), where the error bars include the calibration uncertainties and both Poissonian and non-Poissonian fluctuations of counts, the latter being due to the cosmic variance (Huang et al. 1997). We fit a Schechter function to the data, finding that it is a good representation of the LF ($\chi^2_\nu = 0.78$). Best-fit parameters are $\alpha = -1.30 \pm 0.06$, $M^*_V = -21.25 \pm 0.25$ and $\phi^* = 53 \pm 14$. Lugger (1986) computed the R-band LF of A2151 in a region 4.6 Mpc on a side, but only down to R ~ -19.8 (six magnitudes brighter than our limit). She derived a value of $\alpha = -1.26 \pm 0.13$, similar to ours despite of the differences between both works.

We have also investigated the radial dependence of Schechter parameters as we move outwards the cluster. We divided our area in four different annuli from inside the core radius to outside the virial radius, and computed the LFs in them. The faint-end slope turned out to be almost constant within the errors.

A2151 shows substantial subclustering with at least three distinct subclusters identified by BDB. In our mapping we cover two of them (A2151N and A2151C), while A2151E is out of our survey (see Fig. 1, left). However, a fourth overdensity (A2151S) seems to exist in the southernmost part of the cluster, though it was not identified as a distinct

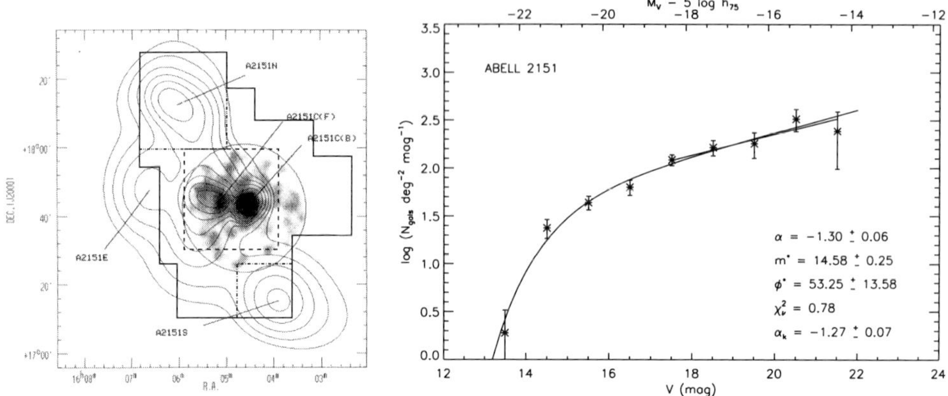

Figure 1. (*Left*) Surveyed area plotted over galaxy density contours and X-ray emission (from Huang & Sarazin 1996). Small squares show the areas where subclusters' LFs have been computed. (*Right*) The global LF of A2151 and its best-fit Schechter function. Error bars also include non-Poissonian fluctuations.

subcluster in the BDB analysis. We have computed the LFs of A2151N, A2151C and (partly) A2151S in the areas shown in Fig. 1 (left) down to V=21. We find that, within the errors, the two identified subclusters have slopes consistent with each other and with the global one ($\alpha = -1.33 \pm 0.06$ and $\alpha = -1.17 \pm 0.16$ for A2151C and A2151N respectively). However, for A2151S we surprisingly find a somewhat steeper slope ($\alpha = -1.64 \pm 0.10$). A detailed inspection of the LF suggests that this increase in the dwarf to giant ratio (D/G) is mainly due to a significant population of faint galaxies. If this increment is not due to a contamination by background galaxies in this region, it would imply the presence of a large population of dwarf galaxies in this southern subcluster.

Acknowledgements

RSJ acknowledges financial support from the *Academia Canaria de Ciencias* and thanks the other authors for their continuous support.

References

Bertin, E. & Arnouts, S. 1996 A&A **117**, 393.

Binggeli, B. & Jerjen, H. 1998 A&A **333**, 17.

Bird, C.M., Davis, D.S. & Beers, T.C. 1995 AJ **109**, 920 (BDB).

Cardelli, J.A., Clayton, G.C. & Mathis, J.S. 1989 A&A **345**, 245.

De Propris et al. 2003 MNRAS **342**, 725.

Giovanelli, R. & Haynes, M.P. 1985 ApJ **292**, 404.

Huang, Z. & Sarazin, C.L. 1996 ApJ **461**, 622.

Huang, J.-S, Cowie, L.L., Gardner, J.P., Hu, E.M., Songaila, A. & Wainscoat, R.J. 1997 ApJ **476**, 12.

Lugger, P.M. 1986 ApJ **303**, 535.

Schechter, P. 1976 ApJ **203**, 297.

Schlegel, D.J., Finkbeiner, D.P. & Davis, M. 1998 ApJ **500**, 525.

Takamiya, M., Kron, R.G. & Kron, G.E. 1995 AJ **110**, 1083.

Outskirts of Galaxy Clusters: Intense Life in the Suburbs
Proceedings IAU Colloquium No. 195, 2004
A. Diaferio, ed.

DOI: 10.1017/S1743921304000948

The dwarf galaxy population in two compact groups

Patricia E. Campos[1]†, C. Mendes de Oliveira[1] and M. Bolte[2]

[1]Instituto de Astronomia, Geofísica e Ciências Atmosféricas, Departamento de Astronomia, Universidade de São Paulo, Brazil email: campos@astro.iag.usp.br

[2]UCSC/Lick Observatory

Abstract. We have undertaken an imaging survey of low surface brightness dwarf galaxies in nearby groups with the main goal of studying the faint-end of the galaxy luminosity function at z ∼ 0. Here we describe the results on the search for LSB dwarf galaxies in two compact groups: HCG44 and HCG68. Our study explores the surface brightness, sizes, magnitudes and colours of these candidate dwarf galaxies. The selection criteria were their central surface brightness, scale factors and diameters at the limiting isophote of 26 R mag/"^{2}. We estimate the faint end of the luminosity function of galaxies through Monte Carlo simulations. The observed magnitude distribution of the candidate dwarf galaxies in the groups (down to $M_R \sim -13$) is compatible with a Schechter function with a slope $\alpha \sim -1.2$.

---

1. Observations, selection criteria and analysis

We have obtained CCD images which cover 1.18 deg^2 and 0.77 deg^2 of the central regions of the groups HCG44 and HCG68, respectively, with the KPNO 0.9m telescope in the B and R bands. HCG44 has 4 bright galaxies (3 late-type and 1 early-type) and HCG68 has 5 bright members (4 early-type and 1 late-type). We adopt a distance of 34 Mpc to both groups. Detection and photometry of sources were done with the program Source Extractor (Bertin & Arnouts 1996). The selection criteria were based on the properties of the Local Group dwarf galaxies, projected to the group's distance, as in Carrasco et al. (2001). We have selected objects with $\mu_0 > 21$ R mag/"^{2}, scale factor $h > 2$" and diameters > 6". We assumed that the dwarf galaxies have exponential profiles (Ferguson & Binggeli 1994). All objects were visually inspected. We have selected a total of 35 and 45 objects from HCG44 fields and HCG68 fields, respectively. Figure 1 shows a mosaic of their images.

2. Colour distribution

HCG44 has only 18 galaxies with measured colours and the group HCG68 has 37 galaxies. The group HCG68 has a more pronounced peak at $B-R \sim 1.1$ while HCG44 has a more diffuse distribution, with two pronounced peaks, one at $B-R \sim 0.9$ and the other at $B-R \sim 2.5$. The red peak is probably due to contaminating background galaxies. It was observed that there are differences in the population of dwarf galaxies in HCG68 and HCG44. The group HCG68 is richer in dwarf galaxies and has a more concentrated colour distribution. This may reflect the fact that HCG68 is in a denser environment than HCG44. HCG68 is early-type dominated and may be a substructure

† Present address: IAG-USP, Departamento de Astronomia, Rua do Matão 1226, Cidade Universitária 05508-900, São Paulo, SP, Brazil.

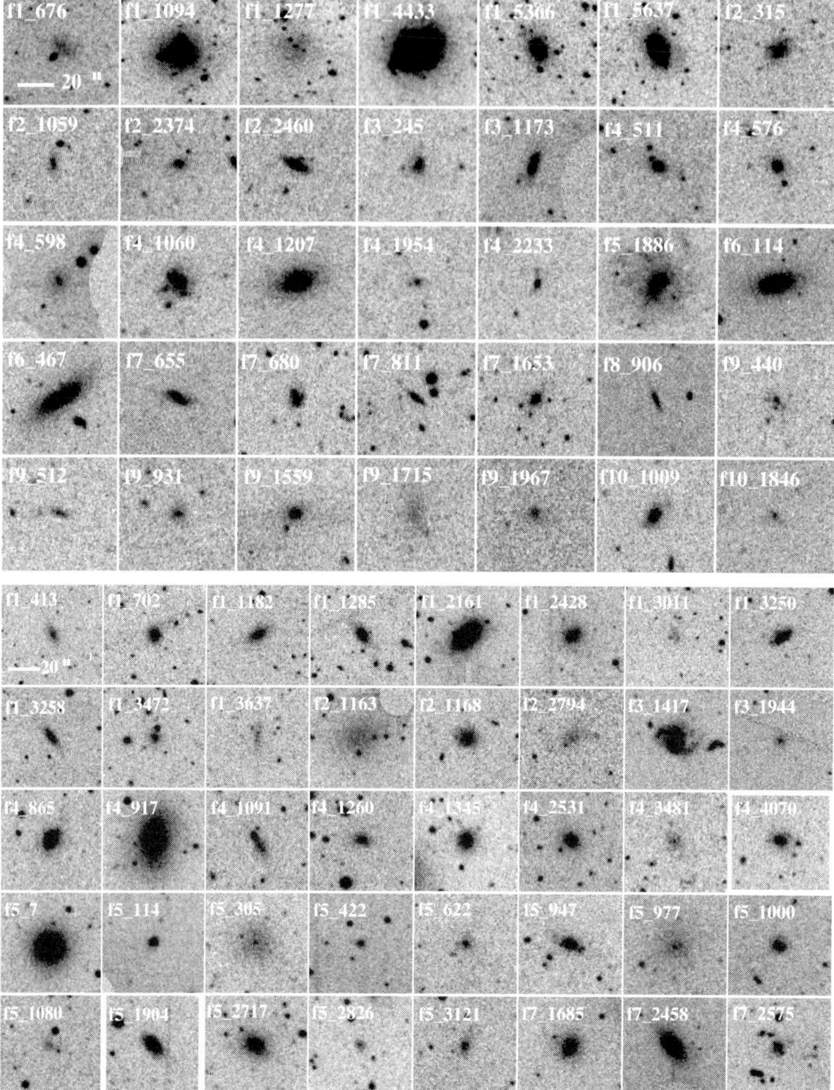

Figure 1. Mosaic of the candidate dwarf galaxies of the groups HCG44 (top) and HCG68 (bottom).

inside a loose group. One of the major problems when studying the faint dwarf galaxy population in groups or clusters is to distinguish group members from giant background galaxies. Figure 2 shows the scale-factor h vs the central surface brightness μ_0 for the candidate dwarf galaxies. One can readily see the reddest galaxies of the sample are also those with the highest central surface brightness, and these are most probably galaxies in the background.

3. Monte Carlo simulations: the faint-end of the LF

In this section we use Monte-Carlo simulations to test the Schechter function parameters which best fit the observed magnitude distribution of the galaxies. For that, we

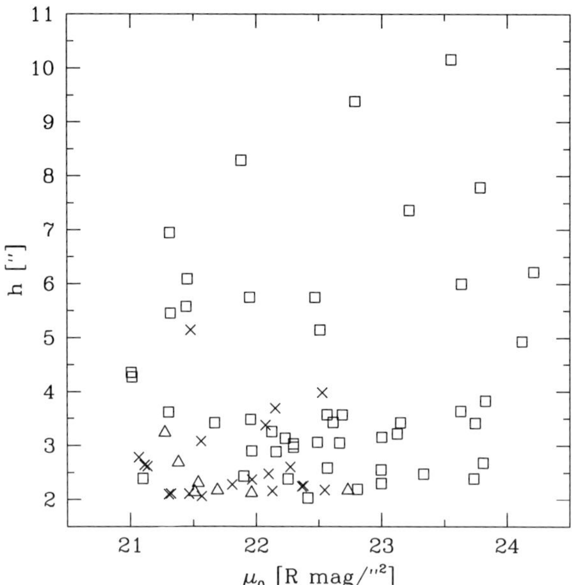

Figure 2. Scale-factor vs central surface brightness for the candidate dwarf galaxies in both groups. The squares represent colours $B-R \leqslant 1.7$ and the triangles represent the $B-R > 1.7$. Crosses indicates the galaxies with no colour information.

simulate 4000 galaxies for each M^* and α pair tested, and then we perform the same process to detect and measure the objects, as done for the real galaxies. The magnitude distribution of the recovered galaxies were then compared with the magnitude distribution of the real galaxies through a Kolmogorov-Smirnov (K-S) test. In addition, a χ^2 test on the binned data was performed in order to confirm the results. The observed magnitude distribution of the candidate dwarf galaxies in the groups (down to $M_R \sim -13$) is compatible with a Schechter function with a slope $\alpha \sim -1.2$. We find that both HCG44 and HCG68 have luminosity functions similar to that seen in the Local Group (van den Bergh 2000), without the enhanced numbers of dwarf galaxies observed in larger groups like Virgo or Fornax. Thus, our results are in agreement with the observed trend of a flatter slope of the faint end of the luminosity function for groups ~ -1.2 (Trentham & Tully 2002, Flint et al. 2001) than predicted from CDM models.

Acknowledgements

P.F. Campos acknowledges the financial support from IAU, PRPG-USP and CNPQ. CMdO acknowledges funding from the Alexander von Humboldt Foundation.

References

Bertin, E. & Arnouts, S. 1996, A&AS, 117, 393
Carrasco, E. R. , Mendes de Oliveira, C., Infante, L. & Bolte, M. 2001, AJ, 121, 148
Ferguson, H. C., & Binggeli, B., 1994, A&ARv, 6, 67
Flint K., Metevier, A. J., Bolte, M. & Mendes de Oliveira, C. 2001, ApJS, 134, 53
Trentham, N., Tully, R.B., 2002, MNRAS, 335, 712
van den Bergh, S., 2000, PASP, 112, 529

Outskirts of Galaxy Clusters: Intense Life in the Suburbs
Proceedings IAU Colloquium No. 195, 2004
A. Diaferio, ed.

DOI: 10.1017/S174392130400095X

The environmental dependence of galaxy properties in the local universe: effects of local and global environment

M. Tanaka[1], T. Goto[1,2], S. Okamura[1], K. Shimasaku[1] and J. Brinkman[3]

[1]Department of Astronomy, Graduate School of Science, University of Tokyo, 7-3-1 Hongo, Bunkyo-ku, Tokyo, Japan email: tanaka@astron.s.u-tokyo.ac.jp

[2]Department of Physics and Astronomy, Johns Hopkins University, 3400 North Charles Street, Baltimore, MD 21218-2686, USA

[3]Apache Point Observatory, P.O. Box 59, Sunspot, NM 88349, USA

Abstract. We investigate the environmental dependence of galaxy properties in the local universe based on the data of the Sloan Digital Sky Survey. We focus on how star formation and morphology of galaxies correlate with luminosity, local environment, and global environment. We find that galaxy properties abruptly change at a critical local density of $\log \Sigma_{\rm crit} \sim 0.4$ galaxies h_{75}^2 Mpc^{-2}. The 'break' at the critical density is found only for faint galaxies ($M_r^* + 1 < M_r < M_r^* + 2$). Bright galaxies ($M_r < M_r^* + 1$) show no break. That is, the star formation-density and the morphology-density relations depend on galaxy luminosity. Next, we focus on global environment, i.e., richness of galaxy groups and clusters. Most galaxies are not forming stars in groups as poor as $\sigma \sim 200$ km s^{-1}. This fact suggests that environmental mechanisms that are effective only in rich clusters, such as ram-pressure stripping of cold gas and harassment, have not played a major role in suppressing galaxy star formation. Our results may suggest that evolution of bright galaxies is not strongly related to galaxy systems such as groups and clusters. On the other hand, evolution of faint galaxies may have a close connection.

---

1. Introduction

It has been widely known that galaxy properties such as star formation and morphology depend on environment (e.g., Dressler 1980). However, it is still unknown how such dependences are established behind galaxy evolution. To further investigate the environmental dependences, we construct a volume-limited sample based on the data of the SDSS (York et al. 2000; Abazajian et al. 2003), and perform a detailed analysis of galaxy properties as a function of local environment and global environment. We define the volume-limited sample as $0.030 < z < 0.065$ and $M_r < M_r^* + 2$ $(= -19.4)$. In what follows, we refer to galaxies with $M_r < M_r^* + 1$ as *bright galaxies* and to those with $M_r^* + 1 < M_r < M_r^* + 2$ as *faint galaxies*. We assume a flat universe with $\Omega_M = 0.3$, $\Omega_\Lambda = 0.7$ and $H_0 = 75$ km s^{-1} Mpc^{-1}.

2. Dependence on local environment

We define local environment by projected local galaxy density which is measured from a projected distance to the 5th nearest galaxy within a redshift sheet of ± 1000 km s^{-1} from the galaxy in question. We characterize star formation and morphology of galaxies by EW(Hα) and the bulge to total luminosity ratio (B/T). Figure 1 shows the star formation-density and the morphology-density relations. It can be observed that faint

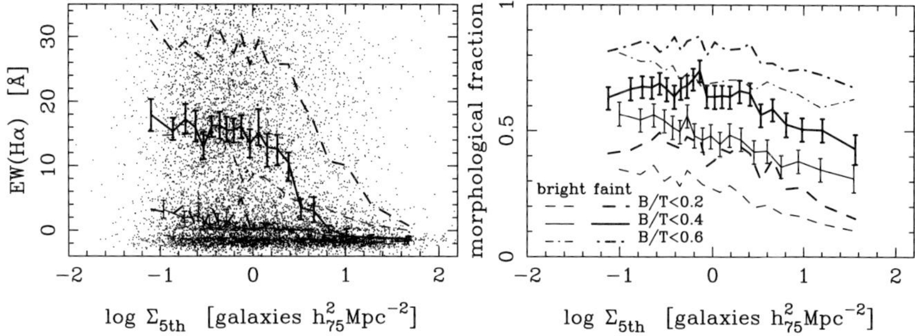

Figure 1. The thin and thick lines represent bright and faint galaxies, respectively. **Left:** EW(Hα) plotted against local density. The solid and dashed lines show the median and the quartiles (25% and 75%) of the distribution. The error bars show bootstrap 90% percentile intervals. **Right:** Morphological fractions ($B/T < 0.2$, 0.4, and 0.6) plotted against local density. The error bars are bootstrap 90% intervals.

galaxies show a 'break' at $\log \Sigma_{5\text{th}} \sim 0.4$ galaxies h_{75}^2 Mpc^{-2} ($\Sigma_{5\text{th}} \sim 2.5$). Below the break, star formation and morphology of faint galaxies do not depend on local density, and faint galaxies show more active star formation and later morphological types than bright galaxies. But, above the break, faint galaxies show weaker star formation and earlier morphological types. In contrast to faint galaxies, bright galaxies show no break, and their properties smoothly change with local density. It is interesting to note that the difference between the star formation activities of bright and faint galaxies becomes small above the break, but the difference in morphology is not strongly dependent on local density. This suggests that star formation and morphology are partly independent quantities.

3. Dependence on global environment

Next, we focus on global environment, i.e., richness of galaxy systems. We identify galaxy groups and clusters via the friends-of-friends algorithm (Huchra & Geller 1982), and measure their velocity dispersions, which are used to characterize richness of galaxy systems. In Figure 2, we plot EW(Hα) against the velocity dispersion. Star formation of bright galaxies do not show any correlation with system richness, and most of the bright galaxies are not forming stars. As for faint galaxies, we find that the median star formation activity is extremely weak in $\sigma > 200$ km s^{-1} systems. It should be noted that the median star formation activity is weak even in systems as poor as $\sigma \sim 100$ km s^{-1} compared with that of field galaxies.

4. Discussion

There are several environmental mechanisms favored in the literature that are expected to transform galaxies into red early-type galaxies. Our finding in §3 suggests that environmental mechanisms that are effective only in rich systems, such as ram-pressure stripping of cold gas (Gunn & Gott 1972) and harassment (Moore et al. 1996), have not played a major role since the dominant galaxy population in poor systems cannot be fully explained by them. Galaxy-galaxy interactions (e.g., Mihos & Hernquist 1996) and strangulation (Larson, Tinsley & Caldwell 1980) are remained to be fascinating candidates since they are expected to be effective in poor systems and/or in the field.

Let us consider implications of our findings in §2. Since the critical density corresponds to *outskirts of galaxy systems* (e.g., Gómez et al. 2003), one possible interpretation of the

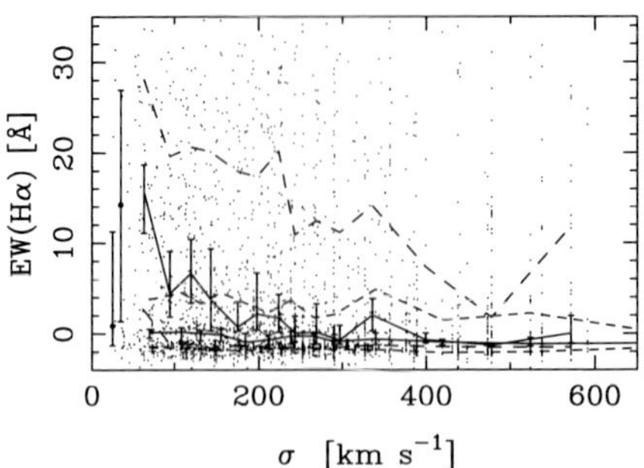

Figure 2. EW(Hα) plotted against velocity dispersion of galaxy systems. The leftmost points and the associated error bars show the median and the quartiles of the distribution of field galaxies. The meanings of the other lines are the same as of the left panel of Figure 1.

fact that faint galaxies show the break at the critical density is that faint galaxies are affected by galaxy systems. That is, evolution of faint galaxies may be closely related to galaxy systems. On the other hand, bright galaxies show no clear break at the critical density and their evolution may not have a strong connection with galaxy systems.

Acknowledgements

Funding for the creation and distribution of the SDSS Archive has been provided by the Alfred P. Sloan Foundation, the Participating Institutions, the National Aeronautics and Space Administration, the National Science Foundation, the U.S. Department of Energy, the Japanese Monbukagakusho, and the Max Planck Society. The SDSS Web site is http://www.sdss.org/.

The SDSS is managed by the Astrophysical Research Consortium (ARC) for the Participating Institutions. The Participating Institutions are The University of Chicago, Fermilab, the Institute for Advanced Study, the Japan Participation Group, The Johns Hopkins University, Los Alamos National Laboratory, the Max-Planck-Institute for Astronomy (MPIA), the Max-Planck-Institute for Astrophysics (MPA), New Mexico State University, University of Pittsburgh, Princeton University, the United States Naval Observatory, and the University of Washington.

References

Abazajian, K., et al., 2003, AJ, **126**, 2081
Dressler, A., 1980, ApJ, **236**, 351
Gómez, P. L., et al., 2003, ApJ, **584**, 210
Gunn, J. E. & Gott, J.R.III, 1972, ApJ, **176**, 1
Huchra, J. P. & Geller, M. J., 1982, ApJ, **257**, 423
Larson, R. B., Tinsley, B. M., & Caldwell, C. N., 1980, ApJ, **237**, 692
Mihos, J. C., & Hernquist, L., 1996, ApJ, **464**, 641
Moore, B., Katz, N., Lake, G., Dressler, A., & Oemler, A., 1996, Nature, **379**, 613
York, D. G., et al., 2000, AJ, **120**, 1579

Outskirts of Galaxy Clusters: Intense Life in the Suburbs
Proceedings IAU Colloquium No. 195, 2004
A. Diaferio, ed.

DOI: 10.1017/S1743921304000961

The relation of the abundance ratios in early-type galaxies with the cluster mass

Conrado Carretero[1], Alexandre Vazdekis[1] and John E. Beckman[1,2]

[1]Instituto de Astrofísica de Canarias, Vía Láctea s/n, 38200 La Laguna, Tenerife, Spain email: cch@iac.es

[2]Consejo Superior de Investigaciones Científicas, Spain

Abstract. We present the values of CN and Mg overabundances with respect to Fe, for early-type galaxies in 8 clusters over a range of richness and morphology. Spectra were taken from the Sloan Digital Sky Survey (SDSS) DR1 spectroscopic database, and from WHT and CAHA observations. Abundances were derived from absorption lines and single burst population models, by comparing galaxy spectra with appropriately broadened synthetic model spectra. We find robust correlations between [Mg/CN] and [CN/Fe] abundance ratios and cluster X-ray luminosity. No correlation was observed for [Mg/Fe]. We also see a clear trend with the richness and morphology of the clusters. This behavior is interpreted given varying formation timescales for CN, Mg and Fe, combined with a different star formation history in early-type galaxies as a function of their environment. This result supports the scenario in which early-type galaxies in more massive clusters are assembled on shorter timescales than those galaxies within less massive clusters.

1. Introduction

Stellar populations offer a fossil record of the formation and evolution of galaxies, providing very strong constraints on the principal galaxy formation scenarios. So, a programme to understand stellar populations as a function of the environment will give many answers to the puzzle. Differences in the abundance of CN, Mg and Fe, as a function of the environment, have been recently suggested by Sánchez-Blázquez et al. (2003). We explore these differences by studying a large number of galaxy clusters over a range of richness and morphology.

2. Abundance estimations and results

To derive mean luminosity-weighted ages and metallicities, we have compared selected absorption line strengths with those predicted by the model of Vazdekis (1999). This model provides flux-calibrated spectra in the optical range at a resolution of 1.8 Å (FWHM) for single-burst stellar populations. The selected absorption indices were CN_2, Mg_2 and Fe2 (Worthey et al. 1994).

Plots of the strengths of the selected indices versus Hβ provide rather orthogonal model grids, allowing us to accurately estimate galaxy mean ages as well as the abundances of these elements. Since the CN_2 index is by far dominated by C and N, the Mg_2 index is governed by Mg and the Fe2 index by Fe, the obtained metallicities can be approximated to [CN/H], [Mg/H] and [Fe/H] abundances. These values are then used to estimate the abundance ratios [CN/Fe], [Mg/Fe] and [Mg/CN] for each galaxy. Finally, we determined the overabundances of each cluster, as the mean values of the overabundances of all the galaxies within it. Errors were computed as the standard deviation of each sample.

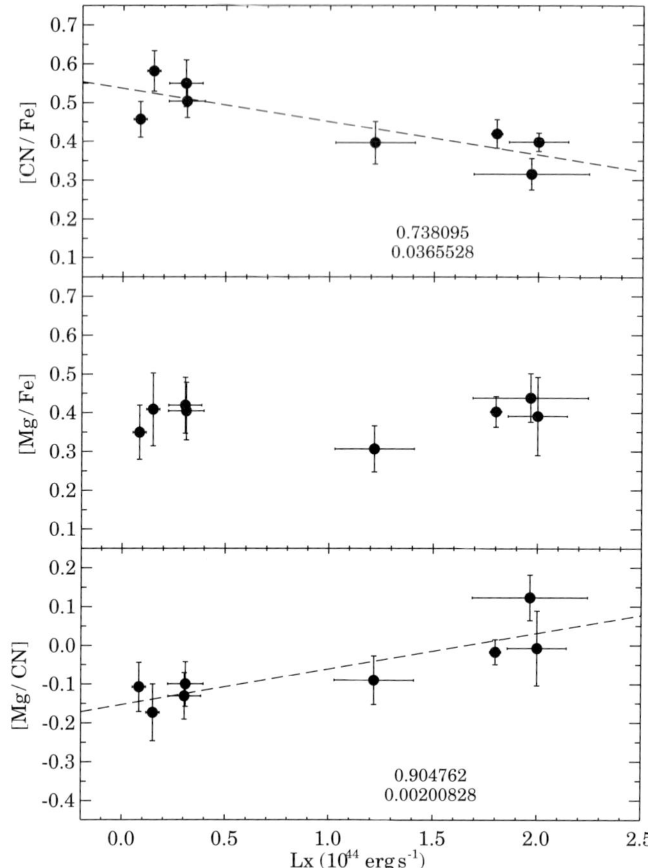

Figure 1. Cluster X-ray luminosity vs. overabundance values of [CN/Fe] *(top)*, [Mg/Fe] *(middle)* and [Mg/CN] *(bottom)*. Each point corresponds to one individual cluster, and is computed as the mean value of the overabundaces of early-type galaxies within each cluster, with $150\ km\ s^{-1} < \sigma < 250\ km\ s^{-1}$. The correlation coefficient and its significance are written in top and bottom panels.

We used X-ray luminosity as a quantitative indicator of the mass of the clusters. We adopted X-ray luminosity values from Ebeling et al. (1998) and Ledlow et al. (2003). Figure 1 shows the values of [CN/Fe], [Mg/Fe] and [Mg/CN] versus X-ray luminosity for each cluster. We found clear correlations between [CN/Fe] and [Mg/CN] values and X-ray luminosity. No correlation was found for [Mg/Fe].

3. Discussion

The correlations can be interpreted in terms of the different formation timescales for each element, and the different star formation history of early-type galaxies as a function of their environment.

Mg is ejected into the interstellar medium (ISM) by Type II supernovae (SNe II) on short timescales (<10 Myr). On the other hand, the iron-peak elements are the products of SNe Ia, which occur on timescales of ∼1 Gyr. Between the two extremes, although there are recent suggestions that most of the C come from massive stars (Akerman et al. 2004), C and N are mainly ejected into the ISM by low- and intermediate-mass stars (e.g.

Chiappini, Romano, & Matteucci 2003), leading to CN formation on timescales longer than Mg but shorter than Fe.

In this framework, the constancy of the [Mg/Fe] overabundance values is explained in terms of the great difference in the formation timescales of the two elements: the galaxies are fully assembled before Type Ia SNe can significantly pollute with Fe the ISM of the smaller galaxies before merging, and right after Mg is fully ejected. Since [Mg/Fe] is found to be constant with the X-ray luminosity of the clusters , which is an indicator of their mass, we conclude that this ratio is independent of the environment. Similar results for the [Mg/Fe] ratio have been obtained by other authors (e.g. Sánchez-Blázquez et al. 2003), by studying field and Coma cluster elliptical galaxies.

However, when considering species with less disparate formation timescales, such as CN and Fe, or CN and Mg, clear correlations are found between abundance ratios and the environment, as shown in Fig. 1. The fact that [CN/Fe] decreases with the cluster X-ray luminosity, and that [Mg/CN] increases with it, suggest that galaxies in more massive clusters are fully assembled on shorter timescales than those in less massive clusters.

The result that there exist relative differences in the assembly timescales of the galaxies due to the properties of the environment is qualitatively in agreement with the hierarchical models. Discrepancies appear, however, when considering the absolute values of such timescales. The fact that [CN/Fe] abundance ratio is not constant implies that early-type galaxies are fully assembled on timescales around the massive release of CN into the ISM. Hierarchical models, on the contrary, predict longer assembly timescales.

Acknowledgements

We acknowledge support from grant AYA2001-0435 from the Spanish Ministry of Science and Technology.

References

Akerman, C.J., Carigi, L., Nissen, P.E., Pettini, M., & Asplund, M. 2004 *A&A* **414**, 931.

Chiappini, C., Romano, D., & Matteucci, F. 2003 *MNRAS* **339**, 63.

Ebeling, H., Edge, A.C., Böhringer, H., Allen, S.W., Crawford, C.S., Fabian, A.C., Voges, W., & Huchra, J.P. 1998 *MNRAS* **301**, 881

Ledlow, M.J., Voges, W., Owen, F.N., & Burns, J.O. 2003 *AJ* **126**,2740

Sánchez-Blázquez, P., Gorgas, J., Cardiel, N., Cenarro, J., & González, J.J. 2003 *ApJ* **590**, L91

Vazdekis, A. 1999 *ApJ* **513**, 224

Worthey, G., Faber, S.M., González, J.J., & Burstein, D. 1994 *ApJS* **94**, 687

Outskirts of Galaxy Clusters: Intense Life in the Suburbs
Proceedings IAU Colloquium No. 195, 2004
A. Diaferio, ed.

DOI: 10.1017/S1743921304000973

Intergalactic HII regions in Stephan's Quintet

C. Mendes de Oliveira[1]†, E. S. Cypriano[1], L. Sodré Jr.[1] and C. Balkowski[2]

[1]Instituto de Astronomia, Geofísica e Ciências Atmosféricas, Departamento de Astronomia, Universidade de São Paulo, Brazil email: oliveira@astro.iag.usp.br

[2]Observatoire de Paris, GEPI, CNRS and Universite Paris 7, 5 Place Jules Janssen, F-92195 Meudon Cedex, France

Abstract. We have discovered four intergalactic HII regions in the Stephan's Quintet, at more than 25 kpc projected distance from the center of the nearest group galaxy, with no apparent optical connection to it. They have M_B ranging from −11.9 to −12.5 mag, colors B−R = 0.7 to 1.1 mag, radial velocities from 6565 to 6651 km/s and they are superposed onto the HI tail east of NGC 7319, with a mean radial velocity of 6610 km/s. In addition, they have metalicities of the order of 12+log(O/H) = 8.58 ± 0.25, which suggests that they were formed from pre-enriched material. We derive a mean age of 4.6 ± 0.6 Myr and a mean stellar mass of $(2.9 \pm 1.4) \times 10^4\ M_\odot$ for the four objects. The masses, ages, colors, velocities, metalicities and location of the objects suggest that they are HII regions that were formed far away from the galaxies through compression of the intergalactic HI gas by galaxy collisions.

1. Introduction

About a decade ago the first convincing examples of new stellar systems formed during galaxy collisions were established. In the high-end of the mass spectrum, the tidal dwarf galaxies were shown to be formed at the ends of tidal tails. In the lower-end of the mass spectrum, young star clusters, possible proto-globular clusters, were identified in gas-rich mergers. The formation of tidal dwarf galaxies and young clusters during galaxy collisions is nowadays an active topic of study. Besides helping in exploring the connection to the host galaxy, the properties of these stellar systems aid at understanding the poorly known processes of star formation during interaction events.

A natural environment where to look for new objects formed in galaxy collisions is in interacting compact groups, small groups of galaxies with large spatial densities, low velocity dispersions, and a high rate of interactions. In this paper, we report on the discovery of another class of objects (different from the tidal dwarfs and the young clusters) which may be formed as a consequence of galaxy collisions: intergalactic HII regions in the Stephan's Quintet.

2. Observations

A 15-min r' image of the quintet (shown in Fig. 1), a 22-min b' image and a 15-min multi-slit exposure (covering from 4000 to 8000 Å, with a resolution of 8 Å) were obtained with GMOS on the Gemini North Telescope. The spectra of the four intergalactic HII regions (IHII) are shown in Fig. 1b. As can be seen, these are typical of HII regions.

† Present address: Universitäts-Sternwarte der Ludwig-Maximilians-Universität, Scheinerstrasse 1, D-81679 München, Germany

This, together with the large projected distance to the closest bright member galaxy (> 25 kpc), justifies the name intergalactic HII regions, IHII.

3. Results and discussion

Radial velocities: The heliocentric velocities of the IHII regions derived from the redshifts of the brightest lines are indicated in Fig. 1a. From Fig. 7 of Williams et al. (2002), we note that the velocities of all IHII's coincide with the HI velocities from the channel maps within $\pm$ 20 km s^{-1}, suggesting a physical association of the knots with the HI cloud.

Metalicities: In order to estimate the metalicities of the objects, we have adopted the N2 calibrator, proposed by Denicoló et al. (2002), which is defined as the logarithm of the [NII]λ6584/Hα ratio. The metalicities inferred for the four objects are very similar, the mean value being 12+log(O/H) = 8.58. This value is close to solar: 12+log(O/H)$_\odot$ = 8.8 $\pm$ 0.1, (Grevesse & Sauval 1998), suggesting that the IHII's have been formed from pre-enriched material.

Star burst parameters: We assume that the IHII are excited by young stars formed in an instantaneous star burst and use Starburst99 (Leitherer et al. 1999), with solar metalicity and Salpeter IMF, to estimate their physical parameters. From a comparison between the observed Hα equivalent widths and those produced by Starburst99, we find ages ranging from 3.2 to 5.6 Myr, with a mean age of 4.6 ± 0.6 Myr for the four regions. The mean mass is $(2.9 \pm 1.4) \times 10^4\ M_\odot$.

The IHII's found in Stephan's Quintet present the following characteristics: (1) they are compact (except for region b), (2) relatively high metallicities, (3) masses of the order of $10^4 M_\odot$, (4) ages ranging from 3.2 to 5.6 Myr, (5) far away from the bright galaxies and with no apparent optical connection to them, (6) they coincide in location and velocity with the HI tail detected 2 arcmin east of NGC 7319 (Williams et al. 2002). The extremely young ages of these objects suggest that they, most probably, are located where they were born. We favour a scenario where the IHII's were formed in the intragroup medium, within the HI tail, through compression of the intergalactic HI gas by galaxy collisions. Further study of these objects is relevant since IHII and intragroup cluster formation might have been more important at higher redshifts, when interactions and mergers happened more frequently. A paper describing the details of this work can be found in Mendes de Oliveira et al. (2004). Similar objects have been recently found in several other systems by Ryan-Weber et al. (2004).

Acknowledgements

The authors would like to thank the Brazilian FAPESP, the PICS program, the Brazilian PRONEX program and the CNPq. CMdO deeply acknowledges the funding from the Alexander von Humboldt Foundation.

Bibliography

Denicoló, G., Terlevich, R. & Terlevich, E. 2002, MNRAS, 330, 69

Leitherer, C., et al., 1999, ApJS, 123, 3

Mendes de Oliveira, C., Cypriano, E.S., Sodré Jr., L., Balkowski, C., 2004, ApJ, 605, 17

Ryan-Weber et al. 2004, AJ, 127, 1431

Williams, B.A., Yun, M.S., Verdes-Montenegro, L., 2002, AJ, 123, 241

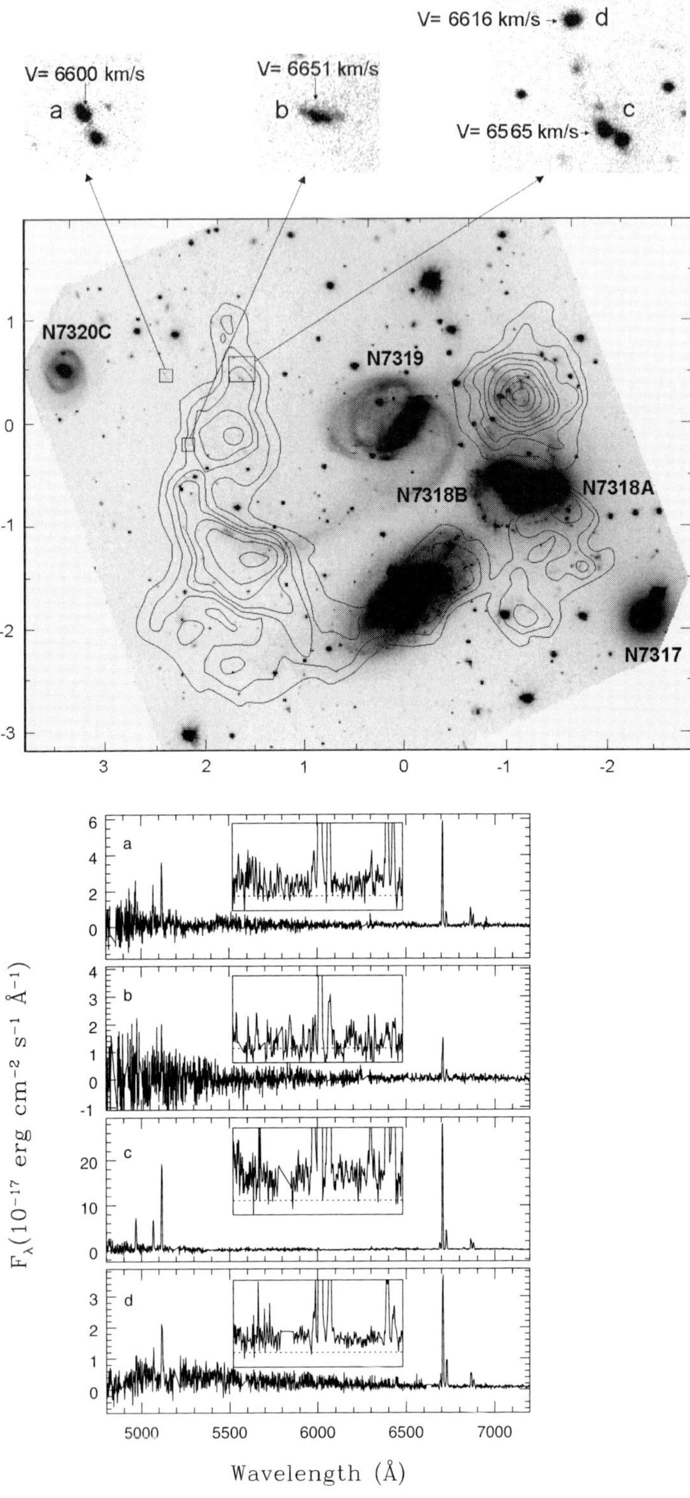

Figure 1. Top – An r′ image of the Stephan's Quintet with the HI contours from Williams et al. (2002) overplotted. The contours are 5.8, 15, 23, 32, 44, 61, 87, 120, 180 × 10^{19} atoms cm^{-2}. The three boxes in the upper left corner (and corresponding zoomed images) indicate the new IHII's. We note that all regions are within the HI tail. Region "a" coincides with the lower-level contour of 1.5 × 10^{19} atoms cm^{-2} (not plotted). **Bottom** – Spectra of IHII "a" to "d". Each inset presents a zoom into the region from 6500 to 6900 Å. Zero flux is indicated with a dashed line, showing that the continuum in each case can be measured and averaged over many pixels.

Outskirts of Galaxy Clusters: Intense Life in the Suburbs
Proceedings IAU Colloquium No. 195, 2004
A. Diaferio, ed.

DOI: 10.1017/S1743921304000985

A new sample of isolated galaxy pairs

V. Zitelli[1], P. Focardi[2], B. Kelm[2] and S. Marinoni[1]

[1]INAF-Osservatorio Astronomico di Bologna, Via Ranzani 1, 40127 Bologna, Italy

[2]Università degli Studi di Bologna, Dip. di Astronomia, Via Ranzani 1, 40127, Bologna, Italy

Abstract. A new volume-limited sample of 84 isolated pairs of galaxies has been selected from the UZC catalogue. Galaxies in pairs are brighter than $M_{Zw} - 5\log h = -18.9$. The projected separation between pair members is $r < 200h^{-1}$ kpc, the isolation criterion requires the pair to have no further companion within a $R = 1h^{-1}$ Mpc distance. Morphological classification for both members is available for 57 pairs. Out of these, 9 are early-type (E+E) pairs, 29 late type (S+S) pairs, and 19 mixed (E+S) pairs. We find a deficit of luminous dominants among E+E and E+S pairs relative to S+S pairs, while the distance between pair members is statistically undistinguishable. However, among the 12 pairs that are not only close but also luminous, the fraction of E+E and S+S pairs is 8% and 83% respectively. This result is consistent with star-formation being enhanced in S+S pairs (but not in spirals in S+E pairs) in extremely close isolated pairs.

1. Introduction

There is observational evidence that galaxy-galaxy interactions/mergers can significantly enhance the star formation rate (SFR) in galaxies (Kennicutt 1998), while it is still unclear whether close galaxy-galaxy interactions/mergers are common and fundamental episodes in the evolution of most galaxies. Galaxy pairs are ideal sites to analyse the effect of interaction/merging on galaxy formation and evolution. The most extensively studied pair sample (Karachentsev 1972) has been selected on plates, based on visual criteria. Reduzzi and Rampazzo (1995) have extended the Karachentsev's selection criteria to the southern sky. More recent flux-limited pairs samples (Barton et al. 2000, Patton 2002, Lambas et al. 2003), selected from large 3D galaxy catalogues, are based on both projected (r) and radial velocity (cz) distance. Results indicate that SF is enhanced in close pairs. Alonso et al (2004) claims that SFR enhacement is observed both in isolated pairs and in pairs in dense environments. On the other hand, on the basis of the analysis of Compact Groups (Focardi & Kelm 2002), we expect truly isolated galaxy pairs to display the largest evidence of galaxy interaction and, as a consequence, SF enhancement. We have thus selected a volumei-limited sample of Bright isolated Pairs of Galaxies (BPG) to investigate the role played by a close companion on galaxy properties.

2. Bright isolated pairs of galaxies: the sample

Isolated pairs have been selected in the UZC catalogue (Falco et al. 1999) applying an automatic neighbour search algorithm. A galaxy pair is defined as a system including a galaxy with one companion within projected radius $r = 200h^{-1}$ kpc and radial velocity difference $|\Delta(cz)| \leqslant 1000$ km/s. Isolated pairs fulfil the additional constraint that no further neighbour galaxy is found within the projected radius $R = 1h^{-1}$ Mpc and velocity difference $|\Delta(cz)| \leqslant 1000$ km/s.

The BPG is a volume-limited sample: it includes 84 isolated galaxy pairs with radial velocity in the range $[2500 - 7500]$ km s^{-1} and galactic latitude $b^{II} \geqslant 30°$. The absolute

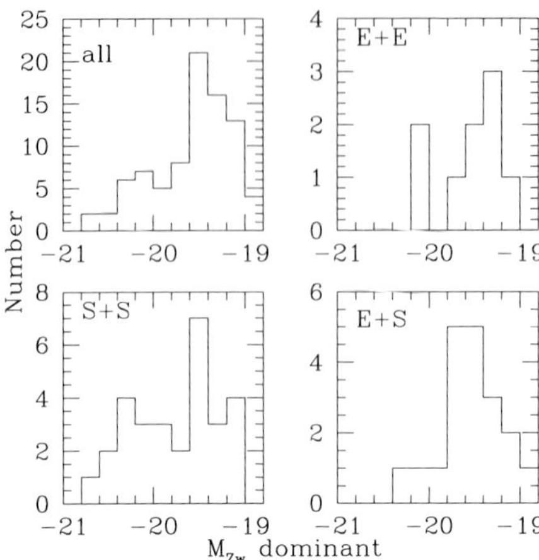

Figure 1. Distribution of the absolute magnitude of the dominant galaxies in pairs, for the whole sample, and for E+E, S+S and mixed pairs.

magnitude limit is $M_{Zw} = -18.87$ corresponding to the luminosity of a m_{Zw}= 15.5 galaxy (the completeness limit of the UZC catalog) at $cz = 7500$ km s^{-1}. The sample includes 77 symmetric systems (both galaxies fulfil all constraints).

3. Morphology and luminosity of BPG members

Only 57 pairs have morphological classification available, in the literature, for both member galaxies. Out of these 57 pairs, 9 (16%) are E+E pairs, 29 (51%) are S+S pairs, and 19 (33%) are E+S pairs. No distinction is made between ellipticals and S0s.

The morphological composition of the BPG sample is actually in excellent agreement with the morphological composition of the visually selected pairs in Karachentsev's and in Reduzzi and Rampazzo's samples.

In figure 1 we plot the luminosity distribution of the dominant galaxy in a pair for the whole sample and for E+E, S+S and E+S pairs respectively. There is evidence that dominant galaxies in E+E and E+S pairs are less luminous than dominants in S+S pairs: while 34% of the dominants in S+S pairs are brighter than $M - 5\log h = -20$, in E+E and E+S pairs the fraction decreases to 22% and to 11% respectively. Conversely, we find the luminosity distribution of the faintest members in pairs to be similar in all 3 pair classes. As a consequence we find the magnitude difference between pair members to be marginally larger in S+S pairs. Along with the observation that isolated E/S0s have fainter characteristic magnitude than E/S0s in Compact Groups (Kelm & Focardi 2004), this suggests that early-type galaxies in low density environments are typically faint, and that the majority of the luminous ellipticals are located in richer environments (groups).

4. Projected separation of BPG members

We next investigate whether E+E and S+S pairs exhibit similar projected separation distributions. In figure 2 we show distributions between pair members in the whole BPG sample and in the 3 morphologically defined subsamples. The fraction of E+E, S+S and E+S in loose pairs (16 with projected distance $r \geqslant 70h^{-1}$kpc) is 33%, 24% and 31% respectively. The fraction of E+E, S+S and E+S in close pairs (31 with projected distance $r \leqslant 25h^{-1}$kpc) is 56%, 66% and 53% respectively.

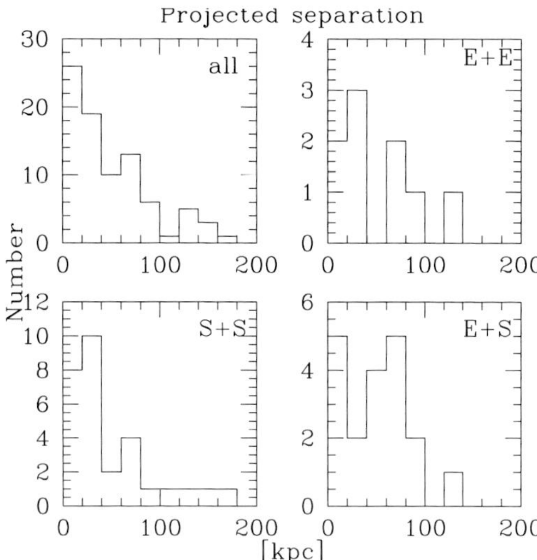

Figure 2. Distribution of the projected separation between BPG members for the whole sample and for E+E, S+S and mixed pairs.

The KS test indicates that there is no statistically significant evidence that the distribution of E+E and E+S pairs are different. However, we find that among the 12 pairs that are not only close, but also luminous ($M_{dom} - 5\log h \leqslant -20$) the fraction of E+E and S+S pairs is 8% and 83% respectively. This is consistent with the luminosity of spirals in close and isolated pairs to be boosted and thus with enhanced star-formation activity, as found by Lambas et al. (2004) and Sorrentino et al. (2003) in 2dF pairs.

Optical observations are in progress, at the Loiano 152 cm telescope, to get complete and homogeneous spectral classification and broad band two colour photometry for the whole BPG sample. These data will allow us to confirm any enhancement in the SFR of late-type pairs and to further check if the fraction of AGNs among isolated pair members is increased relative to the low fractions found in Compact Groups (Coziol et al. 2004, Kelm et al. 2004).

References

Alonso M.S., Tissera P.B., Coldwell G., Lambas D.G. 2004 [astro-ph/0401455]

Barton, E. J., Geller, M. J. & Kenyon S. J. 2000 *Ap.J.* **530**, 660–679

Coziol, R., Brinks, E., & Bravo-Alfaro, H. 2000 *AJ* **accepted**, [astro-ph/0404066]

Falco, E. E., Kurtz, M. J., Geller, M. J. et al. 1999 *PASP* **111**, 438–452

Focardi, P. & Kelm, B. 2002 *A.& A.* **391**, 35–46

Karachentsev, I. D. 1972 *Comm. Spec. Astrophys. Obs.* **7**, 1

Kelm, B. & Focardi P. 2004 *A.& A.* **in press**, [astro-ph/0306414]

Kelm, B., Focardi P., & Zitelli V. 2004 *A.& A.* **418**, 25

Kennicutt R.C. 1998 *ARAA* **36**, 189

Lambas, D.J., Tissera, P.B., Alonso, M.S., Coldwell, G. 2003 *MNRAS* **346**, 1189–1196

Patton, D. R. et al. 2002 *Ap.J.* **565**, 208–222

Reduzzi, L. & Rampazzo R. 1995 *Astron.Lett. & Comm.* **30**, numbers 1–6

Sorrentino G., Kelm, B., & Focardi P. 2003, in Thinking, Observing and Mining the Universe (Sorrento 2003) **in press**, [astro-ph/0312032]

Outskirts of Galaxy Clusters: Intense Life in the Suburbs
Proceedings IAU Colloquium No. 195, 2004
A. Diaferio, ed.

DOI: 10.1017/S1743921304000997

Dominant galaxies in 2dF groups

B. Kelm and P. Focardi
Dipartimento di Astronomia, Università di Bologna

Abstract. We investigate whether the spectral-type of a locally dominant (most luminous) galaxy can be used to select sets of galaxies that are physically associated (groups). We assume that passive dominants trace a group-like potential, and SF-dominants a field-like environment. The group sample includes 988 groups selected in the 2dFGRS applying a maximum magnitude difference criterion. We find that the average number of passive galaxies associated to a dominant is larger when the dominant is passive, a result supporting our assumption that galaxy associations around a passive dominant are reliable groups. Finally we show that, to reduce the contamination by unbound galaxy associations (SF-dominant), a $\geqslant$3 passive-members criterion is more efficient than a $\geqslant$6 all-members criterion.

---

1. Dominant group galaxy selection criteria

Each 2dF galaxy with $\geqslant$4 neighbours (projected separation less than $1h^{-1}$Mpc and line of sight velocity difference less than $1000\,\mathrm{km\,s^{-1}}$), that is more luminous (by $\geqslant$0.2 mag) than all of its neighbours, is a dominant group galaxy and identifies a group. Given the bright and faint apparent magnitude limits of the 2dF, a maximum magnitude difference criterion has been applied providing, for each dominant, complete identification of all neighbours from ~equally-luminous to ~2÷2.5 magnitudes fainter. Dominants are selected in the range $b_j \in [17 \div 17.5]$, neighbours in the range $b_j \in [17 \div 19.5]$. The criteria for selection are $0.03 \leqslant z_{dom} \leqslant 0.12$ and $-22 \leqslant (M - 5\log h)_{dom} \leqslant -19$. The sample includes 988 groups (7281 galaxies): 639 with a passive (Type 1 in Madgwick et al. 2002) dominant galaxy (P-dG), and 349 with a SF (Type>1) dominant (SF-dG). P-dGs are typically richer than SF-dGs.

2. Passive and star-forming dominants

If we assume that P-dGs trace a group-like potential, and SF-dGs a field-like environment (galaxy associations not embedded within a common massive halo), differences in the spectral mix of P-dGs and SF-dGs members should reflect the role of a group environment on the spectral evolution of galaxies. Figure 1 shows the abundance of P-dGs, and SF-dGs (*top panels*) as a function of absolute magnitude of the dominant. Non-luminous dominants are equally likely to be passive or SF, but the former are much more common than the latter among luminous dominants. That SF galaxies represent only a small fraction among luminous dominant galaxies is consistent with X-ray studies that show diffuse hot X-ray halo typically occur in groups hosting a dominant luminous elliptical. It remains unclear whether groups with a faint passive dominant exist at all, or whether these are typically embedded within larger systems (Kelm & Focardi 2004).

The *bottom panels* show the distributions of non-dominants that are passive and star-forming, as a function of the luminosity of the dominant, in P-dGs (*left*) and SF-dGs (*right*). P-dGs with a luminous dominant exhibit more passive than SF non-dominant members, but no passive-member excess is seen in P-dGs with a faint ($M \geqslant -20.5$)

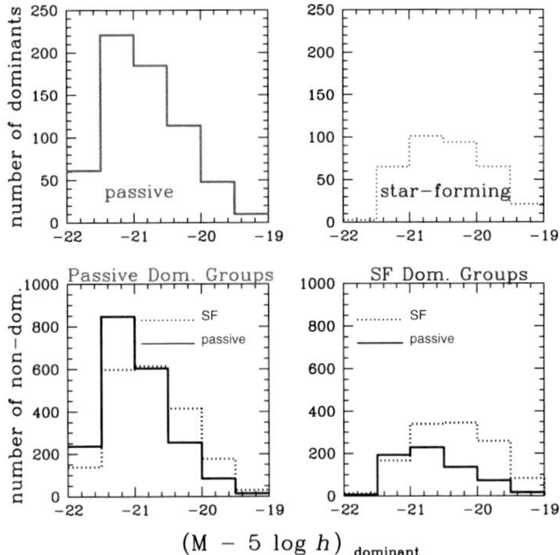

Figure 1. Histograms show the abundance of group dominants (*top panels*) and group non-dominants (*bottom panels*) as a function of the absolute magnitude of the dominant.

dominant. Conversely, SF-dGs exhibit more non-dominant SF members than passive ones, except for very luminous ($M \leqslant -21$) dominants. Being well established that SF activity decreases with increasing local density (Lewis et al. 2002, Gomez et al. 2003) the observed trends in figure 1 might be a further hint that, when luminous, P-dGs are more massive systems than SF-dGs.

In figure 2 (top panels) the average fractional content in passive (*left*) and SF (*right*) galaxies is plotted as a function of the dominant luminosity in P-dGs (*solid*) and SF-dGs (*dotted*). SF galaxies are less frequent in P-dGs, a result that supports our assumption that only passive dominants are key tracer of group-like potentials. It further agrees with recent result (Balogh et al. 2004) that a lower level of SF activity in high density environments is largely due to the smaller fraction of SF galaxies in these environments. However, the fractions of passive and SF galaxies in groups definitely appear to depend on the luminosity of the dominant, rather than on its spectral type. The modest difference observed in the SF-galaxy fraction between P-dGs and SF-dGs suggests that, in group samples, the SF-fraction is generally less efficient than the dominant-spectral-type criterion to segregate a group-like halo from galaxy-size halo associations.

3. Passive dominated groups: is SF depressed?

The slightly lower SF fraction associated to P-dGs might derive from a decrease in the number of SF galaxies or from an increase in the number of passive galaxies. In figure 2 we plot (*bottom panels*) the average number of passive (*left*) and SF (*right*) non-dominants associated to each dominant as a function of the dominant luminosity. At all luminosities, P-dGs exhibit no deficit of SF non-dominants relative to SF-dGs. Conversely, the average number of passive non-dominants per dominant is larger in P-dGs than in SF-dGs (figure 2 *bottom left*). This indicates that the clustering of passive non-dominants is larger in P-dGs than in SF-dGs. Results are clearly consistent with the hypothesis that only P-dGs are tracer of group-like potentials. The excess of passive galaxies in P-dGs further supports a biased galaxy formation scenario in which the formation of massive galaxies (mass assembly and star-formation), from the highest peaks in the initial density fluctuation field, is predicted to occur in an accelerated way (or at earlier time) in

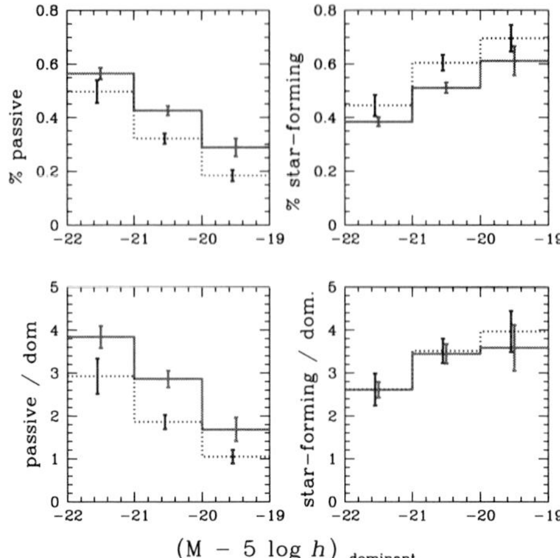

Figure 2. Fractions of passive and SF non-dominants (*top panels*) among non-dominants in groups with a passive (*solid*) and a SF (*dotted*) dominant. The *bottom panels* show the average number of passive and SF non-dominants associated to each dominant. Error bars are multinomial.

systems embedded within larger halos. Projection effects could be responsible for the similar number of non-dominant SF galaxies in the field and in groups. Alternatively, substantial infall on P-dGs from field SF-galaxies is required, thereby matching the prediction that the fraction of mass in the universe bound in group-like systems has undergone a dramatic increase between $z = 1$ and $z = 0$.

Figure 2 (*bottom panels*) also provides evidence that the average number of SF galaxies is less sensible than the average number of passive galaxies, to the luminosity of the dominant (−> mass of the system). This confirms that the relation between density and luminosity is weaker in SF than in passive galaxies (Hogg et al. 2003, Balogh et al. 2004).

4. Group selection criteria: how to find less SF-dominant groups

We have shown that SF-dGs are generally unfair tracer of group-like potentials. Three different selection criteria have been applied to the non-dominant members in the 988 groups to explore which one more efficiently reduces the number of SF-dominant groups:
i) $\geqslant$**6 non-dom** − > 471 groups: 330 P-dGs and 141 SF-dGs
ii) $\geqslant$**3 passive non-dom** − > 426 groups: 330 P-dGs and 96 SF-dGs
iii) $\geqslant$**50% passive non-dom** − > 455 groups: 337 P-dGs and 118 SF-dGs.

Clearly, the request for $\geqslant$**3 passive non-dom** is the most efficient criterion in rejecting groups with a SF-dominant. It is also less biased by projections of field-galaxies and it additionally produces groups with a passive galaxy fraction as high as 50%.

References

Balogh, M., Eke, V., Miller, C.J., et al. 2004 *MNRAS* **348**, 1355–1372
Gomez, P.L., Nichol, R.C, Miller, C.J., et al. 2003 *ApJ* **584**, 210–227
Hogg, D.W. Blanton M. R., Eisenstein D.J., et al. 2003 *ApJ* **585**, L5–L9
Kelm, B., Focardi, P. 2004 *A&A* **in press**
Lewis, I., Balogh M., De Propris R., et al. 2002 *MNRAS* **334**, 673–683
Madgwick, D.S, Lahav O., Baldry I.K. et al. 2002 *MNRAS* **333**, 133–144

Section B
Building galaxies

Outskirts of Galaxy Clusters: Intense Life in the Suburbs
Proceedings IAU Colloquium No. 195, 2004
A. Diaferio, ed.

DOI: 10.1017/S1743921304001000

The mass assembly history of galaxies and clusters of galaxies

Tadayuki Kodama[1], the PISCES team and the SXDS team

[1]National Astronomical Observatory of Japan, Mitaka, Tokyo 181–8588, Japan
email: kodama@th.nao.ac.jp

Abstract. We discuss the mass assembly history on both cluster and galaxy scales and their impact on galaxy evolution.

On cluster scale, we introduce our on-going PISCES project on Subaru, which plans to target ∼15 clusters at $0.4 \leqslant z \leqslant 1.3$ using the unique wide-field (30′) optical camera Suprime-Cam and the spectrograph both in optical (FOCAS, 6′) and near-infrared (FMOS, 30′). The main objectives of this project are twofold: (1) Mapping out the large scale structures in and around the clusters on 10–14 Mpc scale to study the hierarchical growth of clusters through assembly of surrounding groups. (2) Investigating the environmental variation of galaxy properties along the structures to study the origin of the morphology-density and star formation-density relations. Some initial results are presented.

On galactic scale, we first present the stellar mass growth of cluster galaxies out to $z \sim 1.5$ based on the near-infrared imaging of distant clusters and show that the mass assembly process of galaxies is largely completed by $z \sim 1.5$ and is faster than the current semi-analytic models' predictions. We then focus on the faint end of the luminosity function at $z \sim 1$ based on the Subaru/XMM-Newton Deep Survey imaging data. We show the deficit of red galaxies below M^*+2 or 10^{10} $M_{\odot}$, which suggest less massive galaxies are either genuinely young or still vigorously forming stars in sharp contrast to the massive galaxies where mass is assembled and star formation is terminated long time ago.

---

1. Panoramic Imaging and Spectroscopy of Cluster Evolution with Subaru (PISCES)

PISCES is a panoramic imaging survey of distant clusters using the Subaru wide-field optical camera Suprime-Cam which provides $34' \times 27'$ field of view corresponding to a physical area of 16×13 Mpc^2 at $z \sim 1$. This long term project has started since 2003, and we aim to target ∼15 X-ray selected distant clusters in total at $0.4 \lesssim z \lesssim 1.3$, in good coordination with *ACS/HST*, *XMM*, and *Chandra* observations. This unique project is currently underway and some preliminary results on the large scale structures over the entire Suprime-Cam fields are shown in Figs. 1 and 2. These two rich clusters at z=0.55 and 0.83 were imaged in multi optical bands, and photometric redshifts (Kodama *et al.* 1999) have been applied to efficiently remove foreground/background contaminations and isolate the cluster member candidates (c.f. Kodama *et al.* 2001). Many substructures are then clearly seen around the main body of the clusters which tend to be aligned in filamentary structures extending to >10 Mpc scale across. Although these structures should be confirmed spectroscopically later on, these already provide good evidence for cluster scale assembly in the hierarchical Universe.

The large scale structures that we see around the clusters provide us the unique opportunities to look into the environmental effects on galaxies as they assemble to denser regions. Kodama *et al.* (2001) have presented the environmental dependence of galaxy colours along the filamentary structures around the CL0939+47 cluster at z=0.41. They

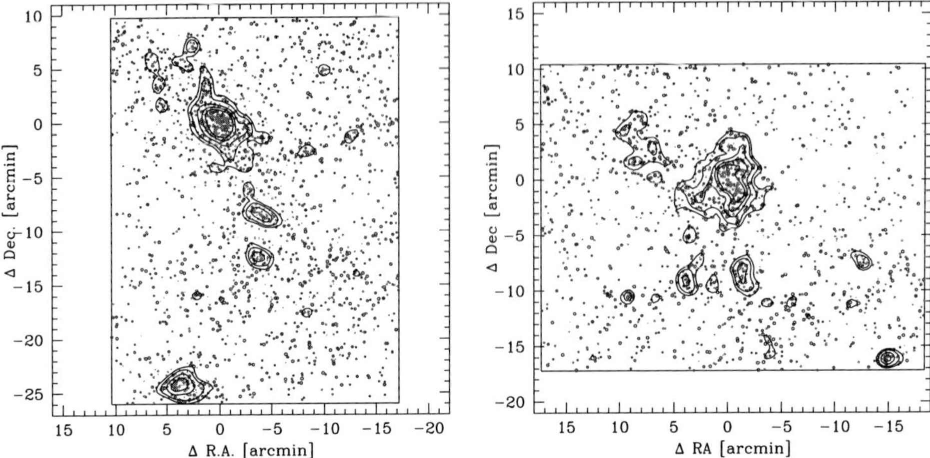

Figure 1. The panoramic maps of CL0016+16 cluster (z=0.55; left panel) and RXJ0152.7−1357 (z=0.83; right panel). 10 arcminutes correspond to physical scales of 3.8 and 4.6 Mpc, respectively. Using photometric redshift technique based on multi-colour data ($BVRi'z'$ and $VRi'z'$, respectively), plotted are the photometric member candidates selected with redshift cuts of $0.50 \leqslant z \leqslant 0.58$ and $0.78 \leqslant z \leqslant 0.86$, respectively. Contours show local 2-D number density of galaxies at 1.5, 2, 3, 4, 5 σ above the mean density. Coordinates are shown relative to the centre of the main cluster. Large scale filamentary structures (>10Mpc) are seen in both clusters.

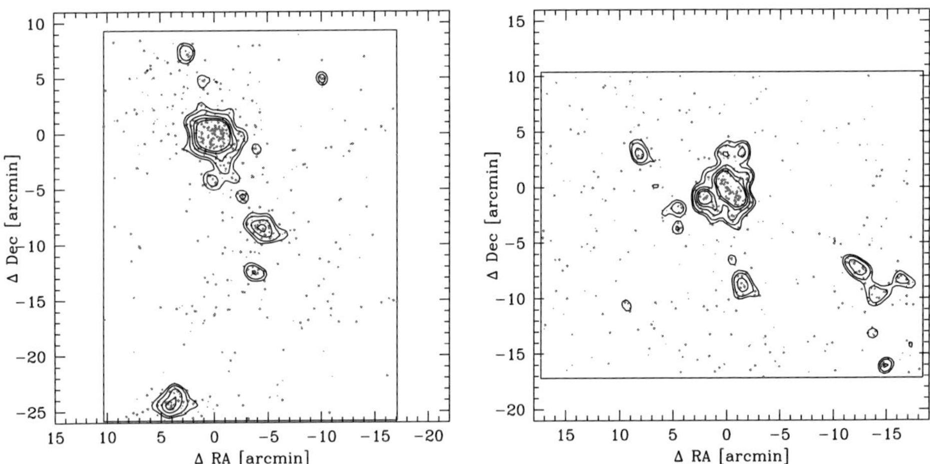

Figure 2. The same as Fig. 1, but only red galaxies on the colour-magnitude sequence are shown rather than phot-z selected galaxies. The red galaxies are selected on the basis of VRi' colours for CL0016+16 and $Ri'z'$ colours for RXJ0152.7−1357, respectively, and some red colour slice cuts are applied to isolate the passively evolving galaxies at cluster redshifts. This technique gives narrower redshift slice ($\Delta z \sim 0.05$) hence has less projection effect, but tends to be biased to the systems dominated by red populations. It is therefore complemental to the phot-z slice technique used in Fig. 1.

have shown that the galaxy colours change rather sharply at relatively low density regions, such as galaxy groups along the filaments, well outside of the cluster core where galaxies have not yet passed the central region of the clusters (see also Gray *et al.* 2004; Treu *et al.* 2003). Together with similar findings in the local Universe (Lewis *et al.* 2002;

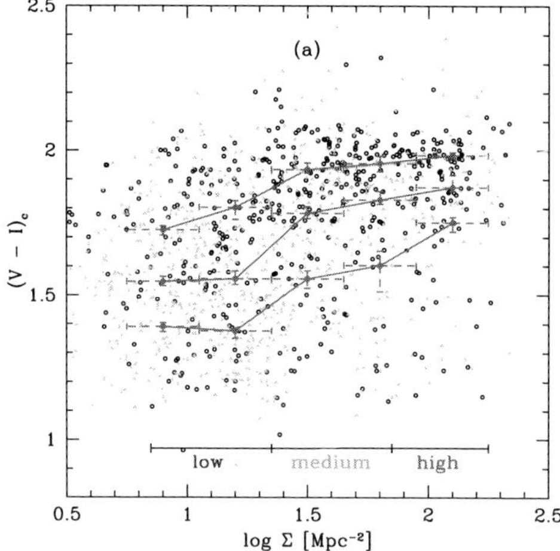

Figure 3. The variation in colour versus local galaxy density, for phot-z selected cluster members brighter than $I = 23.4$ in CL0939+47 cluster at $z = 0.41$ (Kodama *et al.* 2001). The open circles and filled triangles show the galaxies brighter or fainter than $I = 21.4$ ($M_V^* + 2$), respectively. The three red lines represent the loci of the 25, 50, and 75th percentile colours. The local number density is calculated from the 10 nearest galaxies and we correct this for residual field contamination in the photometric members using the blank field data.

Gomez *et al.* 2003), it follows that the environmental effects that truncate star formation are not driven by cluster specific mechanisms, such as ram-pressure stripping (Abadi *et al.* 1999), but by processes that are much more widely spread in low density regions. We should therefore pay greater attention to galaxy groups as the key hierarchy for the environmental effects, and try to identify what is happening on galaxies in this environment (see below). It is also important to extend this analysis to higher redshifts as the galaxy environment should change dramatically during the course of vigorous assembly, which is probably related to the appearance of the morphology-density relation (Dressler 1980).

Obviously, taking as many spectra as possible from the photometrically identified large scale structures is crucial to prove their reality, since our photometric approach may well suffer from the projection effect along the line of sight as we go to lower density regions, due to the broad phot-z slice cuts that we apply. Importantly, OII line and/or the 4000Å break feature are detectable for our PISCES targets in the optical spectroscopy (such as FOCAS) and the Hα line comes in the FMOS window ($0.9\mu m < \lambda < 1.8\mu m$). Besides definitively removing the foreground/background contamination and identifying the physically associated real groups, spectroscopic redshifts of individual cluster members will provide us with two critical pieces of information: (1) the dynamical mass of the systems which can then be compared to the lensing mass and the X-ray mass to address the dynamical state of the systems; (2) 3-D velocity structures, providing the recent and/or near future cluster-cluster/cluster-group merger histories (e.g. Czoske *et al.* 2002).

Also, OII and Hα lines will offer the measures of on-going star formation rate of galaxies. (The latter is preferred since it is much less affected by dust extinction or metallicity variation Kennicutt *et al.* 1984). Therefore we can directly identify when, where and on what timescale the star formation is truncated as the galaxies/groups fall into clusters along the filaments (e.g. Kodama *et al.* 2004b). Moreover, we will combine the information from other spectral indices, such as Balmer lines and colours, in order to resolve the recent star formation histories in galaxies on fine time scales. Different spectral

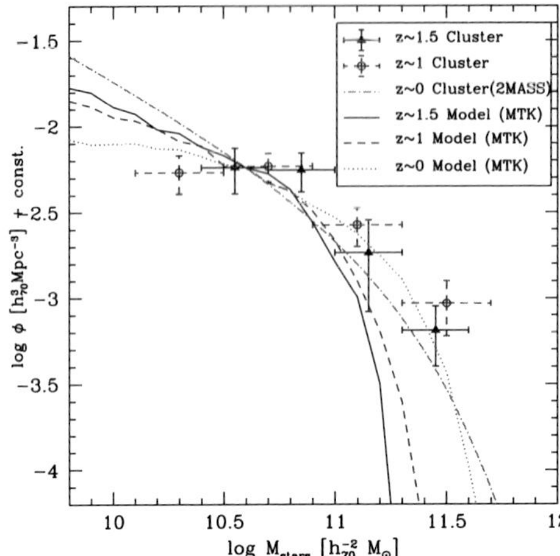

Figure 4. Stellar mass functions of galaxies in clusters as a function of redshift. The Kennicutt's initial mass function (Kennicutt 1983) is used to scale the stellar mass. The open diamonds and the filled triangles show the stellar mass functions for $z \sim 1$ and $z \sim 1.5$ clusters, respectively, which are compared to the local counterpart from the 2MASS survey (dot-dashed curve). All the curves and the data points are normalized at $5{\times}10^{10}$M$_\odot$ so as to have the same amplitude. The theoretical predictions from a semi-analytic model (Nagashima *et al.* 2002) are presented for comparison, which are made for galaxies in the haloes whose circular velocities are greater than 1000 km/s at each epoch. The mass assembly of massive galaxies in the real Universe is much faster than the hierarchical model prediction.

features are sensitive to different stellar ages (Kennicutt *et al.* 1984; Balogh *et al.* 1999; Poggianti *et al.* 1999): the emission lines (OII Hα) will measure the amplitude of on-going star formation (10^7 yrs), while the Balmer absorption lines will give the luminosity (mass) contribution from the stars formed immediately before the truncation (10^{8-9} yrs) (Dressler & Gunn 1992; Couch & Sharples 1987), and the 4000Å break and broad-band colours specify the features of longer-lived populations ($>10^9$ yrs) (Kodama & Bower 2001). Resolving the recent star formation histories in galaxies in the transition regions (groups) is the key to understand the physical processes behind the truncation. In particular, the existence of strong nebular emissions would support the galaxy-galaxy mergers which trigger star-burst, and very strong Balmer absorption lines (E+A or k+a) would follow in the post star-burst phase (Poggianti *et al.* 1999). If the star formation is more gradually truncated ($\gtrsim 1$ Gyr), due to halo gas stripping (strangulation), we would not see any excess of E+A/k+a features (Balogh *et al.* 1999).

It is also important to investigate the morphologies of the galaxies in these groups. The key question here is whether the transformation of morphologies is driven by the same mechanisms as those responsible for the truncation of star formation (Poggianti *et al.* 1999; Treu *et al.* 2003). Furthermore, stellar mass function of galaxies (see §2) in groups compared to other environment will provide information on galaxy-galaxy merger in this hierarchy since the mergers increase the fraction of massive galaxies. On the contrary, strangulation does not change mass, hence it can be distinguished by this test.

2. Stellar mass assembly of massive galaxies in high-z clusters

We now move on to the galactic scale, and firstly we present the stellar mass functions of galaxies in high-z clusters constructed from deep near-infrared imaging in J and K_s (Fig. 4) (Kodama & Bower 2003a; Kodama *et al.* 2003b). The stellar mass function of galaxies derived from the K-band observations is a good tracer of the mass assembly history of galaxies and therefore provides a critical test for the CDM-based bottom-up picture (e.g. Kauffmann & Charlot 1998; Baugh *et al.* 2002).

We have combined two $z \sim 1$ clusters (3C336 and Q1335+28; Kodama & Bower 2003a) and five $z \sim 1.5$ clusters (Q0835+580, Q1126+101, Q1258+404, Q0139–273, and Q2025–155; Hall *et al.* 1998; 2001; Best *et al.* 2003) to increase the statistics. We have subtracted the control field counts taken from the literature (Saracco *et al.* 1999; Saracco *et al.* 2001; Best *et al.* 2003). Applying the same technique described in Kodama & Bower *et al.* (2003), we construct the field-subtracted stellar mass functions of galaxies in high-z clusters, primarily using K_s-band flux and also using $J - K_s$ colour as a measure of the M/L ratio. As shown, little evolution is observed since $z = 1.5$ to the present-day (2MASS clusters; Balogh *et al.* 2001), indicating that the mass assembly on galaxy scale is largely completed by $z \sim 1.5$ in the cluster environment. This epoch of mass assembly of massive galaxies is earlier than the prediction of the hierarchical models as shown by comparison (Nagashima *et al.* 2002).

It is interesting to apply similar analyses to the general field, since the high density regions may be the special places where galaxy formation processes take place in an accelerated way. Recently, Pozzetti *et al.* (2003) and Glazebrook *et al.* (2004) investigated the stellar mass growth of massive galaxies out to $z \sim 1.5$ based on the large (mostly) spectroscopic sample in *K20* (52 arcmin^2) and *GDDS* (120 arcmin^2), respectively, and found that no or little evolution is seen in the stellar mass of massive galaxies, suggesting the early assembly of galaxies even in the general field environment.

It is also important to go to even higher redshifts to firstly identify the epoch of assembly of massive galaxies when they start to break down into pieces. Recent deep NIR observations in fact start to enter such formation epoch (e.g. Dickonson *et al.* 2003; Franx *et al.* 2003; Rudnick *et al.* 2003; Daddi *et al.* 2003), and this aspect will be further extended by on-going and near future space missions such as *SIRTF* (Dickonson *et al.* 2002; Pozzetti *et al.*, 2003) and *Astro-F* (Japanese mission).

3. Down-sizing in galaxy formation seen at $z \sim 1$

It is known that the galaxy properties depend on mass or luminosity as well. Using the SDSS data, Kauffmann *et al.* (2003) and Baldry *et al.*, (2004) have shown an interesting break mass at $3 \times 10^{10} M_\odot$ above which the dominant population is red passively evolving galaxies, while below that mass the contribution of blue active galaxies become dominant. Morphological mix of galaxies is also known to be strongly luminosity (or stellar mass) dependent (e.g. Treu *et al.* 2003). It is therefore indicative that the formation of massive galaxies and less massive galaxies are quite different, in the sense that massive and/or early-type galaxies form early in the Universe, while dwarf and/or late-type galaxies form later on average or still forming stars at present. The mass dependent star formation history is referred to as "down-sizing" (Cowie *et al.* 1996).

Given the down-sizing effect in the local Universe, as we look back into the higher redshift Universe, we expect to go beyond the formation epoch of small galaxies or towards the early stage of their formation, as well as to approach the formation epoch of more massive galaxies. To test this hypothesis, we investigate the galaxy colours at $z \sim 1$ as a function of luminosity (or stellar mass) utilizing the unique Suprime-Cam imaging data-set ($BRi'z'$) on the Subaru/XMM-Newton Deep Survey (SXDS). These data are both sufficiently deep (z'_{AB}=25, 6-10σ) and wide (1.2 deg^2) to enable us, for the first time, to investigate the photometric properties of a statistical sample of galaxies at $z \sim 1$ down to $\sim M^*$+3 with respect to the passive evolution. We first identify five $z \sim 1$

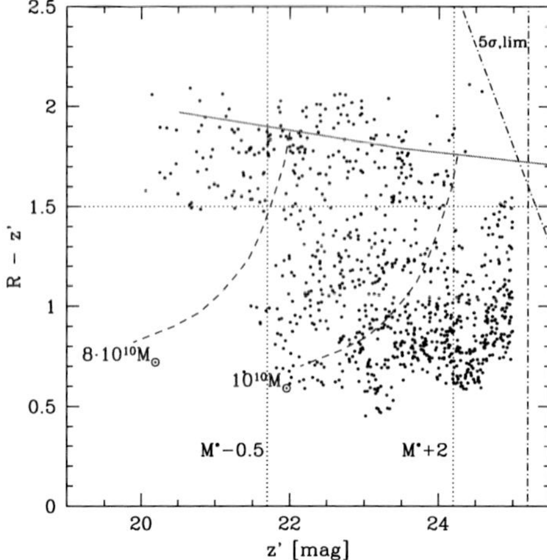

Figure 5. Field-corrected colour-magnitude diagram for the $z \sim 1$ galaxies in high density regions in SXDS (Kodama *et al.* 2004a). The solid line show the expected location of colour-magnitude sequence at $z \sim 1$ assuming a passive evolution with $z_{\rm form}$=5 (Kodama *et al.* 1998). The deficit of both blue galaxies at the bright/massive end and the deficit of red galaxies at the faint/less-massive end are both clearly identified. The stellar masses are scaled using the Kennicutt's IMF (Kennicutt 1983).

high density regions by applying colour cuts at $1.7<R-z'<2.0$ and $0.8<i'-z'<1.1$, which correspond to the colours of passively evolving galaxies at $z \sim 1$ (Kodama *et al.* 1998). We then combine these five regions (amounting to 141 arcmin^2 in total) and subtract off the low density regions at $z \sim 1$ sampled from the same data-set and scaled to the same area. We do this subtraction on the colour-magnitude diagram (Kodama & Bower 2001) and isolate the $z \sim 1$ galaxies in a statistical sense. This method should work since both high and low density regions at $z \sim 1$ are expected to have the same amount of foreground/background contaminations.

The field-corrected colour-magnitude diagram thus constructed for $z \sim 1$ galaxies in high density regions is shown in Fig. 5. The most striking feature in the galaxy distribution on this diagram is that the galaxies are separated into two distinct populations, 'bright+red' and 'faint+blue'. More precisely, we show a deficit of red and faint galaxies below M^*+2, or 10^{10}M$_\odot$ in stellar mass, and a lack of blue massive galaxies above M^*−0.5, or $8{\times}10^{10}$M$_\odot$ in stellar mass. The down-sizing in star formation is therefore also seen at high redshift, where star formation in massive galaxies takes place early in the Universe and is already truncated by $z \sim 1$, while almost all of small galaxies are still forming stars at $z \sim 1$ (Kodama *et al.* 2004a; Poggianti *et al.* 2004; see also De Lucia *et al.* in this volume). Together with the early-assembly of massive galaxies see in § 2, galaxy formation does in fact take place in "down-sizing" fashion as apparently opposed to the "bottom-up" scenario. Some critical physical mechanisms in galaxy formation may be still missing in the current models.

It is important to note that there are two possible interpretations for the blue colours of faint galaxies. One is that they are just forming, and the other is that they have much extended star formation over the long period of time. To separate out these two possibilities, the key quantity to measure is the so-called birth-rate parameter (b) of these faint blue galaxies (Kennicutt 1988). Here b' is re-defined as on-going star formation rate divided by integrated stellar mass of the system ($b'=SFR/M_{\rm stars}$). SFR can be derived from OII and Hα line intensities by spectroscopy or narrow-band imaging of the PISCES clusters and the SXDS field while $M_{\rm stars}$ can be obtained from the deep NIR imaging. This parameter describes what phase of star formation each galaxy is in, and therefore can discriminate between genuinely young galaxies formed recently (large b') and star forming galaxies formed long time ago (small b'). It is also interesting to directly detect the brightening of the break luminosity or mass with increasing redshift by going to even higher redshifts.

References

Abadi, M. G., Moore, B., Bower, R. G., 1999, MNRAS, 308, 947
Baldry I. K., *et al.*, 2004, ApJ, 600, 681
Balogh M. L., *et al.*, 1999, ApJ, 527, 54
Balogh, M. L., Christlein, D., Zabludoff, A. I., Zaritsky, D., 2001, ApJ, 557, 117
Baugh, C. M., Benson, A. J., Cole, S., Frenk, C. S., Lacey, C., 2002, astro-ph/0203051
Best, P. N., Lehnert, M. D., Miley, G. K., Röttgering, H. J. A., 2003, MNRAS, 343, 1
Butcher, H., Oemler, A., 1984, ApJ, 285, 426
Couch, W. J., Sharples, R. M., 1987, MNRAS, 229, 423
Cowie, L. L., Songaila, A., Hu, E. M., Cohen, J. G., 1996, AJ, 112, 839
Czoske, O., *et al.*, 2002, A&A, 386, 31
Daddi, E., *et al.*, 2003, astro-ph/0308456
Dickinson, M., Giavalisco, M., *et al.*, astro-ph/0204213
Dickinson, M., Papovich, C., Ferguson, H. C., Budavari, T., 2003, ApJ, 587, 25
Dressler, A., 1980, ApJ, 236, 351
Dressler, A., Gunn, J. E., 1992, ApJS, 78, 1
Finn, R. A., Zaritsky, D., McCarthey, D. W., 2004, ApJ, 604, 141
Franx, M., *et al.*, 2003, ApJ, 587, L79
Glazebrook, K., *et al.*, 2004, astro-ph/0401037
Gómez, P. L., *et al.*, 2003, ApJ, 584, 210
Gray M. E., *et al.*, 2004, MNRAS, 347, L73
Kauffmann, G., Charlot, C., 1998, MNRAS, 297, 23
Kauffmann, G., *et al.*, 2003, MNRAS, 341, 54
Kennicutt, R. C., 1983, ApJ, 272, 54
Kennicutt, R. C., *et al.*, 1994, ApJ, 435, 22
Kennicutt, R. C., 1988, ARA&A, 36, 189
Kodama, T., Arimoto, N., Barger, A. J., Aragón-Salamanca, A., 1998, A&A, 334, 99
Kodama, T., Bell, E. F., Bower, R. G., 1999, MNRAS, 302, 152
Kodama, T., Smail, I., Nakata, F., Okamura, S., Bower, R. G., 2001, ApJ, 562, L9
Kodama, T., Bower, R. G., 2001, MNRAS, 321, 18
Kodama, T., Bower, R. G., 2003a, MNRAS, 346, 1
Kodama, T., *et al.* 2003b, astro-ph/0312321
Kodama, T., *et al.*, 2004a, MNRAS, in press
Kodama, T., Balogh, M. L., *et al.*, 2004b, in prep.
Hall, P. B., Green, R. F., Cohen, M., 1998, ApJS, 119, 1
Hall, P. B., *et al.*, 2001, AJ, 121, 1840
Lewis, I., *et al.*, 2002, MNRAS, 334, 673
Nagashima, M., Yoshii, Y., Totani, T., Gouda, N., 2002, ApJ, 578, 675
Poggianti, B. M. *et al.* 1999, ApJ, 518, 576
Poggianti, B. M. *et al.* 2004, ApJ, 601, 197
Pozzetti, L., *et al.*, 2003, A&A, 402, 837
Rudnick, G., *et al.*, 2003, astro-ph/0307149
Saracco, P., *et al.*, 1999, A&A, 349, 751
Saracco, P., *et al.*, 2001, A&A, 375, 1
Treu, T., *et al.*, 2003, ApJ, 591, 53

Outskirts of Galaxy Clusters: Intense Life in the Suburbs
Proceedings IAU Colloquium No. 195, 2004
A. Diaferio, ed.

DOI: 10.1017/S1743921304001012

Extended Lyman-alpha emitters in and around a proto-cluster at z=3.1

T. Yamada[1], Y. Matsuda[1,2] and T. Hayashino[2]

[1]National Astronomical Observatory of Japan, Mitaka, Tokyo 181-8588, Japan

[2] Research Center for Neutrino Science, Graduate School of Science, Tohoku University, Aramaki, Aoba, Sendai 980-8578, Japan

Abstract. Here we present our result of deep and wide-field narrow-band imaging of the region of the 'proto-cluster' at $z = 3.1$ around the SSA22 field. We detected 283 highly confident Lyα emitter candidates (LAEs) and discovered that the belt-like high surface density region of these LAEs extends over ≈ 60 Mpc in comoving scale, which is much larger than it was known previously. We then successfully detected the 35 extended Lyα blobs (LABs) which are larger than 16 arcsec^2 in isophotal area and brighter than 0.7×10^{-16} ergs s^{-1} cm^{-2}. The distributions of average surface brightness and morphology are widespread from relatively compact high surface brightness objects to very diffuse low surface brightness ones. For one third of these 35 LABs, simple photo-ionization by massive stars is not sufficient to explain the Lyα luminosities, and other mechanisms, such as gravitational heating or superwind must be considered. From their large size and strong spatial clustering, we consider that these LABs are objects closely related to the massive galaxies in their forming phase in the environment of the proto-cluster or large-scale structure.

1. Introduction

Properties of galaxies in the outskirt of clusters should be considered in the context of biased galaxy formation at an early epoch followed by continuous infall of galaxies from the surrounding region. To understand the whole story of the evolution processes in high-density regions of the universe, it is the most direct way to observe the clusters and their progenitors at sufficiently large scale at various epochs from low to high redshift and trace the properties of galaxies there.

In this contribution, we present the results of Subaru deep and wide-field narrow-band imaging of the region of the 'proto-cluster' at $z = 3.1$ around the SSA22 field (Steidel et al. 1998; 2000). The structure was first recognized as the highest density 'spike' in the redshift distribution of Lyman Break galaxies (LBG; Steidel et al. 1998). The following narrow-band image by Steidel et al. (2000) showed that the overdensity of Lyα emitters (LAE) is as large as that of LBGs. However, there is no conspicuous concentration of LAEs and LBGs *on the sky* in their observed field of view and it was unknown whether the region is a very central part of the proto cluster or just a part of larger-scale structures. It is clearly an important next step to reveal the distribution of star-forming galaxies in a wider area to see the whole extent of the high density region. We therefore conducted the deep Lyα emission-line imaging observation with Subaru Suprime-Cam to study about ten times as large volume as that in the previous one. We here show the distribution of the 283 robust sample of LAEs in the field as well as the 35 extended Lyα emitters, or 'Lyα blobs' (LABs) obtained in the study.

This talk is based on our recent two papers, Hayashino et al. (2004) and Matsuda et al. (2004), and we thank the co-authors of these papers for allowing us to present the results at this conference.

2. Data

We obtained a wide-field and deep narrow-band image centered at $(\alpha,\delta) = (22^h17^m.6, +00°17')$ (J2000.0) with the 8.2 m Subaru telescope equipped with the prime-focus camera, Suprime-Cam. We used our custom narrow-band filter ($NB497$), whose central wavelength is 4970Å and FWHM is 80Å to detect Lyα emission line at $z = 3.05 - 3.12$. The total exposure time was 7 hours for $NB497$ and the limiting magnitudes after correcting for Galactic extinction is $NB497$=26.2 for 5σ in a $2''$ diameter aperture. The FWHM seeing was $1''$. The total size of the field is $32'$ by $24'$, which corresponds to comoving transverse dimensions of 59.3 $h_{70}^{-1} \times 45.5 h_{70}^{-1}$ Mpc at $z = 3.1$ and the total solid angle is 770 arcmin^2. The effective comoving volume covered by the narrow-band image, whose half-power width corresponds to a comoving depth along the line of sight of 61.0 h_{70}^{-1} Mpc, is 165,000 h_{70}^{-3} Mpc^3.

3. Sky distribution of Lyα emitters and absorbers

We selected the 283 robust candidates of LAE at $z = 3.1$ mainly by the criteria (i) $NB497 < 25.8$ (S/N> 7.5) and (ii) $BV - NB497 > 1.2$, which corresponds to the observed equivalent width limit of $EW_{obs} > 160$ Å, or the rest frame EW limit of $EW_0 >$ 40 Åfor Lyα. More detailed description is found in Hayashino et al. (2004).

These 283 strong LAE are distributed in the Suprime-Cam field of view as shown in Fig. 1 with the filled circles. The curved lines delineate the high density region (HDR) where the mean number density of emitters in this field after smoothing with a Gaussian kernel with $\sigma = 1.5$ arcmin is larger than the average. We see a belt-like large structure of LAEs with 60 Mpc long and 15-20 Mpc wide in comoving scale running from NW to SE, together with a rather orthogonal structure with 30 Mpc by 15 Mpc from the field center to NE. The average number density of the emitters in this HDR is found to be three times as large as that of a blank field. We found that the probability of finding such a large-scale high density peak is as small as 0.1% in context of CDM structure formation scenario even if we assume that the distribution of Lyα emitters is biased to mass with linear bias parameter $b \sim 4$. This structure is really a rare peak which may evolve into a massive supercluster at later epoch. The SSA22a area studied by Steidel et al. (2000), which is about $9' \times 9'$ square near the center of this figure, is turned out to be part of this large-scale structure of LAEs.

We also search for Lyα absorbers, namely, narrow-band deficit objects in the field (open circles in Fig. 1). It is remarkable that their distribution is very similar to that of the Lyα *emitters*, which strongly suggests that the HDR of the strong LAEs is a real large-scale structure and these absorbers are associated with it.

4. Lyα blobs

We also detected the 35 highly confident candidates of extended Lyα blobs (LABs) in the field which are larger than 16 arcsec^2 in isophotal area and brighter than 0.7×10^{-16} ergs s^{-1} cm^{-2}. Fig. 2 shows their isophotal area and magnitude distribution and Fig. 3 shows the montage of all the 35 LABs. Eight of these 35 candidates, including the two gigantic 'Lyα blobs' discovered by Steidel et al., are associated with the known Lyman break galaxies at $z = 3.1$. These two Lyα blobs are the most luminous and the largest emitters in our survey volume.

The distributions of average surface brightness and morphology are widespread from relatively compact high surface brightness objects to very diffuse low surface brightness

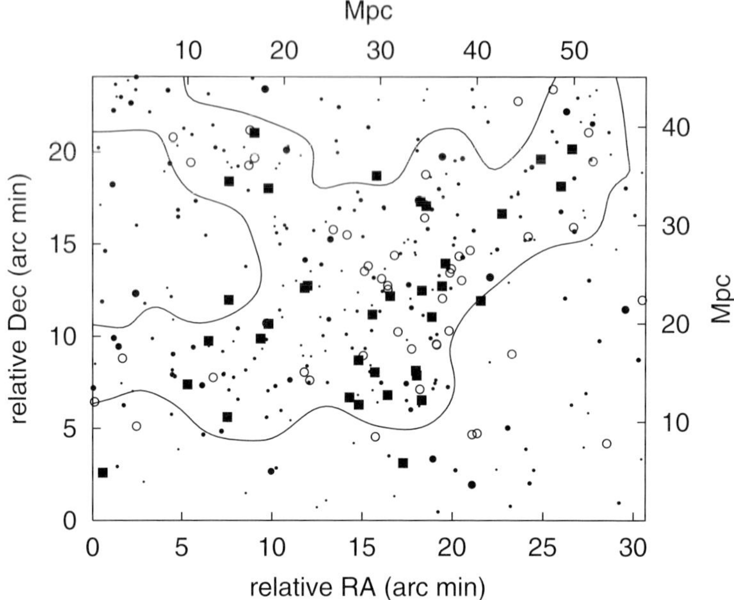

Figure 1. Spatial Distribution of the 283 LAE in the Suprime-Cam field of view (filled circles). The curved lines delineate the high density region of the LAEs. Open circles and filled squares show the distribution of Lyα absorbers and Lyα blobs (LABs), respectively.

ones. The physical origins of these LABs may be (i) photo-ionization by massive stars or active galactic nuclei, or (ii) Lyα cooling radiation by gravitationally heated gas, or (iii) shock heating by starburst driven galactic superwind. Scattering by surrounding neutral gas affects their appearance for all these cases. The case of cooling radiation and superwind represents rather early and late stage of intensive star formation in proto-galaxy while photo-ionization by internal sources reflects on-going events. Sources of photo-ionization may be massive stars or AGNs inside the galaxies, or the background diffuse UV emission. It is interesting to see the objects in the same large-scale structure and may be at different phases of galaxy formation.

In Fig. 4, we plot R versus $NB_{\rm corr}$ for these 35 LABs. If the origin of the Lyα emission is photoionization by massive stars, we may translate the $NB_{\rm corr}$ and R magnitude into the equivalent star formation rates (SFR) at $z = 3.1$, assuming the same Salpeter initial mass function with mass limits 0.1 and 100 $M_\odot$, solar metallicity, no extinction, and case B recombination in the low density limit. Very interestingly, one third of them are apparently not associated with ultra-violet continuum sources that are bright enough to produce Lyα emission, assuming a Salpeter initial mass function. Other mechanisms, such as gravitational heating or superwind activity, may contribute to the excess. We revealed the internal structure of the two gigantic blobs (Steidel et al. 2000) and indeed discovered some bubble-like features, which suggests that intensive starburst and galactic superwind phenomena occurred in these objects in the past.

We also see the sky distribution of the LABs in Fig. 1: 90% of them are located inside the HDR of the 283 relatively compact and strong Lyα emitters. This suggests that LABs very strongly trace the large-scale structure at high redshift where galaxy formation may occur preferentially; in other words, we may be witnessing galaxy formation activity in proto-cluster environment.

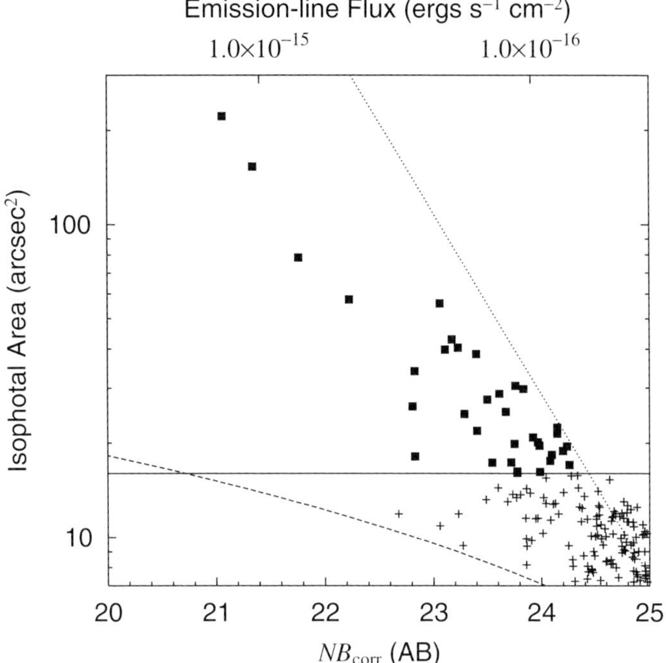

Figure 2. Distribution of isophotal area and narrow-band magnitude of the LABs (filled squares). The dashed line shows the expected value for point sources. The dotted line shows the 8 σ noise level.

Acknowledgements

The co-authors of the two papers on which the contribution is based are as follows: Ajiki, M., Fujita, S., Murayama, T., Nagao, T., Ohta, K., Okamura, S., Ouchi, M., Shimasaku, K., Shioya, Y., Tamura, H., Taniguchi, Y., and Yamauchi, R.

References

Hayashino, T., et al. 2004, *Astron. J.*, submitted.

Matsuda, Y., et al. 2004, *Astron. J.*, in press.

Steidel, C. C., Adelberger, K. L., Dickinson, M., Giavalisco, M., Pettini, M., & Kellogg, M. 1998, *Astrophys. J.*, **492**, 428

Steidel, C. C., Adelberger, K. L., Shapley,A. E., Pettini, M., Dickinson, M., & Giavalisco, M. 2000, *Astrophys. J.*, **532**, 170

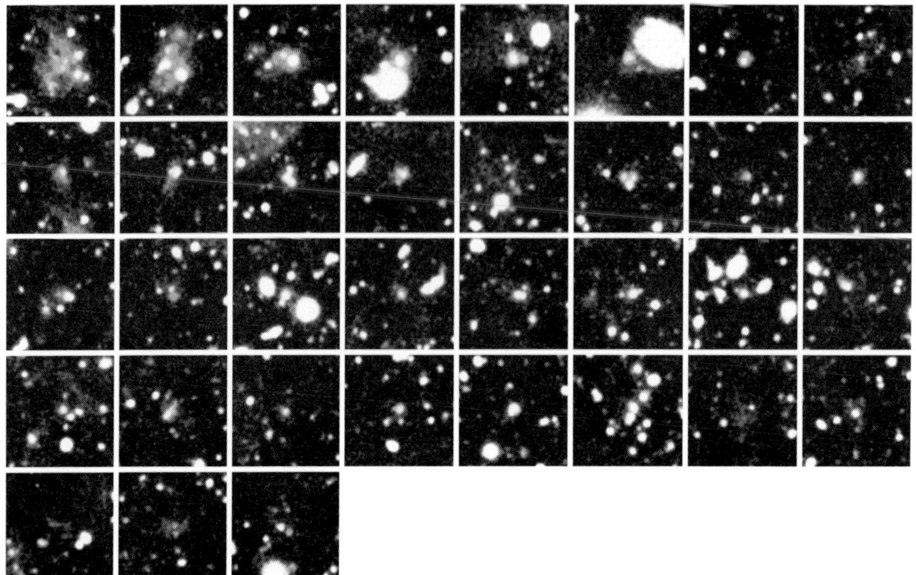

Figure 3. Montage of the 35 LABs. They are B, $NB497$, V composit color images and the green shows the Lα emission.

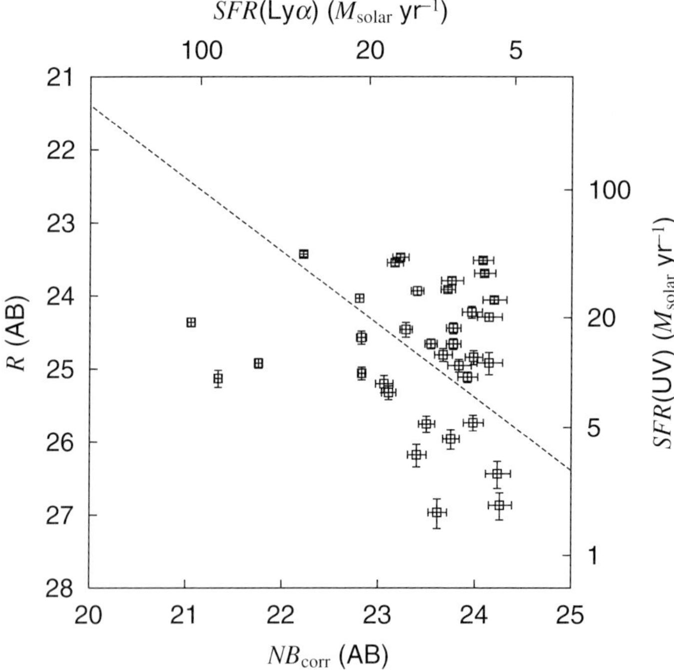

Figure 4. Distributions of R and NB magnitudes of the 35 LABs. R magnitudes are those of continuum sources nearest to the Lyα peak or the known LBGs at $z = 3.1$. The dashed line shows $SFR(\mathrm{Ly}\alpha) = SFR(\mathrm{UV})$.

Outskirts of Galaxy Clusters: Intense Life in the Suburbs
Proceedings IAU Colloquium No. 195, 2004
A. Diaferio, ed.

DOI: 10.1017/S1743921304001024

Observing the build–up of the colour-magnitude relation at redshift ∼0.8†

G. De Lucia[1], B. M. Poggianti[2], A. Aragón-Salamanca[3], D. Clowe[4], C. Halliday[2], P. Jablonka[5], B. Milvang-Jensen[6], R. Pelló[7], S. Poirier[5], G. Rudnick[1], R. Saglia[6], L. Simard[8], S. D. M. White[1] and the EDisCS collaboration

[1]MPA, Garching, Germany – [2]Osservatorio Astronomico di Padova, Italy – [3]University of Nottingham, UK – [4]Steward Observatory, Tucson, USA – [5]OPM, Paris, France – [6]MPE, Garching, Germany – [7]OMP Toulouse, France – [8] HIA, Victoria, Canada

Abstract. We analyse the rest–frame (U−V) colour–magnitude relation for 2 clusters at redshift 0.7 and 0.8, drawn from the ESO Distant Cluster Survey. By comparing them with the population of red galaxies in the Coma cluster, we show that the high redshift clusters exhibit a deficit of passive faint red galaxies. Our results show that the red–sequence population cannot be explained in terms of a monolithic and synchronous formation scenario. A large fraction of faint passive galaxies in clusters today has moved onto the red sequence relatively recently as a consequence of the fact that their star formation activity has come to an end at $z < 0.8$.

1. Introduction

Red cluster galaxies form a tight sequence in the colour–magnitude diagram that, in nearby clusters, extends over a range of at least 5–6 mag from the Brightest Cluster Galaxy (BCG). The existence of a tight colour–magnitude relation (CMR) up to redshift ∼ 1, and the evolution of its slope and its zero–point as a function of redshift, are commonly interpreted as the result of a single formation scenario in which cluster ellipticals constitute a passively evolving population formed at high redshift ($z \gtrsim 2$–3) in a monolithic collapse (Kodama et al. 1998). In this model, the slope of the CMR reflects metallicity differences and naturally arises taking into account the effects of supernovae winds. An alternative explanation has been proposed by Kauffmann & Charlot (1998) and confirmed recently by De Lucia, Kauffmann & White (2004). In this model, elliptical galaxies form through mergers of disk systems and a CMR arises because more massive ellipticals form by mergers of more massive, and hence more metal rich, disk systems.

On the other hand, it is clear that red passive galaxies in distant clusters constitute only a subset of the passive galaxy population in clusters today (van Dokkum & Franx 1996). Distant clusters contain significant populations of galaxies with active star formation, that can evolve onto the CMR after their star formation activity is terminated, possibly as a consequence of their infall onto clusters (Smail et al. 1998; Poggianti et al. 1999).

In this work we present the CMR for 2 clusters at redshifts 0.7 and 0.8 from the ESO Distant Cluster Survey (EDisCS).

† Based on observations obtained at the ESO Very Large Telescope (VLT) as part of the Large Programme 166.A–0162 (the ESO Distant Cluster Survey).

2. The ESO Distant Cluster Survey

EDisCS is an ESO Large Programme aimed at the study of cluster structure and cluster galaxy evolution over a significant fraction of cosmic time. It will provide homogeneous photometry and spectroscopy for 10 clusters at redshift $\sim 0.4 - 0.5$ and 10 clusters at redshift $\sim 0.7 - 0.8$.

In this work we present results for 2 clusters at redshift 0.7 (cl1054.4–1146) and 0.8 (cl1216.8–1201). The clusters have been imaged in V, R, and I with FORS2 at the ESO's Very Large Telescope. The average integration times used were 115 m in the I–band, and 120 m in the V–band. Multi–object spectroscopy was carried out using FORS2 on the VLT. All data and details about their analysis will be presented in forthcoming papers (White et al., Halliday et al., in preparation; see also Halliday et al., these proceedings). Objects detection has been performed using SExtractor (Bertin & Arnouts 1996) in 'two–image' mode using the I–band images as detection reference images. In the following, we will use magnitude and colours measured on the seeing–matched (to 0.8 arcsec – our 'worst' seeing) registered frames using a fixed circular aperture with 1 arcsec radius. At the clusters redshifts, this aperture corresponds to a physical radius of 7.13–7.50 kpc (we use $\Omega_{\rm m} = 0.3$, $\Omega_\Lambda = 0.7$, and $H_0 = 70\,{\rm km\,s^{-1}\,Mpc^{-1}}$).

3. Method and results

We have computed photometric redshifts using two independent codes (see Rudnick et al. 2003) and we have rejected non–members using a two–step procedure: (i) we use the full redshift probability distributions from the codes to isolate objects with high probability to be at the cluster redshift and (ii) we perform a statistical subtraction on the remaining objects using the distribution on the CMR of all the objects at distance from the BCG larger than 1 Mpc.

Fig. 1 (panels a and b) shows the V$-$I diagram for the 2 clusters used in this analysis. At these redshifts, V$-$I approximatively samples the rest–frame U$-$V colour, and is therefore very sensitive to any recent or ongoing star formation episode. A red sequence is clearly visible for each cluster, together with a significant population of the blue galaxies known to populate clusters at high redshift. Thin solid lines correspond to 3 and 5σ detection limits in the V–band. Empty symbols represent galaxies retained as cluster members; filled circles and crosses are spectroscopically confirmed members with absorption and emission–line spectra respectively. Panel (c) of Fig. 1 shows the U$-$V CMR for the Coma cluster including all galaxies in the catalogue by Terlevich, Caldwell & Bower (2001). Here we use magnitudes and colours in a 25.2 arcsec diameter aperture that, at the redshift of Coma, corresponds to a physical size of 11.71 kpc and therefore quite closely approximates our $\sim$ 14 kpc diameter aperture at $z = 0.8$. We have assumed a distance modulus of 35.16 and converted observed colours to rest–frame colours using tabulated K–corrections (Poggianti 1997).

The boxes in Fig. 1 show the location of a model with a single burst at $z = 3$, and the diamonds show the location of a model with a 1 Gyr exponentially declining SFR starting at $z = 3$, calculated with the code by Bruzual & Charlot (2003). Three different metallicities are shown: 0.02, 0.008 and 0.004 (from brighter to fainter). The single burst model provides a remarkably good fit to the CMR of the high redshift clusters, confirming that the location of the CM sequence observed in distant clusters requires high redshifts of formation, and that the slope is consistent with a correlation between galaxy metal content and luminosity.

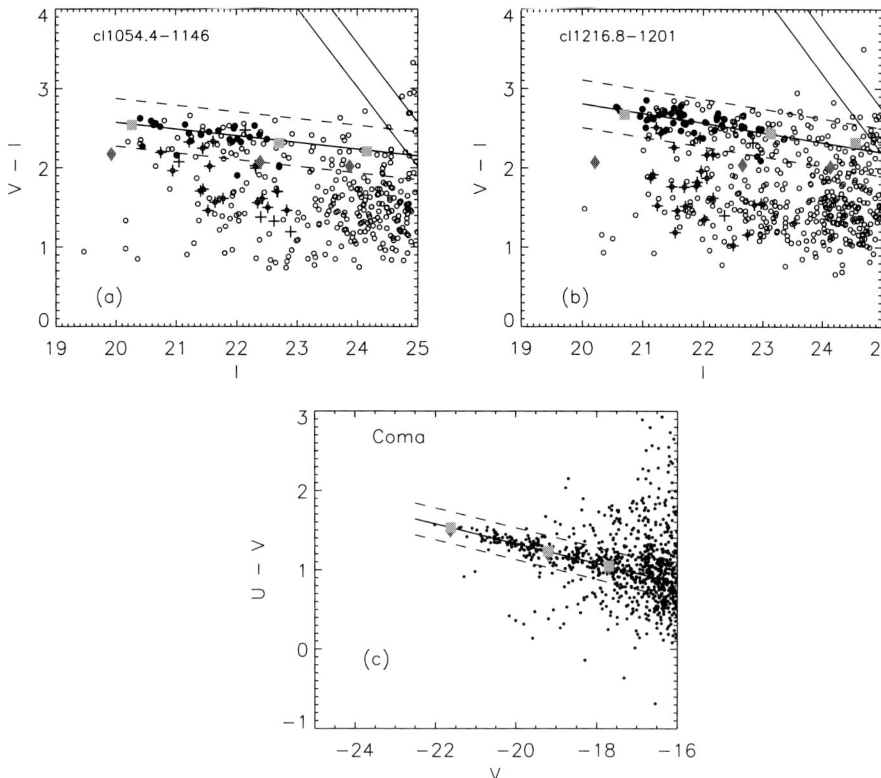

Figure 1. CM diagrams for the 2 EDisCS clusters under investigation (panels a and b) and for the Coma cluster (panel c). The solid thick line represents the best fit relation measured using a bi–weight estimator and only the spectroscopic members with absorption spectra. The dashed lines correspond to $\pm 3\sigma$ from the best fit, where σ is the dispersion of the objects used for the fit. Boxes and diamonds show the location of models with different star formation histories (see text for detail).

However, Fig. 1 also shows that the CMR in our clusters is strikingly 'empty' at magnitudes fainter than ~ 23. In order to quantify this effect, we compute the luminosity function (LF) of red sequence galaxies. Fig. 2 shows the numbers of *bona fide* cluster members within $\sim 3\sigma$ from the best fit relation as a function of galaxy magnitudes. For the EDisCS clusters, we combine the results for the 2 clusters correcting colours and magnitudes to a common redshift $= 0.75$ and average over 100 realizations of the statistical subtraction for each cluster. For Coma, we have corrected the number of red galaxies in each magnitude bin for background and foreground contamination using a redshift catalogue kindly provided by Matthew Colless and the same procedure as in Mobasher et al. (2003).

The histogram of the EDisCS clusters shows a remarkable deficiency of $\lesssim 0.1\mathrm{L}_*$ red sequence galaxies. In contrast, the number of passive red galaxies in Coma tends to increase going to fainter magnitudes, in agreement with the luminosity function of early spectral types in clusters in the 2dF Galaxy Redshift Survey (De Propris et al. 2003). A Schechter fit to the histograms shown in Fig. 2 yields a faint–end slope of -0.82 for the

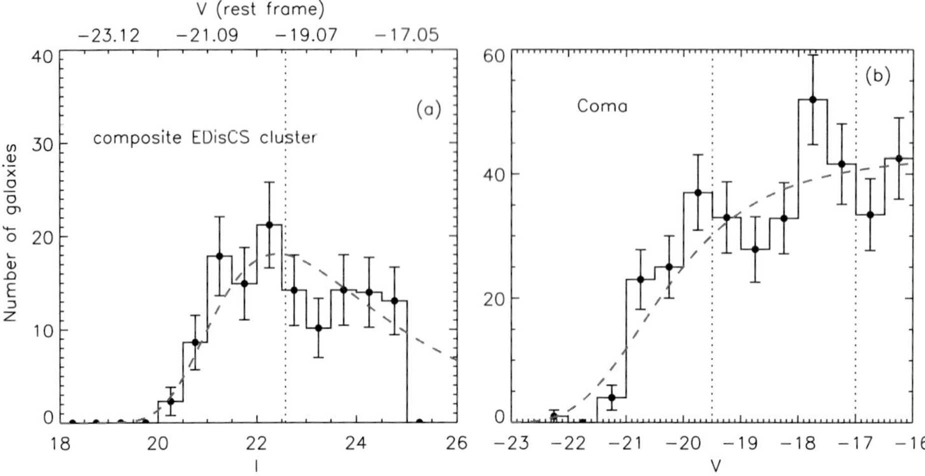

Figure 2. LF of red-sequence galaxies for a composite cluster at redshift 0.75 (panel a) and for the Coma cluster (panel b). Poisson error bars are shown. The dashed lines represent a fit with a Schechter function. The scale on the top of the left panel has been obtained from the observed I–band magnitudes after correcting for passive evolution. Dotted vertical lines show the limits used to define the luminous–to–faint ratio (see text for detail).

EDisCS clusters and −1 for Coma, but the error bars are quite large and therefore the results of the fit are just indicative.

If we arbitrarily classify as 'luminous' galaxies those brighter than $M_V = -19.5$ (this corresponds to an observed I–band magnitude of 22.57 at redshift 0.75), and as 'faint' those galaxies that are fainter than this magnitude and brighter than −17 (this corresponds to the magnitude limit for our high redshift clusters), we obtain a luminous–to–faint ratio for the Coma cluster 0.48 ± 0.06, where the error has been estimated assuming a Poisson statistics. The corresponding value for the composite EDisCS cluster is 0.99 ± 0.17. This ratio therefore differs between the composite high redshift cluster and the Coma cluster at about a 3σ level.

4. Discussion and conclusions

The results presented are robust both against the technique adopted for removing non–cluster members and against the photometric errors. The red galaxy deficit is detected also when rejecting non–members using a purely statistical subtraction or a more stringent criterion for membership based solely on photometric redshifts. In fact, a deficit is evident also in the *full* photometric catalogue, when no field correction is attempted. Photometric errors in the EDisCS catalogue are comparable to the errors for the Coma data, therefore the differences observed in the distributions of Fig. 2 cannot be a spurious result arising from photometric errors.

A decline in the number of red sequence members at faint magnitudes was first observed in clusters at $z = 0.25$ by Smail et al. (1998). Evidence for a 'truncation' of the CM sequence has been noticed in a cluster at $z = 1.2$ by Kajisawa et al. (2000), who suggested that faint early–type galaxies might not have been in place until $z \sim 1.2$. In more recent work Kodama et al. (2004) have come to the same conclusion using deep wide-field optical imaging data of the Subaru/XMM-Newton Deep Survey.

The CMR of the 2 EDisCS clusters at $z \sim 0.8$ also shows a deficit of red, relatively faint galaxies. Our investigation shows that the evolution of the red–sequence population at these redshifts cannot be explained as the result of a monolithic and synchronous formation scenario. A large fraction of the passive $\lesssim 0.1L_*$ galaxies must have moved onto the CMR at redshifts lower than 0.8 as a consequence of the fact that their star formation activity has come to an end. Our analysis enlightens the importance of studying the evolution of the cluster population as a whole, trying to understand how galaxies are accreted from the field, how the dense cluster environment affects their star formation rate, and how these galaxies fade and move on to the red sequence. Only this kind of analysis can unveil the full evolutionary paths of galaxies that lie on the red–sequence today and give strong constraints on the relative importance of star formation and metallicity in establishing the observed red–sequence. We plan to investigate this in more detail in future work.

Acknowledgements

We thank F. La Barbera, S. Zaroubi, I. Smail and M. Colless. G. D. L. thanks the Alexander von Humboldt Foundation, the Federal Ministry of Education and Research, and the Programme for Investment in the Future (ZIP) of the German Government for financial support.

References

Bertin, E. & Arnouts, S. 1996 *A&AS* **117**, 393–404.
Bruzual, G. & Charlot, S. 2003 *MNRAS* **344**, 1000–1028.
De Lucia, G., Kauffmann, G. & White, S. D. M. 2004 *MNRAS* **349**, 1101–1116.
De Propris, R. et al. 2003 *MNRAS* **342**, 725–737.
Kauffmann, G. & Charlot, S. 1998 *MNRAS* **294**, 705–717.
Kajisawa, M. et al. 2000 *PASJ* **52**, 61–72.
Kodama, T., Arimoto, N., Barger, A., Aragón-Salamanca, A. 1998 *A&A* **334**, 99–109.
Kodama, T. et al. 2004 *MNRAS* in press, preprint astro-ph/0402276
Mobasher, B. et al. 2003 *ApJ* **587**, 605–618.
Poggianti, B. M. 1997 *A&AS* **122**, 399–407.
Poggianti, B. M., Smail, I., Dressler, A., Couch, W. J., Barger, A. J., Butcher, H., Ellis, R. S., Oemler, A. Jr. 1999 *ApJ* **518**, 576–593.
Rudnick, G. et al. 2003 *The ESO Messenger* **112**, 19–24.
Smail, I., Edge, A. C., Ellis, R. S., Blandford, R. D. 1998 *MNRAS* **293**, 124–144.
Terlevich, A. I., Caldwell, N., Bower, R. G. 2001 *MNRAS* **326**, 1547–1562.
van Dokkum, P. G. & Franx, M. 1996 *MNRAS* **281**, 985–1000.

Outskirts of Galaxy Clusters: Intense Life in the Suburbs
Proceedings IAU Colloquium No. 195, 2004
A. Diaferio, ed.

DOI: 10.1017/S1743921304001036

Age and metallicity estimates from high-resolution galaxy spectra: application to early-type galaxies

Anna Gallazzi[1], S. Charlot[1,2], S. D. M. White[1] and J. Brinchmann[3,1]

[1]Max-Planck-Institut für Astrophysik, Garching bei München, Germany email: gallazzi@mpa-garching.mpg.de

[2]Institut d'Astrophysique du CNRS, Paris, France

[3]Centro de Astrofísica da Universitade do Porto, Porto, Portugal

Abstract. We present a method for deriving light-weighted stellar metallicities and ages from high-resolution galaxy spectra, based on the new population synthesis code of Bruzual & Charlot (2003). The method relies on the simultaneous fit of several optical spectral absorption features that are sensitive to either age or metallicity, but not to the α-elements abundance ratio. We have constructed a library of stochastic star formation histories, which we have used to derive median-likelihood estimates of ages and metallicities for $\sim 10^5$ galaxies extracted from the Sloan Digital Sky Survey Data Release One (SDSS DR1), spanning the full range in star formation activities, from dormant early type to actively star forming. Here we discuss the results for early-type galaxies. We show that the $g-r$, M_r color-magnitude relation for these galaxies is driven primarily by changes in metallicity and in heavy-element abundance ratios. Changes in light-weighted age contribute mainly to the scatter about the relation. This is consistent with previous interpretations of this relation based on lower resolution models.

1. Introduction

The ages and metallicities of the stellar populations in galaxies are direct signatures of the physical processes that were important in their formation and evolution. If we can interpret the light emitted by galaxies in terms of stellar ages and metallicities, we will therefore be able to better constrain the evolution of galaxies and how it depends on environment. In practice, interpretations of galaxy spectra using population synthesis models have been impaired by the fact that age, metallicity and attenuation by dust all have a similar influence on the integrated colors and low-resolution spectra of galaxies (e.g., Worthey 1994). The expectation is that this degeneracy can be broken by appealing to refined spectral diagnostics involving individual stellar absorption-line features (e.g., Vazdekis 1999).

We present here a method for estimating light-weighted stellar metallicities and ages from high-resolution galaxy spectra, based on the new population synthesis code of Bruzual & Charlot (2003) (hereafter BC03). The method consists in comparing the observed strengths of selected optical absorption features with the predictions of models drawn from a comprehensive library of star formation histories. An important requirement of this method is that the spectral absorption features be measured in the same way in the observations and in the model spectra. The method can be applied to all types of galaxies, from early type to actively star forming. We have used this approach to derive light-weighted stellar metallicities and ages for $\sim 10^5$ galaxies extracted from the SDSS-DR1. We present here the results for a subsample of 5343 high signal-to-noise ratio, early-type galaxies in the narrow redshift range $0.02 < z < 0.08$. We focus in particular on the interpretation of one of the main observational relations for elliptical: the color-magnitude relation.

2. High-resolution spectra

2.1. *Models*

To examine the way in which physical parameters such as stellar age and metallicity can be extracted from high-resolution galaxy spectra, we use the recent population synthesis models of Bruzual & Charlot (2003). The models have a resolution of 3 Å across the whole wavelength range from 3200 Å to 9500 Å for a wide range of metallicities. These predictions are based on a newly available library of observed stellar spectra (Le Borgne et al. 2003).

2.2. *Observations*

We analyze a sample of nearly 10^5 galaxies drawn from the SDSS-DR1. The spectra are obtained with a 3" diameter fiber, positioned as close as possible to the center of the target galaxy. The wavelength coverage is from 3800 Å to 9200 Å with a resolving power $\lambda/\Delta\lambda \sim 1800$. The spectroscopic sample is complete to a petrosian r-band magnitude limit $r < 17.77$. To measure stellar absorption features in galaxies with nebular emission, it is important to first remove emission lines. This is performed using the procedure outlined by Tremonti et al. 2004, which fits the emission-line-free region of the spectrum with a set of template spectra (see also Kauffmann et al. 2003 and Brinchmann et al. 2003 for more details).

2.3. *Stellar absorption-line indices*

Previous studies have made extensive use of the Lick-IDS system of absorption line indices. This system includes 25 spectral features across the wavelength range from 4000Å to 6400Å. These are traditionally parametrized as a function of stellar parameters and observed spectra have had to be degraded to match the instrumental response of the Lick-IDS spectrograph (Burstein et al. 1984, Gorgas et al. 1993, Worthey et al. 1994, Trager et al. 1998). Here we improve on these studies by measuring the indices directly on the model and galaxy spectra using the bandpass definitions from Worthey & Ottaviani (1997), hence circumventing the need to degrade the spectra. As described in Bruzual & Charlot (2003), some of these spectral features are not well reproduced by the models. This is likely due to a discrepancy in element abundance ratios between SDSS galaxies and the models, that rely on the spectra of stars with scaled-solar abundances. The influence of variations in abundance ratios on index strengths has been investigated in several studies (Tripicco & Bell 1995, Tantalo et al. 1998, Trager et al. 2000, Vazdekis et al. 2001, Thomas et al. 2003). Due to the approximately constant (solar) abundance ratios of BC03 models, the spectral indices that can be used to constrain the fit are those that do not show any dependence on element abundance ratio. With this purpose, three 'composite' indices that are sensitive to metallicity but not to α/Fe have been measured: the new $[\mathrm{MgFe}]'$ proposed by Thomas et al. (2003), $[\mathrm{Mg_1Fe}]$ and $[\mathrm{Mg_2Fe}]$ as defined in Bruzual & Charlot (2003).

3. Age and metallicity estimates

We adopt a Bayesian statistical approach to derive estimates of stellar metallicity and light-weighted age for the galaxies of our sample. The goal of the Bayesian statistics is to obtain the likelihood distribution of a given parameter X in the space of all possible values of X. This is obtained by comparing the observational data with a set of models that populate the space of all possible X according to a prior distribution, that represents our knowledge about the relative likelihood of the various X values in absence of any data. We generate a library of 150,000 Monte Carlo realizations of a full range of physically plausible star formation histories (SFHs). These are composed of an underlying continuous SFH with superposed random bursts of star formation (following Kauffmann et al. 2003). The models are distributed uniformly in the space of the parameters that define the models; in particular, they are distributed logarithmically in metallicity from 0.02 to 2.5 solar. All stars in a given model have a fixed metallicity, which we interpret as the light-weighted metallicity.

The probability distribution function (PDF) for a given parameter allows us to derive an estimate of the parameter (the mode or median of the distribution) as well as the error associated with this estimate (the range between the 16th and the 84th percentiles). Each model in the library is assigned a likelihood, P=exp($-\chi^2/2$), where χ^2 is calculated by comparing the strength of a set of spectral features measured from the observed spectra with that measured from the model, normalising by the measurement error on the indices. Since the measure of some absorption indices depends also on the galaxy velocity dispersion, only models with velocity dispersion within 15 km/s of the observed one are considered. The absorption indices that can be chosen to constrain the fit have to be well reproduced by BC03 models and they have to be independent of α/Fe ratio (according to Worthey et al 1994 and Thomas et al. 2003), since the models do not have a variable element abundance ratio. Our preferred choice of indices is the set composed of the 4000Å-break, the Balmer lines Hβ, Hδ+Hγ, and the three composite indices [MgFe]$'$, [Mg_1Fe], [Mg_2Fe].

In Fig. 1 we show the PDF of stellar metallicity (left) and r-band weighted age (right) of four galaxies chosen in different positions along the Hδ – $D_n(4000)$ plane (as indicated in the top panels). The results obtained using only $D_n(4000)$ and Balmer lines to constrain the fit are shown in the bottom panels: while ages are already well constrained, including more metallicity-sensitive indices (upper panels) is required to have good estimates of metallicity. In this plot galaxies are ordered from left to right with increasing value of 4000Å-break: this seems to correspond to a sequence also in metallicity and age. Apart from the choice of indices, the quality of the data also determines the results and their accuracy. We have checked that a median S/N ratio per pixel of at least 20 is required to obtain reasonable estimates of metallicity.

4. The color-magnitude relation of early-type galaxies

Our estimates of metallicity and age allow us to explore some of the well-known observational relations between the physical parameters of galaxies. In this work we focus our attention on the relation between color and absolute magnitude for early-type galaxies. We define our sample of early-type galaxies by taking a concentration parameter $R_{90}/R_{50} > 2.6$ (Strateva et al. 2001) and $D_n(4000) > 1.8$, in the redshift range between 0.02 and 0.08. Here we present results for only 5343 galaxies with S/N > 20. The color-magnitude relation has often been interpreted as a relation between metallicity and mass, by looking at the correlation of spectral features such as Mg_2 or CN with luminosity and color (e.g., Faber 1973, Worthey et al. 1994). Fig. 2 shows the (k-corrected) $g - r, M_r$ color-magnitude relation for our sample. In Fig. 2a the relation has been binned and coded reflecting the average stellar metallicity (derived from our analysis) of the galaxies falling into each bin. The same has been done in Fig. 2b to reflect the average α/Fe

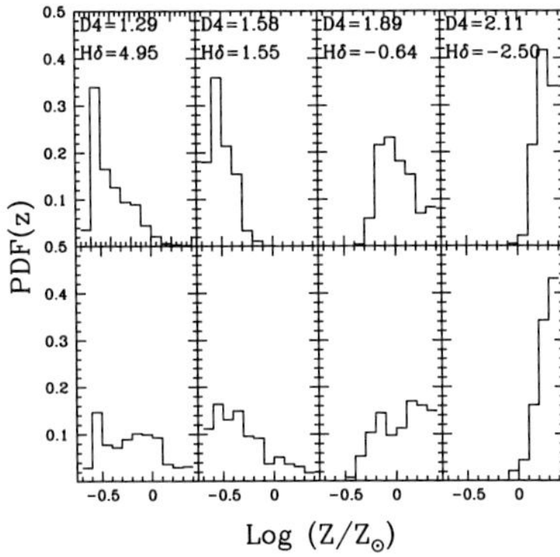

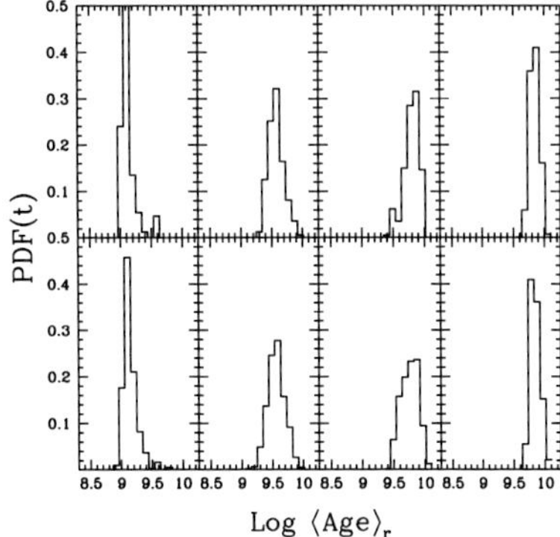

Figure 1. Probability distribution functions of stellar metallicity (left) and r-band weighted age (right) of four galaxies of our sample in different positions along the Hδ-D$_n$(4000) relation (the values of the two indices are indicated in the top panels). Results obtained using only D$_n$(4000) and Balmer lines are shown in the bottom panels, while those obtained by adding Mg+Fe composite indices in the upper panels.

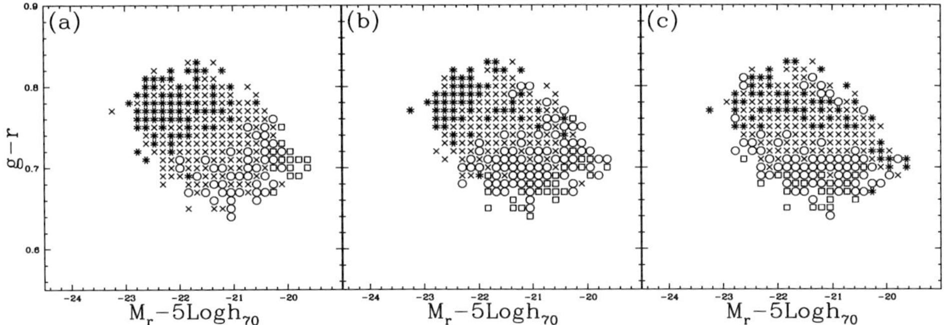

Figure 2. The $g-r$, M_r color-magnitude relation is shown for our sample of early-type galaxies ($R_{90}/R_{50} > 2.6$ and D$_n$(4000) > 1.8) as a function of stellar metallicity (a), $Mg_2/\langle Fe\rangle$ (b) and r-band weighted age (c). The different symbols indicate different average values of the parameter in each bin of absolute magnitude and $g-r$ color. Squares represent $\mathrm{Log}(Z/Z_\odot) \leqslant 0.06$, $Mg_2/\langle Fe\rangle \leqslant 0.08$ and $\mathrm{Log}\langle Age\rangle_r \leqslant 9.75$. Circles represent $0.06 < \mathrm{Log}(Z/Z_\odot) \leqslant 0.15$, $0.08 < Mg_2/\langle Fe\rangle \leqslant 0.09$, $9.75 < \mathrm{Log}\langle Age\rangle_r \leqslant 9.8$. Crosses represent $0.15 < \mathrm{Log}(Z/Z_\odot) \leqslant 0.21$, $0.09 < Mg_2/\langle Fe\rangle \leqslant 0.1$, $9.8 < \mathrm{Log}\langle Age\rangle_r \leqslant 9.83$. Stars represent $\mathrm{Log}(Z/Z_\odot) > 0.21$, $Mg_2/\langle Fe\rangle > 0.1$, $\mathrm{Log}\langle Age\rangle_r > 9.83$.

ratio, as indicated by the $Mg_2/\langle Fe\rangle$ index (where $\langle Fe\rangle$ is the average of Fe5270 and Fe5335 index strengths), and in Fig. 2c to reflect the r-band light-weighted age. There is a marked gradient in metallicity along the relation, with fainter galaxies being more metal-poor than brighter galaxies. In addition the $Mg_2/\langle Fe\rangle$ index ratio also increases with increasing luminosity and color. It appears that, therefore, although stellar metallicity is responsible in generating the observational relation between color and luminosity, element abundance ratio is another important controlling parameter. On the other hand, the light-weighted age of these early-type galaxies varies over a narrow range of about 2 Gyr, with a gradient almost parallel to the color axis. This difference in light-weighted age contributes to the scatter about the observed relation.

Acknowledgements

AG and SC thank the Alexander von Humboldt Foundation, the Federal Ministry of Education and Research and the Programme for Investment in the Future (ZIP) of the German Government for financial support.

JB acknowledges the support of an ESA post-doctoral fellowship. Funding for the creation and distribution of the SDSS Archive has been provided by the Alfred P. Sloan Foundation, the Participating Institutions, the National Aeronautics and Space Administration, the National Science Foundation, the U.S. Department of Energy, the Japanese Monbukagakusho, and the Max Planck Society. The SDSS Web site is http://www.sdss.org/.

References

Brinchmann, J., Charlot, S., White, S. D. M., Tremonti, C., Kauffmann, G., Heckman, T. & Brinkmann, J. 2003 *astro-ph/0311060*

Bruzual, G., & Charlot, S. 2003 *MNRAS* **344**, 1000–1028

Burstein, D., Faber, S. M., Gaskell, C. M., & Krumm, N. 1984 *ApJ* **287**, 586–609

Faber, S. M. 1973 *ApJ* **179**, 731–754

Gorgas, J., Faber, S. M., Burstein, D., Gonzalez, J. J., Courteau, S., & Prosser, C. 1993 *ApJS* **86**, 153–198

Kauffmann, G. et al. 2003 *MNRAS* **341**, 33–53

Le Borgne, J.-F. et al. 2003 *A&A* **402**, 433–442

Schlegel, D. J., Finkbeiner, D. P., & Davis, M. 1998 *ApJ* **500**, 525–553

Strateva, I. et al. 2001 *AJ* **122**, 1861–1874

Tantalo, R., Chiosi, C., & Bressan, A. 1998 *A&A* **333**, 419–432

Thomas, D., Maraston, C., & Bender, R. 2003 *MNRAS* **339**, 897–911

Trager, S. C., Worthey, G., Faber, S. M., Burstein, D., & Gonzalez, J. J. 1998 *ApJS* **116**, 1–28

Trager, S. C., Faber, S. M., Worthey, G., & González, J. J. 2000 *AJ* **119**, 1645–1676

Tremonti, C. A., Heckman, T. M., Kauffmann, G., Charlot, S., Brinchmann, J. 2004 *submitted to ApJ*

Tripicco, M. J. & Bell, R. A. 1995 *AJ* **110**, 3035–3049

Vazdekis, A. 1999 *ApJ* **513**, 224–241

Vazdekis, A., Salaris, M., Arimoto, N., & Rose, J. A. 2001 *ApJ* **549**, 274–280

Worthey, G., Faber, S. M., Gonzalez, J. J., & Burstein, D. 1994 *ApJS* **94**, 687–722

Worthey, G. 1994 *ApJS* **95**, 107–149

Worthey, G. & Ottaviani, D. L. 1997 *ApJS* **111**, 377–386

Outskirts of Galaxy Clusters: Intense Life in the Suburbs
Proceedings IAU Colloquium No. 195, 2004
A. Diaferio, ed.

DOI: 10.1017/S1743921304001048

E+A galaxies in intermediate redshift clusters

Kim-Vy H. Tran[1], Marijn Franx[2], Garth D. Illingworth[3], Daniel D. Kelson[4] and Pieter van Dokkum[5]

[1]Institute for Astronomy, ETH Hönggerberg, CH-8093 Zürich, Switzerland

[2]Leiden Observatory, P.O. Box 9513, 2300 RA Leiden, The Netherlands

[3]University of California Observatories/Lick Observatory, University of California, Santa Cruz, CA 95064, USA

[4]Observatories of the Carnegie Institution of Washington, 813 Santa Barbara Street, Pasadena, CA, 91101, USA

[5]Department of Astronomy, Yale University, New Haven, CT 06520-8101, USA

Abstract. Using extensive ground-based spectroscopy, we isolate the E+A population in three intermediate redshift clusters (z = 0.33, 0.58 and 0.83) and study their physical properties using HST/WFPC2 imaging. Our analysis includes galaxy colors, luminosities, Hubble types, and quantitative structural parameters as well as measured and estimated internal velocity dispersions. We find E+A galaxies make up a non-negligible fraction ($\sim 7 - 13\%$) of cluster members at these redshifts, and their diverse nature indicates a heterogeneous parent population. From their velocity dispersions and half-light radii, we infer that the descendants of the E+A's in our highest redshift clusters are massive early-type galaxies, and we estimate that $\gtrsim 30\%$ of the E-S0 members have undergone an E+A phase. We also find the characteristic E+A mass decreases with decreasing redshift; this is similar to the decrease in luminosity of rapidly star-forming field galaxies since $z \sim 1$, i.e. galaxy "down-sizing".

1. Introduction

Post-starburst galaxies ("E+A"; Dressler & Gunn 1983) in clusters may provide the crucial link in the morphological transformation of spiral galaxies into the elliptical/S0 systems that dominate the cluster population at the current epoch (see Butcher & Oemler 1978). Characterized by strong Balmer absorption and little or no [OII]λ3727 emission, E+A's also are referred to as k+a/a+k (Franx 1993; Dressler et al. 1999, hereafter D99) or Hδ-strong galaxies (Couch & Sharples 1987). Despite the short window of visibility of the post-starburst phase (< 1.5 Gyr; Couch & Sharples 1987; Barger et al. 1996), E+A members can contribute up to 20% of the total cluster population (D99), and have been found in virtually every spectroscopic cluster survey from $0 < z < 1.3$.

Here we summarize results presented in Tran et al. (2003b) on the E+A population in three intermediate redshift clusters: CL 1358+62 ($z = 0.33$), MS 2053–04 ($z = 0.59$), and MS 1054–03 ($z = 0.83$). We have completed extensive spectroscopic surveys (> 130 confirmed members in *each* field) and obtained HST/WFPC2 mosaics of the three clusters ($R \sim 1h^{-1}$ Mpc) to characterize their galaxy populations. By pairing wide-field HST/WFPC2 imaging with deep ground-based spectroscopy, we can study in detail the nature of E+A galaxies in clusters at $z > 0.3$.

First we isolate the E+A galaxies in the three clusters to measure the fraction of members that are also E+A's. This has important ramifications on whether the *majority* of cluster members undergo an E+A phase or if only a small fraction do. Because our analysis includes colors, magnitudes, half-light radii, bulge-to-total fractions, degree of galaxy

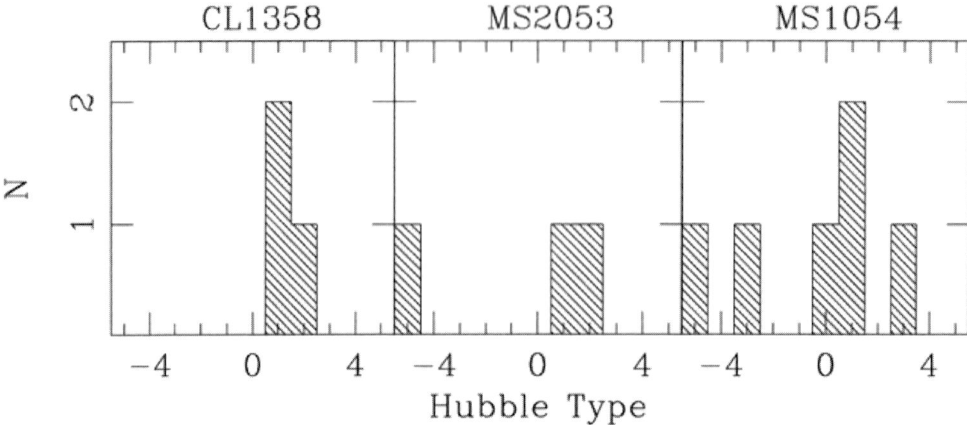

Figure 1. The Hubble types of the E+A galaxies in each cluster; we consider only the E+A's brighter than $M_{Be} = -19.1 + 5\log h$. The morphological types of {E, E/S0, S0, S0/Sa, Sa, Sb, Sc} are assigned values of $\{-5, -4, -2, 0, 1, 3, 5\}$; intermediate values are also used. We do not show the two E+A mergers in MS1054. The E+A galaxies span the range in morphological type to include even early-type members (Type< -2). However, cluster E+A's tend to be systems with disks (Type$\geqslant 1$).

asymmetry, Hubble types, and internal velocity dispersions (measured and estimated), we can determine whether E+A galaxies can include even the most massive members, and if they have a heterogeneous parent population. In addition, we can use their sizes and internal velocity dispersions (σ) to determine what their descendants will be. By analyzing clusters at three different redshifts, we can also determine whether the luminosity and internal velocity dispersion distributions of the cluster E+A's evolves with redshift. We use $H_0 = 100$ km s^{-1} Mpc^{-1}, $\Omega_M = 0.3$, and $\Lambda = 0.7$.

2. Results

From over 1200 redshifts obtained in the three cluster fields, we isolate ~ 500 members. We select E+A galaxies from our cluster sample as members having strong Balmer absorption [(Hδ+Hγ)/2 $\geqslant$ 4Å] and no [OII]λ3727 emission (> -5 Å). We also apply a signal to noise cut ($S/N \geqslant 20$) on the Hδ and Hγ fluxes and use a magnitude limit ($M_{Be} \leqslant -19.1 + \log h$) to ensure we are not including any passive galaxies that are scattered into the E+A regime due to observational errors. For uniformity, we consider only members brighter than this cut when comparing the E+A population between the clusters. We find 14 cluster galaxies that satisfy our strict E+A selection criteria.

2.1. *E+A fraction*

For members brighter than $M_{Be} = -19.1 + 5\log h$, we obtain E+A fractions of $7 \pm 4\%$, $10 \pm 6\%$, and 13±5% at z = 0.33, 0.58, & 0.83 respectively; errors are determined by assuming a Poisson distribution for the E+A galaxies. Considering the low E+A fraction in Coma (Caldwell et al. 1993), these results show the E+A fraction evolves strongly with redshift. They also suggest the E+A fraction continues to increase at $z > 0.5$. However, larger samples at $z > 0.3$ are needed to determine if this increase is real or if the trend flattens at $z > 0.3$.

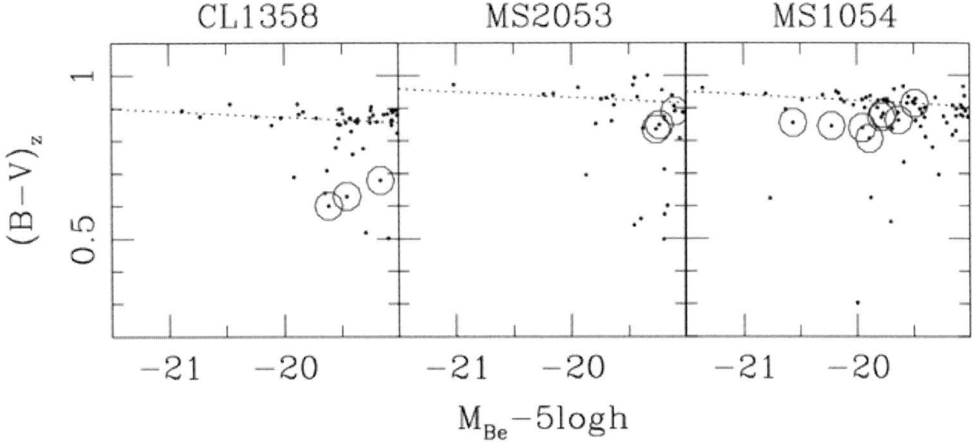

Figure 2. The color-magnitude diagram for all cluster members brighter than $M_{Be} = -19.1 + 5\log h$; the cluster members that are E+A's are circled. The color-magnitude relation (dotted line) is normalized to the E/S0 members and its slope is from van Dokkum et al. (2000). E+A galaxies in the two lower redshift clusters are faint ($M_{Be} \gtrsim -19.5 + 5\log h$) but they can be up to a magnitude brighter in MS1054 ($z = 0.83$).

2.2. *Morphology*

The cluster E+A's in our sample span the range in Hubble type (Fig. 1) and bulge-to-total fraction (B/T; see Tran et al. 2001, 2003a) to include both spirals and E/S0's. However, the majority of E+A's have measurable disks, consistent with results from previous cluster E+A studies (Wirth & Koo 1994; Couch et al. 1994, 1998; Caldwell et al. 1999; D99). Their average bulge-to-total fraction of ~ 0.4 reflects their tendency to be disk-dominated systems.

The observed range in B/T and Hubble type of the E+A population is similar to the heterogeneous morphologies found in studies of lower redshift cluster E+A's ($z \lesssim 0.3$). It also confirms suggestions in earlier studies that E+A's must have a wide variety of progenitors (Zabludoff et al. 1996; D99). We also note that the earliest-type E+A's are in our most distant cluster; it may be that more massive cluster members, i.e. early-types, had their E+A phase at higher redshift.

2.3. *Color-magnitude diagram*

In Fig. 2, we show the color-magnitude (CM) distribution of all cluster galaxies and E+A candidates brighter than our magnitude limit of $M_{Be} = -19.1 + 5\log h$; all members, including the E+A candidates, have been corrected for simple passive evolution (see van Dokkum et al. 1998). In all three clusters, we find bright E+A galaxies ($M_{Be} \lesssim -19.1+5\log h$). Even more striking are the very luminous E+A's at $z = 0.83$: these E+A's are up to a magnitude brighter than their lower redshift counterparts and cover a larger magnitude range. Note the cluster E+A's tend to be bluer than the red sequence. The E+A luminosity range, particularly at $z = 0.83$, indicates that they have a heterogeneous parent population. The fact that the brightest E+A's in this sample are in our most distant cluster suggests that the cluster E+A population evolves with redshift.

2.4. *Internal velocity dispersions*

Having demonstrated that E+A's at higher redshift can be as luminous as the brightest cluster members, we now determine if these brighter E+A's are also *massive* galaxies,

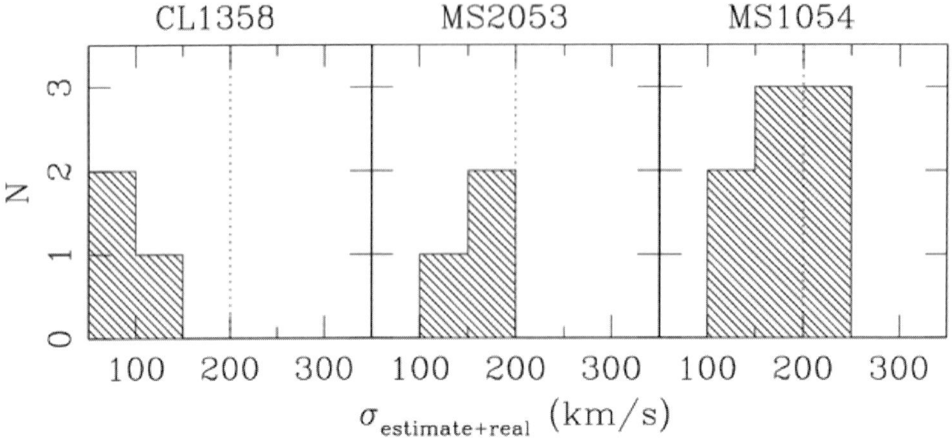

Figure 3. Histogram of internal velocity dispersions (measured and estimated) for E+A members brighter than $M_{Be} = -19.1{+}5\log h$. The characteristic E+A mass (as traced by σ) increases with redshift to include even E+A galaxies with $\sigma > 200$ km s^{-1} in MS1054 ($z = 0.83$). This trend is similar to the observed decrease, since $z \sim 1$, in the maximum luminosity of field galaxies undergoing rapid star formation ("down-sizing"; Cowie et al. 1996).

or whether they are simply low luminosity/mass members that are temporarily brightened. By determining the E+A mass distribution, we can characterize what the E+A progenitors are and also constrain what their descendants at lower redshift can be. We use internal velocity dispersion as a tracer of the mass where we have obtained direct σ measurements for 120 cluster members (Kelson et al. 2000a, 2001; van Dokkum et al. 1998), and we estimate σ for the rest of the sample using the Fundamental Plane (see Kelson et al. 2000b).

Figure 3 shows the distribution of internal velocity dispersions (measured and estimated σ) for cluster members brighter than $M_{Be} = -19.1{+}5\log h$. The range in velocity dispersion for E+A galaxies increases at higher redshift: E+A's at $z = 0.33$ have smaller velocity dispersions ($\sigma \leqslant 150$ km s^{-1}) than at $z = 0.58$ ($\sigma \leqslant 200$ km s^{-1}) and $z = 0.83$ ($\sigma \leqslant 250$ km s^{-1}). Considering the robustness of the spectroscopic data and high quality of the WFPC2 imaging, any E+A's with $\sigma > 200$ km s^{-1} (measured or estimated) in the two lower redshift clusters would have been easily detected *if they existed.*

We find that the half-light radii (~sizes) and dispersions of E+A's at $z = 0.83$ are comparable to those of the massive early-type members in all three clusters. In contrast, none of the E+A's at $z = 0.33$ could be considered a cluster giant. It is apparent that the progenitors of the lower redshift E+A's are very different from those at $z = 0.83$. The observed trend of decreasing internal velocity dispersion with decreasing redshift is consistent with the "down-sizing" of the E+A population.

3. Conclusions

From a sample of ~ 500 spectroscopically confirmed cluster members in three clusters at $z = 0.33$, 0.59 and 0.83, we isolate 14 E+A galaxies and find E+A's make up a non-negligible component of the cluster population ($\sim 7-13\%$) at intermediate redshifts. Although most of them are disk-dominated systems, E+A's span the range in morphological type to include even early-type members. They can be more luminous than L^* and can have internal velocity dispersions in excess of 200 km s^{-1}. We find the E+A's with the highest internal velocity dispersions and luminosities are in our most distant cluster. This galaxy "down-sizing" is similar to the observed decrease in luminosity of rapidly star-forming field galaxies since $z \sim 1$ (Cowie et al. 1996).

Our study indicates the high dispersion ($\sigma > 200$ km s^{-1}) E+A's at $z = 0.83$ are the logical progenitors of massive early-types at lower redshift. We estimate $\geqslant 30\%$ of cluster E-S0's at $z = 0.83$ have had an E+A phase. We consider this a lower limit as evolution in the E+A fraction and characteristic mass as well as the conversion of spirals into early-types can increase the true fraction to 100%. These results show that the E+A phase can play an important role in the transformation of star-forming galaxies into early-type members.

Acknowledgements

K. Tran gratefully acknowledges support from the Swiss National Science Foundation.

References

Barger, A. J., Aragon-Salamanca, A., Ellis, R. S., Couch, W. J., Smail, I., & Sharples, R. M. 1996, MNRAS, 279, 1

Butcher, H. & Oemler, A. 1978, ApJ, 219, 18

Caldwell, N., Rose, J. A., Sharples, R. M., Ellis, R. S., & Bower, R. G. 1993, AJ, 106, 473

Caldwell, N., Rose, J. A., & Dendy, K. 1999, AJ, 117, 140

Couch, W. J., Barger, A. J., Smail, I., Ellis, R. S., & Sharples, R. M. 1998, ApJ, 497, 188

Couch, W. J., Ellis, R. S., Sharples, R. M., & Smail, I. 1994, ApJ, 430, 121

Couch, W. J. & Sharples, R. M. 1987, MNRAS, 229, 423

Cowie, L. L., Songaila, A., Hu, E. M., & Cohen, J. G. 1996, AJ, 112, 839

Dressler, A. & Gunn, J. E. 1983, ApJ, 270, 7

Dressler, A., Smail, I., Poggianti, B. M., Butcher, H., Couch, W. J., Ellis, R. S., & Oemler, A. J. 1999, ApJS, 122, 51

Franx, M. 1993, PASP, 105, 1058

Kelson, D. D., van Dokkum, P. G., Franx, M., Illingworth, G. D., & Fabricant, D. 1997, ApJ, 478, L13

Kelson, D. D., Illingworth, G. D., van Dokkum, P. G., & Franx, M. 2000, ApJ, 531, 159

—. 2000, ApJ, 531, 184

Kelson, D. D., Illingworth, G. D., Franx, M., & van Dokkum, P. G. 2001, ApJ, 552, L17

Poggianti, B. M., Smail, I., Dressler, A., Couch, W. J., Barger, A. J., Butcher, H., Ellis, R. S., & Oemler, A. 1999, ApJ, 518, 576

Tran, K. H., Simard, L., Zabludoff, A. I., & Mulchaey, J. S. 2001, ApJ, 549, 172

Tran, K. H., Simard, L., Illingworth, G. D. I., & Franx, M. 2003, ApJ, 590, 238

Tran, K. H., Franx, M., Illingworth, G., Kelson, D. D., & van Dokkum, P. 2003, ApJ, 599, 865

van Dokkum, P. G., Franx, M., Kelson, D. D., & Illingworth, G. D. 1998, ApJ, 504, L17

van Dokkum, P. G., Franx, M., Fabricant, D., Illingworth, G. D., & Kelson, D. D. 2000, ApJ, 541, 95

Wirth, G. D., Koo, D. C., & Kron, R. G. 1994, ApJ, 435, L105

Zabludoff, A. I., Zaritsky, D., Lin, H., Tucker, D., Hashimoto, Y., Shectman, S. A., Oemler, A., & Kirshner, R. P. 1996, ApJ, 466, 104

Outskirts of Galaxy Clusters: Intense Life in the Suburbs
Proceedings IAU Colloquium No. 195, 2004
A. Diaferio, ed.

DOI: 10.1017/S174392130400105X

The stellar populations of low-redshift clusters

Craig Harrison[1], Matthew Colless[2], Warrick J. Couch[3] and B. A. Peterson[1]

[1]Mount Stromlo Observatory, The Australian National University, Cotter Road, Weston, ACT 2611, Australia email: harrison@mso.anu.edu.au, peterson@mso.anu.edu.au

[2]Anglo-Australian Observatory, PO Box 296, Epping, NSW 1710, Australia
email: colless@aao.gov.au

[3]School of Physics, University of New South Wales, Sydney NSW 2052, Australia
email: wjc@phys.unsw.edu.au

Abstract. We present some preliminary results from an on-going study of the evolution of stellar populations in rich clusters of galaxies. Our baseline sample contains core line-strength measurements from 183 galaxies with $b_J \leqslant 19.5$ from four clusters with $\bar{z} \sim 0.04$, against which observations of higher-redshift clusters can be compared. Using predictions from stellar population models to compare with our measured line strengths we can derive *relative* luminosity-weighted mean ages and metallicities for the stellar populations in each of our clusters. It must be stressed that these ages and metallicities are only accurate when used in a relative sense as the stellar population models, due to differing abundance ratios used in the models compared to those observed in elliptical galaxies, provide inaccurate absolute ages and metallicities. We also investigate the Mgb'-σ_0 and Hβ'_G-σ_0 scaling relations. We find that Mgb' is correlated with σ_0, the likely explanation being that larger galaxies are better at retaining their heavier elements due to their larger potentials. Hβ'_G, on the other hand, we find to be anti-correlated with σ_0. This result implies that the stellar populations in larger galaxies are older than in smaller galaxies.

1. Introduction and observations

The formation and evolution of clusters of galaxies and of their early-type galaxy population is a topic of active debate. Studying the integrated light from stellar populations can help to discriminate between models of giant elliptical galaxy formation: the monolithic collapse model (Eggen et al. 1962), in which giant elliptical galaxies form rapidly in a process that can essentially be considered a single collapse, and the hierarchical merging model (Searle & Zinn 1978), in which giant elliptical galaxies are built up over a long timescale by distinct merger events. If it was found that early-types were largely coeval in clusters, then this would favour the monolithic collapse model, whereas a large spread in ages would favour hierarchical merging.

Increasing a galaxy's age by a factor of two while decreasing it's metallicity by a factor of three results in a spectrum almost indistinguishable from the original (Worthey 1994). This age/metallicity degeneracy can be partially broken by comparing age-sensitive spectroscopic indices with metallicity-sensitive spectroscopic indices. By plotting two such indices against each other and comparing the results to stellar population models (e.g. Worthey 1994; Thomas et al. 2003) we can determine the mean relative luminosity-weighted age and metallicity of a given stellar population.

Previous work in this area has concentrated on individual clusters such as Coma (Jørgensen 1999; Moore et al. 2002) and Fornax (Kuntschner 2000) with sometimes

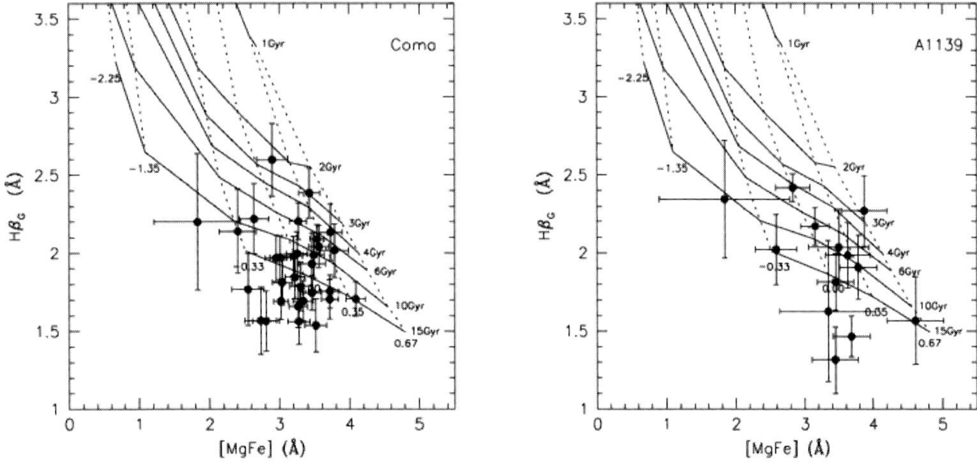

Figure 1. $H\beta_G$ equivalent width versus [MgFe] equivalent width. Thomas et al. (2003) stellar population model grids have been overlaid on data from the Coma cluster (*left*) and Abell 1139 (*right*).

conflicting results coming from the same cluster (e.g. Gonzalez et al. 1993 and Moore et al. 2002). Coma and Fornax represent the extremes of the cluster richness scale and this project will study clusters that fill in this range. We will study, in detail, the stellar populations in several low-redshift ($z \sim 0.04$) clusters from the core out into the surrounding structures. These results will then be used as a comparison for observations of a range of higher-redshift clusters ($0.3 \leqslant z \leqslant 0.55$), allowing us to draw conclusions about the formation and evolution of the stellar populations in rich clusters.

Line-strength measurements, redshifts and velocity dispersions were obtained for 183 galaxies from four clusters over three nights in April 2002 with the 2dF system on the 3.9m AAT. The sample was selected from galaxies with $b_J \leqslant 19.5$, within a 2° diameter FOV centred on each of the clusters and within a 3σ redshift range centred on the cluster redshift. The 300B grating ($\sim$ 9Å FWHM) was used to obtain a wide spectral range covering as many indices as possible, while the 1200V grating ($\sim$ 2Å FWHM) was used to obtain higher resolution spectra yielding precise velocity dispersions.

2. Ages and metallicities

Comparison of the measured equivalent widths of $H\beta_G$ and [MgFe] for two of our clusters, Coma and A1139, and predictions from the Thomas et al. (2003) stellar population models are shown in Fig. 1. Using these models we obtain relative ages and metallicities for our stellar populations. We find that Coma cluster galaxies have a mean age of 9.3 ± 3.8 Gyr with a mean metallicity of 0.15 ± 0.23 dex relative to solar. A1139, we find, has a mean age of 10.3 ± 3.4 Gyr with a mean metallicity of 0.12 ± 0.41 dex relative to solar. Refinements still need to be made to the data set and the other clusters need to be analysed before any conclusions can be drawn. The age/metallicity degeneracy is quite evident from these plots. Also, there seems to be some disagreement between our observations and the models. The data points with low $H\beta_G$ equivalent widths could be star-forming galaxies, which we wish to exclude from this part of the analysis. By examining each galaxy's spectra for emission lines indicative of a star-forming galaxy and by measuring $H\alpha$ equivalent widths we can eliminate these galaxies from our sample.

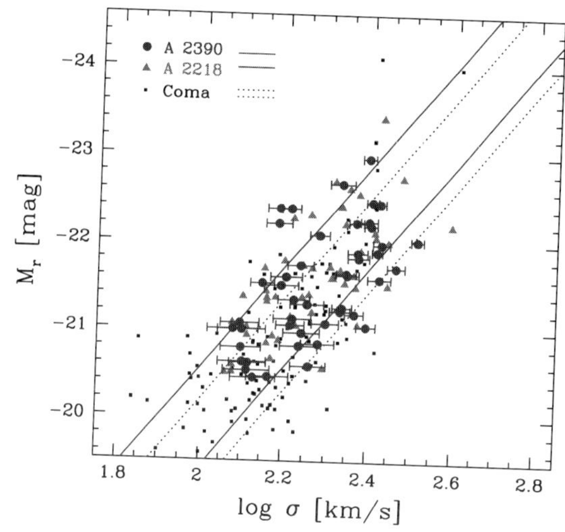

Figure 1. Gunn r Faber-Jackson relation for 94 early-type cluster galaxies in A 2390 and A 2218, compared to the Coma sample (JFK95). The Coma galaxies were restricted to $M_r < -19.5$ and $\log \sigma > 2.0$. The area bounded by dotted lines indicates the mean $\pm 1\ \sigma$ of the local FJ relation using a linear bisector fit, solid lines the respective fits to the distant sample within $\pm 1\ \sigma$.

2. Observations and data reduction for Abell 2390

During two observing runs (09/1999 and 07/2000) we observed three masks using the MOSCA spectrograph at the Calar Alto 3.5 m telescope with total exposure times between ∼8 and ∼12 hours each. In total, we obtained 63 high S/N spectra of 50 different early-type galaxies, out of which 15 are situated within the *HST* field. Only one object was revealed to be a background galaxy, which shows that our sample selection was highly efficient. The instrumental resolution in the spectral regime of Hβ and Mg b was 5.5 Å FWHM ($\sigma_{\rm inst} \sim 100$ km s^{-1}). The average S/N is ~ 41.

The reduction of the spectra was carried out using standard reduction techniques and velocity dispersions were calculated with the FCQ method (Bender 1990). Structural parameters were determined by fitting the surface brightness profile with either an $r^{1/4}$ or a combination of an $r^{1/4}$ plus an exponential law profile (Saglia et al. 1997). Further details concerning the reduction and analysis are outlined in Fritz et al. (2004).

3. A detailed Fundamental Plane of E+S0 cluster galaxies at $z \sim 0.2$

In Fig. 1 the Faber-Jackson (FJ) relation for the complete sample of 94 E+S0 galaxies is shown. As a local reference we use the Coma sample of Jørgensen et al. 1995 (JFK95). Overall, the luminosity evolution is small ($\overline{m}_r \sim 0.32$ mag). The Fundamental Plane (FP) in rest frame Gunn r is shown in Fig. 2. The sub-sample with *HST* structural parameters comprises 34 E+S0 galaxies, splitted into 17 ellipticals, 2 E/S0, 8 S0, 3 SB0/a, 3 Sa and 1 Sab galaxy that enter the FP. With this large sample, possible differences of the galaxies' properties can be explored for various sub-populations. Both clusters have a similar behaviour within and along the FP, with no hint for a slope change. For the 34 cluster E+S0 galaxies we find a combined evolution of $\overline{m}_r \sim 0.43$ mag. Lenticular galaxies (E/S0, S0, SB0/a, Sa, Sab) have a zeropoint offset of 0.30 mag with respect to Coma and seem to inhabit a certain band within the FP (Fig. 2, right). Ellipticals exhibit a similar mild evolution of $\overline{m}_r \sim 0.46$. Both sub-samples have a small scatter (∼0.10 mag).

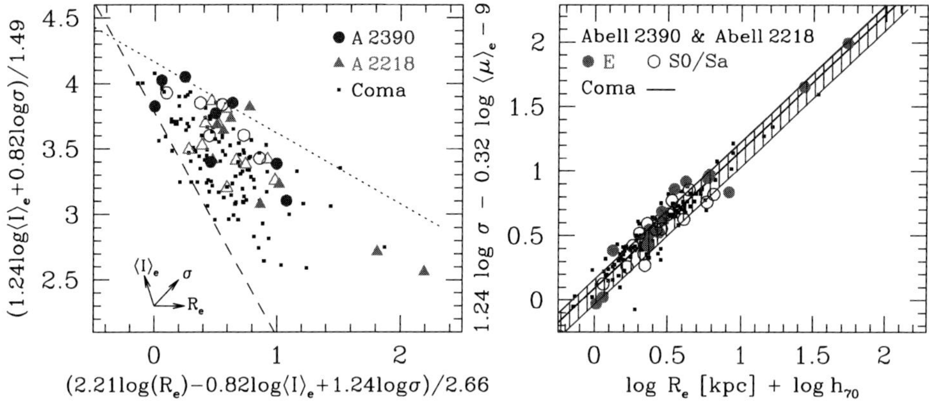

Figure 2. Fundamental Plane for A 2390 and A 2218 in Gunn r. *Left*: Face-on view. The dotted line indicates the exclusion zone for nearby galaxies, the dashed line the L limit for Coma. Filled symbols represent ellipticals, open symbols S0 galaxies. *Right*: Edge-on FP for the combined cluster sample, with a separation into morphologies. The solid line marks the Coma relation of JFK95. Fits along $1-\sigma$ errors for the S0 galaxies assuming a constant slope are indicated (hatched area).

4. Summary

We construct a spectroscopic sample of 48 E+S0 in the rich cluster Abell 2218 ($z = 0.18$) and 50 E+S0 in Abell 2390 ($z = 0.23$). Since both cluster samples are very similar, we combine them and analyse a total number of 98 E+S0 at $z \sim 0.2$. For a subsample of 34 galaxies we explore the evolution of the FP and study their M/L ratios. On average, all 34 cluster E+S0 galaxies show a mild evolution of $\Delta\log(M/L_r) = -0.17$. S0 galaxies show lower M/L ratios of $\Delta\log(M/L_r)_{\rm S0} = -0.12$ with respect to the local reference. For the ellipticals we again deduce a moderate evolution of $\Delta\log(M/L_r)_{\rm E} = -0.18$. Elliptical and S0 galaxies seem to comprise similar sub-populations with no significant differences. An analysis based on the M/L ratios revealed a mean formation redshift of $z_f \sim 3$, which is consistent with stellar population models of passive evolution.

Acknowledgements

This project is a collaboration with R. G. Bower and I. Smail (Durham/UK) and R. L. Davies (Oxford/UK) who have contributed to these results. We thank the Calar Alto staff for efficient observational support. AF and BLZ acknowledge financial support by the Volkswagen Foundation (I/76 520) and the DFG (ZI 663/1-1, ZI 663/2-1).

References

Bender, R. 1990 *A&A* **229**, 441–451.
Bender, R., Burstein, D., Faber, S. M. 1993, *ApJ* **411**, 153–169.
Bower, R. G., Lucey, J. R., Ellis, R. S. 1992 *MNRAS* **254**, 601–613.
Djorgovski, S., Davis, M. 1987 *ApJ* **313**, 59–68.
Fritz, A., Ziegler, B. L., Bower, R. G. et al. 2004 *MNRAS*, submitted.
Jørgensen, I., Franx, M., Kjærgaard, P. 1995 *MNRAS* **273**, 1097–1128. (JFK95)
Saglia, R. P., Bertschinger, E., Baggley, G. et al. 1997 *ApJS* **109**, 79–102.
Ziegler, B. L., Bower, R. G., Smail, I. et al. 2001 *MNRAS* **325**, 1571–1590.

Outskirts of Galaxy Clusters: Intense Life in the Suburbs
Proceedings IAU Colloquium No. 195, 2004
A. Diaferio, ed.

DOI: 10.1017/S1743921304001097

Evolutionary status of early–type galaxies in distant poor clusters

Alexander Fritz and Bodo L. Ziegler

[1]Universitäts–Sternwarte Göttingen, Geismarlandstraße 11, 37083 Göttingen, Germany email: afritz@uni-sw.gwdg.de

Abstract. We introduce our project that investigates the kinematic properties of early–type galaxies in 6 distant poor clusters at $z \approx 0.25$. This study represents a continuation of our efforts to understand galaxy evolution in low-density environments. Higher–resolution MOSCA spectra have been obtained at the Calar Alto 3.5-m telescope with which we can measure absorption line strengths and velocity dispersions. In conjunction with our *HST/F702W* images of all the clusters, we are able to construct the Fundamental Plane of ellipticals and S0 galaxies in poor clusters at a look-back time of ≈ 3 Gyr. For galaxies outside the *HST* field, we concentrate our analysis on the Mg–σ and Faber–Jackson relations. With the line strength diagrams age/metallicity distributions can be derived in densities between the field and rich cluster environments. Comparing with our rich clusters at the same cosmic epochs, the dependence of galaxy formation models on the local environment can be tested more quantitatively.

1. Introduction

Recent analyses of the kinematics and stellar populations of faint distant galaxies up to redshifts of $z \approx 1.0$ have put strong observational constraints on galaxy formation models. The weak evolution seen in the tight correlation of the distance independent parameters Mg–absorption and velocity dispersion, σ, at $z \approx 0.4$ is evidence for a high redshift ($2 < z_f < 4$) of formation for the bulk of the stars in cluster ellipticals (e.g., Ziegler & Bender 1997). Further support gives the small increase of the B-band luminosity derived from the Faber–Jackson relation (L *vs.* σ). The combination of structural (size and surface brightness) and kinematic (σ) parameters in the Fundamental Plane (Djorgovski & Davis 1987) allows us to investigate not only the luminosity but also the mass evolution of early–type galaxies. Most previous studies of the Fundamental Plane (FP) at $0.2 < z \leqslant 1.3$ revealed that the mass-to-light ratio of luminous galaxies evolves only mild in accordance with passive evolution models (Ziegler et al. 2001; van Dokkum & Stanford 2003; see also contribution by Fritz & Ziegler in this volume). However, these observational results apply only to the special environment of dense rich clusters.

Semi-analytic CDM models predict a hierarchical assembly for galaxies through merging of subunits and replenishment of gas from their halos (Cole et al. 2000). Only in the richest clusters galaxies experienced their last major merger at $z \sim 1 - 2$. Thus, the population of early-type galaxies should be more diverse in lower density regions.

For this reason, we analyse the evolutionary status of early–type (E+S0) galaxies in poor clusters at a look-back time of ≈ 3 Gyr with the same quantitative methods as was done in rich clusters. The Mg–σ relation will reveal whether there is more spread in the age/metallicity distribution than in rich clusters at similar epochs, e.g., in Abell 2218 (Ziegler et al. 2001) or Abell 2390 (Fritz et al. 2004). Results on the FP will give insights into whether the mass/structure evolution is more consistent with the monolithic or the hierarchical formation scenario. We use $\Omega_m = 0.3$, $\Omega_\Lambda = 0.7$ and $h = 0.7$.

Table 1. Sample of six poor clusters.

Cluster	R.A. (J2000)	Dec. (J2000)	$< z >$	L_X (0.1–2.4 keV) $10^{43} h_{50}^{-2}$ ergs s^{-1}	$N_{\rm obj/mask}$	$T_{\rm exp}$ (ksec)
Cl 0818+56	08 19 04	+56 54 49	0.2670	1.50	14	29.40
Cl 0841+70	08 41 44	+70 46 53	0.2397	1.22	13	24.54
Cl 0849+37	08 49 11	+37 31 09	0.2343	1.93	17	18.00
Cl 1633+57	16 33 42	+57 14 12	0.2402	0.49	14	30.97
Cl 1701+64	17 01 47	+64 20 57	0.2458	0.40	18	27.90
Cl 1702+64	17 02 14	+64 19 53	0.2233	0.74	18	27.90

2. The low-L_X project

This investigation is an intensive observational campaign to study the evolutionary status of galaxies in 10 poor clusters at intermediate redshifts of $0.2 < z < 0.3$ selected to have very low X-ray luminosities ($L_X < 5 \times 10^{43}$ erg/s), 100 times lower than the CNOC clusters, and poor optical richness class. It comprises ground-based optical and NIR photometry, medium-resolution multi-object spectroscopy, and *HST/F702W* imaging.

The clusters are dominated by passive galaxies with no or little on-going star formation (SF). The fraction of emission-line galaxies (ELG) is as low as in the rich X–ray bright CNOC clusters at similar epochs ($f_{\rm ELG} \approx 0.2$) and at variance with the field value ($f_{\rm ELG} \approx 0.6$). This indicates that processes heavily depending on density like ram–pressure stripping can not be the dominant mechanisms responsible for the supression of SF (Balogh et al. 2002b). Analysis of the *HST* images for all ten clusters revealed that the overall properties of these cluster galaxies are similar to those in more massive clusters with a low fraction of disk galaxies (Balogh et al. 2002a).

3. Sample selection for spectroscopy

The spectroscopic observations have been gained during eleven nights (04/2001 and 02/2002) with the MOSCA spectrograph at the Calar Alto Observatory (Spain). A summary of the cluster sample with the mean redshifts, number of objects on a mask and total exposure times for spectroscopy is given in Tab. 1. Early-type cluster members have been selected based on our low-resolution spectra and from their *HST* morphology. For the FP study, we obtained spectra of 5−7 galaxies lying within the *HST* fields in a single MOS mask. Five masks (one per cluster; Cl 1701 and Cl 1702 only observed with one mask) have been observed to construct a combined FP sample of ∼30 galaxies, comparable to that of a single rich cluster (Fritz et al. 2004). Outside the *HST* field additional spectra of ∼10 E+S0 in each cluster serve for the investigation of the Mg–σ and Faber–Jackson relation. For these scaling relations the number of observed galaxies is sufficient to explore possible variations between the different poor clusters as a function of their σ or L_X. Our total sample comprises ∼80 galaxies.

4. First results

A first test on the formation and evolution of elliptical and S0 galaxies is made with the Kormendy relation (KR), a scaling relation between the effective radius R_e and effective surface brightness $\langle\mu\rangle_e$. As this relation is not based on galaxy kinematics, it can be studied to fainter magnitudes.

The KR for two low L_X clusters, Cl 0841 and Cl 1701, is shown in Fig. 1. We compare these poor clusters with two rich clusters, A 2218 and A 2390 at similar redshift ($z \sim 0.2$)

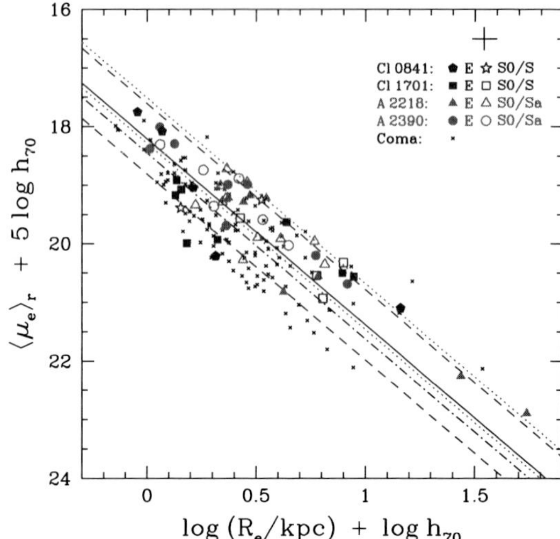

Figure 1. KR in Gunn r for the poor clusters Cl 0841 (z=0.24) and Cl 1701 (z=0.25), compared to the rich clusters A 2218 (z=0.18) and A 2390 (z=0.23) and to the local Coma sample of JFK95. The dot-dashed line is a least square fit to Coma. Dotted lines show the 1 σ errors of the fit for the rich clusters, the solid line indicates the fit for the poor clusters with 1 σ (dashed lines), assuming the local slope. A typical error bar is in the upper right corner.

and to the local Coma sample of Jørgensen et al. 1995 (JFK95). In total, the low L_X sample comprises 19 E+S0 galaxies, divided into 13 E and 6 S0 galaxies. On average, all these objects show an evolution of $\overline{m}_r \sim 0.24$ compared to Coma. The 34 rich cluster galaxies are on average brighter by ~0.47 mag. The small, not significant, difference might be caused by the fact that the two samples comprise slightly different galaxies in luminosity. However, as the scatter of the low L_X galaxies is quite large, a more quantitative analysis must be based on the whole cluster sample.

Acknowledgements

This project is a collaboration with R. G. Bower, M. L. Balogh and I. Smail (Durham/UK) and R. L. Davies (Oxford/UK) who have contributed to these results. We thank the Calar Alto staff for efficient observational support. AF and BLZ acknowledge financial support by the Volkswagen Foundation (I/76 520) and the DFG (ZI 663/3-1, ZI 663/5-1).

References

Balogh, M. L., Smail, I., Bower, R. G. 2002a *ApJ* **566**, 123–136.

Balogh, M., Bower, R. G., Smail, I., Ziegler, B. L., Davies, R. L., Gaztelu, A., Fritz, A. 2002b *MNRAS* **337**, 256–274.

Cole, S., Lacey, C. G., Baugh, C. M., Frenk, C. S. 2000 *MNRAS* **319**, 168–204.

Djorgovski, S., Davis, M. 1987 *ApJ* **313**, 59–68.

Fritz, A., Ziegler, B. L., Bower, R. G. et al. 2004 *MNRAS*, submitted.

Jørgensen, I., Franx, M., Kjærgaard, P. 1995 *MNRAS* **273**, 1097–1128. (JFK95)

van Dokkum, P. G., Stanford, S. A. 2003 *ApJ* **585**, 78–89.

Ziegler, B. L., Bender, R. 1997 *MNRAS* **291**, 527–543.

Ziegler, B. L., Bower, R. G., Smail, I. et al. 2001 *MNRAS* **325**, 1571–1590.

Outskirts of Galaxy Clusters: Intense Life in the Suburbs
Proceedings IAU Colloquium No. 195, 2004
A. Diaferio, ed.

DOI: 10.1017/S1743921304001103

Structure and stellar populations in early-type galaxies

Alexandre Vazdekis[1], **Ignacio Trujillo**[2] **and Yoshihiko Yamada**[3]

[1]Instituto de Astrofísica de Canarias, E-38200 La Laguna, Tenerife, Spain email: vazdekis@ll.iac.es

[2]Max–Planck–Institut für Astronomie, Königstuhl 17, D–69117 Heidelberg, Germany

[3]National Astronomical Observatory of Japan, 2-21-1 Osawa, Mitaka, Tokyo 181-8588, Japan

Abstract. We explore possible correlations between the light profile shapes of early-type galaxies, as parameterized by the Sérsic index n, and relevant stellar population parameters, which were obtained from spectra of extremely high signal–to–noise and stellar population models that provide an age indicator that is virtually free of the effects of metallicity. We do not find any robust correlation between n and mean age or metallicity. However, we find evidence for a strong positive correlation between n and [Mg/Fe] abundance ratio. We speculate that early-type galaxies settle up their structure on short time-scales in agreement with those imposed by their [Mg/Fe] ratios, suggesting that the global structure was already at place at high z.

1. Introduction

Early-type (E/SO) galaxies follow tight relations such as the Fundamental Plane, the Color-Magnitude Relation or the Mg_2–σ relation (Bender et al. 1992). However little is known about the relation between their morphological shapes and stellar populations (Conselice 2003), most likely because these two fields rely on methods that have grown up separately. Aiming at exploring possible correlations between galaxy light profiles and their stellar populations, we combine results obtained from analyzing high-quality spectroscopic data and stellar light profile fitting as modelled by the Sérsic index n (see for details Vazdekis, Trujillo & Yamada 2004).

2. Galaxy measurements

We use the mean luminosity weighted ages, metallicities ([M/H]) and abundance ratios ([Mg/Fe]) derived by Vazdekis et al (2001) and Yamada et al. (2004a,b in preparation) for a sample of Virgo and field early-type galaxies. Flux-calibrated spectra of extremely high signal–to–noise (S/N > 100) from galaxy central regions ($r_e/10$) were used. Relevant line-strengths, were measured and compared to those predicted by the model of Vazdekis (1999). Ages were determined on the basis of the $H\gamma_\sigma$ age indicator of Vazdekis & Arimoto (1999), which is free from metallicity effects.

The Sérsic shape parameters for most of Virgo galaxies were taken from Caon et al. (1993), or by fitting a Sérsic model to published (Peletier et al. 1990) or derived light profiles (HST archive images). We also use the concentration index of Trujillo et al. (2001) $C_{re}(1/3)$, which evaluates the fraction of light enclosed within $r_e/3$ divided by the light within r_e, and is monotonically related to n.

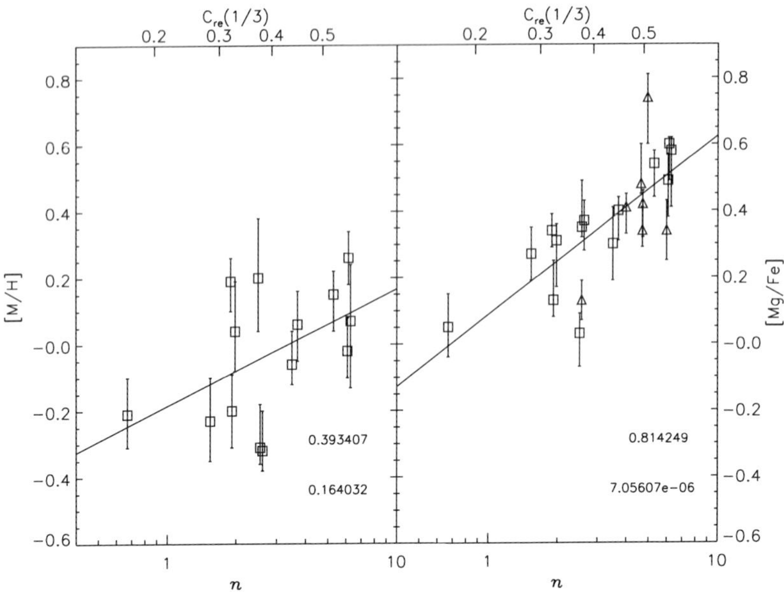

Figure 1. Plot of the shape parameters, n and $C_{re}(1/3)$, versus [M/H], and [Mg/Fe]. Squares mean Virgo galaxies and triangles mean field galaxies. Spearman rank-order correlation coefficient and its significance value are written in each panel.

3. Results

No correlation can be found between galaxy ages and n as most of our galaxies are rather old (Vazdekis et al. 2001). Fig. 1 (left panel) shows a weak correlation (as shown by the Spearman test r_s) between the total metallicity and the shape parameters. However, Fig. 1 shows for the first time a strong positive correlation between [Mg/Fe] abundance ratio and the shape parameter indices. This result is expected since [Mg/Fe] correlates with σ (Trager et al. 2000), and σ correlates with $C_{re}(1/3)$ (Graham et al. 2001). Nonetheless, it is surprising to see that the [Mg/Fe] provides the strongest correlation among other stellar population parameters, reaching a Spearman coefficient 0.8. Finally, similar correlation values are obtained for the already known $C_{re}(1/3) - \sigma$ and [Mg/Fe] $- \sigma$ relations on the basis of the same galaxies, probing the robustness of the new correlation.

4. Discussion

Mg is ejected to the ISM by SN II on short time-scales ($\lesssim 10^6 - 10^7$ yr). If star formation stops before SN Ia release Fe (~1 Gyr), the [Mg/Fe] should be larger than 0. The central σ – [Mg/Fe] correlation suggests that the innermost regions of more massive galaxies stopped their star formation on short time-scales. Next, the fact that n correlates with [Mg/Fe] suggests that early-type galaxies settled up their global structure (and not only the innermost regions) on time-scales according to their [Mg/Fe] (a scenario favored if no [Mg/Fe] radial gradients were present, Mehlert et al. 2003). The fact that we do not find a correlation between n and galaxy mean ages but [Mg/Fe] suggests that the formation time-scale is the main process linked to galaxy global structure, rather than burst peppering as a function of time. Since our galaxies are old and the time scales of most massive galaxies very short, the new correlation suggests that the global structure was already at place at high z, without experiencing a significant time evolution.

Acknowledgements

We would like to thank N. Caon and H.-W. Rix for useful suggestions and J. Blakeslee for providing us with photometric data for NGC 4489 and NGC 7454. This research is based on observations made with the European Southern Observatory telescopes obtained from the ESO/ST-ECF Science Archive Facility.

References

Bender, R., Burstein, D., Faber, S.M. 1992, ApJ, 399, 462

Caon, N., Capaccioli, M. & D'Onofrio, M., 1993, MNRAS, 265, 1013

Conselice, C. 2003, ApJSS, 147, 1

Graham, A., Trujillo, I., Caon, N. 2001, AJ, 122, 1707

Mehlert, D., Thomas, D., Saglia, R., Bender, R., Wegner, G. 2003, A&A, 407, 423

Peletier, R., Davies, R., Illingworth, G., Davis, L., Cawson, M., 1990, AJ, 100, 1091

Trager, S., Faber, S., Worthey, G., González, J. 2000, AJ, 120, 165

Trujillo, I., Graham, A., Caon, N. 2001, MNRAS, 326, 869

Vazdekis, A. 1999, ApJ, 513, 224

Vazdekis, A., Trujillo, I., Yamada, Y. 2004, ApJ, 601, L33

Vazdekis, A., Kuntschner, H., Davies, R., Arimoto, N., Nakamura, O., Peletier, R. 2001, ApJ, 551, L127

Vazdekis, A., Arimoto, N. 1999, ApJ, 525, 144

Outskirts of Galaxy Clusters: Intense Life in the Suburbs
Proceedings IAU Colloquium No. 195, 2004
A. Diaferio, ed.

DOI: 10.1017/S1743921304001115

More evidence of substructure in the motions of nearby halo stars

P. Re Fiorentin[1], A. Helmi[2], M. G. Lattanzi[3] and A. Spagna[3]

[1]Dipartimento di Fisica Generale, Università di Torino, Via Pietro Giuria 1, 10125 Torino, I, email: paola.re.fiorentin@to.infn.it

[2]Kapteyn Instituut, P.O. BOX 800, 9700 AV Groningen, NL,

[3]INAF, Osservatorio Astronomico di Torino, Via Osservatorio 20, 10025 Pino Torinese, I

Abstract. We explore the stellar halo of the Milky Way to search for fossil signatures of past mergers. We use the Beers et al. (2000) catalog of non-kinematically selected metal poor stars in the solar neighborhood to select subsets of halo stars within 1-2 kpc of the Sun. Motivated by the results of high resolution CDM simulations, we look for substructure in the kinematics of the fastest stars. When a two-point velocity correlation function is applied to these subsets, statistical evidence of substructure is found. This appears to be due to a small moving group with dynamical properties similar to the stellar "stream" previously discovered by Helmi et al. (1999).

1. Introduction

Observations of metal-poor stars have often been used to discriminate between alternative galaxy formation and evolution scenarios (e.g. Eggen, Lynden-Bell & Sandage 1962; Searle & Zinn 1978). In a CDM Universe, the first objects to form are small systems which then merge and give rise to the larger scale structures we observe today. This theory predicts the presence of substructures due to the mergers and accretion that galaxies may have experienced over their lifetime.

Several examples of mergers and galaxy interactions have been observed in the Milky Way, such as the disrupted Sagittarius and Canis Major dwarf galaxies (e.g. Ibata, Gilmore & Irwin 1995; Martin et al. 2004), the phase-space stream of halo stars in the solar neighborhood (Helmi et al. 1999), the tidal tails from halo globular clusters and dwarf spheroidal galaxies (Odenkirchen et al. 2001).

2. Evidence of substructures in the Solar neighborhood

The analysis of a high resolution simulation of the formation of a dark matter halo by Helmi et al. (2002) has shown that dark matter particles are smoothly distributed in space in the equivalent of Solar neighborhood. While their kinematics can be represented by a smooth multivariate Gaussian distribution, the motions of the most energetic particles are strongly clumped and anisotropic.

Motivated by this result, we turned to the extensive catalog of metal-poor stars, selected without any kinematic biases, compiled by Beers et al. (2000). This sample provides proper motions, radial velocities, and distance estimates for 1258 stars with a wide range of metal abundances. We identified a set of halo stars defined as those having metal abundances $[Fe/H] \leqslant -1.5\,\mathrm{dex}$, while we excluded those with typical disk kinematics (coplanar and circular orbits, $L \sim L_z$, $V_\phi \sim 220\,\mathrm{km\,s^{-1}}$, $|z| \leqslant 1.5\,\mathrm{kpc}$). Our "clean" halo sample contains 186 and 392 stars within 1 and 2 kpc of the Sun, respectively, and includes subdwarfs, giants, and variable stars (Re Fiorentin et al., in preparation).

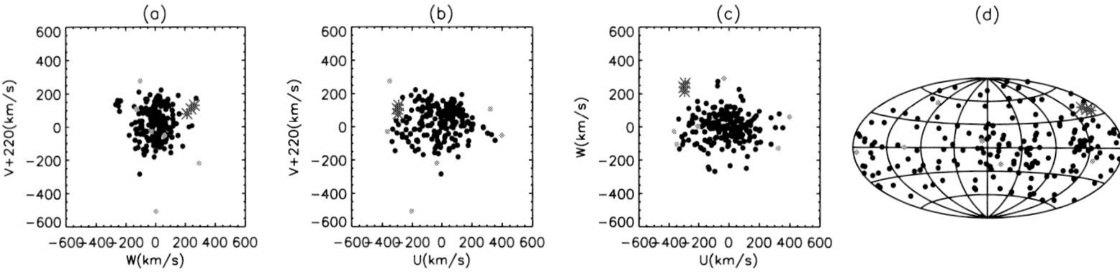

Figure 1. Distribution of the selected sample of halo stars within 1 kpc of the Sun. (a-c) Velocity projections in the (U,V,W) space; (d) Velocity direction: the position of each particle is given by the spherical angular coordinates $\boldsymbol{\alpha} = (\phi, \theta)$ of the velocity vector. Of the 186 stars present in this volume, the 5% fastest are highlighted in gray. Among them, the asterisks identify the objects with velocity difference less than $100\,\mathrm{km\,s^{-1}}$.

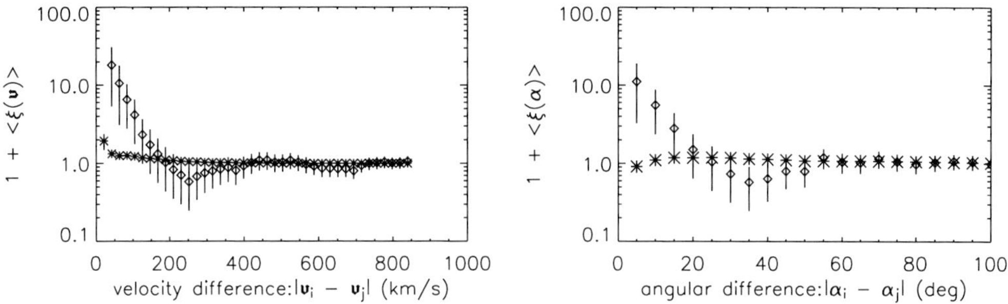

Figure 2. The two-point velocity correlation function $\xi(v)$ (left panel) and angular correlation function in velocity space $\xi(\alpha)$ (right panel) for the 1 kpc sample of halo stars (asterisks) and for the subset of 5% fastest stars (diamonds). Error bars are based on Poissonian counts.

The velocity distribution of the subset of halo stars within 1 kpc is relatively smooth, however this seems no longer to be the case when we focus on the fastest objects, as shown in Fig. 1. The kinematics of the 5% fastest moving stars (gray dots) seem more clumped and anisotropic.

We quantify the deviations from a smooth Gaussian distribution due to kinematic substructures by computing the two-point velocity correlation function, defined as

$$\xi = \frac{\langle DD \rangle}{\langle RR \rangle} - 1 \tag{2.1}$$

where $\langle DD \rangle$ is the number of pairs of stars in our data with velocity difference in a given bin, and $\langle RR \rangle$ is defined analogously for the same number of random points drawn from a multivariate Gaussian distribution convolved with expected observational errors, and averaged over 100 realizations. $\xi(v)$ measures the excess of pairs of stars moving with given velocity difference, above that expected from a random sample. $\xi(\alpha)$ measures the excess of pairs of stars moving in the same given direction range, above that expected from a random sample.

Clumping due to kinematic substructures (i.e. groups of stars moving with similar velocities and directions) appears as an excess at small velocity difference (see Fig. 2). This signal, clearly much stronger for the subset of 5% fastest objects, is due to a moving group formed by 3 stars denoted by the asterisks in Fig. 1.

The distribution in the space E, $\mathrm{L_z}$ and $\mathrm{L_{xy}}$ of the selected set within 2 kpc of the Sun shows that our main kinematic clump and the stars of the stream identified by Helmi

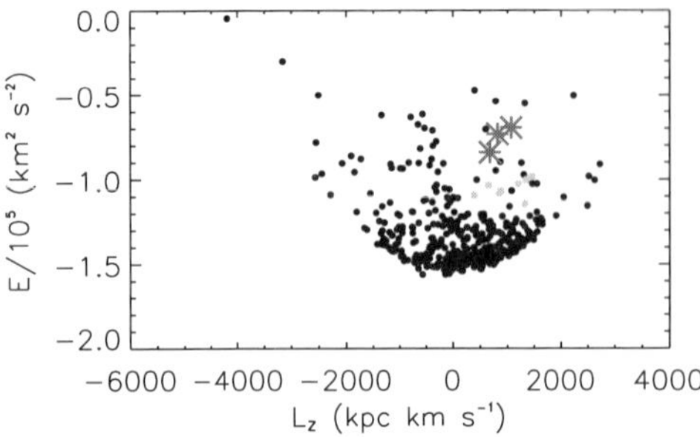

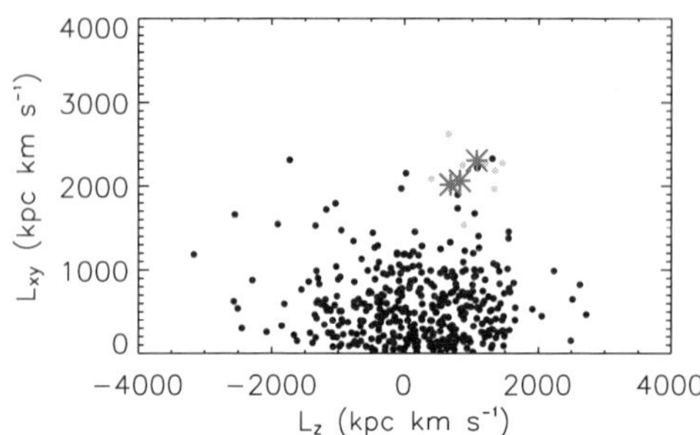

Figure 3. Distribution of the selected sample of 392 halo stars within 2 kpc of the Sun in the space of adiabatic invariants. The asterisks denote the stars of our main kinematic clump, the gray dots denote the stars of the stream identified by Helmi et al. (1999).

et al. (1999) have similar momenta but somewhat different energies (see Fig. 3). This could imply that these groups have different origins. However, it seems also plausible that these groups have been stripped off from the same progenitor in different passages.

3. Conclusions

We have found evidence of substructure in the motions of the fastest stars in the Solar neighborhood, at a level which seems to be consistent with that predicted by high resolution simulations of dark matter halos in a hierarchical universe.

The moving group identified may represent a subsample of a stellar stream spread out along the orbit of a disintegrated body, but the number of stars available is still too small for any definite conclusion.

In the inner halo, where most of the merger activity is expected to have taken place, currently available samples with accurate kinematics are still quite small. Space astrometric missions, such as GAIA, will collect samples of millions of stars with very precise positions and velocities which will dramatically improve the situation.

Acknowledgements

We wish to thank Volker Springel and Simon White for permission to use their simulations; Antonaldo Diaferio for many useful comments. P.R.F. thanks the Sterrekundig Instituut Utrecht for hospitality during her visit. Partial financial support came from the Italian Ministry of Research (MIUR) through the COFIN-2001 program.

References

Beers, T. C., Chiba, M., Yoshii, Y., Platais, I., Hanson, R. B., Fuchs, B. & Rossi, S. 2000 *Astrophys. J.* **119**, 2866–2881.

Eggen, O. J., Lynden-Bell, D. A. & Sandage, A. R. 1962 *Astrophys. J.* **136**, 748–766.

Ibata, R. A., Gilmore, G. & Irwin, M. J. 1995 *Mon. Not. R. astr. Soc.* **277**, 781–800.

Helmi, A., White, S. D. M. & Springel, V. 2002 *Phys. Rev. D* **66**, 063502.

Helmi, A., White, S. D. M., de Zeeuw, P. T. & Zhao, H. S. 1999 *Natur.* **402**, 53–55.

Martin, N. F., Ibata, R. A., Bellazzini, M., Irwin, M. J., Lewis, G. F. & Dehnen, W. 2004 *Mon. Not. R. astr. Soc.* **348**, 12–23.

Odenkirchen, M., et al. 2001 *Astrophys. J.* **548**, L165–L169.

Searle, L. & Zinn, R. 1978 *Astrophys. J.* **225**, 357–379.

Section C

The galaxy melting pot

Outskirts of Galaxy Clusters: Intense Life in the Suburbs
Proceedings IAU Colloquium No. 195, 2004
A. Diaferio, ed.

DOI: 10.1017/S1743921304001127

On the age-radius relation and orbital history of cluster galaxies

Ben Moore, Jürg Diemand and Joachim Stadel

Institute for Theoretical Physics, University of Zürich, Winterthurerstr. 190, CH-8057 Zürich, Switzerland email: moore@physik.unizh.ch

Abstract. We explore the region of influence of a galaxy cluster using numerical simulations of cold dark matter halos. Many of the observed galaxies in a cluster are expected to be infalling for the first time. Half of the halos at distances of one to two virial radii today have previously orbited through the cluster, most of them have even passed through the dense inner regions of the cluster. Some halos at distances of up to three times the virial radius have also passed through the cluster core. We do not find a significant correlation of "infall age" versus present day position for substructures and the scatter at a given position is very large. This relation may be much more significant if we could resolve the physically overmerged galaxies in the central region.

1. Introduction

Are the morphologies of galaxies imprinted during an early and rapid formation epoch or are they due to environmental processes that subsequently transform galaxies between morphological classes? The gravitational and hydrodynamical mechanisms that could perform such transformations were proposed in the 1970's, before the key observational evidence for environmental dependencies was provided - the morphology-density relation and the Butcher-Oemler effect. Many recent numerical simulations support these theoretical expectations. However, until we have self-consistent numerical simulations that can follow the structural evolution of galaxies within a large computational volume, we must resort to semi-analytic treatments or to studying the evolution of galaxies within idealised numerical calculations.

In this paper we study the orbits and infall history of substructure halos within a cold dark matter galaxy cluster. When we observe a cluster today we see a single frame of its entire cosmic evolution. What we would like to know for a given galaxy at a given position is: what is its likely orbit? Is it infalling for the first time? What are the environments that may have pre-processed the galaxy? If it has already passed the pericenter at what epoch did it enter a cluster-like environment? What was its impact parameter and velocity with respect to the cluster center? Are clusters built up in an "onion shell" scenario such that the observed galaxies trace an age-radius relation? We shall use the largest and highest resolution calculations of cold dark matter galaxy clusters to address some of these questions. With over 60 million particles in the high resolution region, up to 25 million particles within the virial radius and high force resolution, we can resolve the orbital histories of many thousands of substructures and halos both within the cluster and in the suburbs. Related studies have been carried out recently: Balogh, Navarro, & Morris (2000) followed *particle* orbits in N-body simulations while Mamon *et al.* (2004) used analytical calculations and the $z = 0$ snapshots of simulations to estimate rebound radii. Shortly after this meeting, other groups (Gill, Knebe & Gibson 2004, Gao *et al.* 2004a) have published results that were also obtained by following *subhalo* orbits and their results are very similar to those presented here.

2. Accretion redshift of cluster subhalos

It is interesting to know how much time today's cluster galaxies spent in dense environments and if the accretion time into a more massive halo is correlated with the current position in the cluster. One could expect that subhalos which fell into the cluster (or one of it progenitors) early have less orbital energy and tend to end up closer to the cluster center. We analyze the redshift of accretion of cluster subhalos in ΛCDM simulations. Note that as accreted structures we count both subhalos of the final cluster and subhalos of the cluster progenitor groups.

We take 20 outputs of run D6h and 10 of run C9, equally spaced in time. The simulations are described in Diemand, Moore & Stadel (2004b) and the properties of their subhalos are presented in Diemand, Moore & Stadel (2004a). Run C9 resolves a $M_{\text{virial}} = 5.0 \times 10^{14} M_{\odot}$ cluster with 10 million particles within r_{virial} and D6h resolves a smaller $M_{\text{virial}} = 3.1 \times 10^{14} M_{\odot}$ cluster with 2 million particles.

The subhalos were identified with SKID (Stadel 2001) and here we consider only structures with at least 32 bound particles. For each snapshot we construct a halo catalogue with FOF using a comoving linking length of 0.164 Δx_0 and trace back in time all subhalos within the virial radius of todays cluster. In Figure 1 the redshift before accretion is plotted, that is the last time a halo is identified as individual field halo. There is a large scatter in the accretion redshifts and no strong correlation with radius.

From the scatter plot and also from the histogram of accretion redshifts in three radial bins (Figure 2) one can see that the accretion rate is not a simple function of time but there are epochs of very rapid or of very slow accretion. Both clusters show very little accretion around redshift 0.4, which seems to be a coincidence.

The inner subhalos were accreted slightly earlier on average, in run D6h the mean and standard deviation of expansion factors at accretion is $a = 0.59 \pm 0.14$ for subhalos that end up in the inner 33 percent of the cluster and $a = 0.80 \pm 0.16$ for the outer 33 percent. For run C9 however all three radial bins give a mean of about $a = 0.7$. More halos must be analyzed to see if there really is a correlation of accretion redshift with cluster-centric radius, but we can already say that such a correlation must be weak and have a very large scatter.

3. Pericenters of halos in the outskirts of clusters

How many galaxies in the outskirts of clusters have passed through the inner, hot dense part of the cluster and how many are approaching the system for the first time? This question is interesting since some spiral galaxies in the outskirts of the Virgo cluster are observed to be deficient in neutral Hydrogen. First attempts to answer this questions include tracing back *particles* in cosmological Nbody simulations (Balogh *et al.* 2000) and analytical, spherical infall and rebound calculations (Mamon *et al.* 2004).

We traced back all subhalos and halos around the cluster D6h and measured the distance to the cluster core going back to the formation epoch of the cluster ($z \simeq 0.6$) with a time resolution of 0.6 Gyr.

Figure 3 shows the pericenter distance of the (sub)halos versus cluster-centric distance today. The points on the diagonal are halos that have their pericenter at $z \simeq 0$, the halos just below the diagonal in the upper right corner of the Figure are orbiting two satellite groups at distances of about 2 r_{virial}. In the lower left corner ($< r_{\text{virial}}$) we see todays subhalos. For ($r > r_{\text{virial}}$) there is a large population of halos that have pericenters well within the cluster. These are halos in the outskirts of the cluster which have passed through the cluster earlier. About half of the halos between r_{virial} and 2 r_{virial} have a

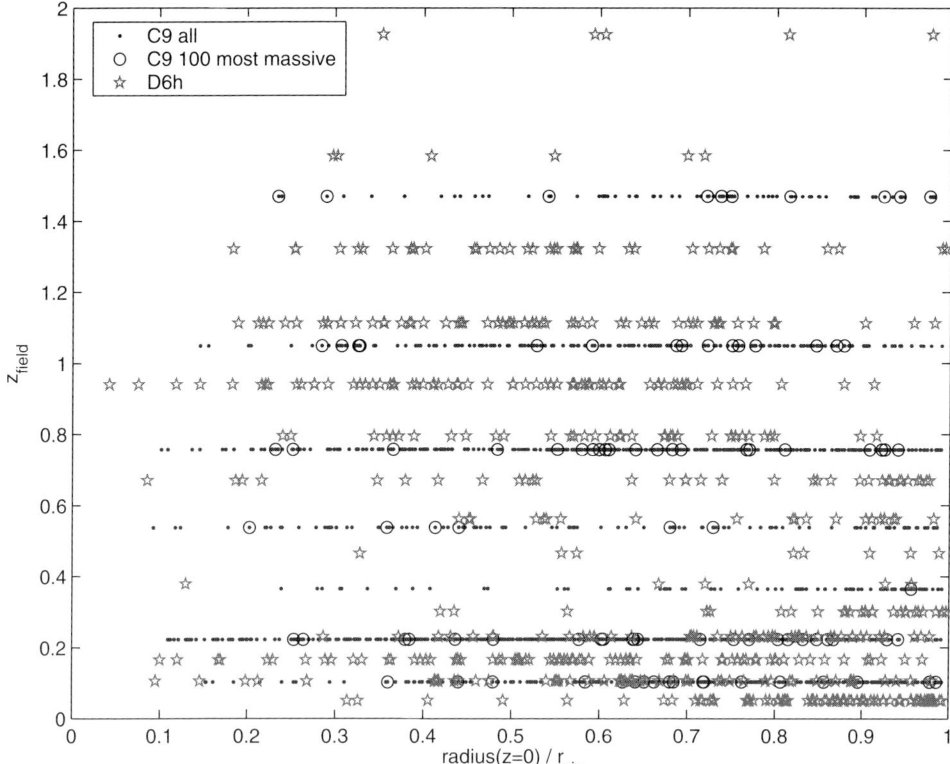

Figure 1. Accretion redshift of subhalos versus distance form the cluster center today. We use the redshift of the snapshot where the halo was identified as a field halos for the last time. The trend that central subhalos are older is weak and there is a large scatter in accretion redshifts at all radii.

pericenter smaller than $r_{\rm virial}$. Most of them (at least 70 percent†) have even passed through the inner part of the cluster ($r < 0.5r_{\rm virial}$). Finally the points in the lower right part of the plot show that in some rare cases halos that passed through the cluster can rebound out to 3 $r_{\rm virial}$, which is a little larger than the maximal distance of 2.5 $r_{\rm virial}$ obtained from analytical, spherical infall and rebound calculations (Mamon *et al.* 2004).

4. Spatial distribution of subhalos

Ghigna et al. (1998) showed that the spatial distribution of subhalos is antibiased with respect to the mass. Diemand *et al.* (2004a) confirmed that this was not a resolution effect but most likely due to physical overmerging of dark matter halos as they entered the central cluster region. In order to reproduce the observed spatial distribution of galaxies (see Figure 4), dissipation is likely to play a key role (Gao *et al.* 2004b). It is not expected that dissipation will greatly alter the internal structure of galactic halos hosting disks with type later than Sb. These galaxies will suffer the same fate as the infalling subhalos and become tidally disrupted by the cluster environment. This immediately leads to the

† Note that in the inner part of the cluster the dynamical times become comparable to the interval between outputs, so the real pericenters will be smaller.

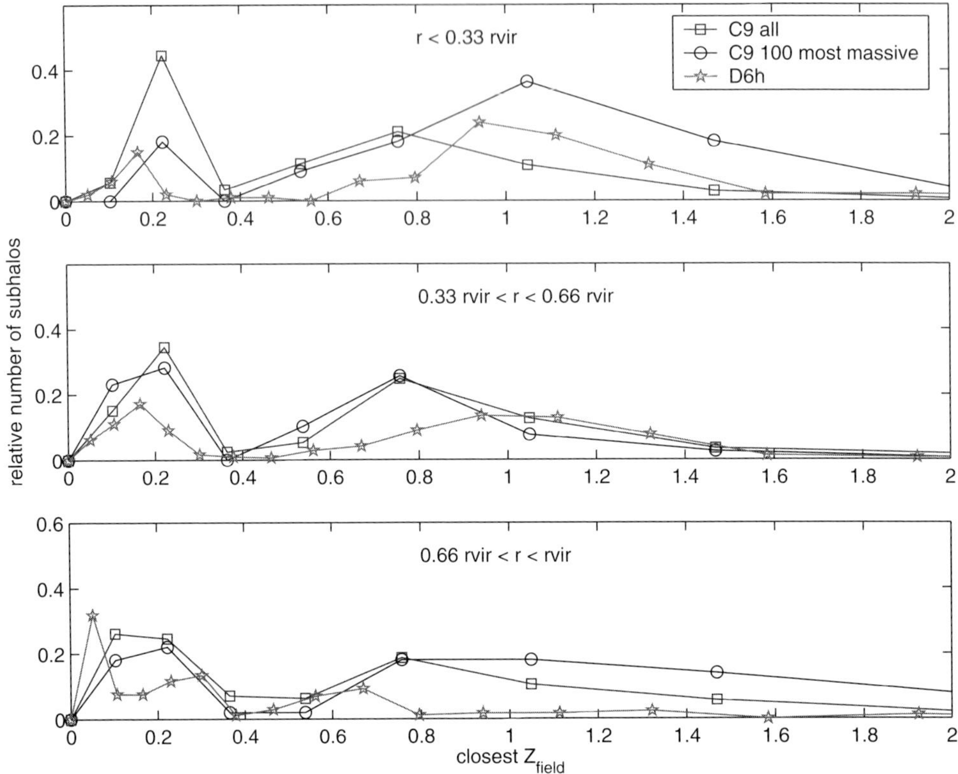

Figure 2. Histograms of accretion redshifts in the inner (top panel), intermediate (middle) and outer part(bottom) of two ΛCDM clusters.

morphology-density/radius relation since only ellipticals and Sa/Sb galaxies can survive near the cluster centre. Ellipticals will be especially dense since the multiple merging of gas rich proto-galaxies will undoubtedly lead to strong gas inflow into the central regions. The inter-galactic medium will rapidly be stripped from the central Sa/Sb galaxies and combined with a moderate amount of disk heating from tides these galaxies will rapidly turn into S0's. However it is hard to distinguish this scenario from one in which disk formation is suppressed within the proto-cluster environment.

5. Conclusions

Numerical simulations that follow only the dark matter component have provided numerous insights into the dynamical evolution of substructures. From these simulations we can infer a great deal about the environmental processes that may have affected galaxies both within the cluster and in the surrounding regions. However it is also clear that dissipation must play an important role in enabling galaxies to survive in the harsh environment within the inner regions of the virialised cluster. We summarise some of the results of our CDM cluster simulations here and look forward to enormous progress over the next decade in hydro-dynamical simulation of galaxy formation in different environments.

• The average accretion redshift of subhalos does not change significantly with the final time cluster-centric distance, i.e. there is no strong age-radius correlation in ΛCDM *subhalos* (however this does not exclude an age-radius correlation for *cluster galaxies*,

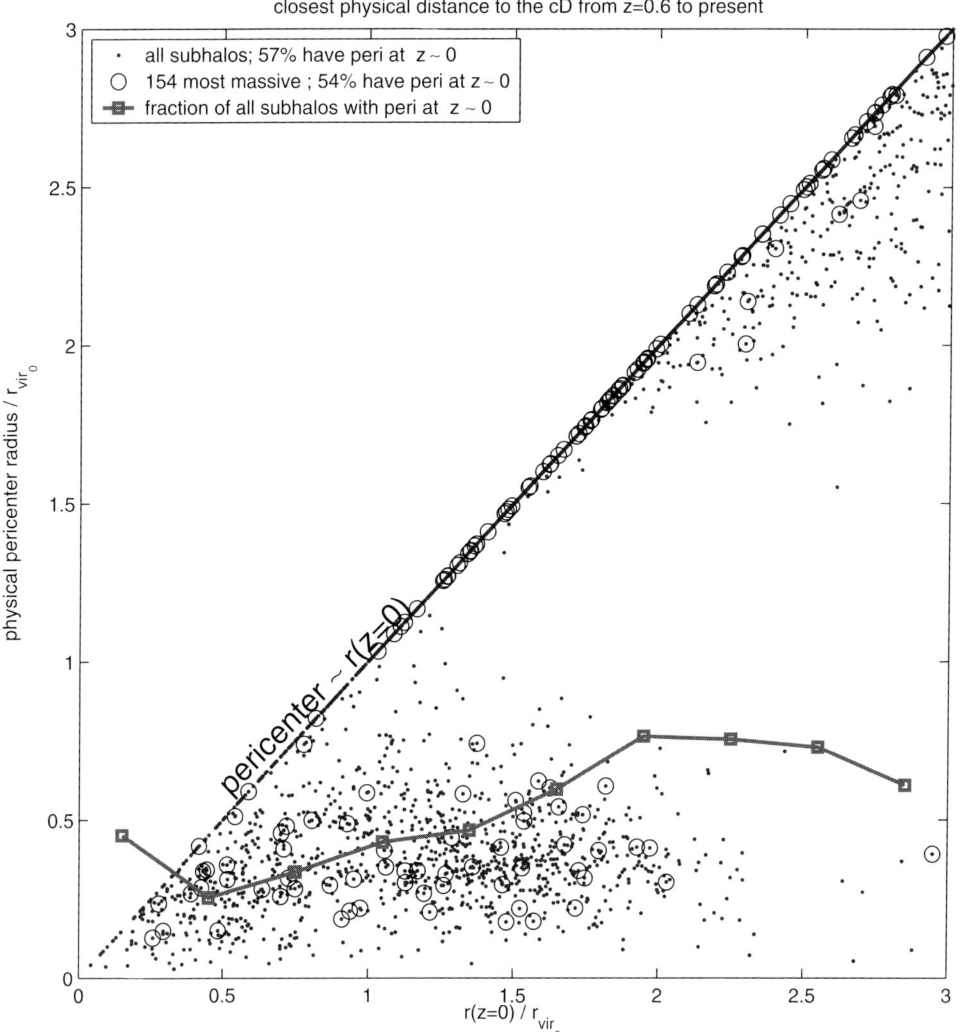

Figure 3. Closest physical distance to the cluster center form $z = 0.6$ to present vs. distance form the cluster center today. The red line gives the fraction of (sub)halos that have their pericenter today (within the time resolution of 0.6 Gyr). The points in the lower right corner show that many halos well outside the cluster today passed deep through it earlier.

since they do not trace the subhalos in a simple one-to-one correspondence: the subhalo number density profile are much shallower).

• About 50 percent of the galaxies that have a distance between one and two virial radii form the cluster center today have passed through the cluster earlier.

• Most of them (at least 70 percent) even approached the cluster center to less than half of the virial radius.

• There are some (rare) cases where a halo passes through the inner part of the cluster and then rebounds out to three virial radii.

• Dissipation must play an important role in enabling galaxies to survive in the central cluster regions. The morphology-density relation may be due to the disruption of disks at the cluster centre.

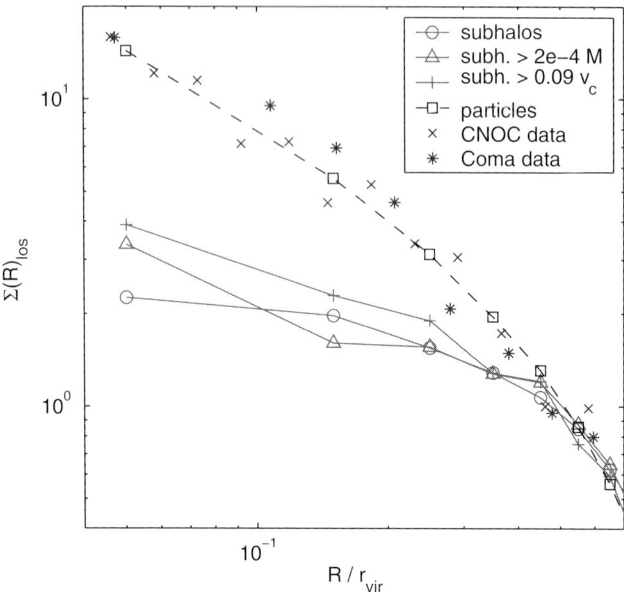

Figure 4. The distribution of subhalos plotted against the smooth dark matter component. The symbols show the radial distribution of galaxies from a sample of CNOC clusters (Carlberg, Yee & Ellingson 1997) and the Coma cluster (Lokas & Mamon 2003).

Acknowledgements

J.D. is supported by the Swiss National Science Foundation. Simulations were carried out on the zBox supercomputer.

References

Abadi, M. G., Moore, B., & Bower R. G. 1999, *MNRAS*, **308**, 947.
Carlberg R. G., Yee H. K. C., Ellingson E. 1997 *ApJ*, **478**, 462.
Balogh, M. L., Navarro, J. F., & Morris, S. L. 2000 *ApJ*, **540**, 113.
Diemand, J., Moore, B. & Stadel J. 2004, *MNRAS in press*, astro-ph/0402160.
Diemand, J., Moore, B. & Stadel J. 2004, *MNRAS in press*, astro-ph/0402267.
Gao, L., White, S.D.M., Jenkins, A., Stoehr, F. & Springel, V. 2004a, *preprint*, astro-ph/0404589.
Gao, L., De Lucia, G., White, S.D.M., Jenkins, A. 2004b, *MNRAS in press*, astro-ph/0405010.
Ghigna S., Moore B., Governato F., Lake G., Quinn T., Stadel J. 1998, *MNRAS*, **300**, 146.
Gill, S.P.D., Knebe, A. & Gibson, B.K. 2004, *preprint*, astro-ph/0404427.
Lokas E. L., Mamon G. A., 2003, MNRAS, 343, 401
Mamon, G. A., Sanchis, T., Salvador-Solé, E., & Solanes, J. M. 2004, *A&A*, **414**, 445.
Stadel, J. 2001 PhD thesis, University of Washington.

Outskirts of Galaxy Clusters: Intense Life in the Suburbs
Proceedings IAU Colloquium No. 195, 2004
A. Diaferio, ed.

DOI: 10.1017/S1743921304001139

Morphological evolution of cluster galaxies

C. Mastropietro, B. Moore, J. Diemand, L. Mayer and J. Stadel

Institute for Theoretical Physics, University of Zürich, CH-8057 Zurich, Switzerland,
e-mail: chiara@physik.unizh.ch

Abstract. The recent discovery of hidden non-axysimmetric and disk-like structures in bright Virgo dwarf elliptical and lenticular galaxies indicates that they may have late-type progenitors. Using high resolution N-body simulations we follow the evolution of disk galaxies within a ΛCDM cluster simulated with 10^7 particles where the hierarchical growth and galaxy harassment are modelled self-consistently. Most of the galaxies undergo significant morphological transformation, even at the outskirts of the cluster, and move through the Hubble sequence from disks to spheroidals. The time evolution of the simulated galaxies is compared with unsharp-masked images from the VLT and we compare the kinematics of our models with the latest high resolution spectroscopic studies from the Keck telescope. We will also discuss the properties of diffuse intra-cluster light formed by tidal debris stripped away from the infalling disks by gravitational interactions.

1. Introduction

Early type dwarfs (dE and dSO) are the most common type of galaxies in the nearby universe, yet their origin and nature are still unknown. Formation scenarios include mainly two different hypotheses: 1) dEs and dSOs are primordial galaxies, 2) they are the result of a morphological transformation of spiral and irregular galaxies accreting into the cluster. This latter idea is supported through observational evidence, i.e. the relative number density of early type galaxies increases with the local galaxy density, suggesting that the enviroment drives galaxy evolution. Moore, Lake & Katz (1998) have shown that galaxy harassment in clusters transforms spirals into spheroidals. Harassment can explain the morphological evolution of the small spiral and irregular galaxies observed in clusters at redshift $z \sim 0.4$ (Dressler *et al.* 1984) into dwarf ellipticals in the nearby universe (Moore *et al.* 1996). Moreover the radial velocities of early type galaxies in the Virgo cluster seem to indicate that these galaxies are not an old cluster population (Conselice, Gallagher & Wyse 2001) but originate from infalling field galaxies. According to this evolutionary scenario we expect to find nearby cluster galaxies that are currently undergoing morphological transformation and retain part of their disk nature. Recent observations from VLT and Keck telescope confirm that early type galaxies have a broad range of photometric and kinematical characteristics. In particular the degree of rotational support is found to vary from zero to a value close to one expected for a galaxy flattened by rotation (De Rijcke *et al.* 2001, Simien & Prugniel 2002, Pedraz *et al.* 2002, Geha *et al.* 2002, 2003). Jerjen, Kalnajs & Binggeli (2000) and Barazza, Binggeli & Jerjen (2002) discovered hidden spiral structures and bar features in five bright dEs in the Virgo cluster, concluding that the fraction of early type dwarfs hosting a disk component could be larger than 20%. Spirals and disks were also observed within dwarf spheroidal galaxies in the Coma (Graham, Jerjen & Guzmán 2003) and Fornax clusters (De Rijcke *et al.* 2003).

The aim of this work is to follow the evolution of disk galaxies orbiting in a cluster enviroment using high resolution N-body simulations and to compare the final harassed remnants with the latest photometric and spectroscopic data.

2. Simulations

All the simulations have been carried out using PKDGRAV (Stadel 2001). The galaxy models are multi–component systems with an exponential stellar disk embedded in a dark matter halo and were built using the technique described by Hernquist (1993). The density profile of the NFW halo is adiabatically contracted in response to baryonic infall (Springel & White 1999). Each galaxy is modelled with 10^6 particles, of which 10^5 are in the disk. The gravitational softening is set to 0.5 kpc for the dark halo and 0.1 kpc for disk particles. The galaxies are low surface brightness models with a central surface brightness $\mu_B = 22\,\mathrm{mag\,arcsec}^{-2}$, assuming a B-band mass to light ratio $\simeq 3$. The peak circular velocity is $100\,\mathrm{km\,s}^{-1}$, the total mass and stellar mass are respectively $7 \times 10^{10}\mathrm{M}_\odot$ and $4.2 \times 10^9\mathrm{M}_\odot$ while the initial disk scale length is 1.5 kpc at $z = 0$. The Toomre stability parameter Q was set to 1.5 at the disk scale length. A cluster with a mass comparable to Virgo ($M_{vir} = 3.1 \times 10^{14}$), at $z = 0$ and a virial radius of 1.8 Mpc was selected from a 300 Mpc ΛCDM cube simulation and resimulated at high resolution (Diemand, Moore & Stadel 2004b). This halo corresponds to the intermediate resolution cluster D9 of Diemand, Moore & Stadel (2004a). The number of particles within the virial radius is $\sim 6 \times 10^6$, while the softening length is 2.4 kpc. At $z = 0.5$ we replaced 20 random cluster particles (10 within the virial radius and 10 within 20% of the virial radius) with the high resolution galaxy models. Disk scale lengths and internal velocities were rescaled according to the redshift and cosmology.

3. Results

The evolution of the galaxies within the cluster is quite violent and is due to a sequence of strong gravitational encounters with substructures and the global cluster potential. Most of the central galaxies lose a significative fraction of stars and undergo a complete morphological transformation from disks to spheroidal systems. Figure 1 represents the different phases of the evolution of a galaxy transforming in a dE. The images are obtained from the simulations applying the unsharp masking technique described in Barazza *et al.* (2002), in order to highlight the presence of hidden structures. The first stage of the evolution is characterised by the formation of a strong bar and an open spiral pattern in the disk. The spiral arms are easily destroyed while the remaining material forms a ring structure around the bar. The bar+ring phase is quite stable and is apparent in the final state of several of the remnant galaxies. Often the ring is not visible directly in the simulation output without applying the unsharp masking. When gravitational heating removes the remaining spiral and ring features, only a naked bar remains. The bar usually buckles, and as a result the central part of the galaxy becomes more spherical; the buckled bar is also subject to tidal heating and loses mass from its edges. Table 1 shows the main characteristics of the harassed remnants within the virial radius of the cluster. The final state is always quite prolate, with the exception of the first two galaxies which are roughly spherical. This is not surprising since most of the galaxies retain part of the radial anisotropy that originated during the bar phase (Mayer et al. 2001). There is a clear correlation between the c/a ratio, which indicates the flattening of the disk, and the final stellar mass. As shown in the top left plot in Figure 3, massive stellar remnants have smaller c/a values and retain more of their initial disk nature, while galaxies that have lost most of their stars tend to be more spheroidal. On average

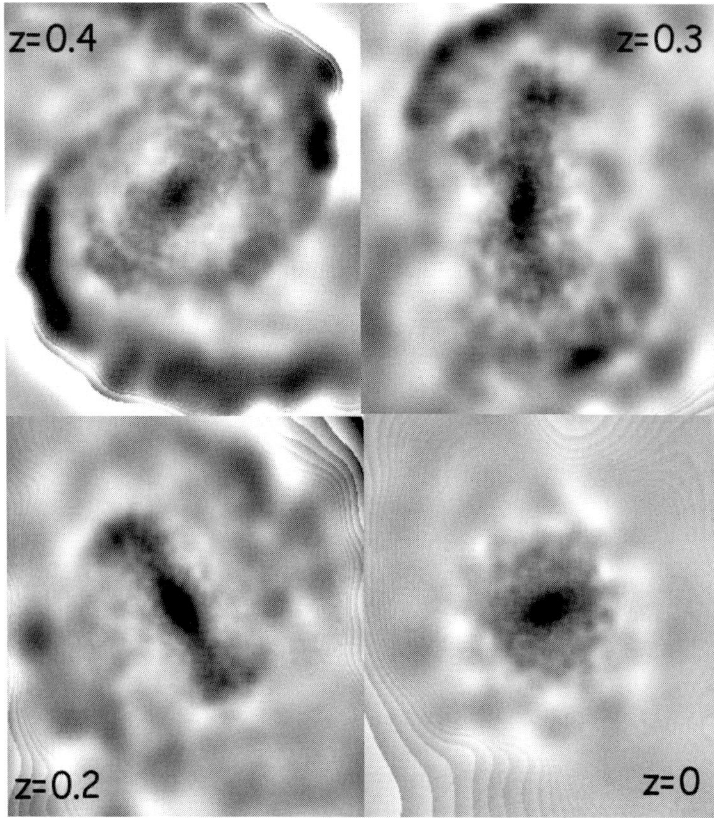

Figure 1. Morphological transformation of a disk galaxy into a dwarf elliptical. For each frame the redshift is indicated.

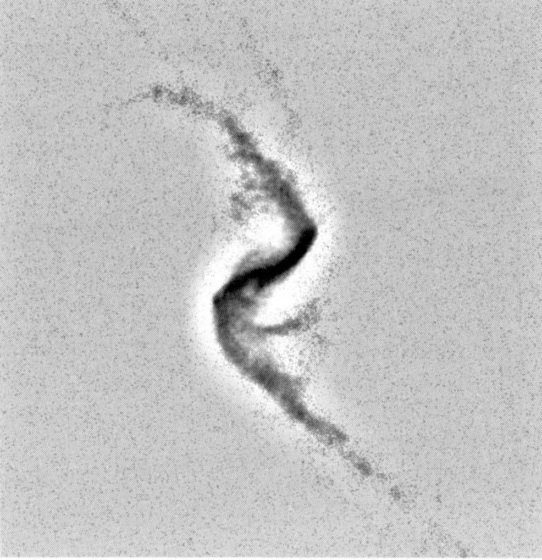

Figure 2. Tidal tails generated during a pericentric passage. The time scale of these smooth features is quite short (tipically about 0.2–0.3 Gyr) since they are perturbed by close encounters with cluster substructures.

r_t (kpc)	r_e (kpc)	M_d ($10^9 M_\odot$)	M_s ($10^9 M_\odot$)	c/a	b/a
4.9	0.6	2.3	0.5	0.8	0.7
7.3	0.9	8.0	1.5	0.8	0.8
12.3	1.1	10.9	1.7	0.5	0.4
14.0	1.1	16.1	1.8	0.5	0.4
8.9	1.3	5.5	2.1	0.7	0.5
13.7	1.3	9.7	2.4	0.6	0.5
15.8	1.4	23.4	2.6	0.6	0.5
11.6	1.7	9.4	3.1	0.6	0.4
33.0	1.9	18.0	3.5	0.4	0.4
12.4	2.7	19.2	3.9	0.3	0.2
33.9	3.2	15.9	4.0	0.5	0.3
18.9	2.4	19.0	4.0	0.5	0.3

Table 1. Final state of the remnants that lie within the virial radius of the cluster at $z = 0$. For each galaxy we list tidal r_t and effective r_e radius, dark matter M_d and stellar M_s mass. The tidal radius is calculated balancing kinetic and potential energy in spherical shells centered on the remnant, while r_e is the radius containing half of the light. The last two columns indicate the average axial ratios (respectively short and intermediate axis to major axis) measured within r_e.

more than 40% of the stars are lost forming irregular tails of low surface brightness ($\mu_B < 30\,\mathrm{mag\,arcsec^{-2}}$) tidal debris. The next two plots in Figure 3 represent the surface brightness and velocity dispersion profiles of the remnants compared to the initial values. Although a significative fraction of the stellar mass has been stripped from the disks, the central surface brightness increases by 1–2 $\mathrm{mag\,arcsec^{-2}}$ as result of the increase in the stellar random motions and conservation of phase space (Moore *et al.* 1999). Within r_e the remnants still have exponential profiles with scale lengths in the range 0.5–1.0 kpc. The radial velocity dispersion rises due to bar formation and impulsive shocks. The degree of rotational support of the two spherical remnants is close to zero (bottom right plot), while the v/σ ratio of the other galaxies depends on the viewing angle. In particular, if viewed down along the short axis, which is always the rotational axis, the galaxies show a $v/\sigma \sim 0$. Along the major axis they have the maximum rotational support, while the intermediate axis projection produces the largest ellipticity values.

4. Conclusions

We have shown that a significant fraction of cluster galaxies undergo a morphological transformation from pure disk systems to dEs and dSOs with photometric and kinematical characteristics similar to observed galaxies. The degree of this transformation is directly correlated with the loss of mass. Up to 80% of the stellar mass is lost due to tidal effects and forms a diffuse low surface brightness distribution of irregular tidal streams.

References

Barazza, F. D., Binggeli, B. & Jerjen, H. 2002 *A&A* **391**, 823
Conselice, C. J., Gallagher, J. S. & Wyse, R. F. G. 2001 *ApJ* **559**, 791
De Rijcke, S., Dejonghe, H., Zeilinger, W. W. & Hau, G. K. T. 2001 *ApJ* **559**, L21
De Rijcke, S., Dejonghe, H., Zeilinger, W. W. & Hau, G. K. T. 2003 *A&A* **400**, 119
Diemand, J., Moore, B. & Stadel, J. 2004a *astroph 0402267, MNRAS* submitted
Diemand, J., Moore, B. & Stadel, J. 2004b *astroph 0402160, MNRAS* submitted
Dressler, A., Oemler, A. J., Butcher, H. R. & Gunn, J. E. 1994 *ApJ* **430**, 107

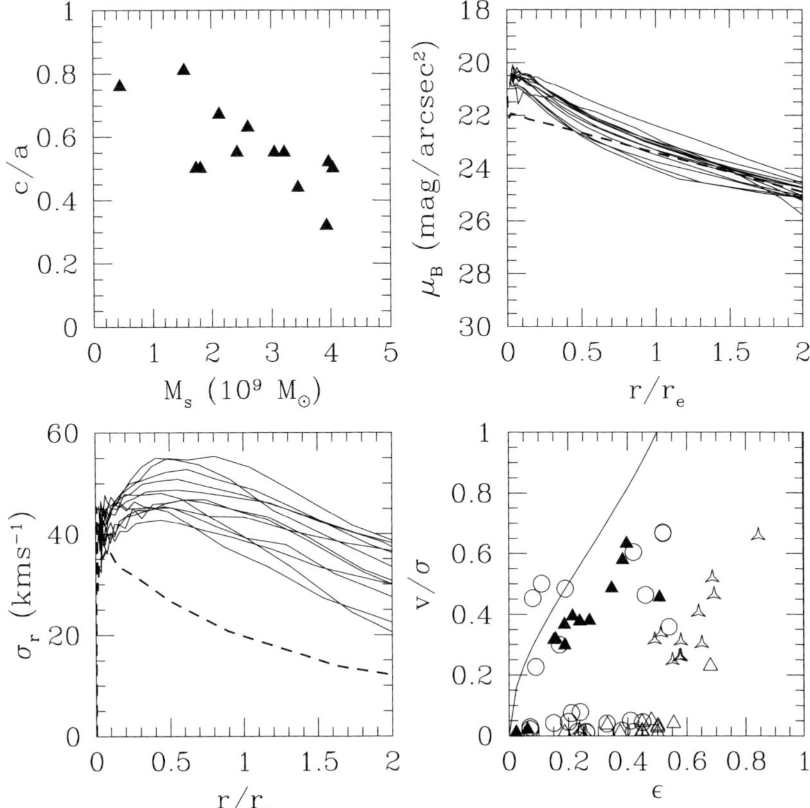

Figure 3. From the top left to the bottom right: final flattening of the remnants plotted against stellar mass, surface brightness and radial velocity dispersion profiles within two scale radii (the thick dashed line represents the initial conditions), ratio of the rotation velocity to the projected velocity dispersion versus the mean ellipticity within r_e. Open circles represent data from Geha *et al.* (2002,2003), Pedraz *et al.* (2002) and Simien & Prugniel (2002). The results of the simulation are indicated for three different projections along the major (solid triangles), intermediate (deformed triangles) and short axis (open triangles). The solid line shows the theoretical relation for an oblate, isotropic galaxy flattened by rotation.

Geha, M., Guhathakurta, P. & van der Marel, R. P. 2002 *AJ* **124**, 3073

Geha, M., Guhathakurta, P. & van der Marel, R. P. 2003 *AJ* **126**, 1794

Graham, A. W., Jerjen, H. & Guzmán, R. 2003 *AJ* **126**, 1787

Hernquist, L. 1993 *ApJ*, **86**, S389

Jerjen, H., Kalnajs, A. & Binggeli, B. 2000 *A&A* **358**, 845

Mayer, L., Governato, F., Colpi, M., Moore, B., Quinn, T., Wadsley, J., Stadel, J. & Lake, G. 2001 *ApJ* **559**, 754

Moore, B., Katz, N., Lake, G., Dressler, A. & Oemler, A. 1996 *Nature* **379**, 613

Moore, B., Lake, G. & Katz, N. 1998 *ApJ* **495**, 139

Moore, B., Lake, G., Quinn, T. & Stadel, J. 1999 *MNRAS* **304**, 465

Pedraz, S., Gorgas, J., Cardiel, N., Sánchez-Blázquez, P. & Guzmán, R. 2002, *MNRAS* **332**, L59

Simien, F. & Prugniel, P. 2002 *A&A* **384**, 371

Springel, V. & White, S. D. M. 1999 *MNRAS* **307**, 162

Stadel, J. 2001 *Ph.D. Thesis*

Outskirts of Galaxy Clusters: Intense Life in the Suburbs
Proceedings IAU Colloquium No. 195, 2004
A. Diaferio, ed.

DOI: 10.1017/S1743921304001140

L.O.S.V.D.'s of elliptical galaxies from mergers of disk galaxies

A. C. González-García and M. Balcells

Instituto de Astrofísica de Canarias, C/ Vía Lactea s/n, 38200 La Laguna, Tenerife, Spain
email: cglez@ll.iac.es

Abstract. Modern kinematic diagnostics from high-S/N spectroscopy provide useful tests on the collisionless merger hypothesis for the formation of elliptical galaxies. Bender, Saglia & Gerhard (1994) show that the skewness of the line-of-sight velocity distributions (LOSVD) has opposite sign to the rotation velocity and follows a distinct pattern. Previous N-body experiments (Bendo & Barnes 2000, Naab & Burkert 2001) disagree as to whether the relation is reproduced by the models.

We present first results from an ongoing effort to explain the skewness properties of the LOSVD's of ellipticals. We use large N-body simulations of disk-disk galaxy mergers (10^6 particles), with mass ratios 1:1 and 3:1. Our 1:1 models fail to reproduce the H3 vs. V/σ relation, in agreement with Naab & Burkert (2001). 3:1 mergers yield trends similar to the observations for large values of V/σ, but with overall lower H3 values in the models than in real galaxies. And we fail to reproduce high H3 values for low V/σ. So far, it is unclear whether the disagreement is related to the lack of rotation in the bulge components of our initial model galaxies, or to a more general shortcoming of dissipationless mergers to reproduce the properties of real ellipticals.

1. Introduction

Bender, Saglia & Gerhard (1994) measured the line-of-sight velocity distributions (LOSVD) of a number of elliptical galaxies up to 1 R_{eff}. Bender et al. fit the LOSVD to a Gaussian plus a combination of Gauss-Hermite polynomials. They find an anti-correlation between the H3 parameter, which gives the amplitude of the 3rd-Hermite polynomial, and V/σ. H3 describes the asymmetry or skewness of the LOSVD from gaussianity. The correlation appears to be more prominent for disky elliptical galaxies. This has been claimed as a signature of the presence of disks inside disky ellipticals (Bender et al. 1994, Naab & Burkert 2001).

In the hierarchical merging scenario, elliptical galaxies can be formed by a major merger between two spiral galaxies (Toomre & Toomre 1972). It has been argued that 1:1 mergers between disk galaxies give as a result a body that resembles a boxy elliptical galaxy, while mergers with mass ratios of the order 3:1 give disky ellipticals (Bendo & Barnes 2000, Naab & Burkert 2001, Gonzalez-Garcia & Balcells 2004).

It is therefore quite straightforward to ask whether the correlation found by Bender et al. (1994) can be found by means of mergers of disks. Naab & Burkert (2001) study this issue by collisionless simulations of mergers of disks. They use galaxy models with mass ratios 1:1 and 3:1 with a disk:bulge ratio 3:1. They find that for 1:1 mergers their systems end up being boxy ellipticals not showing the correlation. Neither the 3:1 mergers, although they show disky deviations, do reproduce Bender's correlation. They argue that this is a proof of the dissipational origin of these structures.

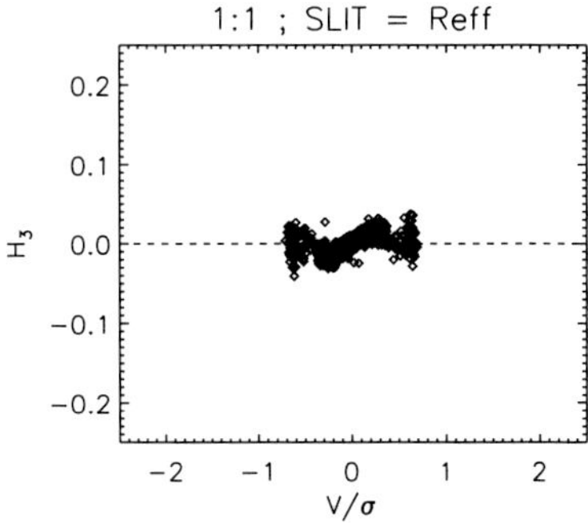

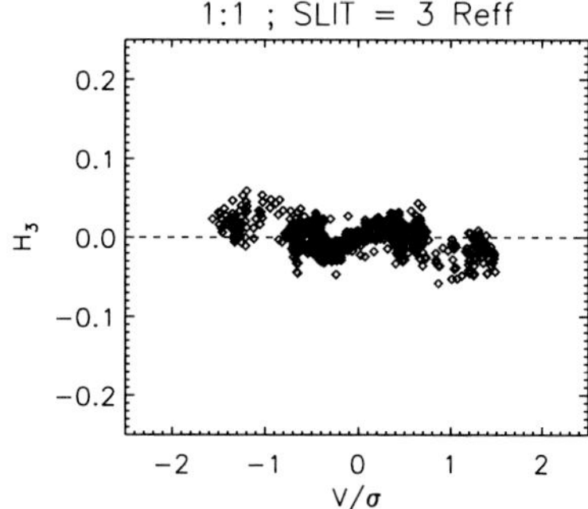

Figure 1. H3 vs. V/σ for the 1:1 merger. Two frames are given: the left one was done with a slit with a length equal to the system effective radius. The second is for a slit with $3R_{eff}$.

2. Models

We build the initial models following Kuijken & Dubinski (1995). Our initial models have a bulge, a disk and a halo. We use non-dimensional units in our calculations, with $G = 1$. The masses of the different components in our initial model are in the ratios: 19:2:1 (halo:disk:bulge). As we can see, the initial mass ratio between disk and bulge is 2:1. Note that our models have a somewhat larger bulge than Naab & Burkert (2001).

We present here results from two simulations of mergers between two of these models. The first one is an equal mass merger. The second one is a merger with mass ratio 3:1. We place the systems on elliptical orbits to ensure merging. Both systems are placed on penetrating orbits with disks orientations: $(\theta_1, \phi_1), (\theta_2, \phi_2) = (70, 30), (10, -10)$. The z-component of the spin angular momentum is then parallel to the orbital angular momentum. The spin-orbit coupling is thus large, although it is not maximum.

Each simulation consists of 10^6 particles in total. We have run these simulations with the parallel version of GADGET (Springel et al. 2001) on a Beowulf cluster with 16 CPU's. Each run takes of the order of 5×10^5 seconds of CPU time. After merger is completed we let the remnants to evolve in isolation for 10 half-mass radius crossing times.

3. Results

We have obtained LOSVD's in the following way. We have binned the slit in ten spatial bins. We find the radial projected velocity and find the number of particles in each bin in velocity for each bin in the slit. In this way we find a line-of-sight velocity distribution. Finally we fit the LOSVD first by a Gaussian and the residuals by a Gauss-Hermite polynomial as given by Bender et al. (1994). We repeated this for 90 different points of view for each model.

Bender et al. (1994) find that for low values of V/σ there is large scatter in values of H3. Most boxy ellipticals are located here. Disky ellipticals show large V/σ with large values of H3.

Figure 1 shows that the inner parts of the 1:1 merger remnant do not show the correlation found by Bender et al., in agreement with Naab & Burkert (2001). However, when inspecting at a larger radius, the general shape of Bender's correlation appears. This may indicate the presence of disk material in the outer parts of the remnant.

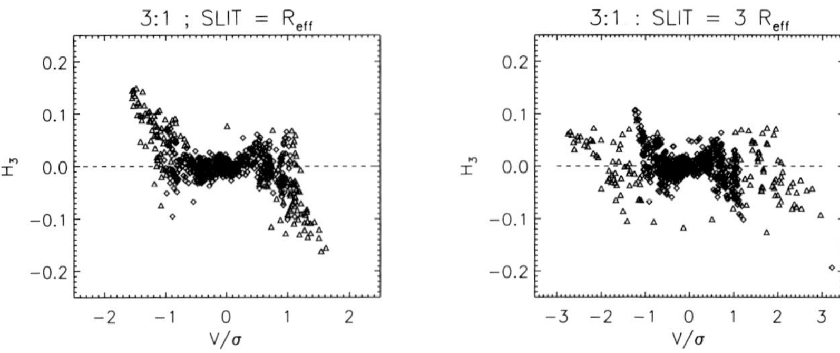

Figure 2. H3 vs. V/σ relation for the 3:1 merger, again for a slit with length equal to R_eff (left) and for $3R_{eff}$ (right).

Figure 2 left panel shows a similar trend to that found by Bender et al. The agreement is best for the outer parts (large V/σ). We fail to reproduce the large values observed by Bender et al. for low values of V/σ. The right panel gives a prediction of how the plot would be for slits including up to $3R_{eff}$ (triangles).

4. Discussion

Collisionless equal mass mergers of disks are unable to reproduce the profile observed in real elliptical galaxies, in agreement with Naab & Burkert (2001). 3:1 mergers on the other hand seem to follow the trend for large V/σ, in disagreement with Naab & Burkert. Our initial models have a larger bulge. This helps to stabilize the disk during the merger (see Gonzalez-Garcia & Balcells 2004).

We are not able to find large values of H3 for the inner parts. Our initial bulges do not rotate. We are carrying out new experiments with rotating bulges to test whether this may help to solve this discrepancy. In the outer parts of the 3:1 merger remnant we can find material from the disk of the large system. Thus, the H3 vs. V/σ relation might be the signature in the LOSVD of the rotation of the initial large disk.

Acknowledgements

We would like to acknowledge Konrad Kuijken and John Dubinski for making their code available.

References

Bender R., Saglia R. P., Gerhard O. E. 1994 *MNRAS* **269**, 785–813.
Bendo G.& Barnes J. E. 2000 *MNRAS* **319**, 315–325.
Gonzalez-Garcia A. C. & Balcells M. 2004 *in preparation.*
Kuijken K. & Dubinski J. 1995 *MNRAS* **277**, 1341–1353.
Naab T. & Burkert A. 2001 *ApJ* **555**, L91–L94.
Springel V., Yoshida N., White S. D. M. 2001 *New Ast.* **6**, 79–117.
Toomre A. & Toomre J. 1972 *ApJ* **178**, 623–666.

Outskirts of Galaxy Clusters: Intense Life in the Suburbs
Proceedings IAU Colloquium No. 195, 2004
A. Diaferio, ed.

DOI: 10.1017/S1743921304001152

Galaxy formation modeling

Andrew J. Benson

Department of Physics, University of Oxford, Keble Road, Oxford, OX1 3RH, U.K. email: abenson@astro.ox.ac.uk

Abstract. Ab-initio modeling of galaxy formation is becoming ever more detailed and is including ever more physical processes in a bid to describe the process of galaxy formation more realistically. We will describe some recent advances and applications of the semi-analytic model of galaxy formation, GALFORM. In particular, we will describe a preliminary study of the effects of environment on galaxy morphology in a $z = 0.4$ cluster.

1. Galaxy formation modeling

Semi-analytic modeling of galaxy formation traditionally follows the birth and evolution of galaxies in the merging hierarchy of dark matter halos predicted to exist in cold dark matter (CDM) cosmologies. The merging hierarchy is computed from the extended Press-Schechter formalism (e.g. Lacey & Cole 1993) or from direct N-body simulations (Kauffmann et al. 1999, Helly et al. 2003a). Galaxy formation is then modeled by considering the rate at which gas can cool within these halos, the rate of galaxy merging (driven by dynamical friction) and the rate and efficiency of star formation and the associated feedback in individual galaxies (Cole et al. 2000). The accuracy of some of these ingredients has now been tested. For example, Fig. 1 shows a comparison of galaxy masses in the GALFORM model and a smoothed particle hydrodynamics calculation (Helly et al. 2003b), which demonstrates good agreement between these two approaches (see also Benson et al. 2001, Yoshida et al. 2002).

2. Recent results

2.1. *The galaxy luminosity function*

We have recently examined the ability of semi-analytic models to reproduce the luminosity function of galaxies (a key observable which all models must strive to reproduce). With only the traditional "reheating" feedback† usually employed in semi-analytic models it is impossible to simultaneously match the rather flat faint-end slope of the observed luminosity function and the very sharp cut-off at bright magnitudes. Fig. 2 shows two possible solutions to this problem. In the left-hand panel we allow thermal conductivity to transport energy into the gas residing in the central regions of galaxy groups and clusters, thereby reducing the rate of gas cooling and hence of galaxy formation. While this produces a good match to the luminosity function, an excessively high conductivity is required. The right-hand panel shows the effects of "superwinds" in which gas ejected from the galaxy by feedback is completely removed from the surrounding dark matter halo and is not allowed to participate in the galaxy formation process again. While superwinds are able to produce a reasonably good match to the luminosity function the

† With reheating feedback for every mass of stars formed in a galaxy a certain mass of cold gas is ejected from the galaxy into the surrounding halo. The amount of ejected gas depends on the depth of the potential well of the galaxy.

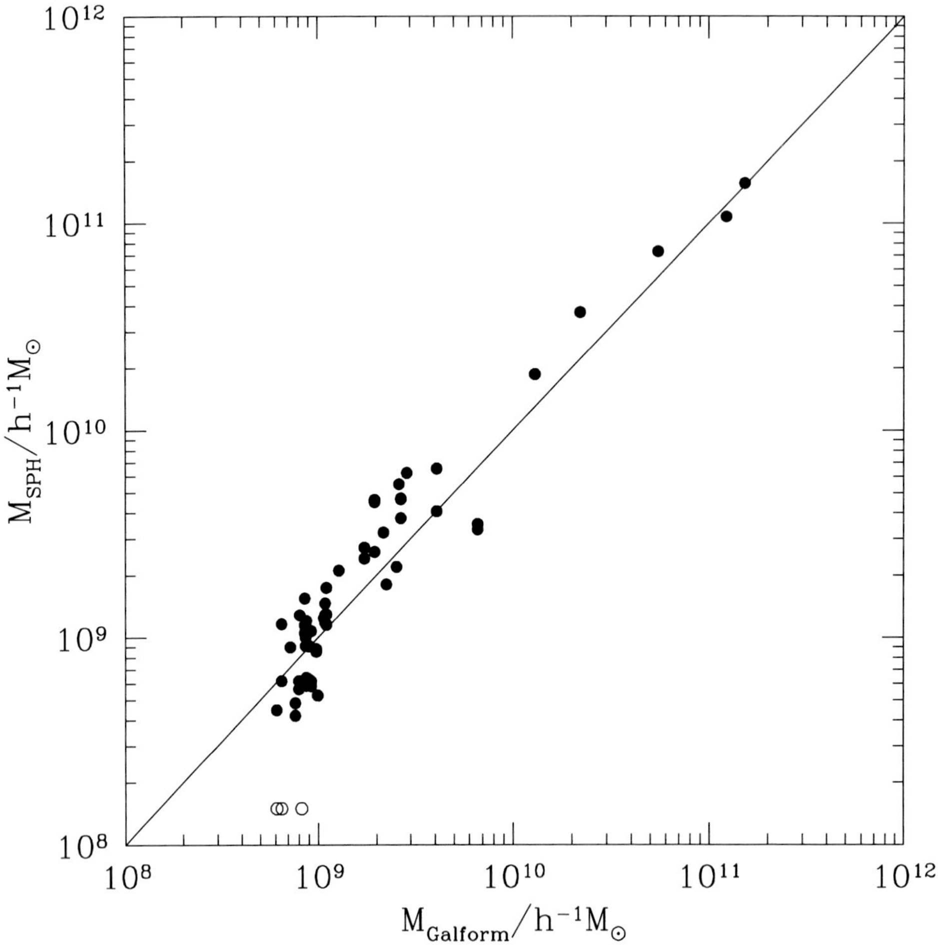

Figure 1. The correlation of galaxy masses as predicted by two different models of galaxy formation Helly et al. 2003b. Semi-analytic and SPH techniques were used to populate a dark matter halo with galaxies. Points show the masses of corresponding galaxies in the two simulations (i.e. those with similar spatial positions). Where no corresponding galaxy was found in the SPH simulation, an open point is plotted at an arbitrary position on the y-axis.

energy requirements are large—larger than is available from supernovae. This hints that perhaps a contribution of energy from AGN is required.

2.2. *Heating of galaxy disks*

Recently, we have developed a much more detailed model of satellite dark matter halo orbits and merging than has previously been possible within semi-analytic models. By tracking the orbit of each satellite halo, accounting for tidal mass loss and dynamical friction, we are able to determine the frequency with which satellite halos interact with the disks of galaxies such as the Milky Way and the consequences of those interactions. Figure 3 shows the initial distribution of orbital velocities for these satellite halos, measured at the time when they cross the virial radius of their host halo. This is used as input to a calculation of the amount of energy transferred from satellite halo orbits to galaxy disks. Figure 4 shows the resulting distribution of galaxy scale-heights (we plot

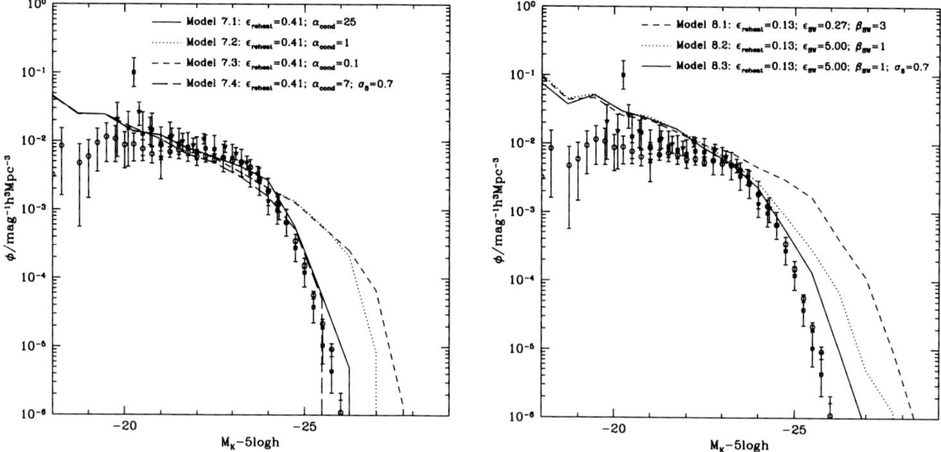

Figure 2. *Left-hand panel:* Lines show K-band luminosity functions predicted by the GALFORM model when the effects of thermal conduction on intra-cluster gas are considered. Model numbers correspond to the models of Benson et al. (2004). The parameter $\epsilon_{\rm reheat}$ specifies the amount of energy (in units of 10^{49} ergs) per solar mass of stars formed that is used to reheat gas from galaxy disks (i.e. to cause feedback), while $\alpha_{\rm cond}$ parametrizes the conductivity of the intracluster gas in units of the Spitzer conductivity. Observational data are taken from Cole et al. (2001) [circles], Kochanek et al. (2001) [squares] and Huang et al. (2002) [stars]. *Right-hand panel:* As left-hand panel except that results are shown for models with superwinds. The parameter $\epsilon_{\rm SW}$ quantifies the energy (in units of 10^{49} ergs) per solar mass of stars formed which is put into a superwind, while $\beta_{\rm SW}$ specifies the mass ejected into the superwind per solar mass of stars formed.

h, the ratio of vertical scale-height to radial scale-length) for a sample of galaxies. These are compared to a recent observational determination by Bizyaev & Mitronova (2002). The thickening of the disks predicted by our model is seen to be in good agreement with the data.

3. Role of environmental effects

The improved model of satellite galaxy orbits described above forms the backbone for a more detailed study of environmental effects than has previously been possible using semi-analytic techniques. As an example, Fig. 5 shows, in the left-hand panel, the projected positions of galaxies in a cluster at $z = 0.4$ with mass similar to that of Cl0024 as studied by Treu et al. (2003). The right-hand panel shows the cumulative mass function of dark matter halos associated with galaxies in the cluster, and demonstrates that the semi-analytic model of these halos reproduces the results of direct N-body calculations extremely well.

Using the positional information predicted by our model of galaxy orbits and the morphological information predicted by GALFORM for each model galaxy we can investigate how galaxy morphology depends upon environment. Fig. 6 shows the fractions of early and late-type galaxies as a function of distance from the cluster centre (left-hand panel) and as a function of local galaxy density (right-hand panel).

GALFORM predicts a strong trend of morphology with radius (namely that the early type fraction decreases towards smaller radii) within the cluster. Beyond about 0.3 Mpc the morphological mix reaches the field value and stays fairly constant. While this radial trend is seen in the real Cl0024, Treu et al. (2003) also demonstrate that morphology

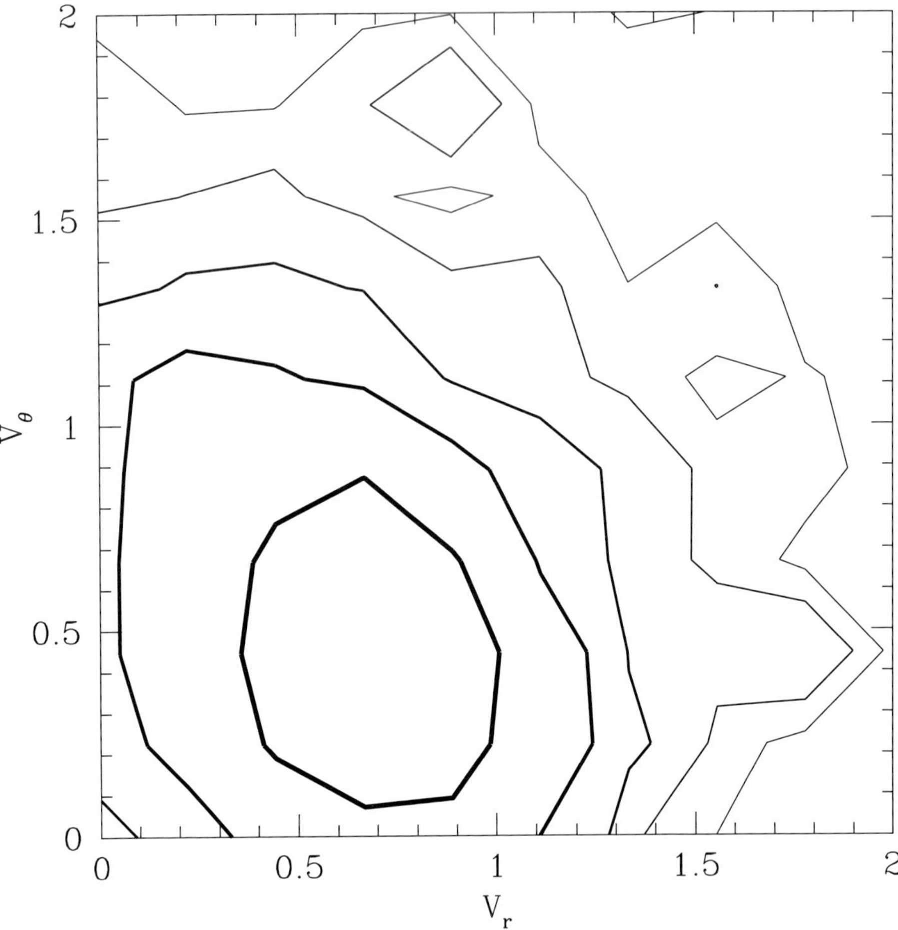

Figure 3. The distribution, $\mathrm{d}^2 f/\mathrm{d}v_\mathrm{r}/\mathrm{d}v_\theta$, of radial and tangential velocities for merging dark matter halos. All velocities are expressed in units of the virial velocity of the more massive halo. Contours indicate the distribution measured from the VIRGO Consortium's ΛCDM GIF simulations. Contours are drawn at levels equally spaced in the logarithm of $\mathrm{d}^2 f/\mathrm{d}v_\mathrm{r}/\mathrm{d}v_\theta$ from 0.01 to 1 (thinnest to thickest line).

correlates with local density (even though there is considerable scatter in the density-radius relation). The right-hand panel of Fig. 6 shows that such a morphology-density relation exists within our model also. Importantly, the figure demonstrates that such a relation exists even beyond 1 Mpc from the cluster centre, where no morphology-radius relation can be discerned.

Our current calculations include three environmental effects which may be driving these relations. These are: 1) galaxy merging, 2) tidal mass loss from galaxies and 3) starvation of galaxies (i.e. galaxies are unable to accrete fresh gas once they fall into a group or cluster and so their star formation rate rapidly declines). Using GALFORM it will be possible to ascertain which of these processes are contributing to building the morphology-density relation. Further processes which may be important (such as ram pressure stripping and tidally triggered star formation) will also be implemented and their importance determined.

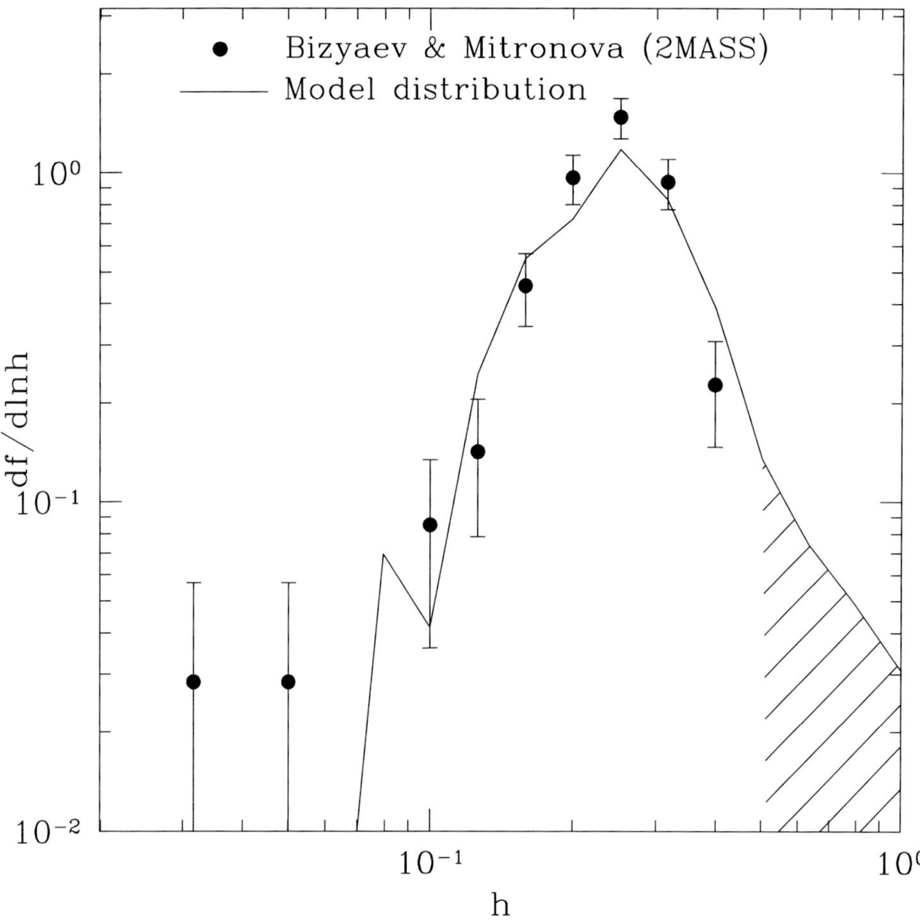

Figure 4. The normalized distribution of galactic disk scale-heights, h, in the observational sample of Bizyaev & Mitronova (2002) compared to the prediction of our model. Error bars on the observational data points indicate Poisson errors. The model prediction is shown by the solid line. The model galaxies have been weighted to match the distribution of absolute magnitudes and morphological types found in the observational sample.

4. Conclusions

Semi-analytic models such as GALFORM continue to be a highly fruitful means of studying the physical mechanisms involved in the process of galaxy formation. Continued development of the GALFORM model has allowed it to be applied to many novel problems (such as the thickening of galaxy disks) and is beginning to permit much more detailed studies of environmental effects than has previously been possible. A preliminary study of simulated clusters similar to Cl0024 shows that a morphology-density relation extending into the field arises naturally from the model. Future work will be directed at understanding the physics which is driving this relation.

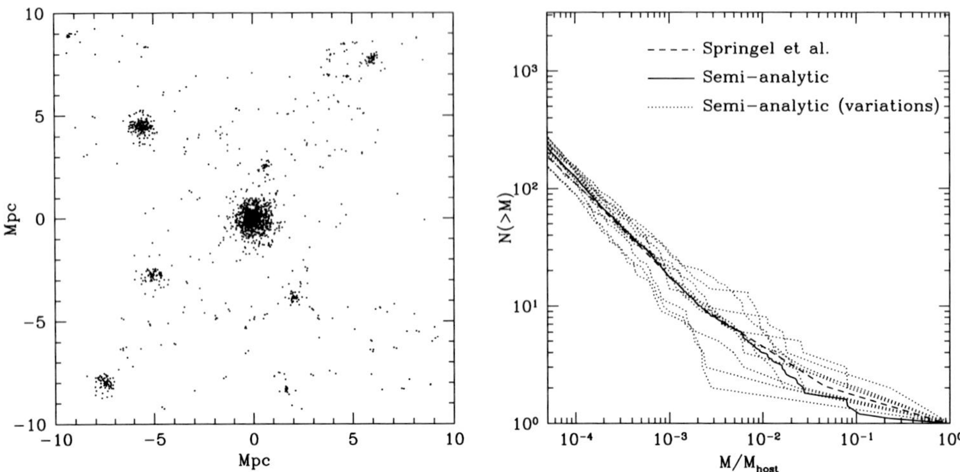

Figure 5. *Left-hand panel:* The projected distribution of model galaxies brighter than $m_{F814} = 21.5$ in a 20 Mpc cube centred on a $z = 0.4$ cluster with mass similar to that of Cl0024 (approximately $8 \times 10^{14}h^{-1}M_{\odot}$). Each dot corresponds to an individual galaxy. Note the presence of several large groups of galaxies near the central cluster. *Right-hand panel:* The cumulative number of sub-halos in the central cluster as a function of their mass (expressed relative to the total cluster mass). The dashed line shows the result computed by Springel et al. (2001) from their N-body simulation, while the solid line shows the average mass function computed from the GALFORM model. Dotted lines show results from ten individual cluster realizations from GALFORM to illustrate the expected cluster-to-cluster variation in this quantity.

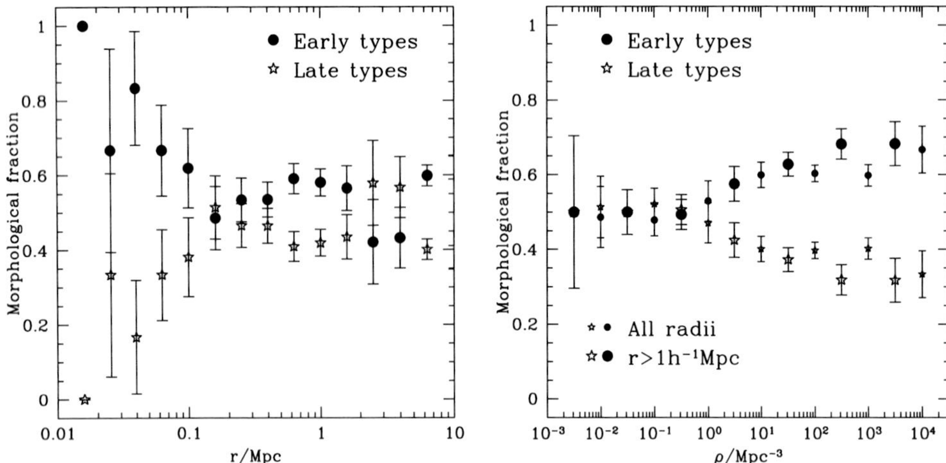

Figure 6. *Left-hand panel:* The morphology-radius relation in a model "Cl0024-like" cluster (and surroundings). Filled circles show the fraction of early-type galaxies as a function of radius, while stars show the corresponding fraction of late-type galaxies. Errorbars are calculated assuming a binomial distribution of early and late-types in each bin. At small radii the fraction of early-types increases as radius decreases. Beyond approximately 0.3 Mpc the morphological mix is constant. *Right-hand panel:* The morphology-density relation in the same model cluster (and its surroundings). Here circles and stars show the early and late-type fractions as a function of local, 3D galaxy number density (estimated from the distance to the tenth nearest neighbour for each galaxy in the sample). Small symbols show the relation including galaxies at all radii in the sample, while large symbols include only galaxies beyond 1 Mpc from the cluster centre.

Acknowledgments

I would like to acknowledge Carlton Baugh, Richard Bower, Shaun Cole, Carlos Frenk, John Helly and Cedric Lacey for developing the GALFORM model of galaxy formation and for permitting me to show results based upon their work.

References

Benson A. J., Pearce F. R., Frenk C. S., Baugh C. M., Jenkins A., 2001, MNRAS, 320, 261

Benson A. J., Bower R. G., Frenk C. S., Lacey C. G., Baugh C. M., Cole S., 2003, ApJ, 599, 38

Bizyaev D., Mitronova S., 2002, A&A, 389, 795

Cole S., Lacey C. G., Baugh C. M., Frenk C. S., 2000, MNRAS, 319, 168

Cole S. et al., 2001, MNRAS, 326, 256

Helly J. C., Cole S., Frenk C. S., Baugh C. M., Benson A. J., Lacey C., 2003a, MNRAS, 338, 913

Helly J. C., Cole S., Frenk C. S., Baugh C. M., Benson A. J., Lacey C., Pearce F. R., 2003b, MNRAS, 338, 913

Huang J. S., Glazebrook K., Cowie L. L., Tinney C., 2002, astro-ph/0209440

Kauffmann G., Colberg J. M., Diaferio A., White S. D. M., 1999, MNRAS, 303, 188

Kochanek C. S., Pahre M. A., Falco E. E., Huchra J. P., Mader J., Jarrett T. H., Chester T., Cutri R., Schneider S. E., 2001, ApJ, 560, 566

Lacey C., Cole S., 1993, MNRAS, 262, 627

Springel V., White S. D. M., Tormen G., Kauffmann G., 2001, MNRAS, 328, 726

Treu T., Ellis R. S., Kneib J.-P., Dressler A., Smail I., Czoske O., Oemler A., Natarajan P., 2003, ApJ, 591, 53

Yoshida N., Stoehr F., Springel V., White S. D. M., 2002, MNRAS, 335, 762

Outskirts of Galaxy Clusters: Intense Life in the Suburbs
Proceedings IAU Colloquium No. 195, 2004
A. Diaferio, ed.

DOI: 10.1017/S1743921304001164

The roles of ram-pressure stripping and minor mergers in the evolution of galaxies

Takashi Okamoto[1,2] and Masahiro Nagashima[3,1]

[1]Department of Physics, University of Durham, South Road, Durham, DH1 3LE UK
email: takashi.okamoto@durham.ac.uk

[2]NAOJ, NINIS, Mitaka, Tokyo 181-8588 Japan

[3]Department of Physics, Kyoto University, Kyoto 606-8502, Japan

Abstract. We investigate environmental effects on evolution of bright cluster galaxies in a Λ-dominated cold dark matter universe using a combination of dissipationless N-body simulations and a semi-analytic galaxy formation model. We incorporate effects of ram-pressure stripping (RPS) and minor merger-induced small starburst (minor burst) into our model. By considering minor burst, the observed morphology-radius relation is successfully reproduced. When we do not consider minor burst, the RPS hardly increases the intermediate B/T population. In addition, the RPS and minor burst are not important for colours or star formation rates of galaxies in the cluster core if star formation time-scale is properly chosen, because the star formation is sufficiently suppressed by consumption of the cold gas. We also find that SF in bulge-dominated galaxies is mainly terminated by starburst induced by major mergers in all environments.

1. Introduction

It has been found that galaxy morphology is a function of environment (Dressler 1980; Whitmore et al. 1993) and redshift (Dressler et al. 1997). Colours and star formation rate (SFRs) of galaxies also show similar dependence on environment and redshift (e.g. Butcher & Oemler 1984; Lewis et al. 2002). To account for these observational trends, several mechanisms that may suppress the star formation (SF) and transform one morphological type into another have been proposed.

Interaction between galaxies is one possible process to promote morphological transformation. Numerical simulations confirmed that major mergers produce galaxies resembling ellipticals as merger remnants (Barnes 1996) and that accretion of small satellites onto their host spiral lead a host spiral to S0 type (Walker et al. 1996). Because the galaxy merger triggers starbursts, the cold gas contained in original galaxies is exhausted in a very short time.

The second mechanism is removal of hot gas reservoir. In denser environments such as clusters, diffuse hot gas reservoir that is originally confined in haloes of non-central galaxies becomes part of the intracluster medium. A galaxy, whose hot gas reservoir is slowly removed, exhausts its cold gas in its SF time-scale (strangulation).

The above two processes, i.e. major merger-induced starburst and strangulation, has been incorporated in most of semi-analytic (SA) models which have failed to reproduce intermediate B/T population (Okamoto & Nagashima 2001; Diaferio et al. 2001). In this paper, we introduce two additional processes in order to solve this problem. One is ram-pressure stripping (RPS) of the cold gas from galactic disks (Gunn & Gott 1972). Another is minor merger-induced small starburst. Hydrodynamic simulations showed that some fraction of disk gas is driven to the galactic centre by a minor merger (Walker et al. 1996). By using a combination of N-body simulations with a SA model, we study the roles of the above four processes in the evolution of galaxies and what process is responsible for the formation of intermediate B/T galaxies.

2. Models

Here we consider a Λ-dominated cold dark matter universe ($\Omega_0 = 0.3$, $\lambda_0 = 0.7$, $\Omega_b = 0.015h^{-2}$, $H_0 = 100\ h$ km s^{-1} Mpc^{-1}, $h = 0.7$, and $\sigma_8 = 1$). We simulate a cluster region to study cluster galaxies and a region whose density is almost the same as the mean density of the universe to study field population. We then construct merger trees of virialised haloes and their substructure haloes. The details of this procedure is given in Okamoto & Habe (1999, 2000).

To clarify the effect of each physical process, here we examine four models. The model in which we consider neither the RPS nor the minor burst is called the *standard* model, because this is the standard prescription in SA models. Model parameters in the standard model are normalised so that they reproduce luminosity functions, cold gas fractions, and sizes of field galaxies. We also show the model using four times longer SF time-scale than the standard model to see the effect of the strangulation. This model is referred to as the *low SFR* model. The model with the minor burst and the model with the RPS are called "minor burst" and "RPS", respectively. In these models, we use the same parameters as used in the standard model. The details can be found in Okamoto & Nagashima (2003).

3. Results

In Fig. 1, we show colour-magnitude relations (CMRs) for simulated galaxies in the standard field model and cluster cores ($r < 0.5h^{-1}$ Mpc). We classify galaxies with B-band $B/T < 0.4$ as late-type and others as early-type galaxies. First of all, the CMRs for early-type galaxies are quite similar in all models, even in the field. It suggests that major merger-induced starbursts mainly truncate SF in early-type galaxies. On the other hand, colours of late-type galaxies are sensitive to the SF time-scale and their environments. Blue population seen in the field model is vanished in the cluster if we use appropriate SF time-scales. Since, in the low SFR model, late-type galaxies have much bluer colours than those in the standard model, it can be said that strangulation is the process that stops SF of late-type galaxies in cluster cores. The suppression of SF by strangulation is so efficient that the effect of the RPS is hardly seen except in the low SFR + RPS model.

It has been known that galaxies in the inner regions of clusters are redder than those in the outer regions. Recent observations have confirmed the existence of colour gradients in the galaxy population, which appear to continue even beyond the virial radius (e.g. Carlberg et al. 1997). In the left panels of Fig. 2 we show the $B-V$ colours of the cluster galaxies at $z = 0.2$ as a function of projected radius. The projected radius is normalised by R_{200} that is the radius inside which the mean density is 200 times the critical density of the universe at a given redshift. The observational data are taken from Diaferio et al. (2001).

For the standard and minor burst models, the effect of the RPS can be seen only at large radii, because, in the central part, the SF is strongly suppressed by the strangulation. By comparing the low SFR + RPS model with the low SFR model, it is found that the RPS effectively suppresses SF at $r < R_{200}$ if there is some cold gas left in the galaxies. While the minor burst makes galaxies slightly redder than those in the standard model, the models with and without the minor burst produces almost the same results.

Recently, some authors have studied the morphology-density (or radius) relations by using a hybrid method of N-body simulations and SA models (Okamoto & Nagashima

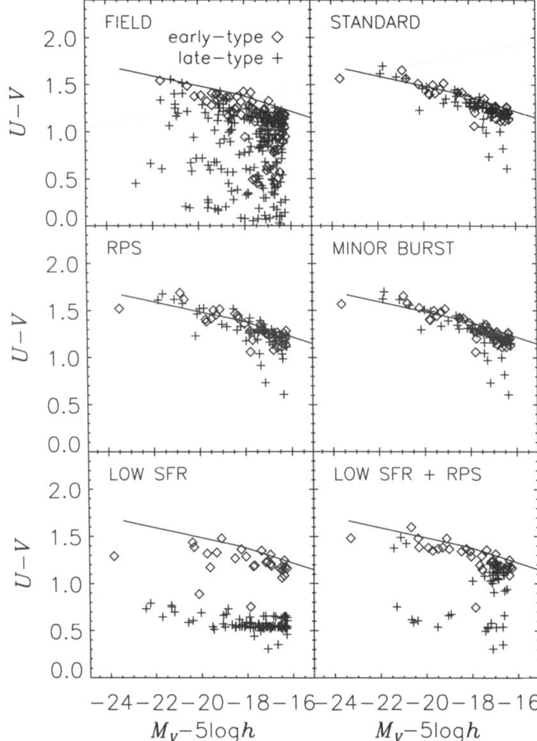

Figure 1. CMRs at $z = 0$. We show $U - V$ colours of galaxies in the field model and galaxies within the cluster core in the standard, RPS, minor burst, low SFR, and low SFR + RPS models as a function of V-band luminosity. Early-type galaxies and late-type galaxies are represented by diamonds and crosses, respectively. The solid lines show the observed CMR for cluster ellipticals by Bower et al. (1992).

2001; Diaferio et al. 2001; Springel et al. 2001). In their work, S0 or intermediate B/T population in clusters is hardly reproduced unless a wider B/T range is adopted to classify simulated galaxies as S0s. Here we examine how the RPS and minor burst affect the intermediate B/T population in clusters.

We show morphological fractions in the cluster as a function of radius in the righ panels of Fig. 2. The simulated galaxies are classified according to their B-band B/T; $B/T > 0.7$ as bulge-dominated, $B/T < 0.4$ as disk-dominated, and $0.4 < B/T < 0.7$ as intermediate. By comparison of the CNOC sample with the standard model, it is confirmed that the intermediate fraction in the standard model is much smaller than the observation, while the bulge-dominated fraction is reproduced. Even when we consider the RPS, the intermediate fraction is still too small. It is because most of the disk-dominated galaxies are almost pure disk galaxies without minor bursts (Okamoto & Nagashima 2003), and then the RPS simply darkens the disk-dominated galaxies rather than increasing their B/Ts. The strange behaviour seen in the RPS models at $r/R_{200} > 0.7$ results from the small number of galaxies at these radii because of the spherical high-resolution region.

The minor bursts increases the intermediate fraction without changing the bulge-dominated fraction because it affects only galaxies with small B/Ts. As a result, the minor burst model reproduces the observed fractions quite well despite the fact that it does not influence other properties of the galaxies. Thanks to minor bursts, now the disk-dominated galaxies tend to be non-pure disk galaxies. Consequently, the RPS can

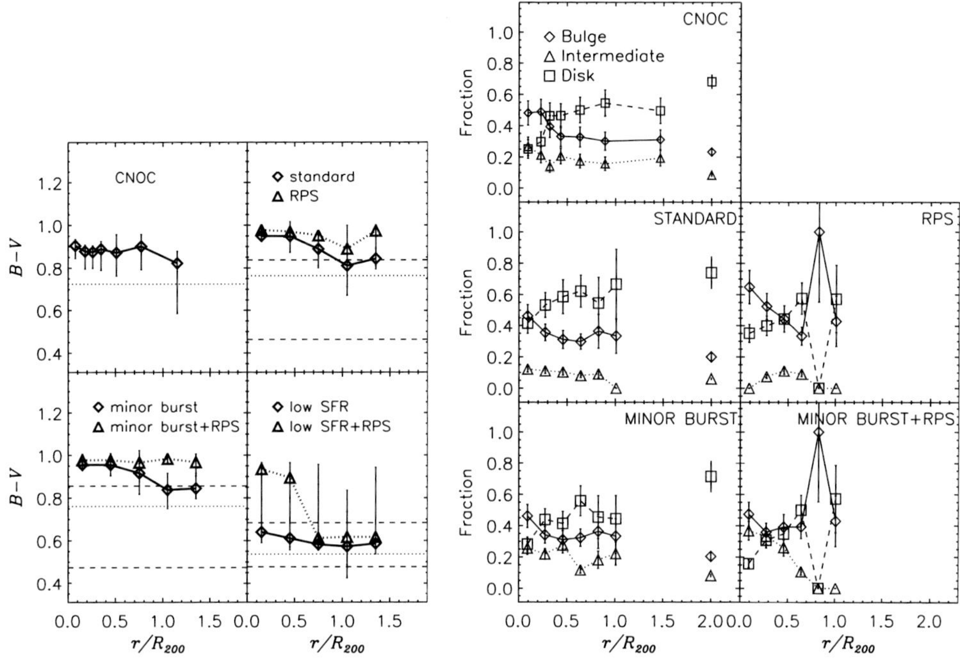

Figure 2. Left: The median $B-V$ colours of cluster galaxies with $M_R < -20.5$ are plotted as a function of projected radius. In the upper left panel, the diamonds denote the observed median colours of the CNOC sample. The median colour of field galaxies at the same redshift range are shown by the dotted line. Other panels shows the median colours for the simulated galaxies at $z = 0.2$ and the corresponding RPS model is indicated by the triangles in each panel. The field model is also shown by the dotted line in each panel. Right: The morphological fractions are plotted as a function of projected radius. In the top panel we show the observed fractions for the galaxies with $M_R < -20.5$ in the CNOC sample. Other panels show the fractions given by our models for the simulated cluster at $z = 0.2$. The symbols at $r/R_{200} = 2$ indicate the fractions in the field in each panel.

change their B/T by the fading of their disks. In the minor burst + RPS model, we can see the increasing intermediate fraction toward the centre.

4. Summary

We have investigated the effects of major mergers, strangulation, RPS of cold disk gas, and minor mergers on the evolution of bright cluster galaxies. We have used a combination of the cosmological N-body simulations and the SA galaxy formation model. This method enables us to study the above environmental effects in a fully cosmological context. We have determined the model parameters of the reference model to reproduce galaxy properties at $z = 0$.

Our results are summarised as follows.

(a) The process that terminates SF in early-type galaxies ($B/T > 0.4$) in all environments is starburst triggered by a major merger.

(b) If we adopt appropriate SF time-scales, so as to reproduce the observed cold gas mass fraction in the field, the dominant process that determines colours of galaxies in the cluster core is the strangulation.

(*c*) Since the strangulation sufficiently suppresses SF in cluster cores, the effect of the RPS is hardly observed.

(*d*) Minor burst does not affect galaxy properties except for morphology.

(*e*) Without minor bursts the fraction of intermediate B/T galaxies in clusters becomes too small. The model with the minor burst can reproduce the observed morphology-radius relations. We conclude that the minor burst is an essential process to form intermediate B/T galaxies.

(*f*) The RPS hardly increases the intermediate population in the cluster without minor bursts. When the minor burst is taken into account, the RPS increases the intermediate galaxy fraction in the cluster cores.

References

Barnes, J. E. 1996 In *ASP Conference Series 92, Formation of the Galactic Halo—Inside and Out*, eds. H. L. Morrison & A. Salajedini (San Francisco: ASP), 415

Bower, R. G., Lucy, J. R., & Ellis, R. S. 1992, MNRAS, **254**, 601

Butcher, H. & Oemler, A. 1984, ApJ, **285**, 426

Carlberg, R. G., Yee, H. K. C., & Ellingson, E. 1997, ApJ, **478**, 462

Diaferio, A., Kauffmann, G., Balogh, M. L., White, S. D. M., Schade, D., & Ellingson, E. 2001, MNRAS, **323**, 999

Dressler A. 1980, ApJ, **236**, 351

Dressler A. et al. 1997, ApJ, **490**, 577

Gunn, J. E., & Gott, J. R. 1972, ApJ, **176**, 1

Lewis, I., et al. 2002, MNRAS, **334**, 673

Okamoto, T. & Habe, A. 1999, ApJ, **516**, 591

Okamoto, T. & Habe, A. 2000, PASJ, **52**, 457

Okamoto, T. & Nagashima, M. 2001, ApJ, **547**, 109

Okamoto, T. & Nagashima, M. 2003, ApJ, **587**, 500

Springel, V., White, S. D. M., Tormen, G., & Kauffmann, G. 2001, MNRAS, **328**, 726

Whitmore, B. C., Gilmore, D. M., & Jones, C. 1993, ApJ, **407**, 489

Yee, H. K. C., Ellingson, E., & Carlberg, R. G. 1996, ApJS, **102**, 269

Walker, I. R., Mihos, J. C., & Hernquist, L. 1996, ApJ, **460**, 121

Outskirts of Galaxy Clusters: Intense Life in the Suburbs
Proceedings IAU Colloquium No. 195, 2004
A. Diaferio, ed.

DOI: 10.1017/S1743921304001176

On the inclusion of environmental effects in the halo model of galaxy clustering

Ravi K. Sheth, Ummi Abbas and Ramin Skibba

Department of Physics and Astronomy, University of Pittsburgh, 3941 O'Hara Street, Pittsburgh, PA 15260, USA email: rks12@pitt.edu

Abstract. Standard models of galaxy formation make specific predictions for how galaxy properties depend on environment. These predictions follow from the assumption that galaxy properties are entirely determined by the formation histories of the halos which host them, and that the distribution of dark halo masses is skewed towards large masses in dense regions. We describe two tests of this assumption which can be made with currently available data.

1. Introduction

We are now in the early years of what is being called the era of precision cosmology; what was a trickle of precise measurements of the large-scale Universe, primarily from exquisitely designed balloon- and space-based measurements of the cosmic background radiation, is growing into a torrent, as data of unprecedented accuracy from large-scale galaxy and cluster surveys floods in. The wealth of data means that, hand-in-hand with precision cosmology, the era of statistical cosmology is upon us.

The start of the year 2003 saw the release of exquisite observations of the microwave sky from the WMAP satellite (Bennett *et al.* 2003). Shortly thereafter, the 2dFGRS (Colless *et al.* 2001, 2003) and the SDSS (Abazajian *et al.* 2003) collaborations released a wealth of data of the optical sky. The auto-correlation functions of the cosmic background radiation temperature, and of the galaxy distribution, and the cross correlation between the two provides strong constraints on the background cosmology. The SDSS dataset has accurate photometry in five bands, so it is well suited to studying a number of galaxy properties, and hence to constraining models of galaxy formation.

Galaxy formation models make specific predictions for how galaxies evolve; in these models, there is a close relation between evolution and environment. The SDSS is not sufficiently deep to allow strong constraints on the models of evolution, but, because it contains a large number of galaxies in a range of environments, it can nevertheless be used to inform galaxy formation models. In what follows, I develop two models of how the environmental dependence of galaxy properties is expected to manifest itself in datasets like the SDSS. Both use the halo model of clustering to illustrate the main features of the argument.

In the halo model (Cooray & Sheth 2002), the nonlinear density field is assumed to be made up of dense objects called halos. At any given time, halos of different masses all have the same density (they are all approximately two hundred times denser than the background). All mass is in such halos, and so all galaxies are also associated with dark halos. Moreover, all physical and statistical quantities of interest are thought of as arising from two types of terms: those which reflect the properties within a single halo, and those which involve more than one halo. For instance, the velocity of a randomly chosen particle is the vector sum of the motion with respect to the center of mass of the halo in which it sits, and the motion of the halo center of mass.

Similarly, the two-point correlation function $\xi(r)$ is determined by the sum of two types of galaxy pairs: pairs in the same halo, and pairs in separate halos. Since the radius of a typical halo at $z = 0$ is less than a Mpc, the one-halo term is negligible on scales larger than a few Mpc. On the small scales where the one-halo term dominates, the shape of $\xi(r)$ is determined by how halo density profiles depend on halo mass, and on how halo abundances depend on mass; on larger scales, $\xi(r)$ is less sensitive to the shapes of halo profiles, and more sensitive to the clustering of the halos themselves.

2. Type-dependent clustering

To illustrate, the power-spectrum, the Fourier transform of $\xi(r)$, is

$$P_{gg}(k) = P_{1h}(k) + P_{2h}(k), \tag{2.1}$$

where

$$P_{1h}(k) = \int dm\, \frac{dn(m)}{dm}\, \frac{g_2(m)}{\bar\rho^2}\, |u(k|m)|^2 \tag{2.2}$$

$$P_{2h}(k) \approx \left[\int dm\, \frac{dn(m)}{dm}\, \frac{g_1(m)}{\bar\rho}\, b(m)\right]^2 P_{\rm Lin}(k) \tag{2.3}$$

with

$$\bar\rho = \int dm\, \frac{dn(m)}{dm}\, g_1(m), \tag{2.4}$$

$$g_1(m) \equiv \langle N_{gal}|m\rangle \qquad \text{and} \qquad g_2(m) \equiv \langle N_{gal}(N_{gal}-1)|m\rangle. \tag{2.5}$$

Here $dn(m)/dm$ denotes the number density of halos of mass m, $u(k|m)$ is the Fourier transform of the halo density profile, $g_1(m)$ and $g_2(m)$ denote the first and second factorial moments of the number of galaxies in m-halos, $b(m)$ is the m-halo bias factor, and $P_{\rm Lin}(k)$ is the power spectrum of the mass in linear theory. The clustering of galaxies is the same as the clustering of the mass only if $g_1(m) \propto m$ and $g_2(m) \propto m^2$; if not, the galaxies are biased tracers of the dark matter distribution. Galaxies of different types cluster differently; for instance, more luminous galaxies cluster more strongly, and the correlation function on Mpc scales is steeper for red galaxies than for blue. In the halo model, this difference arises primarily because the dependence on halo mass of g_1 and g_2 depends on galaxy type. Although, in principle, $u(k|m)$ may also depend on galaxy type (e.g. the red galaxies in a cluster tend to be more concentrated around the cluster center than blue galaxies), accounting for this is only important on scales which are substantially smaller than a Mpc (in most cases of practical interest).

3. Clustering and environment

Most galaxy formation models assume that the formation histories of halos of the same mass are independent of their environments (e.g. the models discussed by Benson in these proceedings). The reason for this assumption is most clearly illustrated by the excursion set approach (Bond et al. 1991; Lacey & Cole 1993; Sheth 1998). The idealization is that objects form from the spherical collapse of initially overdense regions; as the collapse proceeds, initially concentric shells remain concentric. Consider such an initially overdense region in a Gaussian random field. The density of this patch is determined by the superposition of waves of many different wavelengths. Small scale waves determine the structure of the interior of the patch, and hence its formation history, whereas large scale modes determine the environment of the patch. In a Gaussian field, these waves are all independent of one another. So, if the initial fluctuation field was Gaussian (as the CMB suggests), then the assumption that the formation histories patch should be approximately independent of its environment is likely to be accurate (e.g. White 1996). Since these models also assume that the formation history of a halo determines the number of galaxies which form within it, as well as their spatial distribution around the halo center, in these models, the relations $g_1(m)$ and $g_2(m)$ used above, and also the profile shapes $u(k|m)$ are independent of environment. Thus, in these models, all environmental effects arise from the fact that halo abundances depend on environment.

The halo model description of clustering suggests the following test of such models.

- Measure $\xi(r)$ in the full data set.
- Determine the $g_1(m)$ and $g_2(m)$ relations which are required to fit $\xi(r)$.
- Then classify all galaxies by the environment which surrounds them (within 8 Mpc, say). Choose only those galaxies which populate a specified range of environments, and measure their clustering $\xi(r|\delta)$.
- By hypothesis, the relations $g_1(m)$ and $g_2(m)$ do not depend on environment, so they should be independent of δ. Since the spatial distribution of galaxies around halo centers is also assumed to be independent of environment, the only effect of the environment is to modify the halo abundances: $n(m,z|\delta) = [1+b(m,z)\delta]n(m,z)$. Since this modification is known (e.g. accurate formulae for $b(m,z)$ are available), we can use the relations $g_1(m)$ and $g_2(m)$ measured with relatively high signal-to-noise from the full sample to predict $\xi(r|\delta)$.
- Comparison with the measured shape provides a test of the assumption that the only environmental trends come from the halo abundances.

The poster by Abbas elsewhere in these proceedings illustrates that such a test is certainly feasible with current datasets. The remainder of this section formalizes this argument in the context of the halo model.

We imagine sitting on each m-halo, and counting up the mass M in the volume V which surrounds it. We then select only those halos which are surrounded by volumes V which contain a specified range of M. Let $f(M,V|m)dM$ denote the fraction of m-halos which are surrounded by regions V which contain mass in the range dM about M. In the excursion set model, the number density of such halos is

$$n(m)f(M,V|m)dM = n(m)\frac{f(m|M,V)f(M,V)}{f(m)}dM = \frac{\bar{\rho}}{m}f(m|M,V)f(M,V)dM$$
$$= dM\, n(M,V)\, N(m|M,V) \tag{3.1}$$

(Sheth 1998). Thus, the mass density contributed by halos which are embedded in regions of mass $M > M_{min}$ is

$$\bar{\rho}_\delta = \int_{M_{min}}^{\infty} dM\, n(M|V) \int_0^M dm\, N(m|M,V)\, m = \int_{M_{min}}^{\infty} dM\, n(M|V) M. \tag{3.2}$$

In the standard model, the density profile of a halo depends on its mass, but not on the surrounding environment. Therefore, the one-halo term is

$$P_{1h}(k|>M_{min}) = \int_{M_{min}}^{\infty} dM\, n(M|V) \int_0^M dm\, N(m|M,V) \left(\frac{m}{\bar{\rho}_\delta}\right)^2 |u(k|m)|^2. \tag{3.3}$$

The two-halo term has contributions from pairs which are in the same patch V, and from pairs in distinct V-patches. The contribution from pairs in different patches is given by

$$P_{2p}(k|>M_{min}) = \left[\int_{M_{min}}^{\infty} dM\, n(M|V)\, b(M|V) \times \int_0^M dm\left(\frac{m}{\bar{\rho}_\delta}\right) N(m|M,V)\, u(k|m)\right]^2 P_{\rm Lin}(k) \qquad (3.4)$$

where we have again assumed that $u(k|m,M,V) \to u(k|m)$. As a check, notice that when $M_{min} \to 0$ then $\bar{\rho}_\delta \to \bar{\rho}$. A little algebra shows that $P_{1h}(k|>0)$ equals the expression for $P_{1h}(k)$ we had previously, and that $P_{2p}(k|>0)$ reduces to $P_{2h}(k)$.

Galaxies are described simply by replacing the weighting by m in the expression for $\bar{\rho}_\delta$ with a weighting by $\langle N_{gal}|m,M,V\rangle$, which explicitly allows for the fact that the number of galaxies in an m halo may depend on its environment. If it does not, then $\langle N_{gal}|m,M,V\rangle \to g_1(m)$, and the resulting expression reduces to the standard expression for $\bar{\rho}_g$ in the limit that $M_{min}=0$. Similarly, in the expression for P_{2p}, one replaces the weighting by $(m/\bar{\rho}_\delta)$ with $g_1(m)/\bar{\rho}_g$, and in the one-halo term, $(m/\bar{\rho}_\delta)^2 \to g_2(m)/\bar{\rho}_g^2$. These expressions show clearly that the entire environmental dependence of clustering comes from the fact that the halo population depends on environment. Since the most massive halos populate the densest regions which have the largest bias factors, a generic prediction is that galaxies in dense regions are more strongly clustered.

In practice, the analytic calculation is hampered by the fact that while we may observe the overdensity of galaxies δ_g, we do not observe the quantity which affects the halo abundances: the overdensity of the mass δ. We can get a rough idea of the trend if we use the fact that, if V is large, $\delta_g \approx b_{gal}\delta$, where $1+\delta = M/\bar{\rho}V$. In this case, we can replace δ in the expressions above with δ_g/b_{gal}. In practice, it is more accurate to simply Monte-Carlo the predicted dependence on environment by populating halos in a simulation with the desired g_1 and g_2 relations, and then making the same measurement in the mock galaxy catalog as one does in the data.

4. Galaxies and halo substructure

A simple model for the galaxy distribution associates galaxies with the subclumps of dark matter halos which survive the processes of tidal stripping as they orbit within their parent halos (e.g. Kravtsov *et al.* 2004). If m and M denote the masses of the subclump and the parent respectively, then simulations suggest that, on average, the mass function of halo subclumps is a universal function of m/M. (In fact, Kravtsov *et al.* note that estimating subclump masses is difficult, so some estimate of the circular velocity, which is much less sensitive to the precise definition of where the edge of the subclump is, is almost certainly a better indicator of subclump properties. They find that the subclump distribution is a universal function of v_c/V_c.) In particular, a power-law,

$$\frac{dN_{sat}(m|M)}{dm} = N_0\left(\frac{M}{m}\right)^{\mu}\frac{dm}{m} \qquad \text{with } \mu \approx 0.9 \qquad (4.1)$$

provides a good description (e.g. Ghigna *et al.* 2000). The constant N_0 is set by noting that the mass fraction in subclumps is

$$\int dm\left(\frac{m}{M}\right)\frac{dN_{sat}(m|M)}{dm} = N_0\,\frac{1-(m/M)^{1-\mu}}{1-\mu} \qquad (4.2)$$

where we are imagining that the lower limit of the integral $m/M \ll 1$. Simulations show that this mass fraction is approximately ten percent for all M:

$$N_0 = 0.1 \left(\frac{f}{0.1}\right) \frac{1-\mu}{1-(m/M)^{1-\mu}}. \tag{4.3}$$

Moreover, simulations suggest that the cumulative number of massive satellites of M halos follows a Poisson distribution with mean

$$N_{sat}(>m|M) = \int_m^M dm \frac{dN_{sat}(m|M)}{dm} = N_0 \frac{(M/m)^\mu - 1}{\mu} \tag{4.4}$$

(Kravtsov *et al.* 2004).

If there is a monotonic relation between subclump mass and luminosity (or circular velocity), then the number density of subclumps which host galaxies more luminous than L is $N_{sat}(>m(L)|M) \propto \left[M/m(L)\right]^\mu$ when $m \ll M$. In addition to populating halo subclumps, it is likely that there is one galaxy at or close to the center of each sufficiently massive halo. If this galaxy has the same $m(L)$ relation as the other satellite galaxies, then the number of galaxies in a halo can be written as

$$\langle N_{gal}|M\rangle = 1 + N_{sat}\Big(>m(L)|M\Big) \qquad \text{where} \quad M > m(L) \tag{4.5}$$

and recall that N_{sat} is a Poisson variate with mean $\propto [M/m(L)]^\mu$. The luminosity dependence of clustering seen in the SDSS data is well described by this model provided $m(L)$ is chosen to increase with L so as to match the observed number density.

Recent simulations (e.g. Moore in these proceedings) suggest that there may be a problem with this model, because the spatial distribution of subclumps around the centres of their parent halos appears to be shallower than is observed for the distribution of luminous galaxies around cluster centres in the CNOC survey. However, this is a mass-dependent statement, and leads to the following discussion. Since the baryons, and hence the stellar mass, are almost certainly more centrally concentrated than the dark matter, and the stellar mass is almost certainly more closely related to the mass of the subclump before tidal effects stripped-off the mass in the outer layers, it would be interesting to compare the subclump mass function in the simulations before and after tidal stripping. It is the mass function prior to stripping which is perhaps more relevant for predicting stellar mass and/or luminosity functions, and the luminosity dependence of clustering. Such a measurement is not yet in the literature, although simulations are probably up to the task.

5. Weighted or 'marked' correlations

Almost all analyses to date treat galaxies as points without attributes. In fact galaxies have luminosities, sizes, shapes, velocity dispersions, star formation rates, etc. The quality of the data is now sufficiently good that one can imagine measuring the spatial correlations of these attributes. That is to say, rather than measuring clustering as a function of luminosity, one can now measure the clustering of luminosity (or of color, star-formation rate etc.).

The halo-model is the natural language within which to discuss marked correlations. In the standard model, all observables are assumed to be determined by the formation history of the halo, which, in turn, depends on mass. In what follows we will use $W(m|M)$ to denote the observable (W here stands for 'weight'), and we will assume that the distribution of the weight W at fixed m and M is narrow (i.e., we ignore stochasticity in the relation between observable W and m and M). Then the two-halo term is particularly straightforward to model. If we define

$$\langle N_{gal}|M\rangle \equiv \int dm\, N(m|M) \quad \text{and} \quad \langle W|M\rangle \equiv \frac{\int dm\, N(m|M)\, W(m|M)}{\int dm\, N(m|M)}, \tag{5.1}$$

then the average number density of galaxies is

$$\bar{\rho}_g \equiv \int dM\, \frac{dn(M)}{dM} \int dm\, N(m|M) \equiv \int dM\, \frac{dn(M)}{dM}\, \langle N_{gal}|M\rangle, \tag{5.2}$$

the average weight is

$$\bar{W} \equiv \int dM\, \frac{dn(M)}{dM}\, \frac{\langle N_{gal}|M\rangle}{\bar{\rho}_g}\, \langle W|M\rangle, \tag{5.3}$$

and the two-halo term is

$$\mathcal{W}_{2h}(k) = \left[\int dM\, \frac{dn(M)}{dM}\, \frac{\langle N_{gal}|M\rangle}{\bar{\rho}_g}\, \frac{\langle W|M\rangle}{\bar{W}}\, b(M)\, u(k|M)\right]^2 P_{\rm Lin}(k). \tag{5.4}$$

(Notice that when all the weights are equal, $W = \bar{W}$, this expression reduces to the usual expression for the two-halo contribution to the galaxy power spectrum.) If we define $B_W(M) \equiv b(M)\, \langle W|M\rangle/\bar{W}$, then the expression above shows clearly that the large scale power spectrum is simply a linearly biased version of the dark matter spectrum, where the bias factor depends on the particular observable used as the weight.

The one halo term is slightly more complicated. If we assume that the spatial distribution of subclumps does not depend on subclump mass (in fact, it almost certainly does), and that the luminosity of a galaxy in an m-subclump is independent of whether or not there are other galaxies in the same halo, then

$$\mathcal{W}_{1h}(k) = \int dM\, \frac{dn(M)}{dM}\, \frac{\langle N_{gal}(N_{gal}-1)|M\rangle}{\bar{\rho}_g^2}\, \frac{\langle W|M\rangle^2}{\bar{W}^2}\, \left|u(k|M)\right|^2. \tag{5.5}$$

To see what this implies, suppose that the weight is proportional to subclump mass. Then

$$\langle W|M\rangle \to \frac{fM}{\langle N_{gal}|M\rangle}, \quad \text{so} \quad \bar{\rho}_g\bar{W} \to f\bar{\rho}, \quad \text{and} \quad \frac{\langle W|M\rangle}{\bar{W}} \to \left(\frac{M}{\bar{\rho}}\right) \frac{\bar{\rho}_g}{\langle N_{gal}|M\rangle};$$

$$\mathcal{W}_{1h}(k) \to \int dM\, \frac{dn}{dM} \left(\frac{M}{\bar{\rho}}\right)^2 \frac{\langle N_{gal}(N_{gal}-1)|M\rangle}{\langle N_{gal}|M\rangle^2} \left|u(k|M)\right|^2,$$

$$\mathcal{W}_{2h}(k) \to \left[\int dM\, \frac{dn}{dM} \left(\frac{M}{\bar{\rho}}\right) b(M) u(k|M)\right]^2 P_{\rm Lin}(k).$$

If the number of galaxies in M haloes, $p(N_{gal}|M)$, follows a Poisson distribution, then $\langle N_{gal}(N_{gal}-1)|M\rangle = \langle N_{gal}|M\rangle^2$, so $\mathcal{W}_{1h} \to \int dM dn/dM (M/\bar{\rho})^2 |u(k|M)|^2$. If the integral is over all M, then this integral equals that for the dark matter. Thus, if the mass-to-light ratio is independent of galaxy mass (it is not) then the galaxy luminosity weighted marked correlation function will have the same shape as that of the dark matter correlation function. A better approximation perhaps, is to use the dynamical mass estimator $R\sigma^2$ as the mark. Such a measurement is within reach in current datasets. More generally, this analysis reveals an important point: if the subclump mass function is a universal function of m/M, then the marked correlation function term $\langle N_{gal}|M\rangle\langle W|M\rangle$ scales with the same power of M as does the mark with m.

The expressions above have no environmental trends other than those which come from the dependence of halo abundances on environment. In essence, they represent the framework within which to describe the predictions of standard galaxy formation models. The extent to which simple halo-model calculations such as this are able to reproduce measurements of marked correlations in real data provides a test of the standard assumption that galaxy properties are more closely related to the formation histories of their parent halos, rather than to their environments.

6. Conclusions

We presented two tests of a fundamental assumption in galaxy formation models: that environmental effects are primarily a consequence of the fact that the halo population depends on environment (e.g. the most massive halos populate the densest regions). One test uses the clustering of galaxies as a function of environment. The other uses the stastitics of weighted or marked correlations. In both cases, we used the halo model to illustrate what the standard model predicts for measurements which can easily be made in datasets which are currently available. Application of our results to the SDSS is underway. In addition, it would be interesting to apply these tests to the galaxy formation models of van Kampen (which explicitly include environmental effects) and compare the results with the models of Benson (which explicitly assume that the only environmental effects come from the correlation between halo masses and environment).

Acknowledgements

We are grateful to the organizers, and Antonaldo Diaferio in particular, for scheduling the meeting during Torino's chocolate festival. We would also like to thank some members of Torino's police department for their role in an incident involving a lost wallet. RKS is grateful for the receipt of an IAU grant. This work is supported by NASA and the NSF under grants NAG5-13270 and AST-0307747.

References

Abazajian, K., *et al.*. 2003 *A.J.* **126**, 2081.
Bennett, C. *et al.* 2003 *Ap.J.S.* **148**, 97.
Bond, J. R., Cole, S., Efstathiou, G. & Kaiser, N. 1991 *Ap.J.* **379**, 440.
Colless, M. *et al.* 2001 *Ap.J.* **328**, 1039.
Cooray, A., & Sheth, R. K. 2002 *Phys. Rep.* **372**, 1.
Ghigna, S. *et al.* 2000 *Ap.J.* **544**, 616.
Kravtsov, A. *et al.* 2004 *Ap.J.* submitted, (astro-ph/0308519)
Lacey, C., & Cole, S. 1993 *M.N.R.A.S.* **262**, 627.
Mo, H. J. & White, S. D. M. 1996 *M.N.R.A.S.* **282**, 347.
Sheth, R. K. 1998 *M.N.R.A.S.* **300**, 1075.
White, S. D. M. 1996 In *Cosmology and Large Scale Structure, Les Houches Session LX* (eds. R. Schaefer, J. Silk, M. Spiro & J. Zinn-Justin), pp. 349. Elsevier.

Fernando Atrio-Barandela and Rashid Sunyaev *(by G. Murante).*

John Stocke and Gustavo Yepes *(by G. Murante).*

Marc Voit and Sergei Shandarin. In the background, Ravi Sheth and Antonaldo Diaferio gossip *(by A. Ferrari).*

Author Index

Karen Tozzi Colberg, Hans Böhringer and Jörg Colberg *(by A. Ferrari)*.

Howard Yee, Tadayuki Kodama, Brent Tully and Manolis Plionis *(by A. Ferrari)*.